Introduction to Transportation Engineering and Planning

Introduction to Transportation Engineering and Planning

EDWARD K. MORLOK
UPS Foundation Professor of Transportation
Civil and Urban Engineering Department
University of Pennsylvania

McGRAW-HILL BOOK COMPANY
New York St. Louis San Francisco Auckland Bogotá
Düsseldorf Johannesburg London Madrid
Mexico Montreal New Delhi Panama Paris
São Paulo Singapore Sydney Tokyo Toronto

Introduction to Transportation Engineering and Planning

1234567890FGRFGR78321098

This book was set in Helvetica by Bi-Comp, Incorporated.
The editors were B. J. Clark and J. W. Maisel;
the designer was Nicholas Krenitsky;
the production supervisor was Dominick Petrellese.
The drawings were done by ECL Art Associates, Inc.
Fairfield Graphics was printer and binder.

Library of Congress Cataloging in Publication Data

Morlok, Edward K
Introduction to transportation engineering and planning.

Includes bibliographies and index.
1. Transportation. I. Title.
TA1145.M58 380.5 77-16450
ISBN 0-07-043132-9

To O.A.M. and J.A.M.

Contents

Preface

This book is intended as an introductory text in transportation engineering and planning, for use primarily at the junior or senior undergraduate level. It attempts to present the basic principles and methods used by engineers and planners in the planning, design, and operation of transportation systems. It emphasizes principles and methods which are applicable to all modes of transport, passenger and freight, and to all contexts, whether urban or rural, in a mature economy or a developing nation.

This very broad field, which shall be termed transportation system engineering, is still evolving. It represents the union of what is usually termed transportation planning, a field which has developed almost entirely in the last two decades, and the more traditional engineering fields concerned with the components of individual modal systems. Included in the latter category are such areas of specialization as road traffic engineering, railway vehicle design, and airport and seaport design. Out of the particulars of these individual modal fields can be drawn many fundamental principles relating to the motion of individual vehicles, the interaction and flow of numerous transport units on congested guideways, and the operation and performance of terminals. These in effect describe the technology of transportation.

The field of transportation planning has been developed to a large extent by engineers, primarily as a result of the very pressing need to plan better the massive investments which were made in urban transportation systems in the past two decades, to meet the rapidly expanding amount of travel reflective of the rapid population growth of urban areas, both in North America and abroad. Drawing heavily from economics, city planning, and other related disciplines, methods were developed which for the first time permitted the quantitative prediction of future requirements for transport, the evaluation of the performance of entire transport systems in meeting future requirements, and most recently the estimation of impacts on land development and the natural and social environment. These methods are now being used in other contexts, from planning in developing nations to the restructuring of the freight railroad system in the northeast portion of the United States.

The field of transportation system engineering is really composed of these two main streams of development. The technology is critical to the understanding of the functioning of the system, in predicting its performance and costs, and in identifying alternatives in the form of fixed plant investments as well as operational changes. And the methods used in planning are critical, not only for the long-range planning function for which they were developed, but also for dealing with any system problem, whether it be network traffic light timing or the scheduling of an entire airline. This book is intended to bring out the fundamentals of these previously largely unrelated fields, of planning and technology, and to identify the relationships which give unity and coherence to the field of transportation system engineering.

In addition to the integration of technology and planning, this book differs from other introductory transportation engineering texts in numerous other ways. Most other books devote considerable space to a detailed history of transport, usually for one nation or part of the world. In this book, historical references are made within the context of general relationships between the transportation system and the environment it is intended to serve. In this way, the reader gains an understanding of the general principles of interaction and an appreciation of the profound influence of transport in such major changes in civilization as the transition from nomadic tribes to settlements, the industrial revolution, and the recent growth of extremely large urban centers.

Similarly treated in a general way are the principles and primary organizations of transport, governmental and private. Since the book is intended for use not only in the United States but elsewhere, there is a minimum of detailed description of any one nation's governmental structure or other institutions of transport operation, ownership, and regulation. It should be possible for the instructor to fill in details of any particular context chosen for emphasis. Particulars in the United States are provided as examples.

This book was designed for use in a variety of introductory courses, which differ in orientation and duration. Its parts and chapters can be combined in different ways to suit different courses, ranging in duration from one quarter (about 11 weeks) to two semesters (about 30 weeks), the shorter courses focusing on particular aspects of the field, such as engineering, planning, or management and operations. These combinations are shown in the following table. The instructor should be careful to define any concepts necessary for understanding material in the chapter sequence followed but not necessarily covered by it. The engineering course focuses on what has been termed component engineering, which deals with the technology of transportation and culminates in coverage of facility location and design. The planning course omits the detailed component technology chapters, emphasizing system relationships, evaluation, and the planning process. The management sequence is designed to acquaint those in management programs with the characteristics of the transport systems whose operations, marketing, and financing they will be responsible for. Since management students are typically less familiar with the quantitative approach to engineering, a somewhat slower pace is maintained in these courses than in the others, with more emphasis on classroom and team problem solving. The outlines certainly will

WEEKS DEVOTED TO EACH CHAPTER

Chapter	One Quarter (11 weeks)			One Semester (15 weeks)			
	Engineering	Planning	Management	Engineering	Engineering/Planning	Planning	Management
1	1/2	1/2		1/2	1/2	1/2	1/2
2		1	1	1		1	1
3	1	1	1 1/2*	1	1	1	1 1/2*
4	1 1/2†		1	1†	1		1
5	1 1/2†		1	1†	1		1
6				1			
7	1 1/2†		1 1/2*	1 1/2†	1		1 1/2*
8	1/2		1	1	1	1	1
9	1	1	1	1	1	1	1
10		1			1	2†	1
11		1			1	1†	1
12		1		1	1 1/2	2†	1
13		1		1		1	
14	1	1	1	1†	1	1	1
15		2‡			2‡	2‡	
16	2‡			2‡	2‡		
17			2§	1/2¶		1¶	2§
18	1/2	1/2		1/2		1/2	1/2

* Time is allowed here to work some problems in class, since management students may need assistance with engineering concepts.
† Extra time is alloted here to permit classroom discussion of one or more of the lengthy problems, such as Probs. 4-5, 5-8, and 7-12, which can be done in teams.
‡ The time allowed here should permit dealing with one or more major problems which integrate material covered previously, either through one of the problems at the end of the chapter or an original project.
§ For the managers' courses, the sections on facility use and road traffic models can be omitted.
¶ In the courses designed for engineers and planners, the carrier operations planning section can be omitted if necessary.

have to be tailored to each course, reflecting students' backgrounds and the overall curriculum, but it is hoped that they will be of help in structuring courses for particular purposes.

Problems of varying difficulty are provided at the end of each chapter. A companion instructor's guide provides solutions to most of these problems and information which should be helpful in the preparation of lectures (including slides) and problem workshop sessions.

The book should also be useful for self-study, not only as an introductory text but also as an aid for the advanced student of transport in reviewing and integrating material. Numerous references for additional study are presented at the end of each chapter to provide a basis for pursuing particular topics further.

In any enterprise as extensive as writing a book, one is influenced and aided by countless others. My own views of the field, those items which are the basic principles as opposed to the details, the important applications, etc., were influenced by my teachers and colleagues of many years, in particular, Kent T. Healy, Cuyler Professor of Transportation (Emeritus) at Yale University, Donald S. Berry, Murphy Professor of Civil Engineering at Northwestern

University, William L. Garrison, Director of the Institute of Transportation Studies of the University of California at Berkeley, Abraham Charnes, Jones Professor of System Science at the University of Texas at Austin, Paul W. Shuldiner, Professor of Civil Engineering at the University of Massachusetts, the latter three formerly of Northwestern University, and Jack E. Snell, formerly of Princeton University but now with the National Bureau of Standards.

I called upon many persons for information and assistance in preparing this book. In particular, I would like to thank Professor Jarir S. Dajani, who used the text in draft form in courses at Stanford University, for innumerable suggestions for improving both the text and the problems, and for catching many small but troublesome errors; and Professor William J. Dunlay of the University of Pennsylvania for assistance with material on facility design; and Professor Marc Gaudry of the University of Montreal for assistance with the concepts of supply functions; and Professor David E. Boyce of the University of Illinois for assistance with the discussion of land value impacts. Many students in two courses at the University of Pennsylvania used the text as it was in preparation, and I thank them for their many helpful comments on the presentation and the problems. I am indebted to three graduate students, Alain H. Coulon and George F. List, who carefully read the manuscript and solved many of the problems, and John A. Warner, who obtained the photographs used. I am also very indebted to Blythe Kropf, who edited and typed the manuscript.

I am also very grateful to The UPS Foundation, which provided the endowment of the Chair which I hold at the University of Pennsylvania, a situation which enabled me to devote the considerable time necessary to prepare this text.

My greatest debt is to my wife, who insisted that I undertake this book and was a constant source of assistance and morale during the many months I spent writing it.

I would appreciate hearing from any reader who has suggestions for improving this text, and hope that any errors found will be called to my attention.

Edward K. Morlok

Introduction to Transportation Engineering and Planning

1

Introduction

The Field of Transportation Engineering

1.

What is transportation engineering? The dictionary defines transportation as "an act, process, or instance of transporting or being transported," and the verb to transport means "to transfer or convey from one place to another" (Webster's New Collegiate Dictionary, 1977, 1242). Engineering is defined as "the application of science and mathematics by which the properties of matter and the sources of energy in nature are made useful to man in structures, machines, products, systems, and processes" (Webster's New Collegiate Dictionary, 1977, 378). Thus transportation engineering is presumably the application of science and mathematics by which the properties of matter and the sources of energy in nature are utilized to convey passengers and goods in a manner which is useful to mankind. While such a definition may have appeal in its apparent simplicity and conciseness, it really says very little about the profession of transportation engineering or the work which transportation engineers are engaged in. It is the purpose of this chapter to define transportation engineering, to convey the richness and diversity of the subject, and to indicate its evolving nature as one of the newest branches of engineering—one that lies at the interface of engineering, regional planning, and economics.

THE FIELDS OF TRANSPORTATION ENGINEERING

Transportation engineering is by no means at the present time a single unified field with a common discipline and method of approach. Rather, it is a collection of many different fields, each with its characteristic problems and approaches, although these are unified by the use of the scientific method and certain basic fundamentals.

Engineers engaged in many different kinds of activities consider themselves transportation engineers. At one extreme, there is the highway and airport pavement engineer and the transportation bridge design engineer, who, because the facilities which they are responsible for designing are used for transportation purposes, would probably be considered members of the association of transportation engineers. Similarly, those who are engaged in the design and construction of canals and locks, railways, port facilities, and major intermodal terminal facilities would also be included within the field of transportation engineering. Also, many engineers whose origins are in mechanical and aeronautical engineering would consider themselves transportation engineers, including those who design motor vehicles, railway locomotives and cars, aircraft, not to mention the marine engineers and naval architects who are concerned with the design of ships. Even those who are concerned with communication among transport units and with the control of the flow of those units would probably consider themselves transportation engineers. All of these engineers share the characteristic that they are primarily concerned with one major component of the physical transportation system, and each has as a fundamental discipline one of the traditional engineering and scientific disciplines.

But this by no means exhausts the field. In addition, there are many professionals whose primary concern is the planning of improvements to urban transportation systems, whether they be new highway facilities, new transit facilities, novel ways of routing traffic through such systems, or simply the rearrangement of transit routes to serve current travel needs better. Many such professionals work for consulting firms in engineering and planning, which may not only be concerned with such regionwide system planning questions but also with detailed facility design and construction problems. Others might work for agencies which usually include the word planning rather than engineering in their title, yet such people often have their major education within engineering, particularly civil engineering. Professionals who work on similar system planning and large-scale engineering design problems work in other transportation contexts, such as state or provincial highway or railroad planning and the planning of new transportation facilities in developing nations. Many other professionals are primarily concerned with more operational aspects of the transportation system, such as the traffic engineer whose job is to regulate road traffic flow, primarily within urban areas.

Thus the field of transportation engineering is an extremely broad one, encompassing many different kinds of professional activity. It knows no geographic bounds, not being limited to either urban or rural areas or to developed or developing nations. Nor is the province of the transportation engineer one which focuses upon a particular type of movement, such as that for person or freight traffic; nor is it limited to any particular mode of transportation. In fact, those who are attempting to push the technological frontiers of methods of transporting persons and things, whether they be concerned with the development of rocket technology to take us to distant planets or with the development of higher-speed intercity passenger trains, would probably consider themselves transportation engineers.

Although there are many ways of attempting to organize and relate these many fields which fall within transportation engineering, a particularly useful one divides the field into two categories. On the one hand, there are transportation system engineers, or system planners, as they are often called, whose primary concern is with the design or planning of entire transportation systems for a region or market. Such engineers and planners look at all modes and technologies of transportation which might be used to provide the transportation service needed in the particular problem area under consideration. They draw from the available technologies as appropriate. Yet they typically do not deal with the specific design or operational features of the individual components of the transportation system, such as the precise location and geometric design of a new road, but rather focus on the interrelationships among the various parts of the system so that it will properly function as a whole. The other category is what might be properly termed transportation system component engineers, professionals who focus upon a particular component of transportation systems for purposes of analysis, design, or perhaps methods and procedures for its use. Included in this latter category would be such engineers as highway geometric design engineers, automotive engineers, and road traffic engineers.

The relationship of these two broad categories of transportation engineers to one another and to other important professional fields in transportation are shown in Fig. 1-1. It is transportation system engineers or planners and transportation economists, transportation lawyers, and probably other professionals who deal with the broad issues of what new transportation facilities ought to be constructed within a particular region, and what are the proper operation and pricing of all transport facilities and the most appropriate roles for various carrier services. ("Carrier" here is used to indicate that the traveler or shipper would not need to provide a vehicle, but rather the object being transported would be picked up in the carrier's vehicle, such as a bus or truck.)

Since the transportation system is an indispensable part of the infrastructure of almost any region, whether urban or rural, developed or underdeveloped, the planning of transportation is closely allied with broader policies for economic and social objectives. Thus, the broad system planning, which is the main domain of the transportation systems engineer, is usually carried out with close ties to those who form public policies for the region under consideration. Thus, in a developing nation, transportation decisions are closely tied to decisions regarding priorities for economic development and social change. In such a context, the transportation system engineer will usually be found working closely with economists, sociologists, and those who form national policies. In an urban area, the transportation programs are typically closely related to plans for land development, in particular with reference to where people will live relative to where they will work, shop, and amuse themselves. Here the skills of the city planner and lawyer are most often coupled to those of the engineer.

Given the overall system design for new transportation facilities in a developing country or plans for major additions to an existing transportation system elsewhere, the transportation system component engineers' work is in the development, design, production, and use of the components of the system with which they are associated. For example, in an urban area, once plans for particular new transit facilities have been established and accepted at appropriate public and private levels, then it is the job of the transit engineer to design those particular facilities, such as a new rapid transit line. Similarly, those concerned with the design of transit vehicles engage in the design and production of an appropriate number of units, and those involved in the actual operation and maintenance of the system begin formulating detailed plans including precise time tables or schedules for operations, crew rosters, etc. Of

Figure 1-1 The fields of transportation engineering.

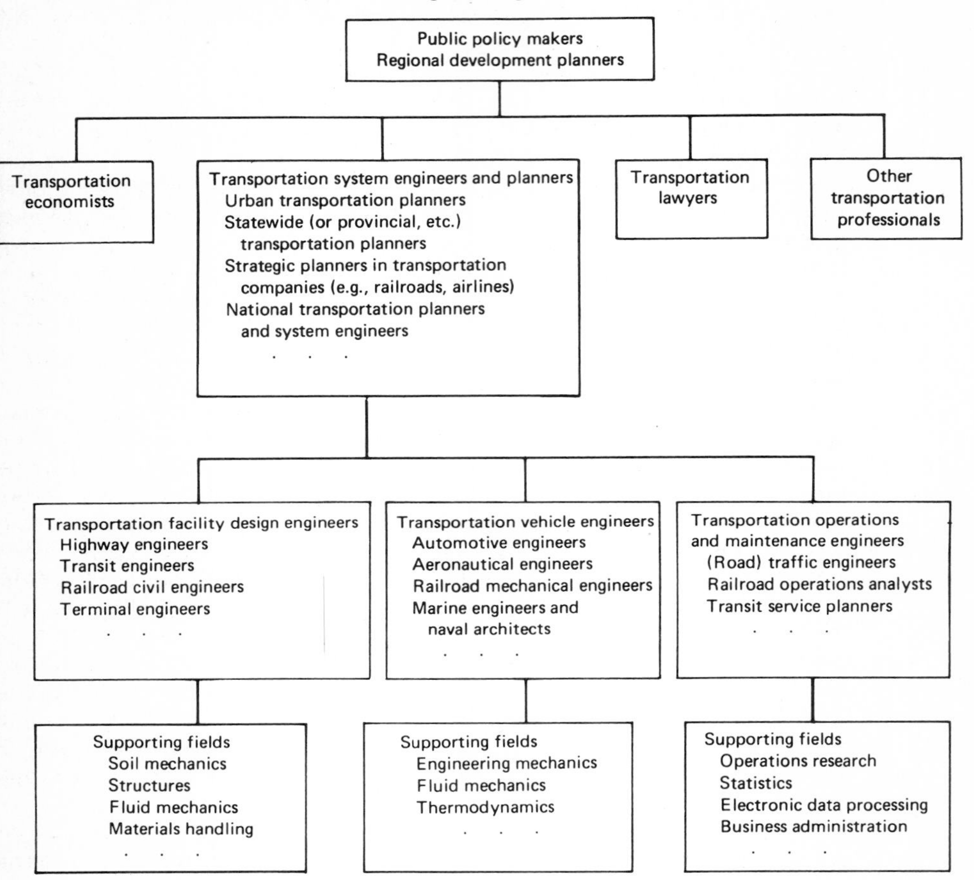

course, many transportation components are so widely used that there is no need for special designs for each particular application. Although individualized designs have traditionally been undertaken in fields such as rail rapid transit and to some extent marine vessels, other transportation vehicles, such as main-line railway locomotives and cars, motor vehicles, and aircraft, are designed for large worldwide markets, and only the selection of number of units to produce and minor design options are dependent upon the particular application.

As shown in Fig. 1-1, the various transportation system component engineering fields are very closely related to what might be considered the more traditional engineering and management fields. For example, highway design engineers draw very heavily on soil mechanics and structural engineering for the methods by which highway pavements and bridges are designed. Similarly, the designers of transportation vehicles typically draw very heavily upon fields which often are associated with mechanical engineering; and those concerned with operations and maintenance draw heavily from some of the more recently developed disciplines such as operations research, systems engineering, and statistics.

Before leaving this section, we will make two more important points. The first is that in our structuring of the various fields of transportation engineering it may appear as though we have neglected a potentially very important component: that of developing new technologies or means of transportation. Here we are not referring to the development of new components which are necessarily compatible with other existing components, for example, the development of an entirely new means of propelling railway trains (such as the linear induction motor), but rather the development of entirely new means of transportation, such as the aircraft–air-control–airport system, which represents an entirely new technology developed within this century. An engineer concerned with the development of such a new form of transportation would necessarily be concerned with questions of facility design, with questions of vehicle design if the system were to involve a vehicle (remember that pipelines and conveyor belts, to mention two, do not typically involve vehicles), and with the control, operation, and maintenance of the system. The interests of such a transportation engineer presumably would cut across the three major fields comprising transportation systems component engineering. He would differ from a transportation system engineer because he would be concerned with developing a new technology of transportation, presumably, largely apart from any particular application. Rather, he would attempt to develop a technology which would be useful in many different sites or locations. Particular applications would, in turn, be planned by transportation system engineers. However, partly because there seems to be so little activity in this area, and because the engineer with such an orientation would have to use many of the methods of the transportation system engineer as they will be outlined in this text, we have chosen not to designate transportation system engineers oriented to such technological development as a separate category.

It is perhaps surprising that in a period of apparently rapid technological advancement in many fields, there is relatively little activity in the development of entirely new transportation technologies. Perhaps the only consciously planned effort in this regard is that concerned with the development of technologies for interplanetary space travel. One might also consider the attempt to develop so-called personal rapid transit systems for urban areas an example of a new technology. However, these systems of small vehicles, operating automatically on exclusive guideways, seem for the most part to use highly advanced and modified forms of conventional rail rapid transit rather than a new technology.

The second major point is that even though most of our discussion has focused upon engineering work at the systems level in the public sector, we do not mean to indicate that it is only performed in the public sector. For many reasons, some of which may no longer be valid, transportation systems in most nations are divided along modal lines, with the entities controlling railways, highway facilities, truck lines, airlines, etc., being independent of one another. There are but a few exceptions: for example, the Canadian Pacific Railway operates railroad, truck, airline, and steamship services, and the White Pass and Yukon Railway operates a highly integrated freight transportation service involving containerized movements by water, over a rail line, with truck services acting as feeders to these, as well as passenger train and water services. Even in nations where almost all of the means of transport are government-owned, the traditional modal division of responsibility usually is perpetuated. Given such a division, the main demand for transportation engineers in the private sector would naturally be for engineers who are primarily concerned with the mode of interest to the particular firm. Hence the need is primarily for a transportation system component engineer of an appropriate type. Thus there is a high demand for civil and mechanical engineers within railways, along with railroad operations analysts. Yet there are many instances in which transportation systems engineers will be needed by the private sector. For example, a firm may be called upon to connect a region without adequate transportation with the remainder of a nation's transportation system, perhaps for making natural resources available. The development of transport connections to bring ore deposits in Labrador (Canada) to the St. Lawrence Seaway required the services of transportation system engineers.

Thus there is a pronounced need for transportation system engineers or planners in government as well as in private industry. This need exists wherever broad transportation problems are being addressed, whether in an urban or rural context, whether the problem involves freight or passenger traffic, regardless of the modes of transportation involved. In the next section we shall attempt to define this field in more detail and to discuss some of its basic principles.

TRANSPORTATION SYSTEM ENGINEERING

The field of transportation system engineering is in reality a relatively new field. The major advances have been made within the past 30 years. The objective of transportation system engineering is to devise the most nearly optimum combination of all transportation facilities and methods of their operation within a given region. If the region were defined rather narrowly and were of limited size, such as within a single shopping center or a single-purpose movement from mine to processing plant, then the problem would not be very difficult. Instead, real transportation problems typically treat large geographic areas and must deal with freight movements as well as person movements of all types and purposes. A typical transportation system engineering problem would be to plan future transportation facilities for an entire metropolitan area or intercity transportation system improvements for an entire state, province, or even nation.

There are basically three characteristics of most transportation system engineering and planning problems which make them difficult to treat and which lead to the requirement for a systematic basis for dealing with them (Lang and Wohl, 1959): (1) The regions typically treated include thousands if not millions of individual person trips and a like number of shipments of freight. (2) Given the large number of available transportation technologies and the different ways in which they may be operated or their use regulated and priced, there exist an almost uncountable number of ways of changing the transportation system within any particular region. (3) The objectives to be achieved by improvements to transportation systems are manifold, often very difficult to measure, and typically not comprehended by such simple notions as minimize the time persons spend in travel. These characteristics, which will be elaborated upon below, all require a systematic and rational basis for planning, designing, and implementing improvements to transportation systems.

The System Planning or Design Process

There has evolved over the past three decades an approach to the design of complex systems which is widely, almost uniformly, used in transportation system engineering. This process has often been called the system planning or system design process. While there are many variations in the name, definition, and number of steps involved in this process, the basic approach is essentially the same. One representation of the major features of this approach follows.

The basic steps in the system planning process seem to be the following: (1) definition of the problem, (2) statement of requirements or objectives to be achieved by the improved design or plan, (3) specification of alternative solutions or improvements to the system, (4) evaluation of the alternative solutions in their meeting the objectives or requirements, and finally (5) selection of the best plan of action. These are depicted in Fig. 1-2 and will be discussed in turn.

Perhaps no part of the system planning process is more important than that of attempting to define the problem and hence specify the requirements or objectives for a solution to this problem. This step really determines to a large extent the nature of the work which will follow in each of the ensuing steps. There are many approaches to defining the problem and specifying requirements for their solution. Within the context of transportation, particularly in developed nations, the problem often is noticed first by users of the system who are dissatisfied with its performance; they expect a higher quality of

Figure 1-2 **The system planning or design process.**

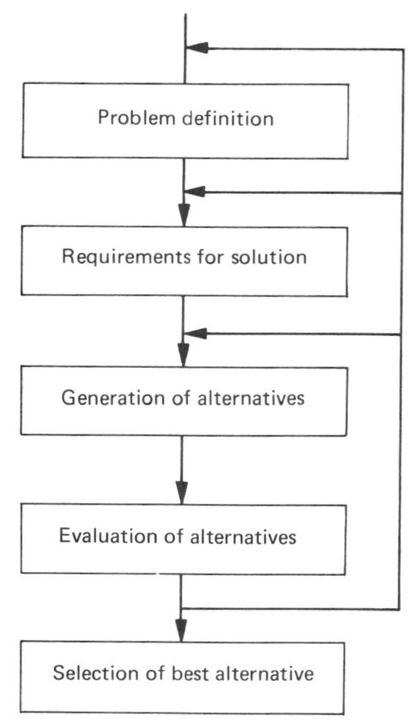

transportation service than they are receiving. Often their criticism carries with it an implicit assumption regarding what changes in the system might make it perform better, such as providing a new expressway or rapid transit line. Another and closely related source of problem awareness is that of a professional who may realize that new technology or improved organization or operation of the system may provide an improvement from the standpoint of various affected groups. Another important source of problem specification may come from potential users of the system, who require new transportation facilities or services in order to take advantage of opportunities for economic or other development. This often happens in developing nations, where valuable resources have been discovered but cannot be extracted and moved to potential points of processing or consumption because of inadequate transportation.

This source of problems reveals an important characteristic of transportation which will appear many times in our work; that is, that transportation is one element of society's process of converting natural resources into things and activities which are of value to humanity. From this stems the concept of transportation as a "derived demand," a service which is carried out because it is desirable to move an object or a person from one place to another to enable some activity at the destination to be engaged in.

The technology of transportation, like our technologies for producing most things of value, has the often undesirable characteristic that it not only produces the desired movements of persons or goods, but also can produce other outputs which are undesirable and sometimes dangerous, such as noise, air pollution, and pollution of ground water. Very recently it has been discovered that such pollutants are reaching levels which are often objectionable and in some cases physically harmful. As a result, changes in the technology of transportation or the way in which particular parts of the system are to be used must be made in order to reduce these undesirable outputs. Such mismatches between the transportation system and its environment, for essentially non-transport reasons, will probably dictate changes in the transportation system to an increasing extent in the future.

Once the problem has been identified, then some means of expressing the direction and perhaps magnitude of the desired changes must be made. These will provide a guideline to the search for ways of improving the system and the evaluation of these alternatives. The usual terms which are used to express such statements of characteristics of desired changes are words such as requirements, goals, objectives, figures of merit, and measures of effectiveness. While there are important differences among these different terms and some may be appropriate in particular situations and not in others, the important facet of all of these is that they identify the desired directions of system change. An extremely important characteristic of these is that they should reflect the desired condition of the system and not be so specific as to exclude some alternatives which would achieve the desired objective but not the overly narrowly stated objective.

For example, in a particular urban situation congestion may be one of the prime problems to be alleviated through a system planning effort. In such a context, some might say that an appropriate objective would be to provide rapid transit service in major corridors, so as to attract people from their automobiles and roads onto the transit system for the journey downtown. Regardless of how such a radial system might be designed and operated, the level of traffic congestion may not be significantly reduced, for the system would be designed primarily for central business district trips, which typically account for less than 20 percent of person trips within large North American metropolitan areas. In fact, one of the major contributions of the early urban transportation studies seems to be that any major improvements designed to reduce congestion would have to concentrate upon nonradial trips (i.e., new freeways should ring cities or provide a grid network rather than a radial network) if they were to be at all effective.

At this stage also any limitations on the alternatives which might be considered feasible or practicable might be introduced into the analysis. For example, in many situations there is an upper limit on the total funds available for new facility construction, such as budget limitations placed by local or national governments. Similarly, requirements related to the degradation of the environment would appropriately be introduced at this point, such as maximum allowable noise levels.

The search for or identification of alternative designs is obviously one of the most important steps in the process, but ironically it is one of the least understood. Modern transportation technology is such that the engineer has a virtually infinite range of alternative designs which might be considered, all of which, at least in principle, should be explored before the best design can be chosen. But engineers are also faced with practical limitations: finite time periods for projects and limited funds to spend. Thus typically they must rely heavily on their intuitive judgments, perhaps on approximate analytical techniques, and hopefully on very structured processes to guide them in selecting the most reasonable set of alternatives.

An important aspect of alternatives within many if not most transportation system engineering problems is that the range of alternatives is extremely large. On the one hand, there are the types of alternatives which engineers—typically civil engineers—gravitate toward: constructing new facilities, reconstructing and perhaps expanding the capacity or improving the quality of service of existing facilities. However, such construction-oriented alternatives by no means exhaust the range. Important transportation alternatives exist in the manner in which the system is operated and the way the system is priced. For example, the provision of high-speed public transportation service in urban areas might be achieved by constructing an entirely new rapid transit line, with attendant high initial cost for land taking and construction. At the same time, similar high-speed service might be provided by simply devoting a lane of an existing freeway to the exclusive use of buses which could be operated in express services at speeds often comparable to or better than those of more typical rapid transit lines. Of course, taking a lane from general use in a freeway is likely to increase congestion on the remaining lanes, unless sufficient people are attracted to the rapid transit line to make the volume per lane even less than it was before the transit service was offered. The evaluation of such an alternative clearly raises many difficult questions, such as the extent of the attraction of automobile drivers to alternative forms of public transit service and the distribution of time savings and possible increased travel time among those affected by such changes. And, of course, this discussion has neglected the possibility of increasing the price of road travel by automobile (such as through increased parking charges or direct road tolls) to induce people to shift from private to public transportation.

The evaluation of alternatives is another critical step in the process. Among the many possible ways, one approach would be to rely primarily upon the judgment of the engineers and planners and other professionals involved in the design activity. On the basis of their experience and knowledge of transportation systems and the interaction of such systems with their environment, these professionals would recommend a particular course of action. This approach is rarely, if ever, used at the system planning level. The reason is simply that the interaction among the various parts of the system and the relationship between design changes and the achievement of the objectives is so complex that human intuition is not a very good guide. It would take a very egotistical engineer to feel that he could specify the best design of a large system by using no more than his own judgment and intuition. In fact, in many planning and system design contexts there are explicit requirements for the use of formal, largely quantitative, design procedures, as in the case of urban transportation planning in the United States.

Another approach is that of experimenting with the real world system. This might be an ideal manner in which to design systems, in that the consequences of various designs would presumably be known with much greater precision after actual implementation than by any other means. However, many types of transportation system improvements do not lend themselves to experimentation, because the cost of instituting significant changes is great and these usually cannot be easily modified or adapted if they do not turn out to be particularly effective. Paramount among these are major new facilities, which typically are extremely costly and from which very little can be salvaged if they are to be abandoned or modified. However, experimentation seems to be very attractive for evaluating alternatives which can be easily modified. Included among these are various pricing options and changes in the regulations for use of the system. Thus, for example, experimenting with one-way streets or changed transit fares is quite possible and has been done on numerous occasions. Similarly, new transit routes or innovative freight services such as high-speed freight trains can be implemented easily and dropped or modified if they do not achieve the desired benefits.

Another possible approach is attempting to ascertain what would happen if a particular alternative were implemented in one place by analogy to another place where the alternative has already been tried out. In principle, this is a reasonable approach, but for many large-scale system design problems, there really is no effective analogy, either because conditions of the area under consideration are unique or because the kind of alternative envisioned has not been tried out in the same form before. However, much of what is commonly termed professional judgment surely is developed by analogy with other situations, and hence this approach is widely used under the name of professional opinion.

There is, however, another more robust means of transferring knowledge gained in one situation to a new situation. This occurs by means of a model of the system under consideration. In some situations, the model may be an actual physical model, as in the case of experimenting with different configurations of aircraft fuselage in a wind tunnel to find those with the most desirable properties. Such small-scale physical models are of course related to the actual aircraft by means of aerodynamic theory, which enables a quantitative extrapolation or extension of characteristics of the small-scale physical model to the actual full-size aircraft. Other examples include the improved design of railroad wheels through experimentation on model railway wheels. The basic approach here is to find relationships in the physical model which will mirror similar relationships in the actual system.

A conceptually similar approach is used in what is termed mathematical modeling, which is by far the most commonly used approach to predicting the effects of various transportation improvements in system planning. Mathematical models consist of relationships between various quantified characteristics of the actual transportation system. Such mathematical models are conceptually identical to the mathematical models which are the common tool and trademark of the engineer and designer. Examples include relationships between the means of support of a beam, the load placed on that beam, and the stress in the beam, which can be extended to use in design problems by varying the material and design of the beam itself. Similarly, in transportation system design problems, mathematical relationships are sought which will help explain the use of various parts of the system. These include relationships between the amount of travel and the location of activities between which people might wish to travel, such as from residences to work places, and characteristics of the transportation service offered, such as the price, travel time, and frequency of departures on a public transit route.

The main difference between traditional engineering mathematical models and those which are commonly used in transportation system planning is that the former deal with inanimate objects, objects which can easily be experimented with in the laboratory, while those in the latter deal essentially with socioeconomic as well as technological systems, and hence laboratory experimentation in the usual sense is not possible. Also, the variability of human behavior enters into many aspects of transportation, from the control of individual motor vehicles to choices regarding the places to which trips will be made, their frequency, and where people live and work. For these reasons and many others the mathematical modeling of transportation systems presents many difficulties which do not appear in the treatment of inanimate systems. Nevertheless, mathematical modeling has been found to be the only effective approach in many contexts to predicting the characteristics of possible alternative transportation systems for purposes of evaluation, and hence it is the most widely used approach.

It is important to point out at this juncture that mathematical modeling or other quantitative treatment of the factors involved is not the only approach to be used in transportation system planning. Many factors are extremely difficult to quantify, and in some cases no effective quantification has been accomplished, and perhaps will not be for many decades, if ever. An example of these are aesthetic properties of transportation facilities. The discussion above is not meant to imply by any means that factors which are not quantifiable are either unimportant or that it is proper to ignore them. It would be folly to think that better decisions are made by considering only those things which can be modeled or conveniently modeled quantitatively; rather quite the reverse is the case. Many factors which are important in transportation system design cannot be quantified and thus must be treated by means other than mathematical modeling. Yet they must be considered if rational decisions regarding what actions to take are to be reached.

To many observers, the "freeway revolt," which resulted in the cancellation of many plans for freeway construction within densely builtup urban areas was really caused by engineers planning facilities which communities would not accept. The reasoning is that the planning was conducted considering only those factors which could be easily quantified, while ignoring such factors as the social disruption of the community which would occur as a result of cutting a transportation system facility through the middle of it. Even air pollution, traffic noise, and possible increases of traffic on streets leading to and from freeway interchanges were largely ignored. Had engineers and planners considered these many factors in their original planning, perhaps such facilities would not have been planned, or alternatives which might have achieved the same transportation objectives while possessing acceptable community impacts might have been called for. While there certainly are many other reasons why extensive freeway construction in the heart of North American metropolitan areas has been terminated, the lesson is certainly a sobering one for those who would either desire to limit analyses and evaluation to only those factors which are easily quantified or to ignore for purposes of evaluation the viewpoint of many of those who will be affected by transportation decisions.

The next step is attempting to select the best among the alternatives. Because transportation projects of a large scale usually affect many different groups which may have conflicting objectives with respect to transportation system improvement, such decisions are ultimately made within a political process, either within the public sector or at the higher levels of management if the project is within the private sector.

A fundamental problem which makes such a choice difficult is that transportation facilities typically have very different impacts upon different groups, these impacts leading to what might be considered a net gain for some groups and a net loss for other. For example, when a new intercity transportation link is built in a developing nation in order to open up a previously undeveloped region, those who will live in that region or are involved in its economic development presumably will benefit from the project. On the other hand, some of the economic activity which might otherwise develop in other parts of the country probably will be shifted to this newly opened region. This in effect reduces the real income of those living in other parts of the nation. Also, by investing in a particular project in a particular location, funds which otherwise might be available for other projects are necessarily consumed, thus precluding opportunities for use of these funds in other ways.

Similar examples can be drawn even more easily within the context of developed nations, particularly urban areas, where the construction of a new freeway from center city to suburb may benefit those who travel long distances to the central business district (CBD) to work, but in the long run it may induce people to continue the outward migration from the central city to the distant suburbs, thereby furthering the decline of the central city as a residential area. For all these reasons, the selection of a best plan of action is extremely difficult and must involve extensive interaction between the technical-professional staff in a systems engineering study and others who are representative of higher-level decision-making units. This gives rise to the development, for such interaction, of complex processes which are still among the most important research areas in system planning and still under extensive development and refinement.

In actual use, the process of system planning is not a simple linear one commencing with the problem definition step and proceeding sequentially to the final selection. Rather, the steps are repeated (in sequence) many times. Such iteration, as it is often called, is indicated by the reverse direction arrows in

Fig. 1-2. Typically, a study will begin with an initial definition of the problem and then continue to the various phases: generating some alternatives, evaluating them, and attempting to at least approximately rank them. At this point a great deal has been learned about the problem through the process of planning. This knowledge can then be used to refine the statement of the problem. Often aspects of the problem which were largely unknown or ignored are brought to light, and perhaps some aspects which were thought to be important turn out to be unimportant. Then perhaps the institution within which the planning is being carried out (such as in urban government units or national planning units) will be able to refine further its statements of requirements and constraints on the final solution. This all then typically leads to modification and further generation of alternatives, which now can be based on greater knowledge of the system, knowledge gained from the evaluation of the first set of alternatives. The process continues, typically cycling many times through the many phases until, in the judgment of all professionals involved in the process, it is logical to terminate the process because the best alternative identified up to that point is likely to be as good as any which could be generated by further work.

The process involves iteration for another reason. In reality the process never logically terminates. What we are saying is that there really never can be a final plan for the future transportation system which is accepted and then does not become changed until that system is fully implemented. Typically, most major new transportation construction projects take one to three decades from first planning to actual implementation and opening for use. During that period, many changes occur in the requirements for transportation in the area, in the opportunities which might be taken advantage of, even in the technology of transportation which might be used. Also, the values of the people and their priorities for transportation service are likely to change. All of these factors combine to make it essential that the planning process be a continuing one, in which the system is monitored and changes taken into account as possible bases for changing the transportation plan. The plan developed at any one point in time for final implementation many decades into the future is likely never to be implemented as originally planned, but will be continuously subject to modification and refinement as time passes.

The Scientific Method

Engineering is differentiated from other disciplines by many factors, including the orientation toward planning and design of facilities and ways of using them which will actually be implemented. Another extremely important characteristic of engineering is that it relies on the scientific method for the principles which guide the discovery and development of the relationships and standards to be used in engineering work.

In almost any field, the engineer must use relationships between variables in order to carry out his work. This is true whether the work be the determination of the thickness of a steal beam needed to carry a load safely or the determination of the number of lanes of a freeway needed to accommodate future traffic. The estimation of the load capacity of a beam must be based upon knowledge of the strength properties of steel. Such knowledge can only be obtained by observing the effect of loading beams, their deflections, and the loads at which permanent deformation or failure occurs. From such observation a theory which explains the deflections, etc., can be developed. In the case of beams, this has been done.

A rectangular beam of length l, height h, and width b simply supported at each end and subjected to a load P at its center will experience a maximum bending moment

$$M_{\max} = \frac{Pl}{4}$$

There is a maximum safe working stress $\sigma_{\max}$ related to $M_{\max}$ by the section modulus Z:

$$\sigma_{\max} = \frac{M_{\max}}{Z}$$

and for this beam

$$Z = \frac{bh^2}{6}$$

Rearranging these equations, we relate P to $\sigma_{\max}$ by

$$P = \frac{2bh^2}{3l} \sigma_{\max}$$

Therefore, if the working stress cannot exceed a value of $\sigma^*_{\max}$, the load P must not exceed

$$P \leqslant \frac{2bh^2}{3l} \sigma^*_{\max} \tag{1-1}$$

This relationship is universally accepted and used in the design of structures in which failure of a beam could bring about serious loss of life or injury and loss of property. Thus it is essential that the relationship accurately represent the strength of beams. In short, this mathematical model of a beam must be a true representation of its characteristics, within some acceptable range of error.

The scientific method provides a procedure for testing the accuracy of such relationships or mathematical models. The first step is to formulate a theory, or model, such as Eq. (1-1). This model must be created by an engineer or scientist, based upon familiarity with beams and the observed effect of loads. Accepted theories about related phenomena might provide guidance: knowledge of the load capacity of a cantilever beam would help here. The search for common patterns of deflection may be facilitated by a carefully planned series of experiments in a laboratory, trying various loads, sizes of beams, etc.

Once a possible theory or relationship has been identified or created, it is treated as an hypothesis. It is tested by comparing the predictions of deflection by the model with the measured actual deflections for identical loads. If the correspondence is close, then the theory may be accepted, as shown in Fig. 1-3. Alternatively, the hypothesis may not predict well, thus requiring reformulation of the hypothesis. Since there will always be unavoidable errors in measurements, the formal methods of statistics are generally used to assist in the test or comparison of predicted and actual values, although the methods of statistics are really the subject of another course.

In order to ensure there was nothing in the initial observations which by chance made an untrue hypothesis appear valid, at least one more test should

Figure 1-3 The scientific approach to model development.

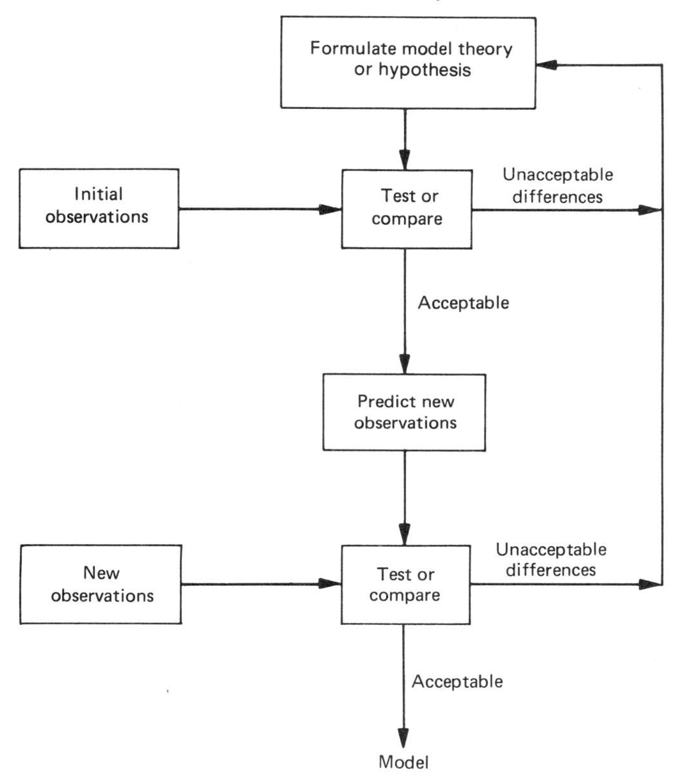

be performed. As shown in Fig. 1-3, this should be with an entirely new set of observations. The predictions of the model are compared with these new data in order to check the model. If the model passes this test and possible additional ones, then it is reasonable to accept it for use. In actual practice, models or relationships (sometimes called "laws" in physical sciences) are continuously subject to testing in use, which often yields ways of refining the models to be better predictors.

Turning back to our traffic problem, the same basic approach is used. In the 1920s and 1930s, there were no methods to estimate the capacity of roads. Engineers hypothesized linear relationships between the speed of traffic on a road (the average speed of vehicles) and the amount of traffic on the road as measured by the density of vehicles per mile of road, of the form shown in Fig. 1-4. Mathematically, this relationship is

$$u = A - Bk \qquad (1\text{-}2)$$

where u = mean velocity of vehicles, mi/h
k = density of vehicles, vehicles/mi
A, B = empirically determined parameters

Given suitable data on actual velocities and densities, the parameters A and B can be estimated.

This linear relationship can be manipulated easily to yield the relationship between the mean speed of vehicles and the amount of traffic passing a point along the road per unit time, this flow rate of vehicles being defined as the volume of traffic. The volume of a steadily moving stream is simply the product of the density and the speed, and hence

$$q = ku = Ak - Bk^2 \tag{1-3}$$

where q = vehicle volume, vehicles/h

Perhaps a more easily understood form of the same relationship is that which expresses speed as a function of traffic volume, this being

$$q = ku = \left(\frac{u - A}{-B}\right) u = \frac{A}{B} u - \frac{1}{B} u^2 \tag{1-4}$$

Equations (1-3) and (1-4) are also graphed in Fig. 1-4.

It can be seen that, starting from a very low volume of traffic traveling at high speed, the speed initially will not drop appreciably as volume increases. But after a certain point is reached, speed drops markedly, as we have all experienced on crowded roads. There is a speed at which the volume or traffic throughput is a maximum; this speed, $A/2$, yields a volume of $A^2/4B$. This volume presumably is the maximum capacity of the road, and could be used for design purposes provided the speed is satisfactory. If not, the relevant capacity is less.

In the 1930s the linear speed–density relationship for road traffic was subjected to considerable testing, in particular, by Greenshields as described in his 1934 paper. It has been widely accepted for traffic flow on two-lane rural roads provided the vehicle density is above about 10 vehicles/mi. As would be expected, the parameters A and B depend upon the road design, including the lane width, grades, curvature, and speed limits. (The speed intercept for zero volume is usually slightly larger than the speed limit.) The comparison of the theoretical relationship with the original data collected by Greenshields is shown in Fig. 1-5. The fit is reasonably good; the relationships were widely used for highway capacity studies.

Relationships such as these for road traffic must be continually checked or tested, however, because they may not remain stable (that is, the parameters or mathematical form of the relationship might change) over time. This is particularly true whenever human behavior is involved and whenever important components of the system being modeled change, as when vehicles become larger and more powerful and roads must be built to higher design standards for width, grades, etc. This linear relationship has been continually tested, and for some road types other mathematical forms for the speed-density relationship are used. As recently as 1968 (see Drake et al., 1967) the linear form and others were tested on data based on a sample of vehicles traveling on the Eisenhower Expressway in Chicago.

We return to the specific case of the two-lane rural roads of the type Greenshields and others studied in the early 1930s. One of their relationships—on a road with 10-ft lanes, based on observations in which the percentage of trucks was very small, about 1 percent, was

$$u = 43.8 - 0.221k$$

Figure 1-4 **Mathematical model of road traffic flow. *(a)* Speed vs. density; *(b)* volume vs. density; *(c)* speed vs. volume.**

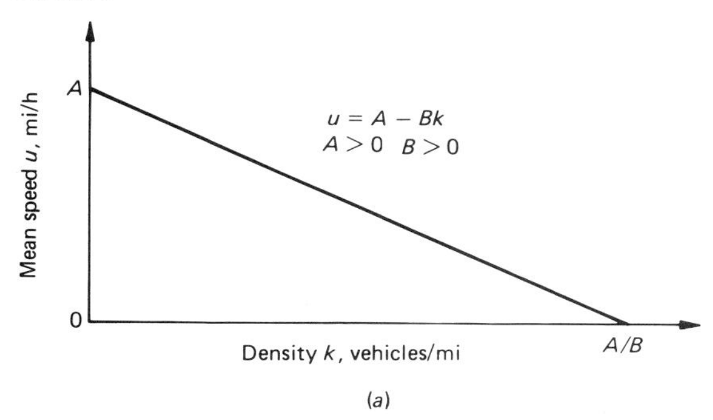

(*a*)

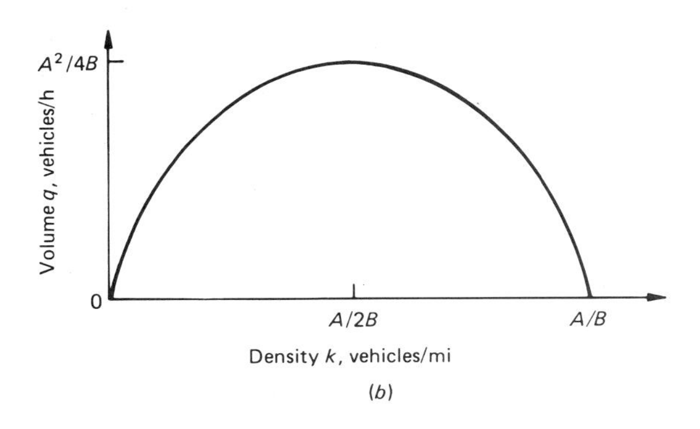

(*b*)

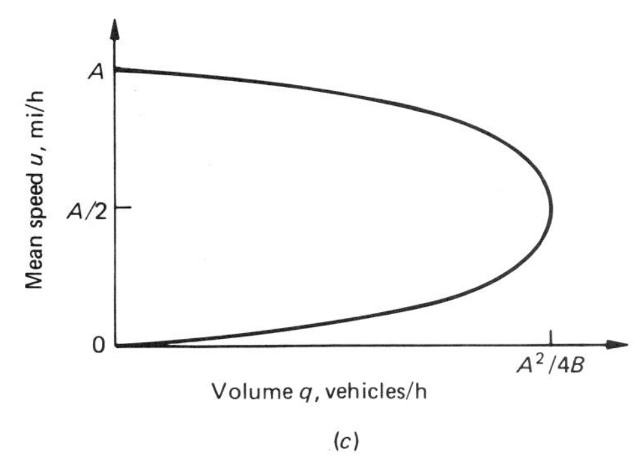

(*c*)

where k is the density of traffic in one lane. This relationship has been superseded. The relationship contained in the 1965 *Highway Capacity Manual* (fitted to the curve appearing in Fig. 10.2b on p. 310), for uninterrupted flow on a two-lane rural road, with a mean free speed of 60 mi/h, a sight distance of 1500 ft, 10-ft lanes with large side clearance, and no trucks or buses, is

$$u = 60 - 0.292k$$

Figure 1-5 **Comparison of linear speed-density curve and Greenshields' road flow data of 1934.** [***Adapted from Greenshields (1934), p. 468.***]

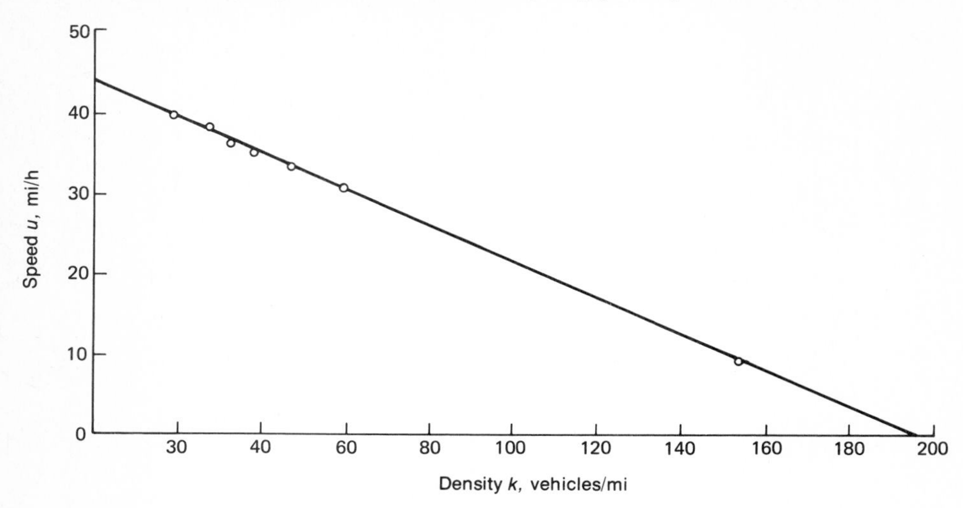

Notice that the mean free speed has increased from 43.8 to 60 mi/h. Also, the slope ($-B$) decreased slightly, revealing that the reduction of speed due to the presence of additional vehicles is slightly greater now than it was in the 1930s. The exact origin of these changes is unknown, but probably reflect a combination of factors, including better-performance motor vehicles (i.e., higher possible speeds and acceleration and deceleration rates), longer vehicles, wider and smoother roads, and changes in drivers' perceptions of safe distances between vehicles.

It is clear from this example that use of the 1934 relationship for speed and traffic flow would yield substantial errors if applied 30 years later. Continual testing and modification as appropriate is much more critical in the case of relationships involving human behavior than those not, for materials such as steel usually do not change in structural properties, but people, their habits and tastes, certainly do.

Thus, the scientific method is at the core of engineering and planning. It provides a guide for developing the models or relationships which are necessary for any quantitative approach to planning and design. Without it or a similar guide, differentiating between what is known or can be accepted based upon prior experience and what is purely conjecture would not be possible.

Science and Professional Judgment

The discussion above is not meant to imply that engineering is composed solely of science; that is far from the truth. Engineering has been defined as

The art of the practical application of scientific and empirical knowledge to the design and production of or accomplishment of various sorts of constructive projects, machines, and materials of use or value to man. (Sarton, 1937, p. 52)

In such application of scientific knowledge, the engineer often must use judgment in order to decide on important options. Examples include (1) decisions on the best materials to use in a particular context, where many may be almost equally good, and (2) which of many different estimates of future traffic to use in the design of a new transport facility, where that future traffic is unknown. In such situations, the engineer will attempt to amass all that is known about the problem and the consequences of selecting any one of the various alternatives, but the final decision must be based upon judgment. Hopefully that judgment will be tempered by experience, his or hers as well as the experience of others who are consulted. Also, it is essential that the information obtained, through the use of whatever engineering methods are applicable, is the best that can be obtained, consistent with the urgency of a decision and the significance of the problem.

In fact, one of the purposes of this book is to convey the methods used by transportation engineers in the solving of typical problems which they face, so that one is capable of using the best available methods in approaching any particular problem. These principles and methods should serve engineers as an indispensable aid in dealing with any problem, giving them insight into the problem which would otherwise not be gained. These methods should identify the factors which should be considered, their importance, and their relationship to the various alternative designs or other options available. Also, they should help predict many, if not all, of the consequences of each alternative, so that the alternatives may be compared to one another and the best selected. As emphasized earlier, these quantitative analyses typically will not treat all of the important aspects of a problem, and therefore they will have to be supplemented by additional information, in order that a balanced and complete portrayal of the alternatives is given. But it is essential that the distinction between that which is deduced from application of scientific quantitative principles and methods and that which is based solely on judgment be maintained, for it is the former which are known to follow from the application of specified principles and assumptions, while the latter are truly opinions. Whether these are in conflict or point to the same course of action, which view is to be given more weight in any final decision and related questions all depend upon the particular context.

That professional judgment can be wrong, just as quantitative models and scientific statements can be misleading or wrong (although the methods of science described above are intended to ensure, as far as possible, that this does not occur), is illustrated by the following two notes from history. Early in 1825, Nicholas Wood, one of the pioneer railroad engineers in England, wrote of railway trains propelled by steam locomotives (instead of horses):

It is far from my wish to promulgate to the world that the ridiculous expectations, or rather professions, of the enthusiastic speculist will be realized, and that we shall see them travelling at the rate of 12, 16, 18, or 20 miles an hour: nothing could do more harm towards their adoption, or general improvement, than the promulgation of such nonsense. (Kirby et al., 1956, p. 277)

Later that year, on October 27, George Stephenson's steam locomotive, named *Locomotion No. 1,* hauled a load of 90 tons, made up of 36 small cars or wagons, carrying about 600 passengers and a small amount of freight, at a sustained speed of 12 mi/h along the Stockton and Darlington Railway. Needless to mention, when Wood published a second edition of his book in 1831, he omitted the previously quoted pronouncement!

Nor were such speculations limited to practicing engineers. In 1835, after steam propulsion had proven itself on the railways, a Fellow of the prestigious Royal Society in England and a popular lecturer on science, Dionysius Lardner, stated (Kirby et al., 1956, p. 259, after Lindsay, 1876, vol. 4, pp. 168–170) "that a voyage on a ship propelled by steam from New York to Liverpool was 'perfectly chimeral'; they might as well 'talk of making a voyage from New York to the moon.'" Regular steamship service linking Britain with the United States became a reality only 10 to 15 years after this statement. And humans began walking on the moon about 140 years later!

There are many ways of approaching problems and attempting to structure and solve them. Engineers have typically drawn on almost all approaches, those which seem to produce useful results being retained and developed further. In addition to the singular characteristic of getting the job done, providing an effective and practical solution to the problem being addressed, the engineering approach can also be characterized as one in which as much of the problem as possible is treated precisely and in quantitative terms which admit to testing of assumptions and relationships. The reason for this is eloquently stated in a summary of what was probably the first organized, concerted multidisciplinary team effort to develop new transport technology. The purpose was not to transport people to the moon, but to design a better public transportation vehicle for urban areas, a new streetcar in 1931. The electric railways of that period pooled their resources to support the research needed to design a faster, more comfortable, and cheaper streetcar, through an organization entitled The Electric Railway Presidents' Conference Committee. As stated by Dr. Thomas Conway, Jr., chairman of the ERPCC

Two courses of procedure were open to the Committee—the first was, by use of empirical methods to endeavor to design a car which, from the viewpoint of the railway executive, is better than any car now available. The second method was to approach the task in a less spectacular but more thorough manner—to apply to this important technical problem the research method, which has produced such remarkable results in many other industries. The Committee unanimously chose the latter course, believing the chances of success were much greater by the use of the scientific method of attack.

The past year has been devoted to the task of planning the details of an elaborate scientific investigation; the recruiting of a competent staff of engineers and technicians; the selection and preparation of the theater in which this research program was to be staged; the mobilization of a representative group of the most modern urban electric railway cars, borrowed for these tests; and the solution of many difficult scientific problems which had to be found before the actual work of test and trial might proceed.

It is an old axiom of research that if a desired result can be accurately and precisely stated in the language of the engineer and the technician, half the battle of attaining that result has been won. We hope and believe that our present fact-finding activities will enable us accurately and clearly to state to what extent each element of the car and of its performance must be improved, and to point out in detail the exact direction which the improvement must take.

When this stage of the Committee's work has been completed—and a number of months will pass before that can be done—we are ready to pass to the second stage of the work, which is, through cooperation with the manufacturers, and indeed with all who may contribute to the desired end, to endeavor to evolve better apparatus and equipment and a railway car which will more nearly meet the exacting requirements of this modern age.

The scientific method permits no short cuts. It is like the siege of a great fortress which goes on grimly and remorselessly day and night, over a long period of time, the victory being attained, if at all, as the culmination of a carefully planned and executed campaign. (Hirshfeld, 1975 reprint)

The effort was a success. The streetcar so designed was a vast improvement over earlier vehicles. The design was used throughout the world. With relatively minor modifications, such cars are still providing reasonably rapid and comfortable transport in many areas today (1977). Many innovations which are commonplace in rail and bus transit vehicles today—rubber suspension systems, unitized all-welded steel body construction, and advanced illumination—were pioneered in this systems engineering effort.

The approach used then is now commonplace, having been used in transportation for such wide-ranging problems as transporting a human to the moon, planning major improvements to urban transportation systems, introducing containerization and other innovations in ocean trade, and recently the planning of intercity transport improvements for both passengers and freight. Not all of these involved the development of new technology; in fact, most did not. Yet the same concern for finding the facts and using them to evaluate alternative courses of action was essential. It is this approach, and the knowledge of transportation phenomena which has been developed by engineers, economists, psychologists, and others—knowledge which is indispensable in the planning and designing of new transport facilities and services—which makes transportation engineering the significant field it is today.

TRANSPORTATION ORGANIZATIONS

Before closing this chapter, it might be useful to describe the various types of organizations which provide transportation facilities and services and their interrelationships. These are the places where engineers and planners concerned with transport find employment, and the character of these organizations influences the type of work such professionals engage in.

The organizations vary widely from nation to nation, and to a lesser extent, region to region, reflecting different economic or political philosophies, as well as chance events of history. Hence any single portrayal will not be completely correct for any particular location. However, there is enough similarity to make a general portrayal informative. Figure 1-6 summarizes the major transportation organizations.

In describing the organization of transport, it is useful to distinguish between three types of functions. One is to provide a carrier service, i.e., to provide a capability to transport passengers or freight, as does a railroad, airline, a private automobile, etc. A second is to provide a facility for others to use, but not the carrier service on it, examples being roads and airports. Finally, there is the

Figure 1-6 Organizations engaged in transportation functions.

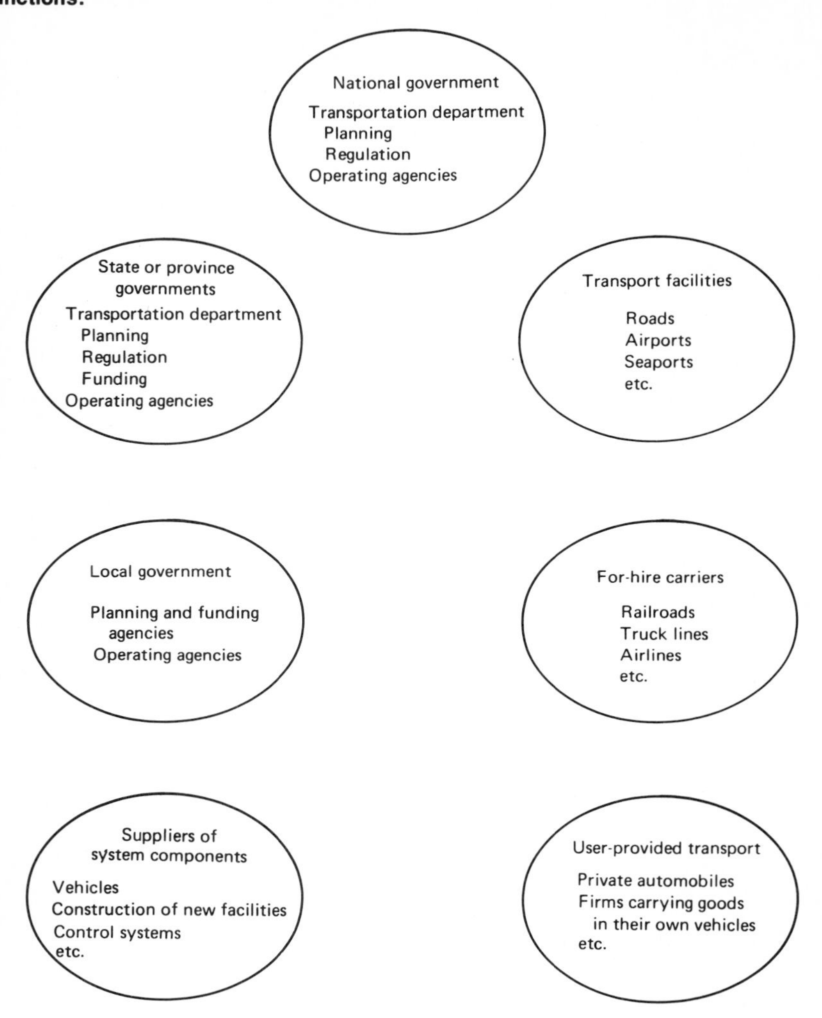

function of planning the entire system in its broadest sense, such as is often done by planning agencies which control to some extent the actions of carriers or providers of facilities. Separate from this classification is the distinction between public or government sector and private sector responsibility in each of these functions.

At the highest level of transportation organizations is the national government, in which is typically located a transportation department responsible for determining national transport policy. It is primarily concerned with formulating policies which specify which organizations will have responsibility for which portions of the transport system, regulations governing such organizations, and possibly some of their funding. For example, in the United States the national

Department of Transportation specifies the state and local agencies which will be concerned with roads and also undertakes some funding of state and local roads and transit systems, while other agencies (e.g., Interstate Commerce Commission and Civil Aeronautics Board) regulate interstate transportation prices and services of carriers. In some nations, the national government may actually own and operate parts of the intercity transport system, such as railroads and roads.

State or provincial and local governments also play an important role. Most undertake planning activities for their regions, and carry out the investments in new facilities under their control, typically roads and often airports and city transit systems. For management purposes, often each major segment of the system is operated by a separate division of government, which may have a nonpolitical management, such organizations often being termed authorities or public corporations. These governments may also regulate privately owned carriers operating within their jurisdiction.

Then there are the transport carriers. One broad class is those carriers which carry other persons or other's freight, often called for-hire carriers. These typically include railroads and bus lines. Ownership is often mixed between the private and public sector. In the United States ownership is almost entirely in the private sector, with varying degrees of regulation depending upon the type of carrier, the area served, and the range of items carried. For example, all railroad freight service is regulated, as are most truck lines, on price and location of service, while truck carriers of agricultural goods are not. Urban transit is the major exception to private ownership, such carriers typically being owned by local or state governments. In many other nations, railroads and perhaps airlines are government-owned, but truck and bus services are private. And in others, all carriers are government-owned.

Also there are the private carriers, users of the system who provide their own carriage, in the sense of movement in their own vehicles (and in the extreme, their own systems, as in the case of some pipelines). Private automobile travel falls into this category, as does much trucking, both local and long-distance. This type of movement is found in all nations and is largely unregulated except for safety. It is dominated by road transport, since that technology lends itself well to individualized movement. In most nations this also appears to be the fastest-growing area of transport.

Then there are the suppliers of the components of the transport system. These include manufacturers of vehicles, control systems, and other largely standardized components. Also there are the builders of facilities. Since these components continually need replacement and often the transport system is expanding, requiring net additions to the stock, this area is very important.

As noted earlier, the actual structure of transportation organizations varies widely among locations, but this description should give some idea of the actual division of responsibilities for the transport activities in a nation.

CONCLUSIONS

This book, then, is designed to introduce the field of transportation system engineering, or as it is often called, transportation system planning. In order to master and understand this field, however, it is not sufficient to learn the techniques and methods of the fields of transportation system engineering alone. Because the field is very closely related on the one hand to the design and operations of components of the transportation system, it is essential that the transportation system engineer know enough about the functioning of the various components of the transportation systems and the processes of design thereof that he can interact with such professionals. On the other hand, transportation system engineering and planning exists within a much broader context of the socioeconomic-political system, not only in the sense that the physical transportation system is part of that system, but also in that the planning and decision making of the transportation system must be related to that higher-level system which it serves. Thus the transportation systems engineer must understand the needs of society which give rise to the demand for transportation and understand the positive and negative nontransportation effects of transportation which will impose requirements on what is done within the transportation system. Only then will the transportation engineer be able sufficiently to understand the system and the context in which it operates in order to design properly that system and to interact effectively with those with whom this work is closely allied.

At the same time it must be recognized that transportation system engineering is a very new field, with an identifiable life of perhaps three decades at the most, and one which is expanding and changing very rapidly. This might be due to the changing nature of the society in which we live, and particularly to the changing emphasis given to different aspects of the quality of life. People's values seem to be changing rapidly, and these have affected transportation, particularly in emphasizing now the environmental and social impacts and implications of decisions regarding this sociotechnological system. But the changes are also undoubtedly due to an increased understanding of the system and its impacts, gained as a result of the intensive research which has been undertaken in this field, primarily during the last two decades. Thus it is necessary to not only equip engineers with knowledge of current practice and current methods, but also equip them to continually learn throughout their professional careers, so that they can assimilate techniques as they are developed. Perhaps even more important is the need to equip engineers to discover new tools for themselves, for that is the way in which scientific fields advance.

SUMMARY

1 The field of transportation engineering consists of many distinct areas which differ in scope and approach. Transportation system engineering, or transportation system planning, as it is often termed, is concerned with the broad questions related to the overall form of the transportation system in a region. It is therefore distinct from the many fields of transportation system component engineering, such as road design or aeronautical engineering, which are much more limited in scope.

2 In very general terms, the objective of transportation system engineering is to determine and implement the most nearly optimum combination of transportation facilities and services in a given region. Thus the field is concerned with all modes or means of transport, with both person and freight transportation, with environmental and economic effects of transport, with transport in any geographic context, whether urban or rural, and with problems of both government and private organizations.

3 Transportation system engineering typically deals with large, complex problems. In order to do this effectively, it makes use of the system planning or design process to structure problems and search for solutions. Like other fields of engineering, it relies heavily on the scientific method as a guideline for the development of generalizable knowledge which can be used to help deal with problems regardless of their context. However, it is important to recognize that many aspects of transportation problems cannot be quantified, and care must be exercised not to ignore these aspects simply because they must be dealt with in a less rigorous manner.

4 Transportation system engineering is a relatively young field, having first been recognized about 20 to 30 years ago. The field is still evolving, as a result of research and also as a result of continuing changes in practitioners' understanding of transportation problems and the process by which solutions are found and implemented.

REFERENCES

Drake, J. S., et al. (1967), A Statistical Analysis of Speed Density Hypotheses, *Highway Research Record,* No. 154, pp. 53–87.

Greenshields, B. D. (1934), A Study of Traffic Capacity, *Highway Research Board Proceedings,* **14**:448–481.

Highway Capacity Manual (1965), Highway Research Board Special Report 87, National Research Council, Washington, D.C.

Hirshfeld, C. F. (1975 reprint), "The Electric Railway Presidents' Conference Committee Streetcar Research and Development Program: Five Technical Bulletins, 1931–1933," Report No. PB-239 996, National Technical Information Service, Washington, D.C.

Kirby, R. S., et al. (1956), "Engineering in History," McGraw-Hill, New York.

Lang, A. S., and M. Wohl (1959), Scientific Research in the Field of Transportation Engineering, *Traffic Quarterly,* **13**:207–220.

Lindsay, W. S. (1876), "History of Merchant Shipping and Ancient Commerce," 4 vols., Sampson Low, Marston, Low and Searle, London.

Sarton, G. (1937), "The History of Science and the New Humanism," Harvard, Cambridge, MA. Copyright © 1937 by the President and Fellows of Harvard College, © renewed 1964 by May Sarton.

"Webster's New Collegiate Dictionary" © 1977 by G. & C. Merriam Company, publishers of the Merriam-Webster dictionaries. Definitions used by permission.

Wood, N. (1825), "A Practical Treatise on Rail-Roads."

Transportation in Society

2.

Transportation is an integral part of the functioning of any society. It exhibits a very close relationship to the style of life, the range and location of productive and leisure activities, and the goods and services which will be available for consumption.

Thus, the introduction of new or improved technologies of transportation has been very closely correlated with the development of modern civilization. While many historians and others have tended to view this as a simple cause and effect relationship, that is, the development of new transportation technology has not only enabled but also caused changes in the societies which use that technology, this is surely an oversimplification. In many instances, new requirements for transport have led to the development of new technology which could meet those emerging needs. Transportation is such an integral part of almost all human activities that it is in principle impossible to differentiate completely between cause and effect. Rather, advances in transportation have made possible changes in the way in which we live and the way in which societies are organized, and thereby have influenced the development of civilizations.

The purpose of this chapter is twofold. The first is to convey an understanding of the importance of transportation in modern society and the potential effects that changes in the transportation system might have on human activities and social structure, primarily through a very brief account of major developments in transportation and the associated changes in the ways in which people live. The second is to present selected characteristics of existing transportation systems, their use, and relationships to other human activities.

ROLES OF TRANSPORTATION IN CIVILIZATION

The movement of people and goods is as old as humanity itself. Neolithic humans moved from place to place in search of food, carrying their few possessions. Such limited and primitive movement has rather steadily given way to a style of living in most of the world in which we travel and ship a great deal. In most developed nations, a large fraction of the working population travels daily in some mechanized vehicle to and from work, not to mention all the travel for shopping and social reasons. Goods are routinely shipped over extremely long distances to provide those material things which are part of the expected standard of living.

At the same time, transport consumes a great deal of resources: the time of many persons in building, maintaining, and operating the transport system, fuel and materials, and land. The expense would not be borne if there were not substantial benefits from such extensive use of transport, benefits in improving the overall quality of life. These benefits, and many of the negative effects of increased transport, can be understood by considering the roles which transport plays in human activities. This is conveniently done by examining its role economically, socially, politically, and environmentally.

ECONOMIC ROLE OF TRANSPORTATION

Economics is primarily concerned with the production, distribution, and consumption of goods and services which are of value to humans—of wealth. A very important role of transport is in this context, and thus much understanding of transport can be gained from this viewpoint.

People must use the natural resources of the earth to satisfy the necessities of life, to provide food, clothing, and shelter. Also, the earth's resources are used to provide much beyond the simple necessities of existence, items which make life more pleasant, comfortable, and rewarding. But the surface of the earth is not uniformly endowed with natural resources, and no location is sufficiently well-endowed to provide the standard of living found in most societies by drawing from only local resources. Thus, there is an almost universal requirement for transportation of things. In addition, since knowledge and skills are not always equal at various locations, there is often a need or an advantage to transporting persons to improve the material well-being of a society. Examples include the movement of doctors to treat the ill and the travel of technicians to assist in overcoming problems falling within their fields.

The Place, Time, and Quality Utility of Goods

A very simple model will serve to illustrate these points. Consider a single commodity which is desired and would be consumed by the people of a certain community if it were available at a sufficiently low price. For purposes of discussion, it matters not whether this commodity is extracted from the ground, grown as a food, or is the result of manufacturing or other processing. On Fig. 2-1, this commodity is produced at point *A*, and it is available there for a price equal to *OC*. The community, which is the point of possible consumption, is located at point *B*, a distance *AB* from *A*, and the maximum price people will pay for a unit of the commodity is shown on the left axis as *OE*.

Transporting a unit of this commodity from *A* to *B* on the original transport system costs an amount *CH*, this cost consisting of a fixed charge reflecting the loading of the vehicle, billing, and other documentation, etc., *CD*, and a charge per mile equal to the slope of the line *D'H'*. With this transport system, the total cost at *B* will be *OH*, which is greater than the maximum price consumers at *B* are

Figure 2-1 **Total cost of a commodity (consisting of price at origin plus transport cost) and its relationship to place utility.**

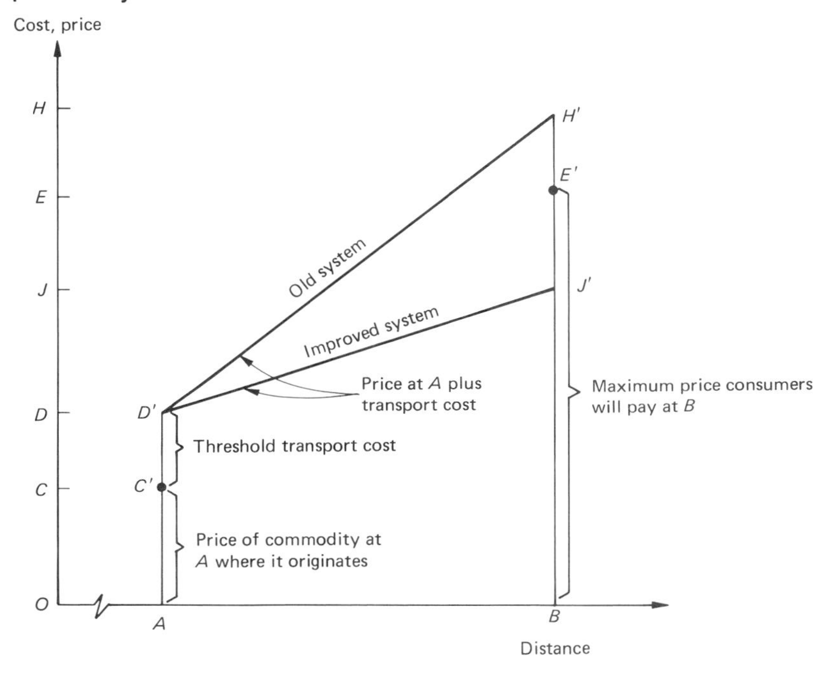

willing to pay of *OE*, and hence none of this commodity will be shipped from *A* to *B*. If *B* were the only point of possible consumption of this commodity produced at *A*, none would be produced at *A* under these circumstances.

Consider the effect of an improvement in the transport system which reduces the cost per unit distance of movement from *A* to *B* to the slope of the line *D'J'*. This might result from upgrading the road or replacing unpaved roads with rail transport, to mention two examples. The resulting total cost at *B* would then be reduced to *OJ*. Since this price is less than the maximum for consumption to occur, the commodity would be sold and consumed at *B*, transported from *A* to *B*, and hence produced at *A*.

In economic terminology, the reduction in the transportation cost between *A* and *B* has given *place utility* to the goods produced at *A*. Whereas these goods would have no value with the high transport cost because they could not be sold in the market, with the lower transport cost they have a value. Specifically, with a total transport cost of *CJ*, the price charged for the good at *A* could be raised by as much as *JE*, and the consumption and hence production and transport still occur. At any price delivered at *B* greater than *OE*, the value at *A* would be zero. Thus transport gives utility to the goods, in the sense that it determines the value of goods at one location in relation to the price at which they might be sold at any other location. It is in this sense that transport gives place utility to goods.

Another closely related concept is that of the *time utility* of goods. The demand for a good may occur during only a particular period of time and perhaps cease thereafter. An example is the demand for Christmas trees. If the trees can be cut and transported to potential buyers before Christmas, then they can be sold. However, if they arrive after Christmas Eve, their value is little or nothing. The transportation must be completed within a certain period in order for the goods to have value, in a manner analogous to the maximum charge for transport in the previous case. While the Christmas tree example is of limited generality, similar considerations apply for many raw materials and manufactured goods. In many manufacturing operations, one stage in the production may occur a long distance from the next, with transportation being a vital link connecting these processes. Since it is costly to maintain a large inventory of parts at each factory, each may rely on regular deliveries of the parts it requires to perform its function, in order to continue production. If those parts are delayed, then that factory may have to shut down until they arrive, halting production not only there, but also at those plants which receive the output it now cannot produce. Again, a very specific requirement is placed on the transport system to deliver goods not only within a specific cost, but also within a specific time.

An example of this arrangement is in the production of automobiles, where one firm manufactures parts such as chassis, bodies, and engines in the Midwest, but assembles these into automobiles throughout the United States. One Eastern assembly plant receives parts daily by train from the Midwest, this train being operated on a daily schedule to meet this auto producer's requirements, as well as to handle other freight. While the railroad makes every effort to keep the train on time, occasionally it is delayed. The auto manufacturer has a contract with an air freight carrier to have certain parts flown to this eastern plant if the rail freight delay were to cause a shutdown of the plant. This seemingly expensive way of moving auto parts being very inexpensive compared to the possible losses due to temporarily halting the vast assembly lines.

There is a third aspect of the utility of movement which is often overlooked but which is quite important. This aspect is transporting the goods so that their essential qualities are not diminished or lost. In the case of perishables, such as food, this is very important. Now technology for maintaining any desired temperature, pressure, or humidity is so readily available and inexpensive that this requirement usually no longer presents a problem, but less than a century ago, refrigeration was very difficult. Instead of shipping cooled or frozen meat, for example, growers had to ship cattle to near the location of final consumption, where the animals would be slaughtered and the meat dressed for use as food. This arrangement included the shipment of much more weight than would ultimately be consumed as food, resulting in added transportation costs and higher prices for the final product. Maintaining the utility of items being shipped depends on the time required for transport and the natural deterioration of some commodities, unless certain environmental conditions are maintained.

The means we use for transport also usually involve some unwanted movement of the items being shipped, in the form of bouncing in the vehicle and rough handling during loading and unloading, etc. Damaged commodities can be a serious problem requiring special packaging and loading. Just as certain commodities require special environmental conditions to preserve their quality, some commodities require special packing in order to avoid damage. This is particularly true in rail and water movement, where the impacts can be very severe. Some particularly fragile goods, such as computers, are almost invariably shipped by truck or air for both this reason and because of the lessened danger of theft in these latter two modes.

Actual Transport Costs: A Discussion

The arguments presented above on the effects of changes in transport costs were made with the assumption that the cost of transporting a given amount of freight was composed of two components: a constant amount which was fixed regardless of distance, plus an amount which varied directly with distance. Mathematically, this would be of the form

$$p = A + Bd \qquad (2\text{-}1)$$

where p = cost or price of transporting a given item
d = distance transported
A, B = constants

A is the cost regardless of length of haul and B is the additional cost per unit distance, the total distance of the movement being d. Of course, both A and B will depend upon the weight and the volume of the shipment. In general, both A and B increase as either weight or volume increases. A reasonable question at this point is: How representative of actual freight costs is this cost pattern we have assumed?

The cost of transporting freight usually does follow the pattern of a cost which is independent of the distance shipped and an additional amount which tends to increase with increasing distance. With shippers operating their own trucks (as many do), that this tends to be the pattern is easily seen. The initial, or threshold, cost, as it is often called, would be primarily the cost of loading and unloading the vehicle, depending on the amount (weight and volume) to be handled but not on the length of haul (ignoring such possibilities as more elaborate stowage on the vehicle if the trip were very long). This threshold cost might also include a portion of the cost of the loading and unloading facilities, and perhaps the packaging of the freight for shipment. There would also be a cost of operating the truck. Some of these operating costs would probably vary very closely with the miles driven, such as fuel, lubricants, wear on the vehicle, and even the driver's wages, assuming that the distance traveled were proportional to the time required. Thus the pattern we assumed seems to be reasonable for private truck transport.

Actual freight rates of carriers also tend to be of the form we assumed, although there are many variations from the precisely linear form in actual practice. An example of an actual railroad's rates for the shipment of freight is shown in Fig. 2-2. These were the rates in effect in 1974 for the shipment of lumber from the Northwest to various locations in the middle and eastern part of the United States, and the rates are shown in units of cents per 100 lb, 100 lb commonly being termed a "hundredweight," abbreviated cwt. The hundredweight is the usual measure of weight in land transport for rate purposes. No rate is shown tor movement to nearby places since such movements probably would be by truck, and as a result the railroads have not established specific rates for the movement of lumber to such nearby points. Lumber could be shipped by rail to these nearby areas, but the rate would be rather high, being the rate for a wide variety of commodities including lumber. (The details of actual rate structures of carriers will be covered in Chap. 11, where

Figure 2-2 **An example of actual freight rates: 1974 rail commodity rates for lumber from the Northwest.** ***(a)*** **Rail freight commodity rates in cents per 100 lb for lumber shipped from western Oregon and Washington origins to various transcontinental destinations. [*From Sampson and Farris (1975), p. 211.*]** ***(b)*** **Lumber rates plotted against airline distance from the origin** ***X*** **along two axes shown in** ***(a).***

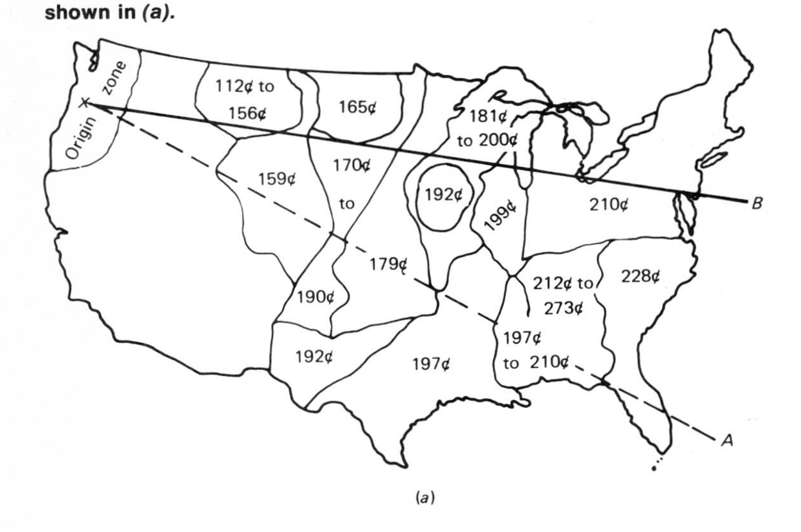

(*a*)

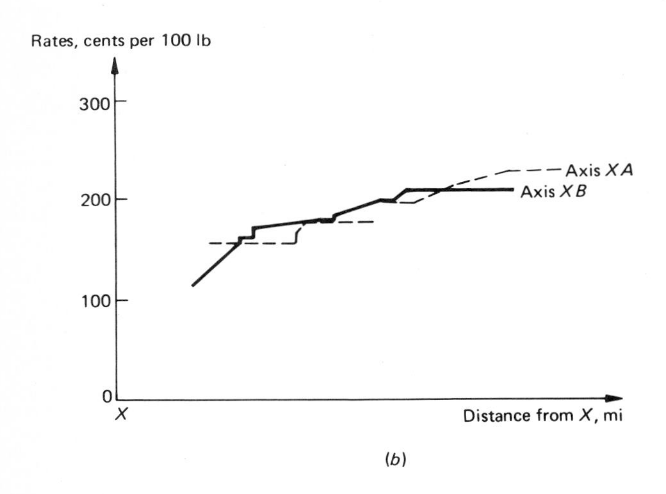

(*b*)

we shall discuss the actual supply of transport services by carriers, including typical price characteristics of carriers.)

As can be seen from the top portion of Fig. 2-2, the rate is the same within rather large areas. This is the case for a number of historical and legal reasons, and is typical of the rates of most carriers, although the size of the area covered by a single rate is often much smaller. Despite the uniformity in the rate to large areas, there is a clear tendency for rates to increase with increasing distance from the origin area. This is shown on the lower graph, in which the rate is plotted against the distance as measured on a straight line from an origin indicated by the *X* in the top figure. The progression of rates along two axes is shown, one passing through New Jersey and the other through Florida. These two progressions show that the rates are somewhat different depending upon the direction traveled from the origin, indicating that there can be, and usually is, some variation in the rate for the same shipment and distance depending upon destination. Even if the distance via the actual route of the carrier rather than the air line distance were used, this variation would still exist. Nevertheless, the charges for hauling freight generally follow the pattern of increasing with increasing distance from an initial threshold value. In fact, the railroad rates shown in Fig. 2-2 can be approximated by the formula

$$\text{Rate in cents per cwt} = 125 + 0.4 \times \text{mi}$$

It might also be noted that these railroad rates are the same per hundredweight, regardless of the total weight shipped. Actually, this is true only to a limited extent. There is a minimum weight which must be paid for regardless of whether or not the actual weight of the shipment reaches that amount. This ensures that shippers will pay for an essentially full car even if they ship a small amount, inducing shippers to use the capacity of the car they have available to them. There is also a maximum weight for which a rate applies, definitely no greater than the maximum capacity of the freight car. There may also be a point at which the cost per 100 lb is reduced, to reflect the fact that it does not cost the railroad twice as much to haul 60 tons of lumber in a single car between two points as it does to haul 30 tons between the same two points.

One might also wonder if actual cost of transporting freight has in fact decreased in recent history, as the preoccupation with the effects of reduced transport costs implies. If a very long period is considered, such as two or three hundred years, then the answer seems rather obvious: Transport costs, particularly for overland movement, have surely decreased. Over such a period there have been numerous new transport technologies introduced and adopted, and without any detailed analysis it seems clear that these, taken together, have probably reduced the cost of transport. Two hundred years ago, overland transport was by crude cart or wagon if the roads were passable by such vehicles or by porter (men on foot carrying goods), pack horse, or other beast of burden. Movement was slow, and maximum loads small. Now since these have been replaced with a wide variety of mechanized means of transport operating over prepared routes, greater speed and carrying capacity require less energy to effect movement.

Professors Young and Hay have attempted to compare the cost of transporting 1 ton of freight 1 mi by various means, ranging from porters to modern freight trains. Their estimates of costs are presented in Table 2-1, and are based on costs of labor and equipment in the late 1950s. While numerous assumptions had to be made in order to generate these values and variations in weather, terrain, and so forth would all affect the productivity values used, they do nevertheless provide an indication of the gains that have been made

Table 2-1 Comparison of Approximate Costs per Ton-Mile of Various Primitive and Modern Forms of Transport†

Type of carrier	Output per carrier, ton-miles per day§	Value of vehicular equipment, dollars	Accessories required	Cost per day circa 1960‡ Accessory Operating (a) Interest (b) Wages, (c) dollars (d)	Total cost per day, dollars	Cost per ton-mile, dollars
Human back (100 lb carried 20 mi)	1	0	Trail and pack rack	0.01 (a) 0.00 (b) 0.00 (c) 0.20 (d)	0.21	0.210
Pack horse (200 lb carried 40 mi)	4	80	Trail and pack saddle	0.02 (a) 0.20 (b) 0.01 (c) 0.40 (d)	0.63	0.158
Wheelbarrow (400 lb moved 20 mi)	4	10	Path	0.04 (a) 0.02 (b) 0.01 (c) 0.30 (d)	0.37	0.093
Cart, best conditions (1000 lb moved 20 mi)	10	10	Pavement	0.08 (a) 0.02 (b) 0.01 (c) 0.30 (d)	0.41	0.041
Team and wagon (3 net tons moved 40 mi)	120	500	Good road	0.44 (a) 0.30 (b) 0.10 (c) 3.00 (d)	3.84	0.032
Motor truck (10 tons moved 240 mi)	2400	8000	Pavement	2.40 (a) 30.60 (b) 1.50 (c) 20.00 (d)	54.50	0.023
Railroad train (2000 net tons moved 40 mi)	80,000	800,000	Tracks and structures	111.74 (a) 424.38 (b) 180.00 (c) 63.92 (d)	780.04	0.010

† Based on a suggestion in unpublished papers of the late E. G. Young, former professor, Railway Mechanical Engineering, University of Illinois.
‡ Costs are those of time and locality where type of carrier was most prevalent. (a) Includes all costs of maintaining and operating facilities, capital costs not included. (b) Fuel (or feed), oil, water, maintenance, etc., except labor. (c) Includes interest on vehicle only, plus simple yearly amortization charge. (d) Direct labor cost of operating vehicle only.
§ Mileage moved is statistical average for all railroads in the United States; obviously a railroad train could run 320 mi ± in 8 h.
Source: Hay (1961, p. 4).

in transport over the centuries. In fact, these figures tend to understate the gains, because they take as costs of such items as labor the values that prevail in the areas where that form of transport is most prevalent. As a result, the cost of a work-day of labor for porter carriage is estimated at $0.20, while that for a truck driver is estimated at $20. Also, the increase in the speed of movement has reduced the cost of transport to the shipper, who now needs to carry his inventory of goods in transit for a shorter period.

A less general but more precise example of the reduced cost of transport over the last century is given in Fig. 2-3. This shows the average revenue per

Figure 2-3 **Average freight revenue per ton-mile of principal United States railroads, 1850–1970.** [***Compiled from Aldrich (1893), Interstate Commerce Commission (1940, 1950, 1960, and 1970), and U.S. Dept. of Labor (1975).***]

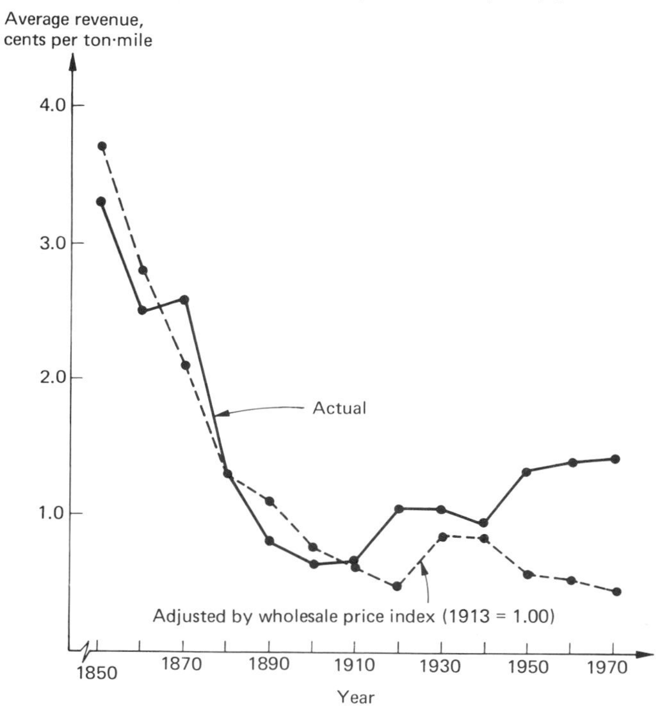

ton-mile received by major United States railroads in the period from 1850 to 1970. The solid line is the actual average revenue. As can be seen, the average dropped rather consistently from 3.3 cents/ton-mi in 1850 to about 0.7 cent/ton-mi in 1900, thereafter increasing to about 1.6 cents/ton-mi in 1970. With the correction for the effects of general inflation, through use of the wholesale price index, the picture remains the same. It is noteworthy that even in the recent past, the average railroad revenue per ton-mile has dropped slightly, after correction for inflation. However, this is somewhat misleading, for in the period since World War I, the railroads have lost much of their former short-distance traffic and the lighter merchandise freight to trucks, so the average revenue per ton-mile has tended to drop partly because of changes in railroad traffic. Nevertheless, the general pattern of decreasing costs of freight transport over the past one or two centuries seems clear.

Now that the appropriateness of the discussion of the effects of reduced freight transport costs and the approximately linear form of those costs has been established, it is appropriate to return to the discussion of the economic roles of transport.

Changes in Location of Activities

Reducing the cost of transport does not have a uniform or necessarily beneficial effect on all locations or sectors of an economy. This can be easily seen by consideration of another example. Following the previous example, imagine a consumption center, located at point *B* (in Fig. 2-4), at which a particular commodity produced at point *A* is consumed. The total cost of this commodity delivered at location *B* is *OJ*. Suppose there is an alternate source of this product, at a point *K*, which is a greater distance from *B* than is *A*. If the cost of the commodity at *K* is *OL*, with the same transport costs as apply between *A* and *B*—a fixed cost plus a cost per unit distance—the total cost of supply from source *K* would be *OT*, as shown in Fig. 2-4. Since this cost is greater than the total cost if it were supplied from *A*, the commodity will be purchased from *A* and no production will occur at *K*.

Again consider the effect of a reduction in transport costs per unit distance. If the reduction were to the slope of the line *V'Q'*, it can be seen from the figure that the total cost from source *K (OQ)* will be less than that from source *A (OP)*. Thus, the demand at *B* will now be supplied from point *K*, and production at *A* will cease. Thus the effect of the transport improvement has been to shift the pattern of production and shipment such that point *K* now experiences an increase in economic activity while *A* suffers a decline.

While this is a very simplified example, it serves to point out the differential effects of transportation system improvements. As a result of a change that usually would be thought of as an improvement, some groups may be adversely affected while others benefit. Transport cost reductions or other improvements in the quality of transport will in general influence the locational pattern of production and the sources of supply of consumption centers. And since production requires workers, changes in the location of production will be accompanied by changes in the location of population, which in turn affects the demand for commodities and may lead to a further shift in production and distribution patterns. Thus the effects of transport system changes

Figure 2-4 Differential effects of transport improvements on two production centers.

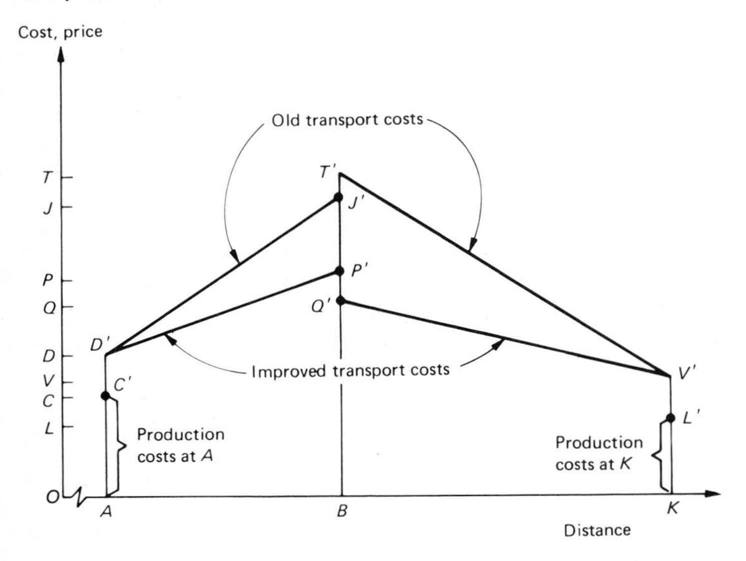

can be very widespread, extending far beyond the immediate positive effects of a reduction in the price of movement.

Another aspect of the reduction in the price of goods which might accompany a transport improvement is what economists term the *income effect*. As a result of the reduced price, if consumers (or firms or other purchasers) continue to consume only the same amount of that good as before, they will spend less money on that good and have some additional money to spend in any way desired. They might spend this money on more of the good which experienced the price reduction or on other goods, perhaps on some which had not been consumed previously. Alternatively, they might choose to maintain exactly the same standard of living as before, and perhaps reduce the total amount of time spent working, and thus in effect consume more leisure time.

Transport system improvements presumably reflect actual reductions in the amount of scarce resources required in the process of transportation. Whether viewed as a price reduction or as simply a freeing of resources to be used for other purposes, the net effect is to make available a larger amount or greater variety of goods and services for possible consumption.

Extent of Freight Transport

The prices of transport, or the resources consumed in transport, are now so low in general that in very few societies do people consume only products produced locally. Most communities consume goods that may be produced thousands of miles away, exchanging for such goods items that are produced locally which are desired elsewhere. As a result, the overall standard of living throughout most of the world has been raised very significantly by reductions in the cost of transport and the increased ability to transport things over long distances.

Examples of the effects of low transport costs and the ability to preserve goods traveling over long distances and for long times are so commonplace that they hardly need mention. However, a few examples will indicate the widespread effect this has had on the range of goods available for consumption. Foods which can be produced only in certain climates and soil conditions are now available almost everywhere, ranging from the common breakfast fruit juices sold in cold climates where they could not be grown to fish and seafoods at places far removed from the oceans and to highly prized delicacies such as caviar from Russia and northern Europe. Natural resources such as ores and oil are transported half-way around the globe from points of extraction to points of processing and ultimate consumption. Often these materials are available locally, but only in low quality or with high costs of extraction, such that it is more economical to transport cheaper or better materials long distances than to consume what is available locally.

As a result of reduced transport costs, in the past hundred years there has been some substantial shifts in the location of points of extraction of raw materials. For example, the extraction of ores and coal from deposits in the western part of the United States rather than from the more costly and inferior quality deposits on the East Coast, where much of the final consumption is located, is economically attractive because of the cheapness of transport. Associated with these shifts in economic activity have been some difficult shifts in employment concentration and associated social problems, such as those found in Appalachia, an area which has seen a reduction in coal mining activity.

An indication of the extent of dependence of modern urban society on goods movement can be obtained from some of the statistics on freight flows in the New York City region. Data were gathered by the Tri-State Transportation Commission (covering the metropolitan area, which is in three states, New York, New Jersey, and Connecticut) on flows into, out of, and within the area in 1962–1963. The results, excluding movements through the region (as in the case of freight on a truck entering the region and continuing on that same vehicle to another destination), are presented in Fig. 2-5. A total of 204 million tons came into the region, and 97 million tons left the region, and another 268 million tons were moved within the region. This represents an average of 10.9 tons inbound and 5.2 tons outbound for each of the 18.5 million people in the region, compared to the United States national averages of 27.1 and 26.6, respectively. The modal distribution of this traffic, the average length of haul, and the revenue received is also presented in Fig. 2-5. The importance of the lengthy movement by water—both domestic (mainly coastal movements) and international—and the lengthy movements by rail are particularly noteworthy. Cities such as New York depend very heavily upon many goods received from very distant points, and similarly reciprocate by shipping many of their products long distances to places of further processing or final consumption.

Summary

In summary then, the economic roles of transport can be seen to be as follows: (1) Transport extends the range of sources of supply of goods to be consumed in an area, making possible the use of cheaper sources or sources of higher quality. In addition, goods which cannot be obtained locally may be made available. (2) The use of more efficient sources of supply results in regional specialization or division of labor, resulting in an increase in the total amount of goods available for consumption. Closely related to this is the opportunity to concentrate production in one or a few locations, but neverthe-

Figure 2-5 **Annual freight flows to, from, and within the New York metropolitan area in 1962–1963. [*From Tri-State Transportation Commission (1966), p. 3.*] Left side of chart shows tons carried as the height of the bar and average length of haul as the horizontal width of the bar for five different modes of freight carriage. The area of the bar, therefore, measures ton-miles. Arrows show direction of movement into, out of, and within the region. Through traffic—that which stays on the same vessel, truck, or rail car as it passes through the region's transportation network—is not included. On the far right, a series of three bars shows the amount of money spent for each mode's freight service—inbound, outbound, and within the region. Note the height of the money bars in comparison with the size of the ton-mile bars.**

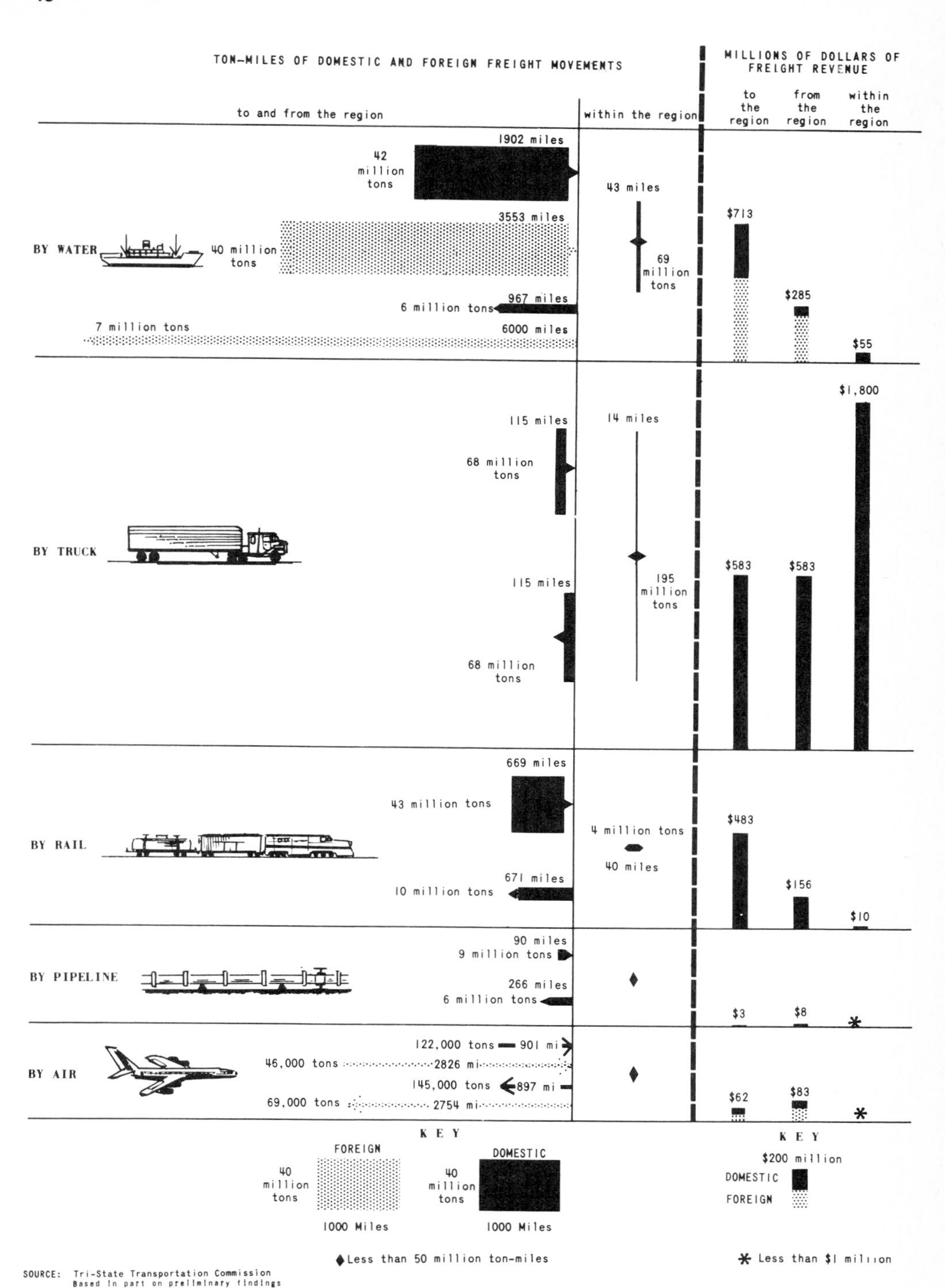
TON-MILES OF DOMESTIC AND FOREIGN FREIGHT MOVEMENTS
MILLIONS OF DOLLARS OF FREIGHT REVENUE
to and from the region
within the region
to the region
from the region
within the region
BY WATER
1902 miles
42 million tons
43 miles
3553 miles
40 million tons
69 million tons
967 miles
6 million tons
7 million tons
6000 miles
$713
$285
$55
BY TRUCK
115 miles
68 million tons
14 miles
115 miles
195 million tons
68 million tons
$583
$583
$1,800
BY RAIL
669 miles
43 million tons
4 million tons
40 miles
671 miles
10 million tons
$483
$156
$10
BY PIPELINE
90 miles
9 million tons
266 miles
6 million tons
$3
$8
*
BY AIR
122,000 tons
901 mi
46,000 tons
2826 mi
145,000 tons
897 mi
69,000 tons
2754 mi
$62
$83
*
KEY
FOREIGN
40 million tons
1000 Miles
DOMESTIC
40 million tons
1000 Miles
♦Less than 50 million ton-miles
KEY
$200 million
DOMESTIC
FOREIGN
* Less than $1 million
SOURCE: Tri-State Transportation Commission
Based in part on preliminary findings

less continue to serve very widespread markets and thus take advantage of economies of scale in production. (3) Because the supply of goods is no longer limited to local areas, items can be supplied from alternative sources if the usual source cannot supply all that is needed. This is especially important in instances of interruption of the supply of the necessities of life, as in the case of food after a drought.

SOCIAL ROLE OF TRANSPORTATION

It is often difficult to differentiate precisely between the economic and the social roles of transportation. However, some distinction is necessary because much of the role of transportation and its effect on the manner in which we live cannot be considered part of a market in which things are transacted for money in any sense. Thus, when we speak of the social role of transportation, we refer to the general organization of society, the style of life in the sense of the range of activities, both economic and noneconomic, that people engage in.

The Formation of Settlements

As mentioned earlier, people originally lived mainly in nomadic tribes and moved from one location to another in search of food. They carried with them what few possessions could be moved with primitive transportation equipment. As a result, they had no need to develop the skills and knowledge required to make things, whether of material, spiritual, or cultural value, for they could not retain them. Partly as a result of this condition, such skills developed only very slowly.

Once the transportation of food and fuel became sufficiently easy in time and effort needed (refer again to Table 2-1) and food could be stored through periods in which game or harvests were not forthcoming, permanent settlements could be established. Then people could devote their attention to things which were not portable. In such settlements people first became extensively concerned with beauty and other aspects of art, and could use written materials to transfer to following generations knowledge and ideas. Also, permanent settlements made it possible to develop and use simple machinery and tools which would make people and the animals which served them more productive, these tools no longer having to be so small and light to be portable. Also, people probably had much more leisure time as a result of the adoption of the settlement pattern of life, for they no longer had to devote time to packing their few belongings and traveling at very slow speeds over long distances to the next temporary settlement. This released time could be devoted to improving material well-being and to other activities which give satisfaction and pleasure, such as addressing questions of religion or morality, developing forms of art and entertainment, etc.

Many of these early settlements were located at points of some transportation significance. Usually, the settlements developed by rivers or other bodies of water, for water transport was by far easiest. Also, goods (and persons) had to be transshipped at these points between water and land transport, and hence there was an immediate need for terminal facilities, warehouses, etc. Many of these settlements have, of course, continued to grow over the centuries to become many of the largest cities throughout the world. Also, settlements grew at points where overland transport routes intersected or crossed, those being natural stopping places and points of greatest accessibility to other places. The advantages of trade with one's neighbors (who might have a relative economic advantage in the production of certain things) tended to reinforce the location of settlements at points of ease of transportation access.

Size and Pattern of Settlements

The original settlements were, of necessity, relatively small developments. In part, this was due to the limited range of territory from which food and other materials could be gathered to support the settlement. As the population increased, improvements in transportation and in the preservation of food were developed and this range was continually expanded.

This can be easily understood by reference to a very simple example. In this example, food and other materials necessary to support a population will be assumed to be proportional to that population, i.e., a certain number of pounds or tons of each material being required per capita. Also, the yield of a unit area of land will be assumed to be constant throughout the region surrounding the settlement, so the number of people which can be supported by a particular land area will be simply proportional to the area of that region. If the maximum distance over which things to support the population is doubled as a result of the transportation improvement, then the area of land which can be used to support that settlement is quadrupled assuming that the region supporting the settlement is a circle with its center at the settlement and a radius equal to this maximum distance. Thus, the additional land area supporting the settlement will increase rather substantially, even with relatively small improvements in transportation and storage capabilities, and hence the added increment in population which could be supported in one settlement resulting from a transport or storage improvement is similarly rather substantial.

An example of this effect is shown in Fig. 2-6. The settlement is located at point *A,* where a price of *OD* is the maximum that can be afforded for the commodity which is supplied from the hinterland. The cost of producing this commodity is *OP* per unit (say, per ton) at the site of production. The original transport cost consists of a terminal cost of δ per ton plus a distance cost of α per ton-mile. Working outward from *A,* it is seen that the most distant point from which this commodity can be supplied and the total cost at *A* still be *OD* or less is at a distance *AC* from *A*. But if the distance related transport cost is reduced to β ($\beta < \alpha$), then this distance is extended, as shown in the figure, to a point *E*. Since *AC* is less than *AE,* the area of possible production for the city at *A* would be greater with the lower transport cost β.

Although the ideal conditions of uniformity in freight charges regardless of direction are not likely to be met precisely in practice, as we have seen earlier in the case of rail rates, this effect of decreasing transport costs on the potential market area or supply area of a city has been recognized to be very important, not only in helping to understand the growth of cities, but also in understanding the effect of rate changes on the total amount of goods shipped. The halving of rates resulted in a quadrupling of the area served from a center is now widely known as Lardner's law of squares, recognizing

Figure 2-6 **The effect of transport costs on the size of the region supplying a settlement or town. *(a)* Maximum distance of supply to *A*; *(b)* area of supply to *A*.**

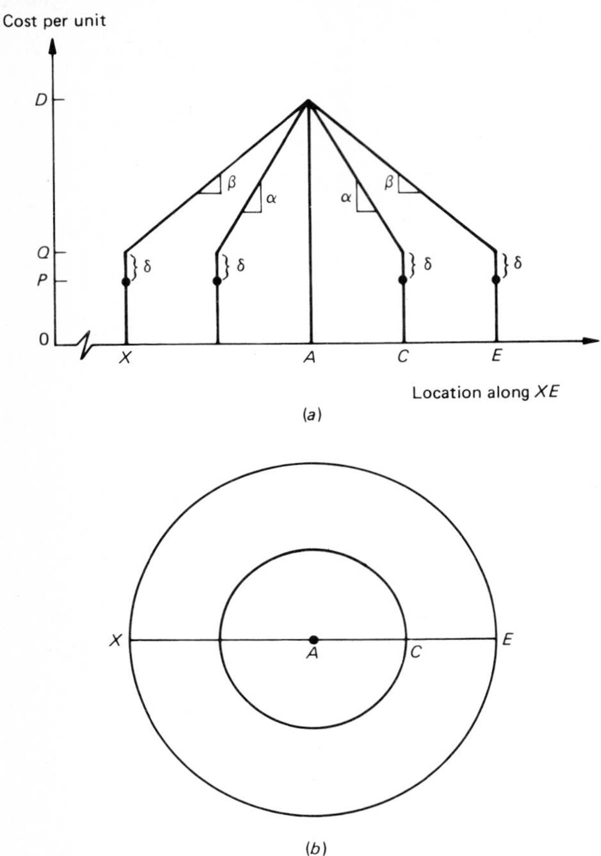

Dionysius Lardner who, in 1850, was perhaps the first person to write about this effect.

The size of settlements is not only limited by the size of the area from which the settlement can obtain food and other necessities, but also by considerations of person movement. A critical component of person travel is the journey to and from work. Any factory, or other business, must be reachable by its employees within a reasonable period of time and expenditure of money. Until the middle of the last century, most persons had to walk to and from work, personal horses being too expensive and in some ways inappropriate for travel within cities, and public transport carriers either not being in existence or costing too much for the average person to afford. Under these circumstances, the area from which a factory or other employment center could draw its workers was limited to probably 2 or 3 mi, the distance a person could walk in about an hour. In these early cities, economic activities were typically concentrated in one or a few places, and hence the maximum feasible size for an urban area was approximately within a 3-mi radius of the city center.

With the advent of early horse cars, small vehicles pulled by horses and operated on rails lying in the middle of streets, travel within urban areas was made much more rapid, probably reaching an average speed (while traveling in the vehicle) of somewhere between 4 and 8 mi/h. With fares relatively low, such that many people could afford them, the maximum size of urban areas with a single central core was expanded considerably.

Since the advent of the horse car, the elevated railways, railway commuter services, electric streetcars, buses, and finally the automobile have all increased the speed of travel in metropolitan areas, so that the distance effect of the time constraint has been relaxed to the point where a single metropolitan area might radiate 30 or 40 mi from its central core, with some people actually traveling such distances to work. Also, the economic constraint was relaxed, partly because of the decreases in average cost of travel per mile as a result of these technological advances, and also as a result of the tremendous increase in personal income, which enabled people to spend more on transport, if they so wished. Thus, it is now possible for a person to work at the very central part of the city and still live in essentially unspoiled rural surroundings of the outer periphery of the metropolitan area. With the advent of larger cities, the pattern of development of these cities has changed considerably.

Instead of a single central core, now most activities are dispersed throughout a metropolitan area into a very large number of small activity centers catering to local needs and so-called higher-order centers which are the sites of major employment concentrations and more specialized shopping and recreational activities. Also, before the advent of the streetcar, residential settlement was largely restricted to that along railroad routes, which provided suburban train services. Now, since the advent of the streetcar, and later the motorbus and automobile, the areas between these original radiating rail routes have been largely filled in, yielding a metropolitan area in which almost all land is being used.

This pattern of growth can be readily seen from examination of Fig. 2-7, which portrays the expansion of the Chicago metropolitan area from 1835 to 1955. The different degrees of shading indicate the periods in which each land area became developed with residences, factories, or other urban activities. The small size of the original city, the development outward along the radial rail lines, and finally, the filling in of the land between these routes since the 1920s, is very evident from this map. Although this will be discussed in much greater detail later in the text, the diffused pattern of residences and activities, and the associated pattern of travel, which involves trips in all directions throughout the metropolitan area, has been made possible by developments in transportation technology. Nevertheless, we emphasize that the size of the population and spatial form of urban areas are matters of social choice, with technology (transport and other) limiting the range of options but not specifying the choice.

The historical trend toward larger cities (or metropolitan areas) in the United States is revealed by the information in Fig. 2-8. In this figure, the population of cities is indicated on the vertical axis, while the horizontal axis

Figure 2-7 **An example of urban growth since the advent of mechanized transportation: Chicago from 1835 to 1955.** [*From Chicago Area Transportation Study (1958), vol. I, p. 15.*]

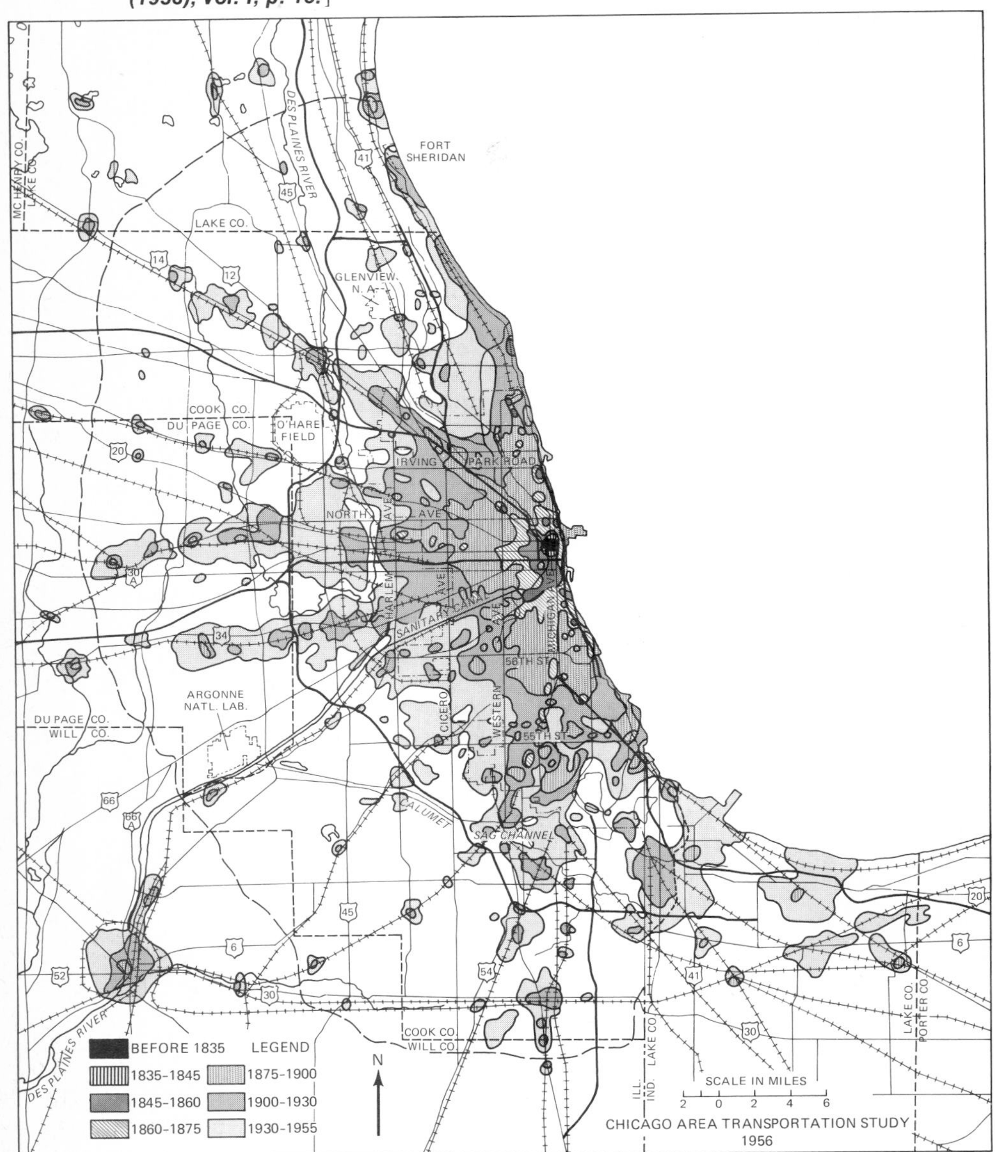

indicates the number of cities, and any point on a line of the figure indicates the number of cities with population greater than or equal to the amount specified. For example, in 1930, there were seven cities with a population of one million or more, while in 1870 there was only one such city (a second city had a population of almost that amount). The trends of the population of the

Figure 2-8 **The number and size of cities in the United States from 1790 to 1930.** [*From Zipf (1949), p. 420.*]

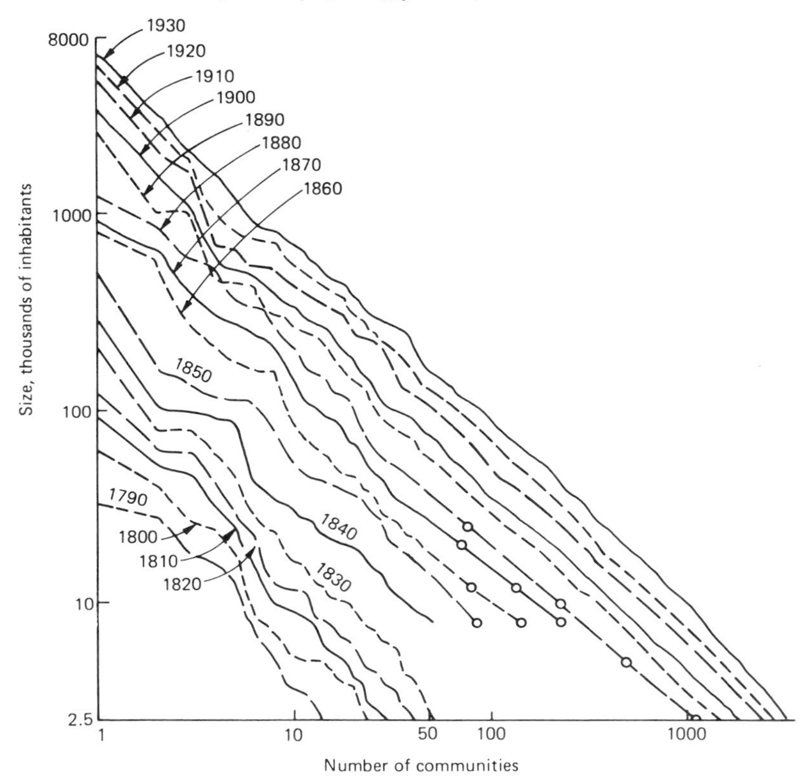

largest city increasing with the passage of time and the increase in the number of cities of at least any particular population level are very evident. Such changes in the size of cities would not have been possible without mechanized transport, both for travel within cities, and for carrying necessary supplies to the cities from farming and other resource areas.

It is not only work travel which has been substantially influenced by the improvements in transportation technology, primarily in the last century. The cost of travel, not only in money, but also in time and energy or discomfort associated with that travel, is now so cheap that there is a great deal of travel for all purposes over distances longer than those which are possible by walking. Suburbanites now typically travel by some mechanized means (usually the automobile, but sometimes public transport) for purposes of shopping, education, and recreation. Inexpensive transport has made possible the elimination of the one-room schoolhouse, with children now being carried by buses over long distances to large regional schools where the specialized teaching of different subjects can be conducted. In the highly developed economies of North America and Western Europe, among others, travel for social purposes is often such that a person is just as likely to have friends 5 or 10 mi away from home as in the immediate neighborhood. Similarly, various kinds of recreational and religious activities might be carried out many miles from where one lives, in marked contrast to the pattern which exists in societies where travel is not so cheap relative to income.

The ease of travel within the metropolitan area has brought with it a number of other changes. One is the wide range of options which this travel now gives to a family in the choice of place of residence relative to the place(s) of work. That range of options is extremely great, and this has enabled many people to move from places fairly close to their places of employment to more distant places, which are desirable for other reasons. While these persons who have moved probably have benefitted, their moves have resulted in the depletion of population in many older areas of central cities, often with very serious problems for those areas and ultimately financial and other problems for the entire central city itself. Similarly, stores and other businesses now can locate almost anywhere within a metropolitan area and expect potential buyers or visitors to travel to the business location. And with increasing amounts of travel, in total for the region and also on a per capita basis, the amount of congestion in the transportation system, particularly on roads, can increase significantly. This often carries with it objectionable or even dangerous levels of noise and air pollution, which can be particularly difficult to deal with in older areas where the road network was not designed for the large amount of current traffic.

The increase in accessibility, or ease of travel, due to mechanized transport has had other, perhaps less obvious, negative effects. One is that a certain level of income is necessary to use the new forms of transport. This is especially true of the automobile, for which a substantial initial outlay must be made. Also, many persons, such as the very young, the elderly, and those with handicaps that preclude driving, are not able to own and operate automobiles. For these persons, the extreme reliance on the auto for most urban travel in the United States, for example, has resulted in such persons being far less mobile than those who own automobiles. Thus a disparity in travel opportunities has been created. Only recently have programs been started to ensure a minimum level of mobility for all, these concentrating on transit service in suburban and rural areas where often none existed before.

The automobile has also become a symbol of wealth and status in society, rivaling homes and other major possessions in this regard. Since each automobile is available only to its owner and others to the extent they are invited, its use (as opposed to some forms of public transit) has undoubtedly reduced interaction among people. It is argued by many social scientists that this is undesirable because it reduces contact and hence creates barriers between different groups, particularly groups of different income levels, just as different classes in rail or air transport do, for example. Of course, the increased mobility due to mechanized transport may offset these tendencies, depending on the nature of any additional travel due to the greater ease of travel. Thus it is difficult to ascertain the effect of increased mobility on human interaction and awareness and understanding of others, the net effect being determined by the way in which people choose to use the increased mobility.

Long-Distance Travel

The discussion above has concentrated largely on travel in urban settlements or within the immediate vicinity of one's residence, in the case of rural families, but the advances in transportation technology have also had a very substantial impact on longer-distance and less regular travel. Traveling distances over 100 mi was virtually unheard of for vacation purposes a century ago, except for a relatively few wealthy families. With the advent of the railway, and later intercity buses, air service, and the automobile, travel at such distances is extremely common, not only for extended vacations, but also for single-day or weekend trips. This has vastly changed the range of experiences of persons within our society, it being very easy to travel to other sections of the country (and in some cases to other nations), which have different cultures, dominant industries, and patterns of living. This has undoubtedly had a very considerable effect on the level of understanding of different groups and the mutual respect of one socioeconomic group for another, and undoubtedly also has helped to bind nations with multiple cultural heritages together.

Intercity travel for business purposes has also been greatly facilitated, primarily in the last 50 years. While travel for the purpose of business contacts has always been a necessary characteristic of economic activities, until the advent of the airplane, such business travel was extremely limited, simply because the time consumed in and largely lost in travel was so great. Since the advent of extensive air service, the range of a 1-day business trip has been extended from perhaps 100 mi via land transport to an upper limit of well over 1000 mi (allowing about $1\frac{1}{2}$ or 2 h for travel to another city and a similar amount of time for the return). In fact, most business travel, which now is such a large component of air travel, simply did not exist before the advent of air service.

The development of air service has also facilitated vacations and other nonbusiness travel over very long distances. Not only has the time been reduced dramatically, but the cost often is less than that of land or water travel.

The effect of transportation improvements has been especially marked on the rural population. It is now possible for persons living in outlying areas to travel very easily to cities or other areas they would like to visit for evenings or for day or weekend trips. Since there seem to be many activities within urban areas which people naturally wish to be involved in, such as going to theaters, plays, restaurants, etc., the advent of cheap and easy travel over distances of up to perhaps 100 mi has enabled people in rural areas to partake of aspects of life from which they otherwise would be effectively separated. Also, persons living in rural areas are no longer isolated from specialized services, such as those of medicine, which enables the benefits of advanced medical care to be had by all.

Although reliable data on intercity travel are less complete than those for urban travel, national surveys by the U.S. Census Bureau provide an indication of the extent of travel for all purposes by United States residents (as reported in Transportation Association of America, 1974, p. 20). In 1972, 458 million trips to a place at least 100 mi from home were taken in the U.S., which, with a population of 209 million persons, averages 2.19 trips (of at least 200 mi round-trip length) per capita. Of these trips, 19 percent were over 1000 mi or more in length. In that same year there was one automobile for every two persons, and each auto was driven an average of about 10,000 mi/year. With over one-half the population possessing drivers' licenses, the degree of mobility via auto and the extent to which it is used are clear. Although the data on total person-miles of intercity travel in the United States are based on many assumptions which cannot be fully verified, it is estimated that in 1972 there were 1.36 trillion person-mi of intercity travel, or about 6500 mi/person, compared to a total of 311 billion person-mi, or 2400 mi/person, in 1939. From these diverse data, the substantial increase in travel, particularly with the advent of the airplane and the interstate freeway network, is evident.

Conclusions

The increased speed of transport and the reduction in the costs of transport have resulted in a much wider variety of spatial patterns for human activities. With the cheapness of transport, any desirable dispersion or concentration of population or of economic activities, is facilitated. In the past few decades there has been a migration from rural areas into urban areas, and within urban areas, a migration away from central cities and dense development to less dense suburban developments. These changes are the results of conscious choices on the part of people as to the manner and location of economic activities. Transport enables other patterns of activities to occur. Although many view the changes that have occurred to be undesirable, it is not the purpose of this chapter to discuss the advantages or disadvantages of any particular patterns or styles of life, but rather to point out the role transportation has had in making these options available.

POLITICAL ROLE OF TRANSPORTATION

The world is divided into numerous political units, within each of which the manner of government and laws are more or less uniform. Such political units are formed for mutual protection from possible enemies, for economic advantages, for enhancement of a common culture, among other reasons. Transportation plays an important role in the functioning of such political units. Also, because of the variety of forms which a transport system can take, given the range of transport technology now available, and the possible long-run effects of transportation systems on a society, the choice of the form of the transport system must be made with full recognition of political consequences.

Rule of an Area

A minimal transport system is a prerequisite for rule of an area by a single government. In order to govern an area, the government must be able to send information, at the very least, to all parts of the governed area, and to receive information from all parts. The information would include such items as the laws which are to be followed, the needs of the local area which are to be met by the central government, the needs of the central government which are to be met by the local area, and the manner in which the laws are to be enforced, and the system of justice to prevail.

Now, with the advent of communication technology and items such as radio and television, it is conceivable that an area could be governed from a distant place to which there were no transportation links (in the sense of a means of moving things and persons) but only communication links. In contrast, a century ago, almost all information had to be transmitted by the physical movement of a written message, or perhaps by word of mouth of a messenger. The only exceptions were message transmission by sound, as in the case of drums or bells, which have a limited vocabulary and limited viable distance, or by visual signals, as in the case of smoke patterns or lights and flags, which are limited in vocabulary and in distance as well as by weather and light. Therefore some means of transport of persons, perhaps carrying written messages, was essential for centralized or uniform government at the time when most nations of the world were being formed.

Any form of representative government is especially dependent upon the transport system for the movement of the representatives from the areas they represent to the seat of the government and back. Providing for their safe passage was a major concern of many governments in the last century and earlier. In the Middle Ages, feudal rulers invariably were accompanied on their journeys by armed escorts. Such escorts are still used in some parts of the world.

If a representative government is to be democratic, then there is an increased requirement for transport, to enable each representative to be in frequent contact with constituents. The more frequent the contact, the more rapidly can their views on any current issues or legislation be obtained. Again, with the advent of modern methods of communication, the transport requirement is lessened, although there is still some advantage to face-to-face contact over even the most sophisticated video or voice communication.

Transport becomes particularly important if it is necessary to send police or troops from the central government headquarters to other areas. This may be to defend the nation from foreign attack or to quell illegal activities. Of course, it can be used to the betterment or the detriment of the people, depending upon the circumstances.

Political Choices in Transport

As the discussion in the previous sections on the economic and social roles of transportation have indicated, the functions performed by a transportation system are many and varied. Since societies can exist in many different forms, each with its own transport requirements, there naturally exist many different forms which the transport system can take. At the most general level, the decision as to what form a nation's transport system will take is necessarily and properly a political one, for it is intimately tied to economic and social issues.

These choices may be described in terms of the various functions which the transportation system is intended to perform. These may be usefully classified for the purposes of this discussion as four: communication, military movements, travel of persons, and the movement of freight. These are in no sense mutually exclusive; a society can choose to develop a transport system which will satisfy any combination of these functional requirements. Each will be discussed in turn, with a few examples of societies which have had transport systems designed to meet that function.

As the above discussion indicated, one of the primary functions of transport systems is to provide for communication—the transfer of messages or information from one place to another, as opposed to the movement of persons or of things of value (for other than their information content). The transport systems of many early civilizations were designed to perform this function. Usually the messages were carried by messengers, sometimes on horseback, thus requiring frequent points at which horses could be changed, and if speed and hence continuous movement of the message were important, transfer of the message from one messenger to another. Examples of such transport systems include the Roman roads and military post houses, the early and very extensive road network in China, the United States pony express system of mail service to the West, and the early air service in the United States, which was for mail only.

A closely related function in many early civilizations was that of providing for the rapid movement of troops. The most famous network probably is that of the Romans, who built such substantial roads on good alignments that some portions still exist today. Their roads were designed to be all-weather routes, with a fairly smooth and hard surface. But they were designed primarily to be used by troops on foot, riders on horseback, and light carts of the chariot type, rather than by wagons. Their roads are noted for being very direct, with very few turns, but it is less well known that these roads followed the terrain very closely, which resulted in exceptionally steep grades and hence very low weight capacities for any wagon. In fact, the Romans had only very crude wagons, probably reflecting a disinterest in carrying goods or passengers for long distances over land. The wagons had axles fixed relative to the wagon body, so a two-axle wagon literally had to be slid and dragged around any turn! Most Roman trade over long distances was by water. The ancient road networks of the Chinese and the Incas similarly were designed more for communication and military purposes than for trade or travel.

Another function of transport facilities is to provide for the movement of persons and goods in general. This is usefully thought of in terms of intratown or intra-urban, intercity, and rural-urban connections.

The provisions for intra-urban movement within settlements or towns has varied widely over the centuries. At one extreme was the Mayan Indian policy of not designating any land for thoroughfares or other public purposes, but rather allowing buildings—including the very large temples for which this civilization is known—to be built in almost any location. The land which happened to remain between buildings was used as the public way. In most other societies, there was some sort of plan, or at least informal agreement, regarding the strips of land which would be public thoroughfares. In the larger and more affluent cities, such as those in the Roman Empire, the streets within towns were paved, with separate areas designated for pedestrians, and specific provisions were made for the movement of rainwater and sewage. In many other civilizations, though, such public ways remained completely unimproved. Congestion, of course, has been a characteristic of urban transportation from very early times, the first known regulations regarding traffic movement being imposed by the Romans and the Chinese, independently, over 2000 years ago.

Decisions regarding the form which intercity transport should take offered many more options for early civilizations. Many such civilizations engaged in extensive maritime transportation activities, primarily for the movement of freight, government officials, and of course, military units. The oceans of the world provide a natural path for movement, requiring little on the part of those using them beyond boats and piers. Undoubtedly, the practice of many early civilizations to engage in substantial freight movement over water, and to improve overland transportation to only a minimal extent—such as in the Egyptian Empire—was due to the fact that little initial expenditure is necessary to begin water trade. Boats and wharf facilities can be easily expanded in number and capacity as the trade warrants.

The wide variation in the forms that intercity transport networks have assumed and the relationship of these to the functions to be performed by the networks is graphically illustrated by comparing the Chinese network of about 1400 (which remained largely unchanged until the late 1800s) and the present United States railroad network. The Chinese network is shown in Fig. 2-9. It consists of two distinct parts. One is the network radiating out from Peking, the city established by Kublai Khan as the capital in the late thirteenth century. Also evident is the radiating pattern of roads from Sian-Fu, the former capital. These roads are primarily for political purposes, establishing communication with the central government. The other part of the network consists of the connections west from Sian-Fu and Sinyang to India and Arabia, which were developed by the first century A.D. to facilitate the supply of silk to Persia and the Roman Empire. With these two primary functions, in addition to purely local transport, there was no need for a more complete network.

In contrast to the earlier Chinese system is the United States railroad network. This system was built primarily by private firms as profit-making ventures in order to serve existing or potential traffic of persons and goods. As the map of Fig. 2-10 reveals, it is not only ubiquitous, serving almost all cities with over 50,000 people, but also is highly interconnected, permitting rather direct travel between almost all points except where physical barriers such as mountains intervene. While this network has played a very important role in wartime in moving people and material and until recently played an important role in postal communication (the truck and airplane having largely replaced it in this function), these functions were accommodated in the basic network, which was designed for a more general purpose. Other kinds of transport networks in the United States have developed similarly, to provide for general freight and passenger movements rather than for more limited purposes.

Figure 2-9 **The road network of China and the route to the west about 1400 A.D.** [*From Goodwin (1937), p. 174.*]

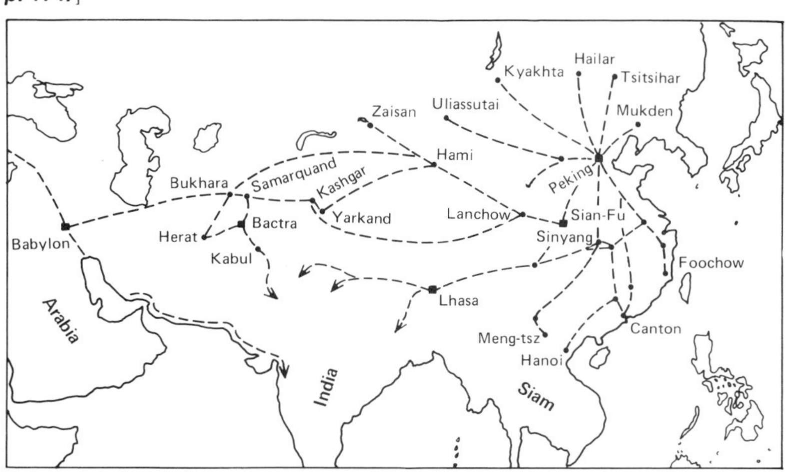

Figure 2-10 **The railroad network of the United States circa 1950–1955.** [*Association of American Railroads.*]

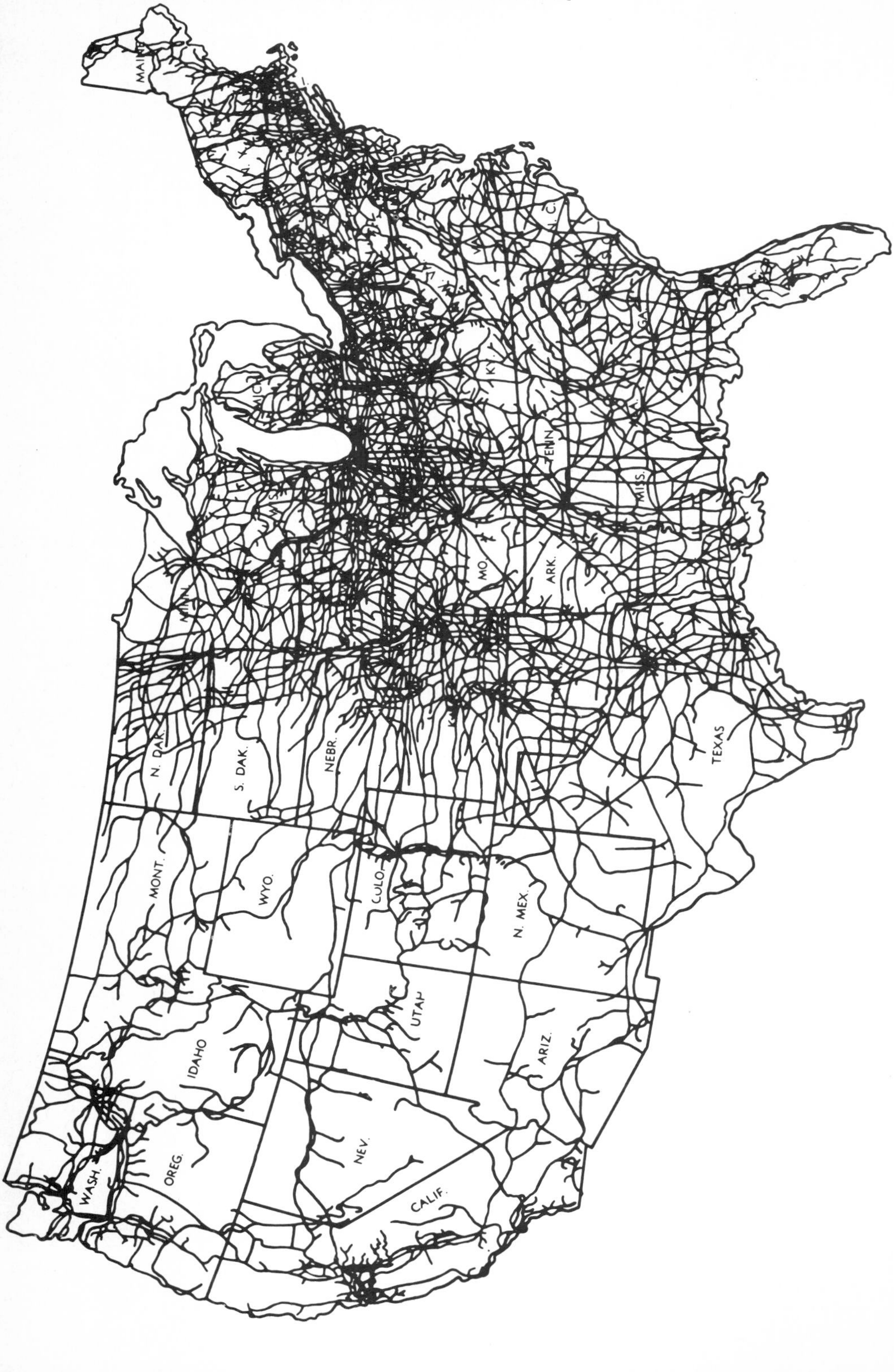

Responsibility for and Financing Transport: Another Political Choice

Extensive overland transportation requires a substantial initial investment in developing some sort of road system on which pack trains or vehicles can be operated efficiently, and then a continuing expense to maintain the system. In addition, expenditure is required on the part of the individuals or agencies which own and operate vehicles on the roadways. Improved roads of any type initially involve national governments in both the construction of the roads and their upkeep. Very often these improved roads were developed primarily for purposes of military movement and communications and the payment of tribute to the central government, the latter primarily between government units. Examples include the roads of the empires of Rome, China, and Central and South America. Thus, the extent to which roads were improved, and the extent to which improved roads would be maintained in their original condition, was largely a matter of government policy and the rise and fall of those governments. In the case of the very elaborate and permanent Roman roads, for example, they fell into disrepair and relative disuse after that empire was dismembered and Europe became governed by relatively small local, feudal units.

In many societies attempts to conscript local residents to maintain roads and other public facilities, primarily under the threat of imprisonment or fine, did not seem to provide a very effective means of ensuring adequate road maintenance. It was not until the development of tolls for the use of roads that an adequate and continuing pool of money was available to maintain and, as necessary, improve, highway facilities.

The political decision to have users pay for the construction and maintenance of road facilities seems to have had very far reaching consequences in the development of transportation systems. Because the financial viability of a road depends on the relationship between its costs and its revenues, such user charge systems become an incentive to construct or improve roads where the expected use will yield revenues consistent with the costs of the improvement. In this manner, economic forces are given a substantial weight in the provision of transportation facilities and in the improvement thereof. This then adds a very important dimension to the considerations of where road improvements will be made, the extent of those improvements, and the price to be charged for use of the improved facilities. Now transportation improvements will be made where there is an economic gain to be obtained from the improvement, a consideration which was not particularly important in the design of transportation networks for many early empires, such as the Roman and the Chinese, where essentially purely political factors, such as military and governmental rule, were the primary considerations. In fact, it is difficult to imagine the existence of such extensive networks of transportation as we have today in most developed areas without some means of charging those who use the facilities for their construction and upkeep. Of course, now the charging of users for most, if not the entire, cost of transport facilities, whether owned and/or maintained by government agencies or private firms, is very well established and almost universally followed.

Primarily in the past three centuries, the transportation systems of the world have evolved from ones which were designed primarily for military and government communication purposes to ones primarily for the general movement of persons and freight. In most developed countries, there is a ubiquitous network serving all parts of the nation with a wide variety of services for both the traveling public and shippers of freight. These transportation systems are primarily paid for by users, and the systems tend to adapt to changing patterns of movement primarily through the operation of a market mechanism or a public enterprise analog which considers benefits from improvements or changes relative to the cost of those changes. These systems may be slightly modified to cater to special needs for military movement, such as particularly large clearances under bridges on the National System of Interstate and Defense Highways in the United States, and airports may be designed for joint use by civilian aircraft as well as military aircraft, as is common in the Middle East today, but these adaptations are rather minimal. Also, with the advent of means of communication which are much cheaper and faster than transporting persons or written messages, the communication role of transport has diminished considerably, although of course its use for the letter mail still remains very important.

Summary

The political roles of transportation are thus essentially two: On the one hand, transportation, along with communication, facilitates the governing of a larger area by a single government and tends to promote uniformity in the application of laws and justice. On the other hand, the broad array of transport technologies and the options they provide as to the form and function of the transport system presents a problem of choice for all societies, one which is likely to have a significant influence on the economic and social structure of that society.

ENVIRONMENTAL ROLE OF TRANSPORTATION

In recent years, it has become increasingly apparent that many of humanity's productive activities have decided effects on the natural environment. These effects must be considered in any overall understanding of those activities. Transportation is among these. While the effect of transportation on the natural environment is primarily in the use of scarce resources, and this might be thought of as part of the cost of transport (and hence included in the section dealing with the economic effects of transport), the environmental effects must be considered separately because many of the resources used are not represented fully, if at all, in economic market transactions. When a price is involved, often it is felt to be an inadequate measure of the value of the resource being used. There are numerous categories into which the environmental effects may be placed; a convenient categorization for purposes of this discussion consists of the following: pollution, consumption of energy, consumption of land and aesthetics, and safety.

Pollution

One of the unwanted by-products of most technologies of transportation is the pollution of the natural environment. Perhaps the most serious form of this pollution, and the one most difficult to deal with, is the contamination of the air by various particles and gases. All forms of transport undoubtedly do this to some extent, but those forms which employ internal combustion (or much less commonly, external combustion) engines on the vehicles seem to emit the greatest amount of pollutants, especially in areas of population concentration where the pollutants are actually injurious to health in some cases. Included among these are highway motor vehicles of all types, and aircraft, the vehicles of other modes tending to be less common or operating in less congested areas. When the output of such vehicles is added to the pollution from other (stationary) sources, the resulting total concentration can be annoying or actually dangerous to health.

An indication of the role of transport in polluting the atmosphere is given by the statistics in Table 2-2. It can be seen that transportation is a major source of many pollutants, and hence that changes in the technology or operation of the transportation system may make a substantial improvement in the pollution level of the air. Since pollutants are diffused in the air, with the passage of time and the circulation of air currents, there is no precise relationship between pollution emissions and the quality of air, so it is not known exactly how much emissions must be cut in order to eliminate dangerous conditions, nor in fact can dangerous conditions be fully specified—research is still being conducted on these questions. However, the need to improve the quality of air in locations of major concentrations of activity (where transport also tends to be concentrated) is clear. This has already led to programs to alter the technology of transportation to reduce emissions, through the introduction of emission control devices on motor vehicles and aircraft, and in some areas, to programs to reduce the amount of vehicle movement.

Another form of pollution which is definitely annoying and may be harmful physically and psychologically is noise. This is an unwanted product of almost

Table 2-2 Sources of Air-Polluting Emissions in the United States in 1969 (in Percentages)

Source	Particulates	Sulfur oxides (SO_x)	Nitrogen oxides (NO_x)	Carbon monoxide (CO)	Hydrocarbon (HC)	Total
Transportation						
Motor vehicles						
Gasoline	0.4	0.2	5.0	2.6	3.2	11.4
Diesel	0.1	0.1	0.5	0.0	0.3	1.0
Total, motor vehicles	0.5	0.3	5.5	2.6	3.5	12.4
Aircraft	0.1	0.1	0.2	0.1	0.2	0.7
Railroads	0.1	0.2	0.1	0.0	0.0	0.4
Vessels	0.2	0.2	0.1	0.1	0.1	0.7
Nonhighway use of motor fuel	0.1	0.2	1.2	0.2	0.8	2.5
Total, transportation	1.0	1.0	7.1	3.0	4.6	16.7
Stationary power sources	10.4	20.2	3.8	0.1	1.1	35.6
Industrial process	20.9	6.2	0.1	0.3	0.1	27.6
Solid waste disposal	2.0	0.1	0.3	0.2	0.2	2.8
Miscellaneous	16.4	0.2	0.1	0.5	0.1	17.3
Total, all sources	50.7	27.7	11.4	4.1	6.1	100.0

Source: U.S. Dept. of Transportation (1972, p. 68).

all movement. Again, it is particularly a problem in the vicinity of roads, where vehicles operate at high speed or accelerate, and in the vicinity of airports. Noise from both of these sources finds its way to noise-sensitive environments, such as residences, schools, and hospitals. Other forms of transport also produce much noise, but usually in areas which are far less sensitive to it. There is currently much research on ways of either reducing noise at its source or in interrupting the path of transmission, and it is expected that a technological solution will alleviate much of this problem.

Another form of pollution or alteration of the natural environment by transportation is the transmission of vibration from the right-of-way of land transport systems to activities which are hampered or made ineffective by such vibration. This is mainly a problem with railroads and rapid transit lines, especially new subways in cities where the proposed line passes near activities which are disturbed by the vibration, such as certain types of laboratories. Also, noise can set up vibrations in fixed objects if it is of the proper frequency. Most problems here seem to have been avoided through relocation of routes and special supports for tracks, which dampen much of the vibration.

The pollution of water, and in effect pollution of the ground through pollution of water near the ground surface, is a problem in certain specific situations. As a matter of routine operational procedure, many vessels still release oil and wastes into the sea, rivers, and lakes, although currently there are efforts to reduce and possibly eliminate these practices. The accidental spilling of cargoes is also a problem, especially since extremely large tanker ships are used in international trade. Here also new regulations, fines, and requirements to pay damages for marine life lost and shores spoiled will force greater concern for safety and fewer accidents. Although land transport modes do discharge some wastes onto the ground, and of course occasionally lose cargoes which may pollute the ground or air, these are infrequent and tend to cause few problems for the natural environment, except in the case of hazardous cargoes, which will be discussed later under safety.

Energy Consumption

The consumption of fuel of various sorts is an example of a cost of transport which is often viewed as being improperly measured by the price of fuel. Hence, it is often treated as a distinct cost in addition to the usual monetary measure. The reason for such treatment is simply the fact that many developed nations, including the United States, and those in western Europe, consume far more oil than they can produce within their own borders, making them dependent upon other nations for a very crucial resource. Also, the world's supply of certain types of fuel, oil among them, is limited; and at present rates of usage is likely to be consumed within the next few decades. It has been estimated by Hirst (1973) that in 1970 one-quarter of all the energy consumed in the United States was used for transportation. Most of the energy used for transport was obtained from oil. About 90 percent of this oil was used in road and air transport alone, where it would not be possible to substitute some other fuel very readily. Thus in many nations there is concern for the amount and type of fuel used in each form of transport and an attempt to gradually shift to fuels which are abundant, as is coal in the United States, and to consume as little as possible.

It might be noted that this special treatment of oil is a prototype for the treatment of any resource with which one is concerned beyond the extent indicated by its price. In many nations, a similar treatment is accorded any items which must be purchased from another nation, in these cases there being a limited amount of foreign exchange currency or credit or a desire to avoid dependence on another nation for the essentials of the economy.

Land and Aesthetics

With the urbanization of population occurring in most nations, there is a pressing need to expand the capacity of urban transport facilities. Such expansion usually requires the use of land. Such additions to the urban transport system are usually in the form of freeways and rapid transit lines, which actually (surprising as this seems) add very little to the total amount of land in a typical city devoted to transport (almost all cities, ranging from the spread form of Los Angeles to the more compact form of New York, devote about 25 to 30 percent of their land surface to transport). Still, the requirement for particular parcels raises some serious problems whenever new facilities are planned. Land for transport must be in a continuous strip, of at least a minimum width, and usually for high-capacity facilities in urban areas it is desirable to depress or elevate it so as to avoid interference with cross traffic. This often results in a barrier to crossing the new facility except at certain places, a barrier which can disrupt neighborhoods and change activity patterns of the residents. Also, there is much displacement of families or firms which were in the path of the new facility, which results in a loss to the area. And of course, the pollution and perhaps the unsightliness of the new facility may make the area less desirable once it is built and in use. For all these reasons, a great deal of care now goes into the location of a new facility, and an attempt is made to integrate it into the communities traversed in as acceptable a way as possible, sometimes putting it underground, as is easily done (albeit at great cost) in the case of rapid transit lines.

Safety

One of the most disturbing by-products of transportation is injury and loss of life. In 1970, for example, over 58,000 persons lost their lives in the United States while engaged in transport. About 91 percent of those fatalities occurred on highway vehicles. Numerous programs have been developed to increase safety in transport, and in most nations there is a regular inspection of transport vehicles and equipment, at least in air transport and other high-speed modes. Also, there is usually some sort of licensing of operators of vehicles. And there is a continuing drive to make vehicle and facility designs that minimize the probability of accidents and minimize the damage done as a result of accidents. As a result of these and other efforts, the actual accident rate of most passenger modes is remarkably low, and is declining. These accident rates for various carriers, the number of deaths divided by the number of miles traveled in that mode measured in units of 100 million passenger-mi, is shown in Table 2-3 for 1972 and, for comparison, for 1947. The fatality rate has declined for most carriers, but general aviation (private aircraft as opposed to commercial service) and motorcycles stand out with the highest accident rates.

As mentioned earlier, a more recent safety problem is that of accidents

Table 2-3 **Fatality Rate per 100 Million Passenger-Mi in Transportation in the United States†**

	1947		1972	
	Number	Rate	Number	Rate
Automobiles and taxis	15,300	2.60	35,200	1.90
Motorcycles	n.a.	n.a.	2700	17.00
Local transit	n.a.	n.a.	n.a.	0.16
Buses	140	0.21	130	0.19
Railroads	75	0.16	48	0.53†
Domestic scheduled air carriers‡	199	3.15	160	0.13
Supplemental air carriers	n.a.	n.a.	0	0.00
General aviation	1352	90.00	1322§	42.06§
Water transport¶	1244	n.a.	n.a.	n.a.

† Includes travelers only, and neither employees or nonpassengers killed by vehicles nor deaths resulting from sabotage.
‡ 1971 rate was 0.24; 1973 rate, 0.07.
§ 1971 number and rate; 1972 not available; rate is per 100 million plane-mi.
¶ Mainly pleasure boating, but includes passenger and crew deaths.
n.a. indicates not available.
Source: Transportation Association of America (1974, p. 17).

involving freight shipments which release cargoes which are dangerous (U.S. National Transportation Safety Board, 1970). While one might think of circus trains releasing wild animals, the problem is actually one of releasing dangerous gases and liquids, and potentially at least, of harmful rays. In the United States in 1970 there were two incidents of communities having to be evacuated because of freight train accidents in which dangerous cargo was released. Also in 1970 a collision involving a barge carrying oil in the New Orleans harbor resulted in 25 deaths, and in 1968 a collision on the Mississippi River resulted in an extensive fire and 17 deaths. A similar problem exists in the case of ships on the ocean being damaged by storms, collisions, or accidental groundings, and releasing oil or other materials which are dangerous (and often deadly) to marine life. Pipelines are also very vulnerable, especially since they often carry highly volatile materials. With the increasingly large volume of cargo movement through areas where accidents can be disastrous, there is increasing attention being paid to ways of minimizing the probability of accidents and the aftermath, including stiff fines, redesigned vehicles, rerouting of movements around sensitive areas, and elaborate clean-up equipment. With the increasing variety of products being shipped and the increasing volume of transportation of things, such problems as these will require increasing attention in the next few years. Hopefully, the results will be as positive as those achieved in recent years in reducing person fatality rates, although even there much work remains to be done.

Conclusions

The environmental role of transport tends to be a negative one, at least as we have defined it relative to the economic and social roles for purposes of this discussion. However, it should be obvious that transportation does enable people to travel, and hence plays an important role in their ability to enjoy the natural environment. And, the ease and cheapness of freight transport presumably also enables society to choose which areas will be used as sources of natural resources, and hence considerations of preserving and enhancing the natural environment can be given greater weight—if the society in question chooses to do so. Nevertheless, there is also the other aspect of transport which has been emphasized in this section—the negative effects of transport on its immediate environment. To minimize this, while maintaining a desired standard of living, requires conscious effort.

TRANSPORTATION TODAY: A BRIEF DESCRIPTION

To reinforce the information on the various roles of transportation in society which were presented in previous sections, it is appropriate to examine a few overall statistics on travel and goods movement and the relationships of these to the remainder of the economy. Although the data to be presented are for the United States, the general conclusions would be the same for almost any other highly industrialized and urban society, such as many nations in Europe and the Pacific. It is recommended that the reader attempt to examine statistics for any area or nation of interest, using the discussion below as a guide.

The Physical System

The transportation system of the United States is very extensive, providing for mechanized movement on some type of improved road to all areas which are inhabited or otherwise used for economic (in a broad sense) activity. The paved road network touches virtually all recognized towns, and the networks of one or more other modes serve most areas of any significant traffic generation. In fact, in some ways the transport system may be too extensive, but this issue will be discussed later.

An appreciation of the extensiveness of the system can be gained from the following statistics. Most of these are measures of the route-miles of types of service or of types of facilities, but the exact measure necessarily differs from one mode or service to another. All data are for 1970 (U.S. Department of Transportation, 1972):

Roads
All types: 3,700,000 mi
Interstate freeway system: 41,400 mi

Intercity road transport
Truck service: on virtually all roads and areas
Scheduled bus service: 268,000 route-mi
Charter bus service: available virtually everywhere
Rental trucks: available virtually everywhere

Urban road transport
Bus and other transit service: 57,500 mi
Taxi service: available virtually everywhere
Households owning one or more automobiles: 79.6 percent

Railroads
Freight lines: 219,000 mi
Passenger routes: 50,000 mi

Airways

Low altitude: 133,000 mi
Jet altitude: 96,000 mi
Domestic scheduled routes: 162,000 mi
Airports: 10,126
Airports with scheduled service: 826

Inland waterways

Total: 26,000 mi

Pipelines

Oil: 219,300 mi
Natural gas: 217,000 mi

Total Expenditures on Transportation

As the prior discussion of the various roles of transportation suggests, transportation is a major component of industrialized, urban societies, as a result of which one expects a substantial amount of resources to be expended on transportation each year. That this is so for the United States is evident from the fact that in recent years (specifically, 1962 to 1975, the last year for which data is currently available) between 19 percent and 21 percent of Gross National Product (the total expenditure for all goods and services) has been devoted to transportation. For example, in 1970, out of a total GNP of $982.4 billion (measured in 1970 dollars), about $190.0 billion was spent on all transportation (U.S. Department of Transportation, 1972, p. 16). There are also other indicators of its importance as measured by the percentage of total national consumption of specific resources used in transportation: 12 percent of all employment, 75 percent of all rubber, 53 percent of all petroleum, 24 percent of all steel, 27 percent of all cement, and 67 percent of all lead, to mention a few (Transportation Association of America, 1974, p. 1).

Figure 2-11 reveals the distribution of the expenditures on transportation in 1970 among various sectors of the transportation system. All the dollar values in this figure are in 1969 dollars, so the total is $181.3 billion rather than $190.0 billion, but the percentages in each sector remain the same regardless of the dollars used. It is interesting to note that only 2.93 percent of the expenditures are for international transportation. The total freight bill (including service trucks) was $90.7 billion, and passenger expenditures were $90.5 billion, each being about 50 percent of the total. Although not shown on this chart, estimates of the split of expenditures between urban and nonurban transport indicates that about one-half of passenger expenditures are urban, as are one-half of the freight expenditures.

It should be pointed out that there are other, perhaps better, measures of the importance of transportation in our economy. GNP data reveal only expenditures, which presumably are less than the value of transport, assuming those traveling and shipping freight would not do so if the value or benefit did not at least equal the cost. However, estimates of the value added by transport are very difficult to make in many cases, and the data presented do provide a picture of current levels of expenditure.

Figure 2-11 Transportation expenditures in the United States in 1970. [*From U.S. Department of Transportation (1972), p. 15.*]

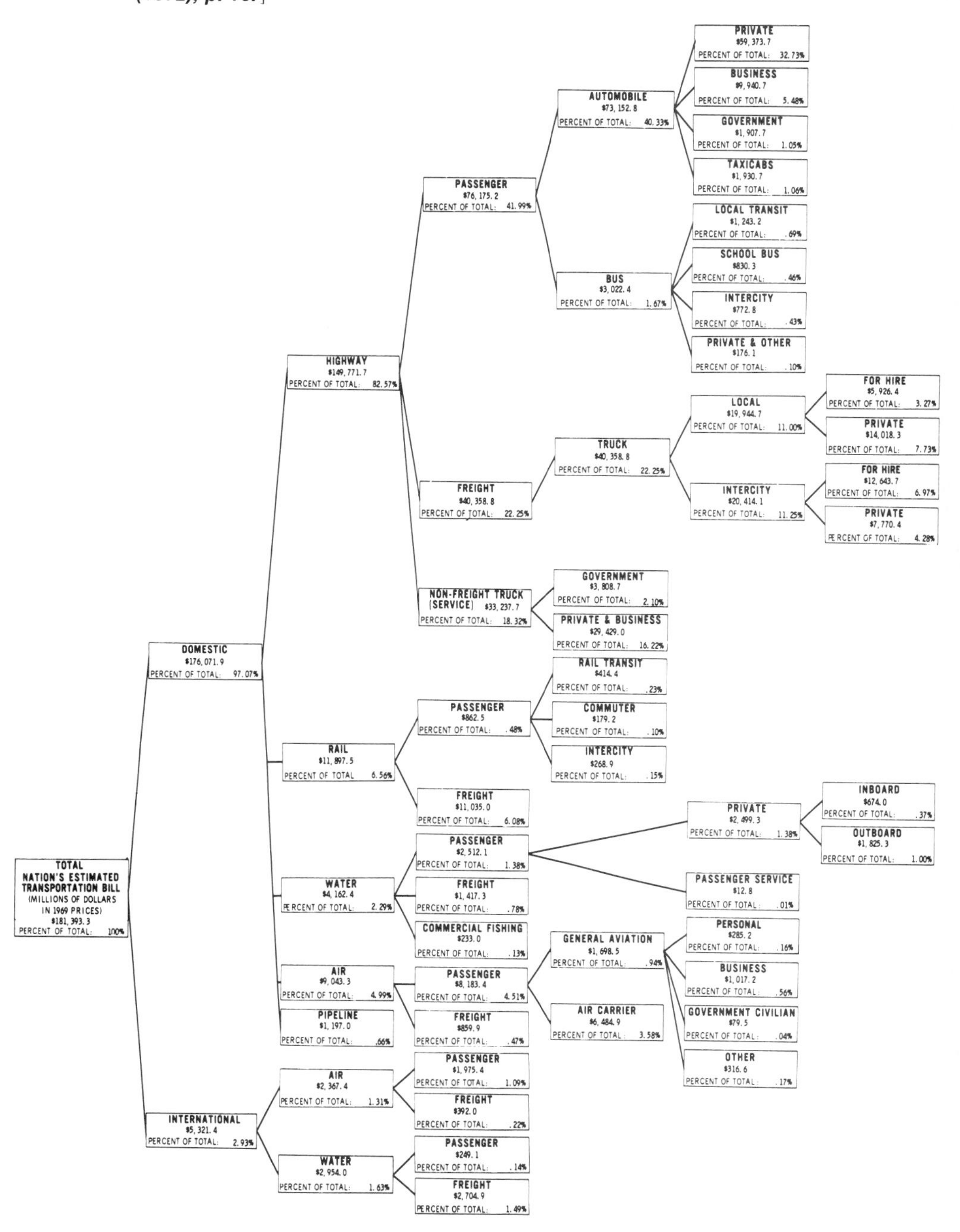

Table 2-4 United States Domestic Person Travel Trends from 1950 to 1970

	1950	1960	1970
Gross National Product (billions of dollars)†	487	673	982
Population (millions)†	152	181	205
Intercity travel†			
Total person-miles (billions)	505	784	1185
Average miles per person	3320	4330	5780
Percent person-miles via			
Air, public	1.8	4.0	9.3
Air, private	0.2	0.3	0.8
Auto	86.8	90.1	86.6
Bus	4.5	2.5	2.1
Rail	6.5	2.8	0.9
Water	0.2	0.3	0.3
Urban travel			
Total person-miles (billions)‡	310	430	710
Average miles per capita	2040	2380	3470
Percentage of passenger-miles via transit (not including taxicabs and school buses)	16‡	10.8§	5.5§
Total transit and taxi trips (millions)§	n.a.	9589	8557
Percentage of transit and taxi trips via§			
Taxicab	n.a.	19.0	27.8
Bus	n.a.	57.5	48.9
Rail (surface, street, and rapid)	n.a.	20.9	20.4
Commuter rail	n.a.	2.6	2.9

† Transportation Association of America (1974, pp. 2 and 18). GNP is in constant 1970 dollars.
‡ Hirst (1970, p. 19), estimate based on numerous assumptions and hence only approximate.
§ U.S. Department of Transportation (1972, pp. 189, 79, and 80).

Person Movement

In the period after World War II the average miles of domestic intercity travel per person has increased substantially. Table 2-4 presents data on total travel and travel per capita in 1950, 1960, and 1970. Over this period, the miles of travel per capita increased from 5360 to 9250, for an increase of 72 percent.

The modal distribution has also changed somewhat, although the automobile has long been the dominant intercity carrier of persons in the United States. Over this 20-year period, the major change in intercity travel has been in the division of the passenger-miles among the public or for-hire carriers—air, bus, and rail—air travel gained a much larger share of the market. However, it is important to notice that all passenger carriers except railroads experienced an increase in the passenger-miles carried over this period, the increase in total (all modes) passenger-miles offsetting declines in modal shares.

In urban areas, the same substantial increase in total movement is also noticed. This was probably primarily due to the spreading out of the growing population and increased land area of urban places, with the consequent tendency toward longer trips, and the increase in income which people choose to spend on travel. But in urban areas there was a substantial shift in the modal allocation of travel, with the absolute number and fraction via public transport carriers of all types except taxi diminishing. By 1970, the taxicab was used for more person trips in urban areas than all rail modes. However, very recently the total number of transit trips has been increasing slightly, but the transit percentage of all trips probably has continued to decline.

Freight Movement

Table 2-5 presents trends in the movement of domestic freight in the United States from 1950 to 1970, divided into intercity and urban components, as were the data on person movements. Intercity movement, as measured by ton-miles, increased by 82 percent over this period, while in the same period population increased by 34 percent and Gross National Product by 103 percent. Thus, intercity freight flows increased on a per capita basis, but not as fast as the GNP. This is typical in developed economies, where the increase in the population and the size of markets allow more production near the place of consumption. Also, other trends, such as the trend toward lighter-weight manufactured items, tend to reduce freight movement as measured by ton-miles.

The modal division of freight has also changed with the railroads' share declining somewhat while other carriers' shares have increased. It is particularly important to note, however, that even the railroads experienced an increase in the total amount of freight carried, measured by ton-miles. Also, the pipeline and water transport have become important in the movement of freight within the United States. Since in the course of daily activities, one typically encounters freight transported only by trains and trucks, it is all too easy to forget these other two very significant carriers.

In urban transport, the data on freight movement are very limited and do not provide a very reliable picture. This is partly because the government has lacked interest until recently, partly because data are difficult to gather, especially on the vast amount of trucking carried out by firms moving their own products (such as household goods), and partly because it is a problem to define adequate measures, since much truck traffic is for purposes in which the movement of freight is rather ancillary, as in the case of household service and repair trucks. Nevertheless, it is clear that most urban freight is via truck,

Table 2-5 United States Domestic Freight Movement Trends from 1950 to 1970

	1950	1960	1970
Gross National Product (billions of dollars)†	548	673	982
Population (millions)†	152	181	205
Intercity shipments†			
Total ton-miles (billions)	1063	1314	1936
Average ton-miles per capita	6990	7260	9440
Percent ton-miles via			
Air	0.03	0.07	0.17
Oil pipeline	12.1	17.4	22.3
Rail	56.2	44.1	39.7
Truck	16.3	21.8	21.3
Water	15.4	16.7	16.5
Urban shipments‡			
Truck vehicle-miles (millions)	n.a.	44,687	80,606
Average vehicle-miles per capita	n.a.	247	393

† Transportation Association of America (1974, pp. 2 and 8). GNP is in constant 1970 dollars.
‡ U.S. Department of Transportation (1972, p. 78).

and hence the measure truck-miles is presented in Table 2-5 as the indicator of total urban goods movement activity. As measured by this indicator, the total amount of urban goods movement has increased by about 81 percent from 1960 to 1970, or on a per capita basis, by about 59 percent. It is interesting that the total expenditures on trucking in 1970 were almost equally divided between urban (49.4 percent) and intercity movements (50.6 percent), but the private trucking share of urban movement expenditures was 34.7 percent, and for-hire, 14.7 percent, in contrast to 19.3 percent and 31.3 percent, respectively, in the case of intercity movements. However, as was revealed by the data on goods movement in the New York metropolitan area (see Fig. 2-5), both railroads and ships are significant carriers of freight within urban areas, also.

Conclusion

This very brief description of the current transport system and its usage has been presented in order to give more meaning to the description of the roles of transportation in modern society. It is important that the engineer and planner keep these in mind throughout their work, for their tasks often involve trying to meet many objectives—each reflecting different roles of transport—rather than simply trying to design and implement the most efficient or technologically advanced system for movement.

SUMMARY

1 The roles of transportation in society can be approached in many different ways. One useful classification is according to economic, social, political, and environmental roles.

2 From an economic perspective, transport makes it possible to move goods from one place to another, the latter presumably being a place where the good is more useful or valuable. Key concepts are those of place utility, time utility, and quality utility. Reductions in the cost of transport have made it increasingly possible to overcome differences in the goods available in different areas.

3 The social role of transport has permitted people to change from nomadic creatures to ones living in permanent settlements. As transport has become cheaper and easier, these settlements have become larger and less dependent upon local resources for support.

4 Transportation has had a substantial influence on political characteristics of society, enabling large areas to be governed from one place and fostering representative government of large nations, for example. Also, it poses some major political questions, such as what functions the transport system will serve (e.g., military versus general travel by citizens) and who will provide and maintain the system.

5 The environmental effects of transportation are usually viewed negatively, as producing such phenomena as air and noise pollution.

6 Most societies have taken advantage of new forms of transport which reduce costs (in resources, time, or other ways) by consuming more

transport. Data for the United States illustrate the very large amount of freight transport and person travel per capita currently characteristic of developed nations.

PROBLEMS

2-1

Calculate the average revenue per ton-mile for each of the modes and movements (to, from, and within the region) of freight traffic of the Tri-State New York Region. The data needed are presented in Fig. 2-5.

a Plot the average revenue versus length of haul for the rail, truck, and ship movements. Is there any evidence of a pattern?

b What is the average length of haul for freight via all modes taken together, into the region, out of the region, and within the region?

2-2

Keep a record of all the things you purchase in a week. Also note where they were manufactured or last processed (usually noted on the package or object). Calculate approximately the ton-miles of transport involved in getting all these things to you. Use airline distances if no others are available.

2-3

Discuss possible effects of the development of an inexpensive and reliable form of videophone and an accompanying facsimile transmitter (in which a photocopy of anything you write can be sent over telephone lines and printed on paper at the other end of the videophone connection).

2-4

What major changes in transport technology do you foresee in the future? To what extent are they likely to reduce the cost of movement or make it more rapid or easy? What effects on where and how we live do you foresee as a result of this new technology being available?

2-5

a Calculate the total ton-miles of freight moved by the various domestic intercity carriers in 1950, 1960, and 1970, using the data presented in Table 2-5.

b Do the same for intercity passenger-miles of movement, using the data contained in Table 2-4.

REFERENCES

Aldrich, Nelson W. (1893), "Wholesale Prices, Wages, and Transportation," Report to the U.S. Congress, U.S. Government Printing Office, Washington, D.C.

Dunbar, Seymour (1937), "A History of Travel in America," Tudor, New York.

Goodwin, Astley J. H. (1937), "Communication Has Been Established," Methuen, London.

Hay, William W. (1961), "An Introduction to Transportation Engineering," Wiley, New York.

Hirst, Eric (1973), "Energy Intensiveness of Passenger and Freight Transport Modes: 1950–1970," Report ORNL-NSF-EP-44, Oak Ridge National Laboratories, Oak Ridge, TN.

Hoover, Edgar M. (1937), "Location Theory and the Shoe and Leather Industries," Harvard Economic Studies vol. LV, Harvard, Cambridge, MA.

Interstate Commerce Commission, "Annual Report on the Statistics of Railways in the U.S. for the Year Ended [Year]," U.S. Government Printing Office, Washington, D.C.

Lardner, Dionysius (1850), "Railway Economy," London.

Sampson, Roy J., and Martin T. Farris (1975), "Domestic Transportation: Practice, Theory, and Policy," Houghton Mifflin, Boston.

Schumer, Leslie A. (1964), "The Elements of Transport," Butterworths, Sydney.

Transportation Association of America (1974), "Transportation Facts and Trends," Washington, D.C.

Tri-State Transportation Commission (1966), "Freight Traffic Regional Profile," vol. 1, no. 1, New York.

U.S. Dept. of Labor (1975), "Handbook of Labor Statistics—1975—Reference Edition," Bulletin 1865, U.S. Government Printing Office, Washington, D.C.

U.S. Dept. of Transportation, Office of the Secretary (1972), "1972 National Transportation Report," U.S. Government Printing Office, Washington, D.C.

U.S. Dept. of Transportation, Office of the Secretary (1975), "A Statement of National Transportation Policy," U.S. Government Printing Office, Washington, D.C.

U.S. National Transportation Safety Board (1970), "Annual Report to Congress," U.S. Government Printing Office, Washington, D.C.

Zipf, G. K. (1949), "Human Behavior and the Principle of Least Effort," Addison-Wesley, Cambridge, MA.

2

The Technology of Transportation

Components of Transportation Systems

3.

Despite the great diversity of components of transportation systems and of manufactured means of transport, transportation systems possess a common set of functional components. In fact, this commonality of components is one of the reasons why there can be a unified field of transportation engineering, in addition to the more specialized fields dealing with individual modes. In this chapter we will attempt to discover these basic components and the ways in which they are assembled to form functioning transportation systems.

TRANSPORT TECHNOLOGY

Prerequisites of a Transport Technology

The function of transportation systems is to provide for the movement of things. The objects to be moved may include inanimate objects, such as natural resources, manufactured items, or foodstuffs, as well as living objects, including persons, animals, and plants. With the exception of persons and animals, most of these items are not suited to movement in their natural state. They require suitable transport technology. In addition, even persons and animals are subject to limitations in their inherent capabilities for movement, in particular, limitations on the maximum speed of travel and distance that can be traversed between periods of rest, so that these capabilities must be supplemented for even the most common types of trips, such as the journey to work.

In order to provide a broad capability for movement of the wide range of animate and inanimate objects which must be transported quickly and over great distances in modern society, humans have developed and perfected many technologies to assist them in transport. A transportation technology must perform the following tasks:

1 Give the object mobility, the capability of being moved without damage. For example, most products of manufacture cannot be safely transported by dragging, rolling, or floating, but rather must be carried in some manner.

2 Provide control of the movement through the application of the forces necessary to accelerate and decelerate the object, overcome the normal resistance to motion, and guide the object—all without damage to it. This property is termed locomotion. In many cases this is done through mechanical forces acting on the object, such as when the object is in a vehicle, which functions to move it in the desired path as well as to protect it. The application of this force must be controlled so that the object is moved without colliding with any items which might be in its path. This need is particularly apparent in the case of vehicular systems, where the danger of collision may be great and the potential damage substantial.

3 Protect the object from any deterioration or damage that may occur as a by-product of its being moved. This is especially important in the case of living things and perishables for which maintenance of the proper environmental temperature, pressure, humidity, etc., may be essential.

Natural Forms of Movement

Over the course of history many different transport technologies have been developed. Most of these are refinements of technologies of movement found in nature. The natural environment abounds in movement, movement which occurs without human involvement. Examples include the walking, running, leaping, and sliding of animals on the earth's surface. Similarly, the swimming of animals in water and the flying of birds in the air are types of natural movement. There is also substantial movement of inanimate things: the rolling of logs and stones down slopes, the floating of wood in streams, the transfer of large quantities of soil and debris by the flow of water and ice, and of course the natural movements of air and water. A more structured examination of natural and manufactured sources of movement is required.

Certain of the objects which must be moved in modern society possess a natural capability for movement which can be taken advantage of. This is obviously the case with persons who can walk and run, as well as all animals—the capability of movement is one aspect of animals which differentiates them from plants.

Similarly, liquids and gases possess inherent capacity for movement. Of the prerequisites for movement identified previously, both liquids and gases generally possess that of mobility without being damaged. The main problems are in providing guidance, as is done by a natural channel in which a liquid might be flowing, and in applying the forces necessary to propel the material, as is done by gravity when the material is flowing downward. Gases are moved above the surface of the earth by various forces, including those due to the rotation of the earth and pressure differences in the atmosphere.

In contrast to liquids and gases, solids for the most part have relatively little inherent capability for movement. Even though solids can be converted into liquids or gases, such transformations often destroy essential properties of objects. Thus, while it is possible to transform metal ingots to liquid at very high temperatures, doing so to a refrigerator would be absurd, even if it were feasible to maintain the high temperature required over long distances. Transforming solids to liquids or gases by raising the temperature, reducing the ambient pressure, or a combination of both is done to facilitate movement only in very rare instances due to the difficulties associated with the maintenance of the required environmental conditions.

However, many solids are buoyant in liquids or gases, and therefore can be moved by placement in a stream of liquid or gas. This is done for the movement of many bulk commodities which are not damaged by such an environment. For example, ores have been moved in water streams for centuries, as have logs. Grains and other light-weight bulk materials are now being moved over short distances in high-velocity streams of air.

The remaining form in which movement occurs in nature is that of humans or animals carrying things themselves. With the exception of limited capabilities for the transport of liquids or gases in body cavities, this form is limited to the carrying of solids. If a liquid or gas is to be moved, it is placed in an appropriate container. Such movement is of course not limited to overland travel, but can occur in water—as a person or other animal swimming on or below the water surface—and in the air by birds. In all cases, the animal or human provides most of the force necessary to achieve the desired movement and to guide it along the desired path.

Technologies of Transport

Despite the wide variety of ways objects are transported in nature, these are not sufficient to meet the transportation requirements of modern society. Therefore, most transport is undertaken by means which are largely man-made. Nevertheless, these means draw very heavily from the natural forms which we have just discussed.

Perhaps the most widespread form of transport is the use of vehicles operating on the land surface. The concept is essentially one of replacing a human or animal carrier with a machine designed to perform the same functions. The most common vehicle has wheels, which give it mobility, and a body which is designed to contain and protect the load. Vehicles have been designed to traverse almost any terrain, military vehicles being perhaps the most versatile in this regard. To spread the weight of the vehicle and its load sufficiently to avoid sinking in the soil, often tracks are used. There are substantial economies to be gained from providing a smooth, hard, prepared path for a vehicle, gains in the form of higher potential speeds, reduced resistance to motion and hence reduced power requirements, capacity to carry heavier loads, and lessened difficulty in protecting the contents from damage. This has led to the development of prepared paths in the form of roads and railroads, in particular. The most common form of propulsion of wheeled land vehicles is through the application of a force to rotate the wheels, with path-wheel friction providing the necessary reaction forces. Guidance so that the vehicle follows the path is provided by friction in the case of road vehicles and reaction of the rail against the wheels in the case of rail vehicles. Examples of off-the-road, road, and railroad vehicles are illustrated in Fig. 3-1.

Rather recently other forms of vehicular ground transport have been developed. These include the provision of mobility by means of an air cushion under the vehicle, with sufficient pressure to raise the vehicle above the

Figure 3-1 **Various land-surface vehicle technologies. *(a)* Off-the-road vehicles: wheeled recreational auto *(Jeep Corp.)* and tracked military tank *(Chrysler Corp.)*. *(b)* Wheeled road vehicles: automobile *(General Motors Corp.)*; truck *(General Motors Corp.)*; bus *(Flxible Co.)*. *(c)* Railroad vehicle: freight train powered by diesel-electric locomotives *(Consolidated Rail Corp.)*.**

pathway, and the use of magnetic forces (in repulsion) to accomplish the same objective. Since there is no longer a physical contact between the vehicle and the path, some means other than mechanical friction is necessary to provide propulsion and guidance. This can be accomplished by propellers or jets acting against the air, or electromagnetic forces between the vehicle and the path, as in the case of a linear induction electric motor. These forms of vehicles are shown in Fig. 3-2.

(a)

(c)

(b)

(c)

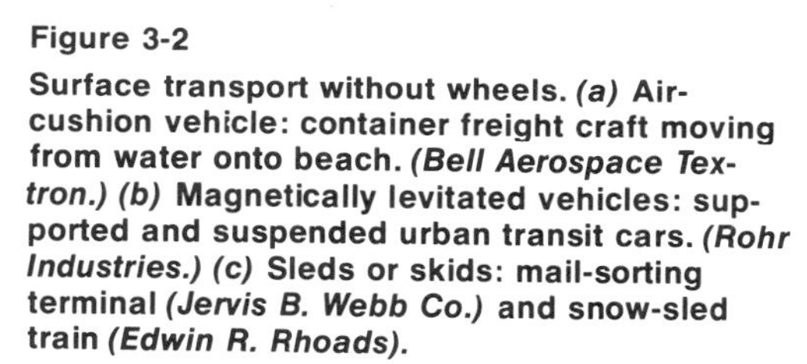

Figure 3-2

Surface transport without wheels. *(a)* Air-cushion vehicle: container freight craft moving from water onto beach. *(Bell Aerospace Textron.)* *(b)* Magnetically levitated vehicles: supported and suspended urban transit cars. *(Rohr Industries.)* *(c)* Sleds or skids: mail-sorting terminal *(Jervis B. Webb Co.)* and snow-sled train *(Edwin R. Rhoads)*.

Vehicle technology is not limited to movement on the land surface, but is also applicable to movement on water and in the air. In either context, the vehicle may be maintained at the proper elevation by buoyancy in the medium (as a ship in water and an airship in air), by reaction forces from the medium (as in the case of helicopters), or by use of an airfoil to create uplift forces as a result of the relative motion of the vehicle and the medium (as in hydrofoil boats and aircraft). Vehicles are controlled by varying the forces against the medium. Examples of these are presented in Fig. 3-3.

In very limited contexts what is in effect vehicular transport is provided with vehicles which may not be thought of as such. Included in this category are sled and skid devices. Their use is rather limited to special situations, such as over smooth paths in factories (which may be greased) and over snow, ice, or mud (see Fig. 3-2).

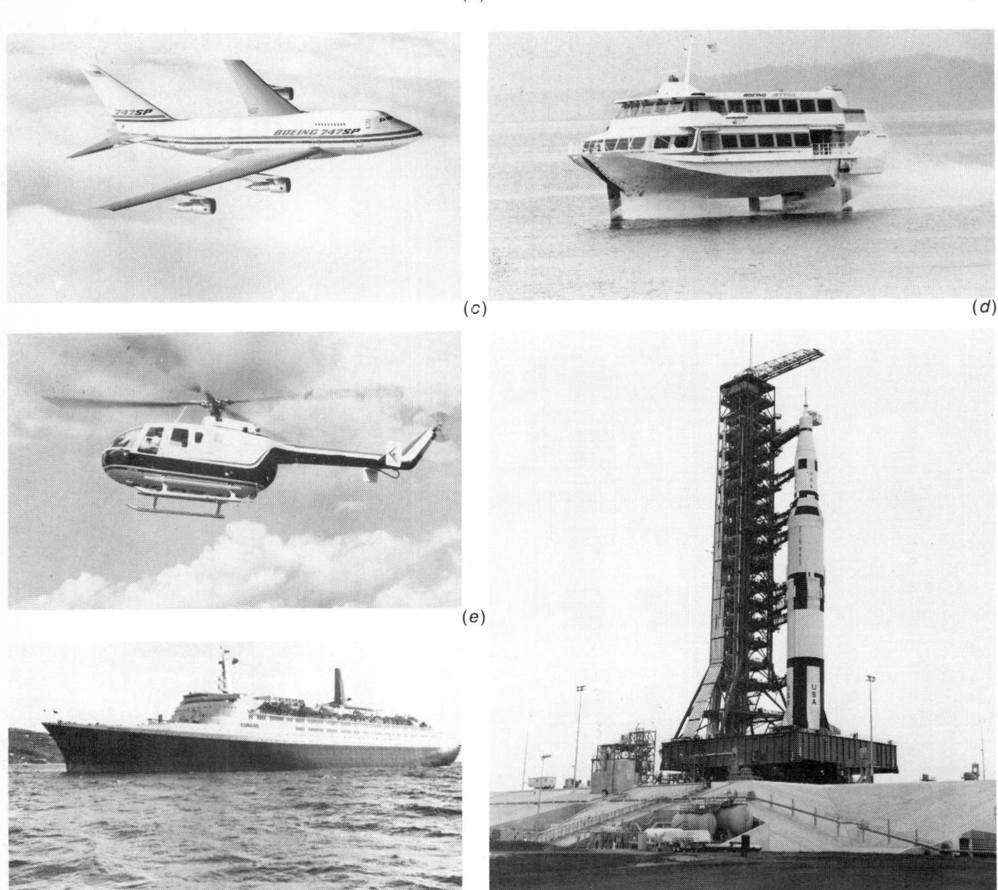

Figure 3-3 **Vehicles for movement in fluids. *(a)* Aircraft: long-distance passenger jet. *(Boeing Commercial Airplane Co.)* *(b)* Hydrofoil craft: passenger carrier. *(Boeing Marine Systems.)* *(c)* Helicopter: small civilian type. *(Boeing Vertol Co.)* *(d)* Rocket: type used in exploration of space. *(Boeing Aerospace Co.)* *(e)* Ship. *(Cunard Lines.)***

Another extreme example is that of objects which can themselves be slid or rolled over land or floated through water without being damaged. An appropriate path must be provided, but the "vehicle" comes automatically with the object to be moved.

The walking of persons and movement of animals on the surface is akin to vehicular transport in that an appropriate path must also be provided. Regular paths of movement which are smooth, level, and free of obstructions have existed since recorded history. Now paths for pedestrians are deliberately constructed with this in mind. Of course, the person or animal provides the necessary control of the movement.

Humans have also adapted the natural means of movement of liquids and gases to their purposes. The main problem in utilizing this form of transport has been that paths do not naturally exist along all routes where such movement is desired. Therefore paths have been constructed, either open channels, which are useful only for liquids for the most part, or pipelines. Gravity is usually used for propulsion in open channels, with mechanical devices used to stop the flow, while pumps of various sorts can be used with pipelines to force the gas or liquid to move in the desired manner. Often solids can be carried in the fluid, although this is most often limited to solids which are buoyant in the fluid. Figure 3-4 illustrates this type of transport.

Between the discrete movement of objects on vehicles and the continuous movement of gases and liquids in pipelines or channels lies a hybrid form of transport. Here, mobility is provided by a device which rests upon the ground and has a moving surface. Conveyor belts represent one such type of system, as shown in Fig. 3-5. Another is a roller line, in which the objects to be moved are transported along a path consisting of rollers placed sufficiently close together to provide a smooth path with almost continuous support. Yet another example is the cable line, or aerial tramway, in which the objects are suspended from a cable which is constrained to move along a predetermined

Figure 3-4 **Forms of fluid flow. *(a)* Open water channel with gravity flow. *(U.S. Army.)* *(b)* Oil pipeline. *(Alyeska Pipeline Service Co.)* *(c)* Transport of logs in water sluice and river. Log sluices and log driving as shown here were last used in the United States in the summer of 1977, when log driving was concluded in the state of Maine. *(Scott Paper Co.)* *(d)* Pneumatic pipes used to unload grain from rail cars. *(CEA · Carter-Day Co.)***

BOSE
FRAGILE-HANDLE WITH CARE
301
DO NOT DROP

RCA
Portable TV

path at speeds governed by rollers which propel it. This form is best known for its application to ski lifts, although it is commonly used in factories and other locations for short-distance movement of items which must follow one another in almost continuous succession.

There are undoubtedly many forms of transport which we have not discussed, at least ones which were not discussed in the sense of using the common name for it and describing the details. But humans are always inventing new ways of doing things, so in a sense any listing would be incomplete. It is more important to turn to the identification of the basic components of all transportation systems, so that the treatment of transport technology can proceed in a more structured manner.

TRANSPORTATION SYSTEMS

Components of Transport Systems

The preceding description of transport technologies has laid the groundwork for the identification of the basic functional elements which are found in all transport systems. The first of these are those which are directly involved in the movement of objects from place to place. In all of the means or technologies discussed above—either natural or manmade—there exist the *object* and a *path* in which it moves. The object is that which is to be moved—the passengers or freight—and the path is the location in space along which it flows.

In addition, in many of the technologies, the object moves in some type of *vehicle,* which gives the object mobility on the particular type of path employed and which can be propelled on that path. It may also serve to protect the object from damage. The most common vehicles are of course those for land, air, or water transport, and typical examples of these were shown in Figs. 3-1 through 3-3. Another example of a vehicle is the capsule used to contain solid materials for transport in a liquid or gas pipeline. In this context the capsule in the fluid has a natural mobility and it is used to protect the solids being carried from damage and contamination. Since vehicular transport is so common, the vehicle is identified as a third functional component. However, it must be realized that some objects possess the inherent property of having mobility relative to paths, and in these instances the vehicle need not be separately supplied because the object itself performs the function of the vehicle.

Akin to the vehicle in its function is the *container*. A container is a device into (or onto) which the objects to be transported are placed in order to facilitate the movement. The container differs from the vehicle in one respect, however, namely that the container does not itself possess either mobility or the capability of propelling itself on the path, or both. A container must be placed on a vehicle or in an appropriate capsule, or combined with additional apparatus in order for it to become part of a vehicle. Thus the primary

Figure 3-5 **Continuous solid transport.** ***(a)*** **10-mi-long conveyor belt as used in construction of large dams** ***(Hennes and Ekse, 1949, p. 581).*** ***(b)*** **Roller line for transferring freight within a warehouse and terminal.** ***(c)*** **Conveyor connecting different levels of a factory.** ***(Jervis B. Webb Co.)***

function of the container is to protect the objects to be moved and to facilitate loading and unloading by keeping the entire shipment together as one unit.

By this definition, a truck tractor plus trailer combination is a vehicle when being operated over the highway. But the trailer alone is not a vehicle, since the trailer alone cannot be propelled in the usual way, with adequate control, without the tractor unit. Similarly, that same trailer is a container when it is being carried on board a railroad flat car or on a ship. It has no inherent mobility with respect to the railroad track or the waterway and must be placed on a vehicle in order to be transported via those technologies. Its function while being so carried is to contain and protect the cargo and to facilitate loading and unloading, exactly the same functions performed by the crates or cartons in which small items might be placed for shipment. This point is further illustrated by a type of container which is designed as a truck body, but which can be attached easily to an over-the-road truck chassis, to a railway undercarriage, to an ocean vessel, or to an aircraft, as illustrated in Fig. 3-6.

In most transport systems, a separate path does not connect each possible origin and destination. This is usually far too expensive in many ways: the use of land, the labor and materials required for maintenance, and initial construction costs. Hence, paths exist at particular locations. However, provision is made for accommodating the traffic between many origins and destinations by linking the paths together, allowing options in the choice of routing and hence in the places reached. This results in an important distinction between two components of paths—the *way link* and the *way intersection*. Links are paths in which the flow is constrained to follow a particular route, as in the case of a railway track, highway (between intersections), a pipe or channel for fluids, and an air lane for aircraft. Flows of two or more links can be merged together at intersections, and a single flow can be separated to follow two or more distinct paths at intersections.

As in the case of vehicles and containers, discretion must be used in defining the links and intersections in particular technologies. For example, in road transport, freedom usually exists to pass from lane to lane or even off the roadway at any point along the road. Yet we will not usually be concerned with such minor variations in the path of movement, so only those sections of roads which are commonly referred to as intersections will usually be termed such for our purposes. Thus even though road links, sections of road between intersections, allow some freedom of choice of path, we usually will treat them as links.

A necessary function in any transportation system is accepting objects to be moved into the system and getting them out of the system at the end of the journey. Also, the movement from origin to destination may entail the movement over more than one technology or mode, requiring a transfer from one mode to another. Even within a single mode, it may be desirable to transfer traffic from one vehicle to another. All of these operations are similar in that the traffic is being transferred from one vehicle or container to another. This function is performed by *terminals*.

For most common technologies of transport, terminals are distinct and often large facilities. Airports, ocean ports, railway stations, are all examples. Yet the same function is performed by the local bus stop at a street corner,

(*a*)

(*b*)

(*c*)

(*d*)

(*e*)

(*f*)

Figure 3-6

Containers for freight in various vehicles. *(a)* Containers on truck chassis. *(Sea-Land Service, Inc.)* *(b)* Containers and truck trailers being loaded onto railroad flat cars. *(Santa Fe Railway.)* *(c)* Containers on ocean vessel with loading cranes in background. *(Sea-Land Service, Inc.)* *(d)* Barges that are floated between docks and ships are carried as containers on this Seabee vessel. *(Lykes Lines.)* *(e)* Roll-on/roll-off trailer-carrying vessel. *(Sun Shipbuilding and Dry Dock Co.)* *(f)* Containers being loaded onto aircraft. *(Flying Tiger Line.)*

(a)

(b)

(c)

Figure 3-7 **Examples of fixed facilities of transportation systems. *(a)* A link: a double track railroad main line with crossover. *(Santa Fe Railway.)* *(b)* Intersections: interchanges connecting freeway and other major highway links. *(U.S. Department of Transportation, Federal Highway Administration.)* *(c)* A terminal: aerial view of large metropolitan airport. *(Chicago Department of Aviation.)***

which may be no more than a place for passengers to stand while waiting and a sign designating the location as a bus stop. In fact, with most land transport technologies, the terminal function could be performed at almost any point along a link where vehicles could stop to load and unload. In wilderness regions this is often the case: Trains, buses, trucks, and water craft stop on call wherever traffic is to be handled. Such is not universally practiced because of the delays to other traffic on the vehicle caused by stopping, and because of the greater economy which can be attained by aggregating traffic at particular points and equipping these for rapid, safe, and, in the case of passengers, comfortable, loading and unloading. Also, if loading and unloading take place only at designated places, the design of the vehicles and the design of the terminals may be jointly optimized so as to achieve lower costs, a higher level of service, or both, for the system as a whole.

The way links, way intersections, and terminals of transportation systems are often referred to as fixed facilities, since they are fixed in location (unlike vehicles or containers). Typical examples of these are illustrated in Fig. 3-7.

A final necessary component of transportation systems is an *operations plan.* Most transport systems are very large, consisting of hundreds if not thousands of components. Thousands of distinct movements of traffic may occur in a single day. It is essential that the terminals be operated in such a manner that the traffic flowing through them can be accommodated, that vehicles (if employed) are available to accept the traffic, that the traffic is routed via the proper links and intersections through the system to the final destination of the traffic. All of this requires substantial coordination of the activities of each of the components. The set of procedures by which this is done is termed the operations plan.

Two other types of components might be separately identified: the maintenance subsystem, and the information and control subsystem. Since maintenance is an activity which occurs everywhere on the system rather than at particular facilities only, we shall treat it primarily as a cost and

Figure 3-8 Basic components of transportation systems.

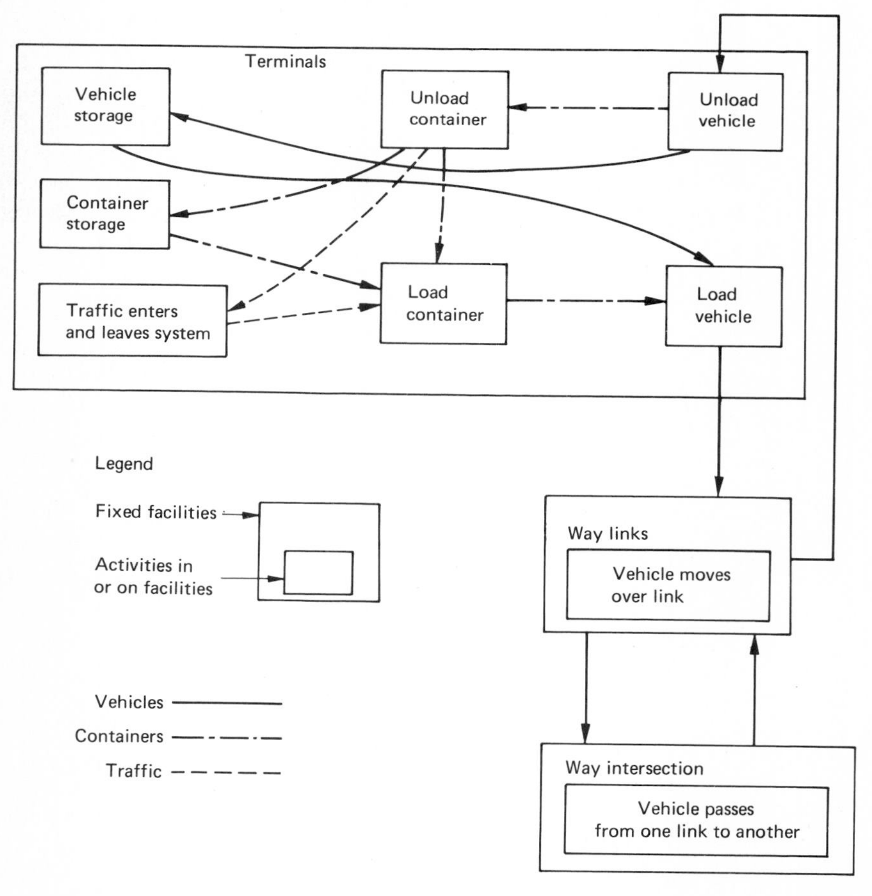

The operations plan governs the flows of vehicles, containers, and traffic through the system.

management function associated with each of the physical components identified previously, not separately. Information and control subsystems will similarly be treated in the context of the components and operations where they apply. For example, the information and control subsystem employed to detect objects in the path of a motor vehicle and cause the vehicle's speed and direction to be modified to avoid a collision will be discussed in the context of vehicle flow on links and through intersections; while the information and control subsystem which monitors the adherence of flights to schedules, and adjusts the assignment of vehicles and crews to future flights, is treated in the context of the operations plan. This is not intended to minimize the importance of these aspects of transportation systems, but rather to facilitate discussion of them in the most appropriate context for understanding transportation systems.

The functioning of these basic components of transportation systems is illustrated in Fig. 3-8. The technology represented is a generalized one, showing all of the functional components. In any particular portion of the transportation system, such as that of a particular mode, some of the basic components may be combined into a single physical component or not be present. For example, a liquid pipeline connecting only two points might consist of two terminals, a link, a control system which might merely regulate flow, and the traffic moving through it. There would be no vehicles as distinct entities, nor containers, even though the functions of these are being performed: the liquid has a natural mobility in the pipe, with forces acting through the wall and the pump causing the liquid to move in the desired path, and the pipe enclosure protects the liquid from loss and contamination by foreign materials. Similarly, no intersections of pipes are present simply because there are no routing options in the system.

To elucidate further the concept of these basic components, Table 3-1 contains the names of the basic components of some of the common modes into which transport technologies are usually divided. This should be read bearing in mind that there are many variations within the usual mode classes, and hence this represents only a sampling of the components.

TRANSPORTATION NETWORKS

Transportation systems exist in order to move traffic from one place to another. A traveler desires to be transported from one particular place, an origin, to another, a destination; similarly, for freight. Since transportation service is not ubiquitous and of uniform type and quality, it is essential that the locational characteristics of the fixed facilities of the system—its terminals, way links, and intersections—be taken into account in all analyses. This is done primarily through the use of the concept of a network. The network representation also serves as a convenient means of arranging information on the characteristics of the various fixed facilities and the flows on them, and hence are useful in describing overall characteristics of the system such as costs and performance.

Network Elements

A network is a mathematical concept that can be applied to describe quantitatively transportation systems and other systems which have spatial charac-

Table 3-1 Examples of Basic Components of Selected Transport Technologies

Component type	Railroad	Airline	Products pipeline	Conveyor
Traffic	Merchandise freight	Passengers	Petroleum	Coal
Terminal	Shipper and consignee docks	Airport	Tank farms	Storage-loading facilities
Container	Box car	Aircraft cabin	Pipeline	Belt and cover
Vehicle	Train	Aircraft	Liquid in pipe and pumps	Belt
Way link	Main track	Air lane	Pipe	Belt support and rollers
Way intersection	Turnout	Air lane junction	Pipe junction	Belt junction
Operations plan	Schedule	Schedule	Batching schedule	Belt speed schedule

teristics. Although the term network often has other meanings, we shall focus primarily on the mathematical concept and its use in analyzing transportation systems. First we shall describe these mathematical concepts; then we cover their application to transportation systems.

Networks consist primarily of two elements—links and nodes. Nodes represent particular points in space. In their graphical representation, nodes are literally points or dots, and links are lines connecting these dots. A link is defined by the nodes which must exist at its ends. Links do not specify direction. In situations where it is important to specify direction (as in the representation of a one-way street), an arc is used. An arc is simply a link with a direction associated with it. On a graph this is indicated by an arrow associated with the line. Often arcs are termed "directed links." The node from which an arc is directed is termed its A-node, and the node to which it is directed is its B-node.

The use of a network to represent the spatial characteristics of a transportation system is illustrated in Fig. 3-9, where the road system shown in the form of a typical map in the upper portion of the figure is represented by links and nodes in the lower portion. There are shown the two types of links, undirected or bidirectional links (often termed simply links) and directed links (often termed arcs) indicated by lines with arrows. A road on which traffic can travel in both directions is represented by either a link or two arcs (one in each direction), while a one-way street is represented by one arc.

This figure also illustrates the usual means of designating the various nodes and links. While on a map names are usually associated with links (street names) or nodes (names of towns), the use of names in mathematical analysis would be extremely cumbersome. Therefore, numbers are usually used to specify the nodes. Less commonly, individual letters such as *a*, *b*, etc., may be associated with the nodes. Numbers can also designate links or arcs.

Figure 3-9 **Example of road transportation system and its network representation. *(a)* Example map of the main road system in a region. Note: Map of hypothetical region to illustrate network concepts; one-way road between towns is very unusual. *(b)* Network representation of that road system. *(c)* The connection matrix. *(d)* The node-arc or node-link incidence matrix.**

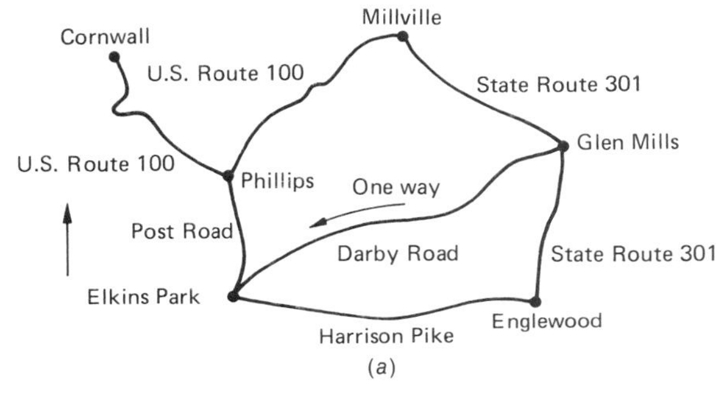

(*a*)

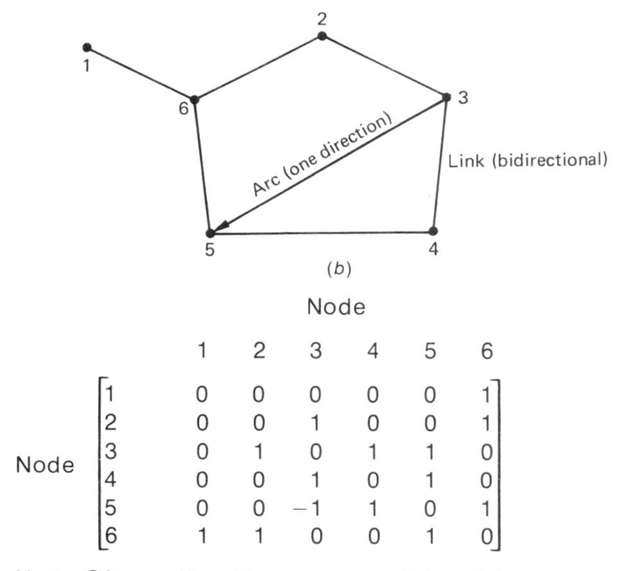

(*b*)

Node

Node	1	2	3	4	5	6
1	0	0	0	0	0	1
2	0	0	1	0	0	1
3	0	1	0	1	1	0
4	0	0	1	0	1	0
5	0	0	−1	1	0	1
6	1	1	0	0	1	0

Note: Often cells with zeros are left blank for ease of writing.

(*c*)

Arc or link

Node	(1,6)	(2,3)	(2,6)	(3,2)	(3,4)	(3,5)	(4,3)	(4,5)	(5,4)	(5,6)	(6,1)	(6,2)	(6,5)
1	1	0	0	0	0	0	0	0	0	0	−1	0	0
2	0	1	1	−1	0	0	0	0	0	0	0	−1	0
3	0	−1	0	1	1	1	−1	0	0	0	0	0	0
4	0	0	0	0	−1	0	1	1	−1	0	0	0	0
5	0	0	0	0	0	−1	0	−1	1	1	0	0	−1
6	−1	0	−1	0	0	0	0	0	0	−1	1	1	1

Note: Often cells with zeros are left blank for ease of writing.

(*d*)

However, in most transportation applications only one link (or two arcs, one in each direction) connects any particular pair of nodes, so in that case the two numbers designating the nodes could be used to specify the associated link or arcs. In Fig. 3-9*b*, for instance, an arc connecting node 3 to node 4 (i.e., 3 is the A-node and 4 is the B-node) would be designated arc (3,4) and the arc in the other direction would be designated arc (4,3). If links are used, as in Fig. 3-9*b*, the link would be designated link (3,4) or possibly link (4,3) since the order of the numbers is unimportant for a link.

In addition to this graphical form, there are other ways of representing networks. Two important ones are the connection matrix and the node-arc or node-link incidence matrix. The connection matrix for the network shown in Fig. 3-9*b* is shown in Fig. 3-9*c*. In this matrix, the rows and columns are the nodes of the network. A zero is placed in the cell corresponding to two nodes if there is no direct connection in the form of a link or an arc between those two nodes. By convention, a zero is also placed in the cells with the same node in the row and column, these cells being called the diagonal of the matrix. If there is a link connecting two nodes, then a +1 is placed in the cell. If an arc connects two nodes, then a +1 is placed in the cell corresponding to its A-node in the row and its B-node in the column. In most transportation applications, a −1 is then placed in the cell for which the row is the B-node and the column is the A-node, although other conventions exist depending upon the application.

Figure 3-9*d* gives the node-arc or node-link incidence matrix for this same network. The columns are the arc or link designations, and the rows are the node designations. If an undirected link connects a node, then a +1 is placed in the corresponding cell. For arcs, a +1 is usually placed in the cell corresponding to the A-node and a −1 in the cell corresponding to the B-node.

In addition to describing the spatial characteristics of a transportation system, the network concept is used extensively to describe such characteristics as capacities, travel times, and flow volumes on the various elements. In almost all applications, these characteristics are associated with links or arcs only. Thus the node specifies no characteristics other than the arcs or links impinging upon it. Figure 3-10 presents the road system originally shown in Fig. 3-9*a* with characteristics associated with the links, including travel times (one direction), actual daily vehicle traffic flows (termed volumes), capacities (maximum possible volumes), and lengths associated with each of the links.

The association of all of these characteristics only with arcs or links and not with nodes, may seem strange, but it is done primarily for mathematical reasons and ease of analysis, and in fact in no way oversimplifies the network. For example, time is, of course, consumed in traveling through an intersection. In order to represent this time in the network, two possibilities exist. One is to include that travel time in the time spent traveling along one of the arcs or links leading into or out of that intersection. This is probably the most commonly used approach in modeling road networks. Another approach is to divide the intersection itself into distinct links or arcs. Travel time would then be associated with each of these links or arcs, and thus the time through the intersection would be so represented. In fact, if it is desirable to differentiate between the travel time for passing straight through the intersection and that for other movements, such as a left turn, then such a dissection of the inter-

Figure 3-10 **Network representation of road system of Fig. 3-9 with data describing link characteristics.**

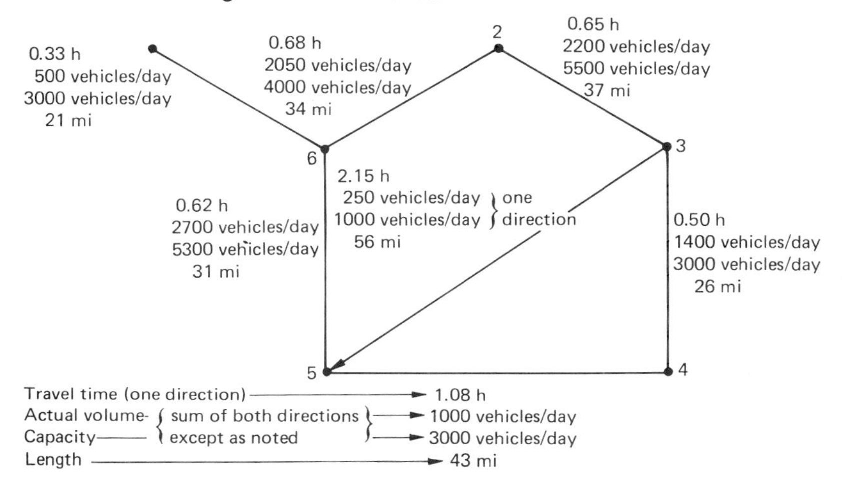

section is essential. One example of this type of network representation is shown in Fig. 3-11. This intersection is the crossing of two two-way streets with all turns permitted. To differentiate among all ways of passing through it (maneuvers) requires the use of directed links, with 8 nodes and 24 arcs. The short arcs at the extremities represent the connections between intersections. If any turns are prohibited, no arc is shown for that turn. Thus it is important to recognize that a single node alone is not the only way to represent an intersection or a terminal in a transportation system.

Network Analysis

A transportation system is represented as a network in order to describe the individual components of the transportation system and their relationships to one another. Some of the most important characteristics of the system are travel time and cost.

This can be illustrated by reference to Fig. 3-12, which is the actual main road network of the San Francisco metropolitan area. Average travel times in minutes are given on all links. The travel time from node 1 to node 8, via links (1,10), (10,24), (24,23), and (23,8) is

$$5 + 10 + 25 + 10 = 50 \text{ min}$$

There are other possible paths, such as (1,11), (11,20), (20,21), (21,22), (22,23), (23,8). Thus in giving origin-to-destination times or costs, it is important to specify the path used. In more general mathematical terms we can express this as follows, designating the path of interest p and the set of links or arcs comprising it L_p:

Figure 3-11 **Detailed network representation of a way intersection: a road example. *(a)* Plan of intersection with all streets being two-way and with left and right turns permitted. *(b)* Detailed network representation using directed links so that each maneuver can be treated separately.**

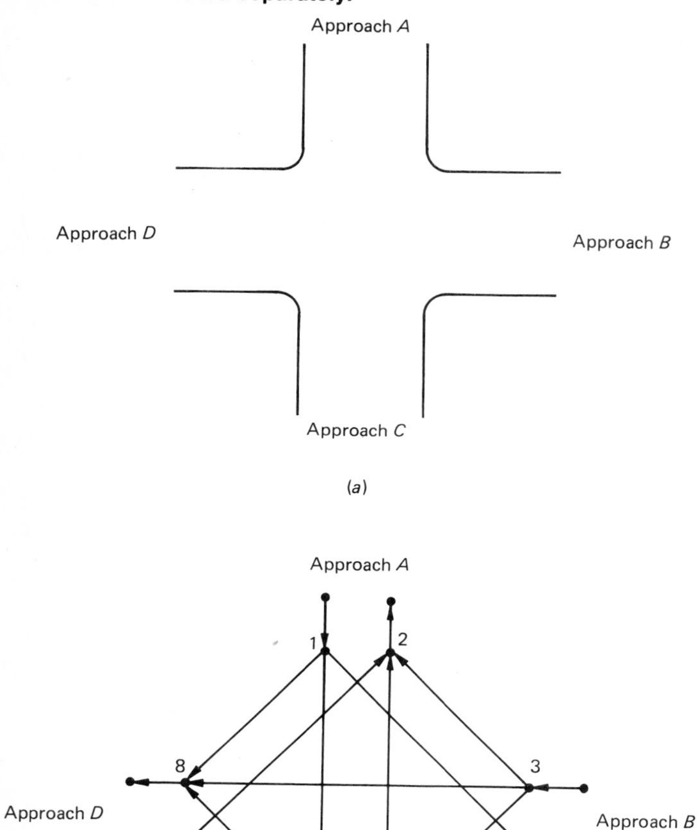

Figure 3-12 **Main road network of the San Francisco metropolitan area showing link travel times. Note: travel times are in minutes.** [***From Potts and Oliver* (1972), p. 56.**]

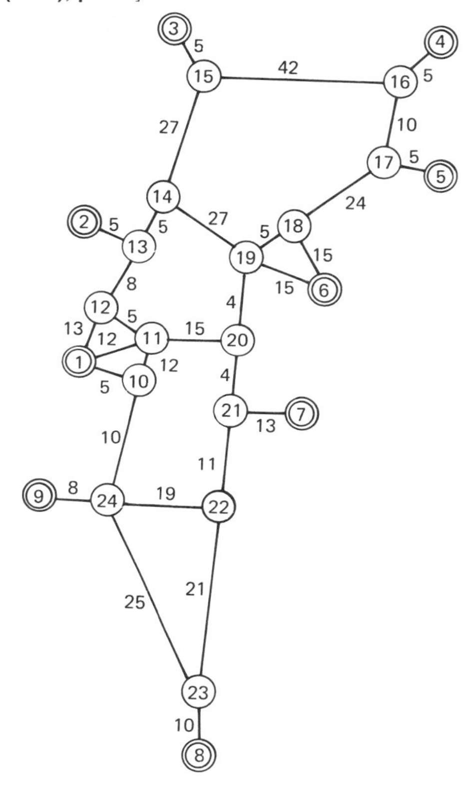

——— Main roads

○ Intersections

◎ Centers of counties

$$t_p = \sum_{ij \in L_p} t_{ij} \qquad (3\text{-}1)$$

where t_p = time from origin of path p to its destination
L_p = set of links or arcs in path p [in above example, (1,10), (10,24), (24,23), (23,8)]
t_{ij} = time on link or arc (i,j)
$ij \in L_p$ means ij is included in the set L_p

In many transportation contexts, the path which is used involves the least total travel time, or in some cases, the least cost. In the context of person travel, most people select the route which minimizes total journey time.

However, in cases where the cost differs among different routes, travelers often modify their selection accordingly. In freight transport, as in the routing of railroad cars over long distances and many different railroad companies, the route is often selected to minimize the total cost. In both of these cases, the problem is essentially one of finding the path through the network having the minimum sum of certain costs (or times) associated with the individual links or arcs which make up the path; such a path is termed a minimum path or best path. Thus these problems are mathematically essentially identical.

A rather simple and elegant procedure has been developed to find these minimum paths through networks. This method, called tree-building, is an application of rather general mathematical procedures called dynamic programming. The procedure is readily explained by use of an example and the performing of the rather simple computations directly on a sample network. For this purpose, we shall use the network in Fig. 3-12. On this network travel times are associated with each link. The problem is to find the path through this network which has a minimum total time from node 1 to node 4. The starting node is termed the home node.

Starting at node 1, compare the cost of traversing the links (or arcs) which emanate from that node, and select the link with the minimum time. In this case, the times are 5, 12, and 13 min, and hence the link to node 10 is selected. The significance of this selection of node 10 is twofold. First, the minimum path from the home node, node 1, to node 10 is via this single link. The reason for this is simply that there is no possible way one could start from the home node and travel on any of the other arcs and reach node 10 with a total time of less than 5 min. This is because all of the other arcs have times greater than 5 min, so even if one could travel from the end of one of the other links to node 10 with zero time (obviously not possible in the case of time on links of length greater than zero), there is no way that the time could be less than 5 min. The second interpretation is that if the best path from the home node to node 4 is via node 10—which it may or may not be—then the portion of that best route from node 1 to node 10 will be via the link (1,10). To indicate that the best path from node 1 to node 10 has been found, next to node 10 place the number of the previous node on the best path, in this case node 1, and the time from the home node, 5 min, as shown in Fig. 3-13*a*. The previous node is often called the predecessor node. The significance of these conclusions will become more apparent later.

In the second step we compare the time from the home node to all nodes which can be reached by traveling over one and only one additional link or arc beyond a node to which the best path has been found (including the home node). Of course, we do not include in the comparison nodes which we have reached, since we have already selected the minimum path to those nodes. In this case, this step would involve comparing the times to node 12 (reached from node 1), node 11 (reached from 1 and 10), and node 24 (reached from node 10). The times from the home node (via the best path to any intermediate node) to those nodes are, respectively, 13, 12, 17, and 15 min. We select the minimum, 12 min, and thus have found the best path to node 11. Node 11 is labeled with 1 and 12 min. The third step involves an extension of the same procedure: we look for nodes which can be reached by traversing one and only one link or arc beyond nodes which have already been reached. Such nodes are 12, 20, and 24 (11 having been reached earlier directly), and the

Figure 3-13 Minimum paths from node 1 to all other nodes. *(a)* Complete network with nodes labeled for minimum path from node 1. [*Adapted from Potts and Oliver* (1972), p. 56.] *(b)* Minimum path tree with node 1 as home node. [*From Potts and Oliver* (1972), p. 58.]

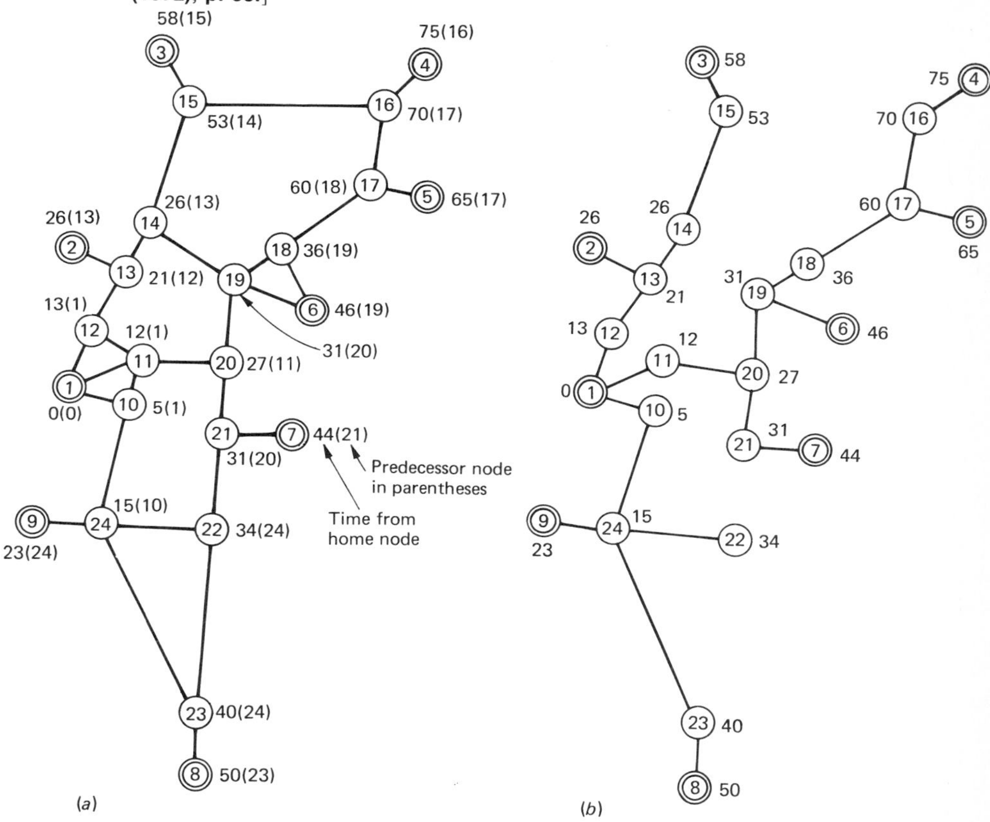

times are, respectively, 13, 27, and 15 min. The least time is 15 min, and we have then reached the best path to node 24. Again this is indicated on the drawing by marking the predecessor node (10) and the time from the home node (15 min).

This process is continued, and the result is that the best path to node 4 is 75 min.

Figure 3-13*a* presents the results of these calculations, with all the nodes being labeled by the predecessor nodes and the times from the home node 1 via the minimum path. The figure also reveals another important characteristic of this procedure, namely, that the best path to many other nodes is found during the process. In fact, this procedure yields the best path to all the nodes which can be reached in a time (or cost) less than or equal to the time to the node desired. While this may seem to indicate an inefficient procedure, actually the procedure is as efficient as any other for finding best paths. It simply yields much additional useful information. The best paths to all other nodes from a home node form what is termed a tree from that node. If only those links or arcs in the tree are shown, the result is as portrayed in Fig. 3-13*b*. To obtain the tree, then, one simply continues the procedure until all nodes have been reached.

In performing analyses for finding minimum paths, it is often desirable to place the information or results in a table rather than on the drawing. For this purpose a rather standardized form has been developed, and it is presented in Table 3-2. The first column is a list of nodes, and every time the best path to a node is found, that node is entered in this table. Entered in the second column of the table is the predecessor node, and thus the node and the corresponding predecessor node indicate the last link selected in the procedure. It is also useful to place in the table the travel time via the best path from the home node, and this is in the third column. Table 3-2 provides the complete list of nodes reached in solving the problem presented in Fig. 3-13. In starting this table, the node which is the home node for the minimum path is initially placed as the first node reached, its predecessor is a fictitious node 0, and the travel time from itself is 0 min. It is useful to note that in this table the list of travel times from the origin to the various nodes should start from 0 (for the home node) and at each successive row the travel time should be at least as great as the travel time in the preceding row. If that is not the case, there is an error in the computations.

With the explanation of this example, it is now possible to formalize the statement of the procedure as a set of rules. In mathematical terminology, a

Table 3-2 Minimum Time Path for Network of Fig. 3-12, With Node 1 as Home Node

Node	Predecessor node	Cost or time from node 1, min
1	0	0
10	1	5
11	1	12
12	1	13
24	10	15
13	12	21
9	24	23
14	13	26
2	13	26
20	11	27
19	20	31
21	20	31
22	24	34
18	19	36
23	24	40
7	21	44
6	19	46
8	23	50
15	14	53
17	18	60
5	17	65
16	17	70
4	16	75

set of rules such as this is termed an algorithm. An algorithm is a procedure which can be proven mathematically to yield a specified result in a finite number of steps. This means, in this case, that by repeated application of the rules shown below, it can be proven that the best path from the home node to all other nodes in the network will be found. Since the algorithm applies to all types of service and cost characteristics, in its statement below the term cost rather than time will be used. [A more extensive treatment of minimum path algorithms is in Potts and Oliver (1972).]

The steps in the minimum path finding algorithm are

1 Compare the costs on all links (or arcs) for which the home node is an A-node. (A- and B-nodes for undirected links result from use of the link in a specific direction in these computations.) Select the one with the minimum cost, and label its B-node with the home node and the cost from the home node. If a table is used, first set up columns labeled node, predecessor node, and cost from home node. On the first row, enter the home node, 0, and 0 min, respectively. In the second row, enter the results of this analysis, the B-node, the A-node, and the time.

2 Compare the costs from the home node to all nodes which are B-nodes of links which have A-nodes in the set of nodes to which the best path has already been found. These are listed in the first column of the table. The times are calculated by adding the link times to the times via best paths from the home node (indicated by the node labels and in the third column of the table). Select the minimum cost, and the minimum path to the B-node of that link has been found. Enter this node in the list, the predecessor node, and the travel cost (from the home node) used in the comparison. Label the node so reached on the figure with the predecessor node and cost. If two or more nodes have equal costs from the origin, treat both as above. Also, if two different paths have equal times, include both in the solution by indicating two alternative predecessor nodes.

3 Continue the procedure specified in number 2 until either the single destination node of interest has been reached or all nodes have been reached if the entire minimum path tree is desired.

Such procedures for finding minimum paths in a network are undoubtedly the most widely used network analysis tools. As has been pointed out above, there is no reason to limit applications to finding paths of minimum time. They can also be used to find the path which has the minimum cost, provided the cost of traversing each link or arc is specified. Also, the measure of cost might be a composite of actual monetary cost to the traveler (or shipper), an imputed monetary value of travel time, plus imputed monetary values of other characteristics of the links such as level of comfort or reliability of service. We will discuss these later, in Chap. 10, when the preferences of travelers and freight shippers for various routes which represent alternative modes, etc., are discussed. The basic limitation of the method is that it applies only to measures of cost which are greater than or equal to zero (for the reason given in our discussion of why this yields minimum cost paths) and for path costs which are the sum of the costs on the links or arcs making up the path. Thus it could be applied to a problem of finding that path containing a minimum number of transfers through a public transportation network, for example, or a path for freight cars through a railroad network which has a minimum number of classification yard handlings. These are illustrated in the problems.

This type of information on the characteristics of a network which relate to movements between all pairs of nodes in the network can be usefully presented in matrix form. The matrix is very similar to the connection matrix, except that the information in the cell is the information desired on the connection between the two nodes specified. Figure 3-14 presents the minimum path travel time from each node of origin to each other node as a possible destination for nodes 1 through 9 in the network of Fig. 3-12. The nodes numbered 1 through 9 are at the approximate centers of the nine counties in the San Francisco area. It is important when presenting such a matrix to be very specific as to the type of information contained in the cells, since it can take many forms, such as travel time, cost, or distance, and also because it is so similar to the connection matrix. (A useful check of your understanding of finding minimum paths in a network would be to calculate the minimum path tree from a node other than node 1 in Fig. 3-12 and check your result with the information presented in Fig. 3-14.)

It is important to realize that the best path through a network, regardless of the criterion used, often depends upon time of day. For example, in the case of highway networks, we saw in Chap. 1 (see Fig. 1-4) that the travel time on any link is influenced significantly by the volume of traffic on that link, the travel time increasing with increasing volume. During the middle of the night or the early morning hours the volume may be quite light everywhere, and thus result in particular times on links and hence a particular minimum path from any origin to any particular destination. When traffic builds up, as in a peak commuter travel period, certain links may be much more heavily used than others, and hence their travel times may increase proportionately much more than the travel times on other links. This may so change the relative travel times among various links that between the same origin and destination another path becomes the minimum travel time path. Thus the best path through a network can in effect be a function of the time of day. A similar phenomenon exists in the case of travel times through the networks of scheduled carriers. For a particular desired time of departure (or arrival), one route

Figure 3-14 **Matrix of minimum path times for nodes 1 through 9 in the network of Fig. 3-12.**

To node

From node	1	2	3	4	5	6	7	8	9
1	0	26	58	75	65	46	44	50	23
2	26	0	42	81	71	52	50	75	48
3	58	42	0	52	62	74	80	107	80
4	75	81	52	0	20	54	65	94	90
5	65	71	62	20	0	44	55	84	80
6	46	52	74	54	44	0	36	65	61
7	44	50	80	65	55	36	0	55	51
8	50	75	107	94	84	65	55	0	43
9	23	48	80	90	88	61	51	43	0

may be best. However, at another time of departure, another series of links may yield the least travel time to the destination.

Another aspect of networks which receives much attention is capacity. Network capacity is a rather elusive concept, because one obviously needs vehicles to move traffic through any vehicular network, and the vehicles themselves may actually be the limitation on capacity, as we shall see in the next section. However, in some cases it is still important to know the maximum volume of vehicles which can be moved from one point to another. If there is only one possible path on which vehicles can flow, then the capacity of that path is simply the capacity of the link or arc with the least capacity. Mathematically this is expressed as

$$c_p = \min_{ij \in L_p} (c_{ij}) \tag{3-2}$$

where c_p = capacity of path p
L_p = set of links or arcs in path p
c_{ij} = capacity of link or arc ij

In the case where the traffic can flow over many possible routes from the origin to the destination of interest, then the problem is much more complex. It has been solved mathematically; the method is presented in Potts and Oliver (1972, pp. 41–43). However, since this is of limited interest to us, we shall not discuss it here.

Characteristics of Transportation Networks

Before leaving this section, we make a few observations on characteristics of actual transportation systems. Perhaps the most important characteristic is that by viewing the transportation system as a whole, rather than as individual modes, we see that virtually all places at which human activity occurs are accessible by some form of transport. Even in the least developed area, footpaths connect communities or settlements. And while there were a few examples of societies completely cut off from the rest of the world, the advent of the helicopter among other means of transport provides a connection with them, at least in principle. Thus transport has become truly ubiquitous.

At the same time there is a great deal of selectivity and specialization of function within the transportation system. While all human settlements may be connected by at least some minimal form of transport, improvements beyond that have been made very selectively. Even though in highly developed nations, as a matter of national policy, roads for at least small motor vehicles reach all separate parcels of land, improved highways are not so ubiquitous. And of course in the least developed countries, roads of any sort for motor vehicles are very limited. Highly improved highways, such as freeways, and other forms of transport, such as railways, canals, and airways are much more limited, serving primarily major traffic generators. (This was discussed in detail in Chap. 2.)

Even within an individual mode or technology of transportation, there is considerable differentiation among the various elements. This is very well illustrated by the classification of highways in the United States. In urban areas there are four classifications of roads: expressways, arterial streets, collector streets, and local streets. The expressway and freeway are designed for relatively long distance through movement, and, of course, provide absolutely no access to land along them since driveways are not permitted. Arterials are major streets for through movement, but also provide some land access. Collectors, too, are primarily for through movement, mainly for travel to and from arterials and expressways, and also provide some land access. At the other extreme is the local road, which exists solely for land access. Table 3-3 provides information on typical percentages of urban road systems which fall in each of these categories and typical distances between parallel roads of each type. Also presented in this table is similar information for rural highways. In this context, there are also four primary classifications: interstate system freeways, primary highways (which include freeway-type facilities), secondary highways, and tertiary roads. Again the principal function and other characteristics are presented in the table.

These classification schemes are not only important in specifying the primary function of the facility, but also influence substantially design standards and the manner in which construction and maintenance of the facility is handled. In the case of roads, for example, as we will see from the chapter on design, there are vastly different standards for different types of roads, which yield very different speeds and capacities. Also, the cost of different facilities is distributed among national and state governments and local governments differently. For example, local governments pay more for tertiary roads because the primary users are local residents.

Table 3-3 Hierarchical Structure of Transportation Networks: Example of Roads in the United States

Classification	Principal function	Typical spacing (assuming grid pattern), mi	Typical percentage of system
Urban			
Freeways and expressways	High volume, rapid movement over long distances; no land access	4–10	0–10
Arterial streets	Lower volume flows and medium speeds over medium distances, often to or from freeways and collectors; some land access	1–3	20–40
Collector streets	Collect trips for arterials, low speeds; land access	$\frac{1}{2}$–1	20–40
Local streets	Land access; low speeds	$\frac{1}{8}$–$\frac{1}{4}$	60–80
Rural			
Freeways	High speed, high volume movement over long distances; no land access	20–100	0–5
Primary roads	Medium speed, lower volume movement; some land access	10–30	10–25
Secondary roads	Feeders to higher type facilities, lower speeds and volumes; land access	5–20	10–25
Tertiary roads	Land access; low speeds	1–10	60–80

Sources: From Hutchinson (1974), pp. 233–234, and unpublished data from U.S. Department of Transportation.

Similar hierarchies of facilities are found in other modes of transportation. There is a substantial difference between branch or feeder railroads and main-line railroads, the latter being designed for much higher speeds and higher volumes of traffic than the former. In the air network, there are distinctions between different types of airports and vast differences in the quality and volume of air service between various pairs of cities. Similar differences exist for bus and truck services.

VEHICLES AND CONTAINERS

There are many distinct transport characteristics of vehicles and containers. Some are determined solely by the vehicle or container design, including the types of objects they are designed to carry, carrying capacity, and degree of operational freedom in creating units of varying sizes (e.g., coupling to form trains). Other characteristics depend upon the paths (links, intersections, etc.) over which the vehicles and containers operate, and the number of other units present, as in the case of congestion at intersections or in terminals. In this chapter, the former type of characteristics will be covered, the flow characteristics being covered in Chaps. 4 and 5.

Vehicle Types

Vehicles and containers can be classified by the degree to which they are divisible into various types of units. At the one extreme, a vehicle may consist of a single unit for performing all needed functions of giving mobility and locomotion and also containing the freight or passengers. Examples of these include automobiles, single-unit buses, and many types of single-unit trucks, aircrafts, and ships. A somewhat more differentiated vehicle might consist of a truck such as the single-unit truck plus a trailer; the trailer gives mobility to the freight and is a container, but lacks its own locomotion (except perhaps brakes). Another possibility is a truck consisting of a flat bed on which containers are placed, the latter actually carrying and protecting the freight. This type of truck is common for the movement of the standardized containers used in international trade between ports and the actual origins and destinations of freight shipments, which are usually removed from the port. An example of a fully differentiated transport vehicle is the typical railroad freight train for carrying containers or truck trailers, in which the power unit is completely separated from the remainder of the vehicle, in which the freight is carried in containers on flat cars which provide the necessary mobility. These different types of vehicles are illustrated conceptually in Fig. 3-15.

The reasons for different kinds of vehicles being used in different situations are manifold. One of the primary reasons why freight containers are often separable from other portions of the vehicle is that freight often takes a long time to load and unload from the container, a few hours to a day being common. In order to avoid detaining railway locomotives and truck tractors, for example, the cargo-carrying portions (containers) are made detachable so that the other portions can be utilized even while the containers are being loaded and unloaded. In local package freight service of a pick up and delivery sort, where the loading is quite rapid, consisting of at most a few parcels, this separation would yield no advantage. Another reason for trains of vehicles is to enable very large units to traverse paths which a rigid single vehicle

Figure 3-15 Different types of vehicles classified by integration or separation of functional components.

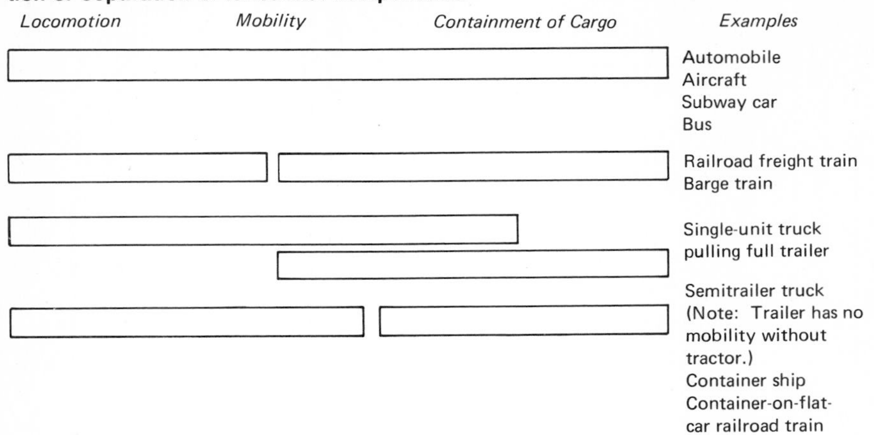

could not, as in the case of railroad trains traversing sharp curves. A further advantage of the multi-unit vehicle is that when one portion of it breaks down, that portion often can be removed (and replaced if necessary) and the vehicle can continue on its way. Of course, the probability of breakdown of a train of units is much more likely than the probability of breakdown of any one unit, and as a result one disadvantage of train operation is, in theory at least, a somewhat lessened reliability of operation. Some idea of the significance of this can be obtained from realizing that the relative frequency of a delay due to mechanical or electrical failures of freight trains (including such items as engine failure, brake failure, failure of couplers and related components) is about 0.19, based on observations on one major railroad, and 0.28 on another. While the total delay time is often small, the rate is high (Belovarac and Kneafsey, 1972, p. 22).

Another advantage of train operation is that very often economies in personnel can be achieved, such as through the operation of the entire train by the same number of crew members as required for a much smaller unit; such economies are particularly evident in the case of railroad trains. Another advantage of operating trains is that very often the capacity of way links and intersections is increased (as we will see from the analyses for railway lines in Chap. 5). On the other hand, there are disadvantages of large vehicle or train operations. One of these is that vehicle movements are necessarily less frequent, often resulting in delays to traffic in terminals while sufficient traffic builds up to warrant operation of a full train. Also, there is often a very considerable expense associated with assembling and disassembling the units to form a train (termed marshaling), since such activities require large terminals and substantial equipment and labor. It has been estimated that in 1973 fully 35 percent of the total freight-service-related operating costs of eight eastern United States railroads was incurred in yard activities (United States Railway Association, 1975, p. 63). Such costs are virtually completely avoided when single units move alone, as is the case for shipments by truck when the vehicle is fully loaded by a single shipment, termed a truck load (TL) shipment. In some cases, single shipments are of sufficient size to load fully a ship (e.g., oil) or train (e.g., coal), which results in direct movement from origin to destination.

On the other hand, it is relatively rare to find a single group of persons traveling together of sufficient size to fill an entire bus, airplane, train, or ship, although there are charter movements of this type. Therefore in person travel by other than private auto (or small private boat or airplane) there usually is much aggregation of persons and groups traveling to diverse places on a single vehicle. This achieves economy in the vehicle movement, but usually requires some deviation from ideal times of travel for some persons and requires the provision of terminal facilities appropriate to such aggregation. While these advantages and disadvantages could be discussed in much more detail, the choice of size of vehicle and whether or not to combine diverse shipments or travel groups is essentially one of attempting to match the service provided with the characteristics of the services demanded by passengers and shippers, including economic considerations best discussed later on.

Container Types

Another aspect of containers is the design of the unit to provide an appropriate environment for the traffic while it is being transported. Specialization of accommodations for passengers is probably well known to everyone through personal experience. On one hand, for very short distance travel merely a place to stand is often acceptable. As a result, much local passenger transport involves carrying passengers who are standing, particularly during peak travel periods when the provision of seats for all passengers would be extremely costly. On elevators, escalators, and conveyor belts, seats are virtually never provided. For longer-distance travel, where seating is a necessity, the types of accommodation vary widely. For short-distance travel, rather narrow seats, close together, are often acceptable, typical seats in commuter trains and buses measuring about 17 in in width and 28 in in pitch, the seat back to seat back distance, usually without package racks or washrooms. In contrast, for longer-distance travel, reclining seats with leg rests are often provided, as well as ample room for baggage and elaborate washroom facilities. Also, for long-distance travel, there usually is provision for meals, as in dining cars on trains, meals during flights, and meal stops or an on-vehicle dining service on bus trips. For trips involving long periods on board the vehicle, sleeping accommodations in the form of private bedrooms or small bunks (termed berths) are provided, as on long-distance trains and ocean vessels. In the extreme, full recreational facilities, movie theatres, small libraries, etc., are often provided on ocean liners. These various types of passenger accommodations are illustrated in Fig. 3-16.

Similar specialization in the type of container exists for freight traffic, and the differences are even more substantial than for person traffic. The many different types of cargo transported in various forms require various types of containers. One of the most important distinctions among types of cargo is whether or not it is in bulk form. Bulk freight is defined as "loose, unpacked freight not subject to the count of pieces. It generally can be loaded or unloaded by gravity, pumping, suction, or other mechanical means" (Grossman, 1959, p. 122). Examples of cargo usually shipped in bulk include oil, coal, sand, gravel, and grain. Nonbulk freight includes such items as machinery, house-

(a)

(b)

(c)

(d)

(e)

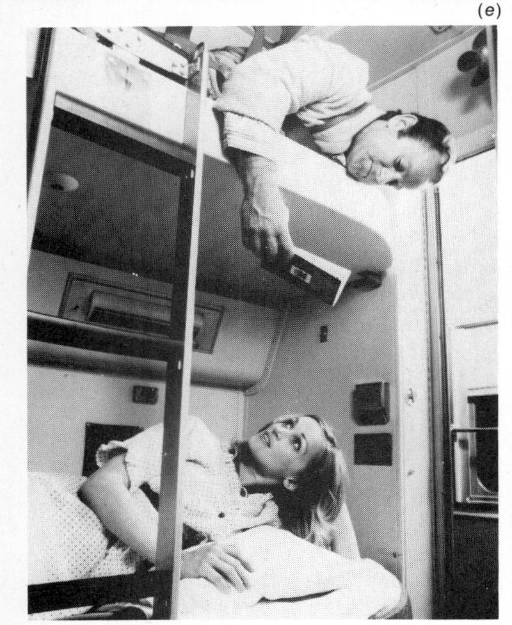

(f)

Figure 3-16

Various types of accommodations on passenger vehicles. *(a)* Urban transport: rail rapid transit car with small seats and places for many standees. *(Pullman Standard.)* *(b)* Suburban commuter transport: high-backed seats, baggage racks, and reading lights, with no provision for standees. *(General Motors Corp.)* *(c)* Intercity transport: reclining seats with leg rests on railroad car. *(Amtrak.)* *(d)* Intercity transport: bar lounge on railroad car. *(Amtrak.)* *(e)* Intercity transport: railroad sleeping car room. *(Amtrak.)* *(f)* Cruise travel: recreational facilities on an ocean liner. *(Cunard.)*

hold goods, automobiles, and in general most large-piece manufactured products. Whether or not something is classified as bulk or nonbulk often depends on the packaging. For example, motor oil can be shipped in a bulk form or in cans packaged in cases. The distinction is important for the type of container required to accommodate the cargo.

In general, nonbulk freight of almost any type can be carried in a fully enclosed container body of a rectangular cube shape. Examples of such containers include railroad box cars, truck vans, and the cargo holds of most vessels. Special devices may be incorporated in the container to hold the cargo securely. An example, in the form of a railroad box car, is illustrated in Fig. 3-17*a*.

Much nonbulk freight does not need the weather protection and the side support of the box type of container. A cheaper and lighter container consists of a flat bed or such a bed with sides and ends only, examples being railroad flat cars and gondola cars. Some weather protection can be provided with tarpaulins. Loading and unloading these is often easier than a box container, which must be loaded through doors in the sides or ends. Railroad flat and gondola cars are illustrated in Fig. 3-17*b* and *c*.

Bulk freight is carried in a variety of container bodies. Gondolas are used for cargo such as coal and ore which require little weather protection. Loading is accomplished by dropping the cargo into the body; for unloading the body is simply turned upside down, obviously by an elaborate unloading device. Often individual cars of trains can be so unloaded without uncoupling one car from another. Similar to these is the hopper body, which is basically a gondola with hoppers or openings in the bottom for unloading. This makes unloading as easy as loading.

Many types of bulk cargo require complete enclosure in transport, usually to prevent spillage and contamination. Solid materials often move in covered hopper bodies, which have a roof, with sealable openings for loading. Typical cargoes are grains, cement, and sugar. Such cargo also is carried in box containers, usually with a paper or other covering on all interior surfaces to allow complete sealing to prevent contamination. A covered hopper is shown in Fig. 3-17*e*.

Liquids and gases are carried in tanks. Variations in loading and unloading devices reflect the nature of the cargo. Special care must be taken with gases under pressure and highly volatile substances; containers are often designed with two walls and special release valves if the cargo pressure should become excessive.

Much freight requires special environmental conditions, particularly temperature, to be maintained. Foodstuffs often require a low temperature, and some, of course, now require maintenance of a frozen state, which can be accomplished in special refrigerated cars. Similarly, certain commodities require a temperature above that of the immediate environment, and there are insulated containers as well as heated analogs to refrigerated containers. Also, there are specialized ventilated containers, usually with a box-type body, for live animals as well as other freight such as fruit requiring fresh air. Examples are shown in Fig. 3-17*g* and *h*.

Often the container for the cargo is an integral part of the vehicle. This is typified by most railroad cars and trucks. On the other hand, the vehicle can be designed to accept containers standardized in size and maximum weight

(*a*)

UNION
PACIFIC
UP
49 10 51

(*b*)

(*c*)

PTLX
70

(*d*)

NW
138 000

(*e*)

ATSF
303681
Santa Fe

(*f*)

ACFX 89394

(*g*)

WESTERN FRUIT EXPRESS
BURLINGTON
NORTHERN
BNFE 3
Temp-Lok

(*h*)

LIVESTOCK SPECIAL BALTIMORE&OHIO
B&O
113000

(*i*)

(including cargo). A standard size of about 8 ft square in cross section, and various lengths up to about 40 ft, is becoming popular for most containers, especially those in international, intermodal service. These are carried on vehicles designed specially for this service, which consist of little more than the devices necessary to give mobility and locomotion to the containers. (Such containers and road, rail, water, and air vehicles to carry them were shown in Fig. 3-6.) In air service, smaller containers are usually used, reflecting the typically smaller shipments carried by air. Also, the shape may be irregular, to fit the aircraft fuselage.

Container or Vehicle Size

An important aspect of transportation system operation is matching of the vehicle or container to the characteristics of the traffic to be accommodated. One aspect of this was discussed above, that of providing a container which maintains appropriate environmental conditions for the commodity being carried. Another aspect is that the container has two limitations on the amount of cargo that it can accommodate. The amount of cargo may exceed neither the weight capacity of the container, nor the container's cubic volume capacity. These are expressed in equation form as follows:

$$S_w \leq Q_w \tag{3-3}$$

$$S_v \leq Q_v \tag{3-4}$$

where S_w = shipment weight
Q_w = container weight capacity
S_v = shipment volume
Q_v = container volume capacity

Any particular container will thus be filled to both its weight and volume capacity if and only if the weight density of the cargo is as given below.

$$D_c \equiv \frac{Q_w}{Q_v} \tag{3-5}$$

where D_c = container design density

If the cargo exceeds this design density, the weight capacity of the container will be reached before the volume capacity, and the container must be moved with unused space in it. Similarly, if the density of the cargo is less than the design density of the container, the volume of the container will be reached before its weight capacity is reached. This condition is termed cubing out. In

Figure 3-17 **Various types of containers for freight as illustrated by railroad cars. *(a)* Box car. *(Association of American Railroads.)* *(b)* Flat car with end bulkheads. *(North American Car Corp.)* *(c)* Gondola car. *(Pullman Standard.)* *(d)* Hopper car (open top). *(Norfolk and Western Railway Co.)* *(e)* Covered hopper car. *(Association of American Railroads.)* *(f)* Tank car. *(ACF Industries, Inc.)* *(g)* Refrigerator car. *(Burlington Northern.)* *(h)* Cattle or stock car. *(Chessie System.)* *(i)* Three-level automobile carrier. *(North American Car Corp.)***

either case there is in effect some unused capacity of the container, and this results in waste in the transportation of that cargo. For this reason, it is important to attempt to match the design density of the container to that of the cargo in addition to matching the total weight capacity of the container to the weight of the shipment.

Because of these considerations, carriers who handle much small parcel freight which they load onto vehicles for long-distance movement will go to great care to try to fill a vehicle to precisely its weight and volume capacity if at all possible. This often takes the form of mixing some very high density cargo with light-density cargo to achieve a full load in both senses. Also, over the long term they will purchase vehicles (containers) of an appropriate weight density for their traffic.

The significance of these considerations can be illustrated through an example. Consider a truck line which carries 500 tons/day of freight between two points. This freight has an average density of 12.5 lb/ft^3. The company is now operating tractor semitrailer units consisting of one tractor and one 40-ft trailer of the standard type, which has a weight capacity of 58,000 lb and a volume capacity of 2600 ft^3. The design cargo density of these containers (truck trailer bodies in this case) is 22.3 lb/ft^3, by Eq. (3-5). Since this is greater than the average density of the cargo, trucks will cube out before the full weight capacity is reached. Thus each truck can haul 2600 ft^3 of freight, which will weigh

$$2600 \text{ ft}^3 (12.5 \text{ lb/ft}^3) = 32{,}500 \text{ lb}$$

To move this load requires 30.7 truck trips per day, or 31 trips rounded to the nearest integer. The truck line is considering replacing this fleet with "double-bottom" units consisting of one tractor and two short 27-ft trailers, such units having a weight capacity of 51,300 lb and a volume capacity of 3500 ft^3. The design cargo density of these is 14.66 lb/ft^3. Hence these will also be filled by volume. Each new truck can carry 43,750 lb, and therefore 22.86 trips per day, or 23 in actual practice, will be required. This is a saving of 26 percent in truck trips. This is likely to yield a substantial savings in operating costs, and since fewer trucks will be required, the added initial cost per vehicle for the larger trucks as compared to the smaller ones is not likely to offset this gain.

CONCLUSIONS

In this chapter the basic components of a transportation system have been identified. Then particular attention was paid to the network concept for describing the fixed facilities, especially the manner in which these elements are connected to one another. Also, some very general properties of vehicles and containers were described, mainly a classification of types of vehicles and a discussion of variations in containers to suit various types of freight or passenger traffic. This background is the basis upon which the characteristics of vehicles (and container) traffic flow over the fixed facilities will be developed in the next five chapters, which include operations plans and system performance, followed by a discussion of methods for cost estimation.

SUMMARY

1 A transportation technology must be capable of performing three functions:

Give the object to be transported, persons or goods, mobility.
Provide control of the direction and velocity of movement (locomotion).
Protect the object from damage.

2 All transportation systems are composed of components which can be classified into the following groups:

Containers
Vehicles
Way links
Way interchanges
Terminals
Operations plan (including a means of control so that the operations plan is followed)

3 These components are related through the spatial arrangement of the fixed facilities and the pattern of flows of vehicles, containers, and objects. The quality, cost, and capacity of the transportation system depend upon all of these elements taken together.

4 Characteristics of the components of the transportation system can be associated with the links of the network. This enables the calculation of characteristics associated with paths from an origin node to a destination node, such as travel time and cost. The specification of the path is important, a path commonly used is the fastest (or cheapest) one. It is usually termed a minimum time (or cost) path.

5 Vehicles and containers may be classified in terms of the degree to which vehicle (or container) size can be modified (i.e., as in forming a train). Important characteristics of containers are volume and weight capacity, the types of accommodations for passengers, or the commodities that can be carried, in the case of freight. These characteristics will determine the suitability of containers for various types of movement tasks.

PROBLEMS

3-1

a Select a particular transportation system (such as that of a modal firm) and describe in detail how the three functions—mobility, controlled locomotion, and cargo protection—are accomplished.

b For this same system, or a portion thereof, identify the six functional elements. Are there any elements which cannot be classified as one of these types of components?

3-2

Find the minimum time path tree from node 6 in Fig. 3-12. Present the results in the form of labeled nodes on the network and a minimum path tree and table.

3-3

Find the minimum path trees for nodes of the network in Fig. 3-12. Enter the origin to destination travel times in a matrix in which the rows and columns are those 24 nodes. (You can check the figures for nodes 1 through 9 with Fig. 3-14.) Can you think of any way to find which node is most central to the region using the data in this matrix? (Hint: Calculate the average time from each node to all other counties.) Note to instructor: This problem can be done by assigning a different home node to each student and constructing the matrix in class.

3-4

Find the route with the shortest travel time from New York, New York (NY on map), to Louisville, Kentucky (LOU on map), via the interstate highway system. The times on the links are given below. Hint: Draw the relevant portion of the map on a sheet of paper and indicate your computations there.

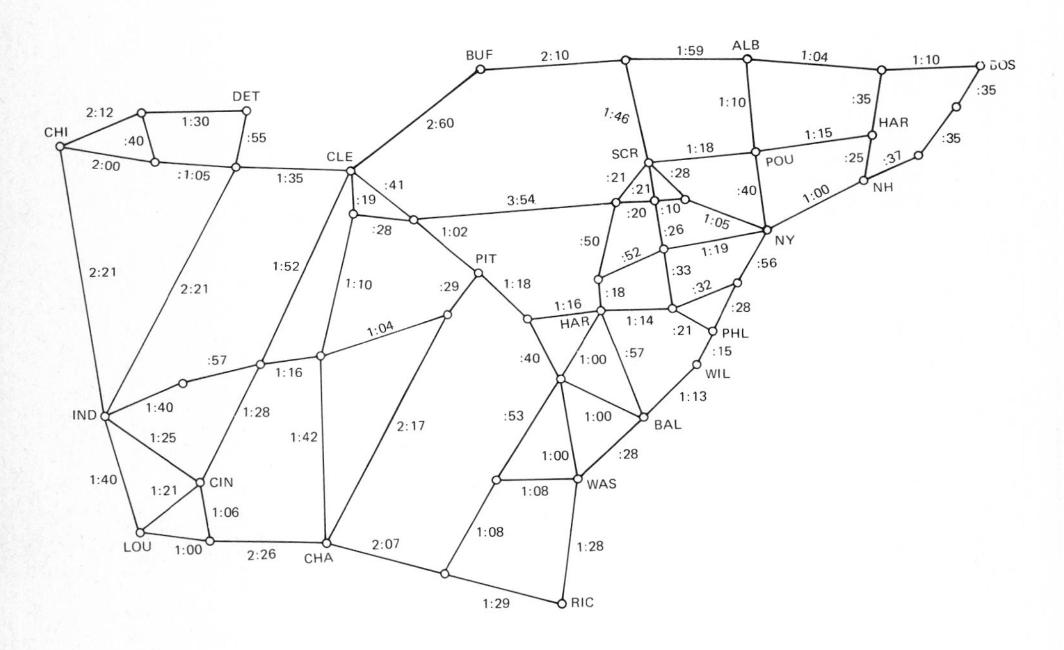

3-5

Obtain a map of transit routes for a city which has some type of high-speed trunk transit routes, such as a rail line or busway. Select a number of different pairs of places between which you might like to travel. To what extent would you make use of these trunk routes, perhaps going out of your way in distance to save time? Discuss.

3-6

Do the same sort of analysis as specified in Prob. 3-5 for auto travel in a region that has a fairly extensive freeway network. It can be a city or a larger area.

3-7

Listed below are the weight and cubic volume capacities of recent models of various types of railroad freight cars. Calculate the design cargo density for each. How does this density compare with the density of typical items of freight which would be carried in each? Such cargo densities are given in Appendix B.

Car type	Weight capacity, lb	Empty weight, lb	Cubic volume capacity, ft³
Railbox box car	140,000	47,000	4,950
PS-1 box car, 50 ft long	100,000	47,900	3,940
PS-1 box car, 60 ft 9 in long	130,000	86,300	5,951
"Hi-cube" box car, 86 ft 6 in long	100,000	117,900	10,000
Refrigerator car	132,000	85,000	4,025
Gondola car	154,000	58,000	1,995
"Pig Palace" cattle car	127,000	91,500	7,654
Covered hopper car	200,000	56,200	4,750
Coal hopper car	190,000	58,700	3,418
Ore hopper car	140,000	44,000	1,000
General service tank car	200,000	69,300	(20,000 gal)
Flat car	150,000	56,500	. . .
TOFC (trailer-on-flat-car) carrying two trailers	100,000	93,340	4,900

3-8

A manufacturer plans to ship 1000 refrigerators from Chicago to New York by rail. Crated refrigerators weigh 180 lb, are 30 in wide, 24 in deep, and 60 in tall, and cannot be stacked (i.e., all must stand upright on the car floor). She can request any of the following types of box cars. Which type would result in the fewest cars required? Which type would result in the least total weight (car plus cargo) to be moved?

Car	Interior dimensions			Empty weight, lb	Cargo weight capacity, lb
	Length	Width	Height		
A	40 ft 6 in	9 ft 4 in	10 ft 5.5 in	45,000	100,000
B	50 ft 6 in	9 ft 4 in	10 ft 5.5 in	54,000	140,000
C	60 ft 9 in	9 ft 4 in	12 ft 10 in	77,200	180,000

REFERENCES

Association of American Railroads Mechanical Division (1966), "1966 Car and Locomotive Builders' Cyclopedia of American Practice," Simmons-Boardman, New York.

Baerwald, John E., Matthew J. Huber, and Louis E. Keefer, eds. (1976),

"Transportation and Traffic Engineering Handbook," Institute of Traffic Engineers, Prentice-Hall, Englewood Cliffs, NJ.

Bekker, Mieczyslaw G. (1956), "Theory of Land Locomotion," University of Michigan Press, Ann Arbor, MI.

Belovarac, Kenneth, and James T. Kneafsey (1972), "Determinants of Line Haul Reliability," vol. 3, "Studies in Railroad Operations and Economics," Research Report No. R72-38, Civil Engineering Dept., M.I.T., Cambridge, MA.

Grossman, William L. (1959), "Fundamentals of Transportation," Simmons-Boardman, New York.

Hay, William W. (1961), "An Introduction to Transportation Engineering," Wiley, New York.

Hutchinson, Bruce G. (1974), "Principles of Urban Transportation Planning," McGraw-Hill, New York.

Morlok, Edward K. (1969), "An Analysis of Transport Technology and Network Structure," Northwestern University Transportation Center, Evanston, IL.

Pignataro, Louis J. (1973), "Traffic Engineering: Theory and Practice," Prentice-Hall, Englewood Cliffs, NJ.

Potts, Renfrey B., and Robert M. Oliver (1972), "Flows in Transportation Networks," Academic Press, New York.

Schumer, Leslie A. (1974), "The Elements of Transport," Butterworths, Sydney.

Taffe, Edward J., and Howard L. Gauthier, Jr. (1973), "Geography of Transportation," Prentice-Hall, Englewood Cliffs, NJ.

United States Railway Association (1975), "Preliminary System Plan," Washington, DC.

Individual Vehicle Motion

4.

The movement of vehicles over a path is central to the capacity and service properties of transportation systems. It lies at the heart of the time required to travel from one point on the system to another. It also is one of the primary factors which determine the number of vehicles required and the associated operating labor, and hence it is an important consideration in the cost of transportation. Thus it is essential that the movement of vehicles over a route be thoroughly understood, not only in its own right but as a basis for much further work.

For purposes of the analysis of movement, transportation systems can be divided into two distinct classes: discrete flow systems and continuous flow systems. Discrete flow systems are those in which the flow of objects, vehicles, etc., past a point on the path is intermittent. Examples include motor vehicles on highways, railroad trains, aircraft, and ships. In these systems there is always some space between the vehicles (except of course when a collision occurs). Continuous flow systems, on the other hand, as the name implies, are characterized by a continuous flow of the objects past any point along the path. Examples include water and gas supply in pipelines. Coal slurry pipelines and conveyors are also usually considered within the continuous flow category. Although precisely speaking, the solids being conveyed do pass any point intermittently, the intervals are so small that the flow is for all practical purposes continuous.

This distinction between discrete and continuous flow systems is important because the principles and mathematical relationships which describe the flows are quite different. Moreover, even though the technologies of, say, rail transport and road transport may seem very different, the engineering methods for dealing with such problems as predicting vehicle running times and power requirements are very similar. Hence in this and the following chapter, we will be concerned with the flow of discrete vehicle systems, and then in Chap. 6 treat continuous flow systems.

The point of departure for describing the movement of a vehicle along its path is to consider the motion of that vehicle as if it were completely unaffected by the movement of other vehicles. In effect, this is a means of treatment of the best possible performance of that vehicle, for the interference of other vehicles can only tend to slow the progress of the vehicle. Once this best performance can be treated, then reductions in speed caused by the other vehicles and other types of interference can be taken into account. Thus this chapter will treat vehicle movement under what might be termed completely uncongested flow conditions, and the next chapter will introduce the effects of congestion on vehicle flow.

EQUATIONS OF MOTION

The basic principles which underlie most of the relationships governing uncongested vehicle movement are Newton's laws of motion, as described in *Principia* published in 1687. These laws are

1. Every body continues in its state of rest or of uniform motion in a straight path unless the application of a force compels a change in that state.
2. An (unbalanced) force causes a proportional rate of change of momentum that takes place in the direction in which the force is impressed.
3. For every action between two contiguous bodies, there is a reaction which is equal, opposite, and simultaneous.

The first law may be regarded as a special case of the second; and the second is the one most crucial to transportation engineering, as will be seen below.

The momentum of a body is simply the product of its mass m and its velocity v, mv; hence the second law states

$$F = \frac{d}{dt}(mv) \tag{4-1}$$

where the force F acts in the same direction and sense as v. If the mass remains constant, then

$$F = m\frac{dv}{dt} = ma \tag{4-2}$$

where a is the rate of change of velocity, or the acceleration. The mass of a body is, of course, equal to its weight in a gravitational field T divided by the acceleration due to gravity, usually termed g. In fact, mass can be thought of as the ratio of the force acting on a body to its acceleration, since, by Eq. (4-2),

$$m = \frac{F_1}{a_1} = \frac{F_2}{a_2} = \cdots = \frac{T}{g} \tag{4-3}$$

In most engineering work, the weight (a force) is measured, and the mass is calculated from that. In the English system, the unit of force and hence of weight is the pound, and acceleration is often measured in feet per second

Table 4-1 Common English and Standard International Units Used in Transportation Engineering

Quantity	English units	Symbol	SI units	Symbol
Length	foot	ft	meter	m
Time	second	s	second	s
Mass	pound-second^2/foot	lb-s^2/ft	kilogram	kg
Temperature	degree Fahrenheit	°F	degree Kelvin	K
Force	pound	lb	newton	N = kg-m/s^2
Energy, work	foot-pound	ft-lb	joule	J = N-m
Power	foot-pound/second	ft-lb/s	watt	W = J/s

Conversion factors

0.3048 m = 1 ft
14.5939 kg = 1 slug = 1 lb-s^2/ft
4.4482 N = 1 lb (force)
$t_K = (t_F + 459.67)/1.8$, t_K in K, t_F in °F
745.6999 W = 1 hp = 550 ft-lb/s
0.003785 m^3 = 1 gal (U.S. liquid)
0.004405 m^3 = 1 gal (U.S. dry)

per second. Dividing the weight by the value of g of 32.16 ft/s^2, the mass is obtained in units of pound-second^2 per foot, a unit sometimes called the slug. In the new Standard International (SI) units, forces including weight are measured in newtons (N) and acceleration in meters per second per second (m/s^2). The unit of mass is the kilogram (kg), 1 kg being equal to 1 N-s^2/m.

By way of an example, consider a railroad train which has a total weight of 10,000,000 lb. It is subjected to a net force of 1,360,000 lb along its direction of motion or path. The gravitational constant or acceleration due to gravity is 32.16 ft/s^2, and hence it will have an acceleration of

$$a = \frac{F}{m} = \frac{1{,}360{,}000 \text{ lb}}{10{,}000{,}000 \text{ lb}/32.16 \text{ ft/s}^2} = 4.374 \text{ ft/s}^2$$

Since very extensive use of both the English units and the SI metric units will be made throughout this book and undoubtedly in one's professional career, it is well to become familiar with them. With this in mind, Table 4-1 has been prepared, containing the more common units of force, time, and length, and important derivatives of these. A more complete discussion of units is presented in Appendix A.

Velocity and Acceleration Defined

Velocity and acceleration are terms in constant, almost everyday, usage in our highly mechanized society, so they often are used without worrying about precise definitions. However, because the correct application of many of the concepts and techniques of this and following chapters depends upon an accurate understanding of these terms, they must be defined precisely.

Transport involves motion in three dimensions. An example is the flight of an aircraft. Not only does it move horizontally, perhaps in a straight line, but usually on any flight it must occasionally turn while in the horizontal flight plane. Even the straight, horizontal movement at a constant elevation above sea level involves movement of a turning variety, because of course the earth's surface is curved. But because most transport motion can be and is analyzed as if it occurs in a plane, we shall devote our attention to this case.

Figure 4-1 represents the motion of a vehicle in a plane. The location of the vehicle at any instant can be represented by the direction and magnitude of a vector, the vector $\mathbf{s}_1$ (a boldface letter indicating a vector rather than a scalar), specifying the location at time t_1, $\mathbf{s}_2$ the location at t_2, etc. The origin of these vectors of location can be chosen in a convenient way in any particular problem, as can the axis directions. These axes define the two rectangular components of $\mathbf{s}$.

Average velocity between two points is defined as the vector difference in locations divided by the difference in times (a scalar difference) as below:

$$\mathbf{v}_{12} \equiv \frac{\mathbf{s}_2 - \mathbf{s}_1}{t_2 - t_1} \tag{4-4}$$

This vector difference is shown in Fig. 4-1*b*.

Often we are not interested in the average velocity between two locations or points in time, but rather in the instantaneous velocity. This is defined as

$$\mathbf{v}_1 \equiv \lim_{t_2 \to t_1} \frac{\mathbf{s}_2 - \mathbf{s}_1}{t_2 - t_1} \tag{4-5}$$

The direction of this vector—velocity is defined as a vector—will be tangent to the path the vehicle is following. Thus the instantaneous velocity of a vehicle moving on a path is a vector, tangent to the path at the point considered, of magnitude equal to the rate of progression along the path. Speed, as it is normally used, is this rate of progression. Thus the odometer in an automobile yields the speed, the rate of progression on the auto's path, but it tells us nothing about the direction of the motion.

In the derivative form, Eq. (4-5) is

$$\mathbf{v} \equiv \frac{d\mathbf{s}}{dt} \tag{4-6}$$

Acceleration is the rate of change of velocity with respect to time. But since velocity is a vector, so must acceleration be a vector. Formally, acceleration is defined as

$$\mathbf{a} \equiv \frac{d\mathbf{v}}{dt} \tag{4-7}$$

and it can be written in the limit form

$$\mathbf{a}_{12} \equiv \lim_{t_2 \to t_1} \frac{\mathbf{v}_2 - \mathbf{v}_1}{t_2 - t_1}$$

Figure 4-1*d* illustrates average acceleration as a vector difference in the velocities between two different points, 1 and 2, in time (and location). But unlike the case of velocity, there is no word to designate the magnitude of

Figure 4-1 Motion of a vehicle in a plane. *(a)* Location. *(b)* Location vector difference. *(c)* Velocity. *(d)* Velocity vector difference.

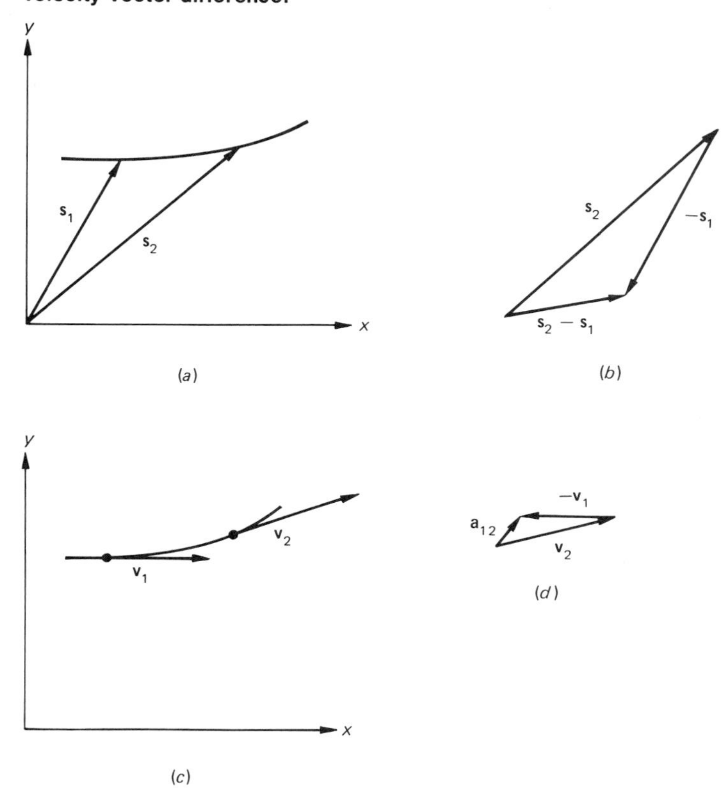

acceleration, as distinct from acceleration in the full sense of a vector with magnitude and direction. Following the usual practice in engineering, we will use the term acceleration in both ways, with the context usually sufficient to reveal whether the scalar or vector concept is meant.

Forces Producing Motion in Transportation

The movement of any vehicle along its path is determined by the forces acting on that vehicle, according to Newton's laws. By consciously applying forces, the direction and magnitude of which are controlled, the velocity and path of the vehicle is determined.

The forces acting on a vehicle can be understood by consideration of the forces shown in Fig. 4-2. The x and y axes are perpendicular to one another and are in the horizontal plane. The z axis is perpendicular to this plane. Acting parallel to the vertical axis is the weight of the vehicle, the force T, which, if not reacted to by an equal and opposite force, would cause the vehicle to fall. If the vehicle is moving, there is an inherent resistance of the vehicle to motion R, acting along the axis of the vehicle's path, which also must be overcome. In addition, if the vehicle is

Figure 4-2 **Forces acting on a vehicle in motion.**

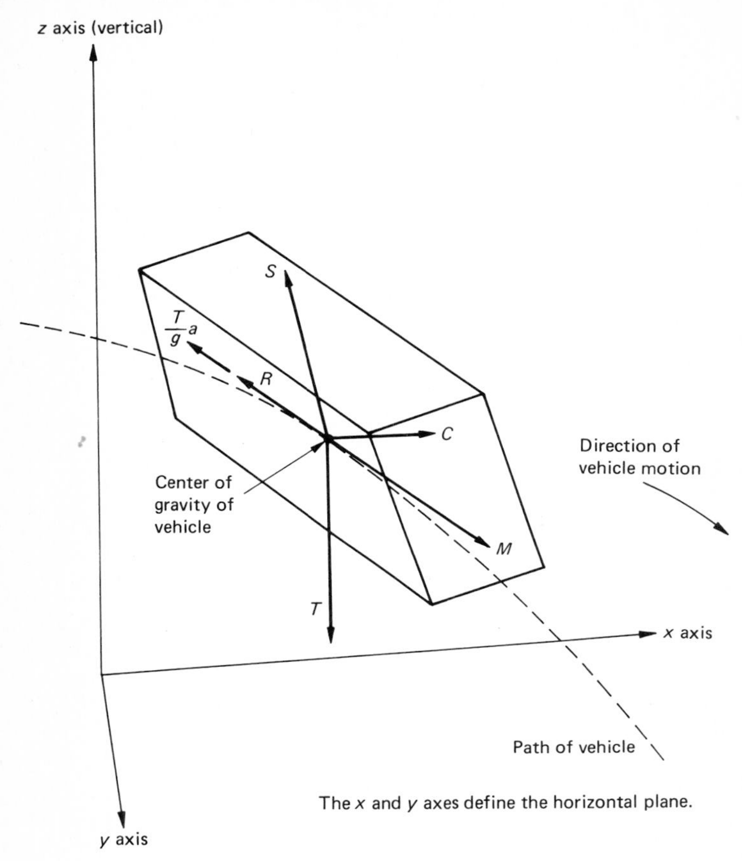

The x and y axes define the horizontal plane.

accelerating, then a net force Ta/g, a being the acceleration and g the gravitational constant, must be the resultant force along its axis. Finally, the vehicle may be changing direction, in which case the centrifugal force C must be overcome, such as by the forces between the wheel flanges of a railroad car and the outer rail on a curve.

The forces to overcome these resistances to the desired motion come from a combination of the propulsive unit, gravity, and the reaction of the surface (or medium) on which the vehicle is moving. The propulsive force M acts along the vehicle's axis, as shown. The reaction force of the supporting medium can include the propulsive forces, as in the case of wheeled vehicles (where the propulsive force is created by wheel to path friction), but after elimination of this component, the resulting support force S acts in a plane perpendicular to the vehicle axis as shown, and will include a component to overcome centrifugal force if the vehicle is turning. Notice that when the vehicle is changing elevation, the weight of the vehicle will have a component along the vehicle's axis, and therefore it will either aid or hinder motion.

Thus there are many forces which must act on a transport vehicle in order that it move in the manner desired. To summarize, these forces may be categorized as those required to

1 Overcome inherent resistance to motion
2 Provide acceleration and deceleration
3 Guide or steer
4 Change elevation

The same principles and general methods apply to all vehicular technologies of transportation, and hence these are most appropriately treated in a unified manner. However, there are differences in the exact manner of treatment among the technologies, which reflect differences in the terminology used and in certain simplifying assumptions, and this makes a perfectly uniform treatment difficult. Nevertheless, the presentation to follow will be of this general type, stressing principles and methods which are applicable to all vehicular technologies.

To provide illustrations of the application of these and their practical significance, examples of the analyses of a railway train will be given, and hence the terminology and some details of this technology will be presented. The railway was chosen because it illustrates most of the principles more fully than could be done with others. It is in this mode that the transportation system engineer is most likely to be required to deal with the widest range of vehicle variations and path variations. In the air mode, choices must usually be made between a limited number of vehicle designs which can be only minimally changed, and the characteristics of the path followed by the vehicle vary only minimally, in ways that are readily dealt with. At the end of this section, the formulas and other information necessary to treat other technologies will be presented, following the pattern established for the railway examples. In this manner, all the vehicular technologies will be treated, but in the level of detail consistent with the purpose of this book—the presentation of the principles and techniques of transportation system engineering rather than all the details of the engineering of any one transport technology. Now we shall turn to the forces underlying the generalized equations of motion of vehicles.

Inherent Resistance

All vehicles traveling on or in the earth or its atmosphere possess an inherent resistance to motion, by which is meant the resistance to motion on a straight, level path at a constant speed. This resistance must be overcome if the vehicle is to maintain that speed, for otherwise this inherent resistance would cause the vehicle to decelerate, ultimately to a stop. In general, this resistance for any given vehicle type and transport technology is a function of its weight and its velocity. It is important to consider weight variations primarily because the load being carried on a vehicle will vary from one assignment to another, and a rapid means of estimating resistance for each load is desired. Also, there is a reduction in the weight of a vehicle which carries its own fuel as a trip progresses, and this often must be taken into account. The inherent resistance will be termed $R(T,V)$ where the parentheses containing the total vehicle weight T (including fuel, load, etc.) and the velocity V indicates that the resistance depends on these two characteristics. The function R, of course, will depend upon the particular vehicle type (size, shape, etc.) and path. Generally, in land transport technology (rail, road, etc.), the inherent resistance increases with increasing speed and also with increasing weight of the vehicle (including its load). The reason for the customary increase with speed is primarily the increase in air resistance, which can be very appreciable at high speeds (say, 50 mi/h or more) for all vehicles. Also, there usually is an increase in other components of resistance, such as the resistance to rolling of wheels on rails or roads, and the resistance of water to the hull of a vessel passing through it. For airborne vehicles, resistance also increases with speed, although not always very rapidly, because of the way in which the "pushing aside" of air is used to provide lift and thereby support for the vehicle. Certain types of water craft, such as hydrofoils, also exhibit unusual characteristics in this regard, and these will be briefly described later when the formulas for these are presented.

The resistances of vehicles are estimated by the use of formulas which have been estimated from numerous tests of actual equipment in conjunction with theories as to the sources and nature of the frictional resistances and the flow of fluids around the vehicle. The resistance formulas for various types of railway equipment are presented in Table 4-2. These are numbered for easy reference. Note that all rail car variables, T, N, and V, are as defined for the locomotive and are in the same units.

Notice that all the resistance formulas for the various types of railway equipment have the same general form: two terms which together depend only on the vehicle design and weight but not velocity, a third term which depends on these two plus the first power of velocity, and a fourth term which

Table 4-2 Inherent Resistance Formulas for Various Railroad Locomotives and Cars

Diesel-electric locomotives

$$R = 1.3T + 29N + 0.03TV + 0.0024CAV^2 \tag{4-8}$$

where R = resistance, lb
T = weight, tons
N = number of axles
V = velocity, mi/h
A = maximum cross section area, ft^2
C = air resistance parameter;
first locomotive, $C = 0.0024$ (less if streamlined, as low as 0.0017);
following locomotives, $C = 0.0024$ (less if closely coupled and streamlined, as low as 0.0005)

Standard freight car

$$R = 1.5T + 72.5 + 0.015TV + 0.055V^2 \tag{4-9}$$

Piggyback car (two truck trailers on flat car)

$$R = 0.6T + 20N + 0.01TV + 0.20V^2 \tag{4-10}$$

Passenger car

$$R = 1.3T + 29N + 0.03TV + 0.041V^2 \tag{4-11}$$

Source: General Electric Company (1969).

depends on the vehicle's shape and size and the square of velocity. The first three terms are primarily rolling resistance of the vehicle, and the fourth term is the additional resistance due to the air. However, this interpretation should not be taken too literally, for there are components of air resistance, especially at low speed when the flow of air past the vehicle is smooth (technically termed laminar flow) that are included in the other three terms. It is unfortunate that much of the literature on the resistance of wheeled land vehicles identifies the term containing V^2 as the only air resistance term.

These formulas are for movement in still air, where the velocity of the vehicle relative to the air is the same as the velocity relative to the path. If the air is moving, then the formula should be modified. In particular, the third term should be appropriately changed, since at high speeds the effect on this term is the greatest. In the case where there is a direct headwind or tailwind acting along the vehicle's path, then the wind velocity is added or subtracted from the ground velocity to yield the velocity value to be inserted into the third term. In the case of a crosswind, the connections are very complicated and will not be treated here. In fact, in practice for rail and road transport any correction for wind is usually not made, partly because often the effect is small, and partly because winds probably would be encountered for only a small portion of the trip anyway.

Other factors besides the wind can affect these formulas. The quality of the track surface influences the resistance. Weather in general affects resistance. In cold weather, oil becomes more viscous, resulting in a higher resistance to rotating parts until the rotation of the parts heats it. Rain or snow can strike a vehicle in such a manner that it increases the resistance, just as wind can. In addition, water on the rail surface will add slightly to the resistance, since it must be "pushed" out of the way; and of course large puddles to be traversed in any flooded areas add considerably to the resistance. The formulas presented above represent typical conditions, and thus care must be exercised in the use of these formulas so as to account for unusual conditions.

To illustrate the use of these formulas, consider a train composed of

4 diesel-electric locomotives, each weighing 136 tons, having 4 axles, and a cross section area of 120 ft^2

20 empty standard freight cars, each weighing 30 tons, with 4 axles

20 standard freight cars, each with a load of 40 tons, weighing empty 30 tons, having 4 axles

10 piggyback cars, of 4-axle design, each carrying 2 empty trailers, total weight of 47 tons (34.5 tons for rail car alone)

10 piggyback cars, of 4-axle design, with 2 loaded trailers each, total weight of 77 tons

1 caboose (car for crew at rear of train), weighing 30 tons, having 4 axles

At a speed of 50 mi/h, on a level, straight route, this train has a total resistance of 31,008.1 lb. This is calculated in the following manner:

4 locomotives, using Eq. (4-8):

$$\text{Resistance of each locomotive} = 1.3(136) + 29(4) + 0.03(136)(50) + 0.0024(120)(50)^2$$
$$= 176.8 + 116 + 204 + 720$$
$$= 1216.8 \text{ lb}$$
$$\text{Resistance of all 4 locomotives} = 4867.2 \text{ lb}$$

20 loaded standard freight cars, by Eq. (4-9):

$$\text{Resistance of 1 car} = 1.5(70) + 72.5 + 0.015(70)(50) + 0.055(50)^2$$
$$= 105 + 72.5 + 52.5 + 137.5$$
$$= 367.5 \text{ lb}$$
$$\text{Resistance of 20 cars} = 7350 \text{ lb}$$

20 empty standard freight cars, by Eq. (4-9):

$$\text{Resistance of 1 car} = 1.5(30) + 72.5 + 0.015(30)(50) + 0.055(50)^2$$
$$= 45 + 72.5 + 22.5 + 137.5$$
$$= 277.5 \text{ lb}$$
$$\text{Resistance of 20 cars} = 5550 \text{ lb}$$

10 piggyback cars with empty trailers, by Eq. (4-10):

$$\text{Resistance of 1 car} = 0.6(47) + 20(4) + 0.01(47)(50) + 0.20(50)^2$$
$$= 28.2 + 80 + 23.5 + 500$$
$$= 631.7 \text{ lb}$$
$$\text{Resistance of 10 cars} = 6317 \text{ lb}$$

10 piggyback cars with loaded trailers, by Eq. (4-10):

$$\text{Resistance of 1 car} = 0.6(77) + 20(4) + 0.01(77)(50) + 0.20(50)^2$$
$$= 46.2 + 80 + 38.5 + 500$$
$$= 664.7 \text{ lb}$$
$$\text{Resistance of 10 cars} = 6647 \text{ lb}$$

1 caboose, using Eq. (4-9):

$$\text{Resistance of 1 car} = 1.5(30) + 72.5 + 0.015(30)(50) + 0.055(50)^2$$
$$= 45 + 72.5 + 22.5 + 137.5$$
$$= 277.5 \text{ lb}$$
$$\text{Total resistance of this train} = 4867.2 + 5550 + 7350 + 6317 + 6647 + 277.5$$
$$= 31{,}008.7 \text{ lb}$$

The total weight of this train is 3814 tons, so the average resistance per ton is 8.130 lb/ton. In terms of the cargo load, however, the average resistance is 28.189 lb/ton for the 1100 tons of cargo being carried. The effect of the weight of the cars (and truck trailers) and the fact that in this train only half are loaded with cargo, is evident.

It is imperative that in the use of any resistance formulas such as Eqs. (4-8) through (4-11) above that you enter values of the variables which are in precisely the correct units. These equations called for velocity in miles per hour, weight in tons, and yielded resistance in pounds. If you desire resistance in any other units, it is advisable to make the change after calculating the resistance in the unit specified in the original formula. Similarly, if you are given speed or weight in units other than those specified in the equation, then you should convert to the required units rather than try to modify the equation. These rules will help avoid errors. As is common in the transportation engineering literature, the resistance equations were presented without specifying the units for all the parameters. This was done deliberately so that you would become accustomed to this. This means, for example, that if you wish to perform a dimensional analysis of the resistance equations, you must first figure out what the dimensions of each parameter are! Many persons who are used to seeing equations only with a complete specification of dimensions for all elements find this difficult to get used to, and therefore it is important that this custom in transportation engineering be remembered, for otherwise a dimensional analysis can lead to very erroneous conclusions.

These calculations of resistance for a railroad train are rather lengthy. This is true of other vehicles also, although often not quite so lengthy unless numerous formulas are involved, as in the case of aircraft. For this reason, such calculations are usually made on a computer, the programming of such repetitive calculations being fairly easy. However, if you are performing them manually, it is useful to note that the railroad train calculations above can be simplified by combining like terms for different cars.

The effect of variations in speed and weight can be ascertained easily with these resistance formulas. The general relationship between resistance and these two variables can be seen from Fig. 4-3, which presents the inherent resistance for a single standard railroad car [using Eq. (4-9)]. The weight of the car is 30 tons (typical North American cars varying between 25 and 35 tons empty) and the maximum load is taken to be 80 tons (typical maxima ranging from 40 to 100 tons).

This figure is significant in that it illustrates two general relationships that

Figure 4-3 Effect of weight and speed on railroad vehicle inherent resistance.

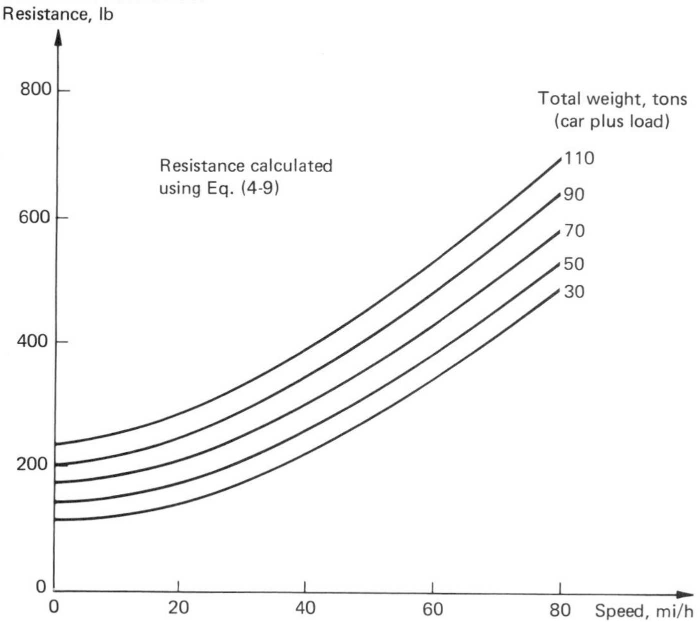

are quite important in transportation. The first is that for any given railroad vehicle, inherent resistance increases with velocity, and it typically increases at an increasing rate. This pattern of increasing inherent resistance with increasing speed applies to all wheeled land vehicles, but it does not apply to some vehicles, such as hydrofoils and aircraft. Second, the inherent resistance increases with increasing total weight, but not proportionately, so that a doubling of weight does not result in a doubling of resistance. This second characteristic seems to apply to all technologies. For example, in the case of this railroad car, at 50 mi/h, an increase in total weight, resulting from an increase in the cargo weight from 40 to 80 tons (a 100 percent increase in cargo), leads to only a 24.5 percent increase in total resistance. In this example, resistance per ton of total vehicle weight drops from 5.25 to 4.16 lb/ton, and resistance per ton of cargo alone from 9.19 to 5.72 lb/ton.

Before turning to the inherent resistance of other technologies, it is appropriate to deal with a question that invariably arises. Often persons using these railway formulas for the first time ask why there is any contribution of the cars following the locomotive to the air resistance. Or, alternatively, should not the parameter of the air resistance term be reduced to reflect the lessened resistance of a following locomotive or car? Part of the answer is that each car of a train contributes to the total air resistance, primarily due to side friction of the car (which often has a rough exterior with many protrusions) moving past the air and causing turbulence in the air. In addition, all cars of a train generally do not have the same height or cross section, and a box car following an unloaded flat car will have to push aside a considerable amount of air as it moves. These formulas represent an averaging of all these effects, and, in fact, each equation represents the added resistance due to the addition of a car (or locomotive) of the type designated, and hence underestimates the resistance of a single car moving along under its own power.

Now that the concepts of inherent resistance have been explained with the use of a detailed railroad example, the resistances of other transport technologies will be described. Any deviations from the railroad form in concepts or manner of application will be explained.

Table 4-3 presents the inherent resistance formulas for wheeled road vehicles—automobiles, trucks, and buses. The form of these is almost identical to the form of the railroad formulas, as one might expect from the similarity of the technologies. The use of these is straightforward, exactly the same as that for the railroad formulas; one need only be mindful of the units, which differ from the trains. In particular, the unit of weight used in motor vehicle transportation is often 1000 lb, termed 1 kip. As in the case of the earlier formulas, many factors influence the actual resistance of a vehicle, these formulas and parameters assume both the vehicle and hard-surfaced road are in good condition. The sources contain more detailed information on variations due to vehicle and road conditions, weather, altitude, etc.

The inherent resistance of many important types of vehicles includes both forces to overcome the resistance to horizontal motion and forces required to support the vehicle—in contrast to the wheeled land transport vehicles discussed above, which are naturally supported by the paths on which they operate. The two forces, support and horizontal resistance, are not so easily represented in equation form as the resistances of wheeled vehicles, partly

Table 4-3 Inherent Resistance Formulas for Wheeled Road Transport Vehicles

Automobiles and buses (Taborek, 1957)

$$R = 0.01T + 0.0001TV + 0.0026CAV^2 \qquad (4\text{-}12)$$

where R = resistance, lb
T = weight, lb
V = velocity, mi/h
C = air resistance parameter
Auto, C = 0.40 to 0.50
Convertible auto, C = 0.60 to 0.65
Bus, C = 0.60 to 0.70
A = maximum cross section area, ft^2

Trucks (Society of Automotive Engineers, 1974)

$$R = 7.6T + 0.09TV + 0.002AV^2 \qquad (4\text{-}13)$$

where R = resistance, lb
T = weight, 1000 lb or kips
V = velocity, mi/h
A = adjusted maximum cross section area, ft^2
= width times height less $\frac{3}{4}$ in or 0.0625 ft

because of the variability of the air and water media in which they operate and partly due to inherent complexity in the underlying physical phenomena. Nevertheless, they can be calculated approximately from appropriate formulas, graphs, or tables, which are the results of considerable theoretical and experimental work which started in earnest as a result of failures or problems with the first large ocean-going steamships of the 1850s.

One large class of such vehicles includes those which are primarily or exclusively supported by their own buoyancy in the fluid medium, air or water or both, in which they operate. In the case of vessels operating on the surface of the water, these must overcome many types of resistances: frictional resistance of the water passing the hull, wave-making resistance, eddy resistance associated with separation in the water flow around the hull, additional resistance due to rough water and the measures used to counteract its adverse or unwanted effects, shallow water effects through increasing the velocity with which water must pass the vessel, and finally air resistance, which is of increasing importance due to the increase in ship speeds. Many years of research has led to the development of relationships for estimating resistances of ships. The results of calculations are shown as lines in Fig. 4-4, with empirical results for actual vessels indicated by specific points.

Following the format of the source (Mandel, 1969), these graphs relate the ratio of weight to resistance to the speed of the ship for various total ship weights. The three weights graphed should be sufficient to allow interpolation of the resistances of ships with other weights. Again the same general pattern as found in land vehicles is observed: an increase of resistance with speed, generally at an increasing rate, and a decrease of resistance per unit weight as the weight increases.

Also shown in this figure are the weight-to-resistance ratios for submarines and airships. The theory of these is identical to that for surface ships but each is traveling in only one medium. It is interesting to note that the

Figure 4-4 **Weight-to-resistance ratios for airships, submarines, surface ships, and planing craft.** [*From Mandel (1969), p. 7.12.*]

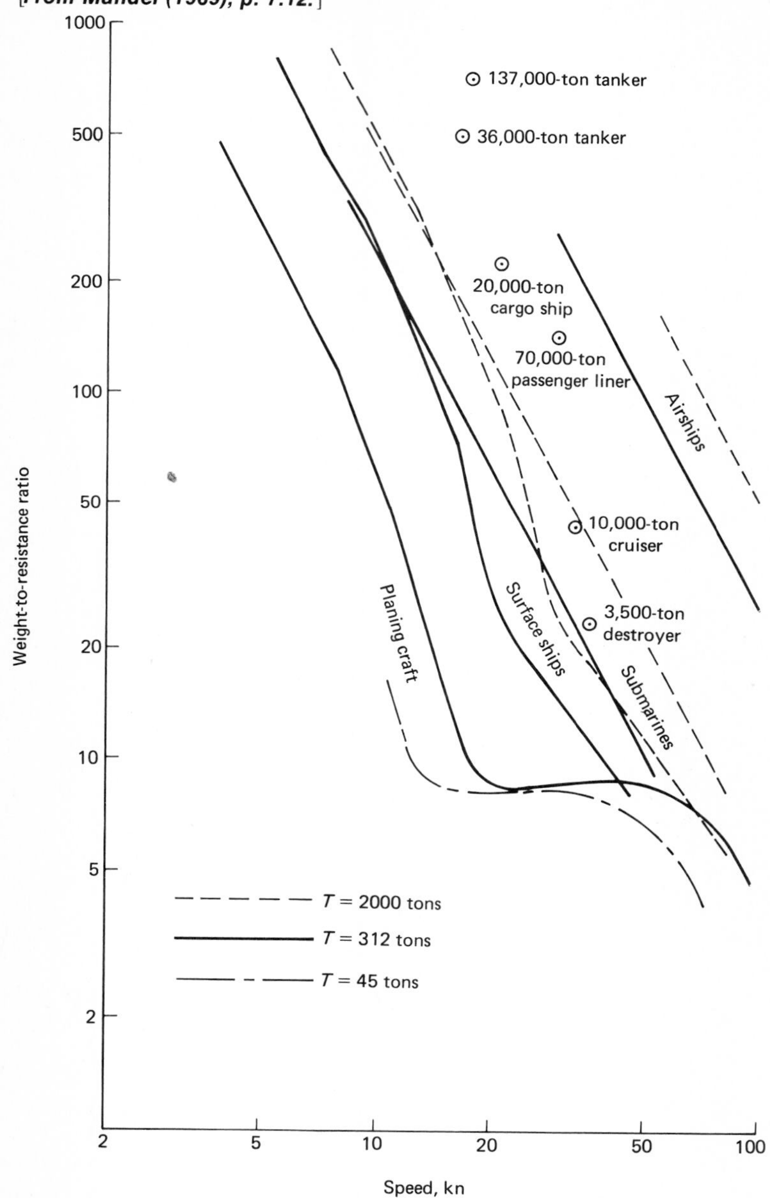

weight-to-resistance ratio of submarines is greater than that for a ship of comparable weight above about 13 knots (kn), indicating that the resistance is less at identical speeds. This explains part of the current interest in underwater technology for the transport of freight and passengers over long distances—the consumption of less power. Also, there are other advantages, such as freedom from surface storms. Of course, many technical problems in the design of strong hulls must be overcome, as must problems of guidance and control, not to mention safety in rescue operations.

The airship curves are primarily theoretical in origin, but give some idea of the magnitude of resistances to be expected. Airships are not now used for commercial service, primarily due to problems of maneuvering these large vessels and their vulnerability to storms due to their size and inability to fly above them as other aircraft can.

Planing craft are vessels which operate in the water supported by buoyant forces at low speeds, but which are designed to be lifted out of the water at high speeds by the dynamic forces on the underside of the hull. While such craft are primarily employed in pleasure boating and in military use, where speed is essential (with weights usually in the vicinity of the 45-ton craft shown or less), they could be used for other purposes. It is interesting to note that with this technology the resistance to motion decreases with increasing speed in the vicinity of 20 to 40 kn (exact speeds depending on craft design), reflecting the change in the relationship between the vessel and the media in which it is traveling.

The resistance of aircraft is also a complex subject which can be treated only in a summary form in this text. However, again the purpose is an understanding of the effect of speed, weight, and other factors on resistance, rather than an ability to design aircraft! Heavier-than-air aircraft are sustained in flight by a difference in pressure above and below the vehicle (mainly the wings) sufficient to support the vehicle. This is achieved through the design of the wings, or airfoils, as they are technically termed. The airfoil is designed so that air flowing above it travels a longer distance than that flowing below it, resulting in greater speed for the air flowing above. According to Bernoulli's law, which states that the pressure of a fluid is less the greater its speed, the pressure above the wing is therefore less than that below. This results in an upward force, which if sufficient will support the aircraft or if even greater will cause it to rise. Thus the elevation or altitude of the aircraft can be controlled by controlling this pressure difference. Examples of curves of weight to resistance for two aircraft are shown in Fig. 4-5. These are based on theoretical calculations, rather than on actual test measurements, but are representative of typical values. The smaller aircraft is typical of private planes and is designed for cruising at a speed of 110 kn at an altitude of 10,000 ft. The other aircraft, a hypothetical one weighing 304 tons, has a design cruise speed of 550 kn at an altitude of 45,000 ft. Resistances for other sizes can be interpolated approximately using this figure.

The principles discussed above for aircraft are identical to those which apply to hydrofoil boats. Thus it would be expected that the resistances of hydrofoils would behave in the same manner as those of aircraft, and this is borne out by the curves plotted in Fig. 4-5. An actual measured resistance curve for a 127-ton hydrofoil used in commercial service is shown. It can attain a top speed of about 60 kn. It can achieve a weight-to-drag ratio as low

Figure 4-5 **Weight-to-resistance ratios for aircraft, air cushion vehicles, hydrofoils, and planing craft.** [***From Mandel (1969), pp. 7.25 and 7.55, except 127-ton hydrofoil from Comstock (1967), p. 369.***]

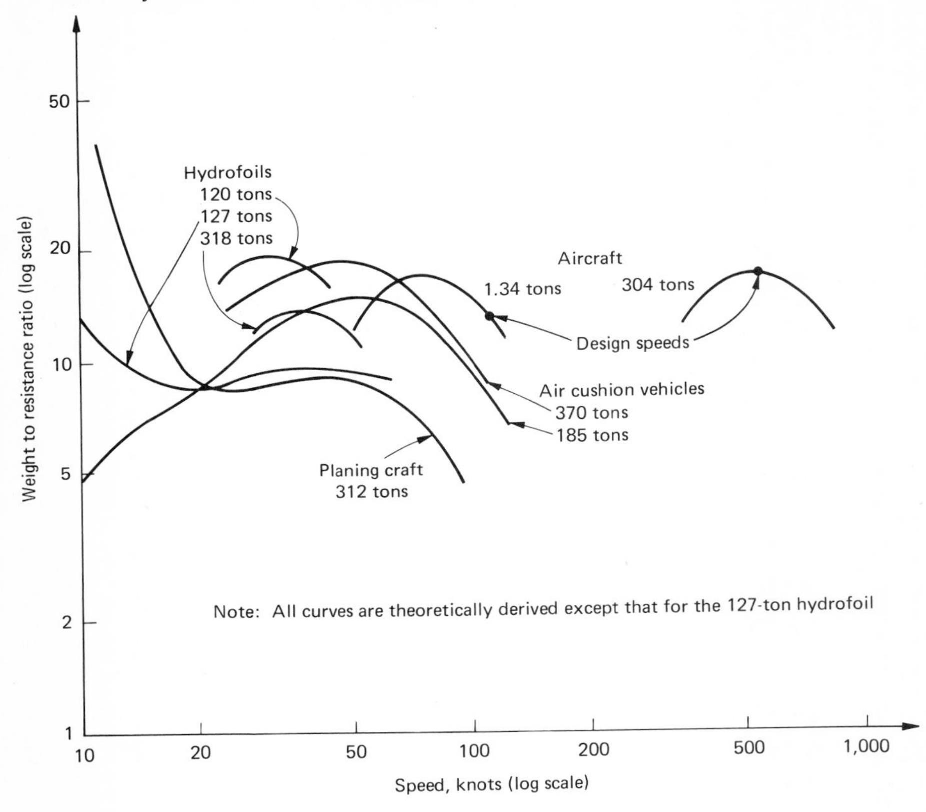

as 7.5 at its typical cruise speeds. Also shown are curves for two other hydrofoils, calculated theoretically, one of 120 tons and the other 318 tons. The variation between theory and measurements reflects differences in design as well as the current inability to predict resistances very accurately in new fields such as hydrofoil technology.

Air cushion vehicles, as the name implies, operate on a cushion of air which can be used to provide mobility over land or water. Such vehicles are in widespread use as commercial carriers of persons and, to a lesser extent, freight in many parts of the world. These vehicles must expend energy to overcome the air resistance of the body (termed profile and trim drag), to impart momentum to the air used to create the cushion, and overcome wave making and wave drag if operating over water. Theoretically these resistances over water sum to the values shown in Fig. 4-5. Actual weight-to-resistance ratios have been found to increase at very low speeds, resistance being approximately a linear function of speed at very low speeds not shown in this figure (Elsley and Devereux, 1968, p. 44).

Gradients

Vehicles of all modes undergo changes in elevation above sea level. Rail and road vehicles ascend and descend grades frequently, and aircraft of course ascend to and descend from the normal elevation of flight. Even water-borne craft ascend grades in the form of movement upstream, where a slight upgrade will exist, or downstream with a downgrade. Because such movement is associated with a change in elevation, the force of gravity is involved and will either reduce or increase any resistance to motion.

The manner in which gravity comes into play can be seen by reference to Fig. 4-6. This shows a vehicle moving on a gradient. There is a component of the force of the vehicle's weight (acting downward) which acts parallel to the surface of the path on which the vehicle is traveling. This force F_g is related to the weight of the vehicle T and the angle θ between the path and horizontal plane (which is by definition perpendicular to the weight force) by

$$F_g = T \sin \theta$$

where F_g = gradient force
θ = angle between grade and horizontal

The most common measure of the severity of a gradient is the so-called percent grade. It is defined as the rise, measured in any unit of length, over a horizontal distance of 100 such units. For example, a 1.0 percent upgrade is a grade which rises 1 ft in 100 ft of movement on the horizontal. Note that if this same grade were measured in meters, the result would still be a 1.0 percent grade, since it would rise 1 m for every 100 m of horizontal distance. Another way of reaching the same conclusion of independence of the measurement units of length is to realize that the rise divided by the horizontal distance is simply the tangent of the angle subtended by the path and the horizontal, as shown on the upper left of Fig. 4-6. Since gradient is so commonly measured in percent, it is appropriate to express the grade force F_g in a form related to the percent grade p. The true relationship is, by the geometry of a right triangle,

$$F_g = T \frac{p}{\sqrt{100^2 + p^2}} \qquad (4\text{-}14)$$

where p = percent grade. However, in almost all applications the value of p is

Figure 4-6 **Forces due to gradients.**

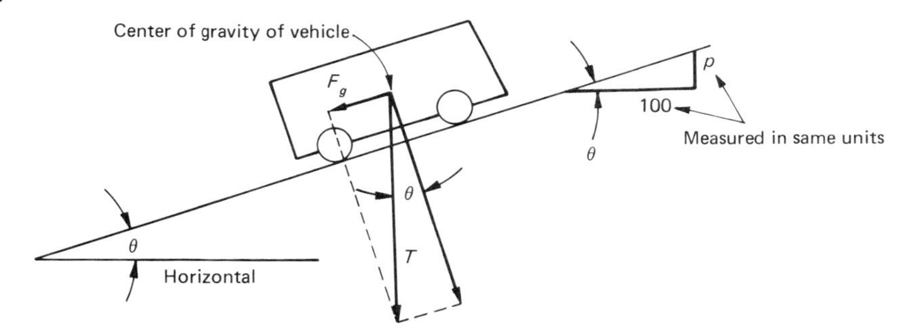

very small relative to 100, and hence this approximation is employed:

$$F_g \doteq T \frac{p}{100} \tag{4-15}$$

This approximation is very good for small gradients, such as the typical railway grades of less than 3 percent and main road grades of less than 5 percent, but for grades larger than 5 percent, it is best to use the true relationship, Eq. (4-14).

Since this force is acting in the downgrade direction, this gradient force will help overcome resistances to motion if the vehicle is moving in the downgrade direction. In fact, in many instances the gradient force is so large that it more than overcomes all resistance forces and causes the vehicle to accelerate. If the speed increases to a dangerous level, then the vehicle must be braked in some manner. Alternatively, with an upgrade the gradient force will add to the vehicle's resistance.

A common grade on railroads is 0.6 percent. If the freight train of the previous example is moving up this grade, then the total resistance of the train consists of the inherent resistance calculated previously plus the resistance due to the grade. The total train weight was 3814 tons. Substituting this into Eq. (4-15), we find that the grade resistance is

$$3814 \text{ tons} \frac{0.6 \text{ percent}}{100 \text{ percent}} = 22.884 \text{ tons}$$

Since this is in tons, we must convert it to pounds in order to add it to the inherent resistance at 50 mi/h calculated earlier as 31,009 lb. Since 2000 lb equals 1 ton, the gradient resistance is 45,768 lb. The total resistance of the train moving upgrade at 50 mi/h is 76,777 lb, which is over twice as great as the resistance on the straight, level line! It is clear why railway engineers attempt to minimize the magnitude of grades, and often lay out a circuitous route in order to pass through mountains at a low elevation. The same principle applies to road vehicles.

If the train were traveling downgrade, then the gradient force would have acted in the direction of motion. The net force acting on the train would have been the grade force of 45,768 lb less the inherent resistance of 31,009 lb, or 14,759 lb, acting in the direction of motion. This force would produce an acceleration of [using Eq. (4-3)]

$$a = \frac{F_g}{T} + \frac{14{,}759 \text{ lb}}{3814 \text{ tons}} \frac{32.16 \text{ ft}}{\text{s}^2} \frac{\text{ton}}{2000 \text{ lb}} = 0.0622 \text{ ft/s}^2$$

if it were not checked by applying the brakes of the train. Thus a substantial braking force would be necessary to hold the train to a constant speed on the downgrade.

Since the effects of gradients, ascending or descending, are so straightforward, additional examples are not warranted.

Curvature Resistance

All vehicles must traverse curves or change direction. There are two important types of curves in the paths of vehicles. Horizontal curves are curves in the horizontal plane and do not result in any change in vehicle elevation. Vertical curves are in the vertical plane and result in a change in elevation. Often the two are combined in actual curves in the path followed by a vehicle, although they can exist independently.

The effect of curves on vehicle resistance to motion can be understood from reference to Fig. 4-2. In the simplest case, the introduction of the curvature or change of direction results in a change, usually an increase, in the total resistance which must be overcome. This change in resistance is due to the centrifugal force which, provided the vehicle is traveling at constant speed on a path of constant radius, is equal to

$$F_c = \frac{T}{g}\frac{V^2}{r} \tag{4-16}$$

where F_c = centrifugal force
r = radius of curvature
V = vehicle speed
g = gravitational constant

A countervailing force must be applied to the vehicle, and it is this which increases the resistance.

This force is applied at the wheel-rail interface in railroad systems, as shown in Fig. 4-7*a*. By banking the track (i.e., elevating the outer rail) as shown, the resultant of the weight and centrifugal forces is made to act more nearly perpendicular to the plane of the track, reducing the force acting parallel to that plane which tends to overturn the vehicle and cause discomfort to passengers and damage to freight. This force can be derived readily. We use the notation of Fig. 4-7:

$$U = F_c \frac{1}{\sin(\theta + \alpha)}$$

$$F_{rc} = U \sin \alpha = \frac{V^2 T}{gr} \frac{\sin \alpha}{\sin(\theta + \alpha)}$$

The force F_{rc} is partly responsible for the increase in resistance of vehicles when rounding a curve. In the case of railroad trains, for instance, this force acts at the wheel-rail interface and increases the resistance of the wheels to rolling.

Also increasing resistance is the increase in the force acting perpendicular to the plane of contact between the wheels and the rails. In Fig. 4-7 this force is shown as X on a curve, in contrast to the usual value of T, on a level, straight path. The equation for X can be derived from that for U in terms of T:

$$U = T \frac{1}{\cos(\theta + \alpha)}$$

$$X = U \cos \alpha = T \frac{\cos \alpha}{\cos(\theta + \alpha)}$$

Since the larger the angle, the smaller the value of the cosine in the range from 0° to 90°, X will be greater than T provided θ is greater than zero.

Figure 4-7 **Curvature forces and terms.** ***(a)*** **Forces on vehicle rounding a curve.** ***(b)*** **Relationship between curve radius and degree of curvature.**

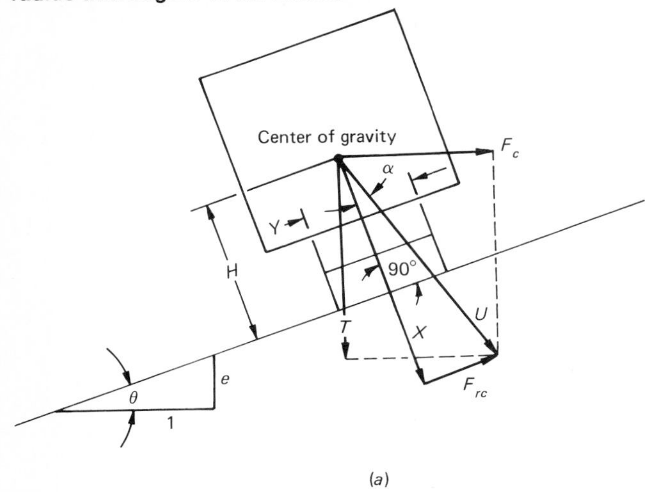

(a)

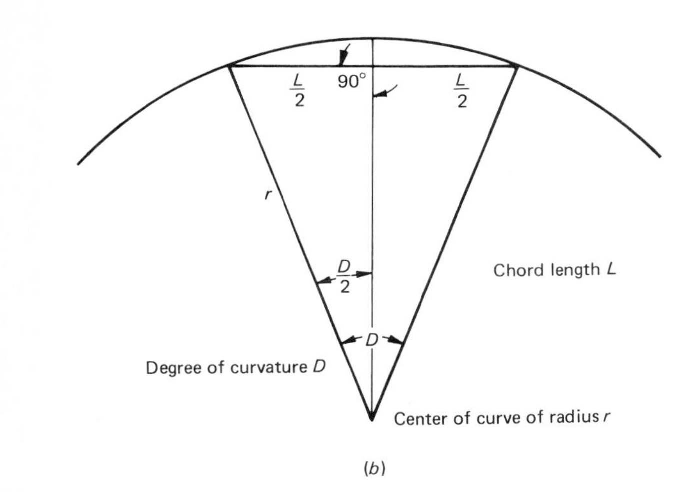

(b)

The general expression relating the centrifugal force F_c to the angles θ and α is

$$\frac{V^2T}{gr} = T \tan(\theta + \alpha)$$

This can be rewritten, using a trigonometric identity and dropping the T from both sides of the equation, as

$$\frac{V^2}{gr} = \frac{\tan\theta + \tan\alpha}{1 - \tan\theta\tan\alpha}$$

This can be expressed in terms of e, the rate of superelevation (equal to $\tan \theta$) and f, the ratio of the side force F_{rc} acting parallel to the wheel-rail plane and the force X perpendicular to that plane (equal to $\tan \alpha$):

$$\frac{V^2}{gr} = \frac{e + f}{1 - ef} \tag{4-17}$$

Since both e and f are small, often the right side is taken as $e + f$ only. Hence to make the side force zero, it is necessary to set the value of e as follows:

$$e = \frac{V^2}{gr}$$

However, if different vehicles operate at different speeds on the curve, then no single value of e will be sufficient, and a compromise value must be used. The selection of this value will be discussed in the chapter on design, Chap. 16.

It is worth noting at this point that the same formulas apply to any vehicle rounding a curve. In fact, in road design Eq. (4-17) is used directly. In that context f has the added interpretation of the coefficient of side friction between the wheels and the road, this friction being the means by which the force F_{rc} is transmitted between the wheels and the road. Of course, on railroads this is accomplished by the wheel flanges acting against the rail.

These additional side forces and forces perpendicular to the plane of the pathway facilities add to the resistance of all wheeled vehicles on curves. However, only in the case of railroad trains, where the power unit is adjusted to what is required to propel the vehicle, is this additional resistance due to curvature usually taken into account in predicting vehicle motion. In the railroad mode, it has been found that the resistance due to curvature varies widely, depending upon factors which are only imperfectly understood. Typically used values range from 0.4 to 0.8 lb per ton of car weight per degree of curvature, the latter value being recommended by the American Railway Engineering Association for calculations. The degree of curvature is a measure of the radius of a curve, but is more suitable for field measurement than the radius. In the manner it is used by railway engineers (in contrast to its use in other fields, such as highway engineering), the degree of a curve is the central angle subtended by a chord of 100-ft length. It is illustrated in Fig. 4-7*b*. As can be deduced from that figure, it is related to the radius of the curve by

$$\frac{L}{2} = r \sin\left(\frac{D}{2}\right) \tag{4-18}$$

where r = radius of curve
D = central angle subtended by chord of length L
L = chord length

Thus in general, for railways, the added curve resistance is

$$R_c = \epsilon DT \tag{4-19}$$

where ϵ equals the curvature resistance coefficient, 0.8 lb per ton per degree being recommended.

An example of the use of these relationships would be useful. We will use the freight train that has been used in earlier computations, this train weighing 3814 tons. If it is traveling around a curve of 3.0°, what is the radius and what is the added resistance due to curvature? The radius is calculated using Eq. (4-18):

$$r = \frac{L/2}{\sin (D/2)} = \frac{50 \text{ ft}}{\sin (1.5°)} = 1910 \text{ ft}$$

The added resistance is calculated on that portion (by weight) of the train which is on the curve. Assuming the entire train is on the curve, this is, by Eq. (4-19):

$$R_c = 0.8 \frac{\text{lb}}{\text{ton-degree}} (3°)(3814 \text{ tons}) = 9153.6 \text{ lb}$$

The added resistance is small, but if the number and power of the locomotives is adjusted very closely to the amount needed to propel the train, this amount is significant. In fact, on most railroad lines grades are reduced on curves by an amount sufficient to compensate for the added resistance due to the curve. In this way, the power requirements of the locomotive are kept at a constant level. Since each percentage of gradient adds 20 lb/ton resistance, the grade must be reduced by a percentage which will reduce the grade resistance by the 2.4 lb/ton that the curve increases it. This is precisely achieved by a gradient reduction of 0.12 percent. In general, the gradient must be reduced by 0.04 percent for each degree of curvature, assuming the value of 0.8 lb/ton-degree for curve resistance.

Propulsion Forces

There are almost as many means of giving locomotion to vehicles as there are different means of giving mobility to them. Forces are required in all transport technologies to overcome the resistance forces described in the previous section, which include inherent resistances (including, where applicable, the force required to support the vehicle), forces required to cause the vehicle to follow the desired path, and forces required to cause the vehicle to accelerate and decelerate as desired. As in the previous sections, the emphasis will be upon principles which apply to all technologies of transport, but nevertheless examples of actual relationships and methods applicable to the various technologies will be given to provide a fuller and more concrete understanding of these principles. In addition, the example application to the railroad freight train used in the previous section will be continued.

In order to discuss the propulsion forces acting on a vehicle it is necessary to understand certain physical terms which are closely related to the forces referred to in Newton's laws that lie at the foundation of the analysis of vehicle motion. Work is defined as the product of a force acting over a distance, specifically:

$$dW = F ds \cos \theta \tag{4-20}$$

where dW = differential unit of work
F = force
ds = differential displacement
θ = angle between force and direction of displacement

and

$$W = \int_s F(s) \cos \theta \, ds \tag{4-21}$$

where $F(s)$ = force as function of location s on path

If the force is constant and the displacement is in the same direction as the force, then

$$W = FS \tag{4-22}$$

where W = work
S = displacement

Power is the rate of performing work, namely,

$$P = \frac{dW}{dt} \tag{4-23}$$

where P = instantaneous power

At any point along the path followed by a force, then, the power is simply

$$P = FV \cos \theta \tag{4-24}$$

where $V = \dfrac{ds}{dt}$

Often in transportation the force and velocity are constant over long distances, and the force is directed along the path, resulting in constant power FV.

Before we leave this general discussion for the specifics of particular transport technologies, some treatment of units is appropriate. Force is now most widely measured in pounds. Distance is often measured in feet, resulting in work units of foot-pounds. Speed is usually measured in miles per hour or feet per second. The latter results in a power unit of foot-pounds per second; then 1 hp, the most common unit of power for engineers, is defined as 550 ft-lb/s. If speed is in miles per hour and force is in pounds, it is easy to derive from this definition the fact that 1 lb force moving at a speed of 1 mi/h is equal to $\frac{1}{375}$ hp. In other words,

$$P = \frac{1}{375} FV \tag{4-25}$$

where P = power, hp
F = force, lb
V = speed, mi/h

The SI units for work and power are much easier to remember, since they are all defined as one unit of the appropriate combination of the force (newton), distance (meter), and time (second) units. This can be seen by referring again to Table 4-1.

The basic relationship which is central to the propulsive aspects of vehicle performance is the relationship between the maximum available propulsive force and the speed of the vehicle. The reason for the connection with the vehicle's speed is partly that the speed is important in the resistance which must be overcome by the propulsive forces and partly that for many types of

Figure 4-8 **Generalized propulsive force diagram.**

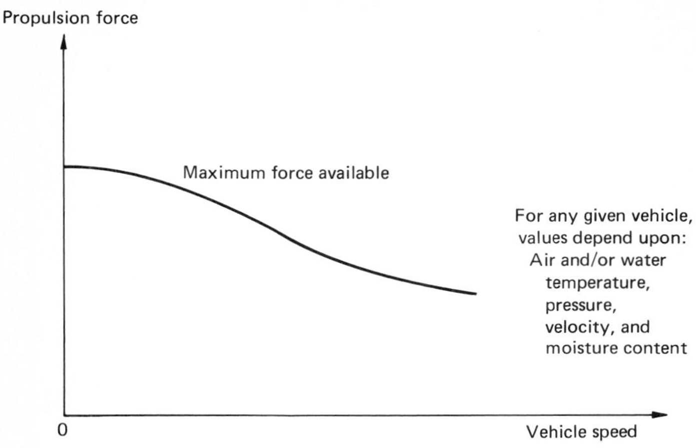

vehicles the maximum propulsive force available is significantly influenced by the speed. The maximum force available for propulsion will also in general vary with various environmental conditions, such as the pressure and temperature of the air (which vary consistently with elevation, for example), possibly wind direction and velocity, and whether or not it is raining or snowing.

A generalized relationship which illustrates how all these factors might interact is shown in Fig. 4-8, which is termed the propulsive force diagram. The maximum force that can be produced on the vehicle to overcome all the resistances is given by the vertical axis; the speed, by the horizontal axis. In this case, the maximum force is shown to decrease with increasing speed. At any speed, the maximum propulsive force depends on the various environmental factors identified above, and on the amount of power used for nonpropulsive purposes, such as air-conditioning passenger compartments and other support functions. Of course, if at any point on a vehicle's trip the maximum force is not needed, then a smaller force may be produced by appropriate control of the engines.

Turning first to railway locomotives, we show an actual propulsive force diagram in Fig. 4-9. This is for a diesel-electric locomotive, the most common type of locomotive in use in North America, and one whose characteristics are almost identical to those of the diesel-hydraulic locomotive commonly used elsewhere. In these locomotives, the diesel engine is the source of rotating power. Since diesel engines, like gasoline and steam engines, produce maximum fuel economy at a given optimal speed, it is desirable to have a transmission to the wheels of the locomotive which would allow the prime mover to operate at its optimum speed regardless of the speed of the locomotive. In the diesel-electric, this is done by having the diesel engine generate electric power, which is then used to supply electric motors which are geared to the locomotive's driving wheels (precisely, the axles). By varying the voltage and current to the motors, as well as certain other characteristics of the motors themselves, the diesel is allowed to operate at constant speed—its maximum power speed if necessary—while the locomotive operates over the

Figure 4-9 Propulsive force diagram for a railroad diesel-electric locomotive. [*Adapted from General Electric Co. (1969).*]

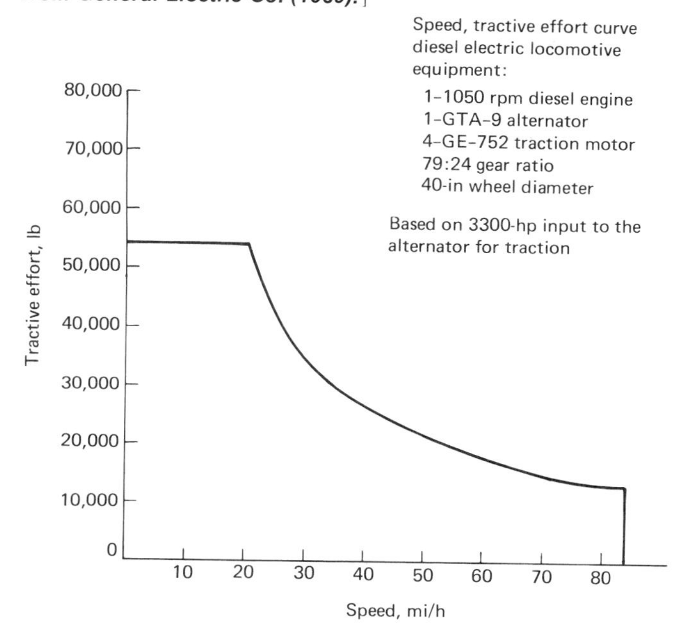

range of actual train speeds. In the diesel-hydraulic locomotive, the same principle applies, but the transmission is hydraulic, like those of large trucks and tanks.

While the locomotive is operating at the maximum horsepower output of the diesel engine, the tractive effort is given, by definition, by the following relationship:

$$M = \frac{375P}{V}\eta \tag{4-26}$$

where M = propulsive force, lb (termed tractive effort in railroad jargon)
η = efficiency of transmission system, usually 83 percent
P = rated diesel power output, hp

This constant horsepower propulsive force is shown as the decreasing portion of the curve in Fig. 4-9, for a 3300-hp locomotive weighing 136 tons. For almost all types of diesel-electric locomotives from various manufacturers, the efficiency of the transmission system is taken as approximately 83 percent, and it does not vary appreciably with speed (Hay, 1953, p. 124). It is about the same for hydraulic transmissions.

The propulsive force of locomotives is limited not only by the maximum power output of the diesel engine, but also by the adhesion between the driving wheels and the rails. Diesel-electric locomotives have been found to have a maximum adhesion of approximately 0.20, although under good rail and wheel conditions this can increase to 0.3 to 0.4. Wheel-rail adhesion limits the maximum tractive effort as shown by the flat portion of the tractive effort curve. This is given by

$$M \leq fT_d \tag{4-27}$$

where f = maximum wheel-rail adhesion for propulsion
T_d = weight on locomotive driving wheels

The electrical transmission system of diesel-electric locomotives also limits the minimum speed at which the locomotive can operate continuously without excessive overheating, this minimum speed usually being in the range of 10 to 20 mi/h, a typical location being shown in Fig. 4-9. However, for a short period (less than 1 h), a lower speed can be safely maintained with tractive effort following the curve drawn.

Propulsive force relationships for all types of locomotives—diesel, electric, and steam—are available from the manufacturers. For examples of these, the reader is referred to Hay (1953).

The braking of railway trains is usually accomplished by an entirely different mechanism from that used to provide for acceleration. In other words, there is one mechanism for producing propulsive forces in the direction of motion, and quite another for producing forces in the opposite direction to retard or stop the vehicle. The primary braking is done through the use of shoes which are forced against the wheel treads (sometimes instead against discs on the axles, termed brake discs). The resulting friction force tends to reduce the speed of rotation of the wheels. Usually all wheels or axles are braked in this manner. Provided the wheels do not slip relative to the rails, the result will be to decelerate the train. The total resistance of the train—the sum of the inherent resistance to motion at the speed the train is traveling, added resistance due to curvature and any gradient forces—will either aid or work against (if total is negative, possible on downgrades) this retardation of motion.

This discussion identifies the main limitation of the braking or deceleration of railroad trains, namely, that the force of braking must not exceed the maximum which can be transmitted at the wheel-rail interface. The coefficient of adhesion, as it is termed, is a maximum if there is no relative motion at the interface, i.e., if the wheels are not sliding, but rolling. The maximum coefficient of adhesion varies widely with the condition of the interface (rain, rust on wheels, etc.), but typical values range up to 0.18. This limits the propulsive force for braking (a negative value of M) to

$$M \leq f'T_b \tag{4-28}$$

where f' = braking maximum coefficient of adhesion
T_b = weight on braked wheels, usually entire train weight

The actual propulsive force for braking is usually considerably less than this, because the design of the brake systems cannot produce greater forces. Although performance varies widely with train design and maintenance, actual freight train braking force values of one-half to one-quarter the maximum are good estimates of normal braking behavior. Somewhat higher values can be achieved in emergency stopping, but that often leads to internal collisions in the train, minor derailments, and damage to the cargo. Passenger trains usually have much more sophisticated systems, which can deliver the maximum braking force allowed by wheel-rail adhesion at all speeds.

The principles of propulsion and braking of motor vehicles are very similar to those for railway trains, as would be expected from the general similarities of the technologies. Most motor vehicles are propelled by a combination of a diesel or gasoline engine and a mechanical (gears and clutch) or hydraulic transmission, with the propulsion force being applied to the vehicle through the wheel-road interface. As in the case of trains, it is desirable to have the engine operate within a particular range of speed, to achieve reasonable fuel economy. The transmission is designed to provide a sufficient number of ratios of engine speed to axle speed (gear ratio) so that this can be achieved. In addition the transmission allows some slippage, in effect giving a continuous range of ratios, to allow a smooth transition from one gear ratio to another. This results in a propulsive force versus speed of vehicle relationship which appears as shown in Fig. 4-10.

The braking of motor vehicles is also similar to that of railroad trains. The main difference is that the available adhesion between the wheel and the road is much larger than that for steel wheels on steel rails. Typical values for various types of road surfaces and weather conditions are shown in Table 4-4. These high values of coefficients of adhesion under normal circumstances, in conjunction with the actual retarding power of the brakes typically found in motor vehicles, result in extremely high possible deceleration rates.

At this juncture, it is appropriate to point out the limits upon the maximum deceleration rate that passengers can withstand before becoming uncomfortable or actually being injured. In a series of tests conducted in the early 1930s, in conjunction with the development of a more advanced streetcar (Electric Railway Presidents' Conference Committee, 1975 reprint), it was found that seated passengers could remain comfortable at deceleration and acceleration rates up to 9 ft/s^2 and a rate of change of acceleration up to 10 ft/s^3. Standing passengers, on the other hand, would remain comfortable only

Figure 4-10 Gasoline- or diesel-powered road vehicle propulsive force diagram.

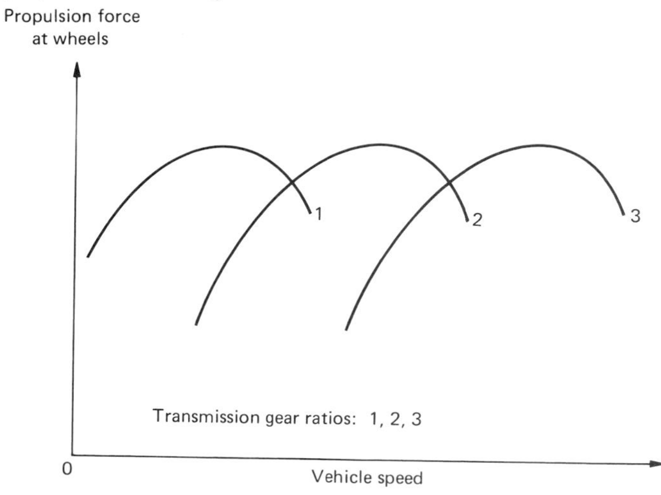

Table 4-4 Maximum Coefficients of Adhesion (Friction) between Rubber Tires and Roads

Road surface condition	Maximum coefficient Rolling	Sliding
Smooth concrete		
Dry	0.75	0.64
Wet	0.60	0.50
Wet and oily	0.50	0.25
Asphalt		
Dry	0.70	0.60
Wet	0.56	0.31
Snow		
Hard-packed	0.20	
With chains	0.30	
Ice		
Dry	0.14	0.10
With chains	0.25	. . .
Wet	0.12	0.06

Source: Fitch (1976, p. 65).

up to values of 5 ft/s^2 and 7 ft/s^3, respectively. These values are still widely accepted as limits in the design of land transport vehicles. However, it should be noted that where seat belts or other restraining devices are used, much higher values can be safely tolerated for seated passengers, and 10 ft/s^2 is commonly accepted in aircraft operation (McFarland, 1953, p. 705).

The propulsion of air- and water-borne vessels is very different from that of wheeled vehicles. The most common form is through the use of a propeller, which is designed so that its rotation causes a force in the backward direction against the air or water medium. This in turn creates an equal and opposite force on the propeller and hence on the vehicle to which it is attached, tending to move the vehicle forward. For stopping, the direction of the forces is reversed, by any one of many possible changes in configuration of the propeller and prime mover system, or reversing of direction of rotation of the propeller.

In the case of most propeller-driven air and water vehicles, a transmission connecting the propeller(s) and the prime mover enables the engine to be operated in reasonably efficient ranges of speed even though the propeller speed may vary greatly. As a result, this system can be described by relationships which are very similar to those for the locomotive. Specifically, the propulsive force, termed thrust in water and air terminology, is given by Eq. (4-26), but with lower efficiencies reflecting the changed means of locomotion. Typically efficiencies for water propellers are 0.5 to 0.7 (Comstock, 1967, pp. 373–439). It is important to realize that in the case of aircraft, where altitude and other environmental conditions can vary so widely, these conditions must be taken into account in adjusting engine horsepower and the efficiency of the propellers. The process of doing this is quite complex and hence beyond this text.

Another common type of propulsion system which has very different characteristics is the jet engine. This is widely used for large aircraft—actually dominating power plants of such vehicles—and it can be used for other types of vehicles also. It is characterized by a propulsive force diagram, too, but a typical curve is essentially independent of speed. An example of a jet engine used in a large commercial aircraft, a Boeing 707, is given in Fig. 4-11 (Boeing, 1975, p. 66). The effect of variations in altitude and temperature are very pronounced.

An Example

Perhaps it is appropriate at this point to bring together the material on resistance and propulsion covered so far. Consider the example freight train used earlier, traveling along a path which is straight but which consists of a 0.6 percent upgrade. If the train is traveling at 50 mi/h, with full power from the locomotives, what will its acceleration rate be?

The propulsive force exerted by the four 3300-hp diesel-electric locomotives is, by Eq. (4-26),

$$M = 4\,\frac{375(3300)}{50}\,0.83 = 82{,}170 \text{ lb}$$

The resistance of the train includes the inherent, grade, and curvature resistances, but the curvature is zero. The inherent resistance at 50 mi/h was previously calculated as 31,009 lb, using Eqs. (4-8) through (4-11). The gradient

Figure 4-11 **Example of jet engine propulsive force diagrams: two engines used in the Boeing 707 aircraft.** [***From Boeing Commercial Airplane Co. (1975), p. 66.***]

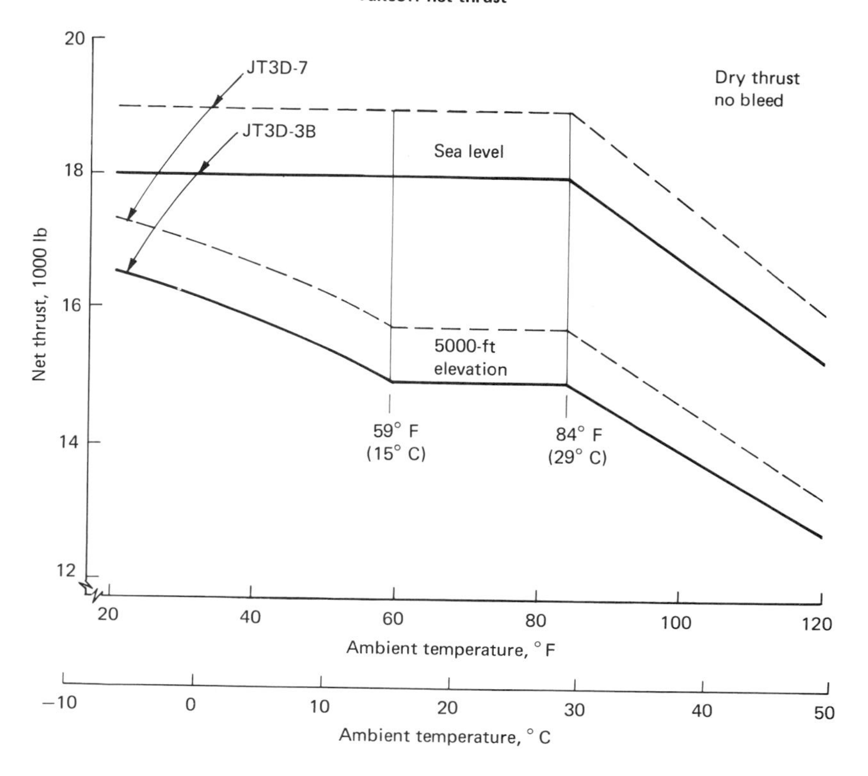

resistance was previously calculated by Eq. (4-15) to be 45,768 lb. Thus the total resistance is 76,777 lb. The acceleration of this train is calculated using Eq. (4-2), Newton's second law, with appropriate changes of units, as

$$a = \frac{F}{T} g = \frac{M - R - F_g - F_c}{T} g$$

$$a = \frac{82{,}170 \text{ lb} - 76{,}777 \text{ lb}}{3814 \text{ tons}} \frac{\text{ton}}{2000 \text{ lb}} 32.16 \text{ ft/s}^2$$

$$= 2.274 \times 10^{-2} \text{ ft/s}^2$$

Converting to the more common mi/(h)(s), the acceleration is

$$a = \frac{2.274 \times 10^{-2} \text{ ft}}{\text{s}^2} \frac{\text{mi}}{5280 \text{ ft}} \frac{3600 \text{ s}}{\text{h}} = 1.550 \times 10^{-2} \text{ mi/(h)(s)}$$

and in SI units is

$$a = \frac{2.274 \times 10^{-2} \text{ ft}}{\text{s}^2} \frac{0.3048 \text{ m}}{\text{ft}} \frac{3600 \text{ s}}{\text{h}} \frac{\text{km}}{1000 \text{ m}} = 2.495 \times 10^{-2} \text{ km/(h)(s)}$$

Alternatively, one may wish to know what the power output of the four locomotives must be in order to maintain the train's speed at exactly 50 mi/h on this gradient. By Eq. (4-2), this means that $a = 0$, and hence that

$$F = M - R - F_g - F_c = 0$$

In other words, the propulsive force must exactly equal the total resistance. Since the latter equals 76,777 lb, M must be 76,777 lb. This is converted into power at the wheels by Eq. (4-25), yielding

$$P = \frac{1}{375} (76{,}777)(50) = 10{,}236.9 \text{ hp at wheels}$$

However, the power output of the diesel engines must be greater, including in addition the power lost in the transmission. This prime mover power output, given by Eq. (4-26), is

$$P = \frac{76{,}777(50)}{375(0.83)} = 12{,}333.7 \text{ hp at prime mover}$$

This is less than the maximum rated prime mover power of 13,200 hp, as it must be if the locomotive is to overcome fully the resistances so as to maintain the steady speed.

PATH CHARACTERISTICS

Before presenting the methods used to predict vehicle performance over a path it is following, it is necessary to specify certain characteristics of the path which influence vehicle motion and performance. These fall into three categories: (1) vehicle exclusion or limitations of size and weight, (2) speed restrictions, and (3) effects of environmental conditions. Since our concern at this point is with vehicle performance prediction, the primary emphasis will be upon the form these limitations take and the manner in which they are presented. In a later chapter, where we are concerned with the design of pathway links, interchanges, etc., we will discuss the underlying principles in detail and the methods used to analyze different path designs which may permit different vehicle performance. The concepts of path limitations and design are not limited to land transport; important options in air transport path choice can also influence significantly travel time and other aspects of vehicle performance.

Exclusion, Size, and Weight Limitations

An important class of limitations are ones which prohibit certain vehicles from operating on a path. Of course, the most fundamental of these is the exclusion of any vehicles which are incompatible with the path with respect to the means used to provide mobility or to give locomotion. Obviously, a road vehicle cannot use a canal. Similarly, a train propelled by an electric locomotive designed to take current from overhead wires (termed catenary) cannot use a path which does not have such electric power supply lines. Similarly, for practical reasons, one would not wish to operate a vehicle over a path lacking the facilities necessary to service the vehicle.

In some cases there are also limitations on vehicles based upon their performance properties. For example, certain airports may be unsuitable for certain aircraft because the aircraft cannot take off in the runway length provided or cannot meet the requirements for steepness of ascent sufficiently to clear obstacles. Also, such requirements may be based not upon physical obstacles but upon considerations of aircraft proximity to noise-sensitive activities. These considerations also apply to other technologies, such as the prohibition of trucks on certain roads at night due to noise or the banning of trucks from certain roads or lanes during peak traffic flow periods when they would impede traffic. In fact, the same applies to railways, but the consideration is entirely internal to the organization, through the preparation of train schedules, which provide for the unimpeded flow of fast and slow trains on the same tracks.

There are also limitations on the size and weight of vehicles that can use any particular facility. All possible paths, whether over land, underground, or in the air, have certain limitations on vehicle size. In the case of paths constructed on the earth's surface (or underground), these limitations are rather obvious, in the form of maximum width, height, and possibly length. These almost invariably take the form of the maximum dimensions of vehicles, and perhaps a few other specifications which are particular to a specific mode or context. It is, then, a simple matter to compare the vehicle's dimensions with the limits to determine whether or not it is acceptable on that path.

An example is given in Fig. 4-12, which is the standard clearance diagram for railroad vehicles in the United States and Canada. Typical limits for other vehicles are given in Table 4-5. It should be noted, however, that such limits do vary among facilities, so it is necessary to check each potential path. Many vehicles are designed so that varying amounts of freight may be placed upon them. Such loading must be done in a way that the vehicle still fits within the maximum limits of size.

Limits on the weight of vehicles are imposed by the path on which it is to travel. This again is most obvious in the case of land vehicles, where weight limits on road bridges are in almost everyone's common experience. Such limits vary, of course, with the design of the facility and the vehicle (e.g.,

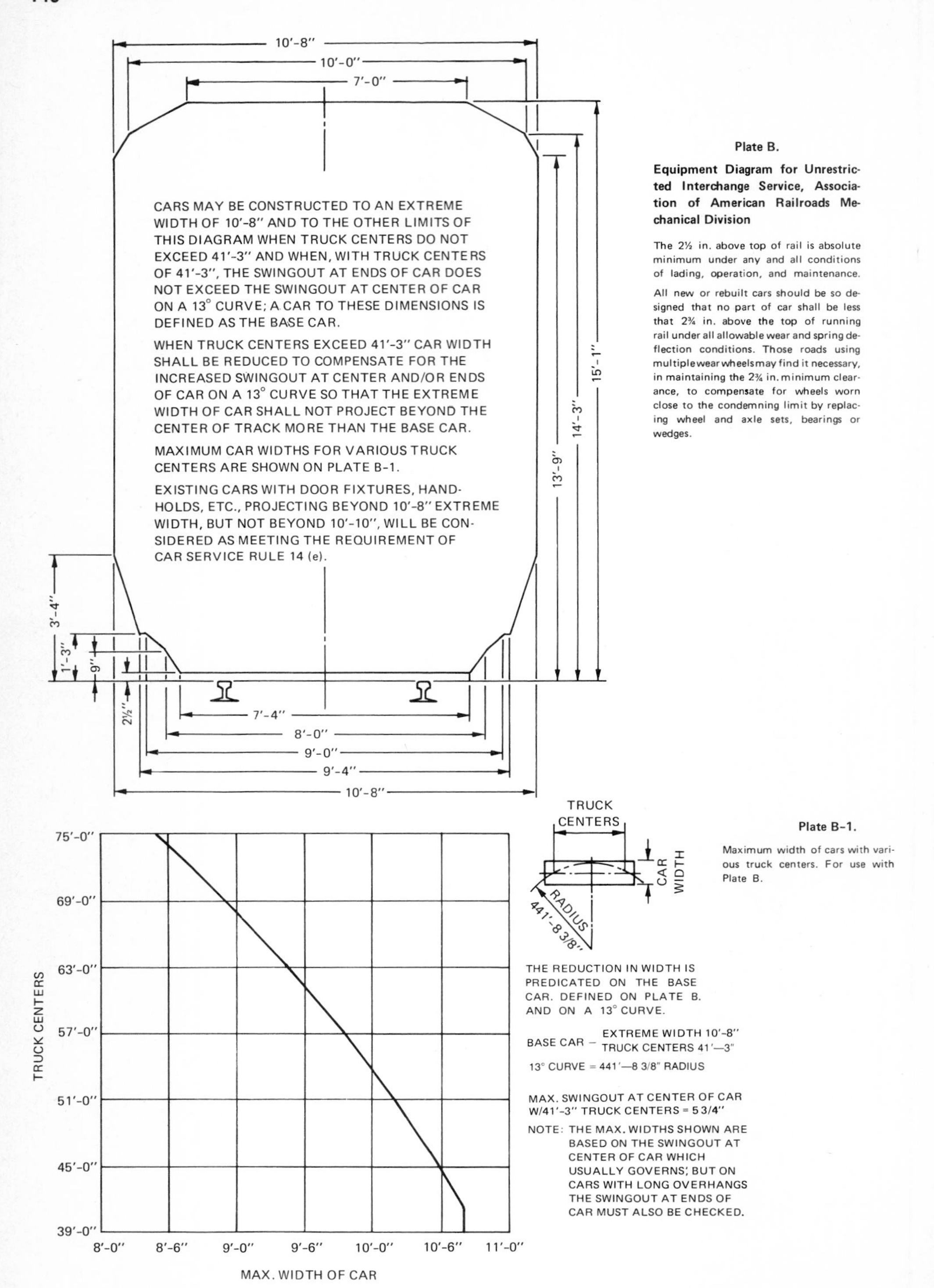

Plate B.

Equipment Diagram for Unrestricted Interchange Service, Association of American Railroads Mechanical Division

The 2½ in. above top of rail is absolute minimum under any and all conditions of lading, operation, and maintenance.

All new or rebuilt cars should be so designed that no part of car shall be less that 2¾ in. above the top of running rail under all allowable wear and spring deflection conditions. Those roads using multiple wear wheels may find it necessary, in maintaining the 2¾ in. minimum clearance, to compensate for wheels worn close to the condemning limit by replacing wheel and axle sets, bearings or wedges.

Plate B-1.

Maximum width of cars with various truck centers. For use with Plate B.

Table 4-5 **Typical Vehicle Size and Weight Limitations in Paths in Canada and the United States. The Most Common Limitation Is Given, Followed by Less Typical Ones in Parentheses, If Applicable**

	Weight	Size, ft
Railroad trains		
(standard gauge, 4 ft 8.5 in)	62,750/axle[a]	See Fig. 4-12
Road vehicles		
All vehicles[b]	20,000 lb/single axle	Width: 8.0 (8.5)
	32,000 lb/tandem axle	Height: 13.5 (13.0)
Single units[b]	80,000 lb	Length: 40
Combination units[b,c]	80,000 lb (127,400 lb)	Length: 65 (98)
Inland and coastal waterways[d]	Depth: 9 ft (8–40 ft)	Width: 110 (56–200)
		Length: 600 (300–1200)
Airports[e]		
Air carrier	Varies widely	Undercarriage width: to fit 75 (50) pavement
General aviation	8000–175,000 lb	Take-off within runway length: 7000–12,000

[a] Association of American Railroads (1966).
[b] American Association of State Highway Officials (1965).
[c] Fitch (1976, pp. 36–42). Largest values primarily on turnpikes.
[d] American Waterways Operators (1965).
[e] Horonjeff (1975, pp. 212–240).

number of wheels) and may also depend upon the speed at which the vehicle is traveling. For any permissible weight, it is important to note any special speed restrictions that might apply. Again, this is straightforward, but nonetheless very important! Typical examples of weight limits are also given in Table 4-5.

Speed Restrictions

Most transport paths possess speed restrictions. Usually these include a maximum speed, and in many cases a minimum speed.

The maximum speed is based upon many considerations. One of the most important is safety. Included here are such considerations as the potential need to stop or diverge from the intended path in order to avoid collision, this being especially important in rail, road, and waterway speed limits. In areas where there is more vehicle traffic, where vehicles may be traveling at widely ranging speeds, and where there may be unforeseen obstacles in the path, speed limits are naturally lower. Speed limits are also posted on the curves of land transport paths and are usually imposed for the sake of safety (to prevent vehicles from sliding outward or overturning), to ensure passenger comfort, and to guard against damage to goods.

There may also be minimum speeds. These are primarily applied to roads and, through segregation of aircraft type, to air lanes. The rationale is to keep vehicles moving at as nearly the same speed as practicable to minimize the chance of rear-end collisions.

Figure 4-12 **Example of vehicle size limitations: United States railroad car and locomotive limits.** [*From Association of American Railroads (1966), p. 72.*]

Environmental Effects

In the discussion above it was noted that the resistances of vehicles to motion and the propulsive power characteristics of vehicles were influenced by many environmental factors. In particular, land vehicle resistance is affected by the weather condition on the surface of the paths, such as rain or snow. Also, all vehicles are affected by the characteristics of the fluid media in which they operate. Thus the temperature of the air, its moisture content, wind direction and speed, and snow or rain, etc., all affect resistance to motion. In addition, for vehicles whose elevation changes appreciably, the effect of the concommitant change in medium pressure must be considered. Similarly, the various prime movers are affected by some or all of these considerations, depending upon the particular energy-conversion-device employed.

Thus it is important to include among the path characteristics the relevant environmental factors which influence the vehicle's performance.

Path Chart

All of these factors can be conveniently included in the path characteristics chart. This chart is a generalization of various types of charts which are used in individual technologies to present the path characteristics relevant to that technology. But since the function of the diagram is the same in all cases, there is no reason to treat them separately.

An example of such a path diagram is shown in Fig. 4-13. It specifies the location of the path to be followed in the three location coordinates and gives information on the location on the surface of the earth as well as on elevation. The location in the horizontal plane is usually called the horizontal alignment, and that in the vertical plane (along the path axis), the vertical alignment. The vertical and horizontal curves may be specified by radii or degrees of curvature, if such detail is required (as it is for horizontal curves for railways if curve resistance is to be calculated).

In addition to this information on the location of the path, it is important to specify any speed restrictions which apply. These typically take the form of maxima and possibly minima which must be adhered to on the various segments of the path (see Fig. 4-13).

As discussed above, in some technologies it is important to specify various environmental conditions. These can easily be associated with the point or section of the path where they apply. For example, atmospheric pressure can be specified. Also, conditions which vary from time to time such as temperature, presence of rain or snow, and humidity can be specified in the appropriate sections.

Given the information appropriate for the particular technology being treated, it is possible to predict the performance of any given vehicle over the path. The methods for doing this are the subject of the next section.

PREDICTION OF VEHICLE PERFORMANCE

With the information on paths given in the path chart as described above and the resistance and propulsive force characteristics of vehicles presented earlier, it is possible to predict individual vehicle performance on any path. Specifically, the time required for a vehicle to travel between any two points on the path can be calculated; and the speed and acceleration rate of the

Figure 4-13 **Hypothetical example of a path chart. *x, y,* and *z* are reference axes. *(a)* Horizontal alignment. *(b)* Vertical alignment. *(c)* Speed restrictions.**

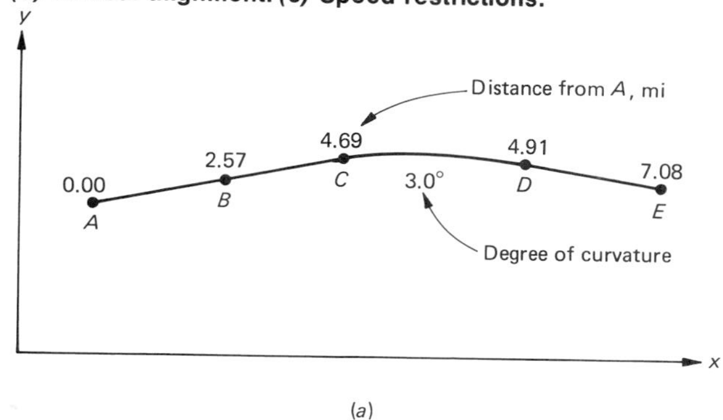

(*a*)

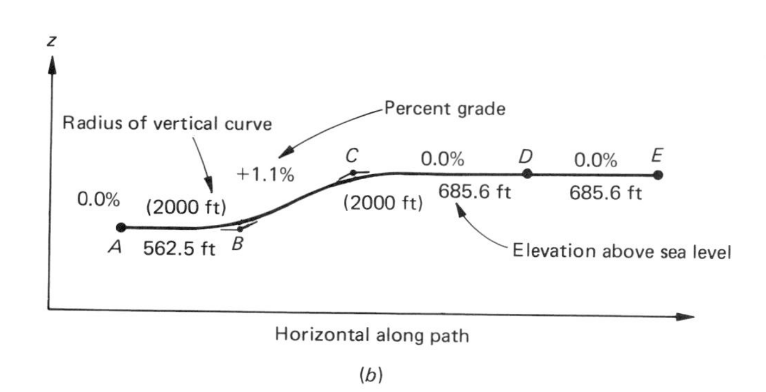

(*b*)

Speed limits, mi/h

Segment	Eastbound Max.	Eastbound Min.	Westbound Max.	Westbound Min.
AB	70	50	70	50
BC	70	50	40	30
CD	50	40	50	40
DE	60	45	60	45

(*c*)

vehicle at any point on the path can be determined. These results also prove useful in estimating fuel consumption, the number of vehicles required, and many other important cost and performance aspects of the transportation system.

A Simple Example

The basic approach to the estimation of vehicle performance can be easily presented through the use of an example, admittedly a highly idealized and simplified one. Consider a vehicle which is to travel from one station to another 2 mi away along a path. The path is level and contains no curves, and the speed limit is 60 mi/h. The vehicle is designed to accelerate at a constant rate of 5 ft/s^2, this value having been selected by the vehicle design engineers as the maximum that passengers find comfortable. It also decelerates at the same rate from any speed.

The movement of the vehicle away from its origin can be treated by application of Eq. (4-2). At a time t after acceleration from rest began at t_0, the vehicle's speed is

$$v = \int_{t_0}^{t} a\,dx = ax\Big]_{t_0}^{t} = a(t - t_0) \tag{4-29}$$

$$v = 5\,\frac{\text{ft}}{\text{s}^2}(t - t_0)$$

Since there is a maximum speed limit, it will stop accelerating after 17.6 s have elapsed. Taking t_1 as the time required to accelerate to the cruise speed V_m, we find

$$t_1 - t_0 = \frac{V_m}{a} = \frac{60\text{ mi}}{\text{h}}\,\frac{\text{h}}{3600\text{ s}}\,\frac{5280\text{ ft}}{\text{mi}}\,\frac{\text{s}^2}{5\text{ ft}} = 17.6\text{ s}$$

The distance d traveled in a time t s after acceleration from rest begins is

$$d = \int_{t_0}^{t} v\,dx = \int_{t_0}^{t} ax\,dx = \tfrac{1}{2}ax^2\Big]_{t_0}^{t} = \tfrac{1}{2}a(t - t_0)^2 \tag{4-30}$$

Alternatively, d can be expressed as

$$d = \frac{V_m^2}{a}$$

At the end of acceleration, this is

$$d = \frac{1}{2}\,\frac{5\text{ ft}}{\text{s}^2}(17.6\text{ s})^2 = 774.4\text{ ft} \tag{4-31}$$

The deceleration of the vehicle will require the same time and distance as acceleration, since the rate has the same absolute value. This can be more precisely determined by application of Eq. (4-2) to deceleration from a speed V_m, this deceleration commencing at time t_2 and terminating at speed 0 at time t_3:

$$v = V_m + \int_{t_2}^{t_3} -\,a\,dx = V_m - ax\Big]_{t_2}^{t_3} = V_m - a(t_3 - t_2) \tag{4-32}$$

For deceleration from V_m to 0, the time required, $t_3 - t_2$, is

$$t_3 - t_2 = \frac{V_m}{a}$$

and hence the time of deceleration is seen to be 17.6 s also. Similarly, the distance traveled in deceleration will be identical to the acceleration distance.

$$d_3 - d_2 = \int_{t_2}^{t_3} (V_m - ax)\,dx = V_m x - \tfrac{1}{2}ax^2 \Big]_{t_2}^{t_3} \tag{4-33}$$
$$= V_m(t_3 - t_2) - \tfrac{1}{2}a(t_3 - t_2)^2$$

But since $(t_3 - t_2) = V_m/a$ and $(t_3 - t_2) = (t_1 - t_0)$

$$d_3 - d_2 = \frac{V_m^2}{a} - \frac{1}{2}\frac{V_m^2}{a} \tag{4-34}$$

Hence deceleration requires 774.4 ft also.

The distance traveled between acceleration and deceleration will be covered at a speed of 60 mi/h, or 88 ft/s. The time required will be

$$t_2 - t_1 = \frac{2 \times 5280 \text{ ft} - 2 \times 774.4 \text{ ft}}{88 \text{ ft/s}} = 102.4 \text{ s}$$

Therefore the total time required to traverse the 2 mi will be 137.6 s. The average speed will be 52.3 mi/h.

Simulation of Vehicle Movement

While the simplified example presented above illustrates some of the relationships between the path, the vehicle, and its propulsion and resistance characteristics, it is highly idealized. Actual vehicles rarely accelerate at a constant rate over an extended range of speed. Inherent resistance varies with speed, as does the propulsive force of a vehicle. Also, the presence of gradients and curves affects vehicle performance. A method is needed which accounts for all of these individual characteristics.

This is accomplished by using Newton's laws and Eqs. (4-1) through (4-6), but in a manner that considers the complexities of path-vehicle interaction which produce motion. Charting the vehicle movement over very small intervals of time as it moves along a path enables us to account for changes in path characteristics, such as gradients or curves and changes in speed and environmental conditions—all of which affect total vehicle resistance. Similarly, changes in speed, elevation, or other environmental conditions which might affect propulsive forces can also be taken into account. While the method may seem difficult at first, it is basically quite simple.

The method is best presented by considering the acceleration of a vehicle from rest. The vehicle will be moving along a path, the characteristics are known or can be determined as they are needed in the resistance and other calculations. Calculations will be made at frequent points along the path, the first point being labeled d_0, this being the point at which movement begins, the next point d_1, and so on. (The location of these points will be explained later.)

The data on the vehicle's resistance and propulsive force characteristics, in conjunction with the path chart data, can be used to estimate the vehicle's acceleration rate when starting from rest at location d_0. If this rate of acceleration is assumed to apply for a small period of time, then the speed and distance traveled in that period of time can be calculated. At the end of this short period of time, the vehicle has traveled some distance and reached a point which we shall term point d_1. At this point the resistance and propulsive force can again be calculated, the acceleration rate estimated, and the increase in speed and distance covered during another short period of time calculated. If the periods of time are short enough so that the forces estimated for one particular point along the line are close approximations to the forces encountered over the entire time interval, then this method should yield an accurate if approximate representation of actual vehicle movement. In fact, this method does yield a good representation, and this procedure is used in almost all technologies of transportation for vehicle performance calculations.

The actual method is as follows. The forces producing motion (propulsion minus resistances) are assumed constant for a period of time and distance corresponding to a small change in speed. The increment of speed is chosen so that the force and hence the acceleration do not change too greatly, and hence the approximation of a constant acceleration is reasonable. The increment of speed over which the acceleration is assumed constant varies with the speed of the vehicle, because in some speed ranges there may be little change in these forces even over substantial ranges of speed, while in another range of speed the change in resistance or propulsive forces, or both, may be very rapid. For example, in the case of a railway train propelled by a diesel-electric locomotive, there is no change in propulsive force with speed in the range in which the propulsive force is limited by the weight on the driving wheels, and at low speed the total train resistance also changes only slowly with speed. But as soon as the prime-mover power limitation controls the propulsive force, this force decreases very rapidly with speed, requiring shorter speed increments to yield the same accuracy. Also, at very high speeds, the resistance increases rapidly because of the square of velocity term. For these reasons, it is customary in manual calculation to choose the speed increments carefully, to minimize the total error. When using computers, it is customary to select very small increments of speed over the entire speed range (e.g., 1-mi/h increments).

The formulas for this simulation of vehicle motion are most easily presented for a vehicle starting from rest. The various points on the path at which calculations are made are designated points 0, 1, 2, etc., as noted earlier. The initial location is d_0, measured from a reference point, and the initial speed v_0 equals 0. If the increment of speed over which the acceleration rate will be assumed constant is Δv for the movement from point d_0, then the most appropriate resistance and propulsive force values are those at $\Delta v/2$, at the middle of the speed range for which these forces will be assumed to apply. These will be termed R_0 and M_0, respectively, and they include all forces (inherent, gradient, curvature, etc.).

$$M_0 = \text{total propulsive force at speed } \frac{v_0 + v_1}{2}$$

$$R_0 = \text{total resistance force at speed } \frac{v_0 + v_1}{2}$$

Therefore a_0, the acceleration rate which applies over this interval of speed, is

$$a_0 = \frac{M_0 - R_0}{T} g$$

If we assume this acceleration rate applies over the time period and distance to reach required speed $v_1 = v_0 + \Delta v_0$, then the time required, the difference between t_0 and t_1, is

$$t_1 - t_0 = \frac{v_0}{a_0}$$

At t_1 the speed is

$$v_1 = v_0 + a_0(t_1 - t_0) = v_0 + \Delta v_0$$

and the location measured from the reference point is

$$d_1 = d_0 + v_0(t_1 - t_0) + \tfrac{1}{2} a_0(t_1 - t_0)^2$$

$$= d_0 + \frac{v_0(\Delta v_0)}{a_0} + \frac{(\Delta v_0)^2}{2\,a_0}$$

In the above equation, if the vehicle is starting from rest, then $v_0 = 0$.

Thus point 1 on the path is located at d_1, and the speed is v_1. Now another speed increment is chosen, Δv_1 (it may be the same value as Δv_0), and the forces are calculated as

$$M_1 = \text{total propulsion force at speed } \frac{v_1 + v_2}{2}$$

$$R_1 = \text{total resistance force at speed } \frac{v_1 + v_2}{2}$$

Then the same approach is applied, i.e.,

$$a_1 = \frac{M_1 - R_1}{T} g$$

$$t_2 - t_1 = \frac{\Delta v_1}{a}$$

$$v_2 = v_1 + a_1(t_2 - t_1) = v_1 + \Delta v_1$$

$$d_2 = d_1 + v_1(t_2 - t_1) + \tfrac{1}{2} a_1(t_2 - t_1)^2$$

$$= d_1 + \frac{v_1 \Delta v_1}{a_1} + \frac{(\Delta v_1)^2}{2\,a_1}$$

The basic formulas are applied repeatedly until the maximum speed is reached. This maximum speed is termed the *balancing speed*—the speed at which the propulsive and resistance forces balance one another. A general statement of the formulas for movement, from one point i to the next point $i + 1$, is as follows:

$$M_i = \text{total propulsion force at speed } \frac{v_i + v_{i+1}}{2} \qquad (4\text{-}35)$$

$$R_i = \text{total resistance force at speed } \frac{v_i + v_{i+1}}{2} \qquad (4\text{-}36)$$

$$a_i = \frac{M_i - R_i}{T} g \qquad (4\text{-}37)$$

$$t_{i+1} - t_i = \frac{\Delta v_i}{a_i} \tag{4-38}$$

$$v_{i+1} = v_i + a_i(t_{i+1} - t_i) = v_i + \Delta v_i \tag{4-39}$$

$$d_{i+1} = d_i + v_i(t_{i+1} - t_i) + \tfrac{1}{2} a_i(t_{i+1} - t_i)^2$$

$$= d_i + \frac{v_i(\Delta v_i)}{a_i} + \frac{(\Delta v_i)^2}{2a_i} \tag{4-40}$$

For deceleration, the same procedure is followed, except that the propulsive force M_i will either be zero, for coasting, or negative, for application of the braking force. Figure 4-14 presents a graphical portrayal of the forces entering these computations for a situation in which the balancing speed is reached after three steps, although typically more are required. As would be expected, the approximation is more accurate the smaller the Δv. But the resulting increase in computations tends to increase the value of Δv that is chosen. Now almost all computations are performed by computer.

Figure 4-15 presents the corresponding vehicle performance curves, in standard form. These are the acceleration-distance diagram, the acceleration-time diagram, the speed-distance diagram, the speed-time diagram, and the distance-time diagram. It should be noted that all of these curves can be derived from any one of them, although in practice of course

Figure 4-14 **Forces used in vehicle performance simulation.**

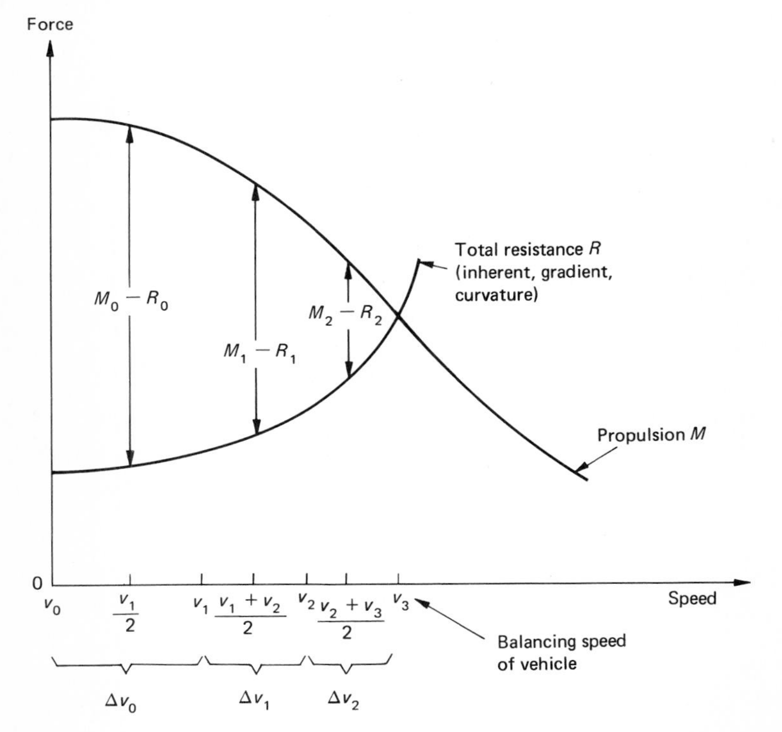

Figure 4-15 **Vehicle performance curves.**

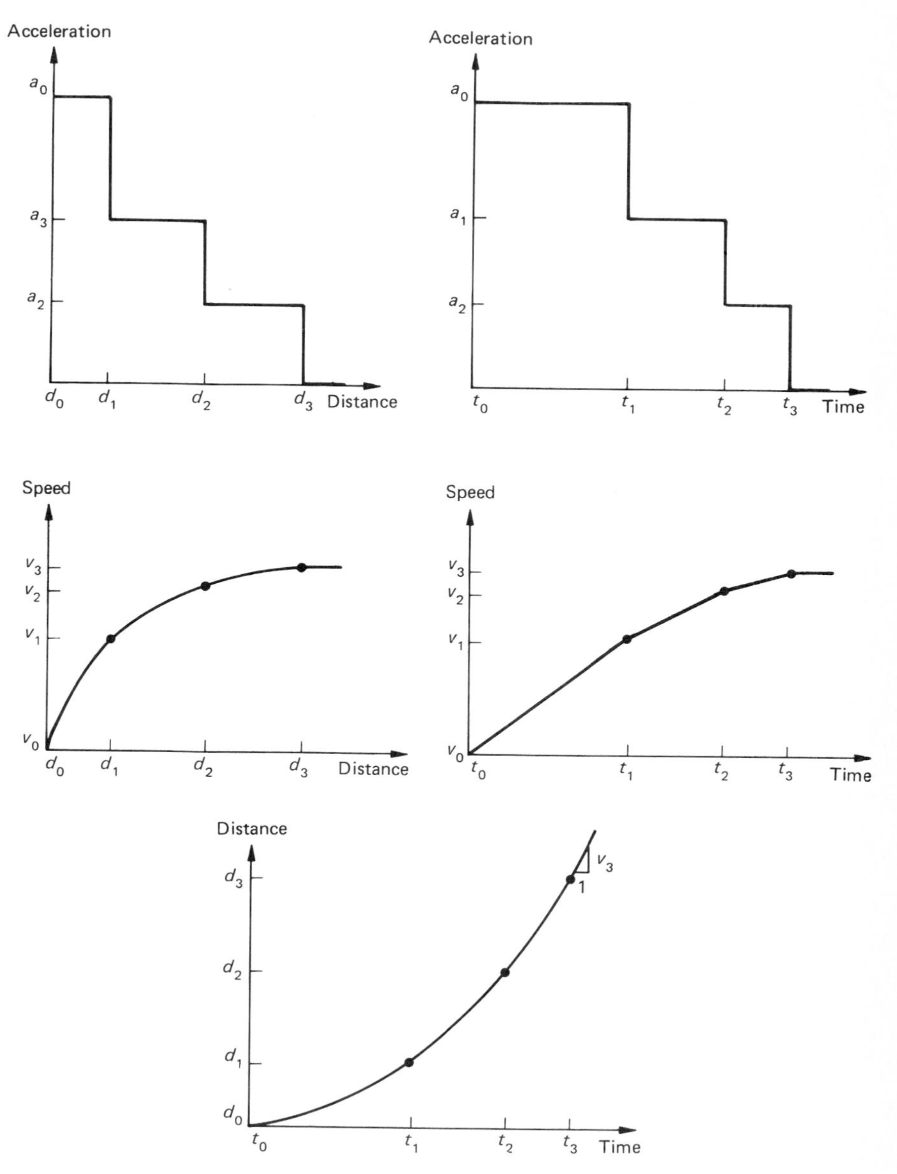

this would involve difficult measurements on the drawings. In practice usually vehicle performance is specified by one or two such diagrams, the distance-time diagram being most common, and the speed-distance diagram the next most common.

An Example

The railroad freight train used in previous examples will be used again to illustrate the application of the vehicle performance simulation method described above. It is a straightforward application of Eqs. (4-35) through (4-40). Table 4-6 presents the results of these calculations, which were performed using a constant 10-mi/h increment of speed (Δv) so that the reader could reproduce the calculations easily as a check on understanding the method. The train is accelerating on a straight level line. For the calculation of the inherent resistances, the resistance formulas for individual cars and locomotives can be summed as suggested earlier:

$$9533 + 59.17V + 7.407V^2$$

yielding total resistance in pounds.

GENERALIZED VEHICLE PERFORMANCE RELATIONSHIPS

Within each technology of transportation a set of generalized vehicle performance relationships has been developed which relate important vehicle performance characteristics to vehicle design, loading, and path characteristics. While the specific form of these varies among technologies, it is useful to review briefly some of them since they illustrate important applications of the principles presented above. In particular, those for aircraft, railroad trains, and road vehicles will be covered, with discussion of the single relationship widely used in water transport.

Aircraft

The use of generalized relationships is probably most fully developed within this technology, because the variations in paths are primarily ones of al-

Table 4-6 Force and Vehicle Acceleration Calculations in Performance Simulation of Example Railroad Freight Train

Speed, mi/h	Resistance, lb	Propulsive force, lb	Acceleration,† mi/(h)(s)	Time,‡ s	Distance,‡ mi
0	9,533	217,600	0.597	16.75	0.023
10	10,865	217,600	0.570	17.54	0.073
20	13,679	203,280	0.442	22.62	0.157
30	17,974	135,520	0.281	35.58	0.345
40	23,751	101,640	0.185	54.05	0.675
50	31,009	81,382	0.113	88.50	1.353
60	39,748	67,760	0.051	196.08	3.540
70	49,969	58,080			

† Based on average of forces between beginning and ending speeds of 10-mi/h increment.
‡ Time and distance required to increase speed by 10 mi/h.

titudes (and of related factors) rather than of gradients, curvature, etc., and because aircraft have an essentially fixed power plant. This permits considerable generalization among diverse application situations of the same aircraft.

A key relationship is between the distance the aircraft can fly, termed the range, and the total weight at take-off, including the load. This curve for the Boeing 707 is shown in Fig. 4-16. As would be expected, as the weight increases, the range decreases. In this drawing the weight of the aircraft is shown at take-off. This weight decreases somewhat with distance of flight due to the burning of fuel. The upper, flat part of the curve represents the weight limit imposed by the structural strength of the aircraft itself. With such a heavy load, the aircraft cannot take off with its fuel tanks completely filled, because that would result in a total weight greater than that which can be sustained by the aircraft structure. The maximum distance the plane can travel with such a heavy load and the maximum fuel load is at the end of this horizontal line. To travel beyond that distance, additional fuel must be carried, which results in a reduction in the load which can be carried, as shown by the next two (downward sloping) portions of the curve. The last point of change of slope reflects the maximum distance the plane can travel with its tanks completely filled. The range can be extended slightly by further reductions in the load, down to zero load, due to the reduction in aircraft inherent resistance as a result of the reduction in weight.

Figure 4-16 **Typical aircraft range-weight performance curve: Boeing 707-320C.** [***From Boeing Commercial Airplane Co. (1975), p. 61.***]

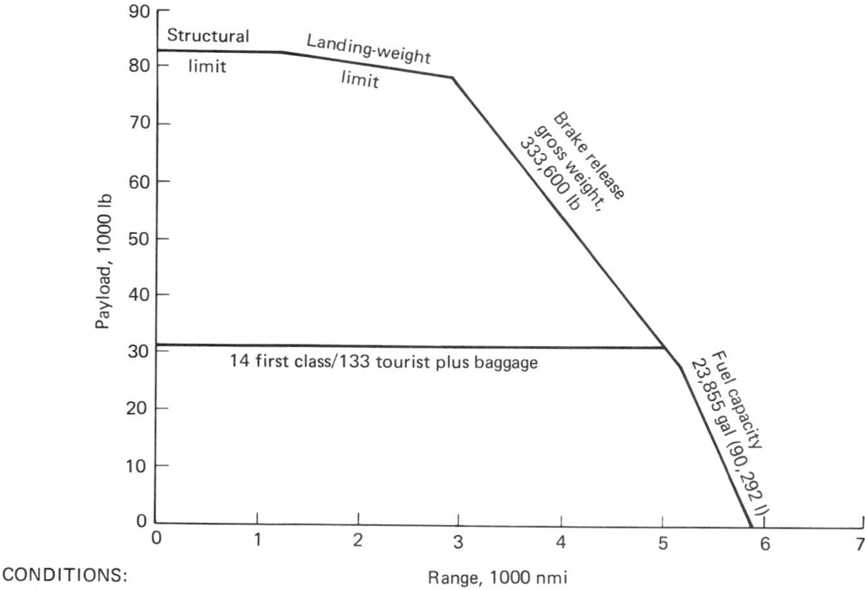

The other important characteristic of aircraft is the runway length required. As the discussion of vehicle performance indicates, this will depend upon the weight of the aircraft, the gradient of the runway (often zero or level, of course), and environmental conditions, especially temperature and atmospheric pressure as specified by elevation above sea level. A typical set of relationships for specifying runway length is presented in Fig. 4-17, again for the Boeing 707 aircraft.

Figure 4-17*a* is the landing performance curve for the Boeing 707-200 jet aircraft described earlier and used for most of our examples. To use this curve, one begins with the landing weight, projecting vertically to the elevation of the airport, then horizontally to the right axis, which yields the runway length required for landing. For example, with a total weight of 180,000 lb, at an airport 2000 ft above sea level, a runway at least 6900 ft long is required. Runway gradient, which is usually very small (less than ±2.0 percent), has little effect on landing requirements.

To use Fig. 4-17*b* for take-off one begins with the temperature and the altitude. A dashed line is shown for a temperature of 75°F and an elevation of 2000 ft. Project horizontally to the reference line marked *RL*. Then move parallel to the curved line to a point above the take-off weight of the aircraft, for this example taken as 220,000 lb. Now projecting horizontally to the right axis yields the runway length required, approximately 8300 ft. If the runway is on a grade, add 20 percent for each 1.0 percent of grade. It is useful to try a few variations to facilitate understanding of the use of this figure. Also try to

Figure 4-17 **Aircraft landing and take-off performance curves: Boeing 707-200.** [***U.S. Federal Aviation Administration.***]

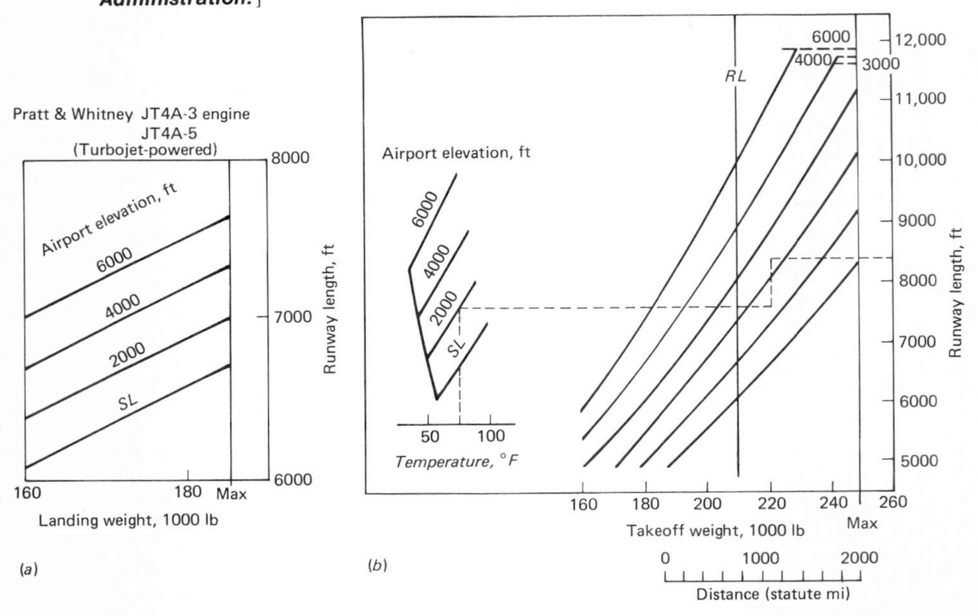

explain the effect of various environmental factors and variations in weight of the runway length. All are a direct result of the principles covered in this chapter!

Road Vehicles

The performance curves of motor vehicles can be specified by the manufacturer from the design only in terms of the maximum performance that can be achieved. But for acceleration, deceleration, and speed performance, ideal values are meaningless for most purposes since so much depends upon the driver of the vehicle, its load, condition of the road surface (e.g., wet or icy), and many other factors. As a result, the actual performance of vehicles is determined by extensive surveys, and the results tabulated in the form of curves for acceleration performance.

These curves for starting are presented in Fig. 4-18*a* for vehicles found in the United States in the 1960s. As can be seen, the acceleration rate depends significantly on the type of vehicle. It should be remembered that these are average values, and any one vehicle's performance may be vastly different from that shown.

Information on deceleration rates is shown in Fig. 4-18*b*. Variations here are so great that a cumulative curve or distribution of values must be shown for each vehicle type. The interpretation of this figure is that the percentage of vehicles (within a class) given on the vertical axis has a deceleration rate greater than or equal to that given on the horizontal axis. Thus, 42 percent of the type-7 vehicles have deceleration rates greater than or equal to 15 ft/s^2.

A closely related type of performance curve is that showing speed performance on grades, this being very important for large vehicles which might not have the power to maintain high speeds in the vicinity of speed limits on such grades. An example of a set of such curves is presented in Fig. 4-19. Such curves can be predicted directly from the methods of vehicle movement simulation presented earlier. The reference contains such curves for many different types of road vehicles.

Railroad Trains

Because the power on trains is typically so closely tailored to the actual speed or travel time requirements of the trains (as imposed by the time table) and the gradients and speed restrictions on railroad lines vary so widely from one to another, generalized speed performance curves for trains have not been widely developed. However, they would appear as shown in Fig. 4-15, with the curve shifting to reflect different ratios of power to total weight of the train.

Urban Rapid Transit

Urban rapid transit lines have characteristics similar to intercity railroads, such as a private right of way and complete control of vehicle movements. As a result, performance curves for transit vehicles are presented in a manner similar to those for railroad trains. Figure 4-20 shows performance curves for a typical diesel-powered city bus (seating 53 passengers, standing about 25 persons) and a typical rail rapid transit train (trains typically up to 10 cars, each seating 60 to 80 and standing 100 to 200 persons). In both cases a straight, level path is assumed. Note that the performance curves for the

Figure 4-18 Acceleration and deceleration performance curves of road vehicles. *(a)* Acceleration curves. [*From Baerwald (1965), p. 27.*] *(b)* Deceleration curves. [*From Baerwald (1965), p. 29.*]

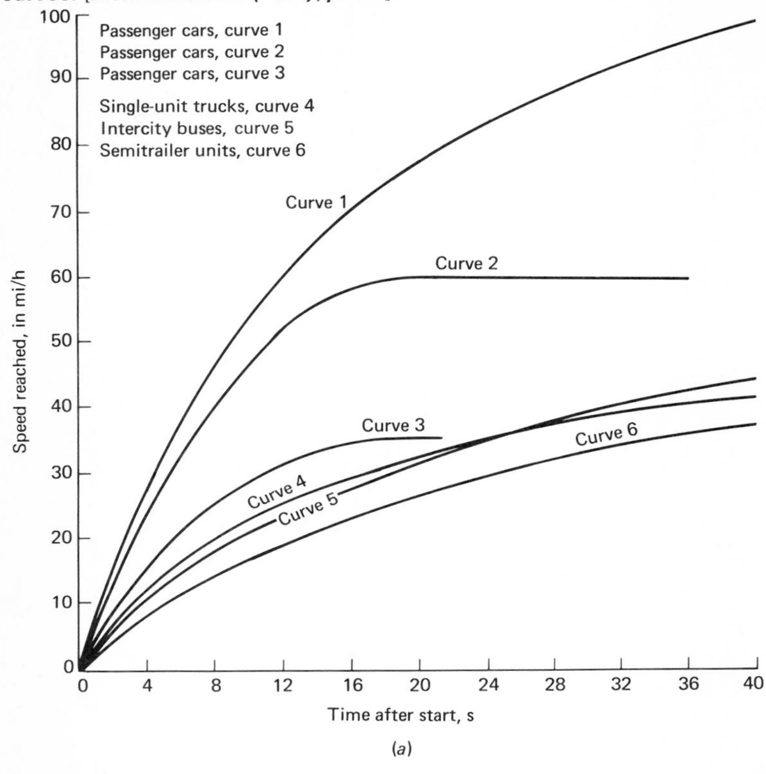

(*a*)

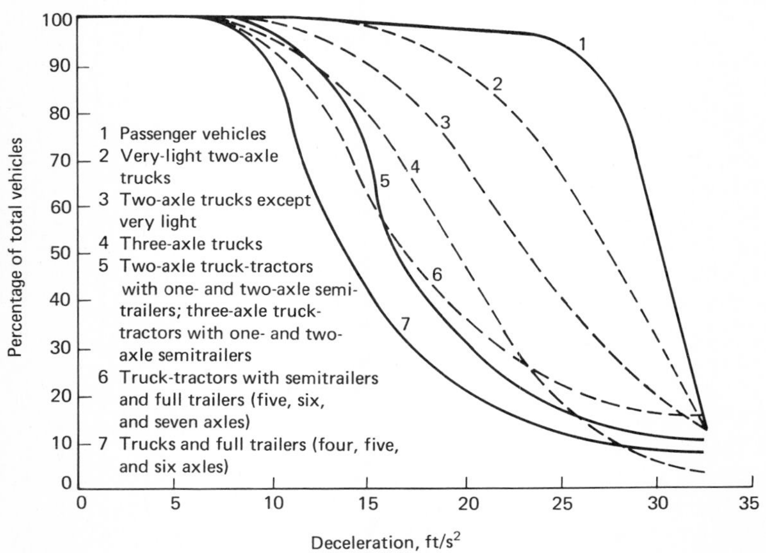

(*b*)

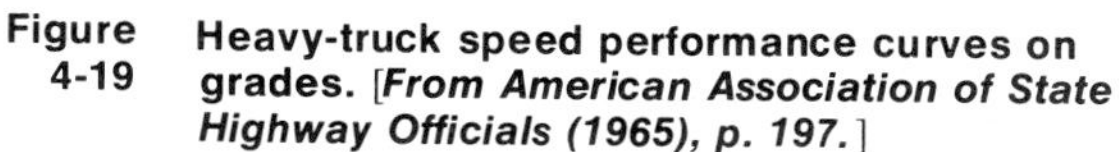

Figure 4-19 **Heavy-truck speed performance curves on grades.** [*From American Association of State Highway Officials (1965), p. 197.*]

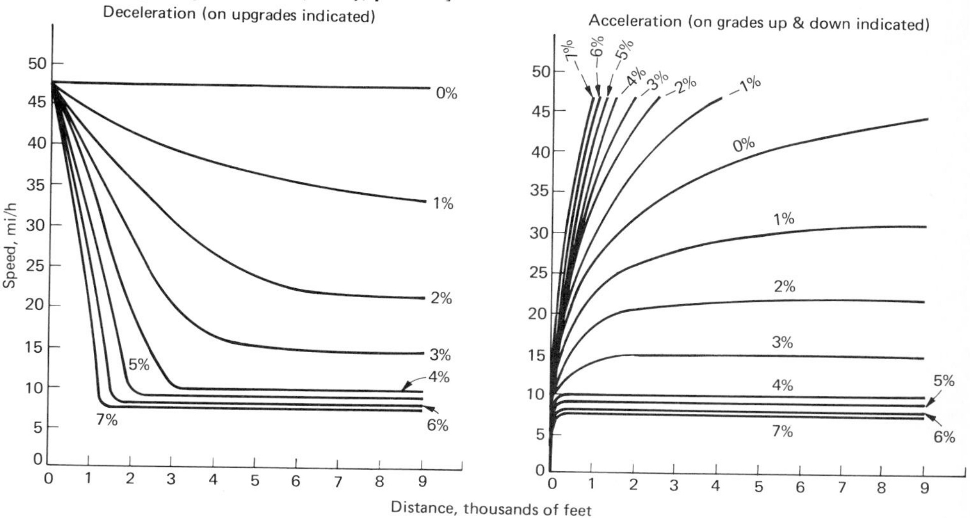

bus are for movement unimpeded by traffic, as would be found on a busway, an exclusive bus roadway similar to a rail rapid transit route.

Ships

There seem to be very few generalized performance curves for water craft. Perhaps the most important is the Plimsoll line, named after the Britisher who brought about the legislation in Britain which required that ships could not be filled to the point of dangerous overloading. A typical line takes the form shown in Fig. 4-21. The horizontal lines indicate the point to which the vessel may be safely submerged, different lines applying to different types of water and temperatures (i.e., salt versus fresh, summer versus winter, etc.).

Since most water vehicles seem to be specially designed for particular services and operators, there is perhaps little need for the generalized relationships of the types described for other types of vehicles which might be used by many different persons or organizations in diverse types of transport services. However, the same types of curves relating speed, acceleration, etc., to the load on the vessel, environmental conditions, etc., can be developed for water craft of all types.

WORK, ENERGY, AND FUEL CONSUMPTION

An increasingly important aspect of transport technologies is their consumption of fuel. As mentioned in Chap. 2, in many developed nations transportation of all sorts currently consumes about one-fifth to one-quarter of the nation's entire fuel consumption. With the realization that the supply of fuel is not inexhaustable, fuel conservation is becoming increasingly important. While the methods used to analyze fuel consumption vary from technology to

Figure 4-20 **Performance curves for urban transit vehicles. *(a)* Urban bus with full seated load (53 passengers); *(b)* typical rail rapid transit train of ten cars with full seated load.**

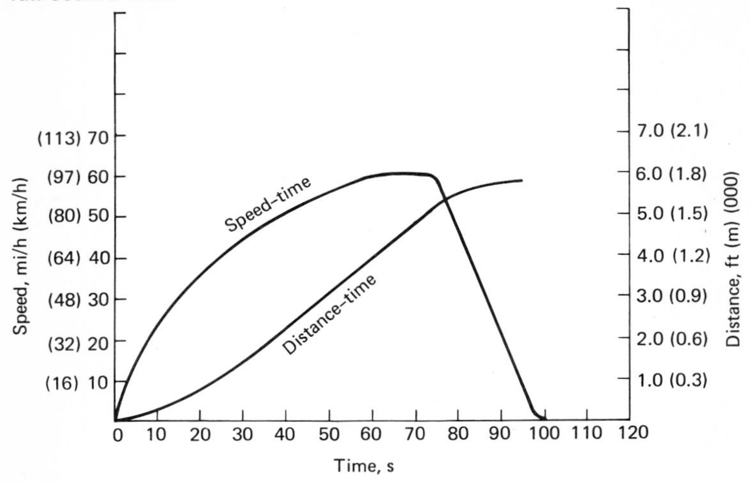

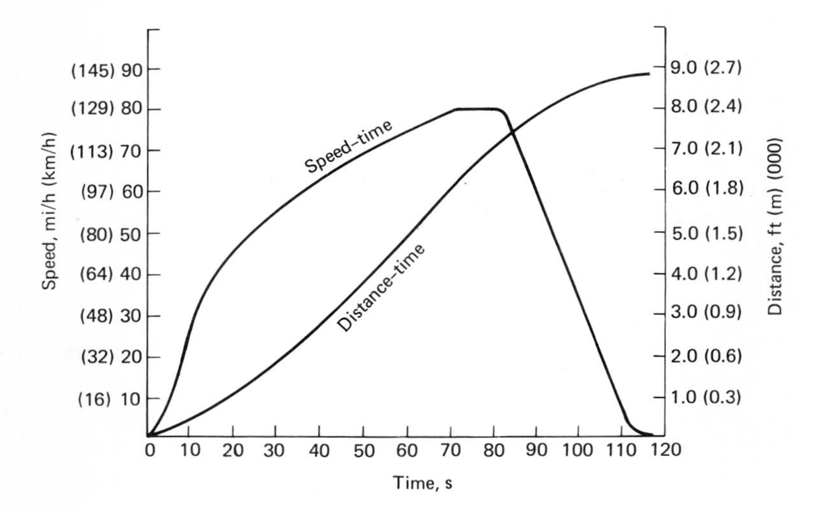

technology, some general approaches which are useful in transportation systems engineering will be presented below.

Fuels are used in transportation primarily to overcome the various resistances to motion which are encountered: inherent, curvature, and gradient. Also, fuel is used to provide and maintain an appropriate environment for passengers and cargo. Some cargoes require refrigeration or heating, while others require extreme pressure conditions. Although these are important uses of fuel in transportation, we will set them aside at this point in order to

Figure 4-21 **Typical vessel markings indicating maximum safe depth: the Plimsoll line.** [*From Hay (1961), p. 164.*]

TF — tropical fresh water
F — fresh water
T — tropical water
S — summer
W — winter
WNA — winter, North Atlantic

concentrate on the use of fuel in propulsion. Typically, propulsion consumes 90 percent or more of the fuel used on vehicles, and hence this will provide a good idea of the fuel consumption of the various carriers, which can be supplemented later when the characteristics of commodities and passengers will be considered.

The Work Method

In many transportation technologies, as well as in many other fields in which fuel is used by machines to create forces, it has been observed that the fuel consumed by such machines is approximately proportional to the total work performed by those machines. The basic relationship is the following:

$$z = Wr \tag{4-41}$$

where z = fuel consumed
W = work performed
r = fuel rate

The fuel rate and the specific measure of work performed vary from application to application, but the approach is identical.

In transportation, the usual approach is to measure the work performed by the propulsion unit or units, termed the propulsive work. In this measure, the work is created by the propulsive force moving along with the vehicle as it travels over its path. Thus, if we specify the variation of the propulsive force at point x along a path as $\mathrm{M}_p(x)$, then the total propulsive work in traveling from d_m to d_n would be

$$W = \int_{d_m}^{d_n} \mathrm{M}_p(x)\, dx \tag{4-42}$$

where $\mathrm{M}_p(x)$ = fuel-consumption-related propulsive force at point x

It should be noted at this point that capital and lowercase italic symbols have been used for all variables and parameters. To differentiate symbols for functions from these, roman letters will be used for functions, followed by parentheses containing the variable or variables on which the value of the function depends.

In using this method, it is extremely important that the force used in Eq. (4-41) is the force created by the propulsion unit solely as a result of consumption of fuel. Thus, for example, the gradient force on a downgrade would not be considered part of this force. Similarly, in many cases the braking force acting on a rail or road vehicle would not be included. On the other hand, in many airplanes and ships, the propellers or other propulsion units are reversed in order to provide a braking force, which consumes fuel in the same manner as if the propulsive force were acting in the direction of motion. In this case, the braking force would also have to be included in the computation of the propulsive work. Thus the user of this method must find the propulsive forces that are created through the use of the vehicle's fuel-consuming power unit and must see that all of these forces, but only these forces, are used in the calculation of propulsive work.

In practice, it would be very difficult to use the formula above for the estimation of propulsive work, because the formulas for the $M_p(x)$ term would be so complex. Instead, an approach similar to that used in the simulation of vehicle motion is employed. The path over which a vehicle travels is divided into many sections, within each of which the (fuel-produced) propulsive force is nearly constant. Then in each such segment the force is multiplied by the length of the segment, and these are summed to yield the total propulsive work. In fact, in most, the same increments of distance which emerged from the simulation of vehicle motion are usually used, which results in an equation for estimating propulsive work from a point d_m to a point d_n (where m and n are specified as the locations of interest).

$$W = \sum_{i=m}^{n} M_{p,i}(d_{i+1} - d_i) \tag{4-43}$$

The freight train used in previous examples of the methods of this chapter can be used again to advantage in illustrating this method. Let us consider a 60-mi trip, including the acceleration of the train to a speed of 40 mi/h, travel for a distance of 50 mi at that speed, and finally deceleration to rest. The forces used in the computation of acceleration are presented in Table 4-6, along with the distances traveled in accelerating over each of the 10-mi/h speed increments used in that simulation. These forces are multiplied by the relevant distances below, which yields the total propulsive work in acceleration to 40 mi/h.

$$\begin{aligned}
217{,}600 \text{ lb} \times 0.023 \text{ mi} &= 5{,}005 \text{ lb-mi}\\
210{,}440 \text{ lb} \times 0.073 \text{ mi} &= 15{,}362 \text{ lb-mi}\\
169{,}400 \text{ lb} \times 0.157 \text{ mi} &= 26{,}596 \text{ lb-mi}\\
118{,}580 \text{ lb} \times 0.345 \text{ mi} &= \underline{40{,}910 \text{ lb-mi}}\\
& 87{,}873 \text{ lb-mi}
\end{aligned}$$

At 40 mi/h, the resistance of the train to motion is 23,751 lb, which remains constant on the level, straight line, resulting in a total propulsive work for this segment of 1,187,550 lb-mi. While the train is decelerating, the locomotion is not providing any propulsive force through fuel consumption, and as a result the work is zero. (Note, though, that the locomotive's brakes, acting in conjunction with the cars' brakes to retard the train, provide a propulsive force in this section.) Thus the total propulsive work is 1,275,423 lb-mi. If the fuel rate of the locomotive is 0.0324 gal of diesel fuel per 1 million ft-lb of work, then the total fuel consumption on this 60-mi trip is

$$1{,}275{,}423 \text{ lb-mi } \frac{5280 \text{ ft}}{\text{mi}} \; 0.0324 \frac{\text{gal}}{10^6 \text{ ft-lb}} = 218.2 \text{ gal}$$

Fuel Rates

This approach is most extensively used in land transportation, where the fuel consumption of the same vehicle can vary widely depending upon the characteristics of the path (portrayed in the path chart), the load on the vehicle, and the speeds at which the vehicle is operated. Approximate fuel rates are presented in Table 4-7 for these technologies. These figures should be used with caution since engine and general vehicle maintenance, fuel quality, temperature, air pressure, and many other factors affect these rates.

Work and Energy

Since the term "energy" is widely used in transportation, it is important to define it. At the outset it should be noted that the term is used in two ways: (1) as a general term referring to the fuel, electricity, etc., consumed, and (2) as a precise term referring to certain types of work that have been performed on a body (such as a vehicle). With the increased concern over fuel consumption of all types, the former usage has become increasingly common, but it is best to use instead terms such as gasoline consumed or energy equivalent of fuel used (as, for example, when the Btu measure of fuel consumed is used). Within the second, more correct, usage of the term, a distinction is made between two types of energy, kinetic energy and potential energy.

The concept of kinetic energy can be explained by considering the acceleration of a vehicle on a level path from rest. If the total resistance to motion of the vehicle (inherent and curvature) is $R(v)$, then the total work performed on the vehicle in acceleration from speed v_0 at location d_0 and

Table 4-7 Fuel Rates of Various Technologies and Vehicle Types

Technology and vehicle type	Approximate fuel rate
Railroad	
Diesel-electric locomotives†	0.0324 gal/10^6 ft-lb
Road	
Diesel trucks‡	0.40–0.50 lb/hp-h
Gasoline autos§	0.003–0.005 gal/mi-lb
Water	
Turbine-electric¶	0.65 lb/hp-h (shaft)
Diesel-electric¶	0.50 lb/hp-h (shaft)
Diesel-geared¶	0.45 lb/hp-h (shaft)

† Poole (1962, p. 137).
‡ Fitch (1976, chap. IV).
§ Winfrey (1969, chaps. 12 and 14).
¶ Comstock (1967, p. 52).
Note: gasoline: 1 gal = 6.0 lb; diesel fuel: 1 gal = 7.1 lb.

time t_0 to speed v_1 at time t_1 and location d_1 will be

$$W = \int_{d_0}^{d_1} \left[R(v(x)) + T \frac{a(x)}{g} \right] dx \tag{4-44}$$

where $v(x)$ = velocity at each point x on path

$a(x)$ = acceleration at each point x on path $= \frac{dv(x)}{dt}$

This work is composed of two parts, overcoming resistances and imparting acceleration. Concentrating on the latter only, and terming this work K, we have

$$K = \int_{d_0}^{d_1} \frac{T}{g} \frac{dv(x)}{dt} dx = \int_{d_0}^{d_1} \frac{Tdv(x)}{g} \frac{dx}{dt} = \int_{d_0}^{d_1} T \frac{v(x)dv}{g}$$

$$= \frac{1}{2} \frac{T}{g} [v(x)]^2 \Big]_{d_0}^{d_1} = \frac{1}{2} \frac{T}{g} \{[v(d_1)]^2 - [v(d_0)]^2\}$$

$$= \frac{1}{2} \frac{T}{g} (v_1^2 - v_0^2) \tag{4-45}$$

This work performed in changing the vehicle's speed from v_0 to v_1 is termed the change in kinetic energy. The kinetic energy of a vehicle (or other moving body) is one-half the product of the mass of that vehicle and the square of its velocity. It is the energy content of the vehicle or body associated with the movement of that body. Recognizing that vehicles have rotating parts which move in addition to the entire vehicle moving at a speed v (e.g., the wheels of an automobile), usually about 5 percent is added to account for the additional energy contained in those parts.

The potential energy is the energy content of a vehicle (or in general, any body) associated with the elevation of that vehicle. It is equal to the work that must be performed on the vehicle to change its elevation, but including only the work necessary to overcome gravitational forces. Thus if a vehicle is being propelled up a gradient of percentage p over a distance (on the incline) of $d_1 - d_0$ at constant speed V, the total work performed will be

$$W = \int_{d_0}^{d_1} \left[R(V) + \frac{pT}{100} \right] dx \tag{4-46}$$

Concentrating on the second portion and defining that portion of the work as P, we have

$$P = \int_{d_0}^{d_1} \left[\frac{pT}{100} dx = \frac{pT}{100} x \right]_{d_0}^{d_1} = \frac{pT}{100} (d_1 - d_0)$$

But $p(d_1 - d_0)/100$ is precisely the change in elevation of the vehicle. If e_0 and e_1 are the elevations of the vehicle at d_0 and d_1, respectively, then

$$P = T(e_1 - e_0) \tag{4-47}$$

The potential energy of a vehicle (or other body) is the product of the weight of the vehicle and the elevation of the vehicle relative to a base elevation, usually taken as sea level. The change in potential energy of a vehicle is simply the product of its weight and the change in elevation; it may be positive or negative.

Both of these concepts are valuable in transportation engineering. The kinetic energy is the work that must be performed in accelerating a vehicle on a level path in addition to that required to overcome inherent and curvature resistance. Similarly, it is the work that must be done in stopping the vehicle. Most of this work must be performed by the propulsion (braking) unit, although some will be performed in the normal course of overcoming resistances to motion. For example, in stopping the sample freight train used throughout this chapter, which weighs 3814 tons from a speed of 40 mi/h, the kinetic energy dissipated is

$$\frac{3814 \text{ tons-s}^2}{2(32.16 \text{ ft})} \frac{2000 \text{ lb}}{\text{ton}} \left(\frac{40 \text{ mi}}{\text{h}} \frac{5280 \text{ ft}}{\text{mi}} \frac{\text{h}}{3600 \text{ s}} \right)^2 = 4.081 \times 10^8 \text{ ft-lb}$$

From the earlier computation, 67.34×10^8 ft-lb of propulsive work were required to propel the train over its entire journey up to the point where deceleration began. Thus the kinetic energy represents about 6.1 percent of the total work put into the train.

If this energy could be recaptured and transferred in some manner to other trains (or otherwise used), then a significant saving in total fuel consumption might be achieved. In fact, various schemes for accomplishing this have been developed and used. Most notable is the braking of electrically propelled (electricity from a central plant) trains by reversing the propulsion motors, so that some of this kinetic energy is translated into electric energy usable elsewhere. It is fed back into the electric lines (overhead catenary or third rail, as on rapid transit lines) for use by other trains. Although problems of train scheduling seem to limit its use on intercity railroads, it is increasingly popular on electric transit lines where both the large number of vehicles operating at one time and the frequent accelerations and decelerations provide an ideal situation.

Potential energy can be used in the same way. As a train descends a grade, if the inherent and curvature resistances are more than overcome by the gradient force, the braking force needed to keep the speed in check can be provided by reversing the electric motors. The electric energy is fed back into the system for use by other trains, which conserves fuel through cutting wasted energy. Lest one be led to believe that this is an ideal propulsion system, it should be mentioned that the expense of these so-called regenerative systems and electric propulsion of railways from a central power station (as opposed to diesel-electric locomotives) has often been found more expensive than diesel-electric propulsion. However, with the increase in oil prices and advances in electrical technology, "straight" electric propulsion for railways, buses (and even trucks operating on fixed routes, as in Moscow) is becoming more attractive.

Another interesting aspect of work requirement and energy is the effect of gradients. One might think that a route that entails many grades would necessarily require more propulsive work than one that does not, assuming they are of equal length and operated at the same speed, but this is not necessarily true. This, and other interesting aspects of work and energy consumption, are covered in the problems at the end of this chapter.

TYPICAL VEHICLES

While the frequent references made throughout the discussions on preceding pages provide some information on the characteristics of vehicles that are important in performance determination, a more organized form of this material is presented in Table 4-8. This information will be useful in many of the problems at the end of this chapter. But the reader is cautioned that this small selection of vehicles in each technology is not meant to be exhaustive.

Table 4-8 Some Characteristics of Typical Vehicles

Road

Type	Weight, lb	Power, hp	Length, ft	Cross section area, ft^2	Comments
Small automobile	3,000	100	14	19	4 seats
Large automobile	4,000	200	18	26	6–9 seats
Bus	38,000	300	40	104	38–53 seats (25 standees)
Small delivery truck	4,000	150	17	33	1–2-ton capacity
Medium truck	7,000	170	22	45	4–5-ton capacity; 2 axles
Tractor + 40-ft trailer	22,000	200	54	104	58,000-lb capacity
Tractor + two 27-ft trailers	29,600	250	65	104	50,400-lb capacity
Tractor + two 40-ft trailers	41,500	300	95	104	95,500-lb capacity

Railroad

	Axles	Weight, tons	Length, ft	Cross section area, ft^2	Comments
Diesel-electric locomotive	4	136	50	120	2500–3500 hp
	6	170	60	120	3000–4000 hp
Standard freight cars	4	25–35	40–60	120–135	50–100-ton capacity
Hi-cube (volume) freight cars	4	30–40	50–89	130–150	50–75-ton capacity
Piggyback car	4	30.5	89	140	Capacity as for truck trailers
with two 40-ft trailers		47			
Passenger cars	4	60–75	85	120–150	44–64 seats, intercity coach; 120–165 seats, urban coach; 30–40 seats, intercity sleeper

Inland waterway

	Length, ft	Width, ft	Draft, ft	Comments
Towboat	140	30	7.6	2000–4000 hp
Tugboat	100	30	12	1000–3500 hp
Small barge	100	26	7	350-ton capacity
Large barge	195	35	9	1500-ton capacity

Air

	Cruise speed, mi/h	Range, mi	Comments
Boeing 747	625	5750	490 seats
Boeing 707-320C	600	4000	200 seats
Douglas DC-9	565	1725	115 seats
DeHavilland DHC-6	204	118	20 seats, 1200-ft runway

SUMMARY

This chapter has focused on the principles of estimating vehicle performance, movement along a path, as governed by characteristics of the vehicle and the path. Although the jargon and specific techniques differ among technologies, identical principles and approaches are used in each. It is this similarity which has been stressed, but with specific examples of individual technologies in order that realistic problems can be addressed. The main points are the following.

1 Newton's laws underlie the prediction of vehicle motion on a path.

2 The forces which result in vehicle motion are those of resistance and propulsion. The resistance force includes inherent resistance, curvature resistance, and gradient resistance. Propulsive forces are characterized by a propulsive force diagram. Path characteristics are specified on the path chart.

3 Because of the complex interaction of these forces, vehicle motion usually cannot be predicted by simple analytic formulas, but must be simulated.

4 There are many limitations on the paths which vehicles can use. Some exclude certain vehicles, others limit speed and/or weight, etc.

5 The fuel consumed by a vehicle can be estimated from the total propulsive work performed on that vehicle trip. Special consideration in such computations must be given to the effects of gradients. The concepts of kinetic energy and potential energy have particular significance in the context of work performed.

PROBLEMS

4-1

A train consisting of diesel-electric locomotives (none streamlined), each of 3000 hp, 4 axles, 125 tons and 120-ft^2 cross section, and 20 4-axle passenger cars, weighing 60 tons on the average, must be able to ascend a 1.0 percent grade with a long 2° curve at 50 mi/h. What number of locomotives must be assigned to this train? What is the maximum speed of this train on level track?

4-2

A tractor plus two trailer combination trucks (often called "double bottoms") must be able to maintain a constant speed of 45 mi/h on a 2.5 percent upgrade. The characteristics of the truck combination are as given in Table 4-7, and it has a full load, by weight. What is the total resistance to be overcome in still air? What would the resistance be on a level road? What horsepower must be delivered to the wheels on the upgrade? If the transmission and auxiliaries result in an efficiency of 85 percent, what must the rated horsepower of the motor be?

4-3
A state which now restricts trucks to one tractor plus one trailer is considering relaxing this restriction to allow combinations of a tractor plus two full trailers. One of the advantages would be a reduction in the fuel consumed in truck movement claim those in favor of the liberalized rules. What would the fuel saving be for a shipment consisting of two 15-ton trailer loads in 40-ft trailers (characteristics in Table 4-7) over the turnpike route shown below from point A to point B? Empirical studies have revealed that tractors consume an average of 0.040 gal of diesel fuel per million foot-pounds of work (measured at the wheel-road interface). Assume the trucks would move at a constant speed of 55 mi/h throughout the run and ignore the acceleration and deceleration at the ends. The profile is

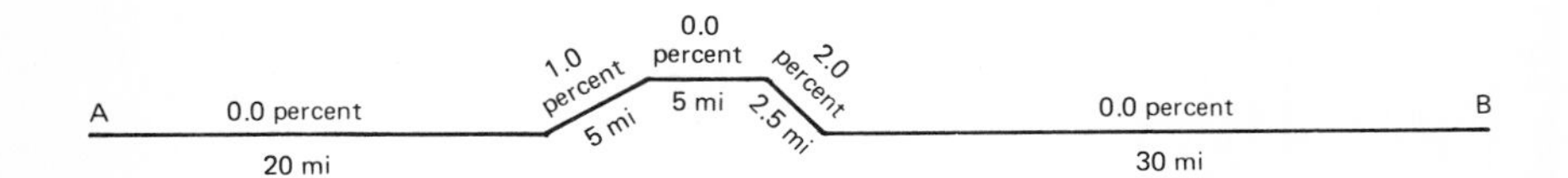

4-4
Assume a motor vehicle driver attempted to avoid an accident by decelerating her vehicle as rapidly as possible by locking the wheels, such that there were skid marks on the pavement from the point that deceleration began to the point of stopping. If the maximum coefficient of friction between the wheels and the road were f for skidding and the road were level, find the relationship between her initial speed V and the length of the skid marks D. What is the relationship for a skid on a grade of percentage G? (Hint: For this problem a reasonable assumption is that the inherent resistance is so small relative to the deceleration braking force that the former can be ignored. The approximation is so good that these relationships are used in courts of law in accident and reckless driving cases!)

4-5
A new trailer-on-flat-car (TOFC) service is to be inaugurated jointly by a railroad and a truck line. The railroad line connecting these cities is 100 mi long and has a profile consisting only of level sections and very slight grades, such that it can be approximated by a level profile. The trains will be composed of one or more diesel-electric locomotives, 20 flat cars (often called "piggyback" cars) each of which carries two 40-ft truck trailers and a caboose. The locomotives are 3300-hp 4-axle units, each weighing 136 tons, with a 120-ft^2 cross section area. The TOFC cars will average 70 tons and have 4 axles. The caboose will weigh 25 tons and have a cross section area of 120 ft^2. Braking will occur at a rate of 1.0 mi/(h)(s) at all speeds. With two locomotives per train and a maximum speed limit of 80 mi/h

- **a** Draw the time-distance, speed-distance, and speed-time diagrams for acceleration to cruise speed, using velocity increments of 5 mi/h.
- **b** Draw the same for deceleration from cruise speed.
- **c** Determine the maximum or balancing speed.
- **d** Estimate the time required for the 100-mi trip.

What is the effect of increasing the number of locomotives per train to four on **a** through **d**?

4-6

Compare the resistances to motion per passenger of three passenger carriers, each operating in its typical speed range: (1) a bus weighing empty 40,000 lb, 8 ft wide and 13 ft high, with 38 seats, (2) a train of 4 4-axle coaches, each weighing 120,000 lb with 88 seats, pulled by one 3300-hp diesel-electric locomotive, fully streamlined, with 6 axles, a cross-section area of 140 ft^2 and weighing 136 tons, and (3) a jetliner weighing 220,000 lb with 157 seats. The speeds are, respectively, 55 mi/h, 70 mi/h, and 375 kn. If average load factors (ratio of passengers to seats) are 0.50, 0.30, and 0.45, respectively, and passengers weigh, on average, 150 lb (with luggage), what are the comparative resistances per passenger?

4-7

A ship designed to carry 120 passengers will weigh about 100 tons. It is intended for a service in which a cruise speed of 40 kn is required. Compare the resistance for this craft as a boat, a hydrofoil craft, an air cushion craft and a planing craft.

4-8

A rapid transit line is to be constructed in a city with numerous hills. One consideration in choosing between a steel wheel on steel rail technology and a rubber tire on concrete technology is the stopping distance on contemplated grades. On a downgrade of 5.0 percent, what is the minimum stopping distance from 60 mi/h for a rail system if the maximum adhesion will be 0.25? What will it be for a rubber-tired system if the maximum adhesion is 0.60? How would considerations of safety and comfort of standing passengers affect these results? Discuss the implications of your results for the choice of wheel and guideway types in hilly terrain.

4-9

Compare the work required to move 2500 tons of freight by rail in conventional rail cars, in trailers on piggyback cars, and on trucks on highways, over the routes shown below. The conventional train consists of 50 cars, each weighing 30 tons empty, a 30-ton caboose, and as many locomotives (3300 hp, 120-ft^2 cross section, 136 tons, 4 axles) as required to propel the train at a constant speed of 35 mi/h. The piggyback train contains 50 TOFC cars, each with two trailers, weighing 47 tons empty, plus caboose and locomotives, and is operated at 60 mi/h. Fifty truck combinations would be used, each consisting of a tractor pulling two trailers, the combination weighing 41,500 lb empty with a 104-ft^2 cross section area. The truck operates at a constant 55 mi/h. Compare the fuel consumption, using the fuel rates in the text. Discuss your results.

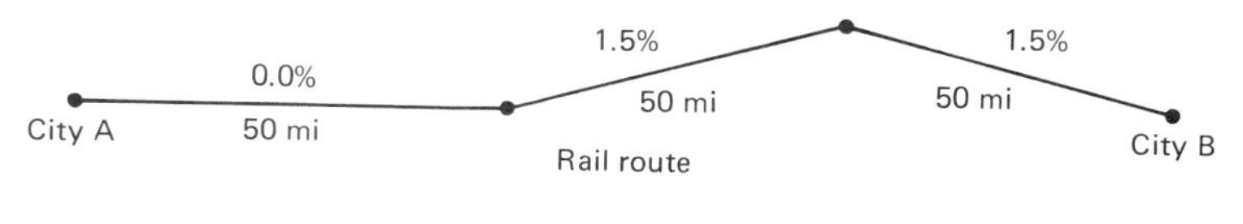

Rail route

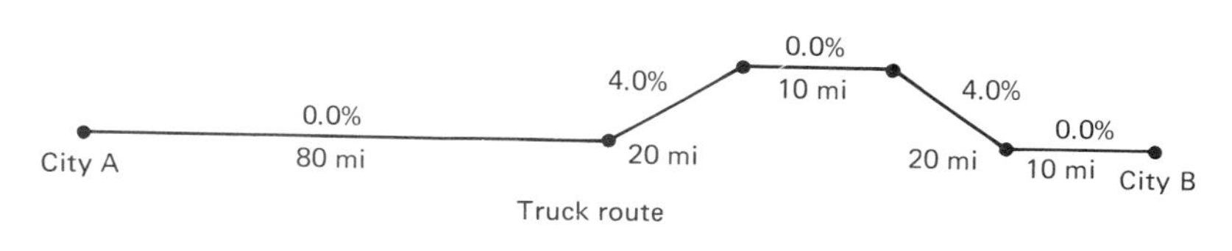

Truck route

4-10
On a rapid transit line, using the rail cars with characteristics as shown in Fig. 4-20, what is the relationship between travel time, start to start at adjacent stations, versus distance between stations (spacing), for distances from 0.5 to 3.5 mi? Assume the dwell time at each station is 20 s. Why is the curve linear above a certain distance, and what is that distance? Also, plot average speed versus station spacing. Even though average speeds increase with increasing station spacing, are there any negative effects of increasing spacing (e.g., from the viewpoint of some potential transit users)?

4-11
Derive the expression for the total propulsive work expended when a vehicle travels over a route with a profile consisting of an upgrade (of constant gradient) and then a downgrade (of constant gradient). The end points are at the same elevation, and the vehicle travels at constant speed. Use the notation of Eqs. (4-41) through (4-48). Under what condition is the total work identical to that required on a perfectly level route connecting the same end points, which is exactly the same length and on which the vehicle operates at the same speed as on the route with grades. What is the implication of this for route design in rolling or mountainous terrain? (Hint: Compare the force due to the gradient with the inherent resistance.)

4-12
Show that the work requirement of technologies with high inherent resistances are least adversely affected by gradients.

REFERENCES

American Association of State Highway Officials (1965), "A Policy on Geometric Design of Rural Highways," Washington, DC.

American Waterways Operators (1965), "Big Load Afloat," Washington, DC.

Association of American Railroads Mechanical Division (1966), "1966 Car and Locomotive Builders' Cyclopedia of American Practice," Simmons-Boardman, New York.

Baerwald, John E. (1965), "Traffic Engineering Handbook," Institute of Traffic Engineers, Washington, DC.

Baerwald, John E., Matthew J. Huber, and Louis E. Keefer, eds. (1976), "Transportation and Traffic Engineering Handbook," of the Institute of Traffic Engineers, Prentice-Hall, Englewood Cliffs, NJ.

Boeing Commercial Airplane Co. (1975), "General Description, 707-320C" Seattle, WA.

Comstock, John P., ed. (1967), "Principles of Naval Architecture," Society of Naval Architects and Marine Engineers, New York.

Davis, W. S., Jr. (1926), Tractive Resistance of Electric Locomotives and Cars, *General Electric Review,* **29** (10):685–708.

Elsley G. H., and A. J. Devereux (1968), "Hovercraft Design and Construction," Cornell Maritime Press, Cambridge, MD.

Fitch, James W. (1976), "Motor Truck Engineering Handbook," 2d ed., San Francisco.

Gabrielli, G., and T. Von Karman (1950), What Price Speed?, *Mechanical Engineering,* **72** (10):775–781.

General Electric Co. (1969), "Application of Transportation Formulae to Diesel-Electric Locomotive Train Haulage," Erie, PA.

Gillman, Thomas C. (1970), "Modern Ship Design," United States Naval Institute, Annapolis, MD.

Hay, William W. (1953), "Railroad Engineering," Wiley, New York.

Hay, William W. (1961), "Transportation Engineering," Wiley, New York.

Hennes, Robert G., and Martin Ekse (1969), "Fundamentals of Transportation Engineering," McGraw-Hill, New York.

Horonjeff, Robert (1972), "Planning and Design of Airports," McGraw-Hill, New York.

Mandel, Philip (1969), "Water, Air and Interface Vehicles," MIT, Cambridge, MA. Figures reprinted by permission of The M.I.T. Press, Cambridge, MA, © 1969.

Margenau, Henry, William W. Watson, and C. G. Montgomery (1953), "Physics: Principles and Applications," McGraw-Hill, New York.

McFarland, Ross A. (1953), "Human Factors in Air Transportation," McGraw-Hill, New York.

Poole, Ernest C. (1962), "Costs—A Tool for Railroad Management," Simmons-Boardman, New York.

Quinn, Alonzo DeF. (1972), "Design and Construction of Ports and Marine Structures," McGraw-Hill, New York.

Society of Automotive Engineers (1975), Truck Ability Prediction Procedure, SAE J688, in "SAE Handbook," Part 2, Warrendale, PA.

Taborek, Jeroslav J. (1957), Mechanics of Vehicles, *Machine Design,* **29** (11):60–65.

Taylor, D. W. (1910), "The Speed and Power of Ships," Wiley, New York.

Winfrey, Robley (1969), "Economic Analysis for Highways," International Textbooks, Scranton, PA.

Vehicle Flow

5.

In the previous chapter, the movement of individual vehicles was described. As a basis we used the engineering relationships which characterize the various forces to which a vehicle is subjected. That treatment ignored any restrictions on the motion of the vehicle other than those imposed by the design of the vehicle and the path on which it operates. In almost all transportation systems, however, the movement of any one vehicle may be limited by the presence of other vehicles, and thus its optimum performance may not actually be realized. Whenever a large (relative to the number for which the facility has been designed) number of vehicles are on a facility, the resultant congestion may cause substantial delays to all vehicles, increased operating costs, and a higher probability of accidents. Most people have undoubtedly experienced these effects while traveling on urban roads, often to the point of considerable irritation.

The understanding of the flow of many vehicles on a path is essential to the rational design of new facilities and to the modification of existing facilities to meet changed traffic conditions. The design characteristics of the physical facility, the manner in which the movement of vehicles is regulated on the facility (e.g., road traffic rules, railroad train schedules), and the characteristics of the vehicles (including their human operators, if any), all interact to determine the performance of that facility in accommodating the vehicle traffic load placed upon it. Hence in the design of facilities, as well as in the determination of the operating plans or procedures, these interrelationships must be taken into account. The main focus of this chapter will be on the determination of characteristics of vehicle flow on way links and interchanges, such as the maximum throughput of vehicles possible and the effect of varying amounts of throughput on travel time and other quality-of-service factors. These results will prove invaluable later when the design of facilities is treated, where the engineer and planner must consider such factors as meeting the vehicle traffic requirements as well as environmental and economic effects.

THE TIME-SPACE DIAGRAM AND FLOW CONCEPTS

Time-Space Diagram

One of the most useful tools for the analysis of the flow of vehicles, and a very useful device for explaining the primary variables and relationships of vehicle flow, is the time-space diagram. The time-space diagram is simply a plot of the movement of all the vehicles on a path in which the location of each vehicle as a function of time is specified. An example is presented in Fig. 5-1. The vertical axis specifies location along the path (space), and the horizontal axis represents time.

Eight vehicles are shown in this figure. The first, labeled 1, is proceeding at a constant speed, and hence the line representing its movement has a constant slope. The width of the line 1 corresponds to the length of this vehicle. Vehicle 2, which follows vehicle 1, initially is traveling at a constant speed, then slows down, eventually stops for a short period, and finally accelerates again. The width of this line is greater than that of vehicle 1, reflecting the greater length of this vehicle (which might represent a bus, while vehicle 1 is an automobile). Other vehicles are shown with varying speeds. These bands are called *vehicle trajectories*.

Variations of the time-space diagram are used in almost all modes of transportation for the analysis of vehicle flow phenomena. Although the time-space diagram is usually applied to a single path or channel for vehicle flow, such as a single lane on a road or a single track on a railway, in some instances the same diagram may include the vehicles on one or more parallel paths. (Such a multipath form may, however, become extremely cluttered

Figure 5-1 **The space-time diagram.**

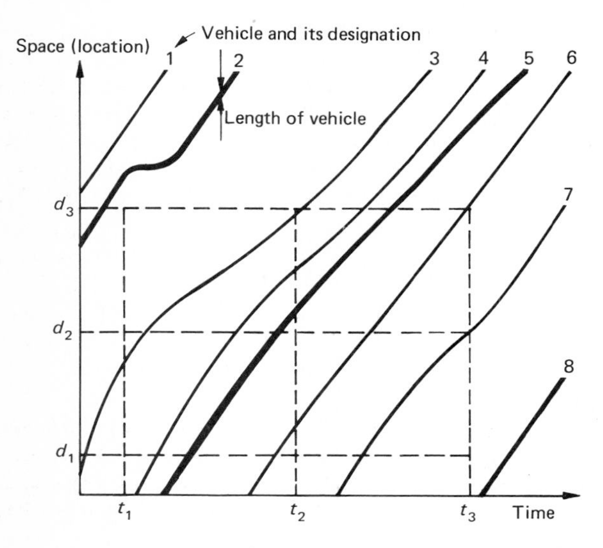

with vehicle trajectories.) It can include vehicles moving in both directions on a path, as in the case of railway trains on a single track line or barges traveling through canal locks. In many applications the vehicle trajectory is shown as a line which indicates the location of one point on the vehicle (usually the front), rather than the entire length of the vehicle. This is usually done in the case of highway traffic applications, for example.

The time-space diagram is obviously very similar to the distance-time diagram of an individual vehicle described and used in Chap. 4. The major difference is that the distance-time diagram presents the movement of one vehicle, such as maximum attainable acceleration from a stop, while the time-space diagram presents the movement of all the vehicles on a given path. Moreover, the distance-time diagram for an individual vehicle is usually used to present the performance of that vehicle under the prescribed conditions, while the time-space diagram usually represents a vehicle's performance as it interacts with other vehicles, possibly a somewhat degraded performance. Also, the time-space diagram is usually used for analysis of particular situations, while the vehicle performance curves are usually generalized for widespread application.

Flow Concepts

The primary variables used to describe or characterize the flow of vehicles on a path are volume, speed, concentration, and headway. This section will define these and related variables, specifying the relationships among them.

Volume is the number of vehicles passing a point on a path per unit time, and hence it is measured in units of vehicles per unit time. It is usually measured by placing a counter of some type at the point where the volume is desired or by manual counting. The count may be of vehicles on one path or on many parallel paths (e.g., the volume on one lane of a street or on all lanes), and it may be of vehicles moving in one direction or in all directions (e.g., all vehicles entering an intersection from one approach or street or all vehicles entering the intersection regardless of the entering or exiting street). Thus the flow or flows included in any volume measure must be specified, as must all the flow measures (e.g., persons per hour, vehicles per hour) to be used. (Although in this general introduction to transportation engineering we will not specify all the details of application of these concepts to each mode, it should be noted that there are often conventions regarding the flows to which the term volume, as well as other terms, applies.)

Returning to the definition of volume, it may be expressed as

$$q = \frac{n}{T} \tag{5-1}$$

where q = traffic volume past a point
n = number of vehicles passing point in time interval T
T = time interval of observation

In Fig. 5-1, for example, the volume past the point d_2 measured over the time interval from t_1 to t_3 is 4 vehicles/$(t_3 - t_1)$. If the time interval were 5 min, the volume would be 0.8 vehicle/min or in the more common units, 48 vehicles/h. The volume is thus an average measure, and it is sensitive to the time at which the measurement began and ended. In this example, if the vehicle count were made from t_1 to t_2 only, a period of 2.5 min, the volume would have been calculated as 1.2 vehicles/min or 72 vehicles/h. In practice, if the flow is somewhat irregular, as it is here, and an average volume is desired, then the count of traffic would be taken over a period sufficiently long that only minor variations in the volume would result from small changes in the time interval of the count.

Another important characteristic of flow is the concentration or density. Both terms are widely used in the literature. Although the term "density" is more commonly used by road traffic engineers, it is also used by professionals in almost all transport carriers (railroads, airlines, etc.) to describe another traffic measure. Hence we shall use the term "concentration." Concentration is the average number of vehicles per unit length of the path at a moment in time, and it is defined by

$$k \equiv \frac{n}{L} \tag{5-2}$$

where k = concentration of vehicles on roadway of length L at a point in time
n = number of vehicles on road
L = length of road

Referring again to Fig. 5-1, we might consider the concentration of vehicles between d_2 and d_3 at time t_2. If the length of that section were 1.5 mi, then the concentration would be 0.5 vehicle/mi. If the measurement were taken at a different moment in time, as at t_3, then the concentration would differ, so again variations in this measure must be expected although making the measurement over a long section of road (so that many vehicles are on it) or taking the average of many measurements, each at a different time on the same section, will reduce the variation. In fact, the average concentration on a length of roadway over a period of time is often calculated by the following formula:

$$k \equiv \frac{n \sum_{i=1}^{n} m_i}{T \sum_{i=1}^{n} s_i} \tag{5-3}$$

where k = average concentration of vehicles over a length of roadway and period of time T
T = period of observation
m_i = time spent by vehicle i on roadway (i = 1, 2, 3, . . . , n)
s_i = distance traveled by vehicle i on roadway (i = 1, 2, 3, . . . , n)
n = number of vehicles on road in period (i = 1, 2, 3, . . . , n)

The term $(\Sigma m_i)/T$ is the average number of vehicles on this section of road in the period T. The term $(\Sigma s_i)/n$ is the average distance traveled by the n vehicles that use this section of road in the period T. Thus these terms are analogous to the two terms which comprise the definition of concentration in Eq. (5-2).

The third primary flow variable is speed. Various definitions of speed may be used to characterize the movement of the many vehicles flowing on a path. The most useful of these seems to be space-mean speed—the mean or average speed of the vehicles obtained by dividing the total distance traveled by the total time they require:

$$u \equiv \frac{\sum_{i=1}^{n} s_i}{\sum_{i=1}^{n} m_i} \tag{5-4}$$

where u = space-mean speed

The space-mean speed is probably not the speed measure which would be devised by someone with no background in vehicle flow phenomena, and therefore its definition should be studied carefully. The mean speed measure which most persons would probably construct is called the time-mean speed, i.e., simply the mean of the speeds of the vehicles passing a point on the road in a particular time interval. The time-mean speed is defined as

$$v \equiv \frac{1}{n} \sum_{i=1}^{n} v_i \tag{5-5}$$

where v = time-mean speed
v_i = speed of vehicle i at a point on roadway

The difference between these two speed measures is shown in Fig. 5-2. This figure portrays in more detail the trajectories of the four vehicles traveling on the road depicted in Fig. 5-1 between the locations d_2 and d_3 in the time interval t_1 to t_3. The space-mean speed considers the movement of the vehicles over the entire section of the roadway in the time interval, and the value of the space-mean speed is calculated to be 41.8 mi/h. The time-mean speed considers the speeds of vehicles at only one point, here chosen to be d_3, and the value is 44.75 mi/h. The disparity between these values reveals the importance of using the correct definition of mean speed. Again it must be stressed that the space-mean speed is by far the most important for vehicle flow analysis.

Figure 5-2 Space-mean speed and time-mean speed. u = **space-mean speed** = {[(1.5 mi)(4)]/(2.3 + 2.5 + 2.0 + 1.8)min} **60 min/h = 41.8 mi/h.** v = **time-mean speed** = [(39 + 48 + 40 + 52)mi/h]/4 = **44.75 mi/h.**

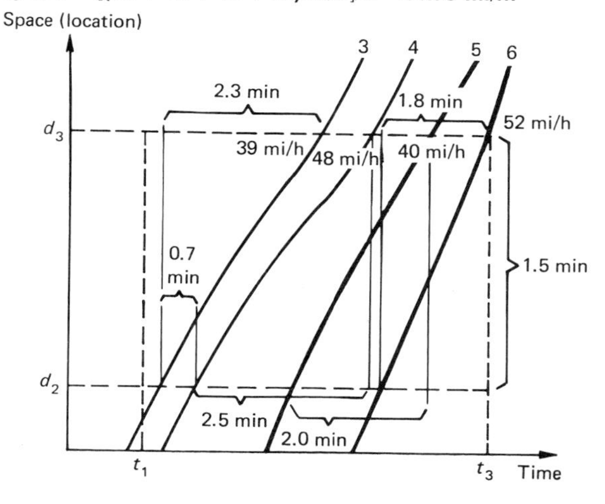

Numerous variations of these three primary flow variables are applied to particular contexts. Two of the most important are measures of headway. The time headway of two vehicles is defined as the time interval between the moment at which the front of one vehicle passes a point to the moment the front of the next vehicle passes the same point. Thus in Fig. 5-2, for example, the headway between vehicles 3 and 4 at location d_2 is 0.7 min. The time headways between other successive pairs of vehicles will in general be different. This gives rise to the concept of average headway. The mean time headway is the mean of the time intervals between successive pairs of vehicles, and it is measured over a period of time at a particular location. Therefore it is closely approximated by the reciprocal of the volume:

$$\bar{h}_t \doteq \frac{1}{q} \tag{5-6}$$

where $\bar{h}_t$ = mean time headway

In fact, if the beginning and end of the period of the volume count both correspond to moments of passage of the front of vehicles, and only one of these two vehicles is counted, the volume is precisely the reciprocal of the average headway. In transport systems where the vehicles pass a point at precisely the same headway, as is often the case on urban transit lines and sometimes on other scheduled carriers, the mean headway term is often used in preference to volume.

Another headway concept often used is the distance headway, i.e., the distance between the front of one vehicle and the front of the following vehicle at a given moment in time. A mean distance headway is sometimes used, especially in situations where the value varies among vehicle pairs in a flow. Here too, the average distance headway is closely approximated by the reciprocal of the concentration, and the two are precisely equal if the distance used for the concentration measurement defines the locations of the fronts of two vehicles at the moment of the measurement, provided only one of these two end vehicles is counted in the calculation.

$$\bar{h}_d \doteq \frac{1}{k} \tag{5-7}$$

where $\bar{h}_d$ = mean distance headway

A final variable often used by those involved with carriers (e.g., railways, airlines) is the average volume (number of vehicles passing a point per unit time) along a length of route consisting of one or more links and intersections. This measure is designated as the density by transportation engineers in those organizations, even though it is very similar to the volume measure defined earlier. The difference is that it is defined in a manner that applies to a section of path rather than to a single point only (as does the volume). It is actually the average volume past all points along the path under consideration. This density measure is defined by

$$\bar{q} \equiv \frac{\sum_{i=1}^{n} s_i}{LT} \tag{5-8}$$

where $\bar{q}$ = traffic density or average volume on a path or paths of interest
L = path length
s_i = distance traveled by each vehicle i on path
T = time interval in which measurements are taken

The vehicle-mile measure includes the sum of the distances operated by each vehicle which used the path in the period of the count. Often the traffic flow unit used in this measure is the vehicle (e.g., train) as specified above. However, it is also frequently used as a measure of passenger or freight traffic density, e.g., in units of passengers per day or tons of freight per year. The only difference is that passengers or tons of freight replaces the vehicle unit in the counting. Since road traffic engineers use the term "density" in a somewhat different way and the dimensions are identical, it is important to determine which definition is being employed in any context where the term "density" is found, although generally outside of the road field, "density" is used as defined above.

Fundamental Flow Relationships

An important relationship exists between the three primary flow variables defined above. Although it has been most widely used in highway traffic applications, it is equally valid in other vehicular modes as well. This relationship, often known as the fundamental flow relationship or fundamental volume-speed-concentration relationship, is

$$q = ku \tag{5-9}$$

The validity of this relationship can be ascertained by simply substituting Eqs. (5-3) and (5-4) for k and u. It might be noted that in road traffic engineering practice the concentration is measured at a moment, following Eq. (5-2), and although this measurement may not be precise, the answers are sufficiently accurate for most engineering purposes. However, particular practices in road traffic engineering or other specific areas need not overly concern us here, as we are primarily concerned with the basic principles and methods rather than the details of any one modal application.

While it might be most interesting to include some examples of how these variables and relationships are used at this point, it is now best to turn to the determinants of the spacing of vehicles on a path. Once this is understood, it is then possible to estimate the capacity of transport facilities and to deal with congestion phenomena and actual speed characteristics of vehicle flows.

CONTROL OF VEHICLE FLOWS

Whenever more than one vehicle is operating in a transportation system, it is necessary to ensure that collisions do not occur or at least that the probability is acceptably low. Means for achieving this, with varying degrees of success, exist in all transportation technologies. In all of the commonly used forms of transportation, except those engaged in for sport (e.g., automobile racing), the accident rates are remarkably low, attesting to the care devoted to problems of vehicle control as well as maintenance of the systems. (This is not to imply that we cannot or should not do better!)

Figure 5-3 **Sequence of events in control of vehicle motion.**

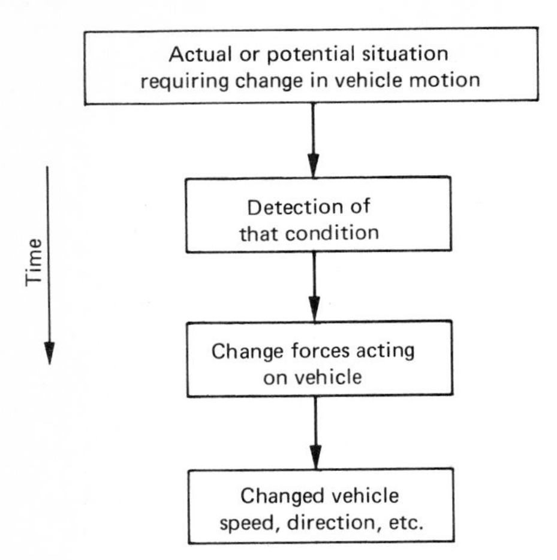

Although the specific means used to effect the desired control of vehicles varies considerably among technologies of transportation, the basic process (shown in Fig. 5-3) is identical. The goal is to enable the detection of any condition that might require a change in the movement of the vehicle, a change in speed, direction, altitude, etc., at a time and distance sufficiently in advance of the point of danger that the danger can be avoided. This involves the four steps in the process shown in the figure.

There are many different types of situations which might require a change in the vehicle's motion. One, of course, is a change in the path of the vehicle, such as a curve in the road or railroad line, which can be negotiated safely only at reduced speed. This type of situation is easily dealt with because the presence of the potentially dangerous condition is known in advance, and appropriate signs can be posted warning the vehicle operator. Temporary situations are more difficult to deal with, since they require unplanned changes in vehicle motion. Persons walking across a street, other vehicles on a roadway, road vehicles crossing a railroad track in front of an oncoming train, and other aircraft in an airspace all fall into this category. Although different technologies of transport and different path designs are subjected in varying degrees to these unforeseen situations, all vehicle movements must be controlled in a manner that takes into account the possibility of unforeseen needs to alter planned vehicle movement.

At one extreme is the local street, which is often used not only for vehicle movement and parking but also as a local playground for children and perhaps as a sidewalk (if none exists) for pedestrians and bicyclists. In this case, the high possibility of potentially dangerous situations is reflected in low speed limits and usually in careful driving. Much less prone to intrusion by persons or objects other than vehicles are airspaces, freeways (which by definition have access only at specific points and are usually completely fenced in elsewhere), railroad lines, subways, and the like. Even in these, however, unexpected persons, animals, or objects occasionally intrude. However, intrusion by another vehicle, requiring a change from the planned movement of some vehicles, is much more likely. Even if other vehicles should not be in the path of the vehicle, a malfunction of the system (e.g., breakdown of a vehicle, delay in a scheduled trip of a train) may create such a condition. Thus there is always a need to detect dangerous conditions in time for adequate response.

Channelization

Probably the most common form of vehicle movement control is channelization. The basic idea is to segregate vehicle movements into categories such that the movements of all the vehicles in a category are as similar as possible. This concept underlies the separation of traffic on a street in two directions, each placed on its own section of the road. It has been observed that the center stripe on highways has probably reduced traffic accidents more than any other single invention of traffic control. Regardless of whether or not this is the case, its contribution to road safety has been immense.

Channelization is used in all technologies of transport. Pedestrians now keep to the same side of walkways as vehicles travel on roads almost by instinct, although in large transport terminals and other crowded places "lanes" for pedestrian movement are often designated. Their use in road traffic is so ubiquitous as to hardly need elaboration—at intersections, on links dividing unidirectional flows into lanes so that vehicles do not sideswipe one another, etc. Similarly, the movement of railroad vehicles is channelized, although this is a natural result of the wheel-rail technology.

Air lanes, space used by aircraft for regular movements over long distances, are divided into lanes, which are defined by altitude along any given route (Horonjeff, 1975, pp. 88–97). For relatively low speed aircraft there are Victor airways, which are 8-mi-wide bands centered on very high frequency radio stations located at frequent intervals along the airway. At higher altitudes, for jet aircraft, are "jet routes," similarly located horizontally by radio stations. The Victor airways extend from 1800 ft above mean sea level up to but not including 18,000 ft, while the jet routes are from 18,000 to 45,000 ft, above which aircraft are controlled on an individual basis. The minimum vertical separation of aircraft is 1000 ft up to and including 29,000-ft elevation, and above that the minimum is 2000 ft. Up to 29,000 ft odd altitudes are used for eastbound flights and even altitudes for westbound flights. Above 29,000 ft the eastbound altitudes are 31,000, 35,000, 39,000 and 43,000 ft. In the vicinity of an airport, usually up to about 5 mi from the runways, all flights are under the supervision of the local air-traffic-control facility.

Even in waterways and in parts of the oceans vessels are guided into paths or channels. Wherever there is so much vessel traffic that separation is advisable, as in some ports and canals, specific channels are designated. Of course, this is also done in order to guide vessels into sufficiently deep water. Lighthouses, lighted and unlighted buoys, and other markers are all used to guide vessels, in effect creating channelization of flows.

Speed Limits and Related Controls

In the land surface and water modes of transport in particular, there is often need to restrict the speed of vehicles in specific locations. For rail and road systems, this is most often due to the presence of curves in the path, gradients

Figure 5-4 **Examples of standard road signs.** [*From U.S. Department of Transportation, Federal Highway Administration (1971) and (1977).*]

Background–red
Legend and border–white

Background–white
Legend and border–black

Background–white
Legend, border and arrow–black
Circle and slash–red

Background and legend–white
Circle–red

Background–yellow
Legend and border–black

Background and diamond–white
Legend and border–black

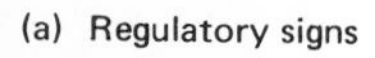

(a) Regulatory signs

Background–yellow
Legend, drawing, and border–black

Background–yellow
Legend and border–black

Background–yellow
Legend, drawing, and border–black

(b) Warning signs

Background–green
Legend, arrows, and border–white
Route numbers–black or white

(c) Guide signs

Background–green
Legend and
border–white

Background–blue
Legend and
border–white

(d) Traffic signal head with circular lights and arrows

which would be unsafe at higher speeds (e.g., because the vehicle could not stop or because other vehicles or objects might be in the path). The latter consideration is also important in waterways, especially in the vicinity of docks and water recreation areas. Such speed restrictions are usually indicated by signs along the route placed sufficiently far in advance of the restriction that vehicles have time to slow down.

Similarly, warning signs indicating the increased probability of dangerous conditions are especially common along roads, but are also present along waterways and railway lines. In the latter modes, where those persons who operate the vehicles may be known and limited (i.e., those employed by the carriers using the facility), often special written instructions are given the operators before they use the facility and they may even be instructed in the operation of vehicles over the route before they are allowed to control a vehicle over it. Examples of the various standard road warning signs and road markings used in the United States are pictured in Fig. 5-4.

Vehicle Control on Way Links

In addition to the controls on vehicle movements provided by channelization and speed control are provisions for ensuring that one vehicle does not collide with another traveling on the same path. On way links (between intersections of links) this is primarily a problem of ensuring that vehicles follow one another in such a manner so as to avoid collision. The second vehicle must follow the first at a distance and speed such that it can decelerate as necessary to avoid hitting the first vehicle, which may have to swerve or decelerate to avoid an object in its path. Thus, the term following control is applied to cover this situation and the term following behavior is used in those cases where human behavior is an element in the vehicle control and decision process. Road transportation illustrates this well. Here the driver of each vehicle is responsible for seeing that his vehicle does not collide with any object, and hence he must take into account the location and speed of preceding vehicles as well as the possibility of a vehicle suddenly entering the path from another lane or road. Of course, there are rules governing such entry into a traffic lane, but they are not always obeyed.

In the United States most state governments rule that drivers are to maintain a distance of at least one automobile length (approximately 20 ft) between their cars and the ones in front of them for every 10 mi/h of speed they are traveling. Thus the distance headway of vehicles is one car length plus one more car length for each 10 mi/h of speed, e.g., six car lengths at 50 mi/h. Actual driving has been observed to be somewhat different from what is dictated by this rule. Specifically, under conditions of approximately constant speed flow (i.e., without traffic lights or stop or yield intersections), at headways of very large distances drivers travel at speeds below those suggested by the rule (presumably as limited by the speed limit and safety considerations other than vehicle collisions), and at headways of very small distances they travel at lower speeds. If speed (space-mean speed) is plotted against vehicle concentration, observations of actual driver behavior on freeways (and other roads with long distances between intersections) yield the relationship shown in Fig. 5-5*a*. There is a maximum speed at one extreme and a concentration at which speed effectively approaches zero or a lock-up condition (termed jam concentration) at the other. Plotted on this figure as a dashed line is the relationship as it would be if drivers always maintained the minimum spacing dictated by the safety rule at each speed.

Given the relationship between space-mean speed and concentration as portrayed in Fig. 5-5*a*, the fundamental relationship of Eq. (5-9) can be used to determine the relationships between volume and concentration and between volume and speed. At values of concentration near zero, the volume is

Figure 5-5 Fundamental diagrams of road traffic flow. *(a)* Space-mean speed vs. concentration. *(b)* Volume vs. concentration. *(c)* Space-mean speed vs. volume.

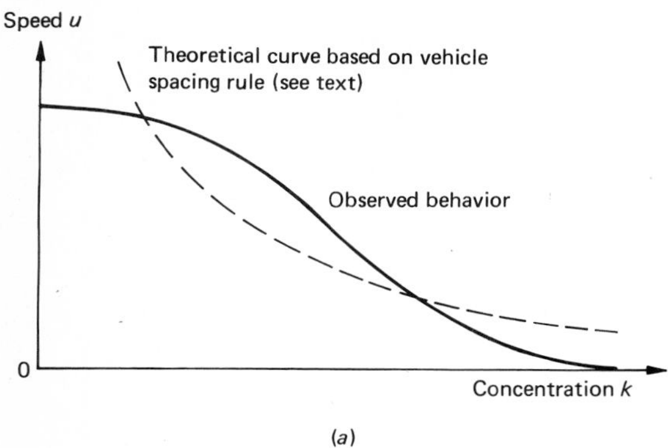

(*a*)

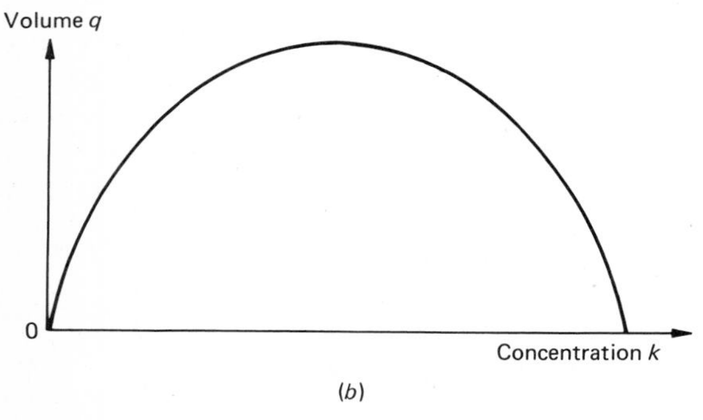

(*b*)

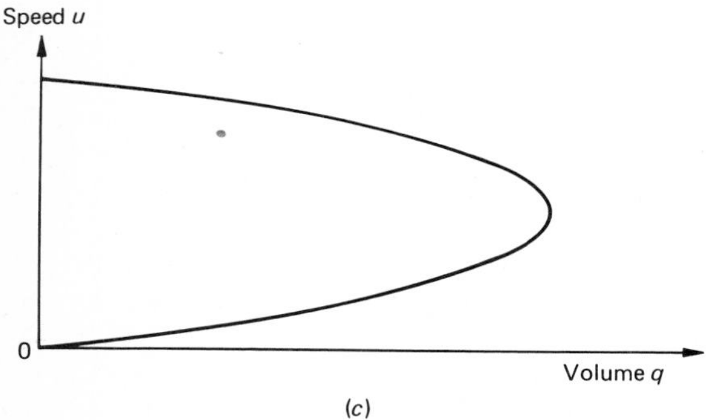

(*c*)

necessarily low despite the high speed because there are so few vehicles. This defines the origin intercept in the volume-concentration curve of Fig. 5-5*b* and the high-speed zero-volume intercept of the volume-speed curve of Fig. 5-5*c*. At the very low speed (approaching zero) of large values of concentration, the volume is also necessarily low, being zero at zero speed. This defines another end condition in each of the two relationships. In the intermediate range of speed and volume, the volume must be different from zero, and it is typically observed to increase to a maximum as shown and then decrease to zero, thus having one maximum value. The three curves of Fig. 5-5 are usually referred to as the *fundamental diagrams of road traffic.* However, it should be remembered that they apply only to flow on roads where the movement of traffic is not interrupted, as it would be by traffic lights or stop signs. Although in the latter case Eq. (5-8) still holds, the relationship of speed to concentration is different as traffic travels through intersections, which results in a form for the curves different from that shown in Fig. 5-5.

The exact values taken by the relationships shown in Fig. 5-5 vary considerably depending upon the road type and features such as speed limit, presence of curves, proximity of obstructions near the running lanes, etc. Also, the presence of trucks, buses, and other nonauto traffic in the stream affects the values. In general, these vehicles have poorer acceleration and deceleration performance than automobiles, and hence the speed at any given level of concentration is lower the greater the fraction of trucks and buses. However, the details of these effects will be covered when relationships resulting from actual traffic measurements are presented in the next section.

The behavior of aircraft is somewhat similar to that of road vehicles since the operators also often rely upon visual detection of other vehicles to avoid collisions. Actually, there are two distinct conditions of flying. Visual flight rules (VFR) are rules pilots follow to avoid other aircraft. VFR are in effect whenever the air traffic controllers determine that atmospheric conditions make such operation safe. In the airspaces described earlier for long-distance flying of the type normally done by commercial air service on jet routes, instrument flight rules (IFR) almost always apply, and often IFR apply to the Victor airways as well. These rules require that the air traffic controllers maintain safe separation between aircraft, and specific rules apply. An abbreviated form of these rules follows (Horonjeff, 1975, p. 96): If radar coverage is available and wake turbulence (relevant for small planes following large ones) is not a factor, at least 5 nautical miles (nmi, 6080 ft in length) must separate two aircraft, except within 40 nmi of the radar antenna where 3 nmi is sufficient. If wake turbulence is a factor, the minimum separation is 5 nmi. The spacing between two heavy aircraft (over 300,000 lb take-off weight) must be at least 4 nmi, and that between a heavy plane following a light one must be at least 3 nmi. When there is no radar coverage, the spacing is also governed by time headway conditions, specifying minima of 3 min if the lead aircraft is 44 kn faster, 5 min if it is 22 kn faster, and 10 min if both aircraft operate at the same speed.

From these distance and time headway rules, the minimum spacing of aircraft on any route can be determined. A useful exercise is to plot the volume-speed and volume-concentration curves for these rules. Since the primary limitation of air traffic is at the airports, relatively little attention has been paid to the actual performance of aircraft routes with respect to the maximum volume of flights which might be accommodated. Of course, the capacity of airports will be covered in the chapter on terminals.

The control of train movements on railroads represents another example of an elaborate transportation control system. This is primarily due to the extremely long distances required to stop railroad trains. As may be recalled from Chap. 4, typical stopping distances of large main-line freight trains are often 1 to 2 mi, depending upon the gradient of the line and the cruise speed. Under these conditions it would be impossible to rely upon visual observation of a train stopped on the track ahead as the basis for brake applications. After a history of many severe accidents in the early years of railroad operation, a system of detecting trains and relaying appropriate information to other trains was devised and is now (in various forms) almost universally used on railways.

The system used is called the block signal system (Association of American Railroads, 1955). All main tracks are divided into sections, termed blocks. The presence of a train or a portion thereof in a block is detected (usually electronically, although manual observation is also possible). Signals along the track are set so as to give approaching trains proper warning of the presence of the train occupying the block, enabling them to stop in time to avoid collision. The application of these concepts can be illustrated by considering the case of a track used for traffic in one direction only, as in Fig. 5-6*a*. A train occupies the block marked "occupied." The signal at the entrance to this block is set at the "stop" indication, usually a red light. If the blocks are sufficiently long that a train can stop in one block length, then the next signal upstream

Figure 5-6 **Railroad block signal systems. *(a)* Three-aspect, two-block system often used on main-line railroads. *(b)* Three-aspect, three-block system often used on rapid transit railroads.**

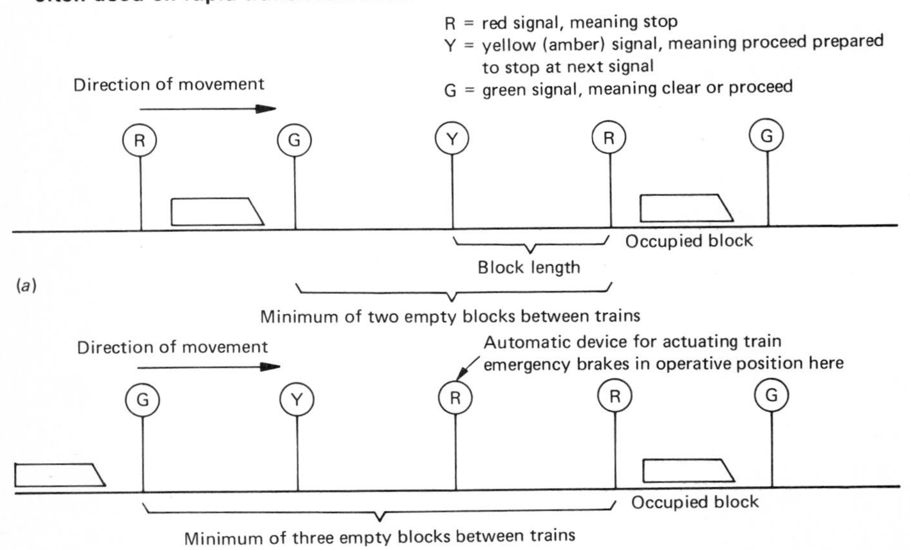

(going in the direction opposite to vehicle movement) can be set at the indication "proceed prepared to stop at the next signal," usually a yellow light. The next signal upstream, at the entrance to the second block removed from the occupied one, can be "proceed," usually a green light. This system will ensure that no train collides with another, provided the block lengths permit stopping (and they are laid out so that this is the case, with longer blocks on downgrades, for example, or reduced speed limits where block lengths cannot be lengthened) and the signal indication is obeyed. Since there are three aspects (different signals, in this case, three different colors) in this scheme, and there are two empty blocks between any clear or proceed signal and an occupied block, this system is called the three-aspect two-block system.

Among the many existing variations, one which is common on rail rapid transit lines includes a simple but effective device for avoiding accidents even if the operator of the train disregards the signals. Wherever there is a stop indication, a device will automatically apply the train's emergency brakes if the train passes the signal. For this to be completely effective there must be an additional stop indication between the successive trains, as shown in Fig. 5-6*b*. This is then termed the three-aspect three-block system. On many main-line railroads other types of automatic stop devices are used which do not require the additional block between trains. Devices for measuring the speed of trains ensure compliance with speed restrictions, with automatic override if the operator does not obey the restriction within a certain period of time. There are many variations of the aspects (e.g., use of two lights simultaneously) which provide more specific indications (e.g., slow to medium speed past next signal) and safely reduce the distances between trains, as described in Armstrong (1957) and Abbott (1975). All of these factors have led to the remarkably low rate of train collisions, although there are a moderate number of accidents due to other causes, such as derailments.

The time-space diagram can be used to analyze the movement of trains under these block signal control conditions. Figure 5-7 illustrates the application of the diagram of this problem. The second train is shown operating as close behind the first train as possible, under the assumption the engineer must observe a clear signal at least 500 ft before passing the signal; otherwise he will apply the train's brakes. The minimum time headway of these two freight trains is 8.4 min, which illustrates why railroads too have capacity problems even though the number of trains operated may seem to be very small. As the problems at the end of this chapter illustrate, this graphical analysis of train following is widely used for the analysis of the capacity of rapid transit lines, where capacity is also often a problem during peak periods, although on such systems trains can often be operated at headways of 90 s to 2 min.

Vehicle Control at Intersections

Almost all vehicluar transport systems have intersections where the channels of vehicles on different way-link paths merge and diverge. The only exceptions are elevators and rather unusual applications of other technologies such as shuttle subway lines. These intersections are very important from the standpoint of capacity and control because they are often the points of bottlenecks and also often present the most difficult accident prevention problems. Most of the discussion will focus upon road and railroad inter-

Figure 5-7 Time-space diagram applied to freight train following with block signal control.

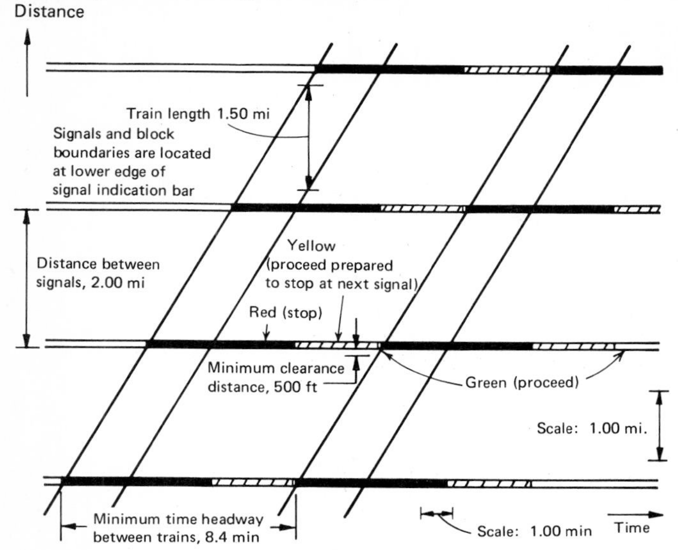

Note: Trains operate at 40 mi/h, from which speed the stopping distance is approximately 3500 ft. Signals are of the three-aspect, two-block type

sections since they present the most serious difficulties. While the intersection of air routes can be treated in a manner analogous to that of rail routes, the most important air intersections occur at airports and will be included in the discussion of air terminal operation in Chap. 7.

There are two basic types of road intersections from the standpoint of vehicle controls: signalized intersections and those which are not signalized, where the driver of the vehicle determines when it is safe to enter the intersection.

The signalized intersection operates in a manner similar to the railway block signal system of control. A traffic light (or lights) with three aspects—green for proceed, yellow for allowing entry into the intersection provided it is cleared before red appears, and red for stop—is operated so as to allow only certain vehicles to enter the intersection at any one moment in time so that collisions of vehicles in conflicting movements are avoided (e.g., turn left on a green light across a stream of traffic only when it is safe to do so). Sometimes additional aspects and indications are used to permit movements while all conflicting movements are prohibited by stop indications, such as a red signal for oncoming traffic which otherwise would conflict with the turn.

The time period required for the traffic lights to pass once through all the different indications displayed is termed the cycle length. The cycle length is divided into a number of phases, within which one or more traffic movements can occur. Most signals are of the fixed-time variety: the duration of each phase is fixed. However, the duration of phases of some lights is set by the amount of traffic or other traffic conditions. For example, in the four-legged intersection of two streets shown in Fig. 5-8*b*, one phase is devoted to movements on street A-C and one to movements on B-D.

Figure 5-8 **Time-space diagram of vehicle movements at a signalized road intersection.** ***(a)*** **Time-space diagram.** **[*From Wohl and Martin (1967), p. 431.*]** ***(b)*** **Intersection layout.**

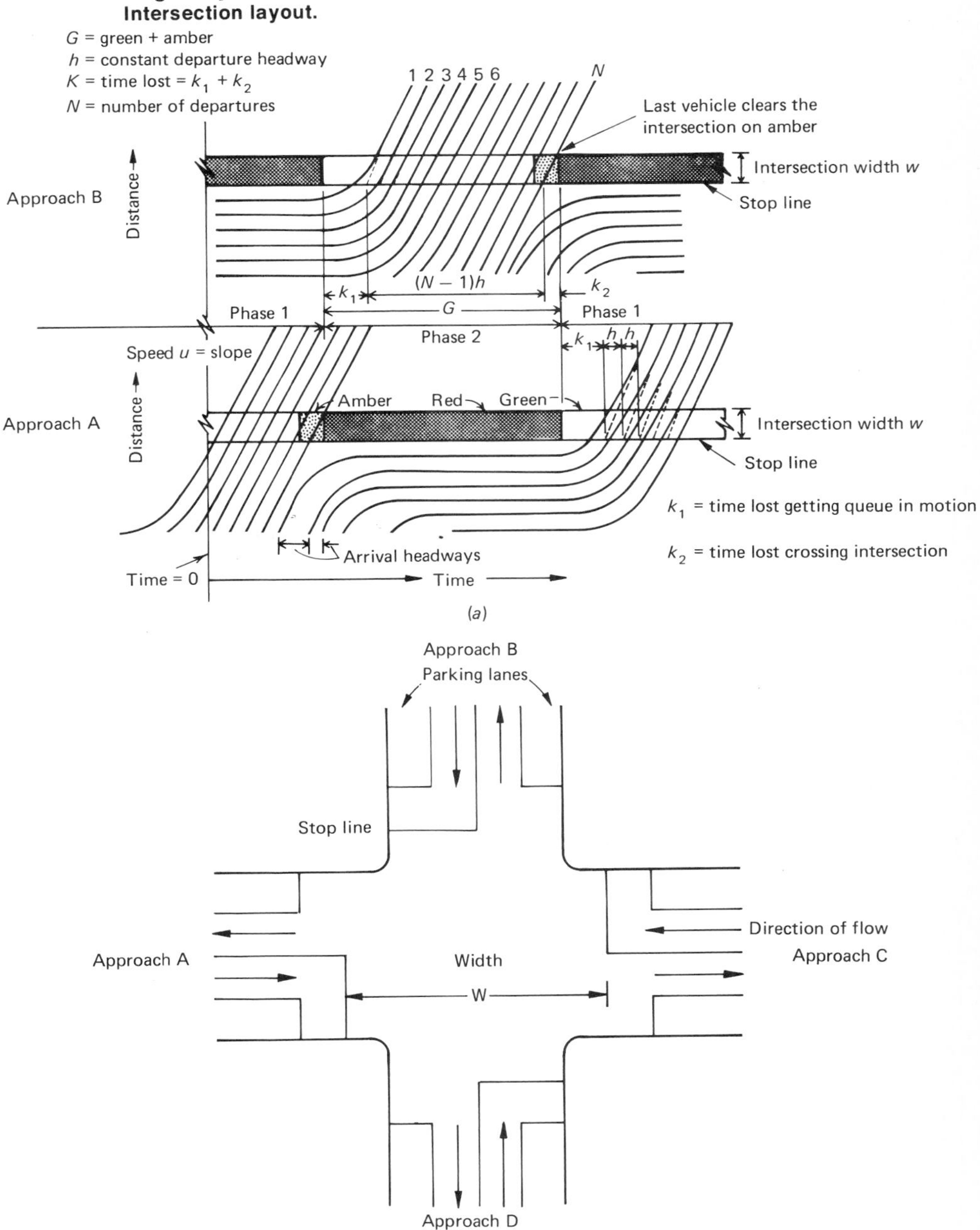

Figure 5-8*a* is the time-space diagram for the traffic moving from two legs of this intersection. The upper part shows the traffic from leg approach B, and the lower part from approach A. An arbitrary time scale is used, with the time $t = 0$ corresponding to the beginning of the green phase (more properly, the green and yellow phases) for traffic on street B-D. This drawing is for an intersection isolated from others, in the sense that the traffic arrives randomly with respect to the time of the green phase. As we shall see in a moment, intersections which are close to one another often can be timed in such a manner that vehicles can travel through them without having to stop for red lights. In the intersection of this figure, many vehicles have to stop. Some green-phase time is in effect "lost" due to the delay of vehicles in accelerating from rest (k_1 in the figure), and, of course, some is lost due to the time required for vehicles to cross the intersection (k_2). In this figure a vehicle is shown passing during the yellow light, a common practice in some parts of the world, especially during high-volume periods.

At isolated intersections such as this one, many vehicles are delayed. Many methods are in use for estimation of this delay. They take into account such factors as the green time, the volume of traffic, the fraction of vehicles making the various possible movements through the intersection (e.g., left or right turns), and the fraction of trucks and buses. Below is presented one of the simpler methods, developed by Webster of the Road Research Laboratory in Britain in 1958. It predicted very well the average delay under various conditions at a few intersections in both the United Kingdom and the United States. The average delay per vehicle is given by the equation

$$d_j = \frac{C(1-\lambda_i)^2}{2(1-\lambda_i x_j)} + \frac{x_j^2}{2V_j(1-x_j)} - 0.65\left(\frac{C}{V_j^2}\right)^{1/3} x_j^{2+5\lambda_i} \qquad (5\text{-}10)$$

where d_j = average delay per vehicle on jth approach during ith phase, s
C = cycle length, s
$\lambda_i = (G_i - k_i)/C$
G_i = green time in ith phase
k_i = lost part of green time ($k_2 + k_1$ in Fig. 5-8)
V_j = volume on jth approach, vehicles/(lane)(s)
s = saturation or maximum possible flow of vehicles during green time, vehicles/(lane)(s)
$x_j = V_j/\lambda_i s$

In the above formula, the notation used by Wohl and Martin (1967, pp. 444–455) has been used, since this reference is more readily available than the original reference (Webster, 1958). Although the formula may appear complex, in application it is rather straightforward, as the following example illustrates. In the intersection of Fig. 5-8, we are interested in the average and total delay of vehicles entering on approach B, i.e., j = B. The cycle length *(C)* is 120 s, and the green time (plus yellow time, assuming both are used) for movements across the intersection from approach B, the second phase in Fig. 5-8 ($i = 2$), is 40 s. The lost part of the green time (k_i) has been estimated as 8 s. The approach volume from B is 300 vehicles/(h)(lane) [and thus $V_j = 0.0833$ vehicle/(lane)(s)], and the saturation flow is estimated as 1 vehicle every 2 s on each lane, so s equals 0.5 vehicle/(lane)(s). Substituting these into Eq. (5-10), we have

$$\lambda_2 = \frac{40 - 8}{120} = 0.267$$

$$x_B = \frac{0.0833}{0.267(0.5)} = 0.624$$

$$d_B = \frac{120(0.733)^2}{2[1 - 0.267(0.624)]} + \frac{(0.624)^2}{2(0.0833)(1 - 0.624)} - 0.65\left[\frac{120}{(0.0833)^2}\right]^{1/3}(0.624)^{2+5(0.267)}$$

$$= 38.68 + 6.22 - 3.48$$

$$= 41.42 \text{ s}$$

Thus the average delay per vehicle is 41.42 s.

As can be seen from inspection of the formula, the average delay per vehicle increases with increasing volume; the delay becomes very large as the volume approaches the saturation flow when all the green time would be required to accommodate the traffic. An example is plotted in Fig. 5-9*a*. In this

Figure 5-9 **Delays to vehicles at intersections. *(a)* Isolated signalized intersection.** [***From Blunden (1971), pp. 78–80.***] ***(b)*** **Intersection with priority traffic stream.** [***From Blunden (1971), pp. 72–73.***]

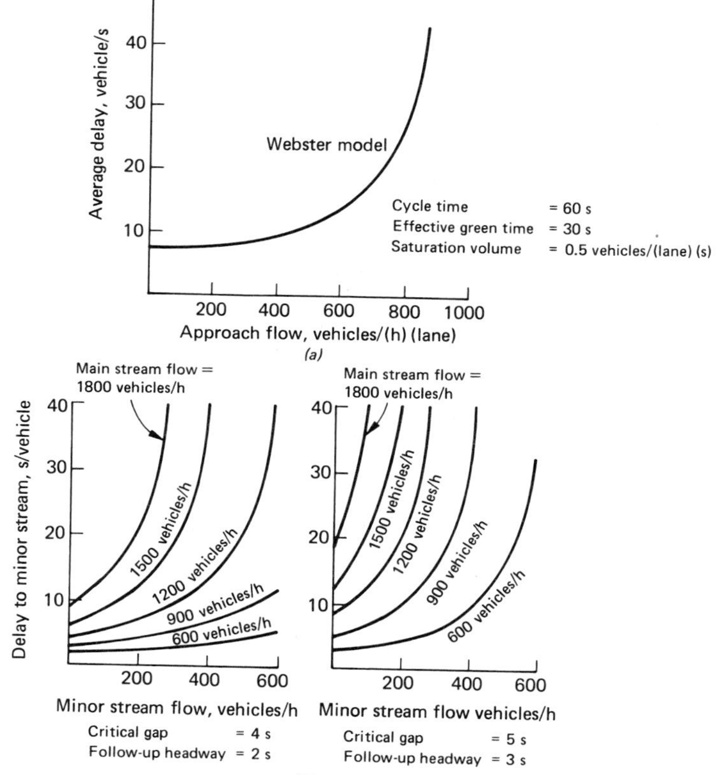

figure, Eq. (5-10) has been simplified by calculating only the first two terms and multiplying the result by 0.9 in lieu of calculation of the third term. This is a reasonable computational simplification for most purposes. In the absence of other data, a saturation flow of 0.5 vehicle/(lane)(s) (one vehicle every 2 s) can be used. For those desiring more detailed information on the delays of vehicles at intersections, Wohl and Martin (1967), Blunden (1971), and Matson, Smith, and Hurd (1955) provide much information on other intersection types as well as on other influencing factors.

Another important type of intersection is that at which there is no traffic signal but at which traffic on one of the roads has the right of way. This characterizes many intersections, not only of roads but also driveways of shopping centers and other major activity centers. The methods used to determine average delays to the minor-stream vehicles (those required to wait until it is safe to enter the major stream) are based on treating the traffic in the probabilistic manner which actually characterizes most road traffic. In particular, the delays are based on the assumption that the time headways between vehicles on the major road vary in a manner consistent with random arrivals of the vehicles at a point on the road. This key assumption will be discussed in more detail in connection with the analysis and design of terminals (Chap. 7). The other assumption is that the driver of a vehicle entering the main road will only enter when he or she finds a headway between vehicles of at least a certain amount—4 to 5 s being commonly observed minimum required headways, or gaps, as they are termed by traffic engineers. Smaller gaps, in the vicinity of 2 to 3 s, will be accepted by a driver following another car entering the stream. For varying volumes of traffic on the main road, the average delay per vehicle can be calculated. The results for a required minimum headway for the first-entering vehicle of 4 and 5 s and for the second or following vehicle of 2 and 3 s are given in Fig. 5-9*b*.

Turning now to the treatment of intersections of railway lines, we first note that virtually all main-line intersections are controlled by some type of signaling, again due to safety considerations and the long stopping distances of trains. Entrance into the intersection is usually governed by block signals which are interlocked in such a manner that the signal is clear (or proceed) on only one approach path and stop on all others. (From this comes the railway terms "interlocking" and "interlocking plant," which are often used to describe railway junctions.) The analysis of these is performed by the use of the time-space diagram in much the same manner that it is used to analyze movement on a single line, as in Fig. 5-7. However, rather than appearing on every block in front of the following train, the "preceding" train now will occupy the block in which is located the turnout (track section in which two tracks converge or diverge) or crossing, and only certain other blocks. It remains in front of the following train if both trains converge from different lines onto the same line, but it vanishes if the two trains continue on different routes, as in the case of the crossing of two lines or the diverging of the two routes followed by trains at a turnout.

The same method can be used in the analysis of intersections of paths of aircraft, vessels, or any other type of vehicle, although as mentioned earlier, the problems in these modes are best treated in the context of the entire terminal facility, where the congestion on merging paths is a problem.

CAPACITY AND LEVEL OF SERVICE

There are two important characteristics of the flow of vehicles through links and intersections. One of these is the *capacity* or maximum volume which the link or intersection can accommodate. As the analysis of vehicle motion in the preceding sections has revealed, the number of vehicles on a path has a profound effect on the speeds at which those vehicles can safely operate, and this in turn, through the fundamental relationship of flow, affects the volume. Thus while there may be a maximum volume which a facility can accommodate, it is essential to know the relationship between speed and volume for any practical transportation work, since speed is such an important characteristic of quality transportation service.

Volume and speed also affect many other important characteristics of transportation service, such as the probability and severity of accidents, which one would expect to increase with increasing speed and increasing volume. The question arises as to what the relationship would be on a given facility where speed and volume are themselves related. Also, as traffic volume on a facility increases and it becomes increasingly difficult to maintain a constant speed, a vehicle driver will experience greater fatigue and scheduled carriers will have possibly greater difficulty in meeting planned running times. Thus there are many other important aspects of vehicle flow in addition to travel time or speed, the aspect on which we have been focusing. The entire set of these factors is usually termed *level of service*. While it would be impossible to list all the aspects of travel on a link or intersection that might be of interest or importance to users of that link, a partial list is

Travel time (or speed)

Reliability (variations in total time)

Comfort

Safety (freedom from damage for freight)

Cost (to be covered in Chap. 9)

Thus it is desirable to know the relationship between volume and other characteristics of the flow of vehicles and the aspects of level of service as discussed above.

The best means for ascertaining such relationships is to observe actual transportation facilities in operation. The variables and relationships would come from preexisting knowledge and careful deduction from theories of vehicle behavior. Such aids to analysis as the time-space diagram might also help. The approach would be that described in the first chapter in the section entitled the scientific method. It is important to test any relationships against actual data on real systems for it is all too easy to assume an ideal flow which could not be achieved in practice except under very unusual conditions. If the relationships used were in error, then substantial errors might be made in the design and operation of transportation facilities. In an activity as important as transportation is to the functioning of society and the well-being of persons in it, the use of idealized or purely theoretical relationships can only be tolerated when there is no other information on the problem being addressed! Fortunately this is not the case for most transportation work.

The amount of research on capacity and the relationship between volume and level of service in general varies considerably among the technologies of transportation. Since poor level of service and capacity problems have probably been most severe in highway transport, much work on capacity has been undertaken in that context. Recently much research has been undertaken on railroad capacity, largely because the traffic on many main lines built many decades (often a century or more) ago is now approaching capacity, and large expenditures may be necessary to alleviate bottlenecks. In water and air transport the capacity problems are almost always located at terminals, and hence most capacity relationships must be discussed in the context of terminal design and operation. Thus the majority of the actual relationships to be presented on capacity, volume, and level of service in this section will be drawn from road transport and to a lesser extent rail transport, but the same principles and methods apply to the other technologies.

In the presentation of actual way-facility capacity and level-of-service relationships to be presented below, the reader will notice that the relationships often describe flow over sections of roads or other modal paths which include both links and intersections. For example, the relationships for freeways describe sections which include all lanes (in one direction) both between and at the locations of the entrance and exit ramps, these being the intersections on that type of facility. This combined treatment is generally necessary, because the functioning of the intersections affects the links connecting them and vice versa. However, if intersections are in fact isolated from one another, then the links and the intersections can be treated distinctly.

Level of Service

Because of the importance of speed or travel time to all users of the transportation system and the ease of measuring it, more work has been done on it than on any other. In particular, since much knowledge has been developed for roads, they will be described first. Most of the concepts are general and can be easily applied to other modes.

Highway engineers have long recognized that speed is not the only important level of service variable. As a result, they have attempted to develop a comprehensive measure of level of service, which includes the following factors (*Highway Capacity Manual,* 1965, p. 78):

Speed or travel time

Traffic interruptions or restrictions (e.g., stops per mile, delays, suddenness of speed changes)

Freedom to maneuver

Safety (accidents and potential hazards)

Driving comfort and convenience

Economy (cost of operating the vehicle)

The determination of all of these factors by quantitative measures is now a practical impossibility, although researchers, mainly psychologists, are trying to quantify such items as comfort and driving tension. Once all of these factors can be quantified and the relative importance of each to drivers (and passengers) measured, then an overall measure of service on roads can be established. In the absence of such a capability at the present time, highway engineers use two measures of the level of service of roads. One is the speed, or travel time, necessarily an average, and usually the space-mean speed. The second measure is the ratio of actual traffic volume to the capacity of the road, the capacity being the maximum volume of traffic which the road can accommodate. The ratio of volume to capacity is felt to be closely related to the many nonquantifiable level-of-service characteristics in the list above, and this feeling is partly supported by surveys of drivers, measurement of driving tension, etc.

Level of service is specified on an interval scale consisting of six levels. These are designated A, B, C, D, E, and F, with A being the highest level of service. As the volume increases, the level of service decreases, a result of generally poorer traffic flow in terms of the characteristics of service specified in the list above. For example, as volume increases, speed typically decreases, freedom to maneuver decreases due to the presence of more vehicles, driving comfort decreases due to the need to watch the movement of more vehicles as well as closer proximity to them. The general relationship between level of service and capacity is shown in Fig. 5-10. The point at which a change is made in the level of service, as from A to B, is arbitrary, but based upon collective engineering judgment. The definitions of each of the levels of service are also given in the figure. In those definitions the term service volume means that for design purposes a facility should be built with suffi-

Figure 5-10 **Generalized relationships among speed, level of service, and volume-to-capacity ratio for roads. *(a)* Graphical portrayal of relationship. *(b)* Descriptive characteristics of level of service.** [***From Highway Capacity Manual (1965), pp. 80–81.***]

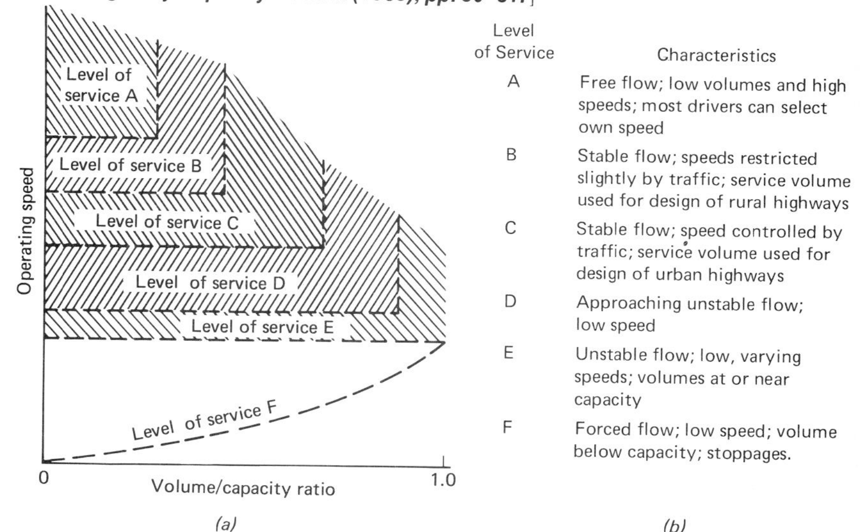

Level of Service	Characteristics
A	Free flow; low volumes and high speeds; most drivers can select own speed
B	Stable flow; speeds restricted slightly by traffic; service volume used for design of rural highways
C	Stable flow; speed controlled by traffic; service volume used for design of urban highways
D	Approaching unstable flow; low speed
E	Unstable flow; low, varying speeds; volumes at or near capacity
F	Forced flow; low speed; volume below capacity; stoppages.

cient capacity to ensure that the volume of traffic always yields a volume-to-capacity ratio such that the desired level of service (e.g., B) is not violated. It is important to notice that the level of service does not necessarily correspond to a particular speed, so one measure is not a replacement for the other.

Freeways and Highways

With these concepts in mind, it is now possible to turn to the actual relationships describing volume, speed, and level of service on roads. Freeways are facilities with separate roadways for traffic in each direction, with no cross streets at grade intersections, which results in complete access control via on- and off-ramps. Expressways can have a limited number of intersections, for very low volume intersecting roads (e.g., farm roads in rural areas). Figure 5-11 presents the relationships for freeways and expressways. Many factors on this figure require explanation. First, the average highway speed refers to the speed for which the road is designed. This is usually the speed at which traffic would move at extremely low volumes, and it is also usually close to the speed limit (sometimes above it). Assuming a facility with a given capacity and average highway speed, the lower axis could be replaced by volume alone with simply a change of scale. Notice that the curve would then be of the shape described earlier for the speed-volume relationship, one of the

Figure 5-11 **Relationship between volume-to-capacity ratio and operating speed, in one direction of travel on freeways and expressways under uninterrupted flow conditions.** [***From Highway Capacity Manual (1965), p. 264.***]

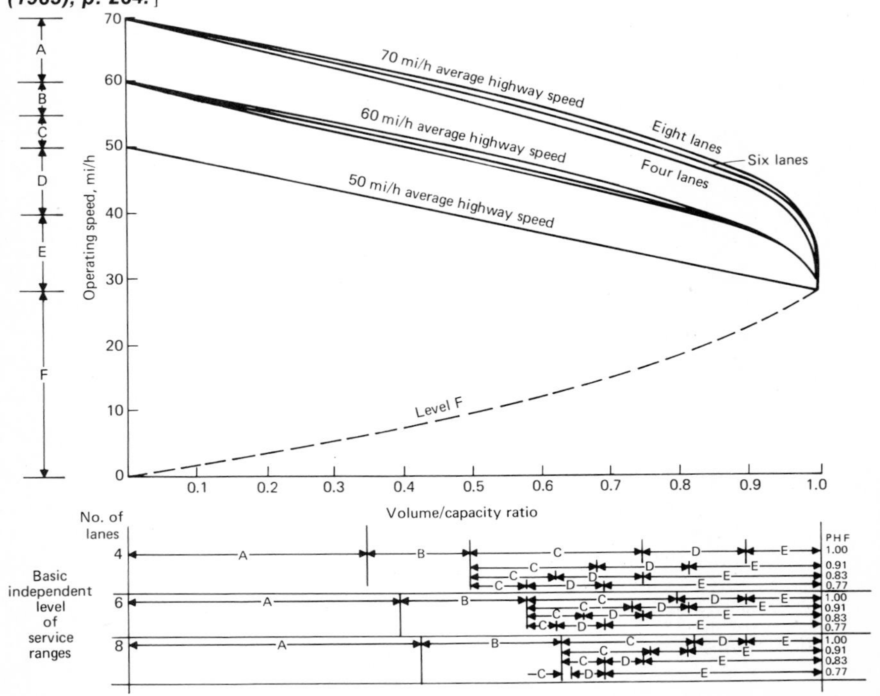

fundamental diagrams of road traffic. The level of service specified under the lower axis varies with what is termed the peak hour factor (PHF). The peak hour factor for freeways is the ratio of the volume during the entire peak hour to the volume during the busiest (highest-volume) 5-min period in that hour. For example, consider a freeway which accommodates 4000 vehicles during the peak hour. During the peak 5 min within that hour, it accommodates 400 vehicles, resulting in a volume during that period of

$$\frac{400 \text{ vehicles}}{5 \text{ min}} \frac{60 \text{ min}}{\text{h}} = 4800 \text{ vehicles/h}$$

The peak hour factor is therefore

$$\text{PHF} = \frac{4000 \text{ vehicles/h}}{4800 \text{ vehicles/h}} = 0.833$$

If the use of a volume measured in vehicles per hour for a period of only 5-min duration seems strange, or even incorrect, remember that the volume is a rate of flow. All we did was to change the units from those implied in the way the flow for the 5 min was given, of 400 vehicles/5 min, to units of vehicles per hour. This in no way implies that the flow occurred for a full hour at that rate. In fact, the peak hour factor would be the same regardless of how we measured the volumes, whether in vehicles per hour, vehicles per minute, or even vehicles per day (although, of course, such units are rare).

The other aspect of the graphical relationship in Fig. 5-11 needing explanation is the determination of capacity. This is actually a very complex task for a road, so only the basics will be presented here. (One book, widely used in the United States, the *Highway Capacity Manual* of 1965, is 397 pages in length, and supplementary books have been written on how to use it. This provides an idea of the depth and complexity of this subject!)

The determination of freeway and expressway capacity begins with the capacity of 2000 vehicles/lane under ideal design conditions and with all vehicles being passenger cars. Ideal design characteristics include lane widths of 12 ft or more, at least 6 ft between the nearest obstruction and a traffic lane (e.g., a bridge abutment), and a level profile. If trucks or buses are present, the capacity as measured in vehicles per hour will be reduced. One bus or one truck is equivalent to more than one automobile on level roads, and the equivalent number of automobiles increases with the severity of the gradients encountered. This is due to the length of trucks and buses and their relatively poor performance in acceleration and braking, as described in Chap. 4. On upgrades such vehicles usually cannot maintain the speeds that automobiles are capable of, and on downgrades they often travel more slowly for safety reasons due to their poorer braking capabilities. Table 5-1 presents the passenger car equivalents of buses and trucks in various generalized terrain conditions.

The capacity of a freeway or expressway can be calculated using the formula below:

$$c = 2000\, NWT_c B_c \tag{5-11}$$

where c = capacity, mixed vehicle types, total for one direction, vehicles/h
N = number of lanes in one direction

W = adjustment factor for lane width and clearance, typically 0.9 to 1.0
T_c = truck adjustment factor
B_c = bus adjustment factor

The estimation of T_c and B_c is by the equations

$$T_c = \frac{100}{100 - P_T + E_T P_T}$$
$$B_c = \frac{100}{100 - P_B + E_B P_B} \qquad (5\text{-}12)$$

where P_T = percentage of trucks
P_B = percentage of buses
E_T = passenger car equivalent of trucks (Table 5-1)
E_B = passenger car equivalent of buses (Table 5-1)

Once the capacity of a freeway is known through use of these equations, then this value can be used in the analysis. The analysis is usually of one of two types. On one hand, with the volume of traffic on a road known, it may be desirable to ascertain the level of service, perhaps in order to determine whether or not the facility is adequate to accommodate the traffic at an acceptable level of service. For example, a freeway of three lanes, with no unusual characteristics (i.e., ideal design), and a 60 mi/h average highway speed, accommodates 2000 vehicles/h during the peak period, including 10 percent trucks and 10 percent buses. It is in rolling terrain. The traffic in the peak 5 min is 200 vehicles. The capacity of this facility is, by Eq. (5-11):

$$c = 2000(3)(1.0)(0.76)(0.83) = 3785 \text{ vehicles/h}$$

using $$T_c = \frac{100}{100 - 10 + 4(10)} = 0.76$$

$$B_c = \frac{100}{100 - 10 + 3(10)} = 0.83$$

Table 5-1 Average Generalized Passenger Car Equivalents of Trucks and Buses on Freeways and Expressways over Extended Section Lengths (including Upgrades, Downgrades, and Level Subsections)

Level of service	Equivalent	Equivalent for: Level terrain	Rolling terrain	Mountainous terrain
A		Widely variable; one or more trucks have same total effect, causing other traffic to shift to other lanes. Use equivalent for remaining levels in problems.		
B through E	E_T, for trucks	2	4	8
	E_B, for buses†	1.6	3	5

† Separate consideration not warranted in most problems; use only where bus volumes are significant.
Source: *Highway Capacity Manual* (1965, p. 257).

The E_T and E_B are trial values, assuming the level of service will be B or less. To determine the level of service it is necessary to know the peak hour factor:

$$\text{PHF} = \frac{2000 \text{ vehicles/h}}{(\frac{200}{5})\ 60 \text{ vehicles/h}} = 0.833$$

The volume-to-capacity ratio is

$$\frac{v}{c} = \frac{2000}{3785} = 0.528$$

Now turning to Fig. 5-11, with a total number of six lanes (three in each direction), the level of service is B. Thus it exceeds the desired standard for level of service for urban freeways, level of service C, as specified in the text of Fig. 5-10. From Fig. 5-11 it can be seen that the speed of this traffic would be about 47 mi/h.

The other use of these relationships and methods is to determine the maximum volume that a given road can accommodate at a given level of service. This volume is often called service volume (SV), as distinguished from the actual volume which might be using the road. The approach is essentially the same as for the previous problem, except Fig. 5-11 is used to determine the volume rather than the level of service or the speed. Taking the same road as in the previous problem and now trying to find the service volume for level of service C, we first calculate the peak hour factor, determined above to be 0.83. From Fig. 5-11 it is determined that the service-volume-to-capacity ratio is 0.67. Given the truck and bus factors of 0.76 and 0.83, respectively, we now use Eq. (5-13) to ascertain the service volume with this mix of vehicles:

$$\text{SV} = 2000\, N \frac{v}{c} W T_c B_c \tag{5-13}$$

where SV = service volume at v/c

Substituting, we have

$$\text{SV} = 2000\ (3)(0.67)(1.0)(0.76)(0.83) = 2536 \text{ vehicles/h}$$

This is the maximum volume that could use the road, with this mix of vehicles and the level of service remaining C.

The calculation of level of service, speed, and service volumes on other types of roads is similar. Figure 5-12 presents the relationship between speed and volume-to-capacity ratio for multilane rural highways. For the multilane highways the automobile capacity of each lane is the same as for freeways, 2000 vehicles/h. This must be reduced if lane widths are below 12 ft or obstructions are closer than 6 ft. Otherwise the method is the same as for freeways, with the exception that now Fig. 5-12, relating speed, capacity, and level of service, must be used. As would be expected for such facilities, speeds are lower at the same v/c ratio than for freeways.

Figure 5-13 provides the relationship for two-lane rural roads, with one lane in each direction. The capacity of such roads, only 2000 vehicles/h (in both directions) is far less than the multilane facilities. The passenger car equivalents of trucks and buses are much greater for these facilities than the multilane facilities because of the reduced opportunity for other automobiles to pass slower vehicles. These are given in Table 5-2. Otherwise the method

Figure 5-12 **Relationship between volume-to-capacity ratio and operating speed, overall for two directions of travel on multilane rural highways under uninterrupted flow conditions.** [***From Highway Capacity Manual (1965), p. 294.***]

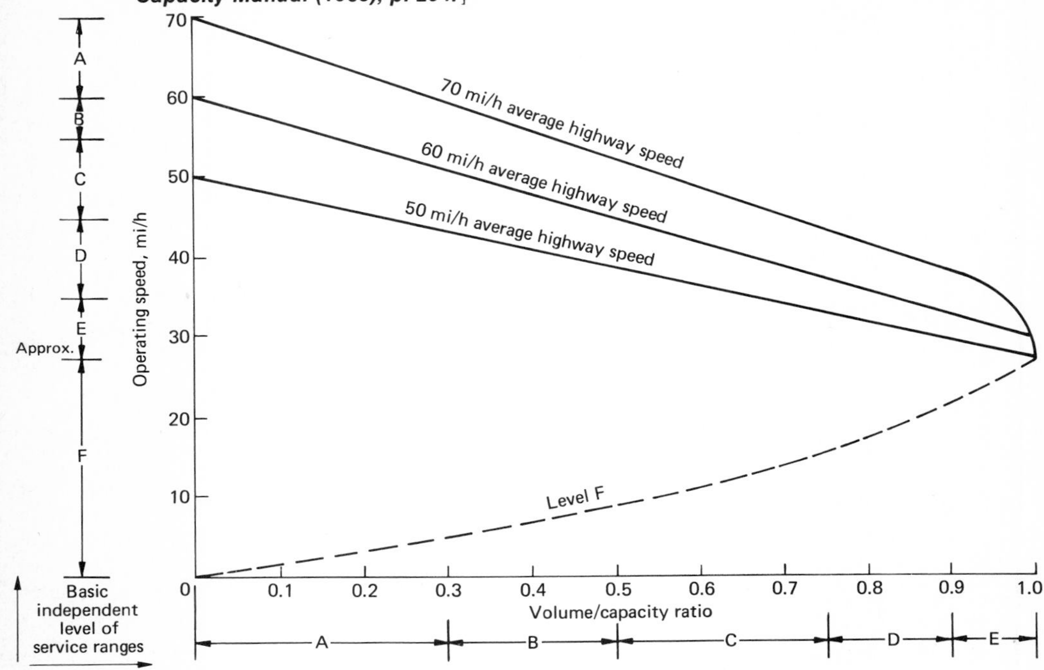

for calculating speed, capacity, service volume, and level of service is identical to that used for the other types of facilities.

Arterial streets are the final type of major road facilities for which capacity and level of service are often a problem. Arterials, the main roads of many areas, are surface streets, usually with no access control (i.e., driveways can be at any location landowners desire) and often intersect with other major and minor streets. Usually all major intersections are controlled by traffic lights, and sometimes all intersections are, especially where volumes are so great that side-road traffic would have little chance to cross the arterial without traffic signals.

Often these intersections with traffic signals are so close to one another that it is only logical to consider them together. The vehicles leaving one signal during the green phase are bunched together in a wave or platoon, as such concentrations of vehicles are often termed. These vehicles then reach the next signal still in such a platoon. Ideally that next signal would be timed in such a manner that the platoon of vehicles would just reach the signal as the green phase appears there, allowing the platoon of vehicles to pass through that intersection uninterrupted. This relationship between the signals at adjacent intersections would be continued to successive signalized intersections. Therefore the traffic would never have to stop, once a green

Figure 5-13 **Relationship between volume-to-capacity ratio and operating speed, overall for two directions of travel on two-lane rural roads with an average highway speed of 45 mi/h under uninterrupted flow conditions.** [*From Highway Capacity Manual (1965), p. 311.*]

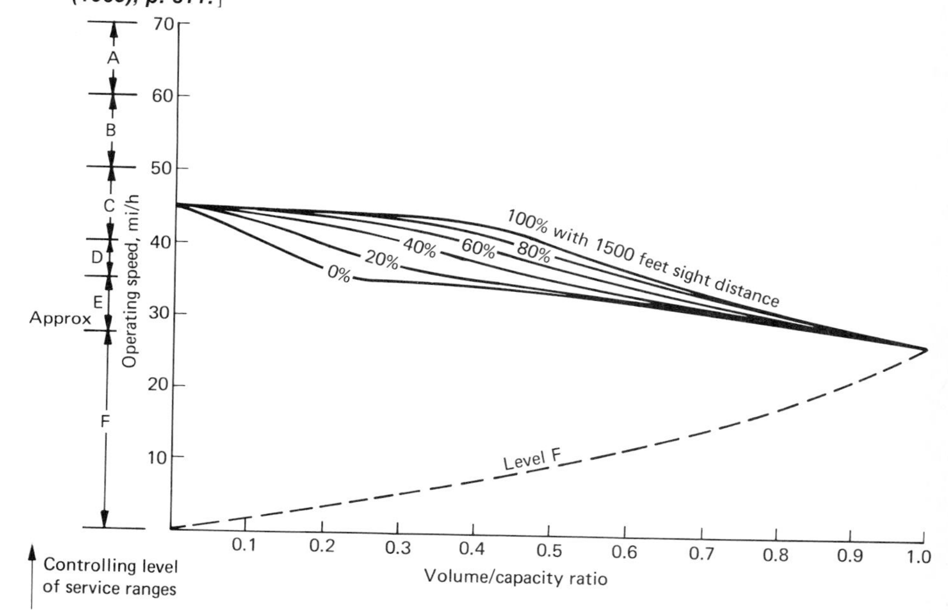

light is obtained, at least in theory. The movement of vehicles in this manner is termed a green wave.

Figure 5-14 illustrates the timing of signals to achieve a green wave (Matson et al., 1955, pp. 343–350). The street shown is one-way, so only movement in one direction need be considered. The key to the formation of a green wave is the relationship of the green phases of adjacent signals. The time from the moment the green phase begins at one signal to the moment it begins at the

Table 5-2 **Average Generalized Passenger Car Equivalents of Trucks and Buses on Two-Lane Highways, over Extended Sections (including Upgrades, Downgrades, and Level Subsections)**

		Equivalent for		
Equivalent	**Level of service**	**Level terrain**	**Rolling terrain**	**Mountainous terrain**
E_T, for trucks	A	3	4	7
	B and C	2.5	5	10
	D and E	2	5	12
E_B, for buses†	All levels	2	4	6

† Separate consideration not warranted in most problems; use only where bus volumes are significant.
Source: *Highway Capacity Manual* (1965, p. 304).

Figure 5-14 **Time-space diagram of coordinated traffic signals on a one-way street.** [***From Matson, Smith, and Hurd (1955), p. 344.***]

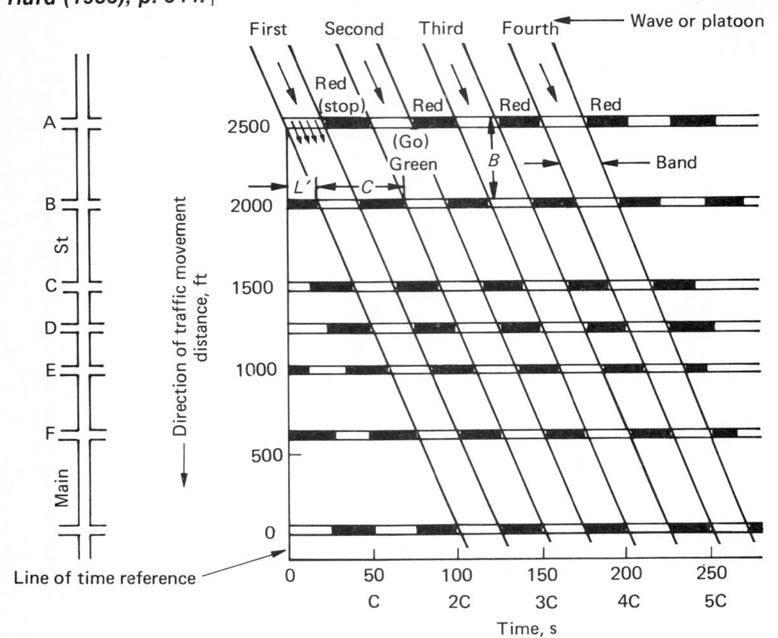

next signal is defined as the *offset*. Thus in order for the green wave, and hence the traffic, to travel at a constant speed V along this street, the offset must be as given by the formula

$$V = \frac{B}{L} \tag{5-14}$$

where V = speed of green wave
B = block length (between adjacent intersections)
L = signal offset

Notice that any desired speed can be achieved by selecting an appropriate offset. Varying block lengths on a one-way street create no problems for the timing.

Actually, for any given offset and cycle length, traffic could move without stopping at any of the speeds given by

$$V_j = \frac{B}{L + jC} \tag{5-15}$$

where C = cycle length
j = any integer $\geqslant 1$

This can be verified by inspection of Fig. 5-14.

In Fig. 5-14 the green phase of all signals is the same length. This is because the volumes are assumed to be the same or at least very nearly so at all points along the street. The duration of the green phase must be adequate to accommodate the volume, a condition that can be checked by use of Eq. (5-16):

$$q_{\max} = \frac{GS}{C} \tag{5-16}$$

where $q_{\max}$ = maximum volume the phase can accommodate
G = effective green time per cycle
S = saturation flow rate

If the street is two-way and the block lengths are uniform, then there is no problem in achieving a constant-speed green wave. An example is shown in Fig. 5-15. Here the speeds are shown as equal in the two directions, although they can be made unequal if desired. For the speeds to be equal, the offset must be a multiple of one-half the cycle length, $\frac{1}{2}C$. The speed is given by the formula

$$V_j = \frac{B}{j(\frac{1}{2}C)} \tag{5-17}$$

If j is an even integer, the pattern is said to be a simultaneous pattern; and if j is an odd integer, it is said to be an alternate pattern. Both are shown in the figure.

In this case, for any given cycle length there is a limited number of possible speeds. However, by varying the cycle length, any desired speed can be achieved, but it is then necessary to vary the green time in each cycle to ensure that the appropriate capacity is being provided.

As an illustration of the effect of poorly timed or uncoordinated signals on the traffic flow, Fig. 5-16 is presented. It shows a two-way street with irregular

Figure 5-15 **Time-space diagram of coordinated traffic signals on a two-way street.** [***From Matson, Smith, and Hurd (1955), p. 346.***]

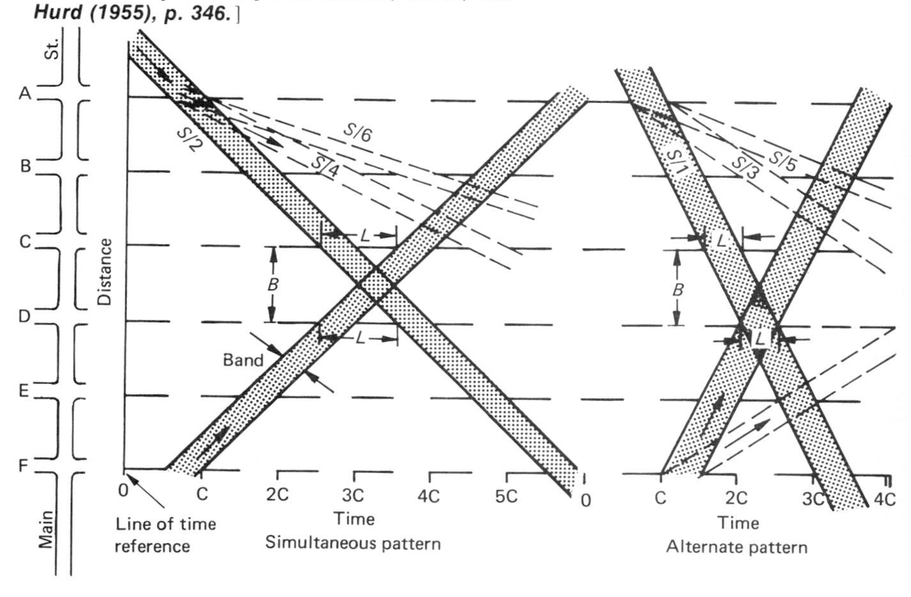

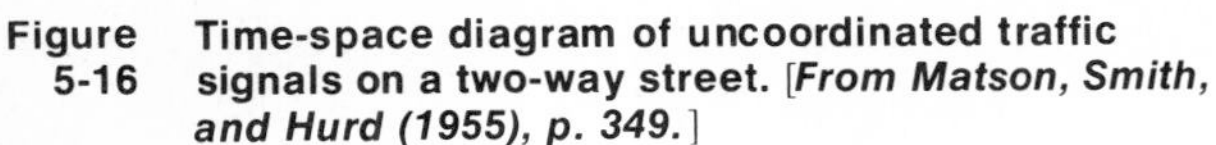

Figure 5-16 **Time-space diagram of uncoordinated traffic signals on a two-way street.** [*From Matson, Smith, and Hurd (1955), p. 349.*]

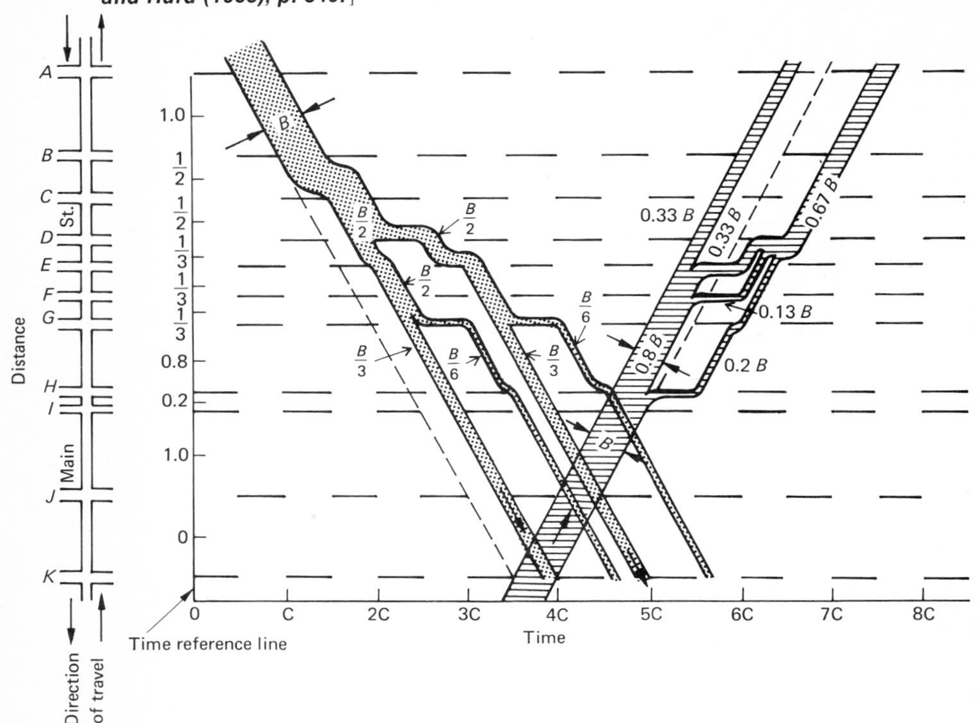

block lengths. The signals are timed and offset in such a manner that any platoon through the first light encountered is broken up at succeeding intersections. This has the effect of increasing substantially the travel time. Also, it may reduce capacity, because of the unused portions of green time at some signals. While coordination of signals on arterial streets is not always possible, it is certainly desirable to make the attempt!

The time-space diagram applied to road intersection coordination assumes an idealized behavior on the part of drivers, but it has been found to give very good results. While there are deviations from ideal platoon flow, they usually are not so severe as to destroy the validity of the results for signal offset and cycle time determination purposes. However, it is useful to examine some empirically based speed relationships.

Generalized relationships among level of service, speed, and the volume-to-capacity ratio have been developed for arterial streets. These are very similar to the relationships for various kinds of links on which flow was uninterrupted, as described in the previous section.

Figure 5-17 presents a curve relating these three factors for arterial streets. The capacity used here is similar in concept but slightly different from that used for freeways and rural highways. The capacity is the maximum volume that can be accommodated during the period when the road in question has a green phase or is free of other predictable interruptions. The calculation of

Figure 5-17 **Typical relationship among speed, level of service, and volume-to-capacity ratio for urban and suburban arterial streets.** [***From Highway Capacity Manual (1965), p. 320.***]

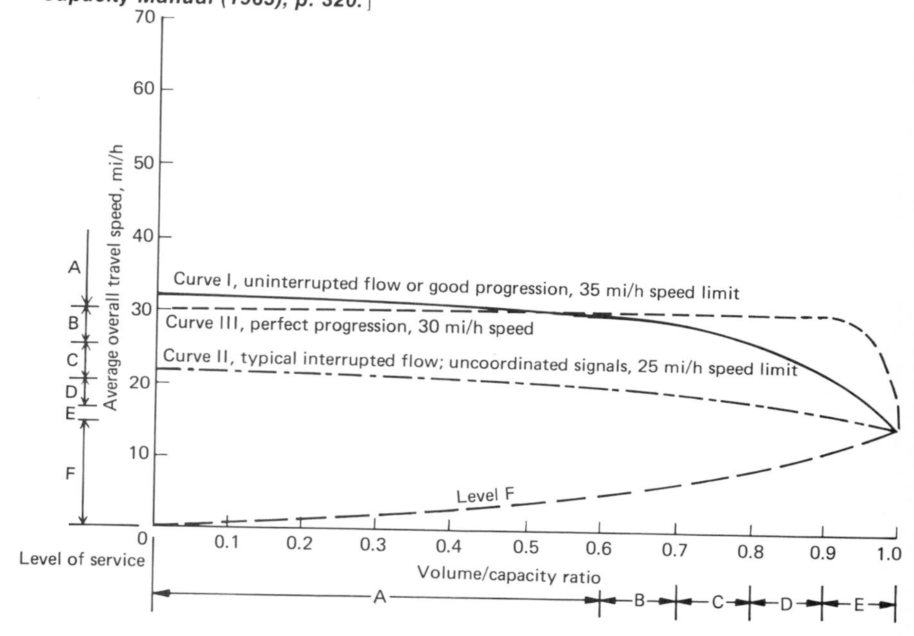

this capacity is much more complicated than that for freeways, so much so that we will only mention the factors to be considered here. (An entire chapter would be required to cover these in the quantitative form necessary to calculate capacity.) The major factors are number and width of lanes, lane assignment to various maneuvers, locations of bus stops, cycle length and allocation to various maneuvers, peaking of traffic, mix of vehicles, size of city and location of intersection within that city, proximity to other intersections or flow-limiting sections, and interaction effects. However, capacity can be approximated by use of Fig. 5-18, in which capacity is related to street width and a few other major influences. These curves are based upon observation of hundreds of intersections throughout the United States, as reported by Normann (1959).

Figure 5-17 indicates that the performance of intersections can vary considerably. Ideal progression yields consistently high speeds, but less than good progression yields a noticeable decrease in speed with increasing volume, as would be expected.

Lest one think that the discussion above includes all or even most of the methods of determining level of service and volume characteristics of roads, consider for a moment what has not been covered. Weaving sections, such as at intersections of freeways, have not been discussed nor have the detailed methods used to treat grades and curves, especially those with restrictive sight distances, nor have the effects of parking and numerous driveways among many other factors. But the rudiments of highway capacity analysis have been presented so that some practical problems can be solved and so

Figure 5-18 **Capacities of various types of street intersections.** ***(a)*** **Two-way streets.** ***(b)*** **One-way streets.** [***From Norman (1959), p. 20.***]

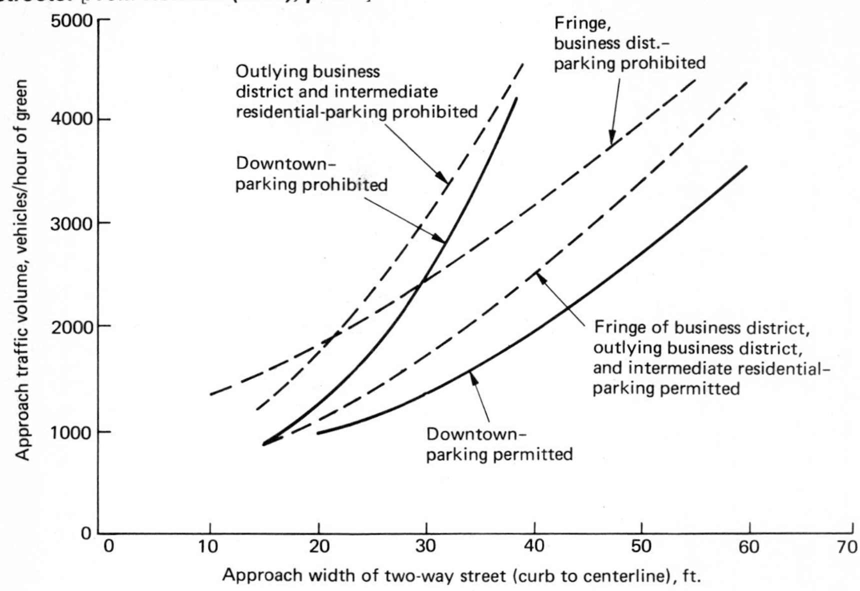

(*a*)

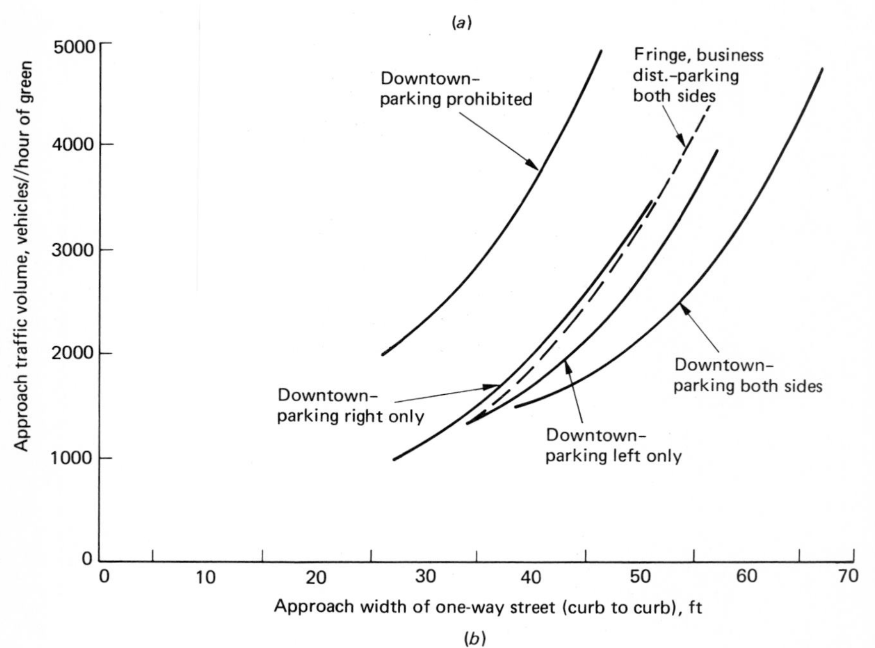

(*b*)

the reader is certainly prepared to read any standard highway capacity manual with facility.

Railroads

The relationships for railroad capacity are surprisingly similar to those developed for road capacity. However, here the primary emphasis is on travel time or speed, to the virtual exclusion of other level-of-service factors. This limitation is probably not very severe, because trains are rarely so close together that there would be a substantial amount of start and stop movement, and, of course, many of the factors are of little relevance, such as the ease of maneuvering and risk of vehicle accidents (although they occasionally do occur). Figure 5-19 presents curves relating the average delay per train in operating over a 100-mi line to the volume of trains using the line per day. The delay is the time required for the train to traverse the line beyond the time that would be required if the train were able to operate at its normal cruise speed. Since trains of many different speed characteristics typically use any single railroad line in a day, these curves were developed with a mix of train types. This mix, given in the figure, specifies the length and weight of trains, their scheduled average speeds from one end of the line to the other, and the delays due to various maneuvers. Faster trains have priority over slower ones, i.e., type 1 has priority over type 2, etc. Of course, the delays will depend upon the mix, but this is a representative one. All of these trains are freight trains.

Figure 5-19 Example of relationship between volume and delay for a railroad line with various track configurations. *(a)* Mix of trains and their characteristics. *(b)* Volume-delay curves. [*From Prokopy and Rubin (1975), pp. 16 and 27.*]

Class	Length, ft	Average Running Speed, mi/h	Time Penalties (Minutes)			Percent of all Trains
			Starting	Stopping	Crossovers	
1	1,500	50	1	2	1	13.9%
2	3,000	40	2	6	3	27.8%
3	5,000	25	4	8	5	52.8%
4	3,000	25	2	5	3	5.5%

(a)

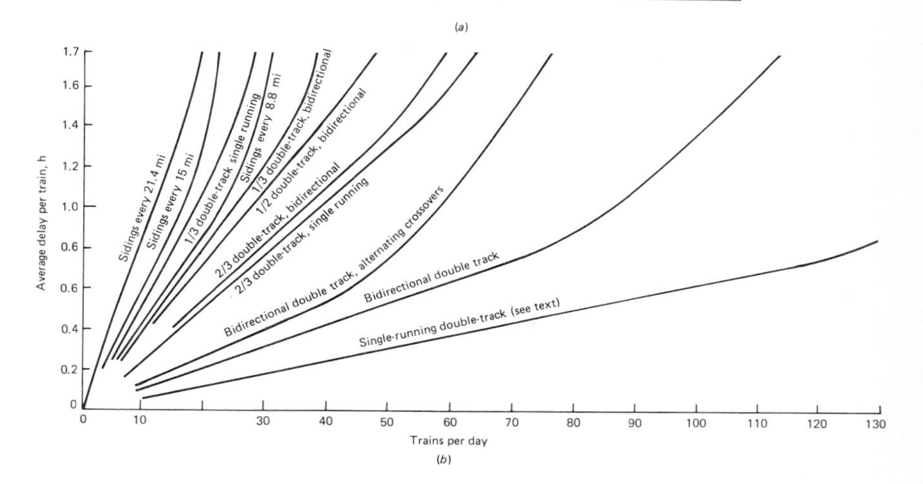

(b)

The relationship of delay to volume is generally as would be expected. The various curves represent different arrangements of the trackage on the link, from a single track line with passing sidings (loops in Continental parlance) to double track with various possibilities for bidirectional running and crossovers between tracks. The only surprising relationship is that the bidirectional double track (on which trains can operate in either direction on either track, crossings being located at frequent points) yields a higher average delay than the single-running double track (one direction on each track only). This occurs because the bidirectional line allows priority trains to be given right-of-way over inferior ones, which results in the delay of the latter trains more than offsetting the reduced delay to the priority trains. If the average delay were weighted more heavily for priority trains, then the relationship might be reversed.

Since railway capacity analysis is only now being studied in detail, generalized formulas or curves have not been developed, but the figure provides an idea of the overall relationships. It is interesting that these curves were developed using a computer program that in effect calculates a time-space diagram for this rail line, and is used to simulate a few days' operations, including breakdowns of trains, etc.

The profound effect that trains operating at different speeds can have on a rail line is revealed by a theoretical analysis of operation of the Northeast corridor route from New York to Washington (Abbott, 1975). A single line in one direction (implying a two-track railroad) can accommodate 7 freight trains per hour if all operate at a steady 50 mi/h. If 80 mi/h passenger trains are introduced, and these have priority over the freight trains, which can enter sidings spaced at 15-mi intervals, and the passenger trains operate at 1-h headways in each direction, the capacity for freight trains is reduced to 5/h, and freight train running times for the 200-mi trip increase from 4 to 5.33 h. This situation is portrayed in Fig. 5-20*a*. If the passenger trains operate at 125 mi/h, as is contemplated on this line, the maximum number of freight trains is further reduced to 3/h, as shown in Fig. 5-20*b*, with overall running times of 6.33 h, 58 percent more than without passenger trains. Considering actual delays in operating freight trains, and the need to remove tracks from active service in order to maintain them, actual capacities would be less. It is therefore no surprise that there are serious capacity problems on railroads, especially where trains of widely varying speeds are to be operated, whether passenger and freight or just high-speed (e.g., piggyback) and lower-speed freight trains.

Since a single main track with passing sidings constitutes so much of the world's railroad mileage, it is of interest to explain methods for analysis of such lines in a little more detail. This is especially true because design engineers must lay out the location of the sidings, and whenever congestion becomes a problem they will be asked to analyze possible locations for additional sidings and to estimate the effect upon capacity and train delays. The time-space diagram is again the primary tool used in this type of problem.

The time-space diagram for a hypothetical railway line is presented in Fig. 5-21. The location of the sidings or loops is shown on the distance axis. Trains operating in both directions are shown. Of course, trains operating in opposite directions can meet and pass one another only at sidings; otherwise a collision could occur, or if not, one train would have to back to a siding so

Figure 5-20 **Time-space diagrams of railroad track used in one direction by freight and high-speed passenger trains. *(a)* 50 mi/h freight trains and 80 mi/h passenger trains. *(b)* 50 mi/h freight trains and 125 mi/h passenger trains. [*From Abbott (1975), pp. 45 and 47.*]**

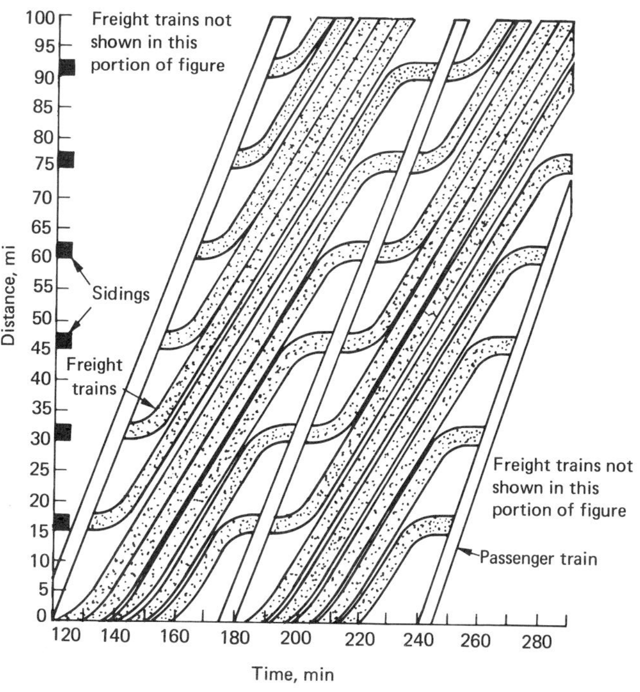

(a)

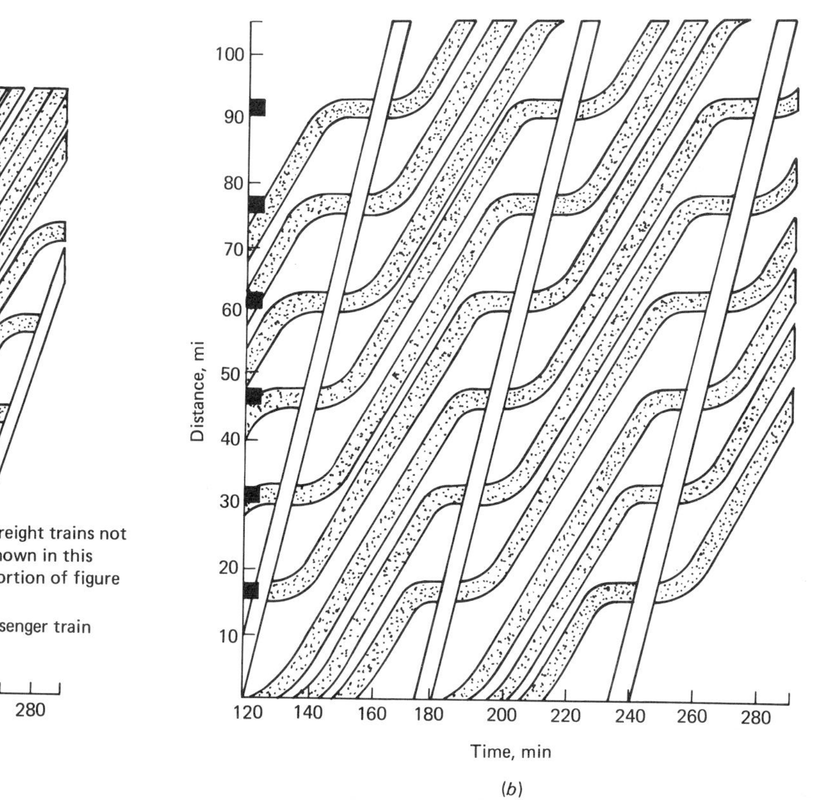

(b)

Figure 5-21 **Time-space diagram or graphical time table of a single-track railroad. The trains in the southbound direction are shown with priority over northbound trains.**

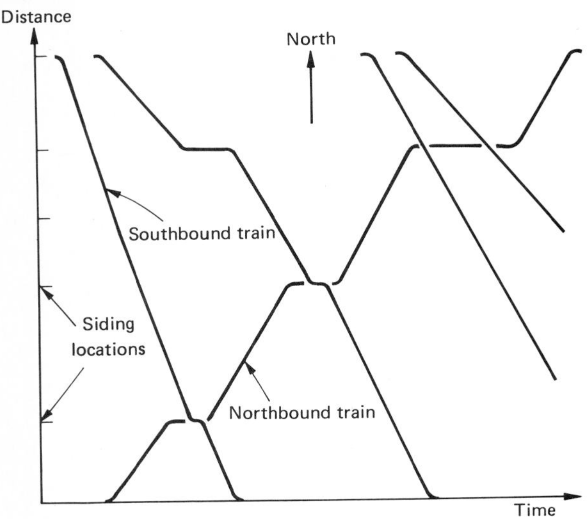

that the passing could occur. In this figure, the siding spacing is assumed given, and the train trajectories simply show the actual movements of trains, assuming ones in the direction marked north are held at sidings for trains in the other direction.

To lay out the location of sidings, the time-space diagram is used in a slightly different way. Consider a situation in which the trains are all of the same type, and hence all trains in one direction will have the same speed at any given point on the line. Also, it is necessary to specify whether sidings will be so long that trains can pass at cruise speed, or if one train must stop on the siding. The latter case is most common outside very high volume lines. Usually trains in one direction are considered inferior and must stop to await the trains in the opposite direction. For example, if trains are heavier in one direction (due to an imbalance in freight traffic) or grades are steeper in one direction, then the trains in the other direction would usually be required to take the siding. We will assume northbound trains must take the siding.

The first step would be to develop the train time-distance trajectories for the southbound trains. The methods for doing this were described in Chap. 4. Then begin the use of the time-space diagram by starting a southbound train (the superior direction) at an appropriate time at its origin, station A in Fig. 5-22. The time of departure may be given in some contexts, in which case that would be used. Given the time headway between trains to be operated in the superior direction, locate the following trains also. The headways may be constant or varying. This yields the odd-numbered trajectories numbered 1 through 15 on the figure.

Now the trains in the other direction must be drawn. The logical time for a northbound train to start is as soon as possible after a southbound train has

Figure 5-22 **Determination of passing siding locations on a single-track railroad line using the time-space diagram. The trains in the northbound direction must stop in the siding at least 10 min before trains in the southbound direction arrive.**

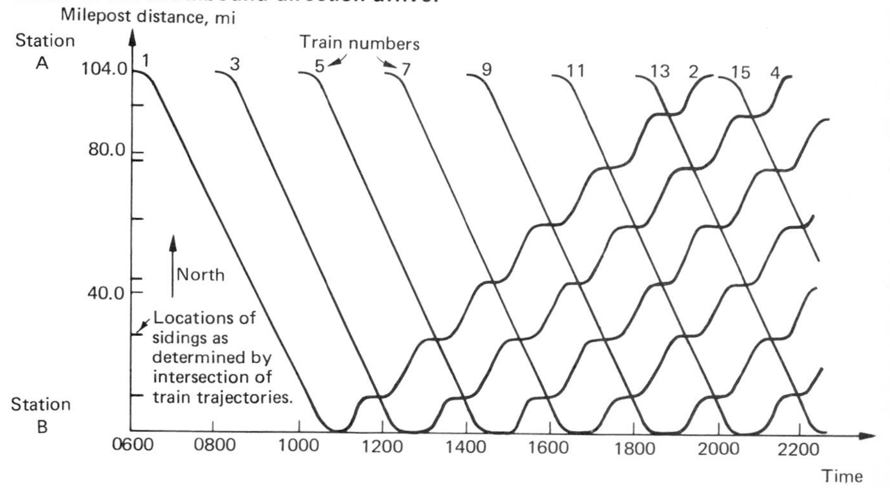

arrived at the terminal B since this will tend to minimize the number of sidings required. A few minutes may be allowed here to account for lateness in the arrival of the superior train, throwing turnouts, etc. Then start the first northbound train, number 2. As it nears the point of passing the next southbound train, number 3 in the figure, locate the trajectory of train number 2 so that it stops before the superior train number 3 passes it. There may be a rule to the effect that the inferior train must be stopped in a siding a few minutes, often 10, before the superior train is due. In this case, train 2 would be stopped 10 min before train 3 arrives at the siding. After train 3 is past, train 2 can continue on its journey. This process is repeated every time train 2 passes another train in the opposite direction until the terminal is reached.

This process has located all the sidings required. A siding must be located at the points where train 2 passed any trains in the opposite direction. As can be seen from the figure, a siding at such locations will be satisfactory for all the other passings of trains required. This is the case provided the trains in both directions operate at the same headways, as generally is the case, since it is desirable to move as many cars and locomotives in one direction as in the other.

The situation is more complex if trains of different types are operated. If a schedule for trains is given, a time-space diagram can be constructed and all necessary sidings located. However, if the number is too great, then judgment must be used in eliminating some, perhaps relocating others, with increases in train delays. If no firm schedule is to be operated, then the problem is even more difficult, sidings being located as a result of compromises among the locations for many different schedules which might be operated.

The capacity and level of service of such lines is not difficult to estimate. The travel times of trains, delays, etc., come directly from the time-space diagram. The capacity is simply the inverse of time headway, again from the diagram. This time headway for trains in one direction (same in the other, of course) is simply the sum of the time for a train to operate in one direction from a siding to the next, plus the time for a train to operate in the opposite direction between the same sidings, plus any time required for clearance of trains. The capacity is the inverse of this sum (in one direction). In equation form this is

$$c = \frac{1}{T_N + T_S + T_C} \tag{5-18}$$

where c = capacity in one direction
T_N = running time between sidings, northbound (i.e., one direction)
T_S = running time between sidings, southbound (i.e., other direction)
T_C = clearance times

This formula reveals a very important characteristic of the capacity of single-track railroads, namely, that the capacity is inversely proportional to the times required by the trains to operate between adjacent sidings. If it is necessary to increase the number of trains operated, then this can be done by decreasing the running time (and perhaps the clearance times, although this raises safety questions) of trains. Also, this provides a way of identifying likely bottleneck points on rail lines: the adjacent sidings between which the sum of the running times in both directions plus the clearance times is the greatest.

The time-space diagram provides a useful tool for analyzing these situations, but likely deviations of trains from scheduled or ideal times must be considered. Also, lost track time due to maintenance, local freight trains which perform switching on the main line, and other capacity-reducing factors, all must be considered.

Figure 5-23 **Pedestrian volume and concentration relationships.** ***(a)*** **Level walkways.** ***(b)*** **Stairways.** [***From Fruin (1971), pp. 44 and 60.***]

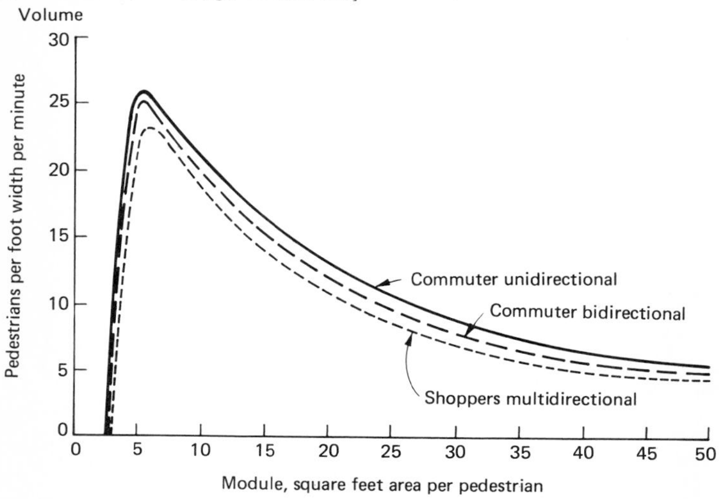

(a)

Pedestrians

Now that the use of the time-space diagram for road and railroad capacity analysis has been explained in detail and actual level-of-service relationships have been presented, it is appropriate to point out that these methods are not applicable to mechanized transport only! Figure 5-23 presents observed volume-concentration relationships for pedestrian flow (Fruin, 1971). Actual concentration as defined earlier is not used, but rather is the area per person, a measure proportional to the inverse of concentration for a given channel width. Since pedestrian traffic is not usually divided into lanes, this measure is more suitable. Speeds vary widely in each case. On the walkways (generally level), the speeds increased with increasing area per person, remaining ap-

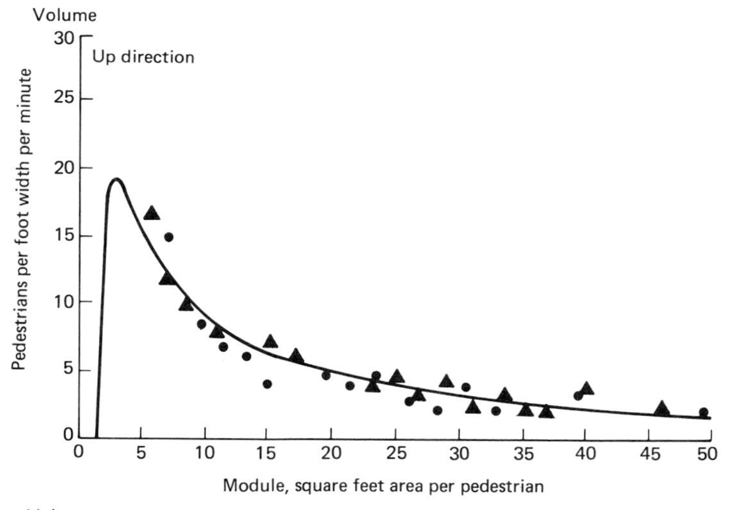

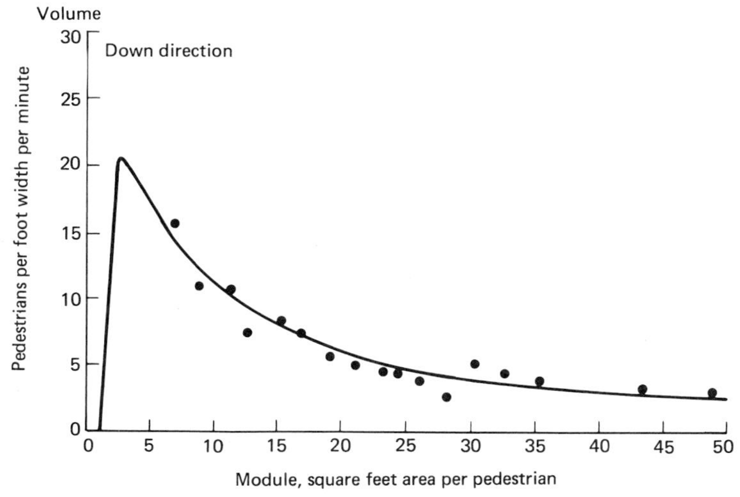

(*b*)

Table 5-3 Accident Rates for Different Road Facility Types in 1958

Area and facility type	Person fatalities per 100 mvm†	Person injuries per 100 mvm†	Property damage only accidents per 100 mvm†
Urban			
Full access control	1.98	88	131
Partial access control	4.64	162	394
No control	4.02	272	350
Suburban			
Full access control	4.03	91	85
Partial access control	5.07	134	253
No control	5.12	338	289
Rural			
Full access control	3.27	125	84
Partial access control	6.12	137	134
No control	8.70	196	217

† mvm is million vehicle-mi.
Source: U.S. Bureau of Public Roads.

proximately constant at 250 ft/min above 20 ft^2 per person. On the stairs, with speed measured along the slope, the speeds were about 100 ft/min in the up direction and 125 ft/min in the down direction.

Examples of the application of these methods to other modes and contexts could be presented, but there is little point. The time-space diagram is perfectly general in its usefulness, and the concepts of flow and the fundamental relationship among speed, volume, and concentration are also applicable to any technology or type of path. The examples of the primary types of application presented above should serve to indicate the usefulness of these methods.

Accidents

One very important level-of-service characteristic in many contexts is the occurrence of accidents. Since the rate of accidents is high and also seems to vary rather consistently with the type of facility and volume, it is a particularly important consideration for road traffic. The rates of fatalities, injuries, and property damage per 100 million vehicle-mi of road travel in the United States for 1973 are presented in Table 5-3. As can be seen, accident rates vary considerably, with full access control facilities yielding fewer accidents per 100 million vehicle-mi in each category than other types of facilities in the same location.

SUMMARY

1 The movement of vehicles from one point to another over a series of links and intersections is affected by the presence of other vehicles on those facilities.

2 The interaction of vehicles sharing the same facility depends upon the form of vehicle control system employed, which in turn varies with the technology of transport and the type of facility. This interaction can be represented and analyzed by use of the time-space diagram, which portrays the following of one vehicle by another.

3 There are three primary characteristics of vehicle flow—space-mean speed, volume, and concentration—and numerous derivatives of these such as time headway. These are related by the fundamental flow relationship: volume equals space-mean speed multiplied by concentration.

4 Many characteristics of flow are important to users of the transportation system besides speed or travel time, and the set of these characteristics is called level of service. The level of service includes such factors as cost, safety, and comfort. The flow performance characteristics of links and intersections are described by relationships among speed, level of service, volume, and capacity, or variations thereof.

5 Actual flow characteristics of links and intersections are best determined by empirical studies, since actual performance may be considerably different from idealized or theoretical performance. In all technologies of transportation, it appears as though congestion phenomena are present, leading to an increase in vehicle travel time as the volume increases, at least above a certain volume.

PROBLEMS

5-1

Intersections on an arterial street are 1000 ft apart, measured from the center line of one intersecting street to the same location in the next. Cross streets are such that the stop lines on the arterial are 80 ft apart at each intersection and the traffic lights are located on the right near-side and left far-side curbs, as seen from the arterial. If it is desired that traffic flow at a speed of 30 mi/h in one direction, what should the offset be? Draw a sketch of two adjacent intersections to be sure that you understand the geometry.

5-2

Suppose it were desired to have traffic move in a smooth progression in both directions on this arterial street. What are the possible speeds, from 10 to 45 mi/h, which will result in equal speeds in both directions? The cycle length is to be 70 s.

5-3

The formula for the capacity of an intersection is $V = 3600W/CD$, where V is volume in vehicles per hour, C is cycle length in seconds, W is green band width in seconds, and D is the average time headway between vehicles in seconds. For a single lane in one direction assume D is 2.5 s and W is $(\frac{2}{3}C - 10)$ s, the 10 s being the amber phase. Plot V as a function of C, for values from 30 to 90 s. What pattern emerges? Does this explain why heavy volume arterials often have very long cycle times?

5-4
Because of proximity to military installations, commercial air service between two cities which are 400 mi apart is restricted to one air lane in each direction. As a result, no passing of aircraft in the same direction is permitted. Passenger air service is provided by two airlines, with airline A operating Boeing 727 jets at a cruise speed of 600 mi/h, and airline B operating slower economy service with Lockheed Electras at a cruise speed of 400 mi/h. Both airlines have the same number of departures each day, flights of one alternating with flights of the other. If the aircraft must stay at least 10 min apart while flying in the same direction, and separate runways are provided for take-offs and landings, what is the minimum time headway of flights of airline A? Of airline B? If the 727 seats 131 passengers and the Electra 98 passengers, how many passengers can depart from one city to the other per hour, on the average?

5-5
Where might a problem of capacity resulting from different speeds of vehicles, such as that for airliners in the problem above, arise in railroad passenger service? In railroad freight service?

5-6
A single-track railroad is to be constructed from Gunnison to a mining area temporarily called Tipple No. 5. The proposed route is 145 mi long. The loaded ore trains will travel in the downhill direction at a speed of 25 mi/h, and will not normally stop except at Midway, a place 75 mi downhill from Tipple No. 5, where they will stop for 30 min. The empty uphill trains will travel at approximately the following speeds:

Milepost 0 to 30, 35 mi/h
Milepost 30 to 75, 20 mi/h
Milepost 75 to 105, 16 mi/h
Milepost 105 to 145, 18 mi/h

The empty trains, operating in the predominantly upgrade direction, must be in each passing siding 10 min before the downgrade train arrives, and hence stop for that period until the downhill train passes. The empties will also stop for 30 min at Midway. Loaded trains will leave Tipple No. 5 10 min after the arrival of empty trains, and a train will operate in each direction every 4 h. Draw a time-space diagram of this operation and locate the required passing sidings. Ignore variations in speed due to acceleration and deceleration. (On your answer sheet, specify the location of the sidings by milepost location.)

5-7
A rapid transit rail line is to be constructed from the present western terminal of the subway line in Philadelphia at 69th Street to the suburban town of Newtown Square. The line will be single-track, with passing sidings at appropriate locations. The distances from 69th Street to the stations contemplated on this line are given on page 219, along with the schedule of typical trains. If trains are to leave 69th St. 4 min after trains arrive and trains will operate at 10-min headways, where should the sidings, each about 1 mi long, be located? (Specify location by distance from 69th St.)

Distance, mi	Station	Time west	Time east
0.00	69th St.	10:00 A.M.	10:26 A.M.
1.09	Highland Park	04	22
2.05	Llanerch	07	19
2.92	Manoa	11	15
4.82	Brookthorpe Hills	16	10
5.63	Broomall	19	07
6.51	Larchmont	22	04
7.34	Ashley	24	01
7.77	Newtown Square	10:26	10:00
		(Read down)	(Read up)

5-8

The mining railroad described in Prob. 5-6 above is faced with a sudden increase in the volume of ore to be transported. It has been suggested that trains be operated in a manner called "fleeting," in which two or more trains in one direction follow one another in close succession, in order to reduce the average time between trains. What would the capacity of the line be if trains were operated in pairs, one train following the other by 10 min? The inferior direction trains (uphill empties) would also operate in pairs, both trains having to pull into a siding 10 min before the superior loaded trains arrived at the siding, and one empty train would leave as soon as both superior trains had passed, the second empty leaving 10 min behind it.

5-9

Is there any similarity between the fleeting of trains as described above and the platooning of motor vehicles through signalized intersections? If so, what?

5-10

A barge canal is operated with locks in each direction. What is the minimum time headway between barges past a lock with the following times required for the locking process? The barges cruise at a speed of 10 mi/h, and acceleration and deceleration occurs at a rate of 0.5 mi/(h)(s).

Close gates after boat enters, 60 s

Raise water level, 540 s

Open gates, upper level, 50 s

Close upper gates after boat clears, 60 s

Lower water level, 300 s

Open lower gates for boat, 50 s

The signal to entering boats is 0.5 mi from the lock gate, and it must be green 30 s before a boat passes it for no time to be lost. It is green only after the lock gate is opened. The upper gate is closed starting when the departing boat has reached a point 100 ft from the gate. Boats are 60 ft long, and the lock is 100 ft long.

5-11

In many of the problems above the operation of vehicles was assumed to be exactly as planned. Such precise operation is often not possible. What are some of the factors which contribute to such unreliable or unplanned operation on railways? In air service? On barge canals?

5-12

Most United States state driving regulations suggest that a safe distance between one vehicle and the vehicle in front of it is one car length for every 10 mi/h of speed. Assuming cars are 18 ft in length, what would the distance headway between cars be at 10, 30, and 60 mi/h? Plot the three fundamental diagrams of road traffic in this range of speed, and give the equation for each.

5-13

Recently the following relationship between the speed of traffic u and the density k was obtained from observations on the center lane of a three-lane (in one direction) freeway in the United States: $u = 53.8 - 0.351k$, where u is in miles per hour and k is in vehicles per mile. Plot the relationships between u and k, q and k, where q is volume in vehicles per hour, and q and u, giving the equations for each.

5-14

Why do the curves and equations in Probs. 5-12 and 5-13 above differ? Which would you use in a consulting job in which you were to determine the maximum volume which a freeway could accommodate? Why?

5-15

Draw a time-space diagram showing four successive vehicles operating at a constant time headway of 10 min over a distance of 3 mi. Each vehicle is 880 ft in length and travels at 45 mi/h. Show by calculation that the mean time headway is the reciprocal of the volume and that the mean distance headway is the reciprocal of the concentration.

5-16

Plot the flow diagrams showing speed versus concentration, volume versus concentration, and volume versus speed for air traffic in the case where the flight rules specify a minimum separation of 10 min. (Assume this refers to time headway, not tail to front time.) Cover a speed range of 200 to 400 mi/h. Also plot the relationship for conditions that require a distance headway of 5 nmi. Discuss.

REFERENCES

Abbott, Robert K. (1975), "Operation of High Speed Passenger Trains in Rail Freight Corridors," PB-247 055, U.S. National Technical Information Service, Springfield, VA.

Armstrong, John (1957), "All About Signals," Kalmbach, Milwaukee, WI.

Association of American Railroads, Communications and Signals Section (1955), "American Railway Signaling Principles and Practices," Washington, DC.

Blunden, W. R. (1971), "The Land-Use/Transport System," Pergamon, New York.

Drew, Donald R. (1968), "Traffic Flow Theory and Control," McGraw-Hill, New York.

Fruin, John J. (1971), "Pedestrian Planning and Design," Metropolitan Association of Urban Designers and Environmental Planners, New York.

Highway Capacity Manual (1965), Special Report No. 87, Highway Research Board, Washington, DC.

Horonjeff, Robert (1975), "The Planning and Design of Airports," McGraw-Hill, New York.

Matson, Theodore M., Wilbur S. Smith, and Frederick W. Hurd (1955), "Traffic Engineering," McGraw-Hill, New York.

Normann, O. K. (1959), Research to Improve Tomorrow's Traffic, *Traffic Engineering,* **29**(7):11–21.

Prokopy, John C., and Richard C. Rubin (1975), "Parametric Analysis of Railway Line Capacity," PB-247 181, U.S. National Technical Information Service, Springfield, VA.

U.S. Department of Transportation, Federal Highway Administration (1971), "Manual of Uniform Traffic Control Devices," Washington, DC.

U.S. Department of Transportation, Federal Highway Administration (1977), "New Look in Traffic Signs and Markings," Washington, DC.

Webster, F. V. (1958), "Traffic Signal Settings," Road Research Technical Paper No. 39. U.K. Road Research Laboratory, London.

Winfrey, Robley (1969), "Economic Analysis for Highways," International Textbook, Scranton, PA.

Wohl, Martin, and Brian V. Martin (1967), "Traffic System Analysis for Engineers and Planners," McGraw-Hill, New York.

Continuous-Flow Systems

6.

Although vehicular transport systems of the types which we have been discussing are undoubtedly the most conspicuous forms of transport, there is another prevalent and extremely important class of systems using a fundamentally different technology, termed continuous-flow systems. In these systems, flow of the objects being transported is continuous, or virtually so, in contrast to the discrete or intermittent flow, which characterizes vehicular systems.

Continuous-flow systems are actually almost as commonplace as discrete-vehicular-flow systems. In fact, they exist in nature and therefore surely predate the vehicular systems. Examples include the flow of water in channels, both natural, as in rivers, and manmade, as in sanitary channels, aqueducts, and pipelines. Closely related to these is the transport of liquid and solid wastes in pipelines and to a lesser extent in open channels. Also, there is the transport of oil and its products in pipelines. As an indication of the importance of continuous-flow systems, oil moved in pipelines represents about one-fifth of all the intercity ton-miles of freight moved in the United States! Gases are similarly transported in pipelines to a large extent, although the significance of this flow is not revealed by a ton-mile measure, for obvious reasons.

Closely related to these liquid and gas systems is the movement of certain solids in pipelines and open channels. Grains, coal, and other material which can be suspended in gases (usually air) or liquids are transported by pipeline. Now experimentation is under way on the movement of capsules in pipelines, in which the capsules would carry and protect freight in much the same manner that vehicles do in other technologies.

Also similar to this entire class of systems are conveyor systems, which can be used for all kinds of freight and passengers. One common form is the belt conveyor, on which the objects to be transported are carried. Closely related are cable-suspended and -propelled systems, in which the cable movement is continuous, as is the movement of objects suspended from the cable. However, these conveyor and cable systems, important in many industrial applications and in mountainous terrain as both recreational and general transport carriers, have elements of both discrete-vehicular and continuous-flow systems.

In the paragraphs to follow we shall first discuss certain general principles applicable to all these technologies. Then we will describe in more detail the most important characteristics of the more significant technologies, in particular the conveyor, the oil pipeline, and the gas and pneumatic pipeline.

GENERAL CHARACTERISTICS

Before describing the characteristics of the many individual types of continuous-flow system technologies, it will prove useful to cover some general characteristics applicable to all.

Fundamental Flow Relationship

The fundamental equation of flow described in the previous chapter with reference to vehicle flow is also applicable to continuous-flow technologies. In fact, the validity of the relationship is perhaps even easier to visualize in the context of continuous flow than discrete flow. As may be recalled, the relationship for discrete flow and typical units is

$$q = ku \qquad (5\text{-}9)$$

where q = vehicle volume, vehicles/h
k = vehicle concentration, vehicles/mi
u = space-mean speed, mi/h

This relationship is applicable in precisely the same form to continuous flow, although the units used to measure flow will not be vehicles but units appropriate for other traffic, such as a fluid or certain types of solids. Therefore in the continuous-flow case the variables of this relationship are modified as follows:

q = volume of traffic
k = traffic concentration
u = space-mean speed of traffic

For liquid traffic gallons or another volume measure is often used, while for solids often a weight measure such as tons is used. In using the relationship, one merely has to be sure the units are consistent throughout.

Since different units may be used for these flow measures, it is useful to have a notation which distinguishes among weight, mass, and other possible measures of the amount of the traffic being transported. A convenient way to do this is to affix subscripts to the volume and concentration measure variables, such as w for weight, m for mass, and c for cubic volume. The term "cubic volume" is used here to denote the amount of (three-dimensional physical) space occupied by a commodity. Usually this is termed volume, but unfortunately the word "volume" is almost universally used in transport to denote the amount of traffic flowing per unit time past a point. Hence we need a means to differentiate between the transport engineer's volume and space occupancy volume. This then yields an additional set of useful relationships, which involve the weight per unit cubic volume of the commodity being transported and its specific gravity, the ratio of its mass per unit cubic volume to that of water at 4°C. These relationships are

$$q_w = gq_m = d_w s q_c = dq_c \tag{6-1}$$

$$q_w = k_w u$$

$$q_m = k_m u$$

$$q_c = k_c u$$

where g = gravitational constant
s = specific gravity of traffic
d = weight per unit cubic volume, of traffic
d_w = density of water, 9.807 kN/m³ or 62.4 lb/ft³ at 4°C

It is useful to remember that water weighs 1 g/cm³ in the traditional metric units. Hence the weight density of any substance in g/cm³ is numerically equal to its specific gravity (Margenau et al., 1953, p. 202).

A continuous-flow system is often used to transport many different commodities, such as the so-called products pipelines which carry the many different types of oil refinery products—materials that cannot be mixed without destroying important properties. One may wonder how the various materials are kept from mixing with one another, particularly in liquid or gas pipelines. Mixing and hence contamination are minimized by sending only large batches of each material through the system. Mixing only occurs at the ends of a batch, so this minimizes the problem. Also, in pipelines a solid ball is often placed at the interface between batches, practically eliminating any mixing. Thus the problem is not really so difficult as it appears at first thought.

Varying-Flow Characteristics

Just as in the case of vehicular systems, the characteristics of the flow in continuous-flow systems may vary along a link. Part of this variation may be due to the merging of flows, which creates a higher-volume flow, or diverging, which reduces the flow. Often it is due to differences in the characteristics of the link along its length. For example, a link may consist of many conveyors arranged end to end. Some may have lower speeds than others. But if the volume transported by one is transferred directly onto another, despite the variation in speed, the volume transported by both must be the same.

What happens in this case can be seen readily from inspection of Fig. 6-1. The graph portrays the fundamental flow relationship, and each of the curves corresponds to a constant volume which, by the relation $q = ku$, can be achieved by different combinations of speed and concentration. If a high-speed portion of a link discharges its flow onto a low-speed portion, then the concentration must increase in order for the volume to remain constant. On a conveyor link, this would mean the amount of the material per unit length would have to increase; e.g., the pile of coal or other material on the belt would be higher on the low-speed end, the belt would have to be wider, as shown in the figure, or a combination of both. The same would be true for

Figure 6-1 **Flow relationships in continuous-flow technologies.**

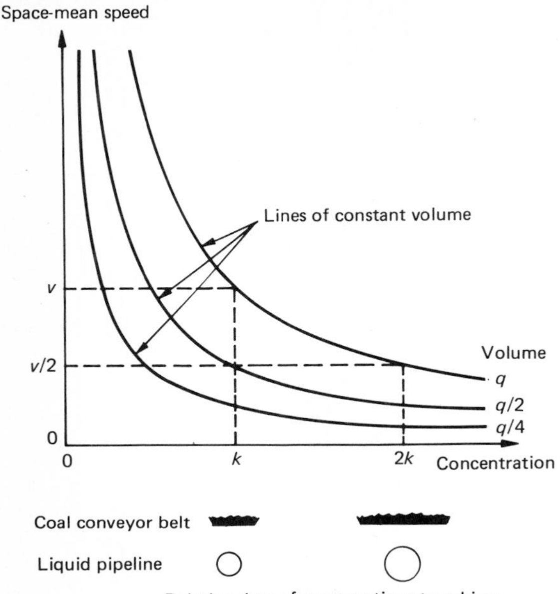

an open channel carrying liquids. In a pipeline, either the density of the material would have to be increased—something that is usually not possible except with gases under pressure—or the cross section area of the pipe would have to increase to permit more material per unit length of the pipe, as shown in the figure.

If the volume cannot be accommodated by a section of a link, then either some must be stored at the point where the change in capacity occurs or the volume over the entire link must be reduced. The maximum volume that a continuous-flow link can accommodate is limited by the capacity or maximum volume on that portion having the least capacity. However, as revealed by Fig. 6-1, if the speed can be varied along a link in order to just compensate for any variations in the concentration of the material being transported, then the volume is the same along the entire length of the link. In fact, this automatically occurs in most fluid flow systems in which the specific gravity of the material usually cannot be changed (i.e., water compressed, etc.) but the speed of flow can be varied.

In those cases where the volumes are different, storage of some material must occur until the input volume is reduced to less than the capacity of the outbound link. This will be discussed in the chapter on terminals, Chap. 7, since the removal of the material from the link and interchange facilities occurs, by definition, at terminals.

Propulsive Work Characteristics

An additional characteristic relating to the propulsive work required differentiates continuous-flow systems from vehicular systems. For the flow to be continuous, all of the material being transported must move together, with no gaps between objects, or at least no substantial ones (i.e., there may be gaps between some lumps of coal, but their size is insignificant). Thus, when a force is exerted on the objects at one point on a link, it is transmitted along the entire link with some reduction due to inherent resistances to motion. This is illustrated for a conveyor belt in Fig. 6-2. An inherent resistance to motion of the belt (and the material being transported on it) acts over its entire length, as shown by the small force vectors. This yields a total force over the entire belt length, $L_1 + L_2$, of $r(L_1 + L_2)$, with r being the inherent resistance of the belt per unit length. On the downgrade portion of the belt, the force due to gravity acts in the direction of motion and hence helps overcome the inherent resistance. This force is

$$\frac{H_2}{L_2} wL_2 - \frac{H_2}{L_2} bL_2 + \frac{H_2}{L_2} bL_2 = H_2 w$$

if w and b are the weights per unit length of the cargo and belt (one strip only), respectively. The forces on the belt cancel one another, since the upper and lower portions of the belt are subjected to equal and opposite forces. On the upgrade portion the force due to gravity, equal to $(H_1 wL_1/L_1)$, is acting to retard the motion. Therefore, the force required to maintain a constant speed u in the belt and the material being transported on it is

$$F = H_1 w - H_2 w + r(L_1 + L_2)$$

Notice that if the end points of the belt are at the same elevation, i.e.,

Figure 6-2 Propulsive and resistance forces in a simple conveyor system.

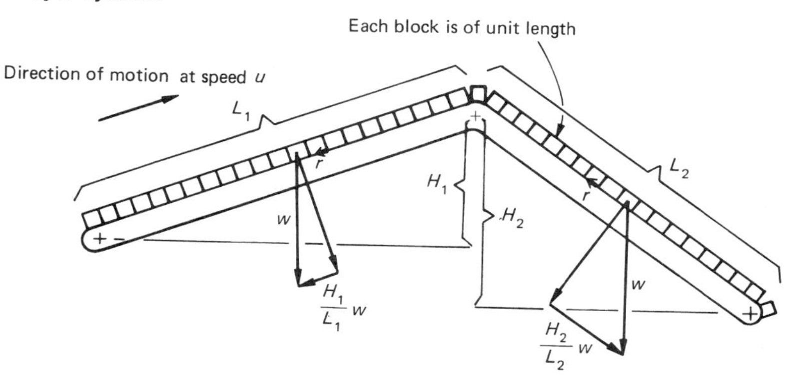

Total inherent resistance = $r(L_1 + L_2)$

Total propulsive force required = $r(L_1 + L_2) + w(H_1 - H_2)$

r = inherent resistance per unit length at speed u

w = weight of cargo per unit length

b = weight of belt per unit length

$H_1 = H_2$, then the forces due to the gradient cancel one another. In this case the total propulsive work required to move any given amount of material, $r(L_1 + L_2)$, is the same as that required to move it at the same speed over a level route of precisely the same length. Despite the undulating profile or gradients, no additional work is required! On a belt transporting material in one direction only, such that the elevation drops, there may be no need for any externally provided propulsive work. In other words, the H_2w term may completely cancel the other terms, if the decrease in elevation $H_2 - H_1$ is sufficiently large.

It is useful at this point to derive the work and power relationships for this type of technology. The basic definitions presented in Chap. 4 apply, of course. The constant force acting on the conveyor to propel it at speed u is $r(L_1 + L_2) + (H_1 - H_2)w$, and for simplicity this will be designated F. Assuming F is positive, i.e., a propulsive force is required, the power required is Fu. This power results in a volume, q_w, measured by weight of wu arriving at the end point of the system. The movement of an amount of this commodity equal to $w(L_1 + L_2)$ from one end to the other requires an amount of propulsive work of $F(L_1 + L_2)$. Thus the average work required per unit distance to move the $w(L_1 + L_2)$ weight of material is work divided by distance, or F. This makes sense intuitively, for we have concluded that the average unit distance work required to move an amount of freight equal to that held at any one moment on the entire length of the belt is the force required to move the belt at the desired speed.

Although the discussion above focused on a conveyor system as an example, the principles and conclusions are precisely the same for pipelines and other continuous-flow systems. Thus the typical small effect of gradients and undulating profiles is a major advantage of continuous-flow systems in mountainous terrain. The links can be constructed over a route with extremely steep grades and even very high intermediate elevations. In comparison to an ideal route, i.e., a direct constant grade route from one end to the other, there will be either little (due to a slightly longer route) or no added work required. Of course, the route may be somewhat longer than such a straight route, due to the presence of major obstacles such as steep cliffs, etc. Nevertheless, the route usually can be much shorter than that which vehicular systems would require, where the maximum gradient may be greatly limited by the means of locomotion and where to meet that limitation the route may be made very circuitous. As was noted in Chap. 4, since fairly shallow downgrades often provide more force due to gravity than the vehicle's inherent resistance, some of the work expended in raising the vehicle to an intermediate summit is lost. For all these reasons, continuous-flow technologies often find application in very mountainous terrain.

However, there are limitations on such technologies. With conveyor belts, there is a maximum grade that can be traversed without the material simply sliding down the conveyor, although some barriers on the belt increase this limit. With pipes, as the liquid is pumped to higher elevations, the pressure differential necessary becomes greater, requiring more substantial piping, with economic and technological (safety) limits on the maximum pressure. With high pressures there is also a possibility of damage to the commodity, especially when it is a solid suspended in a gas or liquid, such as grain or capsules containing general freight. While continuous-flow systems do not provide a panacea for the problems of transport routes over mountainous terrain, they do provide an important option for some commodities.

BELT CONVEYORS

One of the most common forms of continuous-transport technologies is the conveyor belt. While the reader may be most familiar with it in very small-scale installations, as in factories, it is in fact also used to transport bulk materials over long distances, particularly in rough terrain. There have been numerous applications of belts to transporting earth distances up to about 10 mi in major construction projects. Longer belts, up to 100 mi in length, have been proposed for ore and coal, although to date rail and water transport usually have proven more economical.

Physical Principles

Many of the physical principles of conveyor belts have been covered in the previous section as part of the discussion of the work and energy advantages of continuous-transport systems. In this section we will be more specific and present actual numerical values for various parameters and precise limitations of this technology.

As derived earlier, the volume to be accommodated on a conveyor belt limits the combinations of speed and concentration of material or cargo per unit length of the belt. Thus Eq. (6-1) is the first that must be satisfied in designing or operating a conveyor belt system.

The resistances to be overcome and the power required will depend upon the combination of concentration and speed of movement selected. Speeds are typically limited to 600 ft/min (about 6.81 mi/h) or less, so the inherent resistance varies little over the range of speeds. Hence for engineering work a constant inherent resistance which is invariant with speed is used:

$$R = rL(w + b)C_f \tag{6-2}$$

where R = total inherent resistance of conveyor belt, lb
r = inherent resistance coefficient, usually 0.2 to 0.3
L = distance between centers of end pulleys, ft
w = weight per unit length of cargo, lb/ft
b = weight per unit length of upper plus lower belt, including pulleys, idlers, and rolls, lb/ft
C_f = correction factor

For most longer transport applications, a value of r of 0.2 is appropriate. The correction factor C_f (used to include resistances which are independent of belt length), seems to vary with the type of belt and application, but for the longer transport applications in which we are interested it is usually close to 1.00. Note that the dimensions of the variables used in this equation must be exactly as specified, as was also the case for vehicle inherent resistance.

There is, of course, an additional force due to gradients. For the reasons discussed in the previous section, the force due to gradients, regardless of the combination of upgrades and downgrades on any continuous conveyor, is simply the result of the change in elevation from the beginning of the line to the end. If H is the change in elevation and has a positive value, which indicates an increase in elevation, then the net force due to the elevation change acting over the entire belt is

$$F_g = \frac{H}{L} wL = Hw \tag{6-3}$$

Note that the weight of the belt does not appear in this equation because an identical length and hence weight of the belt moves in each direction on each gradient section and the forces of gravity on the belt alone cancel one another.

Since these are the only sources of requirements for propulsive forces applied to the belt, the force required is simply the sum of Eqs. (6-2) and (6-3):

$$F = rL(w + b)C_f + Hw \tag{6-4}$$

The power required and the propulsive work are, respectively,

$$P = Fu \tag{6-5}$$
$$W = Pt \tag{6-6}$$

where t = duration of operation

As noted above, that the force F, the power P, and the work W can be negative indicates that a braking force is necessary to restrain the speed of the belt. But in most transportation applications the force is positive.

Belt System Design

Conveyor systems are usually designed to handle a single commodity in a particular location. Thus they can be designed specifically for that movement and need not, in general, be designed to accommodate other traffic. A number of factors must be taken into account in selecting the belt to be used, its width, the route location and profile, the speed of travel, and the means of propulsion. While these decisions are usually made by experts on conveyor systems with knowledge of the latest technological advances in such systems (such as new belts), it is useful to review some of the more important aspects so as to understand the capabilities and limitations of this important but specialized technology of transportation.

The selection of various critical characteristics of a conveyor system depends greatly on the characteristics of the cargo being transported, primarily because the cargo is being transported largely without the protection of a fully enclosing container as it is in most vehicular transport. An important characteristic of the cargo, its weight per unit cubic volume, is presented for various commonly conveyed materials in Table 6-1. Also presented there is the maximum incline on which the cargo can be carried (the cargo tends to roll or slide down the incline at greater values), which limits the location and design of any conveyor system to lower gradients. Often the gradient limit is met by extending the length of the line to reduce an otherwise excessive grade.

In addition to a gradient limitation, there is a limitation on speed, which, depending in part upon the commodity being carried, arises partly because of the abrasive action on the material against the belt. The higher the speed, the greater the abrasion at the point where the material is being loaded onto the belt. Here the material (usually dropped onto the belt) must be accelerated rapidly to the belt speed. Also, while moving along the belt, the material often

Table 6-1 Weight and Maximum Incline of Materials Commonly Transported by Conveyor Belts

Material	Unit weight, lb/ft³	Maximum incline	
		Degree	Percent
Bauxite, aluminum ore	55–58	17	30
Bauxite, crushed, dry	75–85	20	36
Cement, clinker	88–100	18	32
Clay, wet	95–105		
Coal, bituminous	47–52	18	32
Concrete mix, wet	115–125	12	21
Dolomite, crushed	90–110	17	30
Earth, dry loam	70–80	20	36
Earth, wet loam	104–112	15	27
Granite, broken	96	20	36
Gravel, washed and screened	85	20	36
Gypsum, broken	80–100	17	30
Limestone, broken	95–100	17	30
Ores, sulfides and oxides, broken	125–160	17	30
Sand and gravel, dry	90–105	20	36
Sand and gravel, wet	115–125	12	21
Trap rock, broken	105–110	17	30
Wood chips, dry	15–32		

Source: Hennes and Ekse (1969, p. 582).

bounces and particles or pieces move relative to one another, especially at points of gradient change, resulting in some wear on the belt. And, of course, there is the danger of some material bouncing completely off the belt and becoming lost cargo. Practical limits on speed for various belt widths and types of material are presented in Table 6-2.

Given any maximum speed and a volume to be transported, the weight of cargo per unit length of the belt can be determined from Eq. (6-1). An approximate rule has been developed to specify the relationship between cargo weight concentration, belt width, and speed (Hennes and Ekse, 1969, p. 583):

$$q_w = \frac{5.75}{200{,}000}(B - 3.3)^{1.96}ud \qquad (6\text{-}7)$$

where q_w = weight volume, tons/h
B = belt width, in
u = belt speed, ft/min
d = cargo weight per unit volume, lb/ft³

Table 6-2 Maximum Speeds of Conveyor Belts of Various Widths Carrying Various Commodities

Material to be conveyed	Belt widths, in								
	12	18	24	30	36	42	48	54	60
Light, free-flowing material	300	400	500	600	600	600	700	700	700
Coal and similar lump material	300	400	500	550	550	600	600	650	650
Heavy ore and abrasive material	300	350	450	550	550	550	600	600	600

Source: Staniar (1959).

From the cargo weight and belt weight, both per unit length of belt, the propulsive force and power required can be estimated. Typical belt weights are given in Table 6-3.

Any belt has a maximum tension which it can withstand. As the propulsive force increases, the tension in the belt increases. Hence it is necessary to check the tension level in the belt design assumed, and to modify the speed, width of belt, etc., if the estimated tension is excessive for the belt design. The tension in the belt is a maximum at the point where the propulsive force is applied, which often will consist of more than one drive pulley in a high-force application. The difference in the tension between the slack side and the drive side of the pulley or pulleys must equal the propulsive force required of *F*. From engineering mechanics, the slack side and the drive side tension values are related by the equation (Hennes and Ekse, 1969, p. 587):

$$\frac{T_1}{T_2} = e^{\gamma\alpha} \tag{6-8}$$

where T_1 = drive side tension
T_2 = slack side tension
e = base of natural logarithms, 2.718 . . .
γ = coefficient of belt friction, dimensionless
α = angle of contact, rad

and $F = T_1 - T_2$ (6-9)

Typical values of γ are in the range of 0.30 to 0.35, and typical contact angles are from 150° to 480° for multiple-pulley drives. Steel belts can handle tension levels of up to 1500 lb/in of belt width and nylon belts up to 450 lb/in of belt width.

Table 6-3 Typical Conveyor Belt Weight for Various Widths
Weight in pounds per lineal foot, including belt, all pulleys, and other moving parts

Belt width, in	Light duty conveyors 4-in idlers	Regular duty 5-in idlers	Regular duty 6-in idlers, belts up to 6 ply	Heavy duty 6- or 7-in idlers, belts 7 to 10 ply
14	12	14	. . .	. . .
16	14	15	20	. . .
18	15	17	22	. . .
20	16	18	25	. . .
24	19	24	30	32
30	25	31	38	45
36	29	37	47	58
42	34	43	55	71
48	. . .	48	64	84
54	. . .	. . .	72	97
60	. . .	. . .	81	110
72	. . .	. . .	97	135

Source: B. F. Goodrich Co.

Sample Problem

As an example, consider a belt designed to carry coal over a distance of 2.0 mi. The route is as portrayed in Fig. 6-2, with H_1 equal to 500 ft and H_2 equal to 1550 ft. The belt is to operate at 400 ft/min and will be carrying 20 lb of coal per lineal foot of belt. The belt and accessories weigh 16 lb/ft. Determine the power requirements.

Using Eq. (6-2) and an inherent resistance coefficient of 0.2, we find the inherent resistance is

$$R = 0.2(2.0 \text{ mi})(5280 \text{ ft/mi})(20 \text{ lb/ft} + 16 \text{ lb/ft})(1.0) = 7.60 \times 10^4 \text{ lb}$$

and the force due to gravity, by Eq. (6-3), is

$$F_g = (500 \text{ ft} - 1550 \text{ ft})\ 20 \text{ lb/ft} = 2.10 \times 10^4 \text{ lb}$$

Hence the total force required is

$$F = 7.60 \times 10^4 \text{ lb} - 2.10 \times 10^4 \text{ lb} = 5.50 \times 10^4 \text{ lb}$$

The power required by Eq. (6-5) is

$$P = (5.50 \times 10^4 \text{ lb})(400 \text{ ft/min}) = 2.20 \times 10^7 \text{ ft-lb/min}$$

In the more common units of power this is 667 hp.

It might be interesting to compare the propulsive work required by this conveyor belt with that which would be required by vehicles to transport the same amount of coal. The volume by weight of coal transported is given by Eq. (6-1) as

$$q_w = k_w V = (20 \text{ lb/ft})(400 \text{ ft/min}) = 8000 \text{ lb/min}$$

We will assume that the conveyor operates 8 h/work-day, resulting in a daily volume of

$$q_w = 8000 \text{ lb/min}\ \frac{60 \text{ min}}{\text{h}}\ \frac{8 \text{ h}}{\text{day}}\ \frac{\text{ton}}{2000 \text{ lb}} = 1920 \text{ tons/day}$$

Such a volume is more likely to be handled by truck than rail, since it would correspond to less than ten 100-ton capacity hopper cars per day, so we shall make the comparison with truck movement.

Turning first to the work required for the conveyor movement, this can be estimated from the power required, and the duration of the operation, by Eq. (6-6). The work required to transport the 1920 tons moved in 1 day is

$$W = Pt = (2.20 \times 10^7 \text{ ft-lb/min})(60 \text{ min/h})(8 \text{ h}) = 1.056 \times 10^{10} \text{ ft-lb}$$

Note that the same result would have been obtained by using any of the other approaches to calculating work described in the introductory section.

In order to estimate the work required by the truck, a specific truck design must be used. A typical coal truck consists of a tractor and a single trailer. Although much larger vehicles are often used at mine sites, many of the larger designs exceed the size and weight limits of public roads. As can be ascertained from Table 4-8, this smaller vehicle weighs about 22,000 lb, can carry 58,000 lb, and has a cross section area of 104 ft^2. Hence the daily weight volume of coal, 1920 tons/day, would require a number of truck trips of

$$1920 \text{ tons/day} \frac{\text{truck}}{58{,}000 \text{ lb}} \frac{2000 \text{ lb}}{\text{ton}} = 66.2 \text{ trucks/day}$$

Some trucks might not be fully loaded, so we will use 67 truck trips as the number required, each truck then carrying $\frac{1}{67}$ of the daily volume or 57,313.4 lb on average.

The road could not follow the same route as the pipeline since that would involve grades of about 20 percent, with the summit about 0.5 mi from the coal source end point. The road is 8.0 mi long, and on it trucks operate at 25 mi/h; the summit will be assumed to lie at the midpoint of the route at the same elevation as on the conveyor. Now we can use the methods described in Chap. 4 to estimate the work required. The inherent resistance of the loaded truck, which weighs 79,313 lb or 79.3 kips, by Eq. (4-19) is

$$R = 7.6(79.3) + 0.09(79.3)\,25 + 0.002(104)(25)^2 = 911.1 \text{ lb}$$

and that of the empty truck is

$$R = 7.6(22.0) + 0.09(22.0)\,25 + 0.002(104)(25)^2 = 346.7 \text{ lb}$$

Now we can determine whether or not power must be applied to the truck throughout its run, both loaded and empty in return. The downgrade in the loaded direction drops 1550 ft in 4.0 mi, resulting in a gradient force helping to overcome resistance, by use of Eq. (4-26), of

$$\frac{1550 \text{ ft}}{4.0 \text{ mi}} \frac{\text{mi}}{5280 \text{ ft}} 79{,}313 \text{ lb} = 5821 \text{ lb}$$

Hence no propulsive force is required on the downgrade in the loaded direction, and the total loaded direction propulsive work for one truck is, by Eq. (4-63),

$$W = RL_1 + H_1T = (911.1 \text{ lb})(4.0 \text{ mi})(5280 \text{ ft/mi}) + (500 \text{ ft})(79{,}313 \text{ lb})$$

$$= 5.889 \times 10^7 \text{ ft-lb}$$

In the empty direction, the downgrade force is

$$\frac{500 \text{ ft}}{4.0 \text{ mi}} \frac{\text{mi}}{5280 \text{ ft}} 22{,}000 \text{ lb} = 520.8 \text{ lb}$$

so no propulsive force must be applied there. The total work required to move the empty truck back up to the coal source is therefore

$$W = (346.7 \text{ lb})(4.0 \text{ mi})(5280 \text{ ft/mi}) + (1550 \text{ ft})(22{,}000 \text{ lb})$$

$$= 4.124 \times 10^7 \text{ ft-lb}$$

The total work required for each road trip of a truck is 1.003×10^8 ft-lb, and hence 67 such trips per day would require 0.672×10^{10} ft-lb.

Thus the total work required by the conveyor is about 50 percent more than that required by trucks over their hypothetical routes. Yet the conveyor is often preferred in practice, mainly for economic reasons. It is often much cheaper to build a conveyor line, which usually follows the ground, than to construct a road, which necessitates a smooth surface with much cutting and filling, etc. Also, often the conveyor is cheaper to operate and maintain, usually requiring fewer operating and maintenance personnel.

PIPELINES

Although not so obvious to most persons as vehicular forms of transport, pipelines are quite ubiquitous and are an important form of transportation. As mentioned earlier, they account for a large percentage of the intercity ton-miles of freight moved in the United States and for similarly large fractions of freight ton-miles in most developed nations. Also, they are in widespread use as transporters of water for all purposes, with networks blanketing most developed areas. And, of course, they are widely used for the removal of waste products.

In this section we shall focus on the pipeline as a transporter of liquids. There are three reasons for this selection. First, liquid pipelines are a very large fraction of all pipelines. Secondly, the technology is readily explained in terms of concepts that we have already used in discussing other means of transport. And finally, the basic principles of liquid flow are very similar to those for the flow of gases and of solids in liquids or gases, but the actual methods used in nonliquids are much more cumbersome and difficult, making them extremely difficult to cover in an introductory text. Hence we shall focus on liquid flow and discuss briefly other flow phenomena in the next section.

Physical Principles

As is true of all technologies discussed in this chapter, the relationships among volume, concentration, and speed described earlier and embodied in Eq. (6-1) hold for liquid pipelines. Thus again there are trade-offs in speed and concentration to achieve the same volume. The selection of these will have an important influence on the size of the pipe required and the power required to propel the flow.

Unlike the conveyor belt system, a liquid pipeline may be operated at a rather large range of speeds. The resistance to motion of the liquid depends upon the speed of flow as well as upon characteristics of the fluid and the pipe itself. Specifically, the resistance depends upon the Reynolds number (as you may recall from physics), which is a dimensionless ratio involving the average speed of the fluid, the diameter of the pipe, the density of the fluid, and its viscosity. It is given by the equation

$$R_N = \frac{VD\rho}{\mu} = \frac{VDd}{\mu g} \qquad \text{(dimensionless)} \tag{6-10}$$

where R_N = Reynolds number
V = average velocity of liquid
D = pipe diameter
ρ = mass density or mass per unit volume of liquid (d/g)
μ = viscosity of liquid
g = gravitational constant

Table 6-4 Approximate Density and Viscosity of Liquids Often Transported via Pipeline

Liquid	Weight density, at 68°F, lb/ft³	Viscosity, lb-s/ft², at 32°F	68°F	122°F	212°F
Water	62.31	3.739×10^{-5}	2.089×10^{-5}	1.149×10^{-5}	0.593×10^{-5}
Linseed oil	58.06		7.938×10^{-4}	3.760×10^{-4}	1.358×10^{-4}
Heavy fuel oil	61.81		0.025	0.0036	
Gasolene	42.10	1.042×10^{-5}		0.521×10^{-5}	

Here we have introduced the common symbols in fluid flow analysis, V and ρ, in place of u and d/g. Since R_N is dimensionless, care must be exercised in the selection of the units of the various terms. The density and viscosity of various fluids are given in Table 6-4.

At low values of the Reynolds number, the flow is smooth (termed laminar). In this region of speeds (relative to pipe size, etc.), the resistance to flow is relatively low, and it increases approximately proportionately to speed. Above about 2000 the flow changes from completely laminar to a mixture of laminar and turbulent, and as the speed or Reynolds number increases, the flow becomes increasingly turbulent and at some point becomes completely turbulent. The swiftness of change from laminar to turbulent flow depends upon the roughness of the pipe in which the liquid is flowing, since roughness induces turbulence.

The resistance of a liquid to motion in a pipe depends upon the Reynolds number, and hence upon speed as well as pipe and liquid characteristics. The resistance to motion of a liquid in a pipe is usually expressed as the so-called head loss or the reduction in the total energy of the liquid from one point in the pipe to another. The total energy of a unit weight of liquid at any point along its path is known from thermodynamics to be the sum of the internal energy, pressure energy, kinetic energy, and potential energy. In the movement of a liquid, there is usually so little change in internal energy and in heat energy that these can be ignored. Also, if the pipe is of constant diameter and the liquid is incompressible (as almost all are for practical purposes), there is no change in speed and hence no change in kinetic energy. As a fluid moves from one point in a pipeline to another, energy may be added to or subtracted from the liquid or converted from one form to another. Heat energy and mechanical energy can be added or deleted, but these occur to such a minimal extent that they can be ignored.

As a result, the conservation-of-energy relationship between two points (labeled 1 and 2) on a pipeline can be written as

$$\frac{V_1^2}{2g} + \frac{p_1}{d} + H_1 = \frac{p_2}{d} + H_2 + h_L + \frac{V_2^2}{2g} \tag{6-11}$$

where p_i = pressure at point i in pipe, $i = 1, 2$
H_i = elevation at point i above a fixed datum, $i = 1, 2$
h_L = head loss from point 1 to 2

However, the energy expended in order to overcome the resistance of the fluid

Figure 6-3 **Energy relationships along a pipeline.**

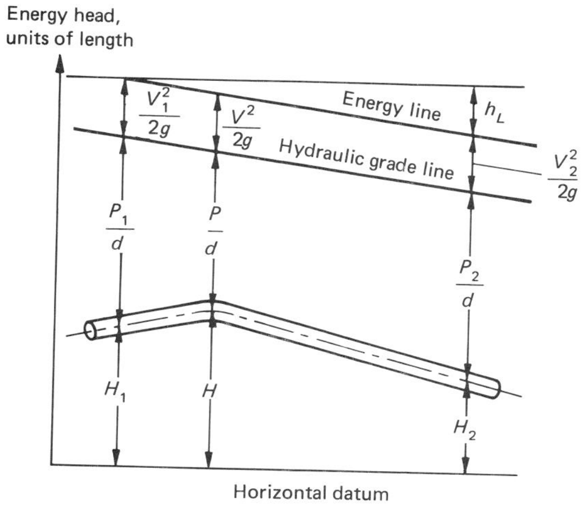

to motion must also be taken into account. The term h_L reflects the resistance to motion of the liquid per unit weight of that liquid, which is termed the head loss. This name arises from the units (of length) and from the similarity of the term to the elevation term. Figure 6-3 illustrates this relationship for the liquid represented in Eq. (6-11).

The head loss has been related to the Reynolds number in order to calculate the force and work required to create flow through pipelines. The Darcy-Weisbach equation, the most common form for estimating head loss, gives head loss as

$$h_L = f \frac{LV^2}{2Dg} \qquad (6\text{-}12)$$

where f = pipe friction factor

L = distance between points over which head loss occurs

Again, care must be taken to ensure that the correct dimension results from this expression.

The pipeline friction factor has been related to the Reynolds number and the roughness of the pipe by numerous experiments. The relationship is presented in Fig. 6-4.

Pipeline Design

Pipelines consist of two major components: pipes and pumping stations. In addition there are the terminals where the liquid is stored, etc., which will be discussed in the next chapter. Since the cargo being transported in pipelines is completely enclosed, there is almost no contamination or loss of cargo (barring leaks, of course) and no inherent limitation on maximum gradient. The primary limitations on this technology are those imposed by pressures in the pipeline, which will, of course, rupture if the pressure is excessive. Modern methods of pipe fabrication and joining have permitted pressures of

Figure 6-4 **Pipeline liquid flow friction factors.** [*From Moody (1944).*]

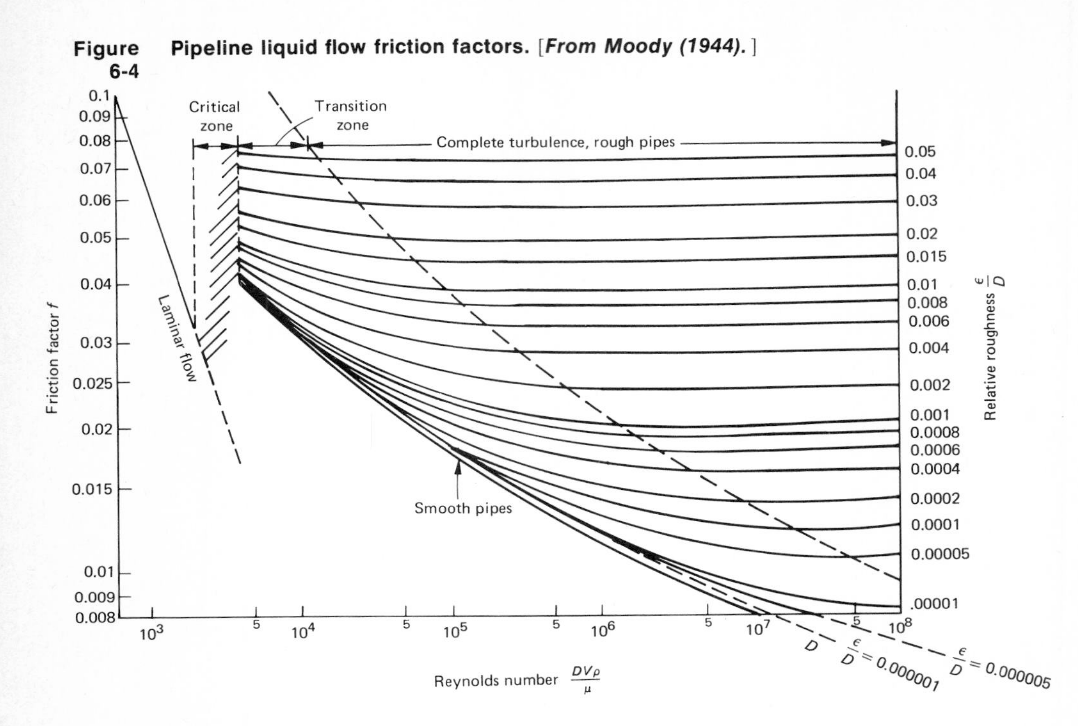

900 lb/in^2 to be commonplace, and undoubtedly higher pressures will be common in the future. It is interesting to note that this pressure corresponds to an elevation change of 3078 ft in a pipe carrying gasoline ($\rho = 42.1$ lb/ft^3) by Eq. (6-11), assuming no head loss and zero pressure at the higher elevation. This means that the elevation change between pumps cannot exceed about 3000 ft. Of course, this rule of thumb would be satisfied automatically in the laying out of a specific pipeline through the use of Eq. (6-11) to assess pressures. Generally the pressure in a pipeline is not allowed to go below 25 to 30 lb/in^2, which provides some slack in case of partial pump failure.

Of the many types of pumps, one of the most common is the reciprocating piston pump, in which the motion of the piston relative to the cylinder forces the liquid out of the pump, and then the vacuum forces more into the cylinder, and so on. The volume of liquid moved by such a pump is the product of the cubic volume of the displacement, the number of strokes per unit time, and an efficiency of the pumping action. This efficiency is usually about 80 to 97 percent, depending upon the design and maintenance, and reflects what is termed "slip" in the pump, literally the slippage of some liquid relative to the motion of the pump. Thus the volume output of a pump is

$$g_c = AzN\eta \tag{6-13}$$

where g_c = cubic volume output
A = piston area
z = stroke length
N = strokes per unit time
η = pump efficiency

The power required to produce this motion (volume of flow) can be calculated readily from these same variables. The force of the piston, its working pressure multiplied by its area, acts through the stroke length, producing work. This is done at a frequency of a given number of strokes per minute, resulting in a power at the piston of

$$P = yAzN \tag{6-14}$$

where y = working pressure of piston

A useful form of this equation is in terms of g_c:

$$P = \frac{yg_c}{\eta} \tag{6-15}$$

The work required to move any given amount of fluid is simply the product of the power so computed and the time period required to pump that amount of fluid, taking into account the pump slip. The volume output and power, etc., calculated above apply for one cylinder only; where more than one is used, it is necessary to multiply by the number of cylinders.

Sample Problem

A pipeline carrying a liquid (weighing 42.1 lb/ft^3, viscosity of 0.000794 lb = s/ft^2 at pipeline temperature) is to be built along a 200-mi level route. The pipe will be 22 in in diameter, with a roughness of 0.00005, and the design volume will be 200,000 barrels (bbl) per day (a barrel being 42.0 gal). If the maximum working pressure is 900 lb/in^2 and the minimum is 30 lb/in^2, how many pumps will be required? Where should they be located? Draw the energy line.

The resistance can be calculated only with the velocity. The cubic volume in the required units is

$$g_c = \frac{200{,}000\ \text{bbl}}{\text{day}}\ \frac{42.0\ \text{gal}}{\text{bbl}}\ \frac{\text{ft}^3}{7.48\ \text{gal}}\ \frac{\text{h}}{3600\ \text{s}}\ \frac{\text{day}}{24\ \text{h}} = 13.00\ \text{ft}^3/\text{s}$$

The cubic concentration per ft of length is

$$k_c = \pi\left(\frac{D}{2}\right)^2 = 3.1416\left(\frac{22}{24}\right)^2 = 2.640\ \text{ft}^2$$

or as it might be expressed, 2.640 ft^3/ft. From the fundamental equation of flow, (6-1), the average velocity is

$$V = \frac{g_c}{k_c} = \frac{13.00\ \text{ft}^3}{\text{s}}\ \frac{\text{ft}}{2.640\ \text{ft}^3} = 4.924\ \text{ft/s}$$

We find the Reynolds number is, taking care that units cancel,

$$R_N = \frac{VDd}{\mu g} = (4.924\ \text{ft/s})(1.83\ \text{ft})(42.1\ \text{lb/ft}^3)\,[(0.000794\ \text{lb-s/ft}^2)(32.16\ \text{ft/s}^2)]^{-1}$$
$$= 1.485 \times 10^4$$

From Fig. 6-5, the friction factor is about 0.027. Hence by the Darcy-Weisbach equation the head loss over 100 mi is

$$h_L = (0.027)(200\ \text{mi})(5280\ \text{ft/mi})(4.924\ \text{ft/s})^2[2(1.83\ \text{ft})(32.16\ \text{ft/s}^2)]^{-1}$$
$$= 5873.1\ \text{ft}$$

This head loss corresponds to a pressure loss of

$$(5873.1\ \text{ft})(42.1\ \text{lb/ft}^3)\frac{\text{ft}^2}{144\ \text{in}^2} = 1717.1\ \text{lb/in}^2$$

or 8.58 lb/in^2 loss per mile of pipe.

Since the pipeline is level, all of the pressure to propel the liquid must be provided by the pumps. Each pump can release the gas at 900 lb/in^2, and the gas pressure can drop to 30 lb/in^2 at the inlet to the next pump. Therefore pumps can be the following distance apart:

$$\frac{900\ \text{lb/in}^2 - 30\ \text{lb/in}^2}{8.58\ \text{lb/in}^2\text{-mi}} = 101.4\ \text{mi}$$

Hence two pumps are required, which could be located at the origin and the midpoint of the line. The energy line is shown in Fig. 6-5.

The power required of each pump, with an efficiency of 0.90, is by Eq. (6-16),

$$P = (900\ \text{lb/in}^2)(13.00\ \text{ft/s})\left(\frac{1}{0.90}\right)(144\ \text{in}^2/\text{ft}^2) = 1.872 \times 10^6\ \text{ft-lb/s}$$

$$= 56.73\ \text{hp}$$

To transport one day's supply, 300,000 bbl, requires propulsive work of

$$W = (2)(1.872 \times 10^6\ \text{ft-lb/s})(24\ \text{h})(3600\ \text{s/h})$$

$$= 3.234 \times 10^{11}\ \text{ft-lb}$$

This movement of liquid could also be performed by rail. The weight of 200,000 bbl is about 23,640 tons. Using standard tank cars of 12,500-gal capacity and an empty weight of 30 tons, this amount could be transported by a daily volume of 675 carloads per day, each car carrying 35 tons of gasoline. Two 3600-hp diesel-electric locomotives could pull 165 such cars (loaded) on level profile at 40 mi/h and ascend a 0.34 percent grade, so we shall use such a train for comparison. Thus this cargo volume would require an average of 4.09 trains/day each way on the line.

The propulsive work required for such a rail movement can be calculated using the methods of Chap. 4. For this the profile will be assumed to require a positive propulsive force over its entire length, the condition leading to

Figure 6-5 **Energy line for example pipeline problem.**

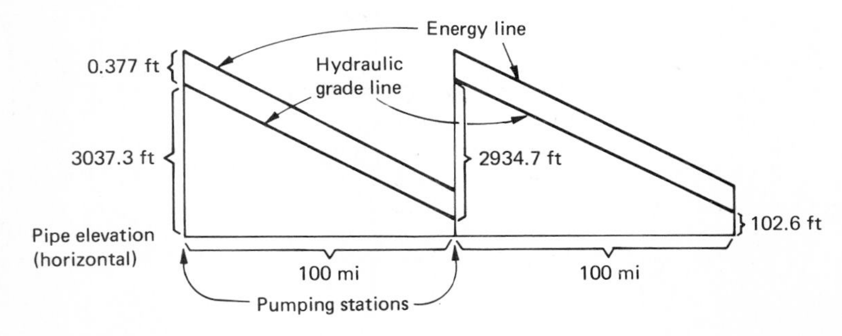

minimum work requirements. The inherent resistance of the train in the loaded direction is 50,655 lb and in the empty direction is 38,775 lb. If the rail route is the same length as the pipeline, the work required for 4.09 trains/day would be

$$(4.09 \text{ trains/day})(50{,}655 \text{ lb} + 38{,}775 \text{ lb})(200 \text{ mi})(5280 \text{ ft/mi}) = 3.863 \times 10^{11} \text{ ft-lb}$$

This is about 20 percent more than the movement via pipeline, under rather favorable conditions. With greater gradients and a longer route, both of which are likely, the difference would be greater.

This reveals one of the reasons that so much liquid moves via pipelines: the propulsive work required is usually less than that via wheeled vehicles. Barges, on the other hand, often are close to the pipeline work requirements. But the pipeline often has other advantages, mainly lower costs of operation and maintenance. However, the initial cost of the pipeline must be spread over a large volume for it to be attractive. These issues will be discussed more in the chapter on costs and planning.

OTHER TYPES OF CONTINUOUS-FLOW SYSTEMS

As was mentioned in the introductory remarks to this chapter, there are many different types of continuous-flow transport technologies. We have covered two of the most important in order to present basic principles.

Gas, Slurry, and Capsule Pipelines

Due to the complexity and rapid change in estimating relationships for resistances and other characteristics, we have chosen not to cover gas pipelines though they are very important. Of increasing importance is the flow of both gases and liquids in the same pipe, termed two-phase flow, which is also very complex and beyond the realm of this text. The reader who is interested in these is advised to peruse the *Journal of the Pipeline Division* of the ASCE and similar professional journals.

There is also much interest in the pipeline transport of solids suspended in liquids or gas. Coal is now transported in water in pipelines, the mixture being termed a slurry. This may become an important technology of transportation of coal in the future, even over very long distances, although most applications to date have been over relatively short distances. One problem with such technologies is that water must often be returned for reuse in the pipeline, much as empty trucks and railroad cars must be returned to places where traffic originates for loading. It is also technologically possible to transport general freight in capsules in pipelines. In fact, this has been done on a very small scale in stores and offices for decades, with pneumatic (propelled by air pressure differentials) pipelines using small capsules for letters, money, etc. As described by Stoess (1970), pneumatic systems can be used to carry a wide variety of light materials, such as grain, wood chips, and cement. Recent experiments indicate that this has potential for moving much larger capsules, perhaps weighing a ton or more. And, of course, if the capsule were sealed properly, it could be carried in a liquid. A very interesting account of these and other developments in pipeline technology appears in the article by Zandi (1974).

Conveyors

Conveyors also have interesting possibilities. In addition to the transportation of bulk materials such as coal and ores over distances of a mile or more, there are numerous installations in factories and warehouses for the movement of general freight. In fact, many urban delivery trucks carry small conveyors to be used in loading and unloading. Usually these are of the roller type, consisting of rollers on which objects with large flat bottoms can roll with ease (see Haynes, 1957).

Another very important application of conveyors is to the movement of people, usually large volumes over short distances, since the speeds are limited to about 3 mi/h so that people may safely enter and leave the conveyor. However, by having end-to-end or side-to-side conveyors, the speeds can be increased, by usually about 2 mi/h for each belt-to-belt transfer. Such conveyors are used in many large airports, for example. There are safety problems with such systems where the speed change occurs, and space requirements would be rather sizable if a large number of speed steps are to be incorporated to achieve moderate line-haul speeds (e.g., five steps to 10 mi/h).

Without a means of increasing speed well above 3 mi/h the role of conveyors would be limited to short movements. This, coupled with the problems of conventional vehicle systems in application to short-distance movements which are nevertheless too lengthy for walking, say 1 to about 5 mi (or shorter instances with luggage, as at airports), has led many engineers to consider possible forms for continuous (or nearly continuous) systems (Tough and O'Flaherty, 1971).

The basic concepts of many of these results from consideration of the fundamental equation of flow, (6-1). To maintain a constant volume of passenger flow at low speeds in terminal areas the concentration of passengers per unit length of path must be higher than that on links where the system operates at a higher speed. This led to the concept of a wide conveyor at terminals, moving at slow speed, becoming narrower and traveling at a higher speed between terminals. Figure 6-6*a* illustrates a specific form that such a conveyor might take (Todd, 1974). The conveyor is composed of rectangular plates, which are accelerated as shown from a speed V_0 to a speed V_{max}. A constant volume is maintained by changing the orientation of the plate relative to the axis of motion. Since in the low-speed direction the length of a plate is B while in the high-speed direction it is L, the speed ratio is L/B. This design has the undesirable feature of spaces being created between plates during acceleration, with potential safety problems. A continuous platform is provided by the design presented in Fig. 6-6*b*, the Dunlop S-Type Speedaway, developed by the Dunlop-Angus Belting Group of the United Kingdom and Battele Institute of Geneva. By limiting passengers to the center section of each trapezoidal plate, a continuous surface is provided. It is termed S-type because in a typical application one terminal would be to the left, the other to the right, yielding an S-shaped path. A prototype of this system has been constructed and tested.

Many proposals have been made for larger-scale uses of conveyors, perhaps with large containers for freight or passenger containers traveling on roller conveyors or belts. The book by Richards (1966) contains photographs

Figure 6-6 **Accelerating conveyor concepts.** ***(a)*** **Original accelerator proposed at Battelle Institute of Geneva;** ***(b)*** **Dunlop S-Type Speedaway accelerator.** [*From Todd (1974), pp. 68 and 71.*]

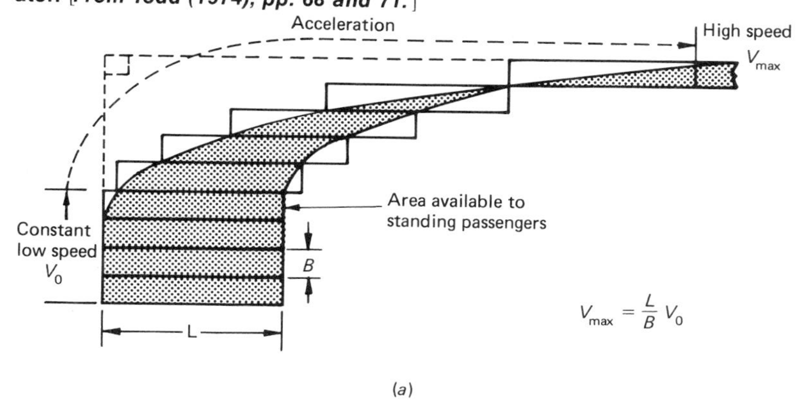

(*a*)

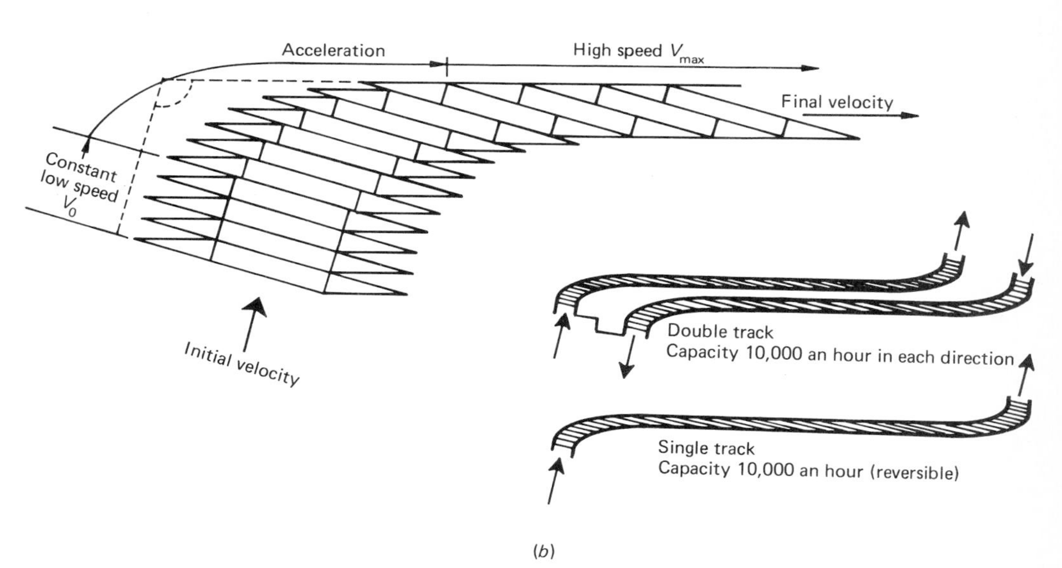

(*b*)

and a brief nontechnical discussion of many of these systems—both built and proposed—primarily for the movement of persons.

Aerial Tramways or Ropeways

Another type of continuous-flow system which is very close to the conveyor in concept is the aerial tramway. Such systems consist of one or more wires from which small containers for passengers or freight are suspended. These containers may be permanently affixed to the wire rope, as is often the case on ski lifts, or the containers may move on wheels on a fixed rope, being pulled by another rope (these two systems being called monocable and bicable, respectively). These containers may travel back and forth on one path or move continuously on a two-path (parallel rope) system, the latter being far more common.

The reasons for using aerial tramways are similar to those for use of other continuous-flow systems (Schneigert, 1966): ease of ascending and descending grades (up to 125 percent), conservation of potential energy in movement over an undulating profile regardless of gradients, ease of construction over any area, since support is necessary only at intermittent points (advantageous not only in mountains, but also over soft ground, water, etc.), continuous flow that may be better matched to actual traffic flow than intermittent flow of conventional vehicular systems, and often low operating costs. However, like conveyors and pipelines, speeds are usually low, typically 1.5 to 3.0 m/s (3.3 to 6.7 mi/h). This is often of little consequence, though, especially for freight and for passengers where no other technology would be economically feasible. Capacities range up to 500 tons/h for freight, about 400 passengers/h for enclosed cabin systems (Schneigert, 1966, pp. 36–48). A capacity of 1200 passengers/h is possible for open-seat ski-lift-type systems (Dwyer, 1975, p. 60). For higher volumes parallel lines are usually constructed. A complete technical presentation of these systems, their planning and design, is presented by Schneigert (1966). Although most systems are short, up to 1 or 2 mi, a monocable ropeway for transporting ore in Gabon is 76 km in length.

SUMMARY

1 Continuous-flow systems are widespread for the movement of many types of freight. The items carried consist mainly of bulk freight, liquids, and gases hauled over long distances. They are also important movers of passengers over short distances. There are also many applications to short-distance goods movement in factories and warehouses, although these applications usually are considered a part of materials handling rather than transportation engineering.

2 Many advances in the technology of continuous systems are being made, especially in liquid and gas flow and in the movement of solids in such media. These advances may result in much more widespread use of these technologies in the future.

3 Continuous-flow systems have advantages over vehicular systems in situations where there are substantial changes in elevation, because they are also easily adapted to steep gradients. Furthermore, they can also use any potential energy gained to overcome inherent resistances to motion. But, as the problem examples illustrated, continuous systems may not always require less work to move a given amount of freight than vehicular systems. Also, in the case of fluid flow (for its own sake, as opposed to use of the liquid purely to provide a slurry), there is no need to return an empty vehicle or container.

4 Continuous-flow systems offer many design options as to the combination of speed and concentration which can achieve the same volume. The options must be carefully considered with respect to the path required (e.g.,

belt width or pipe diameter), resistances to motion, etc., in order to result in a reasonable system. Due to resistance relationships, pipelines are best operated at very low speeds. Belts must be operated at low speeds in order to avoid damage to belts and loss of cargo.

PROBLEMS

6-1

A 15-in pipeline is being designed to transport 75,000 bbl/day of crude oil per day (weight density = 55 lb/ft^3) from a well head to a refinery 350 mi distant. The origin well head is 350 ft higher than the refinery, and the entire loss in elevation occurs at a constant downgrade in the first 55 mi. The pipe friction factor is 0.014. Determine the location of the pumping stations if the maximum and minimum working pressures are to be 700 and 25 lb/in^2, respectively. What is the speed of this flow?

6-2

A pipeline with a diameter of 18 in has a pumping station of 100 hp (efficiency = 0.92) located every 50 mi. The line is level, and it is designed for a maximum safe working pressure of 700 lb/in^2 and a minimum of 25 lb/in^2. What is the maximum volume of gasoline that can be transported through this line per day?

6-3

Ore weighing 125 lb/ft^3 is to be transported by conveyor belt from a mine head to a ship-loading terminal. The mine is 1000 ft above the terminal, and the route is 1.4 mi in length, with a constant downgrade. The belt is 36 in wide, weighs 25 lb/ft with all rolling parts, and travels at 4300 ft/min. The cargo is generally loaded to a concentration as specified by Eq. (6-7). How much power is required?

6-4

Discuss why, for situations such as that in Prob. 6-3, a continuous-flow system has certain inherent advantages in propulsive work required over a vehicular system. Is there any way to give this advantage to a vehicular system? (Hint: Think of ski lifts or inclined plane railways.)

6-5

For the problem on gasoline pipeline flow worked out in the text, calculate the work required if the pipeline were only 18 in in diameter instead of 22 in, or only 12 in in diameter, or 26 in in diameter. Can you generalize these results, basing your statements on the underlying relationships? (Note to instructor: This problem is best worked by many different persons, each performing the calculations for one diameter.)

6-6

Many raw materials transported by pipeline or conveyor are extracted at approximately the same volume throughout the work-day. Does this yield any advantage to continuous-flow systems in contrast to, say, railroad service, which might entail operation of only one or a few trains per day? What additional facilities would be needed at the point of extraction for occasional or intermittent as opposed to continuous movement?

6-7
A passenger conveyor is to be designed with a cruise speed between terminals of 10 mi/h. A design like that in Fig. 6-6 is to be used. If the angle of the longest side of each plate relative to the cruise axis is 20° and a cruise path width of 4 ft is to be provided, what must be the dimensions of the plates? What would the entrance speed be? The plates are 1 ft wide.

REFERENCES

Dwyer, Charles F. (1975), "Aerial Tramways, Ski Lifts, and Tows," U.S. Government Printing Office, Washington, DC.

Hay, William W. (1961), "An Introduction to Transportation Engineering," Wiley, New York.

Haynes, D. Oliphant (1957), "Materials Handling Equipment," Chilton, Philadelphia.

Hennes, Robert G., and Martin Ekse (1969), "Fundamentals of Transportation Engineering," 2d ed., McGraw-Hill, New York.

Margenau, Henry, William W. Watson, and C. G. Montgomery (1953), "Physics Principles and Applications," McGraw-Hill, New York.

Moody, Lewis F. (1944), Friction Factors for Pipe Flow, *A.S.M.E. Transactions,* **66:** 671–684.

Paquette, Radnor J., Norman Ashford, and Paul H. Wright (1972), "Transportation Engineering," Ronald, New York.

Richards, Brian (1966), "New Movement in Cities," Reinhold, New York.

Schneigert, Zbigniew (1966), "Aerial Tramways and Funicular Railways," Pergamon, New York.

Staniar, W., ed. (1959), "Plant Engineering Handbook," McGraw-Hill, New York.

Stoess, H. A., Jr. (1970), "Pneumatic Conveying," Wiley-Interscience, New York.

Todd, J. K. (1974), Dunlop S-Type Speedaway: A High-Speed Passenger Conveyor, *Transportation Research Record,* **522:** 65–75.

Tough, John M., and Coleman A. O'Flaherty (1971), "Passenger Conveyors: An Innovatory Form of Communal Transport," Allan, London.

Zandi, Iraj (1974), Future of Pipeline—Beyond Liquid and Gas. *Transportation Research Forum, Proceedings,* **15:** 187–202.

Terminals

7.

Terminals—the points at which passengers and freight enter and leave the system—are essential components of all transport systems. They not only represent a major functional component of the system but also often are major costs and possible points of congestion. Terminals are especially important to transportation engineers and planners, who are usually responsible for designing them for specific locations and tasks. The correction of design deficiencies is often exceedingly difficult, time-consuming, and costly.

Despite the importance of terminals in all technologies of transport, the level of knowledge of operational characteristics and design guidelines varies greatly among the different types of terminals. In particular, airports and parking facilities seem to be relatively well understood, ports and urban transit terminals are to a somewhat lesser extent, and rail and truck terminals seem to be the least understood. Nevertheless, we shall focus on general principles and engineering methods applicable to all transport technologies, using examples to bring out the nuances of specific applications.

In this chapter we shall focus on the operational characteristics of terminals and their interrelationship with design characteristics such as the overall layout of the terminal facilities. This treatment of terminals will complete the description of the way in which components of transport systems operate, so that we can then deal with the functioning of an entire system in the next chapter, treating such factors as operating plans and the level of service provided by the system as a whole. Then, after discussing the principles of engineering economics (as applied to transport) in Chap. 9, it will be possible to cover the design of facilities, including terminals, links, and interchanges.

TERMINAL FUNCTIONS

The primary function of a transportation terminal, as discussed in Chap. 3, is to provide for the entrance or exit of the objects to be transported, passengers or freight, to and from the system. In vehicular transport systems, this is primarily the loading and unloading of the vehicle or container in which the traffic is to be transported. In continuous-flow systems, where way links and intersections, in conjunction with freight or passenger traffic, may yield mobility and locomotion directly without the need for vehicles, the terminal is located where the traffic enters and leaves the link and intersection system.

Some terminals that only serve a single function, loading and unloading, are extremely simple. Consider, for example, the common bus stop at a street corner, often consisting of only a place for passengers to wait and a sign indicating that it is a bus stop. Similarly simple may be the facilities of a shipper or receiver of freight, which might consist of a single platform (often the height of the truck floor or bed) on which freight is placed before or after shipment.

These extremely simple terminals do not reveal the myriad of other functions which must be performed in addition to the simple physical loading and unloading of passengers or freight and the complexity required at many transport terminals. Of major importance in most systems is a means of collecting the proper charges or fare for the movement, which usually requires a knowledge of the points of both entry and exit to the system. It is also essential to determine the routing of the objects being carried. In the case of passengers, this is often done by the traveler, based upon information (on routes, etc.) provided by the system, although the guidance of others may be sought (e.g., automobile associations, carrier's ticket, or travel agents). But for freight, either the shipper or the carrier (as is often done for long-distance movements where the number of routing options may be very large) must specify the route. In fact, freight movement usually involves a considerable amount of processing at the terminal or origin, including weighing of the shipment (as this forms a basis for charges), determination of any special handling required (e.g., if fragile or requiring refrigeration), and preparation of billing and documents to travel with the shipment to ensure it is directed to the proper destination.

Since it is impractical for all the passengers or freight which might be traveling together in one vehicle to reach the terminal just before departure, it is also necessary for most terminals to provide for the waiting of passengers or storage of freight until it is time to load the vehicle. If the period of wait is long, elaborate facilities may be required. Airports and other passenger terminals provide numerous services for the comfort and convenience of travelers, including comfortable waiting areas, restaurants, places for entertainment, etc. The distribution of airport revenue from airlines (aircraft operations) and other sources may give some idea of the importance of these additional activities. It has been estimated that fully 60 percent of the cost of constructing and operating Cleveland's airport (exclusive of portions owned by the airlines, such as maintenance hangers) will come not from aircraft landing and other fees but from concessions of all sorts—restaurants, stores, parking lots, auto rental firms, etc. (Winchester, 1973, p. 141).

Freight facilities perform a similar function, storing freight to protect it from damage or loss, including covered or enclosed storage of freight susceptible to damage by the elements and refrigeration or heating as required. Parts of freight terminals are often operated as warehouses, where freight can be stored until the owner decides to send it to its destination, perhaps waiting for the demand or price of the commodity to increase. Also, large shipments are often divided up at terminals (e.g., large imports of a raw material broken up into smaller shipments destined for many separate locations) or shipments combined (as different grades or types of coal or grain might be combined to yield a particular blend) if bound for a common destination. In a few cases the shipment may be processed at intermodal terminals, especially ports, where the freight must be loaded and unloaded anyhow.

Terminals are often the sites at which transport vehicles are maintained, since they must stop there anyway and often are completely unloaded. For example, hangars, fueling, and other servicing facilities are a part of almost all large airports. In the case of ground transport, where the terminal may be located in congested parts of cities where the land is quite valuable, maintenance facilities are often located in the outskirts of the urban area, where land is available but close enough so as not to require an undue amount of empty vehicle operation.

Table 7-1 summarizes these functions of terminals. To reinforce the idea of the commonality of functions of terminals regardless of the modal context, Table 7-2 has been prepared. It lists many of the most common names for terminals, and indicates the various functions typically performed at each.

ANALYSIS OF TERMINALS

Terminals can be viewed as processors of freight and passengers and of the containers, vehicles, etc., of the transportation system which will carry the traffic. The processor performs such functions as the loading of vehicles in a certain order. This processing requires a physical plant, labor and equipment, and rules or procedures to govern the operation and ensure that all the functions are carried out properly and in the correct sequence. Processing also requires time—the time of the traveler or shipment of freight and the time of the containers, vehicles, or other transport equipment. These two categories account for much of the cost of terminals, although there are others, such as environmental costs which we shall discuss in a later chapter. Different terminal designs, different loads of traffic to be accommodated, and different procedures, can all be analyzed by considering these two major

Table 7-1 Functions of Transportation Terminals

Functions of Transportation Terminals
Loading of passengers or freight onto transport vehicles (or belt, pipeline, etc.) and unloading
Transfer from one vehicle to another
Storage of passengers or freight from time of arrival to time of departure
Possible processing of goods, packaging for movement
Provision of comfort amenities for passengers (e.g., food service)
Documentation of movement
Freight weighing, preparation of waybill, selection of route, billing
Passenger ticket sales, checking reservations
Vehicle (and other component) storage, maintenance, and assignment
Concentration of passengers and freight into groups of economical size for movement (e.g., to fill a train or airliner) and dispersal at other end of trip

Table 7-2 Common Names for Facilities Performing Terminal Functions of the Various Transportation Modes

Major mode	Facility	Modal interface or other primary function
Air	Airport	Ground access modes and air connections
	Field	Airport with very limited facilities
	Hanger	Repair and servicing
	Helioport	Same as airport
	Seaplane base	Same as airport
Automobile (and other road vehicles)	Parking garage or lot	Vehicle storage, walking access
	Gas station	Vehicle repair and servicing
	Toll booth	Collection of fees
Bus	Bus station	Intercity bus and access modes connections
	Bus stop	Walking access connections
Rail, passenger	Railroad station	Local access modes and rail connections; sometimes includes other intercity modes (e.g., bus)
Rail, freight	Freight house	Local (truck) access
	Team tracks	Local (truck) access (open area as opposed to building)
	Private siding	Loading or unloading by shipper or receiver of freight
	Classification yard	Freight train connections
	Shops, rip track, etc.	Car repair and servicing
	Engine shed, roundhouse	Locomotive repair and servicing
Water	Port	Ground access modes (usually rail, truck and/or pipeline) and sometimes vessel connections
	Dock, wharf, or pier	A single vessel loading-unloading facility
	Dry dock	Repair and servicing

categories of costs—cost to the terminal owner and operator and costs to the user of the system.

Process Flow Charts

One important means for describing and understanding terminals is through the use of a process flow chart. Such a chart portrays the activities which a traveler, vehicle, or other unit of traffic undergoes as it is processed through the facility. It portrays the sequence of activities, possible alternative sequences if they exist, and may be used to determine the time required for processing.

There is no rule to determine when and how finely or coarsely the activities in a terminal should be divided for purposes of process flow chart representation. In fact, the simplest flow chart might portray the terminal as simply a single all-encompassing activity, as Fig. 7-1 does. There the terminal is represented by a single box, with the flows of passenger and freight traffic into and out of the terminal shown, along with the input (arrivals) and outputs (departures) of vehicles.

Figure 7-1 **Simplified process flow chart representation of a transport terminal. (*a*) Transport flows only. (*b*) Extended representation of inputs and outputs.**

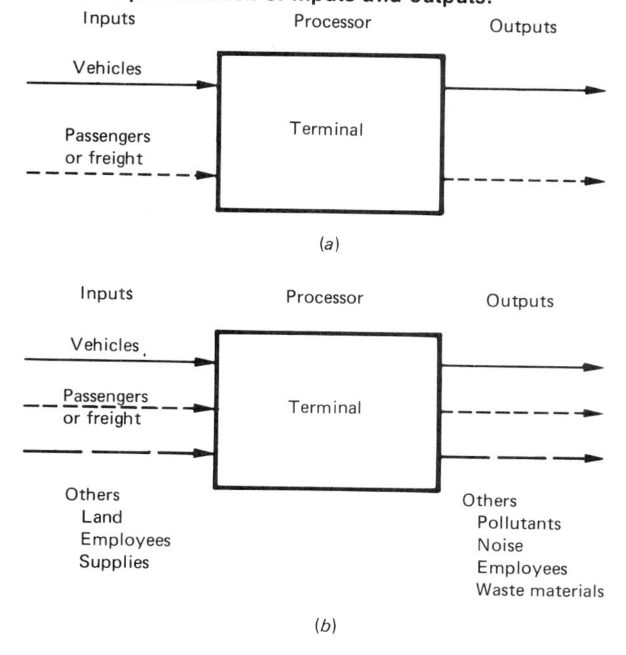

The time required for the processing of passengers and freight originating or terminating at this terminal is the difference between their times of arrival and times of departure. The time of vehicles or containers consumed in terminal processing is the difference between arrival and departure, which may be quite long if servicing is performed. Thus at least conceptually, we can use this representation of a terminal to analyze its operational characteristics.

Cost can also be represented by this method. In addition to the vehicle and passenger or freight traffic inputs to a terminal, many other items pass between the terminal and its environment. Entering are the employees who manage, operate, and maintain the terminal, supplies of all sorts for operation of the terminal, etc. Many of these later leave: the employees after completion of a day's work, the waste products of the terminal, ranging from waste from kitchens of restaurants to used oils and parts, and even pollutants of the air. The cost of operation of the terminal can be viewed as the translation of these inputs and outputs into monetary terms. For example, the labor cost can be determined by calculating the time employees spend at the terminal, by job type, multiplied by the corresponding wage rate, including employee benefits, etc. Even the cost of the land on which the terminal sits can be included, if the land is viewed as an input to the terminal which does not become an output until the terminal is abandoned and the land becomes available for some other use. This more complete representation of the overall cost of the inputs and outputs of the terminal is portrayed in Fig. 7-1*b*.

While the use of process flow charts for complete analysis of inputs and outputs of processes over the entire life of those processes is important for some problems, such as estimating the full costs of terminals over the entire life of the facility, the more limited view of regular transport inputs and outputs such as shown in Fig. 7-1*a* is more useful for our immediate concerns of defining terminal characteristics. A representation such as that in Fig. 7-1*a* can be expanded in detail to provide a fuller indication of the various tasks that are performed at the terminal. The loading and unloading processes may be distinguished, as might any vehicle servicing and storage. Also, the documentation of the movement, ticketing, and perhaps reservation checking for passengers and the weighing and preparation of waybills for freight might be separated from other functions.

In order to include additional details of terminal processing in a process flow chart, it is useful to use different symbols for different types of activities. To facilitate this, a standard set of symbols has evolved for portraying the various components of terminals. While they are not used in all engineering studies, their use is helpful in making flow charts easier to read, in much the same way that standard symbols help in understanding engineering drawings. The standard symbols, shown in Fig. 7-2, portray (1) a source of traffic, (2) a server, (3) a waiting line (presumably waiting to enter a server), (4) a routing decision point, (5) a sink or exit of traffic from the terminal. Arrows connecting these elements indicate the direction of movement of the traffic.

Figure 7-2 **Standard symbols used in process flow charts.**

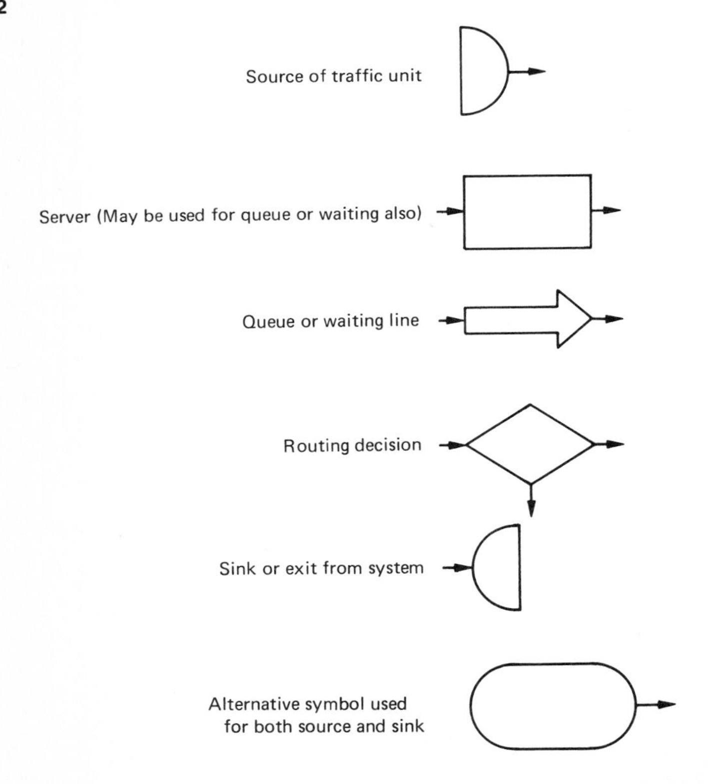

To distinguish between different types of traffic, different types of lines with arrows can be used. From the definitions of these standard symbols, it is apparent that they can be used in the analysis of any technology of transportation, in any context.

Greater detail in representation of a terminal would result in a process flow chart such as that given in Fig. 7-3. It portrays a conventional intercity pas-

Figure 7-3 **Detailed process flow chart of a general passenger terminal.** [***From Consad Research Corp. (1970), vol. II, p. 281.***]

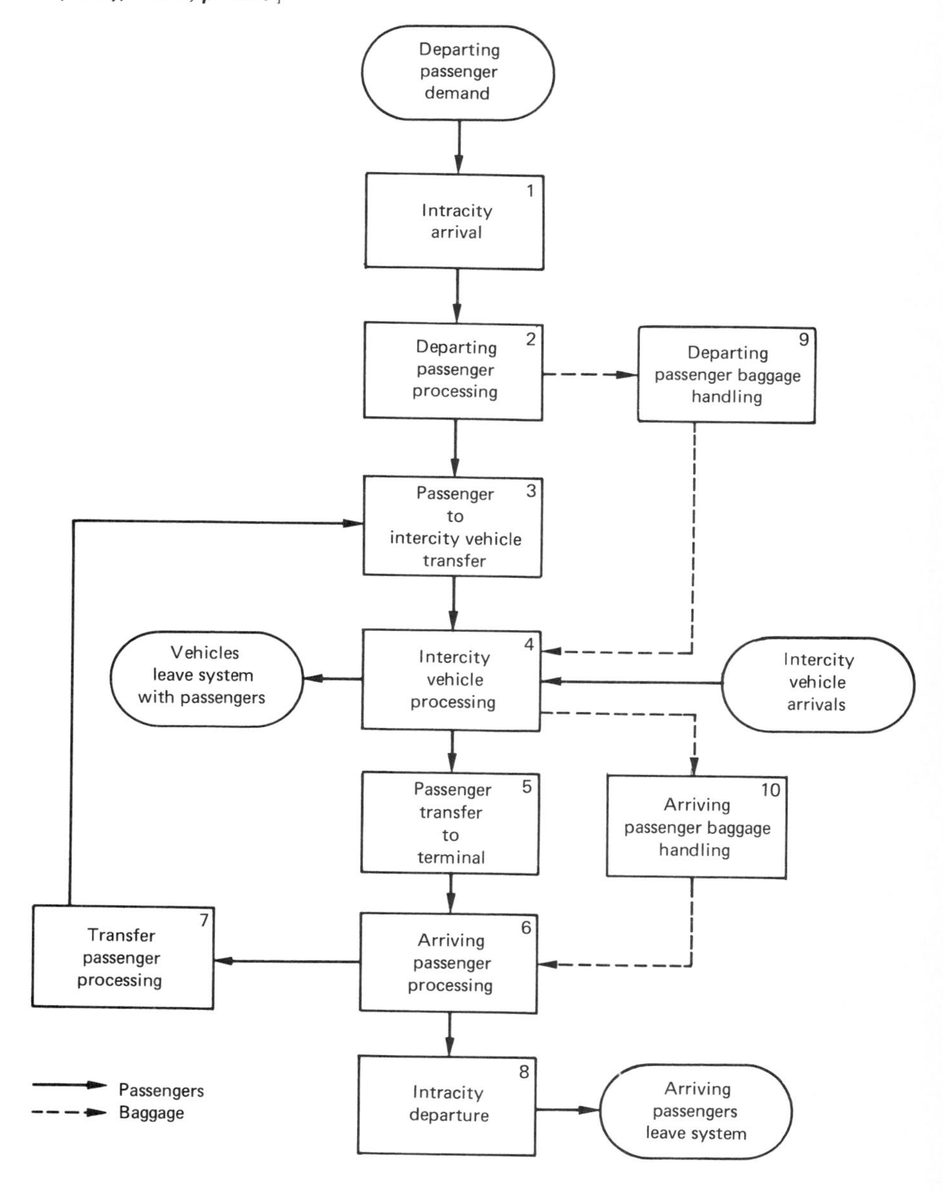

senger carrier terminal, including the processing of the vehicle (e.g., aircraft), passengers, and baggage. This takes the passenger from arrival on an intracity vehicle through the various check-in activities to the boarding of the outbound vehicle and the departure of that vehicle. Since the same vehicle may arrive with passengers, its arrival is also shown, with passengers leaving the vehicle, through to their entry into the intracity vehicle or transfer to another intercity trip. Figure 7-4 portrays an intercity freight terminal in a comparable manner.

In addition to describing characteristics of a terminal, the process flow chart is a powerful tool for the evaluation of alternative designs and operational plans. Different designs can be represented by means of the flow chart, and their advantages and disadvantages can be identified. This then serves as a prelude to more detailed quantitative analysis of each alternative.

Example Application

As an example of the use of the process flow chart in the analysis of transportation terminals, and more generally entire transportation systems, we shall consider two alternative means of transporting freight from a shipper to the receiver (consignee in transportation parlance). Neither the shipper nor the consignee has a railroad siding, but, in this example, movement by rail for the bulk of the journey is preferred. One form of movement would be in containers, which can be transported on flatbed trucks to the rail yard, rapidly and easily transferred onto rail cars, and then moved in trains to the destination city where the reverse terminal and access process occurs. A flow chart of this is shown in Fig. 7-5*a*. The movement could be made in a more conventional manner by loading the cargo into trucks at the origin, moving it to a rail

Figure 7-4 **Process flow chart of a general goods terminal.** [***From Consad Research Corp. (1970), vol. II, p. 282.***]

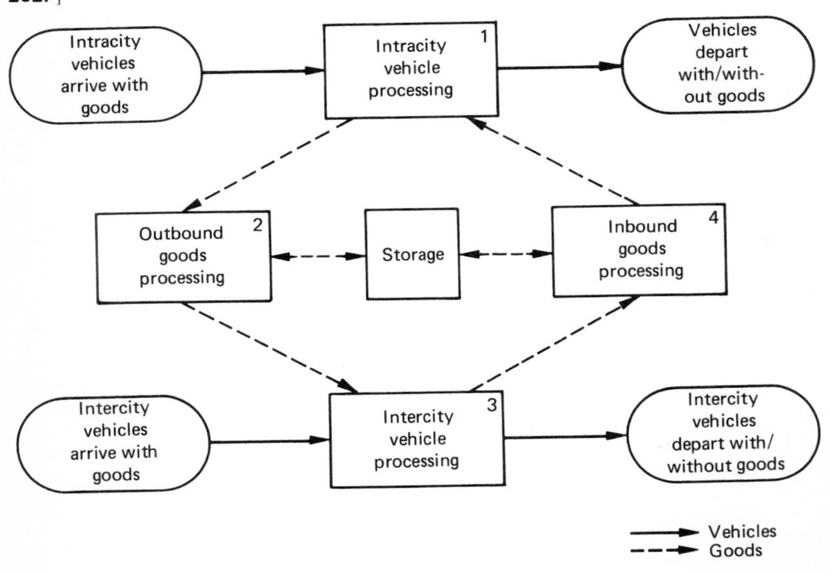

Figure 7-5 A comparison of two methods of carrying freight between nonrail points using a process flow chart. (*a*) Containerized movement. (*b*) Conventional truck-rail movement.

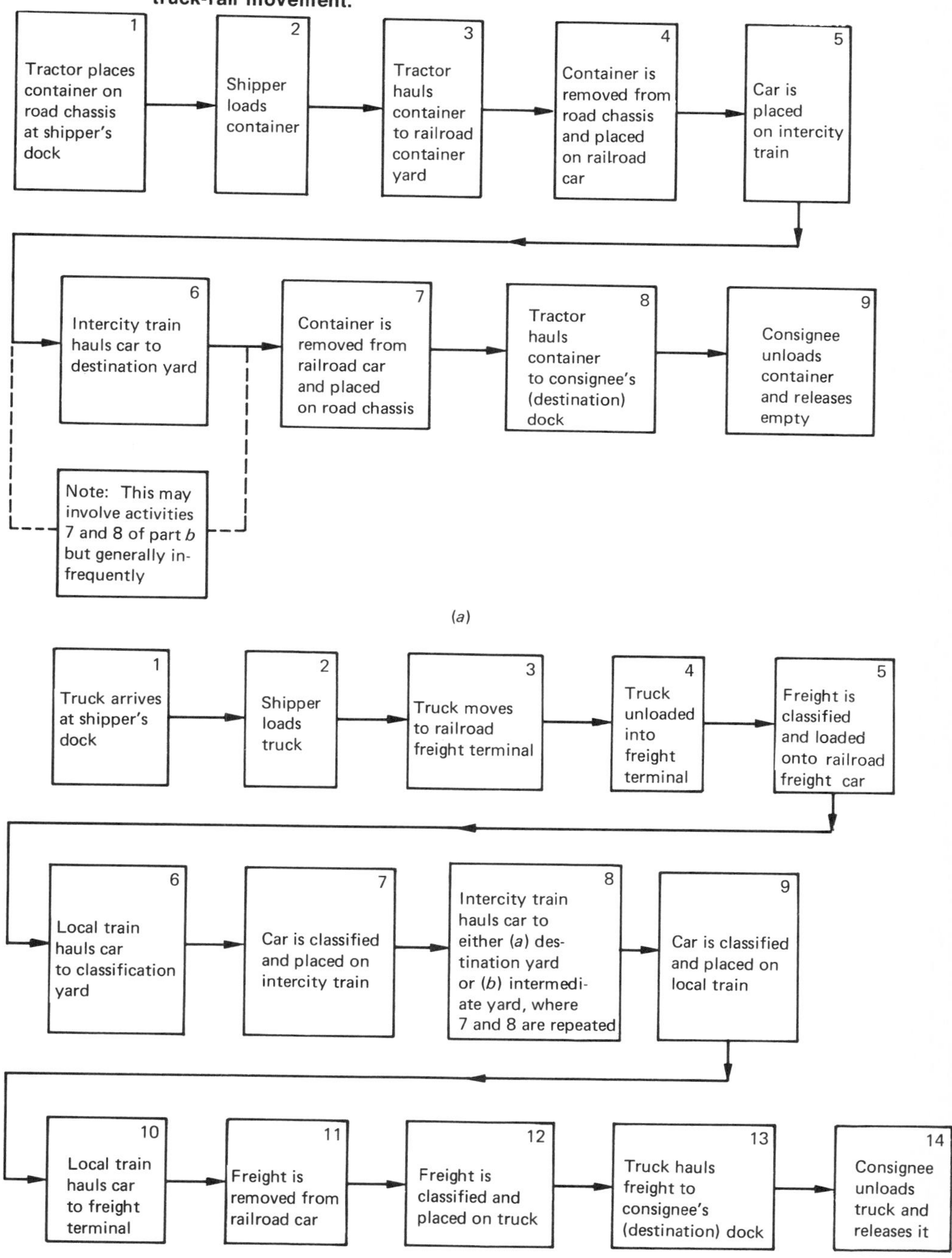

freight house, unloading it piece by piece from the trucks, then loading it onto the rail cars, and then moving it by trains to the destination. This is shown in Fig. 7-5*b*. The reduction in the number of steps and in their complexity resulting from the containerized movement is evident from these charts. As we shall see in subsequent sections of this chapter and Chaps. 8 and 9, such flow charts are very useful in estimating the costs and levels of service.

Terminal Processing Times

The process flow chart may be used to calculate the time required to process vehicles, passengers, and other traffic units at a terminal. The method is basically quite simple, consisting of summing the times required for the individual processes.

In calculating the times of the various activities involved in the processing of a traffic unit, it is important to consider possible deviations from the ideal time which any activity might take. In many of the activities at terminals, the server has a very limited capacity, and if more traffic units wish to use that server in any time period than it can accommodate, some will have to wait. This waiting time must then be added to the actual serving time in order to represent the actual time that activity will consume. Such waiting, or queuing as it is often called (from the British and French term for waiting line queue), is very common in transportation terminals.

One example is the waiting at the check-in counter of an airline terminal, as noted in the airport process flow chart, Fig. 7-4. Another type of waiting, which usually does not involve a clearly defined line and hence may not be thought of as a queue, is the waiting for a subway train at a station. Passengers must wait until the train arrives and its doors are opened. One can imagine the train as a server which can accept passengers only at specified times—when a train is in the station with open doors. During that period, the train can accept passengers at only a certain rate or volume (i.e., passengers per minute) or less, due to the limited number of doors, their width, and the behavior of passengers in following one another, entry presumably occurring after all disembarking passengers have left the train. (In practice, of course, traveler behavior in entering and leaving trains is often not so orderly!) Thus some waiting may be associated with congestion, while some is simply a natural result of the way transport terminals are designed and are operated.

Another complicating factor is that the processing times for different traffic units are usually not the same. This is partly because all travelers intending to use a particular departure do not arrive at the terminal at the same time. Some may deliberately arrive well in advance to be sure they do not miss the departure, while others may not be so cautious and allow less excess time. Thus some of the processing time may in fact be discretionary on the part of the traveler (or shipper in the case of freight), although it is often the perceived or actual unreliability of the access to the terminal that may cause many to allow a great deal of excess time. Also, some may arrive for serving (e.g., airport check-in) when the counter is very busy and may have to wait longer than others who arrive at less crowded times. Then there are usually differences in the times required to serve various traffic units. Again using an airport example, one traveler may be purchasing a ticket for a multileg journey, while another may require a simple one-leg trip ticket. Often this variation leads engineers and others to express the times required in terms of means or distributions showing the fraction of travelers who experience times falling within various ranges.

As an example of the use of the process flow chart to calculate the time required for terminal processing, consider the time consumed in an air journey from Paris to London. The times required for the various activities which comprise the entry journey, according to the breakdown presented in Fig. 7-3, are presented in Table 7-3. These are the actual times taken for a journey in 1969, as reported in an analysis of airport access and terminal problems (O.E.C.D., 1971), and are considered typical for such a trip. This table also gives the times from the origin to the Paris terminal via taxi and from the London airport via taxi to the destination, both origin and destination being in the center of the cities.

Table 7-3 Actual Travel Times Through Terminal Processing for Air Trip from Paris to London

Process activity	Distance, km	Time, min	Cost, francs
Travel from origin to Orly Airport (Paris) via taxi			
Walk to taxi stand	0.25	3.5	
Wait for taxi		4.5	
Taxi trip	20.00	27.0	24.00
Exit taxi		}	
Enter terminal		}	
Walk to ticket counter		} 3.0	
Purchase ticket and check bags		}	10.00 (check-in charge)
Walk to departure area (through passport control)		7.0	
Wait (hold) for aircraft		5.0	
Board aircraft		3.0	
Wait 15 min, then taxi to runway and wait for clearance		25.0	
Flight Paris Orly to London Heathrow	350.00	55.0	166.00
Taxi to terminal and wait		5.0	
Disembark aircraft (and walk to passport control)		5.0	
Wait at passport control		25.0	
Passport control		5.0	
Walk to baggage claim		} 4.0	
Wait for baggage		}	
Pass through customs		6.0	
Travel to destination (London) via taxi			
Wait for taxi		}	
Board taxi		} 45.0	
Taxi trip	17.00	}	45.50
Total, origin to destination	387.25	228.0	245.50
Average speed = 101.9 km/h = 63.7 mi/h			
Total not including taxi to and from airport	350.00	148.0	176.00
Average speed = 141.9 km/h = 88.7 mi/h			

Source: G. Bouladon, The Horizontal Aspect of Air Travel, in O.E.C.D. (1971, pp. 29–39).
Note: The process activities used above correspond to those in Fig. 7-3.

As can be seen from the table, the entire trip of 387 km took 228 min, for an average speed of 101.9 km/h (about 63.7 mi/h). The portion on board the aircraft, in contrast, averaged 381.8 km/h, which included at least 10 min of slack time in the schedule to allow for delays in take-off and landing. From the point of entry at the Paris airport to exit from the London airport, the average speed was 141.9 km/h, somewhat better but still far below the average for the flight. More to the point, of the entire time in the air system—from entry at one airport to exit from the other—148 min, fully 93 min or 63 percent was spent at the terminal! Even allowing for some time for aircraft taxiing, etc., the majority of the time was spent at the airport. Thus the importance of terminals to travelers is obvious. It is conditions such as these which are leading many airport engineers and planners to try to devise less time-consuming and less costly terminals for air travel, such as the remote-parking terminal described by the flow chart earlier may prove to be.

Numerous activities occur simultaneously in many situations. In the processing of passengers at airports, for instance, passengers are debarking from the aircraft and walking toward the baggage claim area while their baggage is being unloaded and taken to the same area. This is clearly shown in Fig. 7-3, where these two operations are in parallel. It is the longer of these two which determines when the traveler is ready to leave the airport; the baggage very often takes longer than the walk, as was the case in the example trip through the London terminal described in Table 7-3.

While an aircraft is at the terminal, basically two separate types of activities are being carried out. One is the loading and unloading of passengers and baggage, and the other is the servicing of the aircraft, including the cleaning of the cabin, the stocking of the galley, and the refueling and maintenance checks on the craft itself. Typical times for these are shown in Fig. 7-6. In this chart the times required by each of these activities is shown, and each one is shown to start only after others which must precede it are completed. For example, the cabin cannot be serviced (mainly cleaned) until all passengers have deplaned, and boarding cannot begin until the cabin service is completed. With such a chart, the longest time from the beginning of any one activity to the beginning of another can be determined. Assuming each activity begins as soon as prior activities are completed, the minimum time for the entire sequence of activities can be identified. This is called the critical time, and the sequence of activities the critical time path, as shown in Fig. 7-6. In this case it is 30 min for aircraft processing at a gate, assuming that each task is completed in the time shown. Conservatively, airlines estimate from 30 to 40 min for gate processing with cabin service. If no cabin or food servicing is required, the time is usually 20 to 30 min. This type of chart can be used in any situation where there are parallel activities to identify the critical time path and hence the minimum time required for the process.

Waiting Lines

As was mentioned earlier, waiting occurs at many terminal activities. The analysis of such waiting is one of the most critical aspects of terminal evaluation, and the prediction of possible queues is an important aspect of terminal design. While the minimization of waiting is often sought, at times waiting cannot be avoided and sufficient capacity must be designed in waiting or storage areas to accommodate the traffic.

Figure 7-6 **Example of critical servicing time: aircraft activities at a terminal gate.** [***From Ralph M. Parsons Co. (1973), p. 6–29.***]

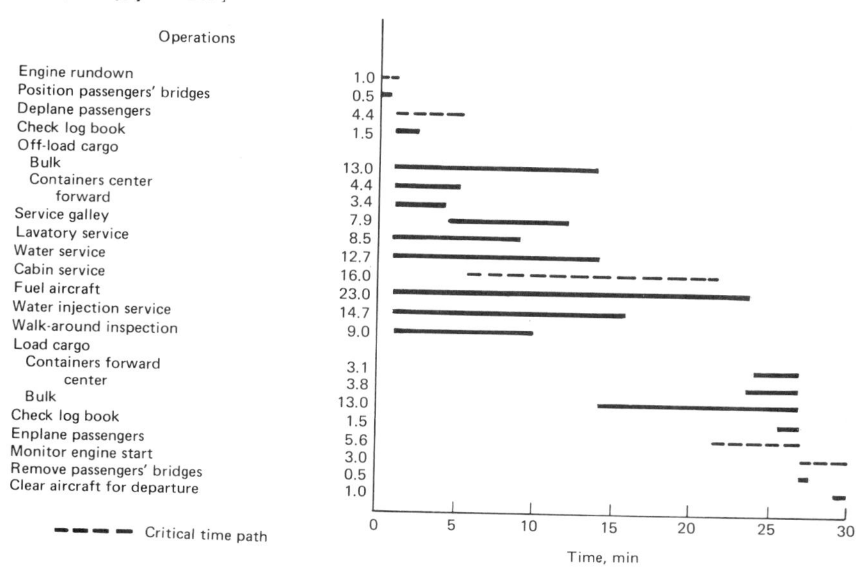

One of the primary methods used to determine the magnitudes of waiting times and the number of traffic units waiting is the cumulative flow-time diagram. This method can be explained by use of the airport example, which will focus upon one gate for aircraft at a terminal for simplicity, although more than one gate can be considered by use of the same method. At this gate, each aircraft requires 30 min for the entire processing, so that 30 min after arrival of one aircraft the gate is available for another.

The use of this gate by aircraft can be portrayed on a diagram such as Fig. 7-7*a*. The horizontal axis denotes time. For convenience the time unit in this figure was chosen to equal 30 min, although any scale can be used. The vertical axis represents aircraft using the facility. At time 1 the first plane to use the terminal that day arrives at the gate. This aircraft occupies the gate for 1 unit (30 min), and then the gate is free for another aircraft. At time 2, the second plane arrives, occupying the gate until time 3. Plane 3 arrives at time 3, holding the gate until time 4. Aircraft 4 arrives at time 5, so the gate remains empty between time 4 and 5. Plane 5 arrives at time 8, leaving at time 9. But the next plane, 6, arrives at time 8.2, so it must wait until time 9 for serving or use of the gate. Planes continue to arrive, all having to wait until plane 11. The waiting occurs even though from plane 8 on the aircraft arrive at a longer time headway than the time required to serve one plane, because a backlog of aircraft remains to be served. In other words, a queue, or waiting line, exists until plane 10 is served.

This diagram shows the time required for each traffic unit (in this example, aircraft) to be served. From it the average waiting time and the average time in the system, as it is called (time waiting plus time being served), can be calculated:

Figure 7-7 **Cumulative flow-time diagrams. (*a*) Showing discrete traffic units. (*b*) Using continuous traffic flow approximation.**

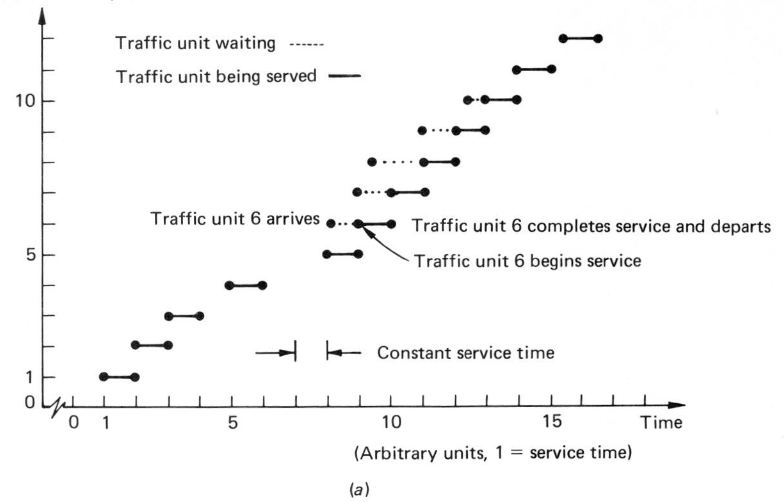

(*a*)

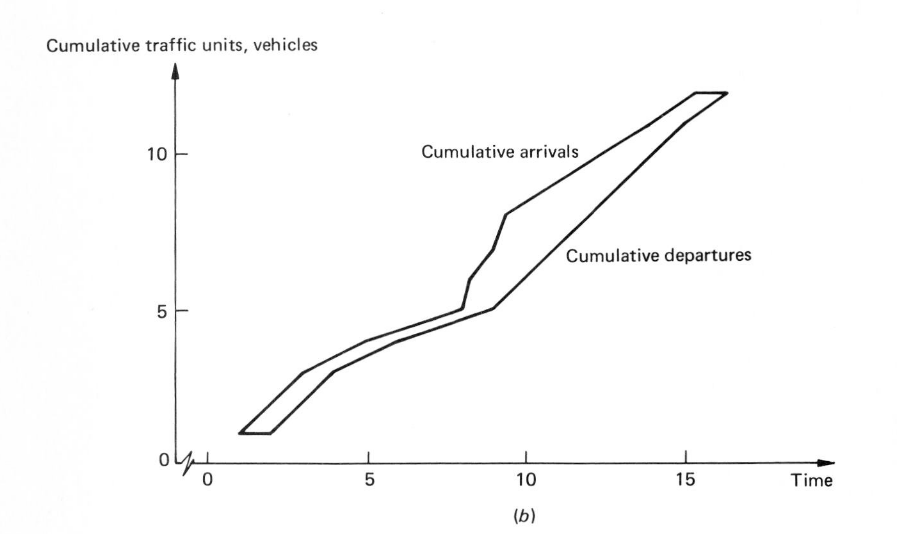

(*b*)

$$\bar{w} = \frac{1}{N} \sum_{i=1}^{N} (E_i - A_i) = \frac{1}{N} \sum_{i=1}^{N} (D_i - A_i - S_i) \tag{7-1}$$

$$\bar{t} = \frac{1}{N} \sum_{i=1}^{N} (D_i - A_i) \tag{7-2}$$

where $\bar{w}$ = mean waiting time
N = number of traffic units, $i = 1, 2, \ldots, N$
A_i = moment of arrival of unit i
D_i = moment of departure of unit i
E_i = moment unit i enters server
S_i = time required to serve unit i
$\bar{t}$ = mean time in the system

These formulas hold even if the serving times are not constant. Also, the figure reveals the number in the waiting line at any point in time. The maximum might be used as a basis for designing the waiting area. At time 9.5, for example, there are three craft in the system, one being served and two waiting. The average number in the system during the period from arrival of 1, A_1, to departure of the last, D_N, is simply

$$\bar{n} = \frac{1}{D_N - A_1} \sum_{i=1}^{N} (D_i - A_i) \tag{7-3}$$

where $\bar{n}$ = mean number of traffic units in system

Figure 7-7*b* portrays the same situation as above, but instead of each traffic unit being represented as a discrete unit, the arrivals and departures are represented as a continuous flow. The cumulative arrival line passes through the points of arrival of traffic units in the discrete diagram; and the cumulative departure line passes through the times of departure. This representation is more commonly used where the number of traffic units is very large, which makes the approximation of a stepped line by a smooth line fairly good. The time of each traffic unit in the system is then approximated by the horizontal distance between the two lines, and the vertical distance is approximately the number in the system. In most applications of this form of the diagram, the lines are drawn as smooth curves rather than short straight-line segments. The average time and average number in the system can be approximated using the area bounded by the two cumulative curves divided by either the number of units or the time interval, as below:

$$\bar{t} = \frac{Z}{N} \tag{7-4}$$

$$\bar{n} = \frac{Z}{D_N - A_1} \tag{7-5}$$

where Z = area bounded by cumulative arrival and cumulative departure curves, the total traffic-unit–time-units spent in the system

Of course, if the cumulative flow curves were drawn as stepped lines, increasing by one traffic unit each time an arrival or departure occurred and the area were that bounded by such lines, then these formulas would give precisely the same results as obtained by Eqs. (7-1) and (7-3), the correct estimates. However, such detail is rarely included in cumulative flow curves.

Capacity and Level-of-Service Concepts

There are basically two concepts of terminal capacity, capacity being a measure of volume through a terminal (or a portion thereof). For the first concept, the maximum possible flow of traffic units through the terminal, to be realized, there must always be a traffic unit waiting to enter a server as soon as it is available. This condition is rarely achieved for long periods, partly because transport flows are usually peaked, such as result from peak-period work movements in urban areas and holiday peak flows to vacation areas. Also, as a practical matter, the sustaining of such large flows would often result in extensive delays to some traffic, delays which are economically and socially intolerable. This leads to the second concept of capacity, the maximum volume that can be sustained with waiting delays that are acceptable.

Any practical measure of capacity must recognize that there are limitations on tolerable delays. First, assume a single activity processor which has a constant service time, as in the aircraft example above. Also, assume traffic units arrive at a constant time headway. Therefore, as long as the headway is greater than the service time, all the traffic units can be served. But if the headway is less than the service time, a queue will form, as in the case of vehicles 5 through 8 in Fig. 7-7. If the volume continued indefinitely at this level, the queue would increase indefinitely in length, resulting in a total time approaching infinity. Of course, in actual systems the volume drops (headways increase) after a peak period, so the system can recover (as illustrated in Fig. 7-7). If the average waiting and serving time were drawn relative to volume (the inverse of headway), the result would be as shown in curve 3 in Fig. 7-8.

In actual systems, time headways usually vary between arrivals of traffic units. Units sometimes bunch together, as do passengers departing from a vehicle. And there are simply random arrivals, resulting from many different people all deciding when to travel almost independently of one another. In the case of varying arrival headways, even if the volume yields an average headway greater than the (constant) service time, some delay is likely. This was shown in Fig. 7-7, where delay occurred even though only 12 vehicles arrived in 15 time units, less than the 15 vehicles that theoretically could be accommodated in that period. As the volume increases, the average headway drops, and the likelihood of delay increases. As a result, actual delay patterns appear as shown by curve 2 in Fig. 7-8, delay increasing with volume and tending toward very large values at volumes near the maximum capacity.

In general, as the traffic is more peaked or bunched in a short portion of the entire period over which the volume is measured, the delay increases, as illustrated by curve 1 in Fig. 7-8. Also, the maximum delay experienced by traffic units is also likely to increase, making it very costly for some units.

Given a maximum tolerable average delay, curves such as 1 and 2 in Fig. 7-8 can be used to specify a capacity, or maximum volume which the processor can accommodate. The maximum volume will, of course, depend upon the choice of that maximum delay. In the processor shown in Fig. 7-8, if the maximum time were selected as t_1 (with a corresponding maximum de-

Figure 7-8 **Typical time-volume curves for terminal processor with constant service times and varying arrival time headway patterns.**

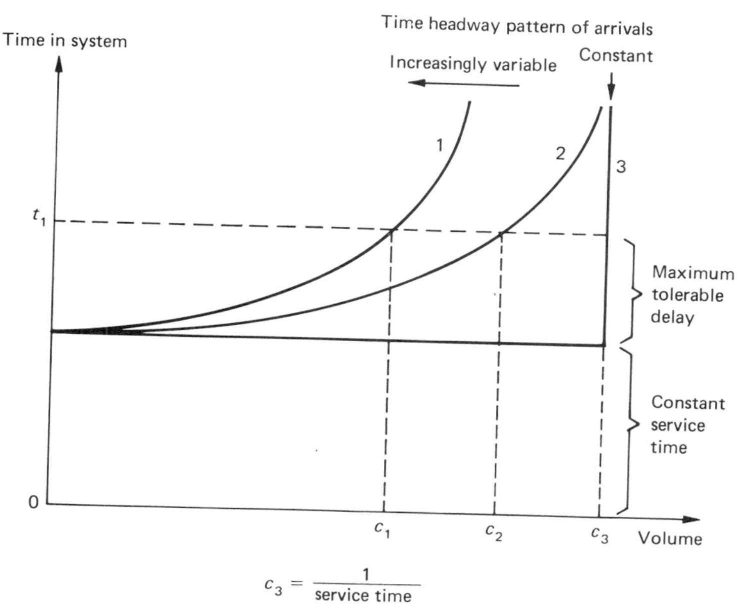

lay), the capacity with curve 1 would be c_1. The capacity will depend upon the peaking of the traffic within the period of analysis, or more precisely upon the entire distribution of the headways. Given any maximum delay, the greater the peaking, the lower the maximum capacity, as illustrated by c_1, the capacity with relatively high variability in headways, and c_2, the capacity with more moderate variability. This is precisely analogous to the effect of the peak hour factor on road link capacity discussed in Chap. 5! As the peaking of road traffic increased (peak hour factor decreased), the capacity at any level of service decreased. Thus the methods and concepts we have been discussing apply not only to terminals but also to any congestion phenomenon.

The service times at most terminal processors are not constant; constant service times were used in the discussion above to facilitate understanding of the operations. If the service time varies, the general form of the time in system versus volume curve remains the same as that illustrated by curves 1 and 2. It begins at zero volume at the mean service time and then increases as volume increases, usually slowly at first, and when the volume approaches the maximum throughput value (the inverse of the mean service time), the delays increase rapidly. Also, the more variable the arrival headways, the greater the delay; a similar relationship applies for the variability of the service times. In cases where both arrival headways and services times vary, the capacity is defined as that volume which has the largest acceptable time. Usually this is based on the average time in the system, but sometimes it is also based on the distribution of the time. However, in all cases it must be based upon judgment, perhaps aided by the use of economic criteria on the costs of the delays and the costs of adding additional servers.

Another useful relationship which has been implied by the foregoing discussion but never made explicit is that among total time in the system, delay, and service time. For any individual traffic unit the total time is, of course, the sum of the delay and service times. As a result, mean times are also related by the sum

$$\bar{t} = \bar{w} + \bar{s} \qquad (7\text{-}6)$$

where $\bar{t}$ = mean time in the processor system
$\bar{w}$ = mean waiting time
$\bar{s}$ = mean service time

In using this formula, it is important to realize that in some situations queue length may affect length of service time since servers tend to be more rapid when the line is long. For instance in an airport check-in line, the agent may pause to exchange pleasantries if the line is short, but faced with a long line, would concentrate on processing travelers as quickly as possible. Thus it is necessary that the mean service time used in the formula applies to the conditions being analyzed. This, merely a word of caution, is not meant to imply that service time will always depend upon the length of the queue. In many mechanized terminal processes, such as automated loading and unloading equipment in freight terminals, the serving times are not affected by the volume of traffic. The critical factor here is that even if the values of the individual serving times vary, the distribution of such times and hence the mean does not vary, so the same value of mean service time can be used for all levels of volume. Thus it is important to check whether or not the mean service times vary with the volume, the same as the delay does. Of course, for much capacity analysis, only the total time versus volume curve is important.

Now that these fundamental concepts have been covered, we shall turn to their application to specific types of terminals.

Capacity and Level-of-Service Example: Airport Runways

The techniques used in airport engineering and planning provide an excellent example of the application of the methods for capacity and level-of-service determination presented above. Although examples could be drawn from other technologies, the methods for airports are generally more developed and refined than those for other types of terminals. This probably reflects the rapid expansion of air terminals in recent decades and the fact that standards for the design of those airports receiving federal aid, including most of those having scheduled airline service, have been set by the Federal Aviation Administration. To ensure the wise expenditure of public monies, methods were developed for the determination of airport capacity and level of service, and, of course, since these are fully public, they may be widely used. While here we shall discuss these as they apply to the determination of the characteristics of an existing facility, it should be obvious that they apply equally well to the evaluation of proposed facilities in the design stage (to be covered in Chap. 16).

The methods begin with the relationship between the time aircraft spend waiting for and then using the runways. Specifically, the average delay experienced by aircraft is related to the arrival rate (or volume) and the average service time formula, of the following type. The formula below is for arrivals on a single runway which is used only for landings.

$$\overline{w}_a = \frac{\lambda_a(\sigma_a^2 + \mu_a^2)}{2(1 - \lambda_a \mu_a)} \tag{7-7}$$

where $\overline{w}_a$ = average delay to aircraft arrivals
λ_a = mean aircraft arrival rate
μ_a = mean service time for aircraft arrivals
σ_a = standard deviation of that service time

This formula is portrayed graphically in Fig. 7-9. Notice that the effect of an increasing arrival rate or volume is as described in the theoretical discussion earlier. Also, as the standard deviation of the arrival rate increases, the average delay increases, as would be expected as a result of more highly variable arrival service times. This model has been validated through testing against the actual waiting times of aircraft at many different airports.

The model for estimating the average delay to departing aircraft for runways used exclusively for departure is identical, except for the definitions relating to departures instead of arrivals and the subscripts, changed from *a* to *d*. The notation used is identical to that in Horonjeff (1975, pp. 116–117) to facilitate more extensive reading on this subject if the reader desires. The model for mixed use of the runway, landing and take off, is too complex to be presented, but it has the same general properties of waiting line models that we have been discussing.

Since the use of these models is rather complex and time consuming, the Federal Aviation Administration has developed approximate methods which are quite satisfactory for most engineering work.

Figure 7-9 Time versus volume relationship for aircraft arrivals at runways based on FAA model.

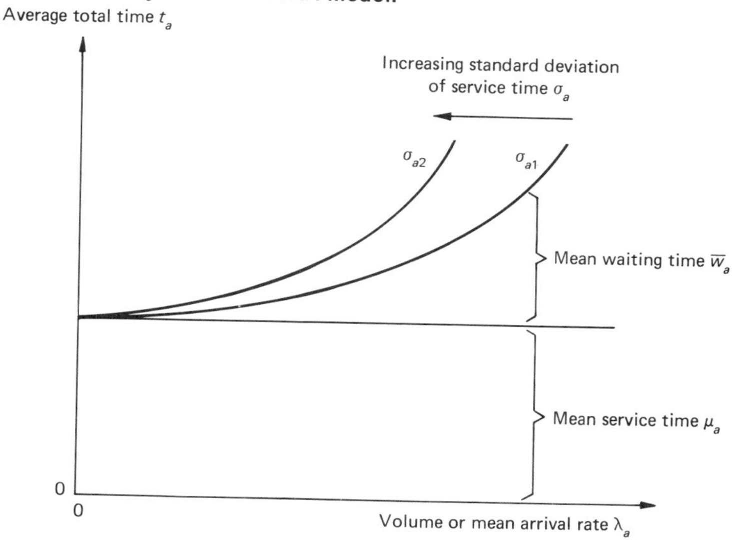

As discussed above, the practical capacity of a runway or other processor depends upon the limits of tolerable delay or, alternatively, total time in waiting plus service. Listed below are the limits of tolerable delay to aircraft set by the Federal Aviation Administration, for both visual flight rules (VFR) and instrument flight rules (IFR) conditions.

1 For departing aircraft, 4 min in VFR conditions for mixed runway operations when more than 10 percent of the aircraft population are small and large jets (classes A and B)
2 For departing aircraft, 3 min for mixed runway operations in VFR conditions when 10 percent or less of the aircraft population are small and large jets (classes A and B)
3 For departing aircraft, 2 min for mixed runway operations in VFR conditions when there is less than 1 percent small and large jets (classes A and B) in the aircraft population
4 For departing aircraft, 4 min for mixed operations in IFR conditions regardless of aircraft class
5 For arriving aircraft, 4 min for mixed operations in IFR conditions regardless of aircraft class
6 For all arrivals, 1 min in VFR conditions.

Since these are averages, some aircraft are delayed more than the specified levels and some less.

When these are applied to runway operations for a single hour, they result in the hourly capacity of that runway, termed in airport jargon PHOCAP, for "practical hourly capacity." The FAA permits airports to be designed in such a manner that the PHOCAP is exceeded for a few hours a year. This overload period must not exceed 5 percent of the time the airport is in operation during the year (invariably 24 hours per day, 365 days per year), nor can such overloaded conditions or periods include more than 10 percent of the flights per year. In addition, the average delays to aircraft during such overload periods cannot exceed 8 min per flight.

Based on these three limitations on exceeding the PHOCAP a "practical annual capacity," PANCAP, is defined. The PANCAP is the largest annual volume of flights such that none of these three limits is exceeded. While the actual computations are somewhat laborious, the method is conceptually rather straightforward.

The capacities of runways depend upon many factors. One is the peaking of traffic, already discussed above. Another is the layout of the runway or runways; the presence of two or more additional runways tends to increase capacity, but depending upon the layout each additional one may not increase the capacity by the same amount because of interference. Also, the layout of the entrances and exits influences capacity. And finally, runway capacity is influenced by characteristics of the aircraft themselves, mainly through the acceleration or deceleration characteristics and time or distance headways between craft, as discussed in Chap. 5. Of course, weather will also affect capacity, through the use of either IFR or VFR flight rules.

Figure 7-10 **Airport runway hourly capacity versus mix of aircraft types.** [*From Horonjeff (1975), p. 122.*]

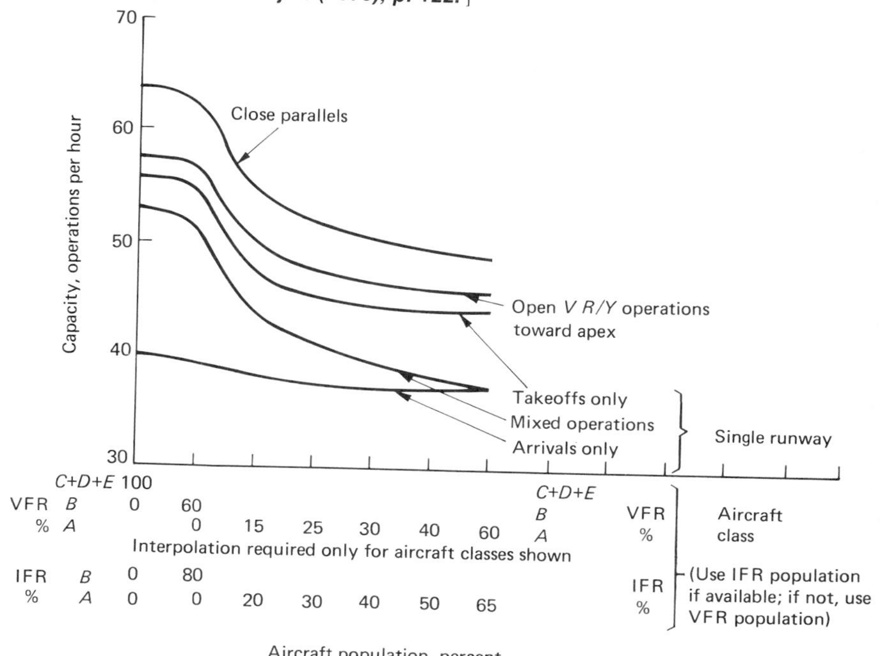

Figure 7-10 presents an example of the combined effect of these factors. On the horizontal axis is the mix of aircraft types. The meaning of the five classes of aircraft—A, B, C, D, and E— is specified in Table 7-4. For example, the sixth vertical line from the left axis corresponds to 65 percent aircraft of type A (large jets) under IFR or 60 percent under VFR, the remainder of the flights being any combination of other types of craft. The curves in the body

Table 7-4 **United States Federal Aviation Administration Classification of Aircraft by Type**

Class	Aircraft type
A	Boeing 707, 747, 720 series; Douglas DC-8 and DC-10 series; Lockheed L-1011 series
B	Boeing 727 and 737 series; Douglas DC-9 series; BAC-111 series. All large piston and turboprop airliners
C	Small propeller-driven airline aircraft such as the Fairchild F-27 and business jets
D	Twin-engine propeller-driven general-aviation aircraft and some of the larger single-engine models
E	Single-engine propeller-driven general-aviation aircraft

Source: Horonjeff (1975, p. 115).

Table 7-5 **Capacities of Various Types of Airport Runways**

Runway configuration: Layout	Description	Mix	PANCAP	PHOCAP IFR	PHOCAP VFR
Single runway	Single runway (arrivals = departures)	1	215,000	53	99
		2	195,000	52	76
		3	180,000	44	54
		4	170,000	42	45
Less than 3,500 ft	Close parallels (IFR dependent)	1	385,000	64	198
		2	330,000	63	152
		3	295,000	55	108
		4	280,000	54	90
3,500 to 4,999 ft	Independent IFR approach/departure parallels	1	425,000	79	198
		2	390,000	79	152
		3	355,000	79	108
		4	330,000	74	90
5,000 ft or more	Independent IFR arrivals and departures	1	430,000	106	198
		2	390,000	104	152
		3	360,000	88	108
		4	340,000	84	90
5,000 ft or more	Independent parallels plus two close parallels	1	770,000	128	396
		2	660,000	126	304
		3	590,000	110	216
		4	560,000	108	180
	Widely spaced open V with independent operations	1	425,000	79	198
		2	340,000	79	136
		3	310,000	76	94
		4	310,000	74	84
	Open V, dependent, operations away from intersection	1	420,000	71	198
		2	335,000	70	136
		3	300,000	63	94
		4	295,000	60	84

of the graph give the capacity as a function of the aircraft mix. Each line corresponds to a different type of runway layout. At 56 percent type-A aircraft under IFR, the capacity of a single runway with mixed operations is about 37 operations or flights/h. If there were two runways intersecting at one end, with operations toward the apex of the "V," the capacity would be 46 flights/h. Inspection of other points on this figure will reveal the effect of the various factors as would be expected.

Further detail on the practical hourly and annual capacities of airport runways is presented in Table 7-5. These capacities are dependent upon the mix of aircraft, with four mixes being used in each case.

Table 7-5 Capacities of Various Types of Airport Runways (continued)

Runway configuration: Layout	Description	Mix	PANCAP	PHOCAP IFR	PHOCAP VFR
	Open V, dependent, operations toward intersection	1	235,000	57	108
		2	220,000	56	86
		3	215,000	50	66
		4	200,000	50	53
Direction of ops	Two intersecting at near threshold	1	375,000	71	175
		2	310,000	70	125
		3	275,000	63	83
		4	255,000	60	69
Directions of ops	Two intersecting in middle	1	220,000	61	99
		2	195,000	60	76
		3	195,000	53	58
		4	190,000	47	52
Direction of ops	Two intersecting at far threshold	1	220,000	55	99
		2	195,000	54	76
		3	180,000	46	54
		4	175,000	42	57

Percent of specified class				
Mix	A	B	C	D + E
1	0	0	10	90
2	0	30	30	40
3	20	40	20	20
4	60	20	20	0

Source: Horonjeff (1975, pp. 137–138 and U.S. Federal Aviation Administration).

To further aid the engineer in the evaluation of airports, the FAA has prepared a simple nomograph to use in the estimation of the delay of aircraft given the ratio practical annual capacity and the number of operations. This nomograph is reproduced in Fig. 7-11. Entering the horizontal axis at the annual capacity, extend a vertical line to the reference line, and then proceed on or parallel to the other set of lines until the above annual volume. This may require movement to the left or right. This point corresponds to the annual delay in aircraft-minutes on the vertical axis. As an example, if the annual capacity were 315,000 operations and the actual volume were 300,000 flights,

Figure 7-11 Annual delay at airport runways versus annual capacity and volume. [*From Horonjeff* (1975), p. 136.]

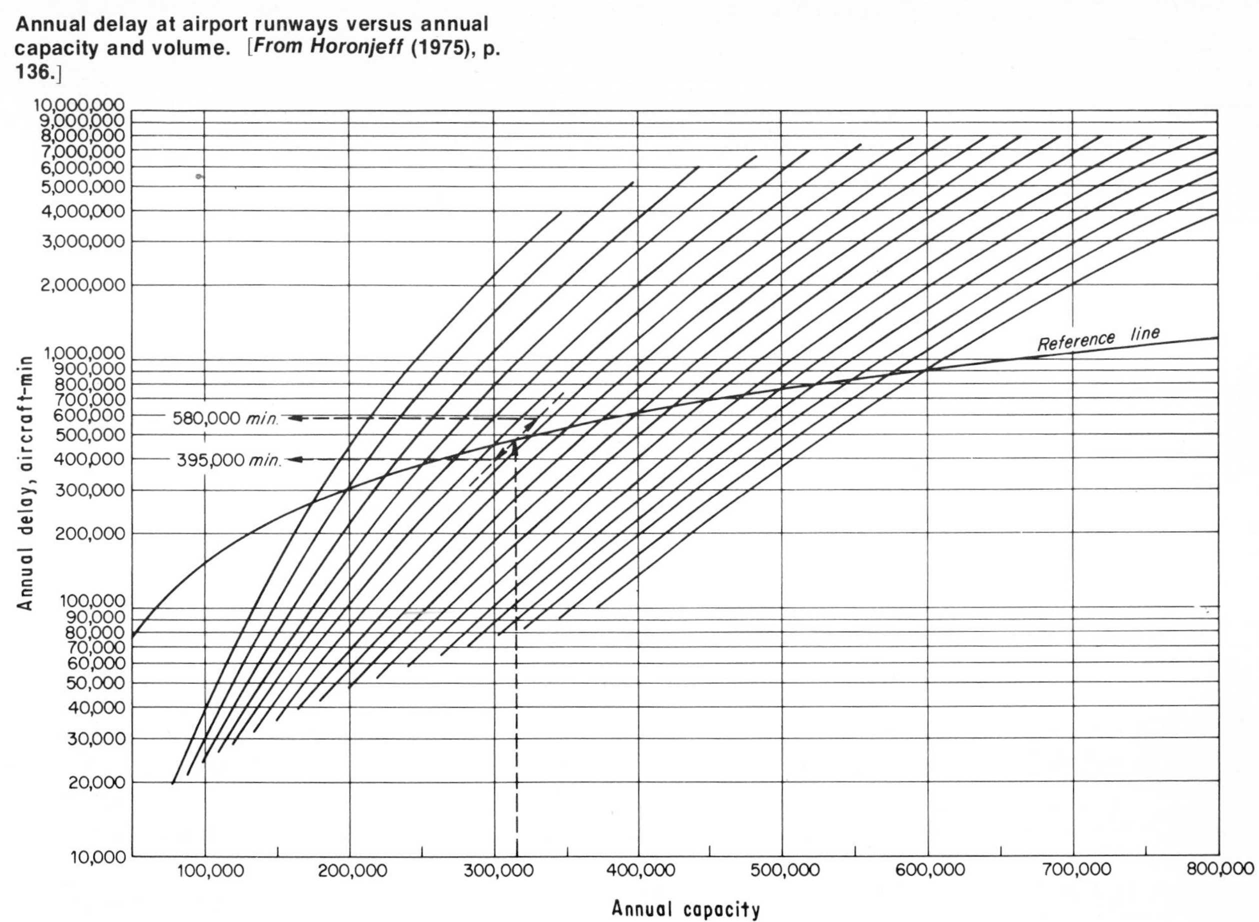

then the annual delay would be 395,000 min as shown. Thus the average delay per flight would be 1.32 min/flight.

Notice that in effect the curved lines extending from the lower origin region toward the upper right corner of the chart are curves of total annual delay versus volume; each curve for that annual runway capacity is given by the intersection of the curve with the reference line. As theory suggests, the delay per flight increases with increasing volume, given a particular runway and hence a fixed practical capacity. At a volume of 330,000 operations/year, the runway used above has a total delay of 580,000 flight-min, for an average delay of 1.76 min/flight.

SIMULATION†

In the above discussion of various types of passenger and freight terminals one ubiquitous characteristic is the irregular processing of traffic. Vehicles arrive at irregular headways, as do travelers and goods to be shipped, which results in periods of congestion followed by periods of free flow, and so on. This characteristic led to the conclusion that the practical capacity of a terminal is far below the maximum possible traffic which can be accommodated; otherwise, the potential of delay to traffic is simply too great. Methods have been developed to treat analytically, to model, the performance of terminal systems in which the flow of traffic is irregular. Of these, the most useful to transportation planners and engineers is simulation. This will be but a brief introduction to simulation, for the subject is very extensive. A book would be required to cover it with any completeness, and as will be clear from this description the applications of simulation extend far beyond the limits of transportation systems.

The basic idea of simulation is to model a process, paying special attention to the various events which occur as the process continues. Thus in the context of transportation, the focus typically is on the arrival of vehicles or other traffic, placement in the appropriate serving facility, loading and unloading, departure of the vehicle, making the server available to another unit, etc. An important distinction is made between two types of simulation: One is deterministic, meaning that all events are characterized by certainty—as to when they will occur, how long each process will take, etc. The other type is stochastic, meaning that there are possible variations in these characteristics of the system as it is represented in the simulation model. In particular, those characteristics which are modeled as variable (such as the time required to load and unload a vehicle) have probabilities associated with each of the possible values, the associated probability indicating the relative frequency or likelihood of each of the possible values.

Similarly, the results of the simulation are characterized by probabilities. A typical result would be that the probability of an aircraft having to wait more than 8 min after arrival at an airport until it is given clearance to land is 0.05. The importance of this approach to terminal analysis and description was revealed in the discussion of airports, where the probability of various delays forms the basis for evaluating the adequacy of airport capacity. A similar

† Those readers unfamiliar with probability concepts can skip this section without any loss of continuity. However, it is recommended that all read the introductory paragraphs.

approach is used in other modes. The term "stochastic" is often used to indicate that the possibility of variations in certain characteristics of the system is being explicitly taken into account and that this is done by means of probabilities of various values. Thus the second type of simulation will be termed stochastic simulation.

Probability Density Functions

Since our attention will focus on stochastic simulation, which involves the use of probability concepts, it is appropriate to review characteristics of probability density functions. (If the reader has not already been exposed to these in a course on statistics, studying an introductory reference is recommended. There are many available as texts for first courses in applied statistics.)

In order to represent characteristics of the system which vary the probability density function is used. This specifies the probability of occurrence of the values which the characteristic can assume. One important condition for this function is that the sum of the probabilities of the various possible values equals unity, since some value must occur.

An example of such a probability density function is given in Fig. 7-12. This presents perhaps the most common probability density function used for traffic arrivals at transport terminals and other transport elements, the Poisson distribution, named after the French mathematician Simeon D. Poisson. Poisson described this function in a book which appeared in 1837 outlining general rules for probability calculation with application to criminal and civil court decisions. The variable is the number of traffic units which arrive in a given interval of time, given a particular mean arrival rate. As the mean arrival rate increases, the value of the number of arrivals associated with the largest probabilities increases, as would be expected. However, it is important to notice that there still is a nonzero probability associated with no or a very few arrivals in the interval of time. Similarly, even with a low mean arrival rate, there still is a possibility of a large number of arrivals in any period, although the probability of this is quite small. Thus this provides the source of variation in the operation of the system by recreating variation in the number of arrivals per unit time much like what might happen at a terminal.

The mathematical formula for the Poisson distribution is

$$\mathrm{p}(n) = \frac{(\lambda t)^n e^{-\lambda t}}{n!} \qquad \text{for } n = 0, 1, 2, \ldots \tag{7-8}$$

where $\mathrm{p}(n)$ = probability of n arrivals in a period t
n = number of arrivals
λ = mean arrival rate (volume in our terminology)
t = time period
$n! = n(n-1)(n-2) \cdots (2)(1)$ and $0! = 1$
e = base of natural logarithms

An important characteristic of any probability density function is the number of parameters required to specify the numerical form of the distribution. The Poisson distribution has only one, the mean, λ, in Eq. (7-8) above.

In Chap. 2 it was emphasized that any mathematical or other model of a system should be tested before it is used in engineering design or planning, for otherwise one can have little confidence that the results of using the

Figure 7-12 **The Poisson distribution of probability of various numbers of traffic units arriving in a given period of time. *(a)* $\lambda = 0.7$ unit/h and $\lambda t = 0.7$ unit. *(b)* $\lambda = 2.5$ units/h and $\lambda t = 2.5$ units. *(c)* $\lambda = 5.0$ units/h and $\lambda t = 5.0$ units. λ = mean arrival rate, units/h; t = time period, 1.0 h; p(n) = probability of exactly n arrivals in the time period t. [*From Benjamin and Cornell (1970), p. 238.*]**

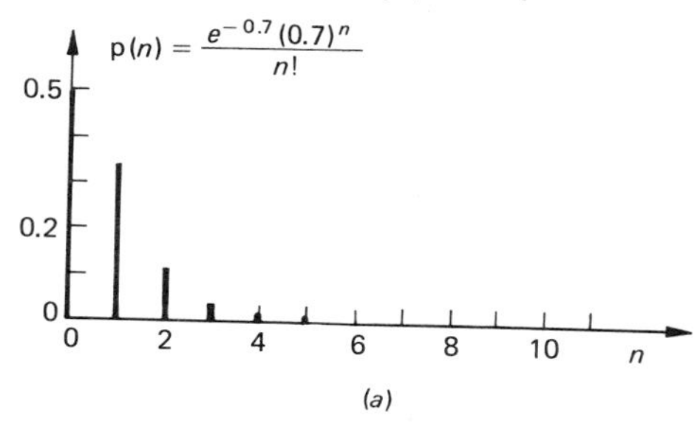

(a)

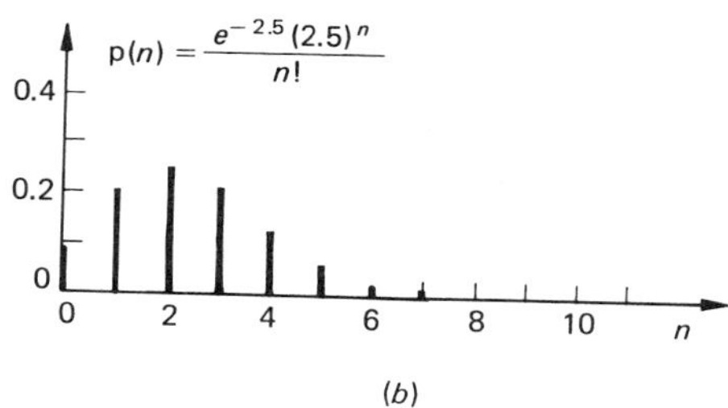

(b)

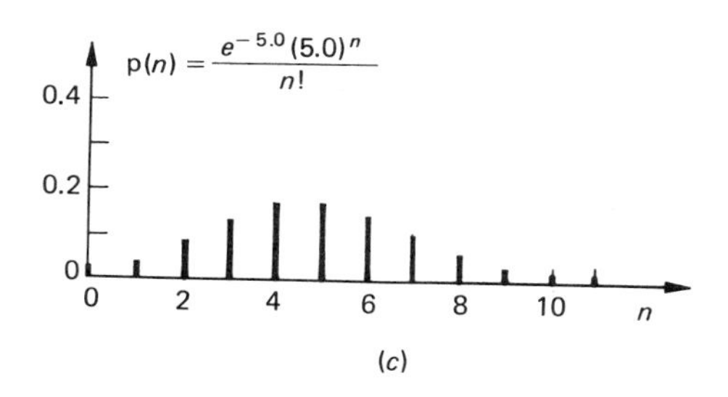

(c)

model will replicate real world phenomena. This is certainly true for simulation models, and thus it is important to be able to test the Poisson distribution against actual data. This is usually done using another form of the same distribution, one for which the gathering of data is easier and less time-consuming. This other form is derived from the Poisson distribution as presented in Eq. (7-8) and is a distribution of time headways between the arrivals of the traffic units rather than a distribution of the number of arrivals in any period of time.

Specifically, the distribution is the cumulative distribution of time headways, which specifies the probability of a headway being greater than or equal to a particular value. First this will be derived from Eq. (7-8), and then examples will be given of the comparison of the theoretical distribution with actual data. Using Eq. (7-8), we find the probability of no arrivals in time t is

$$p(0) = \frac{(\lambda t)^0 e^{-\lambda t}}{0!} = e^{-\lambda t}$$

If no traffic units arrive in t, then there must be a gap or time headway of at least t between successive traffic units. This can occur with a headway between successive traffic units equal to or greater than t; therefore, the probability of a headway h being greater than or equal to t is

$$p(h \geq t) = e^{-\lambda t} \qquad \text{for } t \geq 0 \tag{7-9}$$

where $p(h \geq t)$ = probability of a headway h greater than or equal to t

This is termed a cumulative probability distribution because it specifies the probability which is the cumulation of the probabilities of all values of the variable from that specified (a headway of t in this case) to the upper limit of possible values (a headway of infinity). Because of the mathematical form of this distribution, it is often termed the negative exponential distribution.

Figure 7-13 presents an example of this distribution. Also shown are the results of observations of actual vehicle headways on a road, expressed in terms of the relative frequency of occurrence of values within small intervals of time. As can be seen, the correspondence between the observed relative frequencies and the theoretical probabilities is close. The widespread use of the negative exponential or Poisson probability distribution for traffic arrivals in transport systems is because this correspondence has been found in a wide variety of situations—modes of transport, traffic levels, etc. In fact, Rallis (1967) reports that the negative exponential pattern for arrivals corresponds very closely to travelers arriving at ticket counters, trucks at truck terminals, aircraft at airports, ships at ports, trains at railway stations, etc. Of course, the value of the mean arrival rate will vary by time of day, day of week, etc., for each of these applications.

There are formal methods for the comparison of a theoretical distribution with data in order to determine whether or not the theoretical model can be accepted as a representation of actual systems, but we shall not go into these, as they are more appropriate for a book on statistics. However, it should be noted that an adequate test of a model is not generally simply a visual comparison of a theoretical curve and data on a graph!

An important difference between the Poisson arrival distribution form and the corresponding negative exponential distribution of headways form is that the former is a discrete distribution while the latter is a continuous one. The distinction is made on the basis of the variable involved. The number of arrivals in a period of time is the variable in the former, and it must take on discrete values of 0, 1, 2, 3, and so on. In contrast, the variable is time in the negative exponential distribution, and it is a continuous variable.

Manual Simulation

With this very brief introduction to probability distributions, it is now possible to discuss simulation. This will be presented in a very general form, one suitable for manual use, even though many if not most simulations make use of computers to aid in the computations. Although extension to computer use is conceptually rather straightforward, it will not be attempted here since

Figure 7-13 **The negative exponential distribution of cumulative probabilities of time headways of traffic arrivals and a comparison with observed vehicle arrivals on a freeway.** [*From Gerlough (1955), p. 38.*]

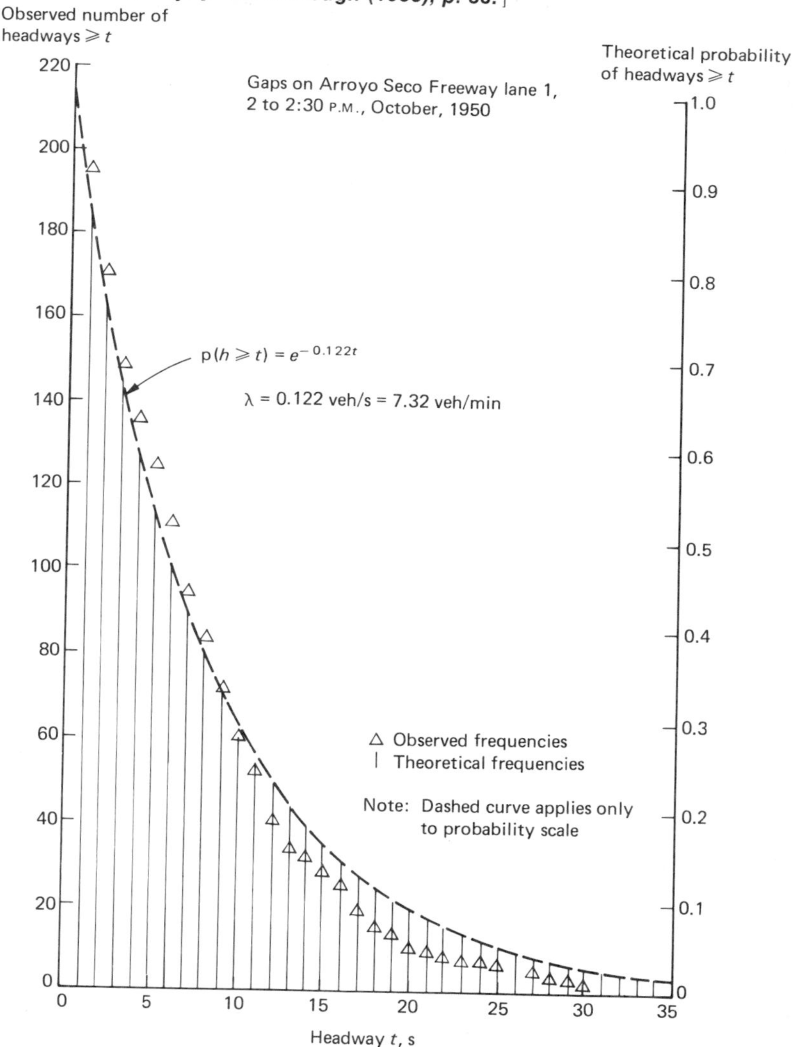

solving problems requires a thorough knowledge of an appropriate computer language. There are many languages which have been written especially for simulation, such as GPSS, Simscript, and Transim, to mention a few.

The key to manual simulation of stochastic systems is the use of a table of random numbers. This table contains a list of numbers, developed in such a way that (ideally) all numbers have equal probability of appearing in any one entry cell. There are many such tables, one being printed as a book with the intriguing title, "A Million Random Digits" (Rand Corp., 1955). The important property of the digits in a random number table for our purposes is that the probability distribution is uniform over all the numbers contained in it.

The significance of this can be seen from the use of the table in connection with the negative exponential distribution of traffic unit arrivals. Suppose we are modeling the arrival of buses at a bus stop and it has been determined that the buses arrive following the Poisson distribution at an average rate or volume of 83 buses per hour. In this case the probability of a headway greater than or equal to t seconds will be, using Eq. (7-9), with $\lambda = 0.02306$ bus/s,

$$p(h \geq t) = e^{-0.02306t}$$

Transforming this, we have

$$\log_e [p(h \geq t) = -0.02306t$$

$$\text{and } t = 43.4 \log_e \frac{1}{p(h \geq t)} \tag{7-10}$$

Equation (7-10) can be used to find the headway t, which will be exceeded by a probability $p(h \geq t)$. Thus we can choose any value of $p(h \geq t)$, substitute it into the equation, and obtain the desired time t.

To obtain a single headway to use in the course of simulating the arrival of vehicles (buses) at a stop, we select a probability $p(h \geq t)$ and then use Eq. (7-10) to find the corresponding time period of no arrivals, t. Each time we do this we wish to have an equal probability that any value of $p(h \geq t)$ will be selected. By drawing a number from a random number table, this is ensured. For example, the first number we take from the table may be 0.32. (If the table gives whole numbers from 0 to, say, 100, we simply place a decimal in front to obtain numbers from 0 to 1.00.) By Eq. (7-10) this yields a headway t of 49 s. If we number buses 1, 2, 3, and so on, this means that bus 2 arrives 49 s after bus 1. Then we draw another random number; suppose this is 0.99. This yields a headway of 1 s, meaning bus 3 arrives 1 s after bus 2. The next random number might be 0.61, resulting in bus 4 arriving 21 s after bus 3. The process is continued until enough bus arrivals are generated for the purposes of the simulation. This number usually must be large enough so that the simulation results which often are expressed in averages and perhaps distributions have settled on reasonably stable values. Twenty bus arrivals, and nineteen headways, are shown in Table 7-6. As can be seen, the mean headway remains fairly stable in this example after about ten bus arrivals, the distribution mean being 43.4 s.

A few remarks should be made about random numbers. First, it is best to use a table of random numbers such as provided in the Rand Corp. book mentioned earlier. In using it, it is best to start at a randomly chosen row and column on a randomly chosen page. These might be selected by opening the book in a haphazard way, without looking, pointing to a place on one of the two facing pages. One can choose to use only two digits, as in the example above, or all those printed in the table. In the absence of such a book, for many problems an accepted substitute is to use the last four digits of the telephone numbers appearing in a city or regional directory. It is best to choose a large directory.

The example of negative exponential arrivals above reveals the general

Table 7-6 Example Simulation of Arrival Times of Buses at a Stop

Bus number	Random number drawn	Time headway, s	Arrival time, s	Cumulative mean headway, s
1			0	
2	32	49	49	49.0
3	99	1	50	25.0
4	61	21	71	23.6
5	16	80	151	37.8
6	66	18	169	33.8
7	10	100	269	44.8
8	49	31	300	42.9
9	83	8	308	38.5
10	12	92	400	44.4
11	36	44	444	44.4
12	31	51	495	45.0
13	92	4	499	41.6
14	8	110	609	46.8
15	74	13	622	44.4
16	84	8	630	42.0
17	33	48	678	42.4
18	16	80	758	44.6
19	28	55	813	44.6
20	64	19	832	43.8

Note: Random numbers are integers drawn from the range 0 to 100, all with equal probability.
Source: Adapted from Blunden (1971, p. 87).

approach to the generation of events so as to correspond to a given or known probability distribution. This approach will be described following the presentation of Blunden (1971, pp. 85–86). If $s(x)$ is the probability distribution of x and $S(x)$ is the cumulative distribution, then

$$S(x) = \int_{-\infty}^{x} s(x)\,dx \tag{7-11}$$

For convenience, let

$$y = S(x) \tag{7-12}$$

$$\text{Hence } x = S^{-1}(y) \tag{7-13}$$

where S^{-1} denotes the inverse function of S. If y arises from a uniform distribution, then for every value of y the corresponding value of x can be found. The variable y has a uniform distribution in the range from 0 to 1. Hence values of y can be generated as random numbers, and through the use of the inverse cumulative function $S^{-1}(y)$ the values of x can be obtained. This holds for any distribution $S(x)$, not just the negative exponential distribution used in the example above.

Now we can turn to its use in the simulation of a waiting line such as those found in terminals. Let us consider the bus stop in the arrivals example, at which buses arrive according to the Poisson distribution, with a mean arrival rate of 83 buses/h. [This example is adapted, with changes, from Blunden (1971, pp. 85–91).] Observations have been made of bus dwell times at the

Figure 7-14 **Relative frequencies and smoothed cumulative probability distribution of bus dwell times at bus stop.** [*From Blunden (1971), p. 86.*]

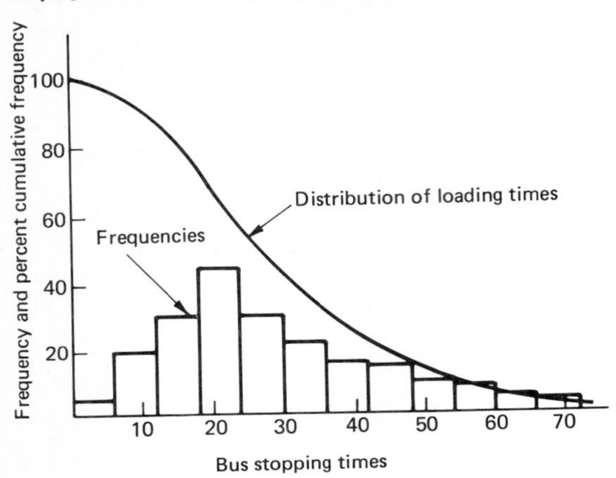

stop, and the observed relative frequency is shown in Fig. 7-14. Also shown is the cumulative distribution function of these times, which was developed by cumulating and then smoothing these relative frequencies. The bus stop is only large enough to accommodate one bus at a time, so in order to simulate the operation of the stop it is necessary to ascertain the minimum time required between the departure of one bus (end of its dwell time) and the arrival at the loading zone of the next bus (beginning of its dwell time). This minimum clearance time was found to be 4 s. With the arrival distribution and the service time distribution both known, the operation of this bus stop can be simulated.

The simulation will be performed using the same notation used in the arrival of buses earlier in this section. Buses will be numbered 1, 2, 3, and so on, and time will be kept track of with a clock starting at 0 s at the time of arrival of the first bus, bus 1, with all times measured in seconds from that instant.

The simulation is really identical to the method used to treat the bus arrivals, only now this must be extended to include the service times, congestion at the stop, etc. This is done by modeling each event as it occurs in time and making sure a new event does not begin until all other events which must occur before it have been completed.

Bus 1 arrives at time 0. This bus must occupy the stop for a period of time to unload and load. We know this time varies according to the probability distribution presented in Fig. 7-14. Hence a random number from zero to unity will be drawn in order to specify a particular time for this bus. Suppose the number drawn is 0.81. On the probability scale of this figure, this value corresponds to a dwell time of 16 s. Thus this bus will wait 16 s, departing from the stop at time 16. Hence the stop is available for another bus at time 20, allowing the 4-s minimum for clearance.

Using the same random numbers and results for a simulation of the arrivals presented earlier, bus 2 arrives at time 49, so it need not wait to enter the stop area. Its dwell time can be determined by drawing another random number: Supposing it is 0.74, the dwell time is 18 s. Thus bus 2 arrives at time 49 and departs at time 67. Bus 3 arrives at 50, so it will have to wait since bus 2 is still pulling out of the stop. It can begin loading and unloading at time 71, 4 s after bus 2 begins to leave the stop. To determine its stop time, suppose the random number draw yields 0.53, resulting in a service time of 25 s. This process can be continued until the desired period of time has been simulated.

An important point here is that the distribution of dwell times is considered independent of the headway between the bus and the preceding bus. In other words, dwell time is not affected by headway. In some cases this is true, while in other cases it may not be. In the latter types of situations, the distribution of dwell times would depend on the headway; presumably the dwell time would increase as the headway increased, reflecting larger numbers of boarding passengers.

Table 7-7 presents the results of a simulation of 20 buses, using the arrival times obtained earlier and presented in Table 7-6. Notice that with only one stopping place, many buses must wait before being able to pull to the stop itself. This is the reason arrival time is often different from the time of entering the stopping zone. Averages can be calculated from the simulation using Eqs. (7-1) through (7-6). The average headway of arrivals is 41.6 s, close to the true distribution mean of 43.4 s. The average stopping time (at the passenger stop) was 30.2 s. The average waiting time was 13.1 s.

A cumulative flow chart presented in Fig. 7-15 portraying this simulation is very useful in understanding the results and hence the operation of the stop.

Table 7-7 Example Simulation of a Bus Stop

Bus number	Arrival time, s	Time entering stop zone, s	Waiting time, s	Random number	Service time, s	Departure time, s
1	0	0	0	0.81	16	16
2	49	49	0	0.74	18	67
3	50	71	21	0.53	25	96
4	71	100	29	0.34	34	134
5	151	151	0	0.45	28	179
6	169	183	14	0.50	26	209
7	269	269	0	0.44	29	298
8	300	302	2	0.14	50	352
9	308	356	48	0.55	24	380
10	400	400	0	0.36	32	432
11	444	444	0	0.10	54	498
12	495	502	7	0.63	22	524
13	499	528	29	0.97	7	535
14	609	609	0	0.42	29	638
15	622	642	20	0.92	10	652
16	630	656	26	0.36	23	679
17	678	683	5	0.50	26	709
18	758	758	0	0.05	72	830
19	813	834	21	0.41	29	863
20	832	867	35	0.14	49	916

Notes: Mean arrival headway = 43.8 s; mean waiting time = 12.9 s; mean service time = 30.2 s; minimum time between departure and arrival, 4 s.

Figure 7-15 **Cumulative flow diagram of bus stop simulation.**

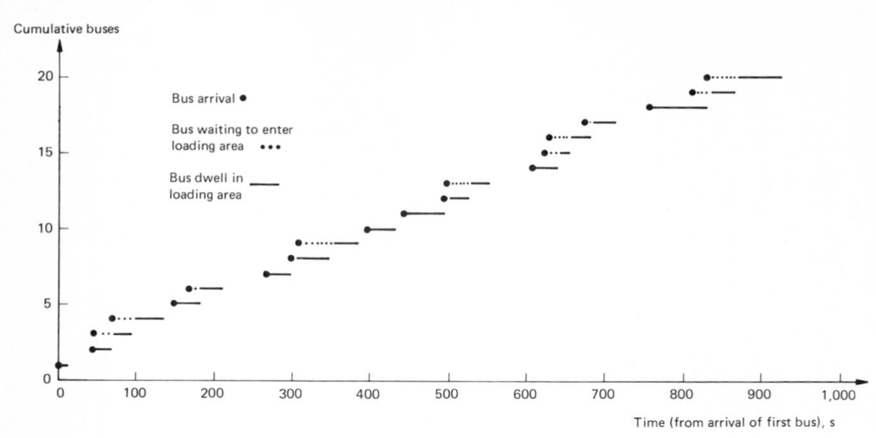

This figure gives the results for the bus stop with only one stopping place, as described above. The horizontal scale is clock time, and the vertical scale represents the arriving buses. The time of arrival of a bus is marked by a vertical line opposite its number (e.g., bus 2 arrives at time 49, 49 s after the start of the clock). Also drawn are the times the bus enters and leaves the loading area. The waiting times and the dwell times are also shown. This chart is very helpful for understanding the sources of congestion and visualizing such characteristics as the number of vehicles at the stop, at any one moment.

While the subject of simulation is much more complex than the small example presented here, hopefully this has provided enough information to convey the role that simulation can play and sufficient amount of technique that at least small manual simulations can be performed.

Queueing Theory

Although simulation is probably the most widely used method for modeling stochastic traffic flows in transportation, such as those occurring at terminals, another important method is the use of queueing theory. Queueing theory deals with queues by deducing characteristics through mathematical analysis, attempting to find formulas which will directly provide information of the type we obtained from simulation. The queueing theory approach thus has the advantage of simplicity and ease of use over simulation, but as one might expect, it is only possible to derive such formulas for certain types of queues. And, of course, the entire problem must be described in mathematical formulas in order to use the equations of this theory. Thus in the bus stop example, we would have had to describe the curve of Fig. 7-14 by an equation which might have been difficult to find, whereas with simulation we could use the graph directly. But for many queueing situations formulas exist for the various elements, and hence this requirement often presents no problem.

Queueing theory formulas provide useful information, for the design and analysis of waiting line systems, quite similar to the information we obtained from the simulation example. For instance, the average number of traffic units in the queue and the average number in the system (queue and servers) is important in ascertaining the adequacy of waiting areas. The distribution of waiting times and the average are important in assessing the adequacy of the entire system in serving the traffic. From this distribution the probability of delays greater than any specified value can be obtained, which permits the application of criteria such as those used for airport runways, described earlier in this chapter.

There are four characteristics of queues which must be specified in order to predict performance in terms of variables such as those mentioned above. One is the distribution of the headways of traffic arrivals, which may be uniform (i.e., with constant headway) or may follow the Poisson or random arrival pattern (i.e., negative exponential probability of headways), or some other pattern. A second characteristic is the distribution of service times (e.g., constant, Poisson, etc). A third important characteristic is the number of serving channels or stations. Finally there is the so-called queue discipline, which specifies the order in which arriving traffic units will be served. In transport, this is usually first to arrive, first to be served, but in other systems the order may be last to arrive, first served, as when mail is stacked in a pile and then sorted from the top down. In queueing theory terminology, the former discipline is usually referred to as, "first in, first out," or FIFO; the second as "last in, first out," or LIFO. Other patterns are also possible, such as traffic being turned away if the waiting line gets too long, but we shall not try to cover all cases in this description.

To illustrate the types of results that are obtained from queueing theory, we shall focus on those which use the Poisson distribution for arrivals and service and the constant service times. Other distributions are also used in transportation, but too much space would have to be devoted to the topic for them to be understood.

One important class of results is for the case of a single-server queue, with Poisson arrivals, negative exponential service times, and a FIFO discipline. Various measures of the performance of such a queue are presented in Table 7-8. Since the Poisson distribution has but one parameter, the mean, the only parameters in the model are the mean arrival rate λ and the mean service rate μ. These would both be expressed in traffic units per unit time, as vehicles per hour. The mean headway of arrivals is $1/\lambda$ and the mean service time is $1/\mu$. It should be noted that the mean time between departures from the servicing channel must be greater than $1/\mu$ since the server is not always in use. The mean departure headway must be equal to $1/\lambda$ since no more units can leave than arrive.

The various results given in Table 7-8 are essentially self-explanatory. These results are termed steady-state results, which means that they are the results that would be observed after the system operated for such a long time that the averages or probabilities did not change as the system ran longer. In fact, they actually are derived from the situation of an infinite period of operation. It has been found useful to express many of the formulas in terms of ρ, ρ being equal to λ/μ, so as to simplify presentation. ρ is called the traffic intensity. ρ must be less than 1.0; otherwise the waiting line would continually build up as time passed and a steady state would not exist. As can be seen from inspection, many of these performance measures would be quite useful in analyzing a facility.

Table 7-8 Single-Station Queuing Relationships with Poisson Arrivals and Exponential Service Times for Steady-State Conditions

	Queueing model†	Description of model
1	$p(n) = \left(\frac{\lambda}{\mu}\right)^n \left(1 - \frac{\lambda}{\mu}\right) = (\rho)^n(1-\rho)$	$p(n)$ = probability of having exactly n vehicles in system
2	$\bar{n} = \frac{\lambda}{\mu - \lambda} = \frac{\rho}{1-\rho}$	$\bar{n}$ = average no. of vehicles in system
3	$\text{Var}(n) = \frac{\lambda\mu}{(\mu-\lambda)^2} = \frac{\rho}{(1-\rho)^2}$	Var (n) = variance of n (no. of vehicles in system)
4	$\bar{q} = \frac{\lambda^2}{\mu(\mu-\lambda)} = \frac{\rho^2}{1-\rho}$	$\bar{q}$ = average length of queue
5	$f(d) = (\mu - \lambda)e^{(\lambda-\mu)d}$	$f(d)$ = probability of having spent time d in system
6	$\bar{d} = \frac{1}{\mu - \lambda}$	$\bar{d}$ = average time spent in system
7	$\bar{w} = \frac{\lambda}{\mu(\mu-\lambda)} = \bar{d} - \frac{1}{\mu}$	$\bar{w}$ = average waiting time spent in queue
8	$p(d \leq t) = 1 - e^{-(1-\rho)\mu t}$	$p(d \leq t)$ = probability of having spent time t or less in system
9	$p(w \leq t) = 1 - \rho e^{-(1-\rho)\mu t}$	$p(w \leq t)$ = probability of having waited time t or less in queue

† λ = average number of vehicle arrivals per unit of time
μ = average servicing rate, number of vehicles per unit of time
ρ = traffic intensity or utilization factor = λ/μ
Source: Wohl and Martin (1967, p. 364).

As an example of their use, consider a truck-unloading platform at a factory where it has been found that trucks require on average 22.5 min to load and unload and that the distribution of service times was negative-exponential. If there is only one space, what would be the average delay if trucks arrived at random at an average rate of one per hour? What is the probability that a truck would have to wait more than 15 min to enter the space? And what is the probability that more than one truck would be waiting to enter the dock space?

In this problem, $\lambda = 1.00$ truck/h; $\mu = 2.66$ trucks/h, and $\rho = 0.375$. Then using the seventh equation, the average waiting time $\bar{w}$ is

$$\bar{w} = \frac{\lambda}{\mu(\mu-\lambda)} = \frac{1.00}{2.66(2.66 - 1.00)} = 0.226 \text{ h/truck} = 13.59 \text{ min/truck}$$

The probability of a wait of 15 min (0.25 h) or less is, by the ninth equation,

$$p(w \leq 0.25 \text{ h}) = 1 - 0.375e^{-(1-0.375)(2.66 \text{ trucks/h})(0.25 \text{ h})}$$
$$= 1 - 0.375e^{-0.415} = 1 - 0.375(0.660) = 1 - 0.247 = 0.753$$

From this it follows that the probability of waiting more than 15 min is

$$p(w > 0.25 \text{ h}) = 1 - p(w \leq 0.25 \text{ h}) = 0.247$$

For one truck to be waiting, there must be one truck at the single dock and one waiting at the queue. For more than one truck to be waiting, there must be three or more trucks in the system. In principle, we could evaluate this probability, using the first equation as follows:

$$p(3) + p(4) + p(5) + \cdots + p(\infty) = \sum_{n=3}^{\infty} p(n)$$

The problem is, of course, that to evaluate the summation through any large-number approximation to infinity would be extremely laborious. But from probability theory we know that

$$\sum_{n=0}^{\infty} p(n) = 1$$

and hence

$$\sum_{n=3}^{\infty} p(n) = 1 - \sum_{n=0}^{2} p(n)$$

Thus the probability of more than one truck waiting in line is

$$p(n \geq 3) = 1 - \sum_{n=0}^{2} (\rho)^n(1 - \rho) = 1 - [(0.375)^0(0.625) + (0.375)^1(0.625) + (0.375)^2(0.625)]$$

$$= 1 - 0.947 = 0.0527$$

Another important set of results is for queues with Poisson arrivals, Poisson service times, the FIFO queue discipline, but with many service channels. Here the FIFO discipline means that an arrival will enter the server which is available first, implying that there is one waiting line. Even though this is unlike many situations where there is a queue for each server, it often approximates such situations well. The formulas for this type of queue are presented in Table 7-9. As might be expected, these are much more complex than the formula for a single-channel server system, but basically they are of the same type. Again these formulas are for steady-state conditions, which in this case require the arrival rate to be less than the average total service rate of all the channels, i.e., $\lambda \leq k\mu$.

The final type of queueing theory model we will cover is the case of Poisson arrivals and constant service times. This is a more difficult system to model mathematically, and as a result performance measures that can be estimated analytically are not as numerous as for the other cases, but they are useful nevertheless. These are presented in Table 7-10.

The performance of these types of queues can be compared to any one of the measures presented in these tables. A very useful one is the mean delay total time in the system, a function of ρ for all of these cases. Figure 7-16 presents the total time, measured in multiples of the mean service time $1/\mu$ versus traffic intensity ρ. Part *a* is for a single-channel queue with various arrival and service patterns. As can be seen, the combination of random arrivals and service is much more prone to delay than the random arrival s

Table 7-9 Multiple-Station Queuing Relationships with Poisson Arrivals, Exponential Service Times, and Leading Traffic Unit in Queue Moving to First Available Station, for Steady-State Conditions

	Queueing model†	Description of model
1	$p(n) = \frac{1}{n!}\left(\frac{\lambda}{\mu}\right)^n p(0)$ for $n = 0, 1, \ldots, k-1$	$p(n)$ = probability of having exactly n vehicles in system for $0 \leq n < k$
2	$p(n) = \frac{1}{k!k^{n-k}}\left(\frac{\lambda}{\mu}\right)^n p(0)$, for $n \geq k$	$p(n)$ = probability of having exactly n vehicles in system for $n \geq k$
3	$p(0) = \dfrac{1}{\left[\sum_{n=0}^{k-1} \frac{1}{n!}\left(\frac{\lambda}{\mu}\right)^n\right] + \frac{1}{k!}\left(\frac{\lambda}{\mu}\right)^k \frac{k\mu}{k\mu - \lambda}}$	$p(0)$ = probability of having zero vehicles in system
4	$\bar{n} = \frac{\lambda\mu(\lambda/\mu)^k}{(k-1)!(k\mu-\lambda)^2} p(0) + \frac{\lambda}{\mu}$	$\bar{n}$ = average no. of vehicles in system
5	$\bar{q} = \frac{\lambda\mu(\lambda/\mu)^k}{(k-1)!(k\mu-\lambda)^2} p(0)$	$\bar{q}$ = average length of queue
6	$\bar{d} = \frac{\mu(\lambda/\mu)^k}{(k-1)!(k\mu-\lambda)^2} p(0) + \frac{1}{\mu}$	$\bar{d}$ = average time spent in system
7	$\bar{w} = \frac{\mu(\lambda/\mu)^k}{(k-1)!(k\mu-\lambda)^2} p(0)$	$\bar{w}$ = average time spent waiting in queue
8	$p(d \leq t) = 1 - e^{-\mu t}\left\{1 + \frac{p(n \geq k)}{k} \times \frac{1 - e^{-\mu k t[1-(\lambda/\mu k)-(1/k)]}}{1 - (\lambda/\mu k) - (1/k)}\right\}$	$p(d \leq t)$ = probability of having spent time t or less in system
9	$p(n \geq k) = \sum_{n=k}^{\infty} p(n) = \left(\frac{\lambda}{\mu}\right)^k \dfrac{p(0)}{k!\left(1 - \frac{\lambda}{\mu k}\right)}$	$p(n \geq k)$ = probability of having to wait in queue

† k = number of service stations or service channels, each having a servicing rate μ
λ_k = mean arrival rate per station
$\lambda = k\lambda_k$
$\rho = \lambda/k\mu$ = traffic intensity
Source: Wohl and Martin (1967, p. 368).

and constant service time situation. Part *b* of this figure compares the times through systems with varying numbers of channels. As would be expected, the greater the number of channels, the lower the average delays.

While there is much more to queueing theory, this will complete our discussion of the topic. For more information on queueing theory the reader is referred to Blunden (1971, pp. 47–100), Haight (1963, pp. 35–66), and for applications to Rallis (1967).

Table 7-10 Single-Station Queueing Relationships with Poisson Arrivals and Constant Service Times for Steady-State Conditions

Queueing model†		Description of model
1	$\bar{q} = \dfrac{2\rho - \rho^2}{2(1-\rho)}$	$\bar{q}$ = average length of queue
2	$\bar{d} = \dfrac{2-\rho}{2\mu(1-\rho)}$	$\bar{d}$ = average time spent in system
3	$\bar{w} = \dfrac{\rho}{2\mu(1-\rho)}$	$\bar{w}$ = average waiting time spent in queue

† λ = mean arrival rate
μ = constant service rate
ρ = traffic intensity = λ/μ
Source: Haight (1963, p. 60).

Figure 7-16 Comparison of total time in system versus traffic intensity for various types of queues. (*a*) Single-channel queue with various arrival and service time distributions. [*From Blunden (1971), p. 65.*] **(*b*) Single- and multiple-channel queues with Poisson arrivals and negative exponential service time distribution.** [*From Blunden (1971), p. 66.*]

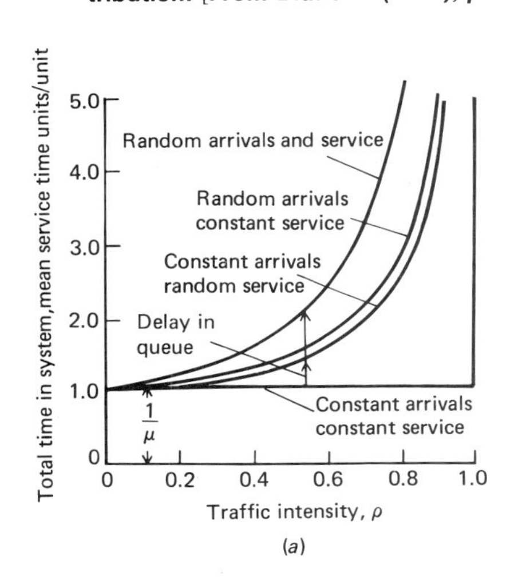

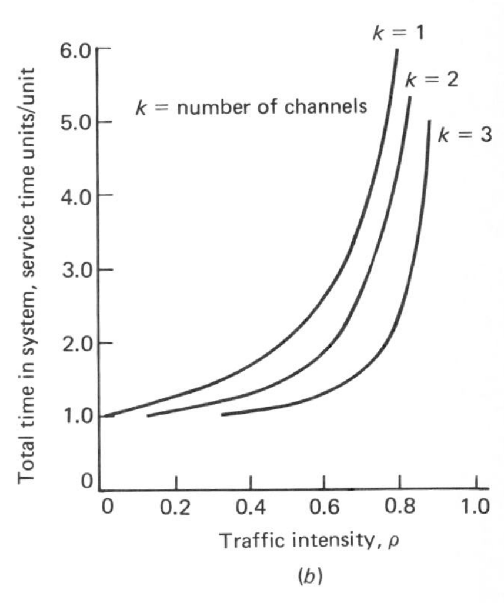

TYPICAL TERMINAL CHARACTERISTICS

Terminals are inherently complex facilities. Many distinct activities are carried out in them, some simultaneously and some in parallel, often with considerable congestion. Also, they are inherently stochastic; they cannot be treated without reference to variations in the volume of arrivals or the times necessary for processing vehicles, passengers, and freight. Furthermore, terminals are often unique, each with a particular design and method of operation found nowhere else. All of these factors make the task of presenting typical terminal characteristics extremely difficult.

However, past studies of terminals of all modes of transport have revealed certain similarities in addition to those revealed by the methods of analysis which we have focused on so far in this chapter. The characteristics of terminals will be presented as a basis for appreciation of their service properties and also as an aid in the solution of problems involving terminal analysis and evaluation, and later, in Chap. 16, on facility design. The presentation is naturally divided into passenger and freight terminals.

Passenger Terminals

The characteristics of terminals are most readily presented in terms of the characteristics of their components. While bus stations are different from airports, many of the components are similar, to the extent that they exhibit similar level-of-service and capacity properties. Thus such a presentation is not only more concise but also facilitates the use of results obtained for one mode or type of terminal in the analysis of another terminal.

Table 7-11 presents the flow characteristics of components of terminals used for the processing of vehicles. The information is divided between capacity and level of service. Again, in theory the capacity of a processor is the inverse of the mean service time, but as was revealed in the analysis of the airport and other terminals throughout this chapter, such a volume of traffic often results in an unacceptably large waiting time. Hence for engineering purposes the capacity is defined as that volume which yields waiting times of an acceptable magnitude. The service times are for the use of one channel of

Table 7-11 Service Times and Capacities of Intercity Passenger Terminal Vehicle Processors

Activity	Mean time or capacity	
Air		
Aircraft service time at gate†		
Intermediate stop (no cabin service)	20–30 min/flight	
End of run (cabin service)	40–60 min/flight	
Gate capacity‡		
Gates assigned to one airline	0.5/G	where G is mean gate occupancy time per flight
Gates shared by airlines	0.7/G	
Bus		
Bus service time at gate§	10–20 min/bus	
Rail		
Train service times at platform		
Day train, pass. only¶	1–4 min/train	
Pass., mail, and express train††	3–20 min/train††	
Pullman train, pass. and baggage only††	3–20 min/train††	
Change locomotives, safety check of train¶	4–15 min/train	

† Horonjeff (1975, pp. 266, 150–153).
‡ Horonjeff (1975, p. 267).
§ Rallis (1967, p. 101).
¶ Droege (1916, pp. 146–208), based on descriptions of terminals and actual train volumes.
†† Rallis (1967, p. 157).

whatever type of processor is specified, such as an airport gate or railroad track at a passenger platform. The interpretation and use of these data are rather straightforward.

Table 7-12 presents similar information for the passenger processors at terminals. There is a very large number of such processors, as the length of this table indicates. Although terminal designs and procedures differ, these processor service times should be fairly representative, since the individual

Table 7-12 Passenger Processing Times at Intercity Terminals

Activity	Mean time or flow rate	Comments
Air		
Airport flight departure		
Pass. ticketing	3.25 min/pass.	Arrivals and service times are reported to be approximately Poisson
Express baggage checking	0.64 min/pass.	
Check-in with seat selection	0.45 min/pass.	
Check-in without seat selection	0.37 min/pass.	
Airport flight arrival		
Deboarding aircraft via jetway	31.9 pass./min	Airline standard capacity is 25 pass./min
Via aircraft stairs	22.1 pass./min	
Via mobile stairs	28.9 pass./min	
Total time from opening pass. doors to baggage available for pick-up	9.40 min	If mobile lounges used, baggage usually arrives at same time as pass.
Minimum time for all air terminal processing		
Departure	24.55–35.10 min	
Arrival	5.39– 7.12 min	
Bus		
Pass. ticketing service	0.70 min/pass.	
Boarding bus	0.065 min/pass.	Based on data for street loading
Total time from opening doors to baggage recovery	4.80 min	
Total time from opening doors to exit of vehicle by pass.		
Without hand luggage	1.56 min	
With hand luggage	2.11 min	
Minimum time for all terminal processing		
Departures	15.37 min	
Arrivals	3.25 min	
Railroad		
Pass. ticketing service	1.06 min/pass.	
Minimum time for all terminal processing		
Departures	19.61 min	Includes times of 3–4 min to walk to end of train platform since terminal is stub end
Arrivals	5.18 min	

All data are based on observations at Washington National Airport, Washington Greyhound Bus Station, and Washington Union Station (railroad), except bus boarding time, which is based on street loading.

Sources: Bruggeman and Worrall (1970, pp. 20–28), except bus boarding times from Kraft and Bergen (1974, p. 17).

processing operations seem to differ little from terminal to terminal. Also presented in Table 7-12 are some typical mean or minimum total times for travelers to use in terminals of various sorts. For example, the mean time for passengers to leave a bus, measured from the time the door was opened, is 2.11 min if the passenger is carrying baggage and 1.56 min if not, presumably reflecting the courtesy of those traveling with luggage in not delaying those without luggage. The minimum time required to go through the processes necessary prior to departure via bus is estimated as 15.4 min. It should be noted that these values were obtained from surveys of the Washington, D.C. terminals in 1968, and values for other terminals may differ, especially the minimum total times, which depend so much upon terminal layout and size.

Characteristics of urban transportation terminals, including those for automobiles (parking lots) and public transportation, are presented in Table

Table 7-13 Service Times and Capacities of Urban Passenger Transport Terminal Processors

Processor	Mean time or capacity	Comments
Automobile parking service times		
Self-park[a]		
Get ticket at entrance	0.1 min/car	Gate
Pay at exit	0.2 min/car	Gate and variable fee
Attendant parking[b]		
Pass. to leave car	0.5 min/car	
Attendant to park or deliver car	3.31 min/car	
Bus stop service times[c]		
Exact fare local bus dwell times		
Simultaneous alighting and boarding		Standard 2-door buses
Morning peak	$y = 0.5 + 1.3A + 2.2B - 0.1AB$	y = dwell time, s
Midday	$y = 0.8 + 1.4A + 2.9B - 0.1AB$	A = no. of pass. alighting
Afternoon peak	$y = 2.4 + 1.1A + 2.1B$	B = no. of pass. boarding
Bus terminal capacity[d]		
Street stops–downtown	160 buses/h	Max. observed with 3–5 berths/stop
Single-terminal berth		
Single-door loading and collect fare	$Q = 1000/B$	Theoretical: Q = buses/h,
Double-door loading, prepaid fare	$Q = 2000/B$	B = boarding pass./bus, approximate for $B \geq 20$
Rail transit service times[e]		
Dwell times		
High-level platforms	$y = 1.1P$	y = dwell time, s
Low-level platforms	$y = 1.8P$	P = no. of pass./door (doors 2.4 ft wide)
Rail terminal capacity (per track)		
Electric rapid transit through station[f]	30–40 trains/h	Lower values are maxima observed; higher values are theoretically achievable
Electric rapid transit stub station[f]	15–20 trains/h	
Commuter railroad through station[g]	15–20 trains/h	
Commuter railroad stub station[g]	6–20 trains/h	
Automobile toll booth service time[i]		
Pay flat toll at booth	0.1 min/car	
Passenger fare collector capacity[h]		
Automatic turnstyle	40–60 pass./min	
Manual barrier	25–50 pass./min	
Exit barrier	50–60 pass./min	Turnstyle or revolving gate
Escalator-per 2-ft-wide pass. lane	60 pass/min	

[a] Blunden (1971, p. 187).
[b] Ricker (1957, pp. 40–41, 147).
[c] Kraft and Bergen (1974, p. 17).
[d] Hoey and Levinson (1975, pp. 31, 37).
[e] American Railway Engineering Association (1936, p. 14-14).
[f] Institute of Traffic Engineers (1965, pp. 20 and 25).
[g] Droege (1916, pp. 146–208), calculated from data on actual terminals.
[h] Leibbrand (1964, p. 226).
[i] Blunden (1971, p. 54).

7-13. Again, the distinction is made between mean service times and the practical capacities of various facilities, for the same reasons as noted above.

In addition to the minimum times necessary for processing at a terminal, it is important to know the actual times taken by passengers. Figure 7-17 presents some typical results for passengers arriving at a terminal preparatory to departure on intercity bus, train, or airplane trips. Although these data were

Figure 7-17 **Time patterns of passenger arrivals at gates before intercity air, bus, and rail departures. (*a*) Washington National Airport. (*b*) Greyhound Bus Terminal, Washington. (c) Union (Railroad) Terminal, Washington.** [***From Bruggeman and Worrall (1970), pp. 22 and 27.***]

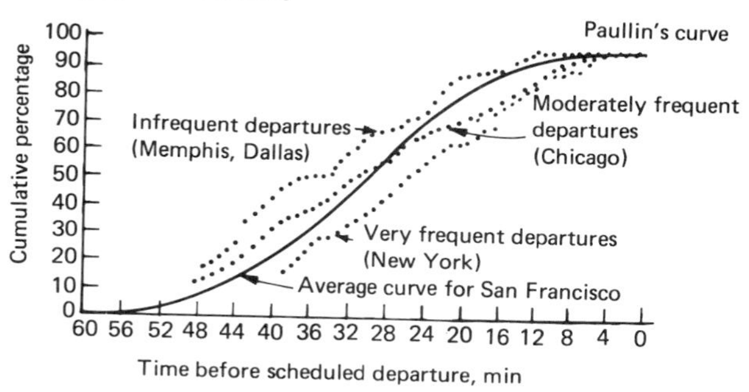

(*a*)

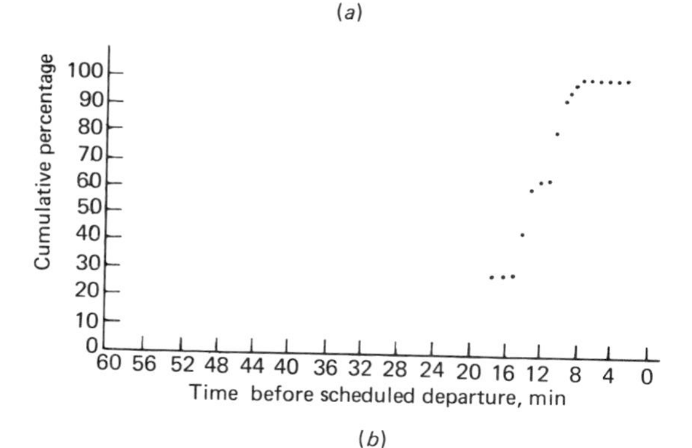

(*b*)

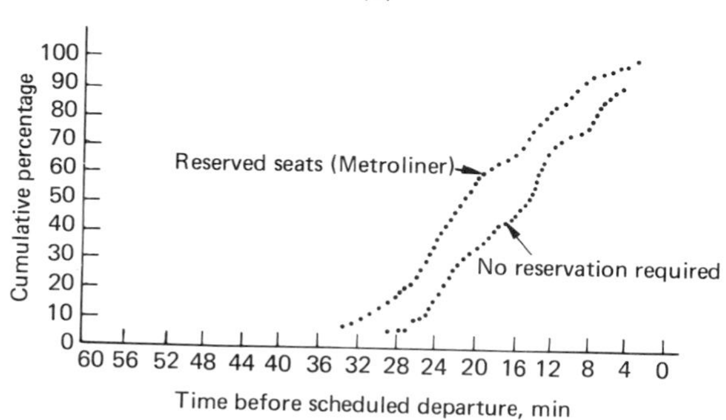

(*c*)

also gathered at Washington, D.C. terminals, they do reveal general characteristics. Many travelers arrive far in advance of the departure, much earlier than necessary for terminal processing, perhaps because they anticipate variability in the travel time from their origin to the terminal and in terminal processing. Given any appreciable variability, to ensure that the desired departure would not be missed, the traveler would allow plenty of time for unexpected delays. There is a noticeable tendency for later (relative to departure time) arrivals if the service is more frequent and if reservations are not required.

Figure 7-18 presents similar results for urban transport. The empirical lines are from studies of bus services in England, these being among the very few actual surveys of passenger arrival patterns. At one extreme is the very reliable bus service, adhering closely to published schedules, as in Leeds during peak periods where passengers arrive on average only 2 to 3 min before departure. At the other extreme is the highly irregular service, with no printed schedule, for which passengers arrive randomly and experience substantial waits. In the middle is the commonly assumed transit situation of random passenger arrivals and adherence to planned headways, yielding an average waiting time of one-half the headway. Also shown is the case of purely random bus arrivals and random passenger arrivals, resulting in average wait larger than observed on any of the routes.

Perhaps an example of the use of these data would be appropriate. Consider an exclusive busway on which a new stop is being proposed. The problem is to determine the additional time required on the bus trip as a result of

Figure 7-18 **Average waiting times for urban transit buses as a function of average time headway. [*London relationships from Holroyd and Scraggs (1966). Leeds and Harrogate relationships from O'Flaherty and Mangan (1970).*]**

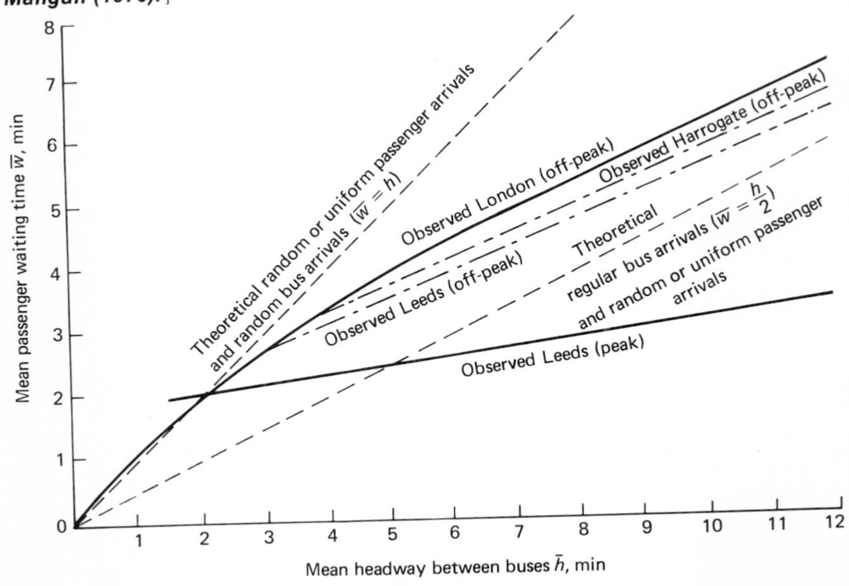

the added stop. Without the stop, buses travel at a speed of 55 mi/h. With the stop buses will decelerate to a stop, requiring 25 s and 0.22 mi; load and discharge passengers; and accelerate to 55 mi/h, requiring 50 s and 0.43 mi. To illustrate the method we shall focus on the morning peak period and the peak direction. Assume that in this period there will be on average 15 persons boarding and 3 leaving each bus. The dwell time, using the morning peak-period formula from Table 7-13, will be

$$0.5 + 1.3(3) + 2.2(15) - 0.1(3)(15) = 32.9 \text{ s}$$

The time required to accelerate and decelerate will be 75 s, and the distance covered will be 0.65 mi. Hence to travel the 0.65 mi the bus will require 107.9 s with the stop. To travel this same distance at 55 mi/h without the stop

Table 7-14 Service Times and Capacities of Freight Terminal Processors

Processor	Mean service time or capacity	Comments
Truck terminal service times		
Align vehicle with dock		
Small single-unit vehicle†	0.5 min	Mean values of Poisson distribution
Medium single-unit vehicle†	2.0 min	Mean values of Poisson distribution
Large trailer truck†	3.0 min	Mean values of Poisson distribution
Loading and unloading times at small shipper and receiver docks†	7 min	Mean values of Poisson distribution
Vehicle loading service times		
Load or unload package goods‡	T = NG/P	T = time, h N = no. of workers G = lb of freight P = productivity rate, lb/worker-h: individual, 5000 to 8000; in team, 6500 to 9500
Load or unload bulk cargo via conveyor§	T = MG/P	T = time, h M = no. of loaders G = tons of freight P = productivity rate, tons per hour: Coal or ore conveyors, 3500–4600 tons/h Sugar conveyor, 700 tons/h
Load or unload oil tanker§	10 h	Current design standard for vessel pumps
Load or unload trailer, container, or pallet on		
Flatcar (side loading)¶	1.0 min/item	
Aircraft¶	1.0–1.25 min/pallet	Pallets on roller bed
Ship§	1.0–2.0 min/item	
Load or unload general cargo on or off ship, per hatch††	30–35 tons/8-h shift	

† Rallis (1967, pp. 101 and 108).
‡ Drake, Startzman, Sheahan, and Barclay (1950).
§ Quinn (1972, pp. 526–536).
¶ Dobson and Stratford (1962, p. 92).
†† Nicolaou (1967, p. 127).

will require 42.6 s. Thus the addition of the stop will add 65.3 s to the bus running time. Of course, if the number of passengers boarding and alighting at other stops is reduced, as it most likely would be, then the time increase would not be so great.

Freight Terminals

Freight terminals are fundamentally different from passenger terminals in one very important respect: the freight must be moved entirely by conscious effort of the terminal operator; it cannot move under its own volition, as can travelers. This difference has far-reaching consequences for the design and operation of the freight terminals. One is that it typically takes much longer to process freight through terminals than passengers because the flow of traffic is smoothed so that more or less full and constant use is made of the equipment and labor engaged in the loading processing. In contrast, a passenger terminal can be designed to accommodate peak loads very rapidly, often with only the provision of additional space and hence at low cost, so traffic volume varies greatly from hour to hour. Also, of course, the value of time savings is often much greater for persons than for freight, making peak-period capacity more desirable in passenger terminals than freight terminals.

The number of processors which can move freight is often limited, and it is essential to consider the capacity of these processors to load and unload and perform other terminal functions. This capacity is usually referred to as a productivity rate, and is usually expressed in tons or other units of freight per processor-hour or -day. Since the operations are usually very labor-intensive, the productivity varies with climate and other environmental conditions, with the quality of the management, culture, etc. Typical values for process times for North America are given in Table 7-14. With the appropriate productivity rate and the number of employees engaged in the operation, the time required to load or unload a vehicle can be calculated easily. The times for other aspects of terminal processing are also given in Table 7-14 and can be added to the actual loading or unloading time to obtain the total time at the terminal.

SUMMARY

1 Terminals are not only an essential component of transportation systems but they also represent a major cost to the system owner/operator and a major point of delay to travelers and freight shipments.

2 The functions of a terminal are numerous. Not only is it the point at which passengers and freight enter and leave the system, vehicles are loaded and unloaded, etc., but it is often the place at which routing decisions are made, reservations are made or vehicles assigned, transport charges are determined and paid, freight is stored, and passengers wait.

3 Terminals can be viewed as processors, in which a certain sequence of activities must be performed in order for the traffic unit (vehicle, freight, etc.) to be processed fully and ready to continue the journey. The activities and the sequence are described by the process flow chart.

4 The flow of traffic through a terminal can be analyzed by use of various types of flow diagrams. These reveal points of congestion and excessive time consumption. They also provide information useful in the design of terminals.

5 Terminal processors operate as queues: As the traffic volume increases, the total time spent waiting increases. As a result, the capacity of terminal processors is the maximum volume that yields an acceptable delay, usually far less than the maximum possible throughput. For many types of terminals, standards specify the maximum acceptable delay and corresponding capacities.

6 The traffic inputs to a terminal and the servicing times of the processors are usually stochastic in nature. Simulation, and to a lesser extent queueing theory, provide means for modeling terminals, taking into account these stochastic elements.

PROBLEMS

7-1

Construct a process flow chart for one of the following types of terminals. Visit the terminal and observe the processing of traffic to ensure an accurate representation. Be sure to include the various types of traffic units also listed.

Airport (aircraft, passengers, baggage, package freight)

Urban transit station (trains or buses, passengers)

Railroad passenger terminal (trains, passengers, baggage, express freight)

Package freight truck terminal (trailers, tractors, single-unit trucks, packages, pallets or containers, drivers)

7-2

Measure the times required for the various activities in one of the terminals listed in Prob. 7-1, following the activity breakdown in Fig. 7-3 or 7-4. Note: This is best done by a large group, permitting the gathering of sufficient data to determine the mean service times and the drawing of the distributions for each processor.

7-3

a Compare the maximum accumulation of passengers at a rapid transit station under two different operating plans. Passengers arrive during the 2-h morning peak period at a volume of 6 persons/min, all to use the same route to the city center. Operating policy 1 calls for 2-car trains to be run at a time headway of 3 min, while policy 2 specifies 10-car trains every 15 min. Draw the cumulative flow diagrams for each of these policies, assuming passengers board the trains at a rate of 3 s/person/car.

b Discuss the results with respect to the passenger's viewpoint, and the size of station waiting area and platform required.

c Can you generalize this result to the effect of operating larger vehicles at greater time headways? What are the advantages to larger vehicles?

7-4

a Plot the delay per aircraft versus annual volume for a runway with a practical annual capacity of 350,000 operations/year. Does the resulting curve follow the general pattern for terminal processor delays?

b What is the additional delay per flight due to an additional 50,000 operations/year if the original volume is 200,000 flights/year? If it is 500,000?

7-5

An automobile parking garage experiences the following average volumes of entering and departing vehicles during the 16-h period it is open each weekday. Draw a cumulative flow diagram showing (a) arrivals only, (b) departures only, and (c) the net flow of cůmulative arrivals minus cumulative departures.

Period	Arrival volume, vehicles/h	Departure volume, vehicles/h
6 A.M.– 8 A.M.	25	5
8 A.M.–10 A.M.	200	5
10 A.M.– 1 P.M.	50	40
1 P.M.– 4 P.M.	30	50
4 P.M.– 6 P.M.	25	190
6 P.M.–10 P.M.	10	50

What is the maximum number of autos in the garage? When?

7-6

Determine the ultimate capacity of one track of a rapid transit station used for trains in one direction only, for two different train lengths, 150 ft and 750 ft. Cruise speed is 55 mi/h, the deceleration rate is 3 mi/(h)(s), dwell time at the station is 30 s, and the acceleration yields the following times from rest to reach the corresponding distances:

Time, s	Distance, ft	Speed, mi/h
10	200	30
20	720	47
30	1500	55

The three-aspect, three-block signal system is used, and a clear signal must be seen 100 ft before one passes it to proceed at normal speed. A signal is at the upstream end of the station platform. What is the ultimate capacity in trains per hour for each of the two trains. If the passenger capacity of the trains is 300 and 1500 persons, respectively, what is the ultimate capacity in passengers per hour on this track?

7-7

In Prob. 7-6, what would be the effect on train and passenger capacity for the longer train of reducing dwell time to 15 s? Increasing it to 45 s?

7-8

An airport runway used for landings has a mean service time of 1.15 min, with a standard deviation of 0.25 min. Draw the volume–mean-delay curve for 0 to 40 operations per hour. If the maximum mean delay can be 1 min, what is the capacity? If it is 2 min?

7-9

Inspection of Fig. 7-10 reveals that the practical capacity, in operations per hour, of a single runway with mixed operations is greater the larger the percentage of small planes (classes C, D, and E) using the runway. Compare the capacity with 100 percent operations of small craft versus 60 percent of class A. If the average passenger capacity of small aircraft is 20 persons and that of class-A aircraft is 250 persons, what are the passenger volume capacities? Assume no class-B aircraft use the terminal.

This problem reveals a rather general relationship: As the size of the vehicles is reduced, the capacity of a facility (terminal, link, intersection) usually increases. But the capacity to carry passengers or freight usually diminishes as a result of the switch to smaller vehicles.

7-10

A city government is under pressure from merchants to increase the amount of street parking space set aside for truck deliveries. On one block, there are presently five spaces. Merchants claim that six spaces should be set aside. The duration of truck dwell times has been measured and found to average 25 min, with a Poisson distribution. During the peak 4 h of deliveries, trucks arrive randomly at a mean rate of 6.5/h. (a) What is the current average delay in looking for a space? What would it be with six parking spaces? If the cost of operating a delivery truck is $20/h, how much will be saved in one 4-h peak period? (b) What is the probability that a truck will be parked for more than 30 min? 60 min? 90 min? What is the average number of trucks that would be in the entire system during these 4 h? (c) Can you think of any way the city could capture some of this saving, which otherwise would be a private gain at public expense?

7-11

Automobiles arrive at a 10-stall parking lot randomly, at a mean arrival rate of one every 5 min. They are parked for an average of 30 min, with a Poisson distribution of parking times. What is the average number of cars in the lot and its entrance, assuming that no cars turn away even if a space is not immediately available? What is the probability of having to wait? What is the mean waiting time?

7-12

For the bus stop used in the example of simulation in the text, compare the average delay to buses and average total time at the stop (delay plus dwell time) for a bus volume of 41.5 buses/h with that obtained in the text with 83 buses/h. Since the time headway between buses is now doubled, on the average twice as many passengers will be boarding each bus as before, assuming the same number of travelers still use the stop. To obtain this doubling of dwell times, for each draw of a random number use a dwell time exactly twice that obtained from Fig. 7-14, e.g., if the random number were 81, the dwell time would be 32 s.

7-13
Draw the cumulative flow diagrams for a terminal handling ore from a mine designed for different line-haul modes. Ore enters the terminal at a rate of 1000 tons/h, 16 h each day. In the terminal the ore is washed to remove dust and dirt, being ready to leave 2 h after arrival. One line-haul alternative is a conveyor belt, which can accommodate up to 1000 tons/h and is operated continuously. Another is a rail connection, with 2 trains/day, each carrying 8000 tons, leaving 10 h and 18 h after the mine starts each day. The other is a truck connection, trucks starting to leave as soon as the ore is ready after mine opening each morning, and trucks departing every 3 min for 16 h, each carrying 50 tons. Contrast the size of the terminal storage and loading facilities with these line-haul alternatives.

REFERENCES

American Railway Engineering Association (1936), ''Manual for Railway Engineering,'' Chicago.

Baerwald, John E. (1965), ''Traffic Engineering Handbook,'' 3d ed., Institute of Traffic Engineers, Washington, DC.

Baerwald, John E., Matthew J. Huber, and Louis E. Keefer, eds. (1976), ''Transportation and Traffic Engineering Handbook,'' (of the Institute of Traffic Engineers), Prentice-Hall, Englewood Cliffs, NJ.

Benjamin, Jack R., and C. Allin Cornell (1970), ''Probability, Statistics and Decision for Civil Engineers,'' McGraw-Hill, New York.

Blunden, W. R. (1971), ''The Land Use/Transport System,'' Pergamon, New York.

Bruggeman, Jeffrey M., and Richard D. Worral (1970), Passenger Terminal Impedances, *Highway Research Record* No. 322, pp. 13–29.

Consad Research Corp. (1970), ''Research Study of Intercity–Intraurban–Interfaces (I^3),'' 3 vols., Report to the U.S. Dept. of Transportation, Pittsburgh, PA.

Dobson, L., and Alan H. Stratford (1962), Operational Experience with Unitized Loading: The Argosy and Rolamat in Operation with British European Airways, ''International Forum on Air Cargo,'' Institute of Aerospace Sciences, New York, vol. 1, pp. 87–98.

Drake, Startzman, Sheahan, and Barclay (1950), ''Manual for Planning and Operating Terminals,'' Washington, DC.

Droege, John A. (1916), ''Passenger Terminals and Trains,'' McGraw-Hill, New York.

Gerlough, Daniel L. (1955), Use of Poisson Distribution in Highway Traffic, in ''Poisson and Traffic,'' Eno Foundation, Saugatuck, CT, pp. 1–58.

Haight, Frank A. (1963), "Mathematical Theories of Traffic Flow," Academic Press, New York.
Hoel, Lester A. (1969), Evaluating Alternative Strategies for Central City Distribution, *Highway Research Record* No. 293, pp. 1–13.

Hoey, William F., and Herbert S. Levinson (1975), Bus Capacity Analysis, *Highway Research Record* No. 546, pp. 30–41.

Holroyd, E. M., and D. A. Scraggs (1966), Waiting Times for Buses in Central London, *Traffic Engineering and Control,* **8**(3): 158–160.

Horonjeff, Robert (1975), "Planning and Design of Airports," McGraw-Hill, New York.

Institute of Traffic Engineers (1965), "Capacities and Limitations of Transportation Modes," Washington, DC.

Jennings, H., and H. Dickins (1958), Computer Simulation of Peak Hour Operations in a Bus Terminal, *Management Science,* **5**(1): 106–120.

Kraft, Walter H., and Terrence F. Bergen (1974), Evaluation of Passenger Service Times for Street Transit Systems, *Highway Research Record* No. 505, pp. 13–20.

Leibbrand, Kurt (1964), Nigel Seymer, trans., "Transportation and Town Planning," MIT Press, Cambridge, MA.

Nicolaou, Stavros N. (1967), Berth Planning by Evaluation of Congestion and Cost, *Journal of the Waterways and Harbors Division, A.S.C.E.,* 93 WW4 (Nov. 1967), 107–132.

O'Flaherty, C. A., and D. O. Mangan (1970), Bus Passenger Waiting Times in Central Areas, *Traffic Engineering and Control,* **11**(9): 419–421.

Organization for Economic Cooperation and Development, Consultative Group on Transportation Research (1971), "Air Transport Access to Urban Areas," Paris.

Quinn, Alonzo de F. (1972), "Design and Construction of Port and Marine Structures," McGraw-Hill, New York.

Rallis, Tom (1967), "The Capacity of Transport Centers," Report 35. The Technical University of Denmark, Department for Road Construction, Transportation Engineering and Town Planning, Copenhagen.

Ralph M. Parsons Co. (1973), "The Apron-Terminal Complex—Analysis of Concepts and Evaluation of Terminal Buildings," U.S. Federal Aviation Administration Report No. FA-RD-73-82, Washington, DC.

The Rand Corporation (1955), "A Million Random Digits," Free Press, Glencoe, IL.

Ricker, Edmund R. (1957), "Traffic Design of Parking Garages," Eno Foundation, Saugatuck, CT.

Sampson, Roy J., and Martin J. Farris (1975), "Domestic Transportation: Practice, Theory and Policy,'' Houghton-Mifflin, Boston.

Whorf, R. P., and D. F. Wilkie (1969), ''On the Generation, Description and Simulation of Terminal Alternatives,'' Ford Motor Company, Transportation Research and Planning Office, Dearborn, MI.

Winchester, James H. (1973), The Business of Airports, in Martin T. Farris and Paul T. McElhiney, eds., ''Modern Transportation: Selected Readings,'' Houghton-Mifflin, Boston, pp. 138–145.

Wohl, Martin, and Brian V. Martin (1967), ''Traffic System Analysis for Engineers and Planners,'' McGraw-Hill, New York.

Operations Plans

8.

In this chapter we shall attempt to bring all of the various components of transportation systems treated in preceding chapters together to describe the functioning of an entire transportation system. A system consists of three basic types of elements: (1) the network of fixed facilities, including way links, way interchanges, terminals, and maintenance facilities, (2) vehicles and containers, and (3) a plan for the operation of the system. As the name implies, the operations plan specifies how these various elements will be used to move passengers and freight from origin to destination. Included in the operations plan, then, must be a specification of how vehicles or containers will be assigned to persons or freight to be transported and how this traffic will be passed through terminals and routed from origin to destination.

Operations plans vary from relatively simple to extremely complex. An example of the former are the plans for the movement of motor vehicles over public highways, which from the standpoint of the highway authorities, are essentially rules for the movement of vehicles such as speed limits, parking restrictions, and the specification of who has right-of-way at intersections, etc. In contrast, the freight operations plan of a railroad is extremely complex, reflecting elaborate procedures for getting empty freight cars to shippers who need them, the routing of those cars from origin to destination including their assignment to various line-haul trains and transfers at freight classification yards from one train to another, the allocation of locomotives to various trains, etc. Furthermore, these procedures may depend upon the type of freight (e.g., perishables receiving higher-priority treatment and traveling on faster trains) and the routing may vary with the time of day or the day of the week.

COMPONENTS OF OPERATIONS PLANS

There are numerous ways in which a transportation system can be operated. Even with the same network of fixed facilities and the same vehicles and containers, a wide variety of possible operations plans usually exists. Although it is often taken for granted since particular types of operations plans are often in practice associated with particular technologies, the selection of an operations plan is an extremely important aspect of planning and managing a transportation system. It is the operations plan which, in conjunction with the network and technology, ultimately determines the capacity, level of service, and cost characteristics of a transportation system.

Operations plans will be discussed in three parts. In this section, we shall present, in rather abstract terms, the primary options with respect to operations plans and their general characteristics. In the next section, the service and capacity properties of a very simple network—a single line with stations (terminals) along it—will be related to the network, vehicle characteristics, and the operations plan. Finally, in the last section, the more generalized relationships for entire networks will be presented. All of this will then provide a basis for the presentation of methods for determining the costs of transportation systems in Chap. 9.

Vehicle and Container Assignments

One of the most critical decisions in the specification of an operations plan is the selection of the manner in which vehicles or containers will be assigned to passengers or freight. Of course, this assignment is limited, as the heading indicates, to transport systems which use vehicles or containers, but these are ubiquitous. There are basically three alternatives with respect to vehicle and container assignment: (1) dedicated assignment, (2) sequential assignment, and (3) shared or simultaneous assignment. These will be discussed in turn.

Dedicated assignment means that a vehicle or container is permanently assigned to a single traveler or shipper, so that in principle no conflicts should arise as to who or what is to use the vehicle or container. Perhaps the best example of this is a bicycle which is used by one and only one person, although occasionally it may be loaned to others. Another example of an assignment which is often but not necessarily completely conflict-free is the family automobile (or increasingly, automobiles), usually used by only members of the family, and often only one or a few members thereof. Although the family car is technically used by more than one person, the group is limited and generally reaches decisions regarding travel collectively, so that it may be considered essentially dedicated assignment. In the case of freight transport, dedicated assignment is perhaps best typified by a truck used to transport a carpenter and his or her tools, supplies, etc. A truck used by a particular firm, such as a manufacturer, is also essentially in dedicated assignment. Again the dedication is essentially to a particular decision-making unit rather than to the same shipment, for obviously on different occasions the truck would be used to move different shipments. Unit trains, often owned by the cargo shipper but operated by the railroad, carrying full trainloads of cargo from a single origin to a single destination, are another form of dedicated assignment.

Sequential assignment involves the exclusive use of the vehicle or container for one shipment or one traveler (or group traveling together) at any one time, but with the vehicle or container being reassigned to another shipment or group of travelers after that movement has occurred. To differentiate it from dedicated assignment, in sequential assignment the vehicle is owned and assigned by an entity other than the shipper or traveler. Examples of this type of assignment are very common, including the usual assignment of railroad cars to freight shippers and shipments, the assignment of trucks owned by transport companies to shippers and shipments, and in the case of passenger transport, the allocation of chartered buses to groups of travelers, as well as taxicabs to individuals or small groups.

The final type of vehicle or container assignment is shared or simultaneous assignment, in which many shipments, perhaps traveling between different origins and destinations, share the same freight vehicle or container, and passengers not traveling together as a group between the same origin and destination share the same vehicle. The vehicle or container is owned by a separate entity, which determines the times of departure, routing, etc. This situation is, of course, most typical in the case of passenger movement. Bus, air, and railroad service all generally fall within this category. In the case of freight movement, shipments of less than a full container typically are consolidated into a containerload by the carrier and then dispersed at various points along the route. Also, freight cars, which are assigned sequentially, are usually moved in a shared vehicle, a freight train, between different origins and destinations. Thus rail freight service (excluding unit trains and less than carload shipments) is a combination of sequential and shared assignment.

Direct comparisons of any but the most general sort are difficult among these three types. Dedicated assignment has numerous advantages, including the freedom to choose times of movement and routing. In a situation where time is very highly valued and the cost of individual vehicle ownership and operation is not excessive, this is the preferred manner of operation, as the popularity of the private automobile and bicycle reveal.

Sequential assignment has some characteristics similar to dedicated assignment, in that the routing of the movement and the times of departure and arrival need not be particularly influenced by the movement of other traffic, except in so far as other movements influence the availability of vehicles or containers. In cases where a shipment is sufficiently large to fill an entire vehicle, as is very often the case with goods movement, this system has considerable advantage. On the other hand, it is not widely applicable to person movement except for taxicabs, air taxis, and rental autos. The shared forms of operation require one person's movement to suit the movement desires of others, but this can often be accomplished without undue hardship, particularly where there are large volumes to be moved. In these situations there is little inconvenience in terms of frequency of service or route to be followed. In the more densely developed parts of many cities, buses and subways may be so ubiquitous and have such frequent service that they are as attractive or more attractive than private automobiles or taxicabs for much travel.

Although the heading of this section indicates these considerations apply to vehicle and container systems only, in principle at least they also apply to continuous-flow systems. In those systems, it is necessary to decide the manner in which the transport capacity is assigned to various shipments or travelers. However, most continuous-flow freight transport systems are in fact designed to be used for one and only one type of shipment. Nevertheless, in the case of products pipelines, many different commodities are carried. As a

Table 8-1 Examples of Different Types of Vehicle and Container Assignments Policies

Dedicated assignment
Family automobile
Privately owned truck (i.e., owned by user)
Sequential assignment
Full unit loads for railroad cars, trucks [termed carload (CL) and truckload (TL) shipments]
Taxicabs and chartered or rented passenger vehicles
Simultaneous assignment
Travelers on regular scheduled carriers, such as aircraft, bus, and rapid transit train
Less than full unit loads for railroad cars and trucks [termed less than carload (LCL) and less than truckload (LTL) shipments]
Movement of freight containers on vehicles which carry many of them, e.g., railroad cars in a train, containers on ships

result, the pipeline must be assigned to these in a sequential or batch manner. Continuous-flow person transport systems are used simultaneously, but service is continuous, and capacity problems are rare, so inconveniences such as queueing delays are minimal. Table 8-1 summarizes these examples of different types of assignment.

Times for Movement

Although various time aspects of movement were mentioned in connection with vehicle and container assignment, this characteristic of transport operations planning is so important that it deserves special attention. Transport systems using separate vehicles inherently provide intermittent service, that is, vehicles depart and arrive at particular moments rather than continuously.

An important consideration is the frequency with which vehicle departures will be made. From the standpoint of the traveler or shipper, generally the more frequent the departures, the better the service quality, since smaller deviations from the ideal departure or arrival time have to be made. On the other hand, generally the larger the transport unit or vehicle, the lower the cost of ownership and operation per unit of carrying capacity. As a result, the carrier tends to favor larger vehicles, which given any particular volume of traffic to be carried, results in fewer departures. Actual scheduling or departure policy is, therefore, in practice, a balancing of these two viewpoints.

If travelers arrive at, or more precisely are ready to depart (i.e., after any processing), a terminal randomly and vehicles depart at a constant headway h, then as we saw in Chap. 7 in the discussion of terminals, the average waiting time will be one-half the time headway. This also applies to a situation of uniform arrivals of passengers, as 1 every min, although this is uncommon. In equation form,

$$\overline{w} = \frac{h}{2} \tag{8-1}$$

where $\overline{w}$ = mean waiting time for vehicle
h = time headway between vehicle departures

Under these circumstances, the maximum wait is the (constant) time headway,

$$w_{\max} = h \tag{8-2}$$

Of course, these formulas are based on the assumption that vehicles (or containers) have available space, but this is usually the case for all passenger carriers except airlines in peak travel periods, although standing may be required during such periods on trains or buses (if permitted). These same formulas apply to freight under the stipulated conditions.

If the headways are irregular, the mean waiting time for travelers on each departure will be one-half the headway interval before that departure. The expected number of travelers is proportional to that period, so if the mean passenger arrival rate is λ, the expected total waiting time for departure i (summed over all passengers) will be

$$\tfrac{1}{2}(t_i - t_{i-1})\lambda(t_i - t_{i-1}) = \tfrac{1}{2}\lambda(t_i - t_{i-1})^2$$

where t_i = time of ith departure
t_{i-1} = time of prior, $(i-1)$st departure
λ = mean passenger arrival rate

Numbering the departures consecutively, $i = 1, 2, 3, \ldots, N$, and taking t_0 to be the time when passengers begin to arrive, the total waiting time (of all passengers) for all departures is

$$\sum_{i=1}^{N} \tfrac{1}{2}\lambda(t_i - t_{i-1})^2$$

Dividing by the total number of passengers $\lambda(t_N - t_0)$ yields the mean waiting time $\overline{w}$:

$$\overline{w} = \frac{1}{2(t_N - t_0)} \sum_{i=1}^{N} (t_i - t_{i-1})^2 \tag{8-3}$$

or, in terms of headways, with $h_i = t_i - t_{i-1}$,

$$\overline{w} = \frac{1}{2(t_N - t_0)} \sum_{i=1}^{N} h_i^2$$

As can be easily shown, this average waiting time is minimized when all headways are equal. This also minimizes the maximum waiting time, which is equal to the maximum headway:

$$w_{\max} = \max_i (h_i) \tag{8-4}$$

Many carriers operate services on a uniform headway, for these reasons. This is common on many intercity bus routes, on most high-speed corridor train services, and on numerous high-volume air shuttle services. Sometimes there are slight deviations because of fluctuations in traffic with time of day. And many urban transit services are operated with a uniform headway in each of three or four periods of the weekday.

However, in many situations travelers do not arrive randomly at the same mean rate throughout the day. Instead, passenger traffic fluctuates, following a regular pattern. In this situation, the average waiting time can be reduced by concentrating more departures in the peak traffic period, although this will increase the maximum possible delay. This is illustrated in Fig. 8-1, which shows the mean waiting time versus the number of departures for a constant headway policy and a schedule for which the departure times were selected to minimize total delay. The latter was developed by a mathematical model based on dynamic programming, which yields the best departure times given any specified cumulative arrival pattern for passengers. This method, somewhat beyond the mathematical level of this book, is described in Bisbee et al. (1969).

There are various forms in which the specification of departure times or other times of movement can be expressed. The most common for passenger carriers is to specify a particular departure and arrival time at each terminal served on a vehicle's trip, e.g., the usual carrier time table made available to the public. Also, some carriers have particular time tables by which they operate vehicle trips, but do not make the schedules available to the public, perhaps because of the expense or the problems that might arise in frequently changing the time table to meet changed traffic conditions. This policy of having a time table but not making it available to the public, at least in most hours of the day, is very common in urban transit systems, where vehicles are usually operated so frequently that the delay caused by waiting for the vehicle is minimal. In such instances often the carrier will have a policy on maximum time headways—that headway being the maximum time which a traveler would have to wait.

Another possibility is to operate vehicles when there is sufficient traffic to warrant a vehicle departure. This has been one traditional way of operating railroad freight services, for example, a typical policy being to operate a freight train whenever a certain number of cars (or total weight of train, or a combination of the two) has been accumulated for that particular destination. This can result in rather irregular service depending upon the variations of traffic volume.

Another possibility is to specify a maximum time of transport, but without specifying the times at which intermediate stages of the movement will occur. This has been tried in many of the new dial-a-ride services in urban areas, where typically either a maximum total time to destination or the maximum waiting time will be supplied to a traveler upon telephoning to request a vehicle. This also seems to be of considerable interest in goods movement at the present time, and probably many carriers will experiment with premium service in which guaranteed delivery times are provided for, possibly with rebates to the shipper if the guarantee is not met. Guarantees are already finding application in certain types of package freight priority movement.

Another important consideration in movements which are characterized by cyclic variations in volume is the possibility of shifting some traffic in time to yield a more nearly uniform volume. This has been tried with considerable success in freight movement, where production is often irregular but consumption is rather uniform, as in the case of some agricultural products. Here warehousing at the origin can be used to advantage. It has also been tried with much less success with person movement by attempting to stagger work hours. Nonetheless, a substantial potential exists for easing capacity problems of transport systems by making more uniform demands on the system. Since this takes us into considerations of the demand for transportation and possible added costs or inconvenience to travelers and shippers, we shall

Figure 8-1 **Comparison of constant headway and minimum passenger waiting time departure scheduling. (*a*) Temporal demand pattern used (actual average daily pattern in New York-Washington air shuttle). (*b*) Departures dispatched so as to minimize total (or average) waiting time, bars showing time of departure and passenger loads. (*c*) Average waiting times with uniform headways and headways varied so as to minimize waiting (as above).** [*From Simpson (1969), pp. 84 and 85.*]

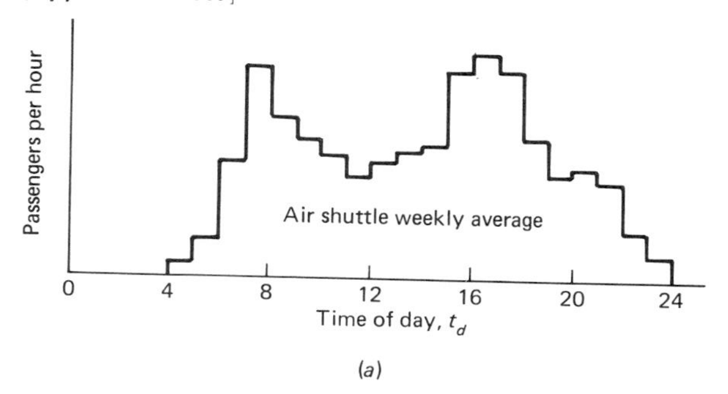

(*a*)

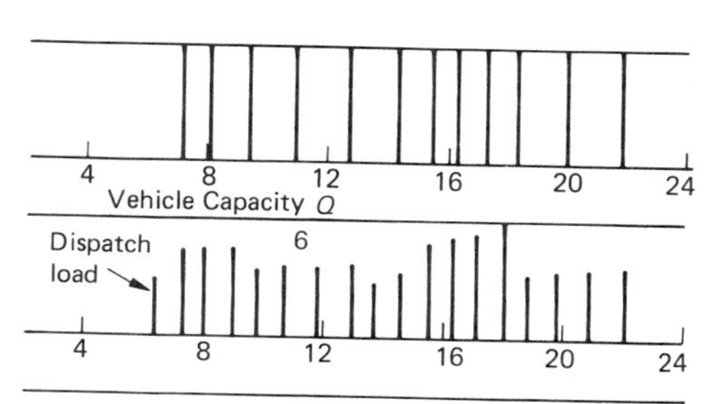

(*b*)

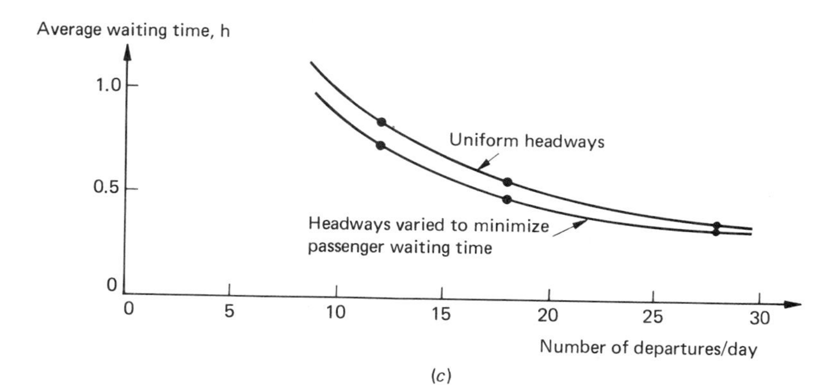

(*c*)

defer detailed discussion of this to Chap. 10. As the name implies, continuous-flow transport systems usually are not characterized by inherent delays in the departure of flows from the origin terminal. Only when the system must be sequentially assigned to different movements, as in the case of a pipeline carrying different liquids, is it a consideration. In those cases, the same options and approaches as for vehicle systems apply.

Spatial Aggregation

The description of transport networks in Chap. 3 revealed that transport service is not available throughout an area. Access to the system and paths for movement are generally provided only at particular locations, due to the scarcity of land and the cost of providing transport facilities. This necessarily leads to a spatial pattern of movement which involves the movement of passengers or freight from origin to destination over paths which may be, to varying degrees, longer than the most direct paths ("as the crow flies"), following a great circle on the earth's surface.

In addition to aggregation of traffic on a particular path due to the limited number of paths in existence, there often is a further aggregation of traffic onto particular main paths. One reason for this is that often by taking a route which is longer than the shortest possible route over the network, the cost or time can be reduced. For example, a traveler may go out of the way to travel on a freeway in order to save time instead of traveling over a shorter route (in distance) on arterial streets.

Similar routing of traffic away from the most direct route in distance is often encouraged or required by the various transport carriers. The traveler's or shipper's time may be reduced if the routes used offer distinctly better performance or cost less than more direct routes. Also, the carrier may wish to concentrate a sufficient amount of traffic on one route so as to utilize fully the vehicles operated or the various fixed facilities, and thereby achieve economies. This is particularly noticeable in railroad, air, and urban transit networks. In intercity networks, traffic traveling to or from small towns is often routed via large cities so that the longer-distance portion of the journey occurs on major or so-called trunk routes. In the case of urban transit, very often feeder buses are operated to and from rail rapid transit stations in order to route travelers via such trunk routes even though local transit routes may be available. As these examples indicate, this spatial aggregation, through operating policies on the routing of traffic, can result in benefits to the traveler or shipper as well as to the carrier, while in other cases the advantage may accrue to the carrier alone.

The networks of carriers are often described as having major gateways, which are terminals that are regional hubs of traffic. In order to travel from one region to another, it may be necessary to travel through particular terminal gateways, which are connected by major trunk routes. In overland networks, particularly railroads and roads, natural barriers such as rivers and mountain ranges are crossed at only a few places. These places form gateways connecting one region to another. The air network, although less subject to natural barriers, is characterized by gateways consisting of major cities which are connected to one another by major trunk routes with high quality of service. Tributary areas around these major hubs are served by local flights. The similarity of this pattern to spokes radiating from wheel hubs gives rise to the term hub. An abstract pattern is shown in Fig. 8-2.

Figure 8-2 **Spatial pattern of operations plan illustrating gateway terminal hubs, major trunk routes, and feeder routes.**

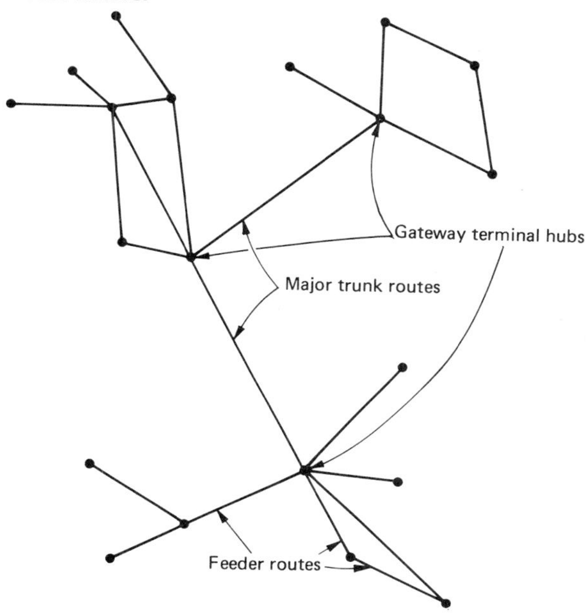

ANALYSIS OF A SINGLE LINE

In this section we shall analyze the effects of various operations plans and other characteristics such as vehicle size on a single line. Although most transportation systems are represented by much more complicated networks, this analysis will illustrate concretely many of the relationships which were discussed abstractly in previous sections. Also, certain transport systems are composed of largely independent individual routes which are essentially single lines connecting two or more terminals. Examples include many urban transit routes as well as some major intercity routes operated as shuttle services. Attention will be directed particularly to the effect of different operating plans on vehicle and crew requirements, the quality of service from the standpoint of the traveler, the capacity of the system, and relationships to the number of lanes or size of the way links and other fixed facilities required.

Size of Vehicle

In operating a transport system of this type, numerous relationships must hold strictly for the system to function properly. One of the most fundamental is that among the capacity of the system for moving passengers (or goods) in one direction, the operating plan, and characteristics of the fleet. On this single line all vehicles will be operated from one end of the line to the other and then return, vehicles shuttling back and forth between the two end terminals. Vehicles will be operated at a uniform departure time headway, and all vehicles are of precisely the same capacity. Under these conditions, the relationship for one direction among passenger capacity, vehicle time headway, and vehicle capacity is

$$q_c = \frac{Q}{h_t} = Qq \tag{8-5}$$

where q_c = passenger flow capacity in one direction
Q = capacity of each vehicle, passengers
h_t = time headway of vehicles (in one direction)
q = vehicle volume in one direction

This is illustrated in Fig. 8-3.

These flows can be related directly to vehicle requirements. Assuming that all vehicles require the same time to make a round trip, measured from departure at the starting terminal to the next departure there—this often termed the cycle time—the relationships are

$$q = \frac{n}{t_r} \tag{8-6}$$

$$q_c = \frac{nQ}{t_r} \tag{8-7}$$

Figure 8-3 Relationships among passenger flow capacity, vehicle volume, vehicle passenger capacity, and time headway, in one direction.

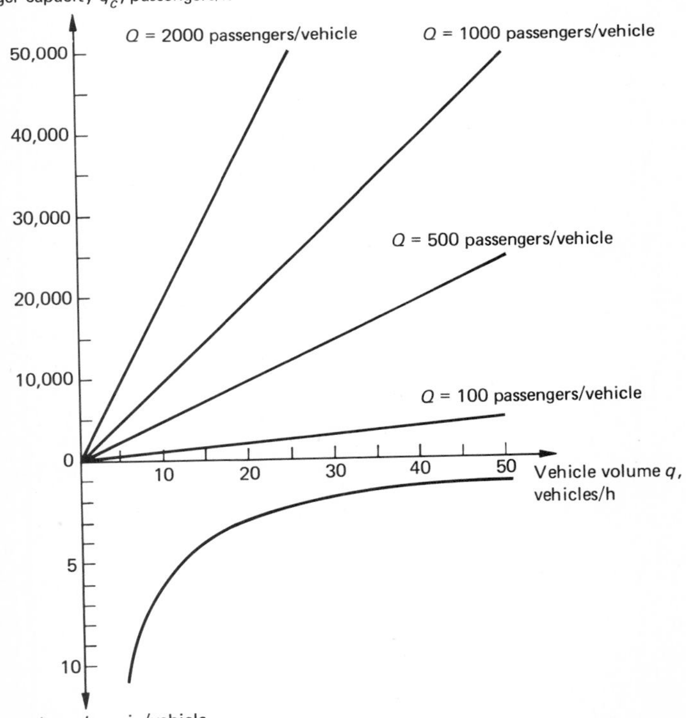

Note: All flow measures are for one direction

where n = number of vehicles
t_r = round trip or cycle time

Equation (8-7) is graphically portrayed in Fig. 8-4, for typical values of vehicle (train) capacity on urban rail transit lines and a cycle time of 1.5 h. Care must be exercised in using Eqs. (8-6) and (8-7): Firstly, n must be an integer. Secondly, t_r is the actual cycle time taken in operation. The minimum possible cycle time may be less, but this merely means that there is slack in the schedule. Finally, this formula applies to situations where the system operates continuously or in which the period of operation in this manner is equal to or greater than the round trip time. To use either equation to find the minimum number of vehicles necessary to provide a given service, simply enter the minimum cycle time in the equation, solve for n, and if n is not an integer, round upward. In equation form:

$$n = \langle q t_m \rangle = \left\langle \frac{q_c}{Q} t_m \right\rangle \tag{8-8}$$

where $\langle \cdot \rangle$ denotes smallest integer containing $\cdot$
t_m = minimum cycle time

Many important interrelationships among the vehicles, network, and operating plan are revealed in this formula. The first is that any given capacity can be provided with a smaller number of vehicles of the same carrying capacity if the round trip time of those vehicles is reduced. This is a particularly important consideration in high-volume transport systems where the cost of vehicles is high. It also reveals the fact that in certain situations an

Figure 8-4 Relationship between passenger flow capacity and number of vehicles on a transit route.

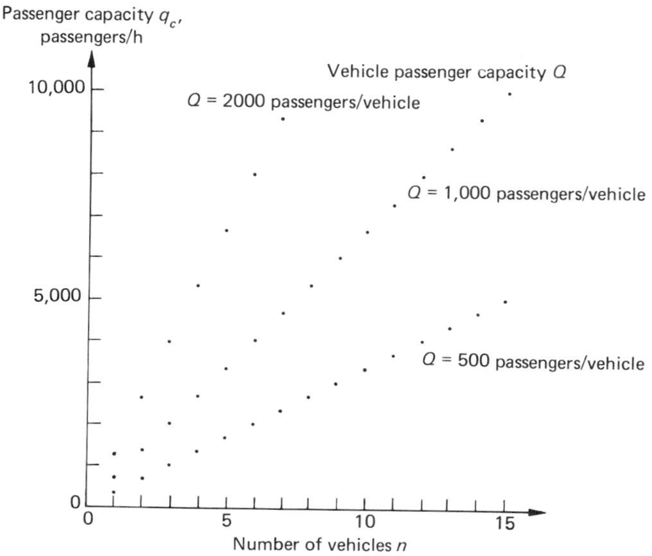

Note: All flow measures are for one direction; all vehicles are available for service; t_r, round trip time, is 1.5 h

increase of the speed of service, a change that benefits the users of the system, might also reduce total system cost, mainly as a result of the reduction in vehicle fleet and crew costs! If it is necessary, increasing the carrying capacity of the system with a fixed fleet size (number of vehicles) can be accomplished by reducing the round trip time, as might be done on transit lines simply by eliminating stops for some or all vehicle trips. On rail transit lines, in particular, this is often done by operating a skip-stop service. Trains are divided into two or more groups, each group serving some stations exclusively, with major stations [as in the central business district (CBD)] being served by all trains. Since train cycle times are reduced, greater capacity is possible; simultaneously most travelers experience a travel time reduction, since usually the reduced running time more than offsets the increased headway and waiting time at skip-stop stations. However, some must change vehicles. Alternatively, for a given capacity, fleet size and costs are reduced slightly. A detailed analysis of application of skip-stop operation to a Philadelphia rapid transit line is presented in Vuchic (1973). These latter relationships are illustrated in Fig. 8-5.

Another important relationship exists between the capacity of the fixed facilities over which the vehicles are to operate and the operating plan. Specifically, the operating plan must not require a headway less than the minimum that can be sustained on those facilities. The relationship which must hold is

$$q < NC' \tag{8-9}$$

or

$$h_t > \frac{1}{NC'} \tag{8-10}$$

where C' = vehicle capacity of each channel
N = number of channels (used in one direction)

In many situations, the required capacity along a line may increase with time, ultimately taxing the facility. One alternative is, of course, to increase the carrying capacity of each vehicle and thus keep the vehicle volume within the line's capacity. However, if the size of the vehicle cannot be increased further, as when the length limitation of terminals is reached, then additional tracks or lanes must be provided to increase capacity. This can result in a step function of channels required, as shown in Fig. 8-6. However, it should be borne in mind that in many transport technologies, there are various ways of controlling vehicle movements which create a smoother relationship between cost and capacity than that implied by this figure. For instance, sophisticated control systems to reduce time headways might be installed in places where there are particularly difficult capacity problems, resulting in a relationship between way facility cost and capacity somewhat smoother than the step function shown in Fig. 8-6.

Another important relationship is that between travel time and the choice of operating plan, including the delay to travelers occasioned by the intermittent departures of vehicles. If we assume that travelers arrive at their origin stations randomly, as has been observed for transit lines with frequent departures, they will face waiting times which on average will be one-half the

Figure 8-5 Relationship between passenger flow capacity and cycle time for a rapid transit line.

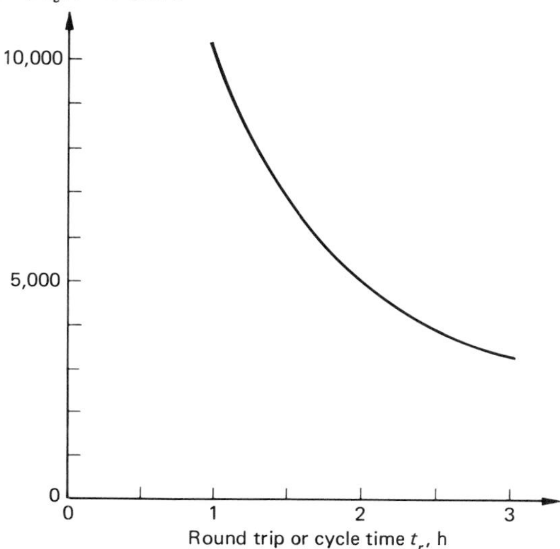

Notes: Relation is for 10 vehicles of 1,000 passengers/vehicle being available continuously; all flow measures are for one direction

headway, given by Eq. (8-1). Thus the travel time is a function of headway:

$$\bar{t}_{ij} = \frac{h_i}{2} + t^r_{ij} + \bar{t}^e_j \tag{8-11}$$

where $\bar{t}_{ij}$ = mean journey time from terminal i to terminal j
t^r_{ij} = vehicle running time from terminal i to terminal j
$\bar{t}^e_j$ = mean time to exit vehicle at terminal j (usually very small)

Figure 8-6 Relationship between number of channels and maximum vehicle volume on a rapid transit line.

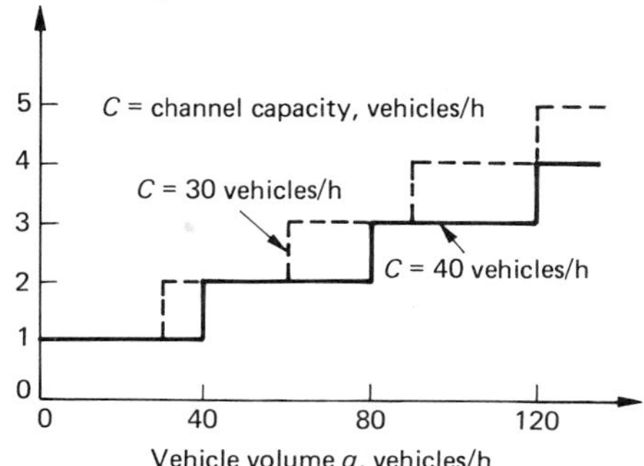

Note: Values of channel capacity of 30 to 40 vehicles/h are typical for rail rapid-transit lines with operation in one direction only on each track

If the carrier operates vehicles of identical capacity and constant headway, as we have assumed for this example, then the mean and maximum waiting times are related to capacity, using Eqs. (8-1) and (8-2),

$$\overline{w} = \frac{h_t}{2} = \frac{Q}{2q_c}$$

$$w_{\max} = h_t = \frac{Q}{q_c}$$

The relationship would be as shown in Fig. 8-7, which is drawn for 100- and 500- passenger vehicles, a 20-min running time, and a 0.5-min average exiting time. The advantage to the traveler of smaller vehicle units—down to that which results in the minimum feasible time headway—is obvious. But to provide greater capacity, a larger vehicle must be operated.

Time Variations in Demand

In most transport systems there are substantial variations in the total demand for movement over time. In the case of urban transport (and many other

Figure 8-7 Relationships among waiting times, vehicle capacity, and passenger flow volume for a rapid transit line.

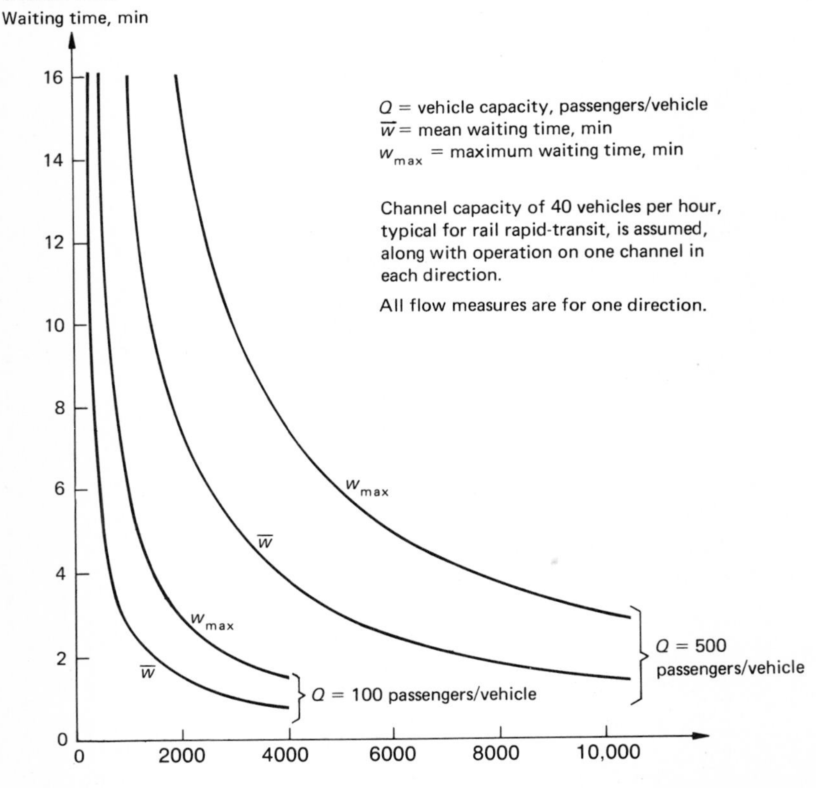

situations), these variations are cyclic in nature; on work days two short periods of high-volume passenger flow separated by approximately 8 to 9 h, with considerably lower volume at other periods. A typical idealized pattern for the capacity provided is shown in Fig. 8-8. Also shown in that figure is an actual pattern of passengers carried past the point of maximum load, the capacity required, which, of course, varies somewhat more smoothly than the step function of capacity provided, which we will use in the analysis to simplify computations.

Under such peak traffic conditions the carrier will usually choose to operate either more frequent service, vehicles of a larger capacity, or a combination of the two during the peak period in order to meet the increased demand. Equation (8-8) can be modified rather easily to cope with this situation of varying traffic. The exact form of the equation depends upon whether the round trip time of vehicles is greater than, less than, or equal to the duration of the peak period. Since with rail service the number of cars per train may vary, it is useful to modify the equations to yield cars rather than trains, or vehicles (as we have been using the term). Car capacity may vary between the peak and base (midday) period, as might headways, etc., so we will use a superscript p to indicate peak and b to indicate base periods. The time headway necessary to provide adequate capacity in any period of constant passenger volume is, following the form of Eq. (8-5),

$$h_t^p = \frac{k^p Q_c^p}{q_c^p} \tag{8-12}$$

where h_t^p = peak period time headway
k^p = number of cars per train in peak period
Q_c^p = passenger capacity of each car in peak period
q_c^p = passenger volume in peak period

If the minimum cycle trip time is less than or equal to the duration of the peak period, then the number of trains and cars required, by Eq. (8-8), is

$$\begin{aligned} n &= \left\langle \frac{t_m^p}{h_t^p} \right\rangle = \left\langle \frac{q_c^p t_m^p}{k^p Q_c^p} \right\rangle \\ c &= k^p \left\langle \frac{t_m^p}{h_t^p} \right\rangle = k^p \left\langle \frac{q_c^p t_m^p}{k^p Q_c^p} \right\rangle \end{aligned} \tag{8-13}$$

provided

$$t_m^p \leq P$$

where t_m^p = minimum cycle time in peak period
P = duration of peak period
n = number of trains required
c = number of cars required

If the minimum cycle time is greater than the duration of the peak period, then the service before or after the peak period must be considered in fleet requirements. The midday period almost invariably requires more cars than the periods on the other side of the peak periods, so the midday period is the period relevant to fleet requirement calculation. This period is often termed the base period, and hence we shall refer to its characteristics with

Figure 8-8 **Typical temporal and directional patterns of capacity required and capacity provided on urban transit routes. (*a*) Inbound toward central business district. (*b*) Outbound from central business district.**

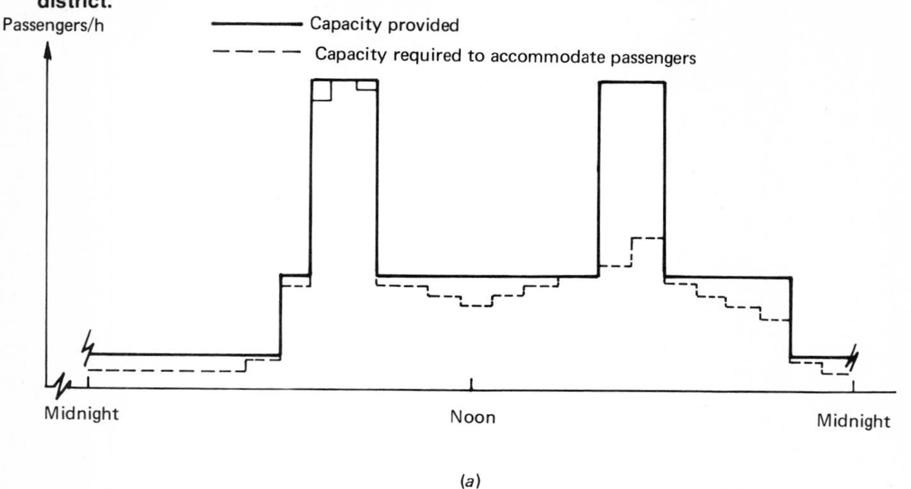

(*a*)

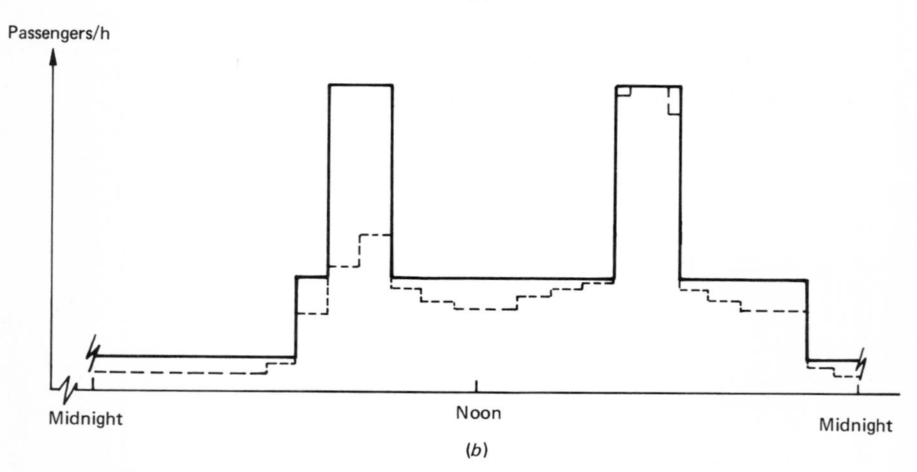

(*b*)

Locations of passenger flow and capacity measurements:

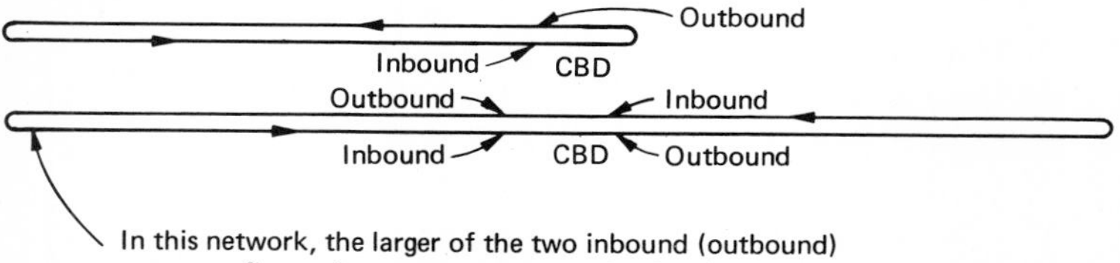

In this network, the larger of the two inbound (outbound) passenger flow volumes at any moment in time is taken as the capacity required.

the superscript b instead of the superscript p used for the peak period. The headway in the base period will be

$$h_t^b = \frac{k^b Q_c^b}{q_c^b} \tag{8-14}$$

The total car requirements will be as follows

$$\begin{aligned} n &= \left\langle \frac{t_m^p}{h_t^p} \right\rangle + \left\langle \frac{P - t_m^p}{h_t^b} \right\rangle \\ c &= k^p \left\langle \frac{t_m^p}{h_t^p} \right\rangle + k^b \left\langle \frac{P - t_m^p}{h_t^b} \right\rangle \end{aligned} \tag{8-15}$$

provided

$$t_m^p \geq P$$

It must be remembered that the expression for the number of trains may include trains of two different sizes, one size for the peak period and one for the base, so care must be exercised in using it. In these expressions, the first term is the number of trains (cars) required to equip the trains run to accommodate the peak flow, and it is assumed all these are run in the period P. The second term is the trains (cars) for the additional trains that must be run before the first of the peak trains returns to the origin station and is available to make up an outbound base period train.

During the base period, the numbers of trains and cars actually operated is less than the numbers in the peak period. Specifically, these are

$$\begin{aligned} n^b &= \left\langle \frac{t_m^b}{h_t^b} \right\rangle \\ c^b &= k^b \left\langle \frac{t_m^b}{h_t^b} \right\rangle \end{aligned} \tag{8-16}$$

These apply regardless of the relationship of minimum cycle time to the duration of the peak period and assume only that the duration of the base period is greater than (or equal to) the base period cycle time.

Equations (8-12) through (8-16) reveal that the effects of variations in vehicle capacity, round trip time, etc., are essentially the same as was discussed above for the case of continuous flow, although reductions in round trip time will yield the greatest reduction in fleet size requirement when that round trip time is less than or equal to the duration of the peak period.

The main problem associated with the substantial peaking of traffic is that many of the vehicles and crews will be required for only a portion of the day, which tends to increase the total cost per traveler or unit of capacity rather substantially over the case where the traffic is uniformly spread throughout the operating period. This can be illustrated easily by considering a system which must have a total capacity sufficient to accommodate 32,000 passengers (in each direction) in the 16-h operating period. At one extreme, all of this capacity might be spread uniformly throughout the period, resulting in a required passenger flow capacity of 2000 passengers/h. Using trains of 5 cars, each car having a capacity of 100 passengers and a minimum cycle time of 1.5 h, would result in the need for 30 cars, or 6 trains of 5 cars each, by Eqs. (8-12) and (8-13):

Table 8-2 The Effect of Peaking of Capacity Requirements on Vehicles Required on a Rapid Transit Line†

Capacity provided			Fleet required	
Total over 16-h operating day, passengers/day	During 2-h peak period, passengers/h	Ratio of peak to average daily capacity	Trains of 5 cars each	Total cars
32,000	2000	1.00	6	30
32,000	4000	2.00	12	60
32,000	6000	3.00	18	90
32,000	8000	4.00	24	120

† Vehicles have a minimum round trip time of 1.5 h and a capacity of 500 passengers, each train being composed of 5 cars. All flow measures are in one direction. Peak to average daily capacity is calculated with both in same units, e.g., in last line the ratio is 8000 passengers/h to 2000 passengers/h.

$$n = \left\langle \frac{2000 \text{ passengers/hr}}{(5 \text{ cars/train})(100 \text{ passengers/car})} \right\rangle 1.5 \text{ h} = \langle 6 \text{ trains} \rangle = 6 \text{ trains}$$

$$c = 5 \text{ cars/train} \left\langle \frac{2000 \text{ passengers/h}}{(5 \text{ cars/train})(100 \text{ passengers/car})} \right\rangle = 5 \text{ cars/train } \langle 6 \text{ trains} \rangle$$

$$= 30 \text{ cars}$$

If greater capacity must be provided in some portions of the operating period than in others, then in general the fleet requirements increase. A typical transit traffic distribution is for about one-half the passengers in each direction to be carried in two peak hours, which results in a similar pattern for capacity required. In this example, this would result in a peak direction capacity of 8000 passengers/h during each of two 2-h peak periods and a capacity of 1143 passengers/h in the nonpeak direction during the peak hours and the same in both directions during the remaining 12 h. Since the minimum cycle time is still less than the peak period, Eqs. (8-12) and (8-13) remain applicable. The requirements increase to 24 trains of 5 cars each, or 120 cars. This is four times the number required with the uniform distribution of traffic!

Table 8-2 presents the fleet requirements for various degrees of peaking of capacity, still providing the same total flow capacity over the entire operating period. From this table one can understand one source of the financial problems that beset urban public transport carriers and others who must accommodate substantial peaking of traffic. Similarly, the possible benefits from flattening peak periods is amply illustrated, although, of course, the traveler may be inconvenienced. Another important characteristic of peaked flow conditions is that usually there is a substantial number of vehicles and crews available during the nonpeak period for the accommodation of any additional traffic which might arise. Thus the added cost of additional traffic during such periods is generally quite small compared to the average total cost of the system, which is often reflected in lower fares for off-peak travel.

Varying Volume Along the Route

Most transport systems consist of more than a single link between two terminals. Usually scheduled vehicle services are designed to connect a number of terminals along a given route. Also, there is often considerable variation of the volume of traffic along such a route. For example, in the case of a transit line operating radially from the CBD of a metropolitan area, the traffic is usually highest in the vicinity of the CBD and drops steadily with distance from it. An illustration of an actual pattern of traffic is shown in Fig. 8-9. This situation presents some interesting problems and challenges for the design of operations plans.

On one hand, all vehicles operating out of the CBD might continue to the end of the line, and thus a uniform capacity would be provided along the line. This capacity would be relatively fully utilized in the vicinity of the CBD and only minimally utilized at the other end of the line, as illustrated in Fig. 8-10*a*. If the traffic drops uniformly along the line, as shown in this figure, the average capacity would be precisely twice that of the demand. This is usually expressed in terms of the load factor, the ratio of actual passenger-miles carried to seat-miles (or passenger-place-miles if standees are allowed) provided:

$$f = \frac{M}{S} \tag{8-17}$$

where f = load factor
M = passenger-miles actually traveled
S = seat-miles (or place-miles) provided

The skip-stop form of service described earlier is illustrated in Fig. 8-10*b*. This also can reduce travel time for most travelers and reduce fleet requirements. It has the advantage of requiring only one track or channel in each direction, but still provides the full capacity all along the route.

Another option is to operate local and express vehicles along the entire route, locals serving all stations and expresses stopping at only a few. This pattern is shown in Fig. 8-10*c*. It still has the characteristics of providing the same capacity all along the route, but the reduced cycle time of the expresses results in a reduction in vehicle requirements, reducing costs somewhat. Also, some passengers will benefit from the faster speeds of the expresses. Even persons originating at or destined to local stations may find it saves time to transfer to the express train. One disadvantage of this type of operations plan is that if time headways are short (as they are on most transit lines), two tracks (or in general, channels) are required in each direction for the time savings of the express to be realized. While the cost may be high, on some subway lines (e.g., in Philadelphia and New York) such four-track lines are provided. On streets wide enough for buses to pass, such an operation can be provided without difficulty.

The so-called zonal service operations plan has the advantage of providing a capacity more consistent with the volume of traffic along the entire route. This is commonly done on radial transit lines by dividing the line into two or more sections or zones, each served by different routes, which might overlap to varying degrees. These can be illustrated for the three-zone case. If most travelers are destined to the CBD, a combination of local and express service is very attractive. One route would stop at all stations in the outer zone, operating nonstop through the middle zone to (or from) the CBD zone, where all stops would be made, this route serving the longer trips to and from the CBD. Another vehicle would then operate as a local in the middle and CBD

Figure 8-9 **Peak hour inbound passenger volume along Cleveland rail transit lines on a typical weekday in January 1960.** [*From Rainville (1961), p. 533.*]

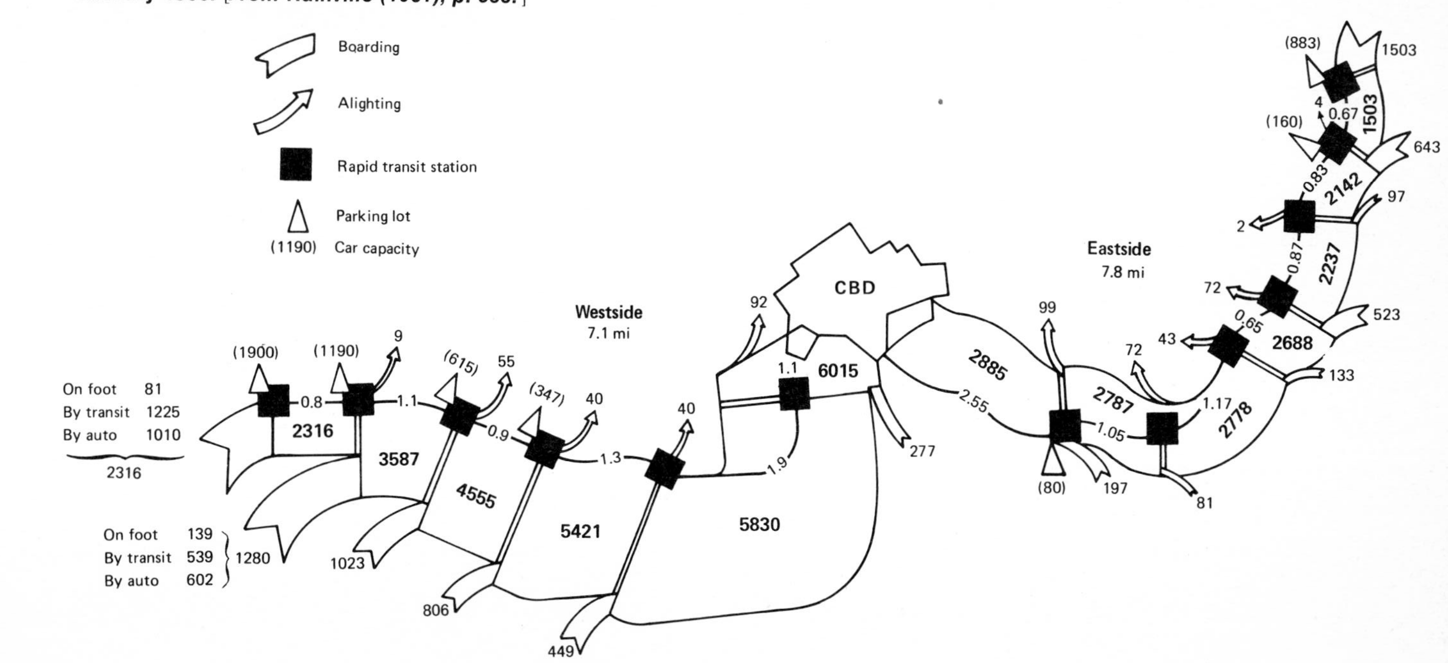

Figure 8-10 **Alternative operations plans and the matching of capacity to traffic volumes along a rapid transit line.**

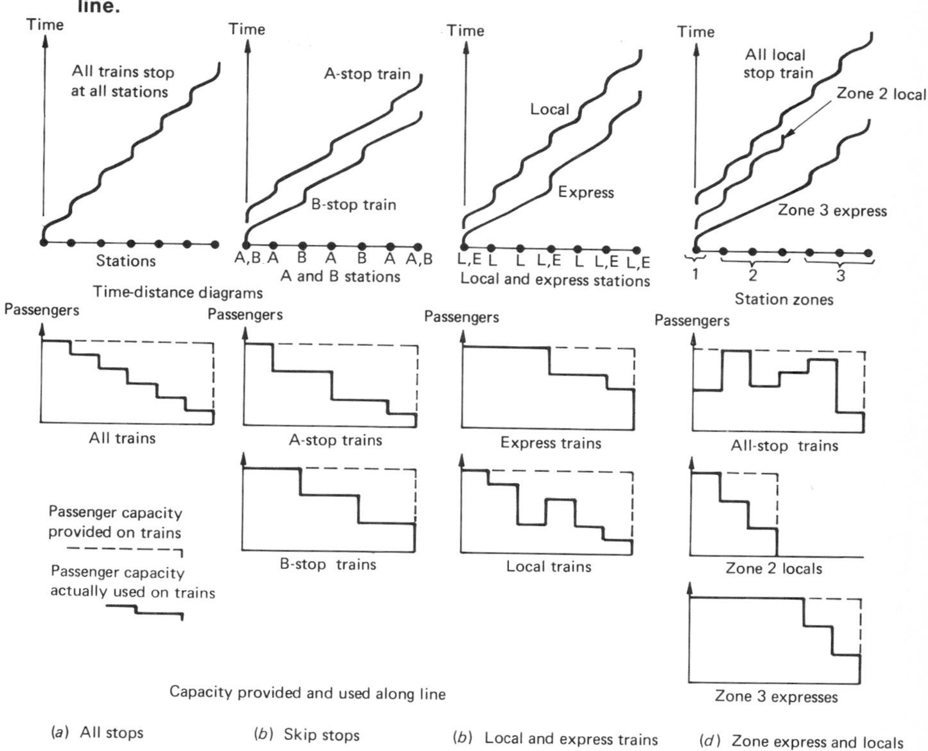

zones, carrying travelers where origins and destinations are along its route. A final route would operate as a local covering the entire three zones, for travelers between origins and destinations that would otherwise require a transfer between the outer-zone express and middle-zone–CBD local. This pattern is illustrated in Fig. 8-10*d*, from which the better matching of capacity and passenger volume is clear.

This type of operation usually not only reduces the cost to the operator in terms of vehicle fleet requirements, but also can reduce the travel time for travelers between the outer zone and the CBD. However, other travelers will experience an increase in travel time, mainly due to the increase in waiting time resulting from fewer vehicle trips being offered. In an application to a Chicago commuter railroad (Morlok and Vandersypen, 1973), the introduction of zone express and local service reduced car requirements by 12 percent and reduced operating costs.

Obviously this type of service is readily provided only on routes where there are two channels for vehicles in the middle zone, for otherwise the express and local vehicles would conflict in that zone. Also, a relatively high volume is necessary to fill vehicles on all three routes, with a reasonable headway. But with vehicles such as typical urban transit buses, seating about 53 passengers and standing up to 25 more, a route with a headway of 10 min would be filled to capacity at 450 passengers/h, for a total three-route corridor flow of only 1350 passengers/h. And often the full-length local can be dropped, requiring travelers between the outer and middle zones to change vehicles at the boundary of the two. The number of such travelers is usually small on CBD-oriented routes, typically about 5 percent to 15 percent of all travelers using the line (Rainville, 1961). Usually such three-route service is provided on heavily used bus routes, multitrack (each direction) rail routes (e.g., New York, Chicago), and often by the combination of city-to-suburb and within-city routes in larger metropolitan areas. The latter often involve different modes and different operating agencies, as with suburban railroads and city bus or rapid transit lines. Also, often bus routes radiate from the outer ends of rail rapid transit lines, resulting in what is in effect multiple zones, within each of which capacity can be adjusted to volume.

Another aspect of urban transit services (as well as many other transport systems) is that the volume of traffic to be accommodated (i.e., capacity required) in one direction at one moment may be very different from the volume to be accommodated in the other direction. As these services are usually designed, with the ability to store vehicles at only one end of the line, the same capacity must usually be provided in both directions, this capacity being offset in time by an amount equal to the time for the vehicle to travel from the point of observation to the end of the line and back to that point. Thus the pattern of capacity provided for these idealized peaking conditions at a point along a route would appear as shown earlier in Fig. 8-8. Also shown is a typical required capacity (passenger flow), which indicates rather full use of capacity in the peak period in the peak direction but substantial underutilization of that capacity in the so-called reverse peak direction. This necessity to provide essentially the same capacity in both directions is an undesirable characteristic of the service, albeit one that our technology does not enable us to circumvent at the present time. However, it also indicates that the empty seats or available capacity in the reverse peak direction is available free of charge, so that any additional traffic in that direction can be accommodated with no increase in system cost. Because of this, many urban transit systems charge lower fares in the reverse peak direction, hoping to induce greater use of the system.

Vehicle-Miles (Kilometers) of Operation

At a number of places in the preceding discussion reference was made to the total vehicle-miles operated with various operations plans, but no precise formula for calculating this factor was presented. Since the total distance of vehicle operation is an important factor in estimating the costs of operating a transport system, such a formula is essential to any presentation of methods of analysis of operations plans. The formula is rather straightforward, the vehicle-miles being simply twice the product of the vehicle volume (in one direction in the period considered) and the one-way length of the line. Mathematically, this is

$$m_v = 2L \sum_{i=1}^{I} q^i P^i \qquad (8\text{-}18)$$

where i = index of periods ($i = 1, 2, \ldots, I$)
m_r = vehicle-miles (or other distance unit) operated
q^i = vehicle volume in one direction in period i
P^i = duration of period i
L = length of route in one direction

In situations where many routes are operated, as in the case discussed above in which a single line was divided into a zone express and local route pattern, the total vehicle-miles, would be simply the sum of those on each of the individual routes, as given by the formula

$$m_r = \sum_{j=1}^{J} 2L_j \sum_{i=1}^{I} q^{ij}P^{ij} \qquad (8\text{-}19)$$

where j = index designating routes ($j = 1, 2, \ldots, J$)

In the case of rail transit operations (as well as other types of transport operations), some costs may depend on car-miles operated. The formula for car-miles is

$$m_c = \sum_{j=1}^{J} 2L_j \sum_{i=1}^{I} q^{ij}k^{ij}P^{ij} \qquad (8\text{-}20)$$

where m_c = car-miles operated

To illustrate the use of these formulas, we shall consider again the rail transit route in which a capacity of 32,000 passengers must be provided each day in each direction during the 16-h operating period, with a capacity of 8000 passengers/h during the peak hours in the peak direction, and a capacity of 1143 passengers/h in the other periods and directions. The line is 15 mi in length, and trains require 1.5 h to make a round trip. During the peak hours, trains with a capacity of 500 passengers are operated, each train consisting of 5 cars, requiring a train volume of 16 trains/h, with a headway of 3.75 min, by Eq. (8-5). During the other periods, the management has decided to operate 3-car trains, each of which accommodates 156 passengers (52 seats/car), at headways of 8 min, thus providing a seating capacity of 1170 passengers/h, sufficient to accommodate all travelers in seats. Since the peak hour trains operating in the peak direction must turn around and run to the other end of the line, they provide capacity in the reverse peak direction during 2 h in the morning and 2 h in the evening also; the shorter trains will operate 12 h each day. Substituting these values into Eqs. (8-19) and (8-20), the total train-miles per day and car-miles per day are, respectively,

$$\begin{aligned} m_r &= 2(16 \text{ trains/h})(2 \text{ h})(15 \text{ mi}) + 2(7.5 \text{ trains/h})(12 \text{ h})(15 \text{ mi}) \\ &\qquad + 2(16 \text{ trains/h})(2 \text{ h})(15 \text{ mi}) \\ &= (960 + 2700 + 960) \text{ train-mi} = 4620 \text{ train-mi} \\ m_c &= 2(16 \text{ trains/h})(5 \text{ cars/train})(2 \text{ h})(15 \text{ mi}) \\ &\qquad + 2(7.5 \text{ trains/h})(3 \text{ cars/train})(12 \text{ h})(15 \text{ mi}) \\ &\qquad + 2(16 \text{ trains/h})(5 \text{ cars/train})(2 \text{ h})(15 \text{ mi}) \\ &= (4800 + 8100 + 4800) \text{ car-mi} = 17{,}700 \text{ car-mi} \end{aligned}$$

Conclusions

Although this discussion has focused on only one type of transport system, rail rapid transit service focused on a central business district, it has highlighted many of the problems and options associated with operations plans. These include: (1) selection of frequency of service, (2) selection of types of connections between terminals (e.g., local, skip-stop, express, or a combination of these), (3) selection of vehicle size, (4) adaptation to peaking of traffic, and (5) adaptation to variations in capacity required along a route. These are common to all transport systems. It is now appropriate to turn to general relationships for more complex networks.

NETWORK RELATIONSHIPS

While the discussion of a rather simple single line revealed many of the important relationships among fixed plant characteristics, vehicle characteristics, and operations plans, it is still necessary to specify the quantitative relationships of these characteristics for the complex networks typical of most transportation systems, particularly freight systems, in which the container (rail car, truck trailer, lighter aboard ship, etc.) is detachable from the remainder of the vehicle. Also, on more complex networks, there is much more opportunity for variation in the manner in which traffic is aggregated temporally and spatially. Although every attempt will be made to be comprehensive in this section, variation in the extent of quantitative data and prior analyses of various carriers necessarily emphasizes certain modes or situations to illustrate particular points. Specifically, data on regulated passenger carriers such as airlines and bus lines are very good, and they are reasonably good on rail freight operations, but there are comparatively few data on truck operations, waterway operations, and international freight services of all sorts.

Operations Plans of Passenger Carriers

The operations plans of passenger transport are logically treated distinctly from those for freight transport primarily because of two characteristics. Firstly, the passenger carrier can take advantage of the fact that persons can be relied upon to transfer themselves from one vehicle to another, along with their luggage, etc. (as noted in Chap. 7), which results in a much smaller fraction of a passenger's total journey time being spent in terminals than is typical in the case of freight. Secondly, passenger travel times are subject to much smaller random variations than freight travel times, because of the relative simplicity of operations plans, the reduced terminal activity, and the scheduled character of most passenger transport services.

Undoubtedly the simplest operations planning in the passenger sphere is that of an individual traveler (or group traveling together) using a dedicated vehicle, whether owned or leased. The main problem is finding the best route, or at least an appropriate one, through the network to the destination, presumably an approximation to a best path as would be found by the best path algorithm covered in Chap. 3. Once the route is known, journey times can be estimated from link travel time characteristics, and other pertinent information can be developed from them. The operation of a rented or chartered vehicle, whether a taxi, auto, bus, or aircraft, can be determined in a similar way from the traveler's standpoint.

The operations plans of scheduled passenger carriers—bus lines, airlines, and railroads—involve many of the same considerations and procedures that were developed for the analysis of the rapid transit line. For routes which are sufficiently heavily traveled to result in the assignment of vehicles and crews to one route only, the methods presented for the single line can be used directly. Running times of vehicles over links and way intersections would be estimated using the methods of Chap. 4, in conjunction with the methods of Chap. 5 to take into account any congestion. The methods of Chap. 7 for the analysis of terminals would be used in estimating the times associated with terminal processing of vehicles (and containers, if applicable), travelers, and their luggage. These would then be brought together to estimate vehicle cycle times on individual routes, which may consist simply of nonstop runs between major terminals, as is common in the case of air and higher-volume bus services, or a chain of stations as is typical of rail services and lower-volume air and bus routes. Once the individual routes have been specified, then the same considerations presented in the analysis of the example transit line would apply. Sometimes very limited options exist as to vehicle size, as in the case of intercity bus services, or there may be limitations on the number of vehicle trips that can be operated due to line capacity considerations.

If traffic on individual routes is very low, it may be possible to use a single vehicle on more than one route. This is particularly common on some air routes, where there may be only one or two flights per day, which may require aircraft for a total of only a few hours. In such cases, one vehicle is assigned to many routes to permit its full utilization, and the precise time table on each route is adjusted so that a vehicle trip on one route directly follows a vehicle trip on a connecting route. Very elaborate computer procedures have been developed by airlines for preparing these schedule adjustments and route combinations to minimize the number of aircraft crews required. The reason for such attention to detail by the airlines is simply that the cost of aircraft and operating crews is extremely large and it is therefore advantageous to minimize the number of aircraft acquired. The procedures used rely heavily on very advanced mathematical methods which are computerized. [For those wishing to pursue this topic, Simpson (1969) provides a rather complete treatment, but one that assumes a knowledge of linear and dynamic programming.]

An innovative class of local public transportation service which many feel has much potential for expanding the types of travel markets which can be served by some sort of public carrier (as opposed to just the private automobile) is paratransit. All of these use essentially conventional transportation hardware but provide novel services because of innovative operations plans. One type of paratransit is the dial-a-bus, in which a van (typically with 9 to 25 seats) provides a door-to-door service, like a taxi does, including use of the telephone for requesting a vehicle, but with the main difference being that many persons traveling between different origins and destinations use the van simultaneously. This reduces costs per passenger, at the expense of some delay for a vehicle and some detour from the best route of any one passenger in order to serve others. Many services are in operation now. Those that are able to control labor costs and achieve high load factors show promise of being economically viable. Some serve special groups only, such as the elderly, and some serve only certain trips, such as to or from a railroad commuter station, while others are general carriers. Other forms of paratransit include allowing taxicabs to offer group rates, and operation of jitneys, small buses, on regular routes, but often with a schedule such as "depart when full or after a 10-min headway, whichever occurs first." These paratransit services may enable transit service to be provided to low-density areas, and areas with short, dispersed trips, for which more conventional technologies and operations plans are inappropriate. Interesting reviews of all of these are in Kirby (1975) and Rosenbloom (1976).

Typical Passenger Carrier Operating Characteristics

There are a number of measures of characteristics of actual carriers which reflect the technology employed (or mode), the operations plan adopted, and the network over which the flows occur. Appropriately, these are called operating characteristics. The measures are designed to reflect particularly the effectiveness in using the vehicles or containers, labor, and other resources under management's direct control, and also the quality of the operations from the standpoint of the user. The exact measures vary from mode to mode, reflecting the ease or difficulty of obtaining certain data, the usefulness of the data to management, etc., but collectively, these data provide a good picture of overall operations.

These data typically include such information as operations plans, size of vehicle, average speeds through system (of vehicles, containers, and passengers or cargo)—often as a function of distance of movement, utilization of capacity (usually load factor as defined earlier for passenger carriers), and some measure of the amount of transport output (e.g., passenger-miles) per vehicle, container, and employee. More or less detail can be provided depending upon the use and the degree to which the carrier wishes to divulge information.

Table 8-3 presents such data for air, bus, and rail intercity passenger services in the United States. All these carriers are scheduled, fixed route carriers, with the bulk of the service being of the simultaneous assignment type, although some charter service is included. Many features of these data are striking. First, only about half the seat-miles provided are used in all cases, reflecting the problems of matching capacity and traffic in rigidly scheduled service and with the fluctuations of passenger traffic. The miles operated by each vehicle varies markedly by mode, air having the largest value by far, reflecting the high average speed achieved in that mode. Speeds, incidentally, are the speeds of scheduled trips between terminals, which in general would be far greater than the average speeds of travelers from origin to destination (as was described in detail for air service in Chap. 7). The measure of employees per million passenger-miles must be used with caution, for all three carriers do not themselves maintain many of the facilities they use and the other employees required are not included.

Similar data for intra-urban carriers are presented in Table 8-4. Again, the exact measures used differ slightly from mode to mode.

Operations Plans of Freight Carriers

Freight carriers exhibit a wide variety of operations plans, and as a result, substantial variation in operating characteristics. At one extreme is the individual vehicle transporting a single shipment from origin to destination, either a dedicated or a sequential assignment. The most prominent example of this is truck movements of truck loads, which accounts for about 5 percent of all intercity truck movement by class I and II common carriers subject to In-

Table 8-3 Selected Operating Characteristics of Intercity Scheduled Passenger Carriers in the United States

	Air (scheduled)—1973	Bus—1973	Rail (Amtrak)—1974
Average vehicle capacity, seats	126.7[a]	42[b]	93 (per coach)[c,d]
Average load factor	0.521[a]	0.471[b]	0.46[e]
Average passenger trip length, mi	801[a]	116[f]	233[d]
Average train size,			
Cars/train	. . .	. . .	8.80[d]
Locomotives/train	. . .	. . .	1.95[d]
Average number of employees/million passenger-mi	1.92[a]	1.83[f]	1.90[d,g]
Average vehicle-mi/vehicle-yr	1,037,336[a]	56,635[f]	140,725 (passenger-car mi/passenger-car year)[d] 111,175 (locomotive-mi/locomotive-year)[d]
Average speed, mi/h	415[a]	~45[h]	50.8 (train-mi/train-h)[d]
Percentage of on-time arrivals	~80[h]	u[i]	31 percent (1973, long haul)[j] 72 percent (1973, short haul)[j]

[a] Air Transport Association of America (1974, pp. 17 and 20).
[b] Transportation Association of America (1974, p. 15).
[c] Averages for all United States railroad cars.
[d] Association of American Railroads (1975a, pp. 10 and 16).
[e] Private communication.
[f] National Association of Motor Bus Owners (1974, pp. 16–20).
[g] Does not include non-Amtrak railroad employees engaged in joint passenger and freight work.
[h] Estimated, based on studies of particular routes.
[i] u indicates unavailable.
[j] U.S. Department of Transportation (1974, p. 307).

terstate Commerce Commission regulation in the United States (ICC 1975a). For these movements, the ICC states that the average speed of trucks in traveling from shipper's dock to consignee's dock is 42 mi/h exclusive of time the driver is off duty for eating or sleeping. Nonregulated truckload movements and trucks carrying small shipments which have been consolidated into truckloads presumably travel at the same speed in intercity service.

At the other extreme are the various forms of simultaneous or shared assignment of vehicles and possibly containers. This type of service is exemplified by conventional railroad freight service, excluding unit trains. Also, it is very common to operate barge trains on inland waterways, one tug boat propelling up to 30 or 40 barges which may be carrying different shipments moving between diverse origins and destinations. Similarly, ocean vessels carry many different shipments, often in containers traveling between diverse origins and destinations but being consolidated for the line-haul movement on to a single vessel. The recently developed lighter-aboard-ship (LASH) system and sea barge carrier (SEABEE) systems are ocean analogs to railroad trains, in that barges are used to carry the freight. Barges are un-

Table 8-4 Selected Operating Characteristics of Urban Passenger Carriers in the United States

Auto and mass transit—1971[a]	Auto	Bus	Rapid rail	Commuter rail
Average vehicle capacity, seats	u[b]	47.6	50.7	104.6
With standees (estimated)	na[c]	75	150	na
Average occupancy, passenger-mi/vehicle-mi				
Weekday	1.56	11.52	24.51	41.33
Peak hour	1.28	16.62	34.17	57.45
Average load factor, passenger-mi/seat-mi	u	0.27	0.50	0.41
Average vehicle-mi/vehicle-year (1972)	u	32,435	39,790	u
Average passenger trip length				
In distance, mi	5.8	3.7	6.6	21.6
In time, min	12.6	18.6	20.3	39.8
Average peak hour speed, mi/h	u	12	20	33
Average peak hour headway, min	na	20.7	4.6	18.3

Taxicabs—1973	
Average trip length, mi	6.05[d]
Average paid trip length, mi	2.60[e] (1971 est.)
Average paid passengers/cab trip	1.60[d]
Average no. of employees/cab	2.84[d]
Average mileage per cab/year	43,460[d]

[a] U.S. Department of Transportation (1974, pp. 178, 262, 262, 261, 143–144, 184, 184, 266, 266).
[b] u indicates unavailable.
[c] na indicates not applicable.
[d] International Taxicab Association (1975, p. 6-2).
[e] U.S. Department of Transportation (1974, p. 58).

loaded and loaded at port facilities, then towed by tug boats into deep water where they are loaded aboard an appropriately designed ocean vessel for transocean movement. When the vessel nears the port of destination, barges, or lighters, as they are termed, are removed, with a tug boat propelling the lighter to the port for loading and unloading, while the ocean vessel continues on to other ports. Trailer truck operation is often quite similar, with the tractor being used to haul other trailers while any given trailer is being loaded and unloaded.

In general, the greater the number of separate containers consolidated to form a vehicle unit for movement, the greater the likelihood of delays before actual movement on the line-haul vehicle. Also, if a vehicle must be stopped frequently for the removal, addition, or exchange of containers, it is more likely to be subject to delay and to fluctuations in travel time. Also, the greater the number of times a container must be transferred from one vehicle to another in its journey from origin to destination, the greater the likelihood that the container will miss a planned connection at a particular terminal and thereby be subject to a substantial delay, perhaps having to wait an additional day or two for another vehicle moving in the direction it is to travel.

Some idea of the travel time and its variation for systems having an extreme degree of shared vehicle assignment is illustrated by the travel time of railroad cars. Martland (1972) analyzed rail car movements in the United States and developed a relationship for estimating the average number of days required to travel from origin to destination. This time depends upon the distance to be traveled, the number of times the car must be classified (marshaled) in yards (differentiating between flat yards switched by locomotives and hump yards, in which cars roll down a "hump" to the appropriate track), the number of times the car must be transferred from the lines of one railroad company to another, and the railroad's policy regarding the reliability of service. This latter policy is reflected in the "3 day percent," the maximum percentage of cars arriving at their destination within 3 consecutive days, the maximum percentage being obtained by examining the distribution of journey times (days) to the destination and selecting that consecutive 3-day period having the largest number of cars. This is illustrated in Fig. 8-11. The model (developed using regression) is (Martland, 1972, p. 87):

$$t = 0.036m + 0.15y_h + 0.12y_f + 0.17i + 0.07r \tag{8-21}$$

where t = mean time, days
m = distance, mi
y_h = no. of hump yardings
y_f = no. of flat yardings
i = no. of interchanges from one railroad to another
r = [100 − (3 day percent)]/10

Figure 8-11 **Illustration of the 3 day percent rail service reliability measure.**

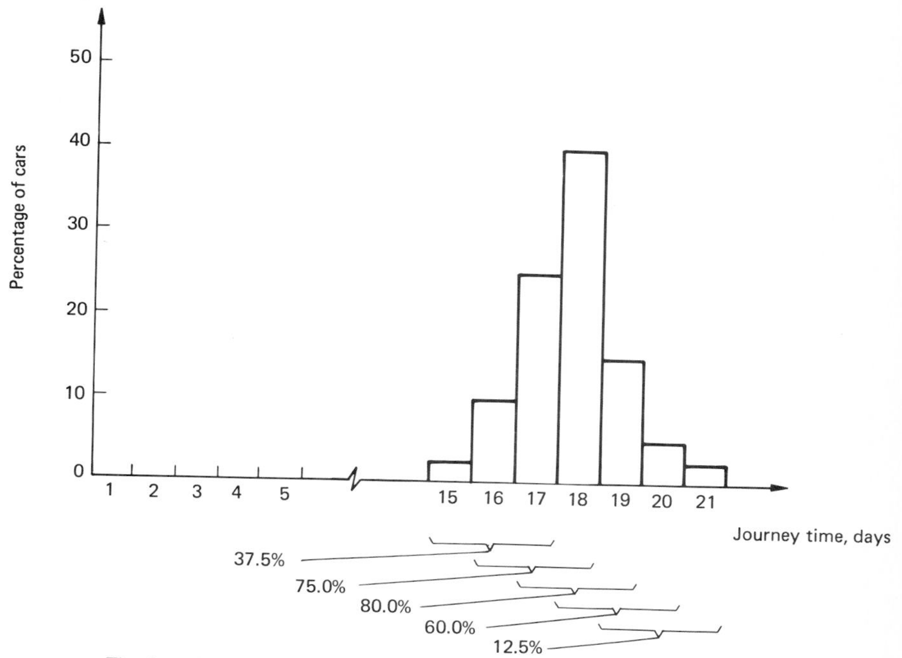

The three-day percent is 80.0 percent, since this is the largest percentage of cars' journey times in any consecutive three-day period.

Figure 8-12 **Typical journey times for rail, TOFC, and truck shipments.** [*Railroad car from Martland (1972), p. 58; TOFC from Interstate Commerce Commission (1975b), p. 143; truck from Interstate Commerce Commission (1972), p. 51.*]

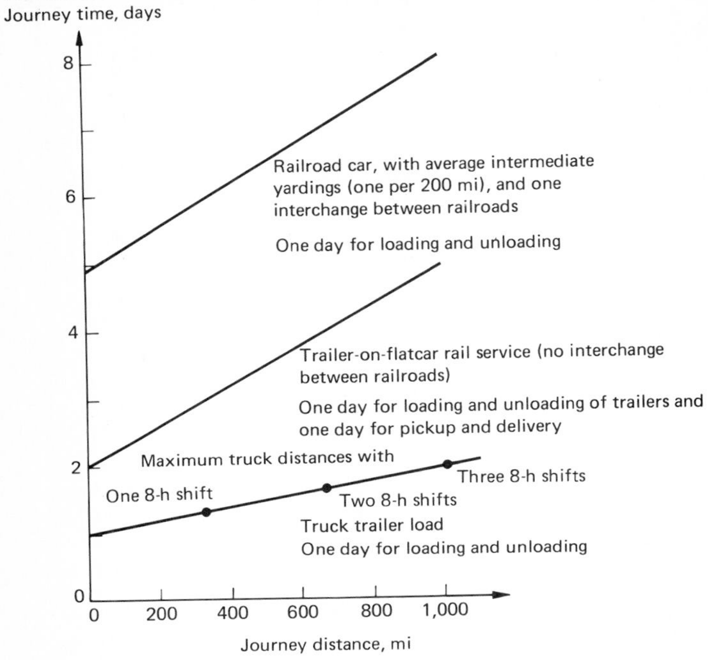

Figure 8-12 illustrates travel time as a function of distance for rail and truck movements under various conditions. Truckload movement is by far the fastest. Also shown is less than truckload movement, in which 1 day is allowed at each end of the journey for consolidation of traffic and pickup and delivery. The railroad movement is for a situation of typical yardings, one every 200 mi, and one railroad to railroad interchange. Unfortunately, data on actual travel time for air and water traffic are not available, but presumably air would be faster than trucks over long distances, above perhaps 300 mi, and water shipment would probably be somewhat slower than rail for all distances. Figure 8-13 is an interesting pie chart indicating the fraction of time rail cars spend in various activities. There is considerable time lost standing still, indicating that rail travel time could be improved substantially. The cycle of time included in this chart covers the period from loading of the car with one shipment to loading with the next, this time period often being referred to as the turnaround time.

Because of the rather low average speed of freight cars, due primarily to the long periods spent in yards, railroads are now considering and experimenting with many changes designed to speed service and make it more reliable. Not only may they make service better from the shipper's standpoint, but some costs may be reduced. For example, if the mean cycle or turnaround time of freight cars is reduced by 20 percent, the same amount of freight can be handled with 20 percent fewer cars! One approach is to oper-

Figure 8-13 **Breakdown of average turnaround time of containers in combined sequential and simultaneous assignments: example of United States rail freight cars in 1971.** [*From United States Railway Association (1975), p. 58.*]

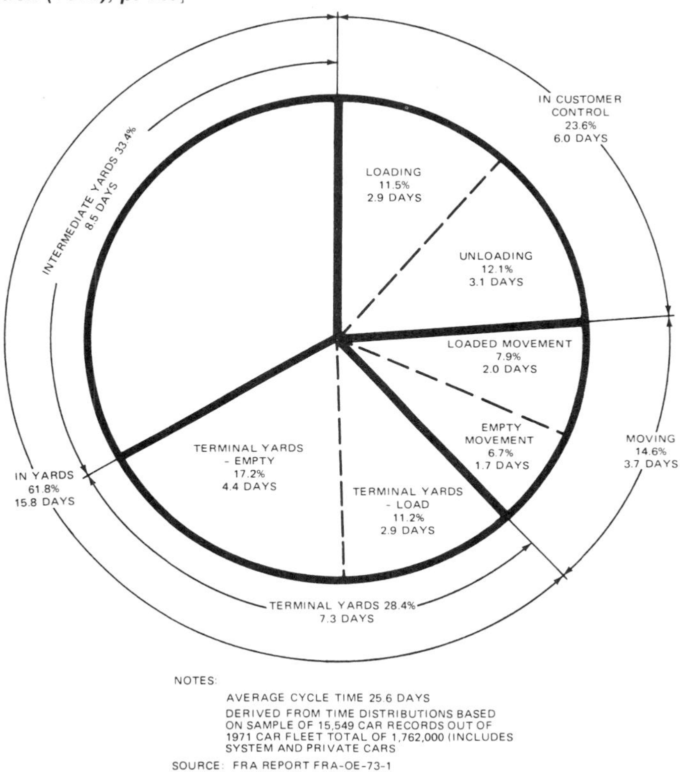

ate shorter, more frequent freight trains, thereby reducing average time waiting for outbound trains in yards. It also tends to increase on-time train performance, which increases the probability of achieving scheduled or planned train to train transfers at a yard. Also, greater use of scheduled trains is being tried as opposed to "operate the train when 125 cars (or some other number, or tonnage) are assembled for it." Much attention is being paid to placing the proper amount of locomotives (power and/or tractive effort) on each train so it can achieve the desired schedule. So-called "preblocking" of cars is being used more and more, i.e., early in their trip, cars are assembled into "blocks" which contain only cars for a particular destination yard, so that they never need to be sorted at a yard again until the destination yard is reached. Such blocks can be transferred from one train to another quickly, if necessary, also. This, coupled with shorter trains that meet connections more reliably and require fewer transfers of blocks, promises to speed freight considerably. Also, the costs of much yard operations and yard facilities are eliminated.

Much attention is also being devoted to the problem of management of the car fleet: assigning empty cars to shippers that need them, inducing shippers to load and unload them quickly, and routing cars via the fastest path through the network. Similar attention is being given to containers of other carriers: barges in the LASH and SEABEE systems, truck trailers, and intermodal containers of all carriers.

Various types of relationships can be used to estimate vehicle requirements, of which the simplest is one that simply relates vehicle and container requirements to some measure of total transport activity on the system, such as total ton-miles or passenger-miles. This would be of the form:

$$n = \frac{1}{\bar{P}_n}(M) \tag{8-22}$$

where n = number of vehicles required
$\bar{P}_n$ = average productivity per vehicle, total output/total no. of vehicle days
M = total output per unit time (e.g., ton-mi/day)

The units vary with the application to locomotives, containers, persons versus freight, etc., but the form remains the same.

However, as revealed by the discussion of rapid transit examples in the preceding section, the number of vehicles required can be influenced by such factors as the time between one loading and another, the round trip time in the transit example, and the average load carried by the vehicle. These and related factors can be taken into account in the freight situation by the following formula:

$$n = \frac{M}{\bar{V}_n(\bar{T})24(\bar{E})} = \frac{Z(\bar{D})}{24(\bar{V}_n)(\bar{T})(\bar{E})} \tag{8-23}$$

where $\bar{V}_n$ = average vehicle speed (e.g., total vehicle-mi/vehicle-h, mi/h)
$\bar{T}$ = average vehicle load (e.g., total cargo ton-mi/loaded vehicle-mi, tons/vehicle)
D = average haul (e.g., total cargo ton-mi/total cargo tons originated, mi)
Z = total cargo originated (e.g., tons)
$\bar{E}$ = ratio of loaded to total vehicle movement, dimensionless (e.g., vehicle-mi/vehicle-mi)

To be correct, the formula must use variables calculated as indicated in the parentheses. Explicit consideration is given to the fact that most freight transportation vehicles must be moved empty from the point of unloading to the point where a load is available, although in some cases a receiver of freight will also use the same vehicle for loading outbound freight. Thus the ratio of loaded miles to total miles must be included in the relationship. With minor modification, it can be used for passenger vehicles and traffic also, using averages of the type presented in Tables 8-3 and 8-4.

The use of this type of relationship can be refined further by considering individual types of shipments and different types of vehicles. Their characteristics often differ in terms of average length of the movement, as well as average load carried per container, average speed (reflecting priority to perishables, for example). The total number of containers required of any type can be estimated by the following formula, in which the index i is used to designate different movements which might have different characteristics:

$$c = \sum_{i=1}^{N} \frac{M_i}{24\bar{V}_{ni}(\bar{T}_i)(\bar{E}_i)} \tag{8-24}$$

These formulas and methods are primarily applicable to estimates for the number of containers required and their miles of operation. Specifically, this includes such items as railroad cars, truck trailers, barges, and containers (in the narrow definition). The operation of other portions of freight vehicles, primarily the power and control units, such as locomotives, truck tractors, and vessels, is quite different, primarily because they are comparatively expensive to purchase and operate. As a result, carriers are much more careful to ensure their maximum possible utilization, in much the same manner as passenger vehicles, and the methods discussed above in the passenger section (and presented in the rapid transit example) can be used directly. Alternatively, the number of such power units required can be estimated on the basis of average relationships to some measure of total output.

Typical Operating Characteristics of Freight Carriers

Typical values for the factors that enter into Eqs. (8-22) through (8-24) are presented in Table 8-5. These represent averages observed over many different movements, and care should be taken to use them only in situations where the characteristics are likely to be similar to existing averages. As can be seen from this table, data are most readily available for rail service, somewhat less available for truck service, and much less available for other freight carriers, for the reasons discussed earlier.

As an example of the application of these formulas, consider a proposed railroad freight service. The line will be 350 mi in length, carry about 10^9 ton-mi of cargo annually, at an average haul of 300 mi. Trains are expected to average about 20 mi/h. How many locomotives, cars, and employees will be needed, assuming the operating characteristics are similar to current United States patterns?

Using data from Table 8-5, the number of locomotives can be calculated if the number of car-miles is known. Assuming the same average load and ratio of loaded to total car-miles as in the United States, the car-miles per year would be

$$\frac{10^9 \text{ ton-mi/year}}{50.1 \text{ ton-mi/loaded car-mi}} \, \frac{\text{total-car mi}}{0.572 \text{ loaded car-mi}} = 34.90 \times 10^6 \text{ total car-mi/year}$$

Hence the locomotive requirements would be, by Eq. (8-24):

$$n_{\text{road}} = \frac{34.90 \times 10^6 \text{ car-mi/year}}{1.475 \times 10^6 \text{ car-mi/year/road locomotive}}$$
$$= 23.66 \text{ road locomotives} \approx 24 \text{ road locomotives}$$

$$n_{\text{yard}} = \frac{34.90 \times 10^6 \text{ car-mi/year}}{4.769 \times 10^6 \text{ car-mi/year/yard locomotive}}$$
$$= 7.32 \text{ yard locomotives} \approx 8 \text{ yard locomotives}$$

The number of cars could be estimated in various ways. The simplest would be to use the same formula and average cargo ton-miles per year per car.

Table 8-5 Selected Operating Characteristics of Intercity Freight Carriers in the United States

Railroads—1974[a]

Freight cars	
Average weight capacity, tons	72.8
Average load carried when loaded (total cargo ton-mi/total loaded car-mi)	50.1
Average length of haul, mi[b]	531
Average loaded to total car-mi	0.572
Average route circuity (actual to minimum distance path)[b]	1.16
Average car-mi/day	57.4
Average turnaround time, days	23.8
Average cargo ton-mi/car-year	657,740
Trains and locomotives	
Average cars/train	65.4
Average locomotives/train	2.93
Average speed of trains, mi/h	19.9
Average gross trailing ton-mi/train-h	77,919
Average car-mi/year/road locomotive	1,475,000
Average car-mi/year/yard locomotive	4,769,000
Network	
Typical minimum path circuity (relative to great circle distance)[c]	1.25
Average employees/million cargo ton mi (estimate)	0.58

Trucks—1972

	Single Units	Combinations	Both
Average load carried when loaded, tons[d]	2.95	14.16	11.38
Average loaded to total truck-mi[d]	0.549	0.693	0.629
Average length of haul (class I intercity common carriers), mi[e]			278
Average truck capacity (class I intercity carriers), tons (1969)[f]			12.2
Average speed—intercity service, mi/h[g]			42.2
Average cargo ton-mi/truck-year[h]	3116	395,394	22,753
Average vehicle-mi/truck-year[h]	4088	41,667	5969

This yields

$$c = \frac{10^9 \text{ cargo ton-mi/year}}{657{,}740 \text{ cargo ton-mi/year/car}} = 1520 \text{ cars}$$

Operations Plans for Roads

So far we have neglected operations plans for roads and, in fact, for other types of facilities where the facility management is quite distinct from the decision-making unit associated with the operation of vehicles through the facility. This group includes roads, airports, and ports. In fact, much of the field of road traffic engineering is really concerned with operations plans in the sense of rules, signs, signals, channelization, etc., all for the efficient operation of the road network.

To a large extent, we have already covered much of these facility operations plans: traffic flow in Chap. 5 and airports, ports, and terminals in general

Table 8-5 Selected Operating Characteristics of Intercity Freight Carriers in the United States *(continued)*

Inland and coastal water—1972		
Number of units (1972)[i]		
Barges		20,947
Towboats		4278
Ships (approximation)		800
Average loaded barge-mi/total barge-mi (1970)		
Tank		0.50
Dry cargo		0.57
Average length of haul (1972), mi[i]		612.8
Average power per towboat and tug, hp[i]		1006
Average capacity per barge, tons[i]		
Tank		2189
Dry cargo		1298
All cargo air carriers—1973[k]	**Domestic**	**International**
Average aircraft capacity (available ton-mi/vehicle-mi), tons	46.8	56.8
Average load factor	0.612	0.694
Average ton-mi/aircraft-mi	28.63	39.4
Average flight length, mi	1135	1698
Average flight speed, mi/h	473	484

[a] Association of American Railroads (1975a, pp. 11–14, and 1975b, pp. 26, 41–53), except as noted.
[b] Interstate Commerce Commission (1975b, p. 131).
[c] Hannon (1974, p. 364).
[d] U.S. Dept. of Transportation, Federal Highway Administration (1973, p. 78).
[e] Transportation Association of America (1974, p. 14).
[f] U.S. Dept. of Transportation (1974).
[g] Interstate Commerce Commission (1972, p. 51).
[h] Calculated as follows: Average cargo ton-mi/truck = total cargo ton-mi by truck type/total trucks by type, data in U.S. Dept. of Transportation, Federal Highway Administration (1973, p. 78), and Motor Vehicle Manufacturers Assoc. (1974, p. 41); average vehicle-mi/truck = total vehicle-mi by truck type/total trucks by type, data sources same as for average cargo ton-mi/truck.
[i] U.S. Department of Transportation (1974, pp. 325, 321, and 327).
[j] U.S. Department of Transportation (1974, p. 43).
[k] Air Transport Association of America (1974, pp. 23 and 25).

in Chap. 7. However, it is useful to recall a few salient points from those discussions as they relate to operations plans.

One of the most important aspects is channelization. To accommodate peak flows in one direction for short periods (e.g., weekday work travel peaks, holiday traffic peaks) traffic engineers have developed reversible lanes. Through the use of informational aids ranging from simple signs to overhead lights, they allocate a different number of lanes to each direction at different times. Also, parking (and even stopping) may be prohibited during peak travel periods. By allocating more lanes to a flow in this manner, capacity may be dramatically increased exactly when needed. The same principle is used on commuter railroads (e.g., three of the four tracks into Grand Central Terminal in New York are used in one direction during peak hours) and for runway use at airports where more than two can be in use simultaneously. Signal phases on roads are also adjusted accordingly. The conversion of narrow streets to one way can also speed traffic flow, although travelers may experience slight increases in driving distances as a result.

The development of regional road operations plans, or traffic management schemes, as they are often called, for roads is a very complex task. It must be done considering the actual pattern of vehicle trips, so that adequate capacity and level of service are provided where needed. Major system alternatives, such as auto-free zones and road pricing, are primary options, along with such alternatives as new freeway and public transportation facilities. Discussion of this more elaborate type of road management scheme must follow our discussion of travel demand (Chap. 10) and pricing (Chap. 11), and will appear in Chap. 17, where transportation system management will be covered.

SUMMARY

1 A transportation system consists of many components in addition to the hardware components described in Chaps. 4 through 7; it includes (1) the fixed facilities, (2) vehicles and containers, and (3) an operations plan.

2 The movement of persons and freight and containers and vehicles (if used) over this network is determined by an operations plan. The main features of this plan are rules for (1) the assignment of containers and vehicles to traffic, (2) the routing of these through the network, including any aggregation of those into trains.

3 Numerous options exist as to container and vehicle characteristics, such as size and speed capabilities, and as to the operations plan. These, closely related, together largely determine the number of vehicles and containers required, vehicle-miles and container-miles of movement, utilization of capacity provided (e.g., as measured by the load factor), labor requirements, etc.

4 Because of the complexity of the relationships, and varying patterns of traffic, generalizations are often misleading. However, typically vehicles required (and operating employees required) are increased by increased peaking of traffic, especially for passenger carriers where traffic peaks cannot be smoothed easily (e.g., by staggered work hours). Larger vehicles typically have lower costs per unit of capacity and increase the flow capacity of guideways as measured by traffic unit (passenger or freight) flow capacity. However, flows of larger vehicles are often more difficult to match with traffic variations, and often the resulting poorer load factors offset some of the cost gains and result in poorer (e.g., less frequent) service. In freight movement, aggregation and disaggregation of containers onto vehicles usually results in less reliable service than that provided by single units.

PROBLEMS

8-1
Discuss the advantages and disadvantages of large versus small vehicles in a transit system. Consider all the factors you can think of, based on information not only in this chapter but also in earlier ones.

8-2

Compare the advantages and disadvantages of so-called free-wheeled vehicles (meaning a change of lanes, turn, etc., can be made at the will of the driver) such as buses, with those of fixed-guideway vehicles such as rail transit trains, in urban transit service. Consider not only operations plan aspects, but also such factors as congestion and terminal capacity.

8-3

Calculate the average load factor on the two Cleveland transit lines emanating from the downtown station (east side and west side). Assume that a total capacity of 7500 passengers/h is provided on the west, and 3000 passengers/h on the east. How do you think this compares with the average daily load factor? Why? Passenger volumes are given in Fig. 8-9.

8-4

In the discussion of operations on a single-rail rapid transit line used as an example in the text, the trains and cars required, and the total train-mi/day and car-mi/day were calculated for an all-stop operating pattern. The route was 15 mi long, operated 16 h/day; trains took 1.5 h for a round trip; passenger flow capacity required was 8000 passengers/h in the peak period, peak direction and 1143 passengers/h at other times and directions, resulting in 16 trains/h of 5-car trains (500-passenger capacity, including standees) in the peak period and 7.5 trains/h of 3-car (156-passenger capacity, all seated) otherwise, yielding a fleet requirement of 120 cars and 4620 train-mi/day and 17,700 car-mi/day. Calculate train and car requirements, and train-mi/day and car-mi/day with changed peak period operations. This is a zone express and local service, with half the trains being CBD zone–middle zone locals taking 0.80 h for a round trip, and the other half being CBD–outer zone expresses, taking 1.25 h for a cycle. The locals terminate 8.5 mi from the CBD end of the line. Traffic is divided equally between the two services.

8-5

A transit authority wishes to operate trains on a rapid transit line at headways not to exceed 10 min throughout the 16-h period of operation each weekday. How many cars should trains consist of in each of the periods listed below to provide adequate passenger volume capacity? What will the corresponding time headways be? Cars consist of permanently coupled pairs, each pair seats 100 and has room for about 50 standees. By policy, standees are desired only in peak periods. Station platforms limit trains to 8 cars.

Morning inbound peak period: 19,000 passengers/h

Midday base period: 1600 passengers/h

Afternoon outbound peak period: 21,000 passengers/h

Evening period: 750 passengers/h

8-6

Compare the average travel time (including one-half the headway where a wait is required) between all pairs of nodes in the two networks on page 336.

In plan A, buses operate on all major roads, at a headway of 10 min. In plan B, the network is consolidated, concentrating traffic on a subway, with few bus lines. In plan B, the subway operates on a 5-min headway and the buses on a 10-min headway.

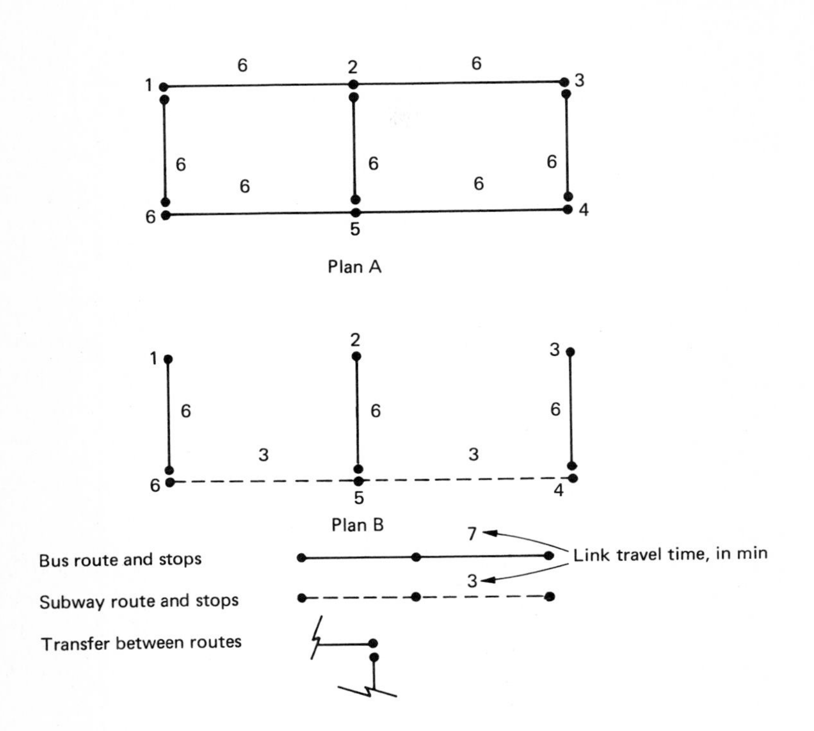

8-7
A truck line carries package freight between the major terminals shown on the map below, from which it also operates a local pickup and delivery service. One truck per day is operated between each pair of terminals, in each direction. All departures occur at 11:00 P.M. What would be the time taken for a parcel from arrival at terminal B, at 5:00 P.M., to arrival at terminal H, via the fastest route? What links make up that route? How could you improve the service? Running times between major terminals are shown below.

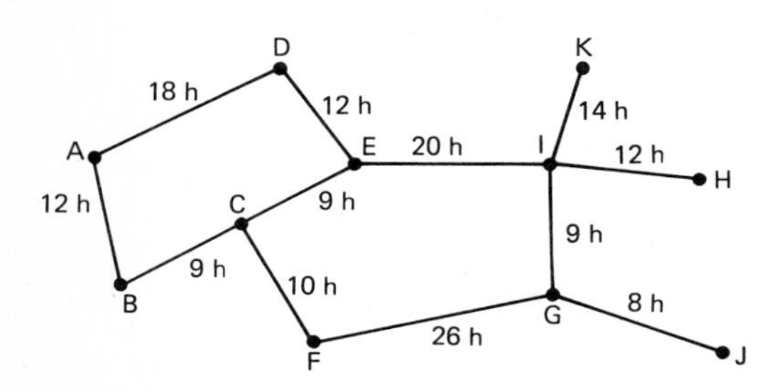

8-8
What would be the effect of reorganizing freight train schedules so that a car making a 1500-mi trip passes through only one hump yard and two flat yards instead of two and four, respectively. The 3-day percent is 70 percent. What types of changes in operations might achieve this?

8-9
Would you expect that an increase in the cruise speed of freight trains from 40 to 60 mi/h would affect the travel time of freight from origin to destination significantly? Why?

8-10
In what ways are the operations plans of railroad freight service, truck LTL (less than truck load) freight service, LASH or Seabee service, and dial-a-bus service similar? Different?

8-11
Obtain a time table for a public transportation route in your city. Calculate the average speeds of the vehicles from one end to the other during weekday peak periods and midday. Do they differ? Calculate the mean travel time for a passenger traveling from one end of the line to a major stop in the CBD during both periods. What is the travelers' average speed? Would a doubling of the frequency of departures or an increase of 20 percent in average vehicle speed reduce that traveler's time by a greater amount?

8-12
The April 1975 time table of Metroliner trains (high-speed, reserved seat) from New York to Washington is reproduced below, including the stop at Philadelphia (other stops were also made). Assuming random passenger arrivals, with arrivals starting 1 h before the first morning departure, what is the average journey time, terminal to terminal, including terminal processing and waiting time, from New York to Philadelphia (90 mi), Philadelphia to Washington (134 mi), and New York to Washington (224 mi). Use terminal processing times of 10 min at origin and 5 min at destination. Compare the average train speed

Leave New York	Leave Philadelphia†	Arrive Washington
6:30 A.M.	7:48 A.M.	9:34 A.M.
7:30	8:45	10:34
8:30	9:48	11:34
9:30	10:41	12:29 P.M.
10:30	11:43	1:29
11:30	12:45 P.M.	2:29
12:30 P.M.	1:45	3:34
1:30	2:42	4:29
2:30	3:44	5:29
3:30	4:41	6:29
4:30	5:42	7:33
5:30	6:42	8:33
6:30	7:46	9:39
7:30	8:39	10:29
8:30	9:40	11:30

† Trains arrive 1 min earlier.

and average speed of travel (including waiting) for these same city pairs. Compare these terminal times with those observed for Metroliners, as presented in Table 7-12 and Fig. 7-18. What would be the effect of reducing train running times by 5 min from New York to Philadelphia and 25 min from Philadelphia to Washington? Compare this with the effect of doubling the frequency of trains, with current speeds. If only one of the two changes could be instituted, which would you recommend?

REFERENCES

Air Transport Association of America (1974), "Air Transport 1974," Washington, DC.

Association of American Railroads (1975a), "Statistics of Railroads of Class I in the United States, 1964 to 1974," Washington, DC.

Association of American Railroads (1975b), "Yearbook of Railroad Facts," 1975 ed., Washington, DC.

Association of American Railroads Mechanical Division (1966), "1966 Car and Locomotive Builders' Cyclopedia of American Practice," Simmons-Boardman, New York.

Baerwald, John E., Matthew J. Huber, Louis E. Keefer, eds. (1976), "Transportation and Traffic Engineering Handbook," Institute of Traffic Engineers, Prentice-Hall, Englewood Cliffs, NJ.

Belovaroc, Kenneth, and James T. Kneafsey (1972), "Determinants of Line Haul Reliability," vol. 3 of Studies in Railroad Operations and Economics, Research Report No. R72-38, Civil Engineering Dept., MIT, Cambridge, MA.

Bisbee, E. Farnsworth, J. W. Devanney, D. E. Ward, R. J. vom Saal, and S. Kuroda, Dispatching Policies for Controlled Systems, in Wilhelm Leutzbach and Paul Baron, eds. (1969), "Beitrage zur Theorie des Verkehrflusses," Strassenbau und Strassenverkehrstechnik Heft 86, Bundesminister für Verkehr, Bonn, West Germany, pp. 255–263 (note: most papers in English).

Grossman, William L. (1959), "Fundamentals of Transportation," Simmons-Boardman, New York.

Hannon, Bruce D. (1974), A Railway Trust Fund, *Transportation Research,* **8**(4/5): 363–372.

International Taxicab Association (1975), "An Analysis of Taxicab Operating Characteristics," Report No. PB-251 147, U.S. National Technical Information Service, Springfield, VA.

Interstate Commerce Commission (1972), "Costs of Transporting Freight by Class I and II Motor Common Carriers of General Commodities, 1971," Statement No. 2C1-71, Washington, DC.

Interstate Commerce Commission (1975a), "Costs of Transporting Freight by Class I and II Motor Carriers of General Commodities 1973," Statement No. 2C1-73, Washington, DC.

Interstate Commerce Commission (1975b), "Rail Carload Cost Scales, 1973," Statement No. 1C1-73, Washington, DC.

Kirby, Ronald J., Kiran U. Bhatt, Michael A. Kemp, Robert G. McGillivary, and Martin Wohl (1975), "Para-Transit: Neglected Options for Urban Mobility," The Urban Institute, Washington, DC.

Leibbrand, Kurt (1970), trans. by Nigel Seymer, "Transportation and Town Planning," MIT Press, Cambridge, MA.

Martland, Carl D. (1972), "Rail Trip Time Reliability: Evaluation of Performance Measures and Analysis of Trip Time Data," Report No. 72-37, Dept. of Civil Engineering, M.I.T., Cambridge, MA.

Moore, E. F. (1957), The Shortest Path through a Maze, "Proceedings of the International Symposium on the Theory of Switching," Harvard University, April 2–5, 1957, pp. 285–292.

Morlok, Edward K. (1965), "Rail Transit Costs," Northwestern University Transportation Center, Evanston, IL.

Morlok, Edward K. (1969), "Transport Technology and Network Structure," The Transportation Center at Northwestern University, Evanston, IL.

Morlok, Edward K., and Hugo L. Vandersypen (1973), Schedule Planning and Timetable Construction for Commuter Railroad Operations, *Transportation Engineering Journal of ASCE,* **99**(TE 3): 627–636.

Motor Vehicle Manufacturers Association (1974), "1974 Motor Truck Facts," Detroit, MI.

National Association of Motor Bus Owners (1974), "The Intercity and Suburban Bus Industry 1973–1974," Washington, DC.

Pignataro, Louis J. (1973), "Traffic Engineering: Theory and Practice," Prentice-Hall, Englewood Cliffs, NJ.

Potts, Renfrey B., and Robert M. Oliver (1972), "Flows in Transportation Networks," Academic, New York.

Rainville, Walter S. (1961), Preliminary Progress Report of Transit Subcommittee, Committee on Highway Capacity, *Highway Research Board Proceedings,* **40**: 523–540.

Rosenbloom, Sandra, ed. (1976), "Paratransit," Special Report 164, Transportation Research Board, Washington, DC.

Schumer, Leslie A. (1974), "The Elements of Transport," Butterworths, Sydney.

Simpson, Robert W. (1969), "Scheduling and Routing Models for Airline Systems," Report No. PB 196 528, U.S. National Technical Information Service, Springfield, VA.

Taafe, Edward J., and Howard L. Gauthier, Jr. (1973), "Geography of Transportation," Prentice-Hall, Englewood Cliffs, NJ.

Transportation Association of America (1974), "Transportation Facts and Trends," Washington, DC.

Ullman, Edward L. (1949), The Railroad Pattern of the U.S., *Geographic Review,* **39**: 242–256.

U.S. Bureau of Public Roads, Department of Commerce (1965), "The National System of Interstate and Defense Highways: Status of Improvements as of Dec. 31, 1965," Washington, DC.

U.S. Department of Transportation, Federal Highway Administration (1974), "Highway Statistics," No. 5001-00088, Washington, DC.

United States Railway Association (1975), "Preliminary System Plan," Washington, DC.

Vuchic, Vukan R. (1973), Skip-Stop Operation as a Method for Transit Speed Increase, *Traffic Quarterly,* **XXVII** (2): 307–327.

The Transportation System and Its Environment 3

Transport Costs

9.

The importance of costs for the transportation engineer can hardly be overemphasized. The engineer or planner must be concerned not only with the estimation of costs of existing and proposed facilities and operations, but also with the correct interpretation and use of those costs in any analyses or evaluations. In this chapter we shall focus upon methods for the estimation of costs, building upon the characteristics of transport technology and the resources required to provide various transportation services which have been the focus of the preceding chapters. Later we shall be concerned with the evaluation of various transportation projects, including considerations of their costs as well as other characteristics relevant to public and private choices.

COST CONCEPTS

Although by virtue of living in a modern civilization everyone is familiar with many concepts of costs and economics in general, the casual use of many concepts leads to a great deal of confusion and ambiguity whenever a detailed analysis of costs is required. Therefore in this section we shall introduce many cost concepts which may seem familiar, but will attempt to do this with precision so that there will be no ambiguity in the future use of these terms in our quantitative engineering and planning analyses.

Costs to Whom?

The ease with which the term costs is often used is very misleading, in that it implies that there is a single cost associated with providing a good or service. It is true that in principle it may be possible to identify a single total cost to society resulting from producing a product or service such as transportation, but the term usually refers to the cost borne by a particular person, group, or organization, and thus may be very different from the total cost to society.

This multifaceted characteristic of cost arises because in general different costs are borne by different persons or groups, and usually such persons or groups are interested in only those costs which accrue to them. For example, in traveling to work on public transportation, the traveler typically perceives the fare charged to ride and also the travel time consumed. This time is a cost in the sense that it is a scarce resource of the traveler which is given up in order to travel. Also, the traveler may be aware of other costs such as a psychological cost associated with discomfort and perhaps loss of some physical energy in traveling, perhaps as a result of having to stand for a portion of the trip. But in general these costs—the fare plus travel time and related items—are only a small fraction of the total cost of transporting this traveler. The cost to the transit system operator will not include the traveler's time or discomfort, but it will include operating costs which usually are substantially greater than the revenue derived from fares paid by the travelers. Also, the operator is usually responsible for some or all of the costs of the vehicles; and in the case of transit operating on its own right-of-way (such as subways and rail rapid transit lines), the cost of providing those facilities may be borne by the authority. Very often, the costs of the facilities and vehicles are partly borne by government, often a combination of the federal, state, and local governments, and any operating deficits might be made up by subsidies from local and perhaps state governments. Thus these primarily monetary costs of providing service will in general be perceived as quite different among different groups, because they are responsible for different portions of those costs.

Furthermore, there may be other costs, such as costs perceived by persons residing near transit lines, who may experience noise and air pollution and loss of aesthetic qualities as a result of proximity to the facility. While these costs may be only indirectly measured in any market transaction, such as a possible reduction in the value of property as a result of these environmental impacts, the residents themselves may perceive a much more substantial cost. And on the other side of the coin, properties near the transit line may increase in value as a result of the greater accessibility given those properties by existence of the line, and this may in fact outweigh any negative effects due to adverse environmental impacts. In addition to these rather direct effects and costs, other effects occurring over a long time period may result in changes in the costs of providing all of the goods and services consumed by persons in the region. More specifically, for example, the existence of the system may result in persons tending to live close to the stations rather than dispersed throughout the region, and this may have a profound effect on the cost of housing as well as the cost of providing the various public and private services necessary for the maintenance of homes within the region. Any changes in the cost of maintaining a given quality of life in the region, as well as changes in the quality of life itself, which occur as a result of the existence of the transit system, thus may be considered costs (or cost reductions) associated with that system. Even though such long-term effects may be considered very indirect costs at best and are certainly extremely difficult to identify and quantify, it is important to realize that the notion of cost is in fact much more subject to variations in interpretation than a casual user of the term might believe.

In this chapter we shall be primarily concerned with much more direct costs, and primarily with those which are reflected in identifiable market transactions where money changes hands and places a value on resources used. In later chapters we shall consider the effects of transportation systems on the pattern of human activities and the quality of life, partly in terms of costs and partly in other ways.

The example discussed above provides a very widely used classification of groups in terms of their perceptions of transportation costs. These groups can be identified as follows: (1) users of the system, (2) owners and/or operators of the system, (3) affected nonusers of the system (such as those living in residences near facilities), (4) government at various levels, and (5) the region as a whole. This list is in no way meant to be exhaustive or to imply that the groups are mutually exclusive. On the contrary, a person may experience a certain set of costs associated with transport as a result of her or his use of the system and a quite different set of costs as a result of living near a link in the system, thereby experiencing nonuser costs as identified in the example above. Thus, while the classification is useful and in fact will reappear in later chapters when we discuss the benefits of transportation improvements, there is a very real possibility of double counting, which must be carefully guarded against. With these caveats, the groups are presented in Table 9-1, along with typical examples of the types of costs experienced by each group. Of course, other breakdowns could be used, and others may be more appropriate than that in Table 9-1 for particular kinds of analyses. Also, these five groups may be subdivided into finer divisions, such as into different users of the system or different levels of government, and such subdivisions may be important in particular situations.

The final important point regarding costs in general, is that many of the costs which we have been discussing are not associated with prices of items exchanged in the marketplace. Although most of the costs are quite real, such as the time travelers spend in traveling, very often the traveler is not directly paid (or charged for) the time consumed in travel. (Some obvious exceptions are the operators in the employ of transport agencies.) Thus we may have difficulty associating a money value with many of the costs which we would like to consider. Also, as indicated in the example above, money

Table 9-1 Groups Which Experience Different Transport Costs

Users
Direct prices (fares, tolls, etc.)
Time consumed
Discomfort of travelers (fatigue, etc.)
Loss and damage of freight
System owner-operator
Direct costs of construction, operation, maintenance
Nonuser
Changes in land value, productivity, etc.
Environmental degradation (noise, pollution, aesthetics, etc.)
Government
Subsidies and capital grants
Loss of tax revenues (e.g., when road or other publicly owned facility replaces tax-paying land use)
Region
Usually indirect, through reorganization of land uses, altered rate of growth, etc.

prices paid for transport services may or may not reflect the costs of providing that service. There may even be no relationship between the price paid or the total revenue received from the transportation service and the direct money cost of providing that service. Also, other costs might not be reflected in money transactions. Thus we must be careful to distinguish between the term cost and the term price.

Fixed and Variable Costs

Fundamental to any discussion of costs is the cost-output curve. This curve relates the total cost of providing a particular good or service—in our case a transportation service—to one or more measures of the amount and characteristics of the good or service provided—in this case, again, presumably a measure of the quantity and quality of transportation service. The cost is usually the total monetary cost experienced by the firm or agency providing the service, although the cost may be defined differently, perhaps as only a portion of that monetary cost or perhaps including other costs (such as a value of traveler's time, for example). Thus it is important to specify clearly what costs are included. The measure of output varies considerably among applications, but at the minimum includes a measure of the quantity of transportation service provided (e.g., the ton-miles of freight carried).

More complicated specifications of output might involve many variables, such as the tons of freight carried, perhaps broken down by commodity and distance carried, and the number of passengers carried, perhaps broken down by their origin and destination. The choices of the costs to be included and the measures of output depend very much upon the particular application.

An example of such a cost-output curve, involving a single measure of cost and a single measure of output, quantity, is presented in Fig. 9-1. In cost-output functions such as this, where there is only a quantity measure of output, it is presumed that all other characteristics of the output remain fixed; that is, the nature of the product, such as the quality of transport service, remains unchanged with quantity of output. This is often described in economic analyses as the product being homogeneous. The upper portion of Fig. 9-1 presents the total cost as a function of quantity of output. Total cost can be divided into two components: fixed cost and variable cost. As the name implies, the fixed cost remains the same regardless of the total amount of output, they are written with x, the output variable, appearing in parentheses after the designation of the two costs, as follows: $\mathrm{TC}(x)$ and $\mathrm{VC}(x)$. The fixed cost is designated FC. These and many other common symbols in economics consist of more than one letter. To differentiate these from the product of two or more symbols, we shall use roman capital letters for such multiletter symbols, in contrast to the capital and lowercase italic letters used for symbols consisting of single letters.

The lower portion of this figure presents two other important ways of presenting costs as a function of output. One of these is the average total cost, which is simply the total cost of any given amount of output. In equation form, the total cost, by definition, is

$$\mathrm{TC}(x) = \mathrm{FC} + \mathrm{VC}(x) \tag{9-1}$$

Figure 9-1 **General cost-output relationships. (*a*) Total cost and its components. (*b*) Average and marginal costs.**

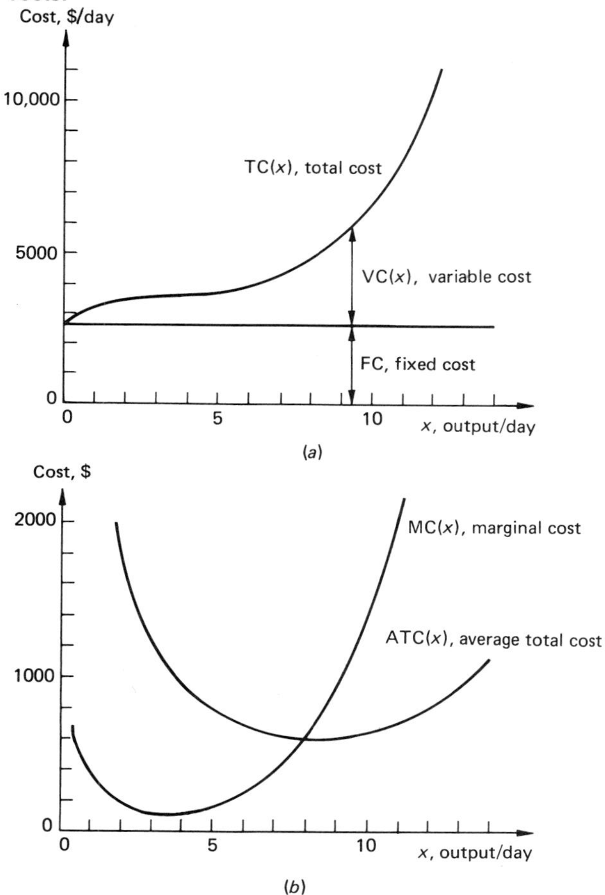

The average cost (average total cost) is

$$AC(x) = \frac{TC(x)}{x} = \frac{FC}{x} + \frac{VC(x)}{x} \tag{9-2}$$

where $AC(x)$ = average cost

The other curve presented in Fig. 9-1*b* is the marginal cost. The marginal cost is defined as the additional cost associated with the production of an additional unit of output. In the equation form the marginal cost of the xth unit of output is

$$MC(x) = TC(x) - TC(x - 1) \tag{9-3}$$

The equation above is for marginal cost in situations where the output must be produced in integer quantities. In most cases the quantity of output is treated as a continuous variable. For example, in Fig. 9-1 it might be trains per day, over a given line, and the output could be measured as an average over a fairly long period of time; as a result, a noninteger output is possible. In cases where the output can vary continuously, the differential form of marginal cost is used, in which the marginal cost is the rate of change of total cost with respect to a change in output. In this form, the equation is

$$MC(x) = \frac{dTC(x)}{dx} = \frac{dVC(x)}{dx} \tag{9-4}$$

One additional concept often used is that of average variable cost, which is defined as follows:

$$AVC(x) = \frac{VC(x)}{x} \tag{9-5}$$

In this example, the average costs decrease over a range of volume from 0 to 8 trains/day, and increase thereafter. The reason for this can be understood from the geometry of Fig. 9-1*a*. The average is proportional to the slope of a line connecting the origin of the total cost curve with a point on that curve corresponding to the total output. The slope of such a line begins at (plus) infinity at zero output and then decreases with increasing output up to a level of 8 trains/day, at which point the line is tangent to the total cost curve. Beyond that output level, the slope is increasing again; as a result, it is shown in the total cost curve.

The marginal cost curve, as the equation above reveals, is the slope of the total cost curve. As can be seen from the total cost curve, its slope is decreasing with increasing output over a range up to approximately 3 trains/day, beyond which it increases. Furthermore, the slope of the total cost curve is less than the slope of the average cost (line from the origin to the total cost curve at the output level under consideration) up to the output level of 8 trains/day, at which point the two coincide since the average cost line is tangential to the total cost curve at that point. At this point then the marginal cost and the average total cost must be equal, which is shown in Fig. 9-1*b*. In fact, since the marginal cost is less than the average total cost for all output levels up to this point, the average total cost decreases over this range, for each additional unit of output adds an additional cost (a marginal cost) that is less than the average up to that point. As a result, including the unit and its associated cost reduces that average. Beyond the output level of 8 trains/day, where these two cost curves cross; the marginal cost is greater than the average total cost. As a result, increasing output tends to pull up the average total cost, again as shown in the figure. The marginal cost curve always intersects the average total cost curve at the lowest point on that total cost curve.

The total cost function shown in Fig. 9-1 has the shape it does for illustrative purposes only. In many but not all situations cost is found to increase slightly with increasing output and then increase very substantially after output exceeds a certain level. The form depends upon the particular process being considered and the technology being used, as well as the way in which the costs of the various resources used vary with different amounts of those resources used.

The form of the total cost function also varies considerably depending on the period allowed for adjustment of production to a changed output level. In the short run, most costs tend to be fixed, for there are at least short-run commitments to continue operations as they are, regardless of changes in output. But as the time period allowed for adjustments of the manner of production increases, the range of possible changes increases and hence the proportion of total costs that can be varied increases. At the extreme, in the very long run, virtually all components of a transport system must be replaced and hence even major so-called fixed plant items such as railway tracks and terminals become elements of variable costs, since the choice may be made to replace them or not, or to replace them but in a modified form. A good example is provided in the context of railroad passenger service. In the short run, unexpected changes in volume from one day to the next often cannot be met with corresponding changes in the number of trains run or the capacity of those trains, and hence little if any variation in cost is observed. However, if a trend of, say, reduced traffic is observed and is expected to continue, the management might reduce the size of trains operated and therefore experience some cost reduction due to reductions in the number of cars owned and maintained, the number of ticket collectors on trains, etc. If the trend continues, there may be adjustments in the schedules, e.g., eliminating certain lightly used trains, with further reductions in costs. If these reductions in service continue for a long period, it may be possible to reduce the number of tracks at terminals, in storage yards, etc. In the very long run, if the reduction in traffic is very substantial, the type of train operated may be altered, long, locomotive drawn trains might be replaced with self-propelled rail cars, which often are much cheaper to operate if the train has only one or two cars. Ultimately, if traffic is expected to be extremely light, the service might be discontinued entirely, with no replacement of terminals and other passenger facilities or of cars and locomotives. Conversion of some facilities with useful life remaining to freight uses, etc., might occur. Thus, depending upon the time period over which the change can be planned for, virtually all expenses can become variable.

One might wonder at this point how the fixed and variable costs are determined or more generally how the cost versus output relationship is developed. The best approach is to use a combination of theory or deduction from basic principles underlying the process, usually engineering principles, and observations of the variations in costs with variation in output in actual operations. One must take great care to be sure that any operations being observed are operated efficiently and that the cost-accounting methods lend themselves to development of these cost relationships. However, cost-estimating methods will be discussed later.

Joint Costs

Joint costs are costs incurred in the production of one product which is produced using a technology that simultaneously yields another product. The costs are called joint because they are associated with the production of both products, and the joint costs cannot be separated into components due to each product. An excellent transportation example is the following: A truck line provides service between town A and town B. The amount of freight to be hauled from A to B may be much greater than that from B to A, and as a result even though trucks operate full from A to B, many may operate completely empty from B back to A. The cost of operating these trucks (which, of course, must be operated from A to B and then back to A again in order to be available for use again), necessarily involves the production of transport capacity not only from A to B, but also from B to A. If all the trucks operate loaded rather than empty from B to A, it would add only a very small amount to the total cost of operation, such as in additional fuel and wear and tear on the vehicles (remember the analyses of fuel consumption based on propulsive work, which depends in part upon the weight of the vehicle and its load, found in Chap. 4). Thus many if not most of the costs involved in movement of freight between these two towns are joint costs associated with movement of vehicles in both directions, regardless of the traffic. It might be noted that joint costs are also often termed common costs in some of the older economics literature.

It should be noted that the existence of a joint cost situation depends solely upon the necessity, due to the technology used, or perhaps institutions or regulations, of producing two or more products simultaneously, regardless of whether both or all of them are desired. It in no way relates to the difficulty of allocation of costs between two services or products. There may be many situations where it is difficult to identify exactly what costs are associated with what products, but the costs are not joint. An example in the transportation sector is the provision of a railway track maintained to standards appropriate for both freight and high-speed passenger trains. Separate maintenance is not performed on the same track for the two types of trains. As a result it may be difficult to decide exactly what additional maintenance is required to accommodate passenger trains at higher speeds. This does not mean that the maintenance is entirely a joint cost. In principle, it would be possible to conduct experiments to determine exactly what cost is incurred in maintaining the track for freight trains and then to continue that experiment with the operation of freight as well as passenger trains to determine the additional cost associated with the higher level of maintenance.

Indivisible Costs

Although in the minds of many persons fixed costs are associated with indivisible costs, the two concepts are really quite different. Fixed costs are incurred regardless of the level of output of a particular product or service. Indivisible costs refer to costs which cannot be reduced if a certain range of output is to be provided. This indivisibility presumably reflects the technology of the processes involved. All indivisible costs are not necessarily fixed, because an indivisible cost is not fixed until a commitment has been made actually to incur that cost. Furthermore, certain indivisible costs may be incurred at higher levels of output, resulting in no change in cost over some range of output, above and below which range that particular indivisible cost is not incurred, so that that indivisible cost is a variable rather than a fixed cost.

As an example, consider a railroad line between two cities. At certain ranges of traffic to be carried, a single track may be quite adequate, the cheapest possible single track representing the fixed cost of providing any railroad service whatsoever between these two points. As traffic increases, costs increase, reflecting in part additional locomotives and cars, and in part probably more sophisticated train control systems, perhaps some double track or passing sidings, etc. At some point it may be absolutely necessary to install a complete double track in order to accommodate any additional traffic, and this double track may be adequate to accommodate a considerable increase in traffic. The least expensive double-track installation then represents an indivisible cost in order to enable the provision of transport capacity over that range. However, as this discussion has clearly indicated, the cost of that double track is in no sense a fixed cost until an actual commitment has been made to install it and incur the expenditure—presumably a commitment that would not be made until there is a reasonable expectation that the double track will be necessary because of the volume of traffic.

It might be noted here that there is often considerable indivisibility in transport costs. Very often this seems to arise largely because of institutional constraints on engineering designs and the way in which systems are operated, rather than because of the technology. An excellent example is provided by the cost of road capacity. In many cases, there are design standards for lanes which require a particular lane width, such as at least 13 ft on the United States Interstate and Defense Highway System. Such lanes, at typical land and road construction costs, yield a cost versus capacity relationship as shown in Fig. 9-2. This relationship portrays capacity levels at level of service B, the most common level of service standard for traffic on that system in rural areas. However, if one were to allow lane widths to be varied, at low

Figure 9-2 **Indivisibilities in transport costs and their reduction through design changes. Notes: Capacities are based on the relationships in Committee on Highway Capacity (1965). Costs are based on those for freeways in fringe areas of metropolitan areas of over one million population, of $1.20 million per lane-mi for construction and $1.11 million per lane-mi for land cost, assuming 13-ft lanes, taken from Sanders et al (1974, pp. 4-12 and 4-14).**

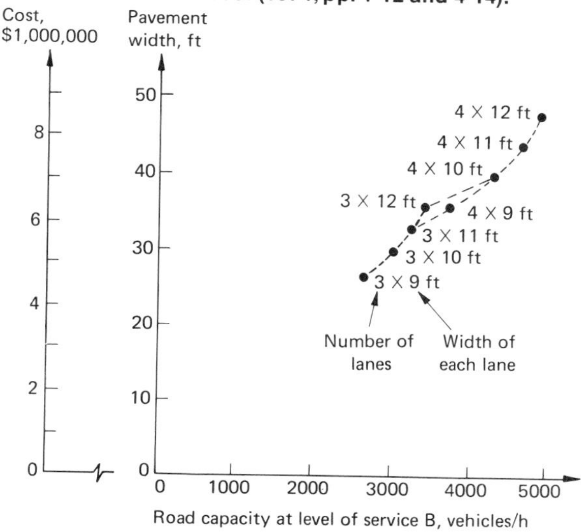

traffic volumes (and hence lower design capacities), a narrower pavement could be provided, at a reduced cost. At these low volumes, that narrower pavement would still permit traffic to flow at the same speed and with the same other characteristics as level of service B. As volume and capacity increased, the width of the lanes would be increased appropriately, ultimately reaching 13 ft where required. By use of the methods presented in Chap. 5, derived from the *Highway Capacity Manual* (1965), the capacity of various combinations of lane widths and number of lanes up to and including a total pavement width of 39 ft (three 13-ft lanes) can be calculated. Using the same freeway cost-estimating relationships, the costs would then be as shown in Fig. 9-2. As can be seen, the costs for roads in which the widths can be varied in increments of 1 ft rather than increments of 13 ft increase much more gradually. Thus, much of the lumpiness or indivisibility of highway costs can be avoided. It is not the purpose here to argue that such a wide latitude in the design of road or transport facilities should be provided, for there are many other considerations (such as the expectations of drivers of standard lane widths and types of facilities), but merely to point out that much of the lumpiness or indivisibility in transport facilities is really due to regulations and design standards, rather than inherent characteristics of the technology.

Future Costs and Present Value

The engineer and planner must be concerned with future costs as well as present costs because so much of professional practice is devoted to planning and design, which by definition relates to future actions. It is essential that the costs as well as the benefits of transportation systems be considered through all phases in the life cycle of such systems, including planning, construction, design, operation (including maintenance, etc.), possible renovation, rehabilitation or replacement, and finally possible abandonment. While our attention in this chapter is devoted to the costs over the life cycle, similar considerations will arise in the benefits of transportation systems, so many of the concepts introduced for cost analysis will reappear when we consider benefits in future chapters.

Since in general different costs will occur in different years as a facility or system passes through its life cycle, it would be useful if there were a way to combine these costs of different years into a single total cost for the system. In general, costs vary considerably from one stage in the life cycle to another. For example, in the case of a new highway facility, a relatively small cost would be incurred during the planning and design phases, and this cost would increase to a very large amount during the construction period (involving purchase and clearing of right-of-way, as well as direct facility construction), and then typically a somewhat lower cost would be incurred during the period of use, this cost increasing as traffic increases and as maintenance becomes more intensive with age, and then this entire cycle of costs may reappear with substantial renovation or replacement of the facility. If the facility were ever abandoned, then there would be a negative cost in the form of revenue accruing from the sale of land and positive costs associated with possible demolition and readying of the land for alternative uses.

It would be inappropriate simply to sum the costs occurring in different years. The reason this is incorrect can be understood by considering a hypothetical situation in which the value of $1 now is compared with its value in the future. Specifically, that dollar will be invested in a savings account or other suitable investment medium so that it earns interest with the passage of time. As a result of the earning of interest, the $1 deposited now, earning an interest of $(100I)$ percent per year, will yield the following amounts:

1 year hence: $\$1(1+I)$

2 years hence: (Value of $1 one year hence)$(1+I)=(1+I)(1+I)=\$1(1+I)^2$

3 years hence: $\$1(1+I)^3$

***N* years** hence: $\$1(1+I)^N$

It is thus apparent that the future value FV_N of an amount of money N years from now is related to its present value PV by the equation:

$$FV_N = PV(1+I)^N$$

where FV_N = future value N years hence
PV = present value
I = interest rate per year, decimal fraction

Rearranging this relationship, we can specify the present value of any future amount of money:

$$PV = \frac{1}{(1+I)^N} FV_N \tag{9-6}$$

The factor $(1+I)^{-N}$ is the factor which relates the future value to the present value. This factor is called the present worth factor and it is given by the following equation:

$$PWF_{I,N} = \frac{1}{(1+I)^N} \tag{9-7}$$

where $PWF_{I,N}$ = present worth factor at interest I for N years

When multiplied by the appropriate present worth factor the future value is converted to its present value, or more generally, a given amount of money is converted to its value N years earlier given the interest rate I. For ease of reference to other books, it might be noted here that two other common symbols for the present worth factor are $pwf_{i,n}$, which differs from our notation only in the use of lowercase symbols, and *(P/F,i,n)*, where P is present value, F is future amount, i is interest rate, and n is years in the future, the latter notation having been suggested by the Engineering Economy Division of the American Society of Engineering Education.

The effect of the present worth factor is to decrease the value of an amount of money exchanged in the future to its equivalent present value. The extent to which future values are reduced, or discounted, to present value is revealed by Fig. 9-3. The present value of $1 twenty years from now is only 31 cents at an interest rate of 6 percent, and only 15 cents at an interest rate of 10 percent. Thus the greater the interest rate, the lower the present value of any future amount of money. Similarly, the greater the number of years in the future when $1 is received, the lower the present value. Table 9-2 presents the present value factors associated with various numbers of years into the future and various interest rates.

Figure 9-3 **Present worth of $1 spent or received in the future at various discount rates.**

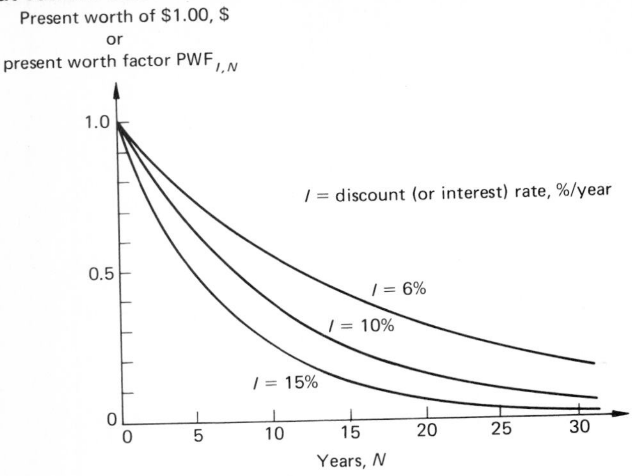

Since the effect of the interest rate on the present worth factor is to discount future monetary values to their present value, the interest rate is often termed a discount rate rather than an interest rate. The distinction is also important because the term interest rate can be used to refer to many different interest rates, such as those on stocks and bonds or on personal loans, and the value of these interest rates are often different. Moreover, they often will be quite different from the interest rate or the discount rate which should be used for the purpose of discounting future money values for engineering and planning purposes. Hence we shall usually use the term discount rate.

Given the obvious importance of the discount rate in determining the present value of any future costs (or other money values), a very important question is, What is the proper discount rate? Unfortunately, there is no "scientific" method for determining precisely what the discount rate should be (Wohl and Martin, 1967, pp. 217–222). Part of the problem is that the discount rate can reflect at least three reasons for the existence of interest. One is that people in general have a preference for receiving a dollar now rather than at some time in the future. Secondly, the discount rate can reflect the existence of some risk, that is, whether or not the full dollar of value will actually be received in the future, so that one is willing to in effect pay less for the receipt of a dollar in the future than for the receipt of a dollar now. Finally, the discount rate can be interpreted so as to reflect a type of profit associated with the expenditure of money now, the same type of profit which enables banks and other institutions to which money might be lent to pay the interest. In this interpretation, a discount rate would reflect the profit attainable from the expenditure of money now. Despite these differing views and the difficulties of selecting a single rate, it is essential that a particular discount rate be used in the calculations involving costs in various years. In most analyses in the United States at the present time, a discount rate of 10 percent/year seems to be very widely used, this being slightly greater than the interest which one would obtain in rather secure investments. To account for

Table 9-2 **Present Worth Factors† for Various Time Periods and Discount (or Interest) Rates**

Year N	Discount rate I, percent per year 6	8	10	12	15	Year N
1	0.943396	0.925926	0.909091	0.892857	0.869565	1
2	0.889996	0.857339	0.826446	0.797194	0.756144	2
3	0.839619	0.793832	0.751315	0.711780	0.657516	3
4	0.792094	0.735030	0.683013	0.635518	0.571753	4
5	0.747258	0.680583	0.620921	0.567427	0.497177	5
6	0.704961	0.630170	0.564474	0.506631	0.432328	6
7	0.665057	0.583490	0.513158	0.452349	0.375937	7
8	0.627412	0.540269	0.466507	0.403883	0.326902	8
9	0.591898	0.500249	0.424098	0.360610	0.284262	9
10	0.558395	0.463193	0.385543	0.321973	0.247185	10
11	0.526788	0.428883	0.350494	0.287476	0.214943	11
12	0.496969	0.397114	0.318631	0.256675	0.186907	12
13	0.468839	0.367698	0.289664	0.229174	0.162528	13
14	0.442301	0.340461	0.263331	0.204620	0.141329	14
15	0.417265	0.315242	0.239392	0.182696	0.122894	15
16	0.393646	0.291890	0.217629	0.163122	0.106865	16
17	0.371364	0.270269	0.197845	0.145644	0.092926	17
18	0.350344	0.250249	0.179859	0.130040	0.080805	18
19	0.330513	0.231712	0.163508	0.116107	0.070265	19
20	0.311805	0.214548	0.148644	0.103667	0.061100	20
21	0.294155	0.198656	0.135131	0.092560	0.053131	21
22	0.277505	0.183941	0.122846	0.082643	0.046201	22
23	0.261797	0.170315	0.111678	0.073788	0.040174	23
24	0.246979	0.157699	0.101526	0.065882	0.034934	24
25	0.232999	0.146018	0.092296	0.058823	0.030378	25
26	0.219810	0.135202	0.083905	0.052521	0.026415	26
27	0.207368	0.125187	0.076278	0.046894	0.022970	27
28	0.195630	0.115914	0.069343	0.041869	0.019974	28
29	0.184557	0.107328	0.063039	0.037383	0.017369	29
30	0.174110	0.099377	0.057309	0.033378	0.015103	30
31	0.164255	0.092016	0.052099	0.029802	0.013133	31
32	0.154957	0.085200	0.047362	0.026609	0.011420	32
33	0.146186	0.078889	0.043057	0.023758	0.009931	33
34	0.137912	0.073045	0.039143	0.021212	0.008635	34
35	0.130105	0.067635	0.035584	0.018940	0.007509	35
36	0.122741	0.062625	0.032349	0.016910	0.006529	36
37	0.115793	0.057986	0.029408	0.015098	0.005678	37
38	0.109239	0.053690	0.026735	0.013481	0.004937	38
39	0.103056	0.049713	0.024304	0.012036	0.004293	39
40	0.097222	0.046031	0.022095	0.010747	0.003733	40
45	0.072650	0.031328	0.013719	0.009595	0.003246	45
50	0.054288	0.021321	0.008519	0.008567	0.002823	50
55	0.040567	0.014511	0.005289	0.007649	0.002455	55
60	0.030314	0.009876	0.003284	0.006830	0.002134	60
65	0.022653	0.006721	0.002039	0.006098	0.001856	65
70	0.016927	0.004574	0.001266	0.003460	0.000923	70
75	0.012649	0.003113	0.000786	0.001963		75
80	0.009452	0.002119	0.000488	0.001114		80
90	0.005278	0.000981	0.000188			90
100	0.002947	0.000455	0.000073			100

† $PWF_{I,N} = \frac{1}{(1+I)^N}$

the uncertainty of whether or not this is precisely the proper rate, usually important analyses are also done with lower and higher rates, such as 7.5, or 8.0 percent and 12.5 or 15.0 percent, these aiding in the determination of the effect of different discount rates on total cost (or other outcome) of the analysis.

An example may clarify the use of the present worth factor and the importance of the discount rate. The problem is to determine which of two types of motor vehicles would be less expensive, type A, which costs $20,000 and has a useful life of 6 years, or type B, which costs $12,000 and has a useful life of only 3 years. The cost of operating each of these vehicles in the years of its life are given below:

Total Operating (and Maintenance) Cost, $

Year	Type A	Type B
1	4000	4000
2	4200	4250
3	4400	4550
4	4600	...
5	4900	...
6	5400	...

Each vehicle will be purchased in the first year of its use.

The present value of the type-A vehicle, for its full life of 6 years, using Eq. (9-6) and obtaining the present worth factors for 10 percent from Table 9-2, is

$$\$20{,}000 + \$4000 + \$(0.909091)4200 + \$(0.826446)4400 + \$(0.751315)4600 + \$(0.683013)4900 + \$(0.620921)5400 = \$41{,}610.33$$

By similar methods, the present value of all the costs of the two type-B vehicles which must be purchased, one in the first year ("present") and one in the fourth year, is

$$\$12{,}000 + \$4000 + \$(0.909091)4250 + \$(0.826446)4550 + \$(0.751315)(12{,}000 + 4000) + \$(0.683013)4250 + \$(0.729021)4550 = \$41{,}373.00$$

Thus the type-B vehicles represent the least-cost choice. Notice that this is a conclusion different from what would be reached if total cost over the lifetime of the vehicles were simply summed, as if the discount rate were zero! Also, if the discount rate were reduced to 6 percent per annum, the present values of the type-A vehicle expenditures would be less than that of the type-B vehicle, as can be verified easily by the reader. From this it is apparent that the relative magnitude of costs of two different alternatives can differ, depending upon the discount rate selected; therefore one must be careful to use the discount rate appropriate for the economy and time. As this example has illustrated, simply reducing the discount rate can often reduce the costs of alternatives involving a large immediate expenditure relative to those involving lower immediate expenditures but higher future expenditures. Unfortunately, there are many examples of deliberate choices of low discount rates in order to make projects involving large initial expenditures, such as major construction projects or massive expenditures on new vehicles, look attractive relative to alternatives involving a more uniform distribution of expenditures.

A minor variation of the present value concept which is important in many engineering studies involves the calculation of present values of costs which are subject to growth in the future. Various formulas have been derived for many different types of steady growth. A particularly useful formula in the transportation engineering context is one in which the expenditures grow at a constant percentage rate from year to year. These expenditures are assumed to begin 1 year in the future, at a value of $1, and then grow at a constant rate of $100 \cdot G$ percent per year (i.e., G is a decimal fraction) through a terminal year, N years hence. The formula can be derived readily from the following considerations, which present the present value of the expenditure in various years in the future;

Present value of $1 cost in year 1: $\$1 \dfrac{1}{1+I}$

PV of $\$1(1+G)$ cost in year 2: $\$1(1+G)\dfrac{1}{(1+I)^2}$

PV of $\$1(1+G)^2$ cost in year 3: $\$1(1+G)^2\dfrac{1}{(1+I)^3}$

. . .

PV of $\$(1+G)^N$ cost in year N: $\$1(1+G)^{N-1}\dfrac{1}{(1+I)^N}$

The present value factor for a uniform series growing at a rate G and discounted at a rate I is

$$\mathrm{PVF}_{G,I,N} = \frac{1}{1+I}\left[1 + \frac{1+G}{1+I} + \left(\frac{1+G}{1+I}\right)^2 + \cdots + \left(\frac{1+G}{1+I}\right)^{N-2} + \left(\frac{1+G}{1+I}\right)^{N-1}\right] \tag{9-8}$$

where $\mathrm{PVF}_{G,I,N}$ = present value factor of the steady growth rate series
G = growth rate in expenditure which begins in year 1 with discounting to year 0, as a decimal fraction

Table 9-3 presents values of this uniform series present value factor for various discount rates and growth rates. The use of Eq. (9-8) and Table 9-3 is rather straightforward. Consider a railroad in which the construction was performed in 1 year, 1975, costing $13 million. The cost of operation and maintenance then began in 1976 at a level of $2 million and these grew at a constant rate of 6 percent per annum. What is the "present" value in 1974, the year before construction occurred, of all of these expenditures, assuming the line operates for a period of 20 years? Use a 10 percent/year discount rate. From Table 9-3, the present value factor for this situation is 13.0820. This discounts the costs to the year before the first of these costs is incurred, to year 1975. From Table 9-2, the present worth factor for the expenditures in 1975 discounted for 1 year to 1974 is 0.909091. Thus the present value is

$$0.909091\,(\$13 \times 10^6) + 0.909091(13.0820)(\$2 \times 10^6) = \$35.6036 \times 10^6$$

Table 9-3 Present Value Factors for a Uniform Series Experiencing Compound Growth at Various Growth Rates and Discount Rates†

All items accrue at the end of each year and are discounted to year 0.

Year	Rate of growth of benefits, percent per annum								
	2	3	4	5	6	7	8	9	10
	6 percent per annum discount rate†								
1	0.9434	0.9434	0.9434	0.9434	0.9434	0.9434	0.9434	0.9434	0.9434
2	1.8512	1.8601	1.8690	1.8779	1.8868	1.8957	1.9046	1.9135	1.9224
3	2.7247	2.7508	2.7771	2.8036	2.8302	2.8570	2.8839	2.9110	2.9383
4	3.5653	3.6164	3.6681	3.7205	3.7736	3.8273	3.8817	3.9368	3.9926
5	4.3742	4.4574	4.5423	4.6288	4.7170	4.8068	4.8984	4.9916	5.0867
6	5.1525	5.2747	5.4000	5.5285	5.6604	5.7956	5.9342	6.0763	6.2220
7	5.9015	6.0688	6.2415	6.4198	6.6038	6.7936	6.9896	7.1917	7.4002
8	6.6222	6.8404	7.0671	7.3026	7.5472	7.8011	8.0648	8.3386	8.6229
9	7.3157	7.5902	7.8772	8.1771	8.4906	8.8181	9.1604	9.5180	9.8916
10	7.9830	8.3188	8.6720	9.0434	9.4340	9.8447	10.2766	10.7308	11.2083
11	8.6251	9.0268	9.4517	9.9015	10.3774	10.8810	11.4139	11.9779	12.5747
12	9.2431	9.7147	10.2168	10.7514	11.3208	11.9270	12.5727	13.2603	13.9926
13	9.8377	10.3831	10.9674	11.5934	12.2642	12.9829	13.7533	14.5790	15.4640
14	10.4098	11.0327	11.7039	12.4274	13.2075	14.0488	14.9562	15.9350	16.9909
15	10.9604	11.6638	12.4265	13.2536	14.1509	15.1247	16.1818	17.3294	18.5755
16	11.4902	12.2771	13.1354	14.0720	15.0943	16.2108	17.4305	18.7632	20.2199
17	12.0000	12.8730	13.8310	14.8826	16.0377	17.3072	18.7028	20.2376	21.9263
18	12.4906	13.4521	14.5134	15.6856	16.9811	18.4138	19.9990	21.7538	23.6971
19	12.9626	14.0148	15.1829	16.4510	17.9245	19.5309	21.3198	23.3129	25.5347
20	13.4169	14.5615	15.8399	17.2689	18.8679	20.6586	22.6654	24.9161	27.4417
21	13.8540	15.0928	16.4844	18.0494	19.8113	21.7969	24.0365	26.5646	29.4206
22	14.2746	15.6091	17.1168	18.8225	20.7547	22.9459	25.4334	28.2599	31.4742
23	14.6793	16.1107	17.7372	19.5883	21.6981	24.1058	26.8567	30.0031	33.6053
24	15.0688	16.5981	18.3459	20.3469	22.6415	25.2766	28.3068	31.7956	35.8168
25	15.4435	17.0718	18.9432	21.0984	23.5849	26.4584	29.7843	33.6389	38.1118
26	15.8041	17.5320	19.5292	21.8427	24.5283	27.6514	31.2896	35.5343	40.4934
27	16.1512	17.9792	20.1041	22.5801	25.4717	28.8557	32.8234	37.4834	42.9648
28	16.4851	18.4137	20.6682	23.3105	26.4151	30.0713	34.3861	39.4876	45.5295
29	16.8064	18.8360	21.2216	24.0339	27.3585	31.2984	35.9783	41.5486	48.1910
30	17.1156	19.2463	21.7646	24.7306	28.3019	32.5371	37.6005	43.6679	50.9529
31	17.4131	19.6450	22.2973	25.4605	29.2453	33.7874	39.2534	45.8472	53.8191
32	17.6994	20.0324	22.8200	26.1637	30.1887	35.0496	40.9374	48.0881	56.7934
33	17.9749	20.4088	23.3328	26.8603	31.1321	36.3236	42.6532	50.3925	59.8799
34	18.2400	20.7746	23.8360	27.5503	32.0755	37.6097	44.4014	52.7621	63.0830
35	18.4951	21.1301	24.3297	28.2338	33.0189	38.9079	46.1825	55.1988	66.4068
36	18.7406	21.4754	24.8140	28.9108	33.9623	40.2183	47.9973	57.7044	69.8562
37	18.9768	21.8110	25.2892	29.5815	34.9057	41.5412	49.8463	60.2810	73.4356
38	19.2041	22.1371	25.7555	30.2458	35.8491	42.8765	51.7302	62.9304	77.1502
39	19.4228	22.4540	26.2129	30.9038	36.7925	44.2243	53.6496	65.6549	81.0049
40	19.6332	22.7619	26.6617	31.5557	37.7358	45.5850	55.6053	68.4564	85.0051
41	19.8358	23.0611	27.1021	32.2014	38.6792	46.9584	57.5978	71.3373	89.1562
42	20.0306	23.3518	27.5341	32.8410	39.6226	48.3448	59.6280	74.2996	93.4640
43	20.2182	23.6343	27.9580	33.4746	40.5660	49.7443	61.6964	77.3459	97.9344
44	20.3986	23.9088	28.3739	34.1022	41.5094	51.1570	63.8039	80.4783	102.5734
45	20.5722	24.1756	28.7819	34.7239	42.4528	52.5830	65.9512	83.6994	107.3875
46	20.7393	24.4347	29.1822	35.3397	43.3962	54.0224	68.1389	87.0116	112.3832
47	20.9001	24.6866	29.5750	35.9497	44.3396	55.4755	70.3679	90.4176	117.5675
48	21.0548	24.9313	29.9604	36.5539	45.2830	56.9422	72.6390	93.9200	122.9474
49	21.2037	25.1691	30.3385	37.1525	46.2264	58.4228	74.9530	97.5215	128.5303
50	21.3470	25.4002	30.7095	37.7454	47.1698	59.9174	77.3106	101.2249	134.3239

† Heggie (1972, p. 254).

Table 9-3 (*Continued*)

Year	2	3	4	5	6	7	8	9	10
	Rate of growth of benefits, percent per annum								
	8 percent per annum discount rate‡								
1	0.9259	0.9259	0.9259	0.9259	0.9259	0.9259	0.9259	0.9259	0.9259
2	1.8004	1.8090	1.8176	1.8261	1.8347	1.8433	1.8519	1.8604	1.8690
3	2.6263	2.6512	2.6762	2.7013	2.7267	2.7521	2.7778	2.8036	2.8295
4	3.4063	3.4543	3.5030	3.5522	3.6021	3.6526	3.7037	3.7555	3.8079
5	4.1430	4.2204	4.2992	4.3795	4.4613	4.5447	4.6296	4.7162	4.8043
6	4.8388	4.9509	5.0659	5.1837	5.3046	5.4285	5.5556	5.6858	5.8192
7	5.4959	5.6476	5.8042	5.9657	6.1323	6.3042	6.4815	6.6643	6.8529
8	6.1165	6.3121	6.5151	6.7259	6.9447	7.1717	7.4074	7.6520	7.9057
9	6.7026	6.9458	7.1997	7.4650	7.7420	8.0313	8.3333	8.6487	8.9780
10	7.2562	7.5501	7.8590	8.1836	8.5246	8.8828	9.2593	9.6547	10.0702
11	7.7790	8.1265	8.4939	8.8822	9.2926	9.7265	10.1852	10.6701	11.1826
12	8.2727	8.6762	9.1052	9.5614	10.0465	10.5624	11.1111	11.6948	12.3157
13	8.7391	9.2005	9.6939	10.2217	10.7863	11.3905	12.0370	12.7290	13.4696
14	9.1795	9.7004	10.2608	10.8637	11.5125	12.2110	12.9630	13.7728	14.6450
15	9.5954	10.1773	10.8067	11.4878	12.2252	13.0238	13.8889	14.8262	15.8421
16	9.9883	10.6320	11.3324	12.0947	12.9248	13.8292	14.8148	15.8894	17.0614
17	10.3593	11.0657	11.8386	12.6846	13.6114	14.6270	15.7407	16.9625	18.3033
18	10.7097	11.4794	12.3260	13.2582	14.2852	15.4175	16.6667	18.0455	19.5682
19	11.0407	11.8738	12.7954	13.8158	14.9466	16.2007	17.5926	19.1385	20.8565
20	11.3532	12.2500	13.2475	14.3580	15.5957	16.9766	18.5185	20.2416	22.1687
21	11.6484	12.6088	13.6827	14.8851	16.2329	17.7453	19.4444	21.3550	23.5051
22	11.9272	12.9510	14.1019	15.3975	16.8582	18.5070	20.3704	22.4786	24.8663
23	12.1905	13.2774	14.5055	15.8958	17.4719	19.2615	21.2963	23.6127	26.2527
24	12.4392	13.5886	14.8942	16.3801	18.0743	20.0091	22.2222	24.7573	27.6648
25	12.6740	13.8854	15.2685	16.8511	18.6655	20.7498	23.1481	25.9124	29.1031
26	12.8958	14.1685	15.6289	17.3089	19.2458	21.4836	24.0741	27.0783	30.5679
27	13.1053	14.4385	15.9760	17.7540	19.8153	22.2106	25.0000	28.2549	32.0599
28	13.3032	14.6960	16.3102	18.1868	20.3743	22.9308	25.9259	29.4425	33.5796
29	13.4900	14.9415	16.6321	18.6075	20.9229	23.6444	26.8519	30.6410	35.1273
30	13.6665	15.1757	16.9420	19.0166	21.4614	24.3514	27.7778	31.8506	36.7038
31	13.8332	15.3990	17.2404	19.4143	21.9899	25.0519	28.7037	33.0715	38.3094
32	13.9906	15.6121	17.5278	19.8009	22.5086	25.7459	29.6296	34.3036	39.9447
33	14.1393	15.8152	17.8046	20.1768	23.0177	26.4334	30.5556	35.5472	41.6104
34	14.2797	16.0089	18.0711	20.5423	23.5173	27.1146	31.4815	36.8023	43.3069
35	14.4123	16.1937	18.3277	20.8976	24.0078	27.7894	32.4074	38.0689	45.0348
36	14.5375	16.3699	18.5748	21.2430	24.4891	28.4580	33.3333	39.3474	46.7947
37	14.6558	16.5380	18.8128	21.5788	24.9615	29.1205	34.2593	40.6376	48.5872
38	14.7675	16.6983	19.0419	21.9054	25.4252	29.7768	35.1852	41.9398	50.4129
39	14.8731	16.8511	19.2626	22.2228	25.8803	30.4270	36.1111	43.2541	52.2724
40	14.9727	16.9969	19.4751	22.5314	26.3269	31.0712	37.0370	44.5805	54.1663
41	15.0668	17.1359	19.6797	22.8315	26.7653	31.7094	37.9630	45.9192	56.0953
42	15.1557	17.2685	19.8768	23.1232	27.1956	32.3417	38.8889	47.2703	58.0600
43	15.2396	17.3950	20.0665	23.4068	27.6179	32.9682	39.8148	48.6339	60.0611
44	15.3189	17.5156	20.2493	23.6825	28.0324	33.5889	40.7407	50.0102	62.0993
45	15.3938	17.6306	20.4252	23.9506	28.4392	34.2038	41.6667	51.3991	64.1752
46	15.4645	17.7403	20.5946	24.2113	28.8385	34.8130	42.5926	52.8010	66.2896
47	15.5313	17.8449	20.7578	24.4646	29.2304	35.4166	43.5185	54.2158	68.4431
48	15.5944	17.9447	20.9149	24.7110	29.6150	36.0146	44.4444	55.6437	70.6365
49	15.6539	18.0398	21.0662	24.9305	29.9925	36.6070	45.3704	57.0849	72.8705
50	15.7102	18.1306	21.2119	25.1834	30.3630	37.1940	46.2963	58.5394	75.1459

‡ Heggie (1972, p. 256).

Table 9-3 (*Continued*)

Year	Rate of growth of benefits, percent per annum 2	3	4	5	6	7	8	9	10
	10 percent per annum discount rate§								
1	0.9091	0.9091	0.9091	0.9091	0.9091	0.9091	0.9091	0.9091	0.9091
2	1.7521	1.7603	1.7686	1.7769	1.7851	1.7934	1.8017	1.8099	1.8182
3	2.5337	2.5574	2.5812	2.6052	2.6293	2.6536	2.6780	2.7026	2.7273
4	3.2586	3.3037	3.3495	3.3959	3.4428	3.4903	3.5384	3.5871	3.6364
5	3.9307	4.0026	4.0759	4.1506	4.2267	4.3042	4.3831	4.4636	4.5455
6	4.5539	4.6570	4.7627	4.8710	4.9821	5.0959	5.2125	5.3321	5.4545
7	5.1318	5.2697	5.4120	5.5387	5.7100	5.8660	6.0269	6.1927	6.3636
8	5.6677	5.8435	6.0259	6.2151	6.4115	6.6151	6.8264	7.0455	7.2727
9	6.1646	6.3807	6.6063	6.8417	7.0874	7.3438	7.6113	7.8905	8.1818
10	6.6253	6.8837	7.1550	7.4398	7.7388	8.0526	8.3820	8.7279	9.0909
11	7.0526	7.3548	7.6738	8.0107	8.3664	8.7421	9.1387	9.5576	10.0000
12	7.4487	7.7958	8.1644	8.5557	8.9713	9.4127	9.8817	10.3798	10.9091
13	7.8161	8.2088	8.6281	9.0759	9.5542	10.0651	10.6111	11.1946	11.8182
14	8.1568	8.5955	9.0666	9.5724	10.1158	10.6997	11.3273	12.0019	12.7273
15	8.4726	8.9576	9.4811	10.0464	10.6571	11.3170	12.0304	12.8019	13.6364
16	8.7655	9.2967	9.8731	10.4989	11.1786	11.9174	12.7208	13.5946	14.5455
17	9.0371	9.6142	10.2436	10.9307	11.6812	12.5015	13.3986	14.3801	15.4545
18	9.2890	9.9115	10.5940	11.3430	12.1656	13.0696	14.0640	15.1584	16.3636
19	9.5225	10.1898	10.9252	11.7365	12.6323	13.6223	14.7174	15.9297	17.2727
20	9.7390	10.4505	11.2384	12.1121	13.0820	14.1599	15.3589	16.6940	18.1818
21	9.9398	10.6945	11.5345	12.4706	13.5154	14.6828	15.9888	17.4513	19.0909
22	10.1260	10.9231	11.8144	12.8129	13.9330	15.1914	16.6071	18.2018	20.0000
23	10.2987	11.1370	12.0791	13.1396	14.3354	15.6862	17.2143	18.9454	20.9091
24	10.4588	11.3374	12.3293	13.4314	14.7232	16.1675	17.8104	19.6823	21.8182
25	10.6072	11.5250	12.5659	13.7491	15.0969	16.6357	18.3957	20.4124	22.7273
26	10.7449	11.7007	12.7896	14.0332	15.4570	17.0910	18.9703	21.1359	23.6364
27	10.8725	11.8652	13.0011	14.3044	15.8041	17.5340	19.5345	21.8529	24.5455
28	10.9909	12.0192	13.2010	14.5633	16.1385	17.9649	20.0884	22.5633	25.4545
29	11.1006	12.1635	13.3900	14.8104	16.4607	18.3840	20.6322	23.2673	26.3636
30	11.2024	12.2985	13.5688	15.0463	16.7712	18.7918	21.1662	23.9649	27.2727
31	11.2968	12.4250	13.7377	15.2715	17.0704	19.1883	21.6904	24.6561	28.1818
32	11.3843	12.5434	13.8975	15.4864	17.3588	19.5741	22.2052	25.3410	29.0909
33	11.4654	12.6543	14.0485	15.6916	17.6367	19.9494	22.7105	26.0197	30.0000
34	11.5407	12.7581	14.1913	15.8874	17.9044	20.3144	23.2067	26.6923	30.9091
35	11.6104	12.8553	14.3264	16.0744	18.1624	20.6694	23.6938	27.3587	31.8182
36	11.6751	12.9463	14.4540	16.2528	18.4111	21.0148	24.1721	28.0191	32.7273
37	11.7351	13.0316	14.5747	16.4231	18.6507	21.3508	24.6417	28.6735	33.6364
38	11.7908	13.1114	14.6888	16.5857	18.8816	21.6776	25.1028	29.3219	34.5455
39	11.8423	13.1861	14.7967	16.7409	19.1040	21.9955	25.5555	29.9644	35.4545
40	11.8902	13.2561	14.8987	16.8890	19.3184	22.3047	25.9999	30.6011	36.3636
41	11.9345	13.3216	14.9951	17.0304	19.5250	22.6055	26.4363	31.2320	37.2727
42	11.9756	13.3830	15.0863	17.1554	19.7241	22.8980	26.8647	31.8572	38.1818
43	12.0138	13.4404	15.1725	17.2943	19.9160	23.1826	27.2854	32.4767	39.0909
44	12.0491	13.4942	15.2540	17.4173	20.1009	23.4595	27.6983	33.0905	40.0000
45	12.0819	13.5446	15.3311	17.5347	20.2790	23.7288	28.1038	33.6988	40.9091
46	12.1123	13.5917	15.4039	17.6467	20.4507	23.9907	28.5019	34.3015	41.8182
47	12.1405	13.6359	15.4728	17.7537	20.6161	24.2455	28.8928	34.8988	42.7273
48	12.1667	13.6772	15.5379	17.8558	20.7755	24.4934	29.2766	35.4906	43.6364
49	12.1909	13.7160	15.5995	17.9533	20.9291	24.7344	29.6534	36.0770	44.5455
50	12.2134	13.7522	15.6577	18.0463	21.0772	24.9690	30.0233	36.6582	45.4545

§ Heggie (1972, p. 258).

Table 9-3 (*Continued*)

Year	2	3	4	5	6	7	8	9	10
	Rate of growth of benefits, percent per annum								
	15 percent per annum discount rate¶								
1	0.8696	0.8696	0.8696	0.8696	0.8696	0.8696	0.8696	0.8696	0.8696
2	1.6408	1.6484	1.6560	1.6635	1.6711	1.6786	1.6862	1.6938	1.7013
3	2.3249	2.3460	2.3671	2.3884	2.4099	2.4314	2.4531	2.4750	2.4969
4	2.9317	2.9707	3.0103	3.0503	3.0908	3.1319	3.1734	3.2154	3.2579
5	3.4698	3.5303	3.5919	3.6346	3.7185	3.7835	3.8498	3.9172	3.9858
6	3.9471	4.0315	4.1179	4.2064	4.2971	4.3899	4.4850	4.5824	4.6821
7	4.3705	4.4804	4.5936	4.7102	4.8303	4.9541	5.0816	5.2129	5.3481
8	4.7460	4.8824	5.0237	5.1702	5.3219	5.4790	5.6418	5.8105	5.9851
9	5.0791	5.2425	5.4128	5.5902	5.7749	5.9674	6.1680	6.3769	6.5945
10	5.3745	5.5650	5.7646	5.9736	6.1926	6.4219	6.6621	6.9137	7.1773
11	5.6365	5.8539	6.0828	6.3237	6.5775	6.8447	7.1261	7.4226	7.7348
12	5.8689	6.1126	6.3705	6.6434	6.9323	7.2381	7.5619	7.9049	8.2681
13	6.0750	6.3444	6.6307	6.9353	7.2593	7.6042	7.9712	8.3620	8.7782
14	6.2578	6.5519	6.8660	7.2018	7.5608	7.9447	8.3556	8.7953	9.2661
15	6.4200	6.7378	7.0789	7.4451	7.8386	8.2616	8.7165	9.2060	9.7328
16	6.5638	6.9043	7.2713	7.6673	8.0947	8.5565	9.0555	9.5952	10.1792
17	6.6914	7.0534	7.4454	7.8701	8.3308	8.3808	9.3739	9.9642	10.6062
18	6.8045	7.1870	7.6028	8.0553	8.5484	9.0861	9.6729	10.3139	11.0146
19	6.9049	7.3066	7.7451	8.2244	8.7489	9.3233	9.9537	10.6453	11.4053
20	6.9939	7.4137	7.8738	8.3785	8.9338	9.5445	10.2173	10.9595	11.7790
21	7.0729	7.5097	7.9903	8.5198	9.1042	9.7501	10.4650	11.2572	12.1364
22	7.1429	7.5956	8.0955	8.6485	9.2613	9.9414	10.6976	11.5395	12.4783
23	7.2050	7.6726	8.1907	8.7660	9.4060	10.1194	10.9160	11.8070	12.8053
24	7.2601	7.7415	8.2768	8.8733	9.5395	10.2850	11.1211	12.0605	13.1181
25	7.3089	7.8033	8.3547	8.9713	9.6625	10.4391	11.3137	12.3009	13.4173
26	7.3523	7.8586	8.4251	9.0608	9.7759	10.5825	11.4946	12.5286	13.7035
27	7.3907	7.9081	8.4888	9.1424	9.8803	10.7159	11.6645	12.7445	13.9773
28	7.4248	7.9525	8.5464	9.2170	9.9767	10.8400	11.8241	12.9492	14.2392
29	7.4550	7.9922	8.5985	9.2851	10.0655	10.9554	11.9739	13.1431	14.4896
30	7.4819	8.0278	8.6456	9.3473	10.1473	11.0629	12.1146	13.3270	14.7292
31	7.5057	8.0597	8.6882	9.4040	10.2227	11.1629	12.2468	13.5012	14.9584
32	7.5268	8.0883	8.7267	9.4558	10.2922	11.2559	12.3709	13.6664	15.1776
33	7.5455	8.1138	8.7615	9.5032	10.3563	11.3424	12.4874	13.8229	15.3873
34	7.5621	8.1367	8.7930	9.5464	10.4154	11.4230	12.5969	13.9713	15.5878
35	7.5768	8.1573	8.8215	9.5858	10.4698	11.4979	12.6997	14.1119	15.7796
36	7.5899	8.1756	8.8473	9.6218	10.5200	11.5676	12.7962	14.2452	15.9631
37	7.6014	8.1921	8.8706	9.6547	10.5663	11.6325	12.8869	14.3715	16.1386
38	7.6117	8.2068	8.8917	9.6847	10.6089	11.6928	12.9720	14.4913	16.3065
39	7.6208	8.2200	8.9107	9.7121	10.6482	11.7490	13.0520	14.6048	16.4671
40	7.6289	8.2319	8.9280	9.7372	10.6845	11.8012	13.1271	14.7123	16.6207
41	7.6361	8.2424	8.9436	9.7600	10.7178	11.8498	13.1976	14.8143	16.7676
42	7.6424	8.2519	8.9576	9.7809	10.7486	11.8951	13.2639	14.9110	16.9082
43	7.6481	8.2604	8.9704	9.8000	10.7770	11.9371	13.3261	15.0026	17.0426
44	7.6531	8.2680	8.9819	9.8173	10.8031	11.9763	13.3845	15.0894	17.1712
45	7.6575	8.2748	8.9923	9.8332	10.8272	12.0127	13.4393	15.1717	17.2942
46	7.6614	8.2809	9.0018	9.8477	10.8495	12.0466	13.4908	15.2497	17.4118
47	7.6649	8.2864	9.0103	9.8610	10.8699	12.0782	13.5392	15.3236	17.5244
48	7.6680	8.2913	9.0180	9.8731	10.8888	12.1075	13.5847	15.3937	17.6320
49	7.6708	8.2957	9.0250	9.8841	10.9062	12.1345	13.6273	15.4601	17.7350
50	7.6732	8.2996	9.0313	9.8942	10.9222	12.1602	13.6674	15.5230	17.8334

¶ Heggie (1972, p. 259).

In many situations it is desirable to discount a present value calculated using either the present worth factor, the present value factor, or both, to another year, as illustrated above. In general, this can be accomplished easily because the present worth factor involving discounting between year 0, year N, and a year M lying somewhere between 0 and N, is the following:

$$\text{PWF}_{I,N} = \text{PWF}_{I,M}\,(\text{PWF}_{I,N-M}) \tag{9-9}$$

That this, Eq. (9-9), is true can be seen from the following substitution:

$$\frac{1}{(1+I)^N} = \frac{1}{(1+I)^M}\,\frac{1}{(1+I)^{N-M}}$$

Thus, if in the above example, the construction actually took 3 years, with expenditures of \$4 million in 1973, \$5 million in 1974, and \$4 million in 1975, the present value in 1974 of all the costs would be as follows:

$$(1+0.10)^1(\$4 \times 10^6) + \$5 \times 10^6 + (1+0.10)^{-1}(\$4 \times 10^6) + (1+0.10)^{-1}(13.0820)(\$2 \times 10^6) = \$35.8218 \times 10^6$$

Another important concept in engineering and planning is that of the annual cost occurring over a period of years, which is exactly equivalent to an expenditure of a certain amount in 1 year. This concept is particularly important in the transportation context because it can be used to specify an annual cost over the life of the transportation facility (say, 30 years) which is equivalent to the initial expenditure for construction of that facility. By using this equivalent cost, one can then speak of annual cost including an appropriate portion of the original capital expenditure for the facility. The factor by which the original single expenditure is multiplied to yield an equivalent annual cost is called the capital recovery factor (CRF) and it is derived in the following manner.

Assume the actual expenditure in the first year is \$$Z$ and the value of the uniform equivalent annual expenditure over a period of N years of \$$Y$ is desired. The equivalency factor or capital recovery factor can be derived by equating the future value of the \$$Z$ at the end of the N years with the future value of the Y yearly payments made in each of these N years. The future value of the actual expenditure of Z is

$$\text{FV of } \$Z = Z(1+I)^N$$

The future value of the N yearly payments of \$$Y$ is as follows:

FV_N of \$$Y$ in year 1: $Y(1+I)^{N-1}$
FV_N of \$$Y$ in year 2: $Y(1+I)^{N-2}$
. . . .
FV_N of \$$Y$ in year N: Y

Total FV_N of the N payments of \$$Y$ each: $Y \sum_{i=0}^{N-1} (1+I)^i$

By equating these two we have

$$Z(1+I)^N = Y[1 + (1+I) + (1+I)^2 + \cdots + (1+I)^{N-2} + (1+I)^{N-1}] \tag{9-10}$$

If this is then multiplied by $(1+I)$, then we have

$$Z(1+I)^{N+1} = Y[(1+I) + (1+I)^2 + (1+I)^3 + \cdots + (1+I)^{N-1} + (1+I)^N] \tag{9-11}$$

Subtracting Eq. (9-10) from (9-11), we obtain

$$Z(1 + I)^N[(1 + I) - 1] = Y[(1 + I)^N - 1]$$

and hence

$$Y = \frac{I(1 + I)^N}{(1 + I)^N - 1} Z$$

Thus the capital recovery factor which converts a single expenditure of $\$Z$ in year 0 to an equivalent amount paid over N years is

$$\mathrm{CRF}_{I,N} = \frac{I(1 + I)^N}{(1 + I)^N - 1} \tag{9-12}$$

where $\mathrm{CRF}_{I,N}$ = capital recovery factor at interest rate I per annum for a repayment over N years

Values of the capital recovery factor are given in Table 9-4. As can be seen from the table as well as from Eq. (9-12), the capital recovery factor decreases with an increasing number of years (life of the repayment period) and increases with an increase in the interest rate, as would be expected. Other common symbols for the capital recovery factor are $\mathrm{crf}_{i,n}$, differing from our notation by the use of lowercase letters only, and *(A/P, i,n),* where A is the annual or annuity payment, P is the present amount, i is the interest rate, and n is the period.

The significance of the equivalent annual expenditure derived using the capital recovery factor is that if that expenditure is made in each year of the N years of the life of a facility, then at the end of the N years the original expenditure will have been paid in the sense that those N payments will yield a total amount of money identical to the amount that would have been obtained by retaining the original expenditure. The use of the capital recovery factor is similar to the concept of depreciation used in business financial practice, but there are differences (Winfrey, 1969, pp. 176–197). In accounting, depreciation is an allocated portion of a prepaid cost. In this sense, the use of the capital recovery factor can be considered one method of depreciation. However, depreciation has also been used by accountants and others as a means of accounting for a loss or a decrease in value of an item of property. Depreciation in this sense is quite different from the capital recovery factor. Since various methods of calculating depreciation result in different amounts of charges in each of the years of a facility's life, it is not possible to generalize precisely how the constant annual amounts obtained by use of the capital recovery factor will differ in all cases with an amount calculated by one of the depreciation methods. However, for purposes of engineering and planning, the use of the capital recovery factor is preferred to the use of any of the methods of depreciation.

In most engineering projects there is at least in principle some value of the facility or project at the end of its N-year life. This value, the terminal or salvage value, usually reflects the price that can be obtained for any salvageable or usable portions of the facility, less costs associated with any required dismantling, demolition, or return of the land to its original purpose. Such salvage values have received careful consideration in many engineering contexts, such as in connection with machinery whose salvage value can repre-

Table 9-4 Capital Recovery Factors† for Various Project Lives and Discount (or Interest) Rates

Life, N	Discount rate, percent per year					Life, N
	6	8	10	12	15	
1	1.060000	1.080000	1.100000	1.120000	1.150000	1
2	0.545437	0.560769	0.576190	0.591698	0.615116	2
3	0.374110	0.388034	0.402115	0.416349	0.437977	3
4	0.288591	0.301921	0.315471	0.329234	0.350265	4
5	0.237396	0.250456	0.263797	0.277410	0.298316	5
6	0.203363	0.216315	0.229607	0.243226	0.264237	6
7	0.179135	0.192072	0.205405	0.219118	0.240360	7
8	0.161036	0.174015	0.187444	0.201303	0.222850	8
9	0.147022	0.160080	0.173641	0.187679	0.209574	9
10	0.135868	0.149029	0.162745	0.176984	0.199252	10
11	0.126793	0.140076	0.153963	0.168415	0.191069	11
12	0.119277	0.132695	0.146763	0.161437	0.184481	12
13	0.112960	0.126522	0.140779	0.155677	0.179110	13
14	0.107585	0.121297	0.135746	0.150871	0.174688	14
15	0.102963	0.116830	0.131474	0.146824	0.171017	15
16	0.098952	0.112977	0.127817	0.143390	0.167948	16
17	0.095445	0.109629	0.124664	0.140457	0.165367	17
18	0.092357	0.106702	0.121930	0.137937	0.163186	18
19	0.089621	0.104128	0.119547	0.135763	0.161336	19
20	0.087185	0.101852	0.117460	0.133879	0.159761	20
21	0.085005	0.099832	0.115624	0.132240	0.158417	21
22	0.083046	0.098032	0.114005	0.130811	0.157266	22
23	0.081278	0.096422	0.112572	0.129560	0.156278	23
24	0.079679	0.094978	0.111300	0.128463	0.155430	24
25	0.078227	0.093679	0.110168	0.127500	0.154699	25
26	0.076904	0.092507	0.109159	0.126652	0.154070	26
27	0.075697	0.091448	0.108258	0.125904	0.153526	27
28	0.074593	0.090489	0.107451	0.125244	0.153057	28
29	0.073580	0.089619	0.106728	0.124660	0.152651	29
30	0.072649	0.088827	0.106079	0.124144	0.152300	30
31	0.071792	0.088107	0.105496	0.123686	0.151996	31
32	0.071002	0.087451	0.104972	0.123280	0.151733	32
33	0.070273	0.086852	0.104499	0.122920	0.151505	33
34	0.069598	0.086304	0.104074	0.122601	0.151307	34
35	0.068974	0.085803	0.103690	0.122317	0.151135	35
36	0.068395	0.085345	0.103343	0.122064	0.150986	36
37	0.067857	0.084924	0.103030	0.121840	0.150857	37
38	0.067358	0.084539	0.102747	0.121640	0.150744	38
39	0.066894	0.084185	0.102491	0.121462	0.150647	39
40	0.066462	0.083860	0.102259	0.121304	0.150562	40
45	0.064700	0.082587	0.101391	0.121163	0.150489	45
50	0.063444	0.081743	0.100859	0.121037	0.150425	50
55	0.062537	0.081178	0.100532	0.120925	0.150369	55
60	0.061876	0.080798	0.100330	0.120825	0.150321	60
65	0.061391	0.080541	0.100204	0.120736	0.150279	65
70	0.061033	0.080368	0.100127	0.120417	0.150139	70
75	0.060769	0.080250	0.100079	0.120236		75
80	0.060573	0.080170	0.100049	0.120134		80
90	0.060318	0.080079	0.100019			90
100	0.060177	0.080036	0.100007			100

† $CRF_{I,N} = \dfrac{I(1+I)^N}{(1+I)^N - 1}$.

sent a significant amount of money. However, in the transportation facilities context, it is rather difficult to speak of a salvage value. This is partly because most of the facilities in common use at the present time either are not abandoned, so that we have no real knowledge of their salvage value (or for that matter of the cost of demolition and returning the area to some other use), or where facilities such as railroad lines and canals have been abandoned, the land and fixed plant have often been of so little value that their salvage value is of little importance. Therefore, in most transportation analyses, extensive reference is not made to the salvage value. If that value is believed to be significant in a particular context, then the appropriate capital recovery factor for the salvage value should be calculated from the following equation:

$$\mathrm{CRF}_{I,N,S} = \frac{I(1+I)^N}{(1+I)^N - 1} - \frac{I}{(1+I)^N - 1}\left(\frac{S}{Z}\right) \qquad (9\text{-}13)$$

where S = salvage occurring at end of year N
Z = initial capital expenditure

A rather detailed discussion of the inclusion of salvage value in transportation projects and the small effect that these typically have on projects with long lives and current rather high discount rates is presented in Winfrey (1969, pp. 164–165).

Treatment of Inflation

Inflation, the increase in the price of items from year to year, has been a characteristic of most economic activity for many years, and recently the rate of increase in prices has been very pronounced. This raises an important question in estimation of future cost, namely, how should possible future changes in cost be taken into account in estimating future costs?

To answer this question, it is necessary to distinguish between two types of inflation. In one case, the rate of change in prices for different items is essentially the same, with the result that the percentage increase in the price of any one item from one year to the next will be about the same as for any other item. This type of inflation might be termed general price inflation. The other type of inflation is one in which the prices or costs of particular items are increasing more rapidly than others, with the result that the percentage change in price of one item from one year to the next is quite different from the percentage change for another item. As a result of this latter type of inflation, the relative prices of goods change over time.

If the inflation is a general inflation, then usually it is best not to attempt to include inflation or price increases expected in the future in the analyses. While this may seem very strange at first reading, the reasoning is quite sound. By doing this, the prices of all items will be equally understated for any year into the future, and hence any decision regarding which items to purchase or any other type of decision depending upon cost will still be made on the basis of the same relative prices of the various items in the future, and presumably the decision will be the same. Of course, to be consistent, any other monetary measure entering into such decisions, such as a measure of revenue or profit for private firms or a general measure of benefits in the public sector, would also have to be based upon prices in effect in the same year, rather than prices increased by an expected inflation. The basic reason for not attempting to include inflation (since to include it or not under these circumstances would not change the decision), is that it has been found extremely difficult to predict inflationary rates in the future. Thus, the problems associated with such prediction and the uncertainty it brings can simply best be avoided.

On the other hand, if the inflation is selective, and results in changes in the relative price of items, then consideration should be given to including inflation in the analysis. However, for inflation to be included, two conditions must hold. First, it must be possible to predict the different rates of inflation associated with different items of cost. Secondly, an item which has experienced an unusual inflation, either high or low, should be a sufficiently sizable component of the total variable cost in the analysis to make a difference in the final conclusion based on the analysis. Otherwise, there is no need to add the complication of different inflation rates.

Because it is so difficult to predict inflationary rates, and small differences in the rate of inflation between the prices of different items could result in different conclusions or decisions, it is essential to be very explicit regarding the assumptions made. Also, it is very helpful to vary the assumed inflationary rate slightly, to ascertain the effect on cost. In practice, in transportation engineering and planning, it is usually customary to use prices of the most recent year for which they are available or some other agreed upon year's prices for all analyses and to not include inflationary adjustments to these

Table 9-5 Recent Inflation in Selected Transportation Costs

	1950	1960	1967	1970	1973	1975
Consumer price index[a]	72.1	88.7	100.0	116.3	133.1	161.2
Wholesale price index[a]	81.8	94.9	100.0	110.4	134.7	174.9
Highways						
Automobiles, new[c]	83.4	104.5	100.0	107.6	111.1	127.6
Composite federal aid construction[d]	66.6	80.1	100.0	125.6	152.4	203.8
Roadway maintenance and operation[e]	51.31	78.35	100.0	116.78	145.75	u[b]
Roadway maintenance and operation labor[e]	43.58	71.02	100.0	122.02	148.04	u
Railroads[f]						
Wage rates	44.8	74.3	100.0	125.6	173.9	207.2
All materials	69.5	95.1	100.0	109.7	126.7	227.2
Fuel (coal and oil)	88.4	98.4	100.0	110.5	136.5	321.9
Airlines[g]						
Total cost	u	85.9	100.0	122.0	143	u
Fuel	u	u	100.0	105.8	123.1	232.7
Rail transit cars[h]	40	75	100	105	155	u

[a] U.S. Department of Commerce (1976, pp. 437 and 441).
[b] u indicates unavailable.
[c] U.S. Department of Commerce (1976, p. 440).
[d] Federal Highway Administration, (1976, p. 2).
[e] U.S. Dept. of Transportation, Federal Highway Administration (1973, p. 117).
[f] Association of American Railroads (1976, p. 61).
[g] Air Transport Association of America (1975, p. 11).
[h] Estimated from Boyd et al. (1973, pp. A-36 and A-37).

prices. However, with rapid changes in certain prices, the use of inflationary adjustments may become more common.

To see the relative magnitude of inflation in selected transportation costs, Table 9-5 has been developed. It shows the price index of various items from 1950 to 1974, with the price in 1967 for each item being the base, with an index equal to 100 in all cases. The price index for consumer goods and for wholesale items increased approximately 35 percent from 1967 to 1973, while almost all the items portrayed in the transportation sector increased more rapidly. In particular, the cost of construction and of fuel seemed to have increased the most rapidly, as would be expected. It is noteworthy that the cost of labor does not seem to have increased much differently from the cost of other factors of production, contrary to widespread belief. Perhaps it is merely the ease with which labor costs per unit of work (or per unit of output) can be determined, often simply by an agreed upon wage rate with fringe benefits, that focuses attention on this one component of cost. As can be seen from this table, the types of items subject to the greatest inflation vary from one sector to another.

COST-ESTIMATING METHODS

Cost estimates are used in almost all aspects of engineering and planning for the evaluation of alternative designs and other options, and therefore it is important that professionals have at their command means for estimating costs. There are basically two approaches to estimating costs, although in practice a combination of both is often used.

The distinction between these can be understood by reference to Fig. 9-4, which portrays the manner in which costs, or the consumption of various resources, are incurred. As we saw in Chap. 8, the transportation capacity or output, such as passenger-miles, of a system can be related to the number of its constituents, such as vehicles, containers, the size and design of links, interchanges, etc. Each of these items represents a scarce resource which must be purchased. The price of each of these various resources, multiplied by the number consumed, summed over all resources yields the total cost of the transportation system.

One of the two types of cost estimation, the so-called engineering unit cost method, actually traces through this process, first estimating the amount of physical resources needed and then applying prices to yield the total cost. The other approach, termed the statistical cost or cost-output method, relates costs and transport service (output) provided, bypassing the need to develop an explicit model of the particular resources used.

Statistical Cost Models

Statistical cost models are developed with the aid of data on the costs incurred in actual transport systems. The usual procedure is to specify an expected mathematical relationship between cost and output, in which the functional form of the relationship is specified but the numerical values of the parameters are not. Then data on the actual costs incurred for the types of systems being considered are examined, and the parameter values are esti-

Figure 9-4 **Components of cost estimation.** [*From Morlok (1967), p. 8*]

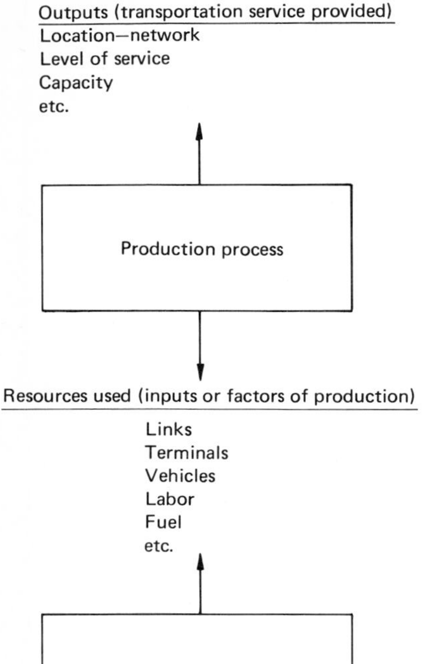

mated, usually using statistical regression or related methods. These methods also involve the development of measures of the degree to which the model with its parameter values determined actually reproduces the costs incurred in actual operation. Often if the original hypothesized model with the initial estimate of parameter values does not adequately predict or reproduce these costs, the model is modified or refined until a satisfactory degree of correspondence is achieved.

These rather abstract ideas can be illustrated by consideration of a specific example. The problem was to develop a model for the estimation of total cost of intercity bus service. Total cost was to be divided into four components, each of which would be estimated separately, these components being the cost of owning the vehicles, the cost of owning and operating terminals, the cost of vehicle operation (termed transportation cost in that industry), and other, primarily administrative, costs. It was hypothesized that these costs depend upon the number of bus-miles operated. This was felt to be a plausible hypothesis, particularly since the range in size of buses operated by intercity carriers is rather minimal, and there seems to be little difference in quality of service or other factors which might make particular carriers incur abnormally high costs. The number of buses which had to be owned to provide a given amount of service was hypothesized to vary with the bus-miles simply because the amount of peaking and characteristics of traffic other than the total volume seem to be fairly uniform throughout that industry, and most services seem to use freeways and other high-quality highways to a large extent, resulting in about the same average vehicle speeds. Hence about the same number of bus-miles would be operated per vehicle regardless of the firm or location. Also, it was expected that the cost of an individual bus was about the same regardless of the firm, location, etc. The cost of stations or terminals was assumed to essentially depend on the amount of traffic, bus companies typically tailoring the size of terminals to suit the volume of traffic, to the extreme that at very low volumes the terminal functions would be performed on a commission basis by a drugstore, restaurant, or hotel in a community. If it can be assumed that the average distance traveled per passenger and average load per bus is fairly constant, then the station costs should depend essentially upon vehicle-miles also. Administrative and other costs probably vary with the size of the organization, and an appropriate measure of size was hypothesized to be vehicle-miles. Thus the form of this model of cost is linear, a very simple form, as indicated below:

$$\mathrm{TC} = \sum_{i=1}^{4} \mathrm{TC}_i \qquad (9\text{-}14)$$

$$\mathrm{TC}_i = B_i m_r$$

where TC_i = total cost of ith category/year
B_i = parameter, $/bus-mi
m_r = total operated bus-mi/year
i = index of cost category (1 = vehicle operation, 2 = vehicle ownership, 3 = terminals, 4 = other)

In order to estimate the parameters of this equation for each of the four cost components, data on the cost of intercity bus operations in the United States as reported by the Interstate Commerce Commission for the years 1968 and 1969 were used. The two years were used in order to provide enough data points for a test of the model. During this period inflation in this industry was rather minimal, so price changes should not affect costs abnormally. The data contain the cost in each of these four categories incurred by all bus companies in each of nine regions, resulting in 18 data points for each cost category. The single parameter for each of the four cost-estimating relationships was estimated by finding that parameter which minimized the sum of the squares of the deviations between the predicted costs and the actual costs incurred. The deviation for each data point is the difference between the actual cost and the cost predicted by the estimating equation at the same actual output (bus-miles). A deviation is calculated for each data point, then squared, and added to the sum. This is simply a minor variation of ordinary least squares regression methods. The resulting four equations are presented on page 372 along with information on the degree to which the predicted values corresponded to the actual values.

$TC_1 = \$0.3293 m_v$ (Bus operation)
$D_{10\,\text{percent}} = 50.3\%$ $r^2 = 0.9794$
$D_{20\,\text{percent}} = 82.9\%$ $SD_{\text{percent}} = 14.64\%$

$TC_2 = \$0.0429 m_v$ (Bus ownership)
$D_{10\,\text{percent}} = 42.8\%$ $r^2 = 0.9757$
$D_{20\,\text{percent}} = 75.0\%$ $SD_{\text{percent}} = 17.41\%$

$TC_3 = \$0.1022 m_v$ (Station)
$D_{10\,\text{percent}} = 34.7\%$ $r^2 = 0.9682$
$D_{20\,\text{percent}} = 62.7\%$ $SD_{\text{percent}} = 22.48\%$

$TC_4 = \$0.1607 m_v$ (Other)
$D_{10\,\text{percent}} = 34.0\%$ $r^2 = 0.9476$
$D_{20\,\text{percent}} = 61.6\%$ $SD_{\text{percent}} = 22.97\%$

The percent of actual data points which lay within ±10 percent of the line and those which lay within ±20 percent of the line are presented ($D_{10\,\text{percent}}$ and $D_{20\,\text{percent}}$, respectively). Also given is the coefficient of determination, r^2, and the percent of standard deviation (expressed as a percentage of the mean), SD_{percent}, two common measures of statistical correspondence which those familiar with regression methods can interpret. Basically, a value of r^2 close to 1 tends to indicate good correspondence and a percent standard deviation approaching 0 (as opposed to 100 percent) tends to indicate a good correspondence also. As can be seen from this information, the models seem to perform well in estimating total costs.

The correspondence between this model and the actual costs incurred can

Figure 9-5 Example of statistical cost model: intercity bus service costs in the United States [*From Bolusky et al. (1973), vol. III.*]

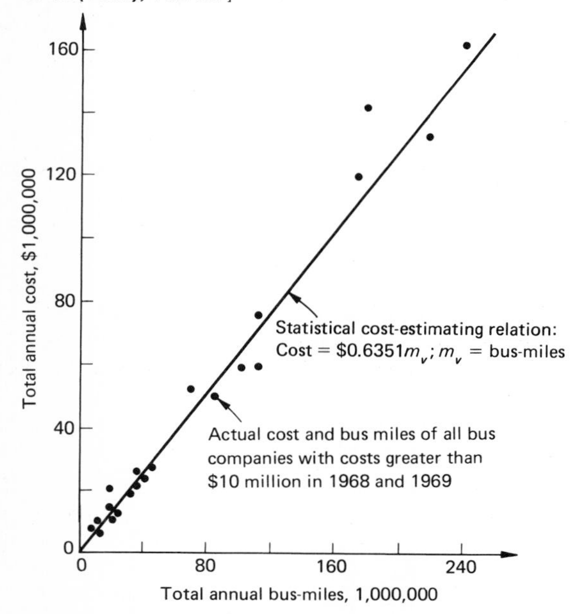

also be visualized from Fig. 9-5, which presents total costs as a function of vehicle-miles. The total cost is estimated by simply summing the estimates of the four component costs described above. It is interesting to note that 78 percent of the bus companies had actual costs which were within ±10 percent of the value estimated by this total cost equation, and 98.6 percent had costs which fell within ±20 percent. For the purpose which this model was intended, a general comparison of the costs of bus service with the cost of other types of intercity short-haul transport service (along with a comparison of other characteristics such as level of service and capacity), this model was felt to be sufficiently accurate and was used. However, if one wished to examine certain types of variations in bus service, this type of model relating costs to output would be inappropriate. For example, if you wish to estimate the cost with a radically different type of bus, such as an articulated bus, seating many more passengers than conventional buses, then certainly the vehicle ownership and vehicle operation portions of this cost model would have to be modified. Also, if you wish to estimate the costs of a luxury bus service, with wider seats (perhaps two on one side of the aisle and one on the other), with refreshment service, etc., then one would expect the operating costs and perhaps other costs to increase. Also, if you wish to provide terminal facilities comparable to the usual terminal facilities for air travelers, then that component of cost would have to be modified. Thus, this type of model is useful for predicting costs of services which are essentially identical to those on which the model is based, but the model is not usable in this form in situations with deviations from that type of service.

Not all statistical cost models have as simple a functional form as the model discussed above. Also, statistical methods might be used to estimate only a portion of total costs in a particular transportation system, perhaps a portion for which development of an engineering cost model would be very difficult because of the many different factors of production involved and perhaps because of uncertainties in the amount of particular factors required.

An example of this is the cost of highway construction, which depends upon numerous characteristics of the highway itself as well as the area through which it is to be constructed. One particular model estimating the total cost of constructing highways includes the cost of purchasing right-of-way, demolition of any buildings, preparation of the land, and actual construction of the freeway (Kane and Morlok, 1970). Numerous factors influence this cost, including the number of lanes of the facility and the location, a distinction being made between three areas of the United States—area 1 being the northeast and northcentral states, area 2 being the southeast and southcentral area, and area 3 being the western area. Also influencing costs is the terrain, which can be described by a relief measure which compares the actual distance along the surface of the earth between two points and the short line or great circle distance between these two points. This measure is commonly used by geographers to describe the terrain characteristics of an area. A further variable is the value of the land, for which the measure of average property tax revenues per unit area was used. Finally, a measure of the intensity with which structures would have to be demolished was incorporated, in the form of gross residential density, the number of persons per square mile, although it should be noted that a preferred measure would include all types of activities, although none was available.

The models were developed with various functional forms and tested against data using standard statistical regression methods. The models which appeared to perform best are presented below for each of the three regional areas along with the multiple r^2 value and the standard error of the estimate.

Area 1:

$$TC = -3.948 + 1.11814(NL) + 0.01092(DEV) + 0.00012(GRD) - 0.10074(RLF) \qquad r^2 = 0.6962 \qquad SE = 2.060$$

Area 2:

$$TC = -6.518 + 1.12534(NL) + 0.00014(GRD) + 0.22255(RLF) + 0.03546(DEV) \qquad r^2 = 0.6387 \qquad SE = 1.485 \qquad (9\text{-}15)$$

Area 3:

$$TC = -13.82 + 1.63440(NL) = 0.01250(DEV) + 0.22283(RLF) + 0.00074(GRD) \qquad r^2 = 0.7626 \qquad SE = 2.993$$

where TC = total construction cost, $ million/mi
NL = number of lanes (sum of both directions)
DEV = annual property tax revenue density, $ thousand/mi^2
GRD = gross residential density, persons/mi^2
RLF = relief or terrain measure (defined in text), dimensionless
SE = standard error

Although r^2 values may seem rather low, they are typical of r^2 values obtained in models of complex processes and, it should be noted, are considerably higher than the r^2 values of other road cost models tested against the data which had r^2 values of the order of 0.12 to 0.31. For purposes of approximate estimates of costs of new freeway facilities, based on a minimum of variables, such as might be desired in the context of initial freeway planning in a region, these models would be quite adequate. However, if very accurate cost estimates are required, many additional factors would have to be taken into account, and perhaps engineering unit cost methods would be preferred.

Engineering Unit Cost Models

As mentioned above, engineering unit cost models differ from statistical models in that the technology is explicitly considered. Specifically, the first step is to develop a relationship between the various scarce resources to be used and the nature of the transportation capacity and service to be provided. Such a model would be developed on the basis of the kinds of relationships presented in Chaps. 4 through 8. In fact, the relationships developed for the single rapid transit line in the previous chapter are good examples of the types of relationships needed. The additional item required is a relationship or price for the various factors of production, such as the cost per vehicle, the cost of providing a mile of track, etc.

A rather common classification of factors of production included in this type of cost model has been developed, although there are some variations from mode to mode and from context to context depending on the purpose of cost estimation. Typically factors such as those presented in Table 9-6 are used. In some cases, individual factors which are purchased as complete units are considered, such as vehicles and containers. On the other hand, some items which are really made available as a result of purchasing many different factors are treated as single items, such as the cost of a way link. In

Table 9-6 Typical Composite Factors Used in Engineering Costing of Transportation Systems

Factors	Characteristics
Fixed facilities	
Way links	Length, number of channels, design standards (e.g., freeway vs. arterial street)
Way interchanges	Number, design standards (e.g., at grade vs. cloverleaf)
Terminals	Number, size
Land for right-of-way	Area, by location (related to value)
Maintenance facilities	Vehicle capacity, amount of fixed plant to be maintained
Vehicles and containers	
Vehicles	Number by type (e.g., capacity, speed)
Containers	Number by type (e.g., capacity)
Operation and maintenance	
Vehicle and container movement	Vehicle-mi and container-mi, by type and location; less commonly vehicle-h or container-h, by type
Maintenance of way	Vehicle movement, usually including weight (e.g., gross ton-mi)
Administration	Varies: constant percent of all other costs, or a function of total activity (e.g., total vehicle-mi or passenger-mi)

practice, many different contracts are let for different components of way links, and each of these might be treated separately. However, for ease of construction and application of the cost model, often all of these are combined into one single cost of the way link which might include purchase of the right-of-way, preparing the land, construction, provision of any appropriate control devices, etc. To avoid confusion with more indivisible factors, these are often termed composite factors. Also shown in Table 9-6 are the types of characteristics of each of these factors which are often found to influence significantly the cost and hence which must be taken into account in any cost-estimating relationship.

The basic approach of engineering unit cost methods, then, is to develop relationships which enable the estimation of the number and type of each of the factors listed in Table 9-6. Then costs of each are estimated. An example of such a cost-estimating relationship is the model of cost of a freeway mile as described earlier—one developed by statistical methods. Such a cost-estimating relationship could be used to estimate the cost per mile of freeways in a particular metropolitan area, which could be multiplied by the miles of such freeways (taking into account number of lanes, terrain, etc.) to yield the total cost of way links of that type in that region. This cost would then be added to the cost of various other factors to yield total cost.

The advantages of the engineering unit cost approach are essentially that it enables exploration of changes in the technology and also examination of particular components of costs. Since the technological relationships are explicitly taken into account, any change in the technology can be treated, as long as the price of any new or modified items can be estimated or ascertained. Also, it is very easy with this type of cost model to estimate marginal cost, variable costs, etc., since this usually is simply a matter of combining the costs of certain portions of all of the factors of production used. These concepts can be illustrated by considering an example.

Example of Rapid Transit Costs

In order to illustrate the use of engineering cost models, we will develop the total cost and various cost components of the rail rapid transit line which served as an example for describing operations plans and integrating the various relationships from Chaps. 4 through 7, in Chap. 8. As you may recall, this line is 15 mi in length and is to be operated with a required capacity of 32,000 persons/day in each direction. During each of the two 2-h peak periods, 8000 passengers/h in the peak flow direction are to be accommodated; this is done by operating 5-car trains, each with a capacity of 500 passengers at a headway of 3.75 min, for a train flow volume of 16 trains/h. During the other periods of the 16-h operating day, 3-car trains, seating a total of 156 passengers, are operated on an 8-min headway for a capacity of 1170 passengers/h, somewhat in excess of the required 1143 passengers/h. To operate this service requires a fleet of 120 cars. During each weekday, 4620 train-mi are operated, and 17,700 car-mi are operated. On this line there are 30 stations.

The typical form in which unit costs are used for the estimation of transit costs is shown in Table 9-7. Fixed facility costs are estimated on the basis of the costs of two-track lines recently constructed. In 1973 the cost of surface, grade-separated line construction averaged $4.22 million/mi. Using a life of 30 years and an interest rate of 10 percent, which results in a capital recovery

Table 9-7 Unit Costs for Example Rail Rapid Transit Line†

Category	Daily cost		Cost		
	$/route-mi	$/car	$/car-mi	$/crew-day	$/station-agent-day
Capital stock					
Guideway (at grade)	1226.45				
Stations (2/mi)	581.26				
Cars		105.64			
Yards and shops		5.40			
Subtotal	1807.71	111.04			
Operating and maintenance					
Maintenance of way and structures	28.77		0.239		
Maintenance of cars			0.211		
Vehicle operation			0.443		
Train crews (2 people)				121.92	
Station agents					37.54
Administration and other	2.88		0.089	12.19	3.75
Subtotal	31.65		0.982	134.11	41.29
Total expenses	1839.36	111.04	0.982	134.11	41.29

† Adapted from Morlok (1965) and Sanders et al. (1974), both in 1973 prices.

factor of 0.106079, the cost per day (365 days/year) of such a guideway is \$1226.45. Stations for 5-car trains cost \$1 million each, which, with the same life as the guideway, results in the cost shown. The cost of vehicles of the type used (50 seats, with room for about 25 standees) was about \$300,000, which, with a 25-year life, a 10 percent interest, and a maintenance reserve of an additional one-sixth of the fleet, yields a daily cost of \$105.64. Closely related to this cost is the cost of yards and shops for the cars, which, at \$15,000/car and a life of 15 years, results in the costs shown. Thus the cost of fixed facilities and of vehicles is directly related to the miles of double-track route, the number of stations, and the number of cars required.

Operating and maintenance costs are divided into a number of components. In transit, usually the costs of operating labor—train operators, guards, and station attendants—are separated from the other costs, primarily because these costs depend very much upon the peaking of traffic and the size of trains operated, items which may vary from year to year and with management policies. The maintenance of track, structures, and stations has been found by statistical analysis to depend in part upon the passage of time and in part upon the traffic on the line. Thus this cost consists of two components, \$28.70/day/double-track mi and \$0.239/car-mi. The maintenance of cars is a function of their usage, as shown. Other costs of vehicle operation, mainly electricity and the costs of various train control activities, is \$0.433/car-mi. Train crews, consisting of a motor operator and a guard, cost \$121.92/day, with the requirement that once called on duty, a crew must be paid a minimum of 8 h. Also, in this particular case, crews are allowed to work for any length of time up to and including 8 h within a period of 12 h, enabling one crew to cover both morning and evening peak operations with a rather lengthy break in between. On some transit properties, the fraction of crews that work such split or "swing" shifts is limited to 30 to 50 percent. The cost of station agents depends upon the number of such agents, each of whom works an 8-h day. Finally, there is the administration cost, which is taken as 10 percent of all other operating costs. The resulting costs, then, are unit costs which are multiplied by the number of the various units involved, such as double-track route-miles or car-miles. The exact form of the resulting cost equation is

$$\text{TC} = \$1839.36L + \$111.04c + \$0.982m_c + \$134.11d_c + \$41.29d_a \qquad (9\text{-}16)$$

where TC = total cost of this rail transit line, \$/day
L = one-way line length, mi
c = number of cars required
m_c = number of car-mi/day
d_c = number of paid train-crew-days/day
d_a = number of paid station-agent-days/day

In order to use these unit costs and the above model, all of the factors or units are known from the analysis previously conducted in Chap. 8 with the exception of the number of crew-days and station-agent-days. Since there are 30 stations, each of which is open 16 h/day, a minimum of 60 station-agent-days/day of operation will be required if fares are to be collected at the stations (although on some rapid transit lines fares are collected on board the train at lightly used stations during low-traffic periods). Here we shall assume 60 station-agent-days are required per day; 12 trains are in operation throughout the 16-h operating period at any one time, with an additional 12 trains being in operation during each of the two peak periods. Given the possibility of swing shifts, a reasonable way to staff these trains is to employ two groups of 12 train crews each, one for the first 8 h of operation and one for the second 8 h of operation, plus additional 12 crews on a swing shift covering both peaks. This results in a need for 36 train-crew-days/day.

With these computations of the number of units or factors required, Eq. (9-16) can be entered and the resulting total costs estimated, as below.

$$\begin{aligned} TC = {} & \$1839.36/\text{route-mi-day}(15 \text{ mi}) + \$111.04/\text{car-day}(120 \text{ cars}) \\ & + \$0.982/\text{car-mi}(17{,}700 \text{ car-mi/day}) \\ & + \$134.11/\text{crew-day}(36 \text{ crew-days/day}) \\ & + \$41.29/\text{station-agent-day}(60 \text{ station-agent-days/day}) \\ = {} & \$65{,}601.96/\text{day} \end{aligned}$$

Thus the total costs of operation of this system are $65,601.96/day. This results in a cost per seat-mi of exactly 7 cents, where the seat-miles are estimated on the basis of the required levels of capacity, some of which are exceeded for operating convenience. It is interesting that of this total cost, $40,915.20 is due to the cost of fixed facilities, the minimal maintenance required to keep the line ready for use, and the cost of ownership of the cars, this being about 62 percent of total cost. Only 37.6 percent of the total cost is devoted to the actual operation of these trains, labor costs, etc. For more typical rapid transit lines, which involve some elevated or subway construction at a much higher cost than $4.22 million/double-track mi and $1 million/station, the ratio of fixed costs to variable costs would be even greater.

Another interesting analysis is that of the marginal cost of accommodating additional traffic in the peak and base periods. In the midday period the cost of operating an additional train trip is only $88.38, this being the additional cost associated with the additional car-miles operated, since both vehicles and crews are available and stations are fully staffed in that period. This results in a cost of 29.5 cents/seat in each direction. In the peak period, however, to operate an additional train requires the purchase of the cars for that train and the hiring of an additional crew. If a 3-car train were added, the additional cost for two round trips, one in the morning and one in the evening, would be $643.99. The cost per seat provided in the peak direction would be $1.073, all the cost being associated with the peak direction since excess capacity is provided in the other. This graphically illustrates the disparity in marginal costs of accommodating additional traffic between the peak and the midday period, which is often the reason why transit lines charge much lower fares in the midday period.

Figure 9-6*a* presents the relationship between total cost of this transit operation and the daily volume of traffic. Trains are operated at the same headway as assumed for the preceding example, with train length being varied to achieve variations in peak capacity. Also, where possible, trains during the off-peak period are reduced in length, under the assumption of the same ratio of required capacity in the midday period to that in the peak as was used in the preceding example. The cost relationship takes the form of a step function because of the discrete variations in train length possible. Also shown in Fig. 9-6*b* is the cost per seat-mile as required by capacity considerations. The high portion of fixed cost is very evident, but it results in a cost

Figure 9-6 **Example rapid transit line total and average daily costs and capacity.** ***(a)*** **Total daily cost.** ***(b)*** **Average cost per seat-mi or place-mi (including standees) in peak periods. Note: Capacity is measured past the peak-load point, and is the capacity required by traffic, which may be exceeded by actual capacity due to inability to adjust train capacity to exactly that required and maintain desired headways.**

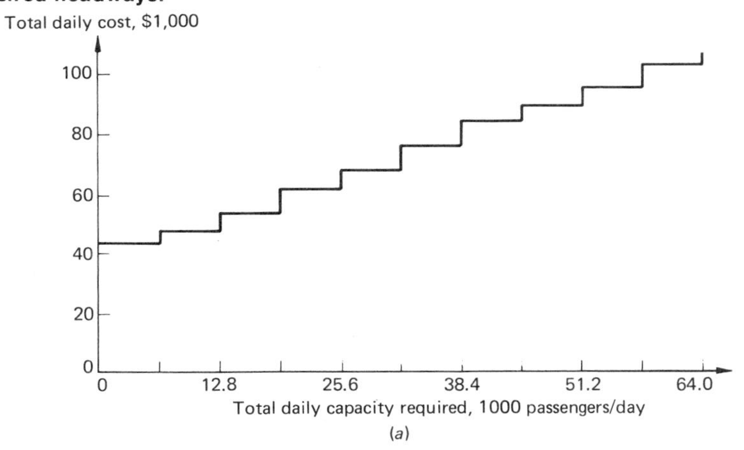

(*a*)

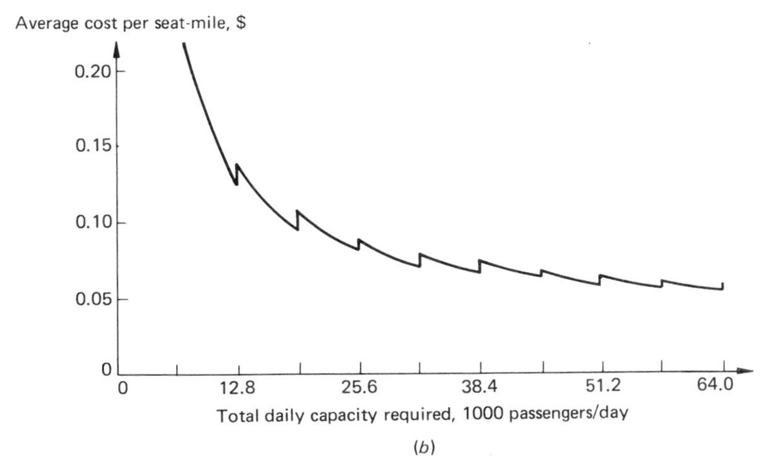

(*b*)

per seat-mile which drops markedly with increasing traffic. It should be noted that in general the utilization of seat-miles on urban public transit is quite low, because of the inherent peaking and directional split of traffic, as discussed in Chap. 8, which results in a cost per passenger-mile often three or more times the cost per seat-mile. The magnitude of these costs and the effect of increasing required capacity (or traffic) graphically indicates why rapid transit lines are generally limited to routes with fairly high volumes of passenger traffic.

Now that the two approaches to cost estimation and the examples have been introduced, it is appropriate to turn to typical cost models.

STANDARDIZED COST MODELS

In almost all aspects of transportation where engineers and planners are engaged in analyses of alternative policies, plans, and designs, it has been found useful to have standardized cost models which can be used readily for different types of analyses. These models are standardized in the sense that they are developed for future application to different problem situations and are designed to be used with ease, requiring a minimum of data gathering on the part of the user. In some cases the standardized models or approaches to cost estimation are developed with a view toward their mandatory use by certain agencies, e.g., the United States federal government requires the use of largely standardized cost-estimating methods if states and other governmental organizations are to receive federal aid for road construction and maintenance. In other cases, the use of certain models is not required, but they have become widely used in the calculation of certain types of costs. Nevertheless, by providing the engineer and planner with cost-estimating tools which are relatively easy to use, the problems of cost estimation are reduced considerably.

There are many different types of standardized cost models, even within a particular mode or context. Many of the differences reflect the various types of problems to which the models might be applied, ranging from situations requiring very detailed analyses of costs to reflect subtle differences between alternative designs, to aggregate analyses designed to yield only very approximate estimates. Typically the models designed to estimate costs with great precision are of the engineering unit cost variety and take explicitly into account the technological relationships underlying the provision of transport services and the resources required. In many cases, the cost-estimating relationships for some of the factors of production in these models are based upon statistical cost-estimating methods. On the other hand, models designed to yield more approximate results are usually of the statistical variety, estimating costs directly from characteristics of the transport service provided with only minimal reference to the technology (e.g., specifying only the mode and not more detailed design characteristics of the system). The engineering unit cost models of course require much more effort on the part of the user than the statistical models. They take more time as well as require more information on the transport systems being analyzed. Therefore the engineer or planner must take care to choose a cost-estimating method appropriate to the problem. To use a more detailed and exacting model than required represents a waste of professional time, even though it may yield in some sense a better estimate.

To illustrate the nature of the standardized cost models and their use we shall concentrate upon a road example. The reason for this is partly that since the field of road transportation engineering has probably made more use of such models than any other, they are well developed and understood. Also, the models for road transport reflect various technological characteristics of the transportation system, and hence our discussion of these models serves to relate costs to various factors discussed in previous chapters and to illustrate how engineering principles are used to aid in the estimation of costs. However, the emphasis will be upon principles applicable to all modes of transportation and all contexts, rather than to the details of road transport. At the end of this section, we shall present briefly methods which have been developed for use in freight transportation operations, mainly rail and truck, since these incorporate a few additional factors not found in the road transport model.

Road Costs

Very extensive standardized cost models have been developed in the highway field, primarily under the influence of national governments which wish the various local governments to whom they allocate funds for road construction and maintenance to use those funds in the most economical manner. To ensure that the benefits of road improvements are accurately and consistently calculated for purposes of allocation of available monies among different construction and maintenance projects, standardized models have been developed in many nations. In the paragraphs below we shall focus upon a set of models developed for application in the United States, which were selected primarily because they relate costs to characteristics of road design and vehicle flow.

One of the primary criteria upon which the decisions regarding road improvements have been made is the total cost of transporting persons and freight via highway. Total transportation cost, as the term is used in the highway field (and to a lesser extent in other ares involving substantial public expenditures such as urban public transportation and air transport) is an approximation to the total cost to society for transportation via road. It includes the expenditures for right-of-way and construction and maintenance of the highway (often together termed system costs because they are incurred by the owner and maintainer of the highway system, one or more government agencies). It also includes user costs, which include the cost of operating and maintaining the motor vehicles, paying drivers of commercial vehicles, and similar items for which an actual monetary transaction occurs. In addition, user costs include certain other items representing scarce resources used in highway transport but for which there may not be direct monetary payment, such as the time consumed by highway users. Also, typically total transportation cost includes the cost of highway accidents accruing to users and nonusers, for many of which there is a direct monetary payment of hospital bills, repair to property, compensation of injury and loss of life, legal fees, etc. For some of these there may be no such monetary payment, or perhaps more importantly, the monetary payment may not truly reflect the loss incurred. While in some cases certain additional items are included in total transportation costs, these three—highway construction and maintenance costs, user costs, and accident costs—are the most typical components. It was emphasized earlier, and merits repeating, that these costs by no means represent the total costs of highway transport, for obviously they do not include damage due to air pollution or noise from the highway vehicles, reduction of property value due to aesthetic considerations, etc. Nor, of course, do they include some of the possible benefits from highway improvement which go beyond gains for the system owner, users, or those involved in accidents, such as general benefits to the regional economy from improved transport. Thus the term total transportation cost, although widely used, is somewhat misleading. We now turn to discussing its components.

Table 9-8 Unit Costs of Road Right-of-Way and Construction in $ million/lane-mi†

Construction	Location	Construction (excluding right-of-way)					Right-of-way				
		Urban area population (1000s)					Urban area population (1000s)				
		0–100	100–250	250–500	500–1000	Over 1000	0–100	100–250	250–500	500–1000	Over 1000
New freeway	CBD	1.04	1.07	1.10	1.14	1.24	0.36	0.43	0.54	0.72	1.11
	Fringe	0.71	0.75	0.81	0.94	1.20	0.36	0.39	0.43	0.54	0.72
	Residential	0.62	0.62	0.65	0.71	0.84	0.32	0.36	0.36	0.46	0.64
Major freeway widening	CBD	1.08	1.10	1.20	1.27	1.48	0.28	0.30	0.34	0.39	0.51
	Fringe	0.78	0.81	0.88	1.04	1.33	0.14	0.17	0.20	0.24	0.34
	Residential	0.62	0.65	0.70	0.79	0.98	0.04	0.06	0.09	0.16	0.30
New arterial street	CBD	0.33	0.36	0.39	0.46	0.56	0.32	0.36	0.41	0.52	0.75
	Fringe	0.29	0.30	0.33	0.36	0.46	0.29	0.31	0.33	0.39	0.49
	Residential	0.26	0.27	0.29	0.33	0.38	0.17	0.18	0.19	0.21	0.27
Major arterial widening	CBD	0.33	0.36	0.39	0.42	0.55	0.16	0.19	0.23	0.33	0.54
	Fringe	0.31	0.33	0.34	0.39	0.45	0.16	0.19	0.22	0.24	0.32
	Residential	0.31	0.33	0.34	0.39	0.45	0.09	0.10	0.11	0.16	0.26

† Sanders et al. (1974, pp. 4-10 to 4-17), 1973 prices.

Various models have been developed for estimating the costs of constructing highways. One such set of models, developed by statistical regression, was described earlier. Another set of cost-estimating relationships has been developed on the basis of an analysis of data from numerous urban highway construction projects and road improvement projects throughout the United States. These costs are presented in a rather useful tabulation of costs and other characteristics of urban highways and transit systems developed jointly by the Urban Mass Transportation Administration, the Federal Highway Administration, and the Office of the Secretary of Transportation, entitled *Characteristics of Urban Transportation Systems* (Sanders et al., 1974). The cost model is of a very simple mathematical form, it estimates costs on the basis of lane-milès constructed or improved, the costs depending on the size of the metropolitan area and the location within it (central business district, fringe, or residential area), these being indicators of the cost of land for right-of-way and of land preparation (e.g., building demolition) costs. The cost for new roads and major widening in millions of dollars per lane-mile, for land (right-of-way) and construction, are presented in Table 9-8 for two types of facilities, freeways and arterial streets. These costs conform to 1973 price levels. These parameters are used in the equation below to estimate the total cost of any improvement. To obtain the equivalent annual expenditure, this cost is then multiplied by the capital recovery factor, as detailed earlier.

$$C_i = \mathrm{UC}_i NL \qquad (9\text{-}17)$$

where C_i = road cost of type i
UC_i = unit cost per lane-mile
N = number of lanes
L = length of road segment

Typical lives for new roads are of the order of 20 years for the pavement, 30 years for bridges and other major structures, and an infinite life for the right-of-way. If land costs cannot be separated from the constructed portions of the facility, a life of 25 to 30 years is often used.

The cost of road maintenance, although typically small compared to the total transportation cost (as used here) for roads, is nevertheless often a very significant expenditure for states and agencies which must maintain these facilities. Typical values for the maintenance cost of roads per lane-mile are presented in Table 9-9, again in terms of 1973 price levels.

The costs accruing to road users are somewhat more complex in the models usually used. Perhaps the most complete analysis of road user costs

Table 9-9 Unit Costs of Road Maintenance in $/lane-mi/year†‡

Facility type	General	Lighting	Total
Expressways and freeways	2500	2150	4650
Arterials	1250	500	1750
Residential and CBD streets	900	850	1750

† Data expressed in terms of 1973 costs.
‡ Sanders et al. (1974, p. 4-18).

resulting in a specific standard cost model was that developed for the American Association of State Highway Officials about 1959, printed in their book *Road User Benefit Analyses for Highway Improvement* (1960), a book which is widely used for the purpose indicated in the title and which has been nicknamed "the AASHO red book." Although the costs reflect 1959 conditions, they can be updated using approximations without too much difficulty, and more importantly for our purposes, they illustrate the basic principles involved in road user cost estimation and a methodology widely used by professional engineers. A portion of the total user cost estimated according to this procedure represents estimates of dollar expenditures actually made by motor vehicle users, including fuel, oil, tire replacement, repair of the vehicle, and depreciation, some of which is due to the wear and tear on the vehicle and some of which is simply a result of age. Also, monetary values are associated with the time of drivers and passengers in vehicles and with levels of comfort and convenience on a highway, which depend on the quality of the road surface, traffic conditions, etc. Also, some of the costs of accidents which accrue to users might be included in this, but this is done separately.

The cost of operation of a passenger car depends upon many characteristics of the vehicle, the road on which it is operated, and the traffic flow on that road. These factors are taken into account in the cost relationships developed by AASHO.

AASHO has developed a cost-estimating relationship which considers the following six characteristics: (1) type of area, (2) type of highway, (3) type of operation or traffic flow conditions, (4) running speed, (5) gradient (divided into ranges or classes), and (6) alignment. The area is divided into two classes, rural and urban. The basic cost-estimating relationships have been developed for rural conditions. For analyzing urban conditions, if the traffic flow is very smooth, the same costs as for rural operation should be used. However, on major streets, where traffic flow is somewhat interrupted, the costs should be increased by from 10 to 30 percent. With stop and go operation, these costs should be increased further by consideration of the number of stops and average standing time, as provided for in specific estimating relationships to be discussed below.

Costs also depend upon the type of highway, the classification being between divided highways, such as freeways, and two-lane roads. The cost for two-lane highways is used for undivided highways of other types and widths.

Traffic flow is categorized as free, normal, or restricted. These levels are similar to level of service concepts described earlier. (They are somewhat dissimilar because they predate the 1965 *Highway Capacity Manual*.) The type of operation is related to the ratio of traffic volume to the practical capacity, where the traffic volume used is that of the thirtieth highest hourly volume occurring in the year for which the analysis is being performed. The relationship between actual capacity as used by AASHO and the concepts of capacity or service volume of various levels of service in the current *Highway Capacity Manual* is somewhat unclear, but most professionals equate practical capacity to the service volume at level of service B or C for rural highways and level of service C or D for urban highways (Pignataro, 1973, p. 68). If the ratio of the thirtieth highest hourly traffic volume to this capacity is greater than 1.25, operation is restricted; if it ranges from 0.75 to 1.25, it is normal; and if it is less than 0.75, the operation is considered free.

Figure 9-7 **Example profile for calculation of average gradient used in road user cost computation.**

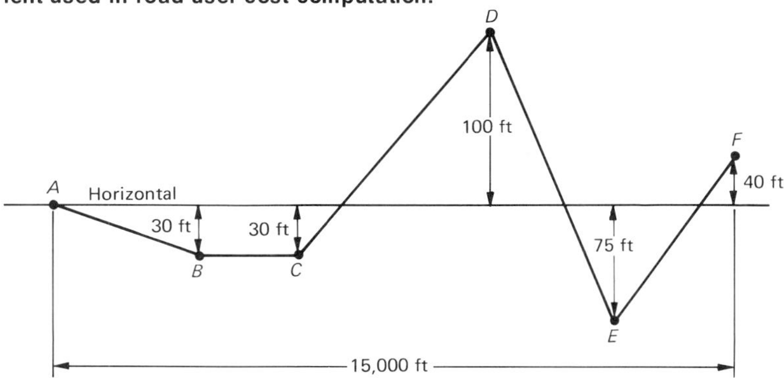

Running speed is measured by the average speed of vehicles over the section of highway being analyzed. The section should be chosen so that speed variation is not excessive. If vehicles stop, then the time stopped is not included in the calculation of average running speed, average running speed being equal to the total time spent moving divided by the distance. Since it would be impractical to develop such a relationship for each different type of profile that might be encountered, various gradient classes are used. Each class represents a range of average gradient, where that average gradient is defined as the total rise and fall of the highway divided by the entire length of the section under consideration. For example, in Fig. 9-7, the total rise and fall is

$$\begin{array}{ccccccccc} AB & & BC & & CD & & DE & & EF \\ 30\text{ ft} & + & 0\text{ ft} & + & 130\text{ ft} & + & 175\text{ ft} & + & 115\text{ ft} = 450\text{ ft} \end{array}$$

Since this section is 15,000 ft in length, the average grade is

$$\frac{450\text{ ft}}{15{,}000\text{ ft}}\,100\text{ percent} = 3.00\text{ percent}$$

The surface of the highway is specified as either paved, which may be either rigid (e.g., concrete) or flexible (e.g., asphalt), loose (primarily all weather gravel), or unsurfaced. The alignment of the road is tangent or curved, and if curved, the actual radius or degree of curvature is used in conjunction with information on the superelevation.

The cost model can be understood by reference to the specific components of costs included. Motor fuel costs obviously depend upon the fuel consumed, which is determined primarily by the total propulsive work which must be overcome by the vehicle, as was described in Chap. 4. This depends upon speed, including variations in that speed (as related to restricted, normal, and free operation), grades to be encountered, and curvature. Using a price of gasolene of 32 cents/gal, AASHO developed relationships between these factors and total fuel cost per vehicle-mile. In a similar manner, esti-

Table 9-10 Road User Costs for Passenger Cars in Rural Areas on Tangent Divided Highways with Pavement in Good Condition in 1959†

User costs, cents/vehicle-mile for

Free operation

Running speed, mi/h	Gradient class, percent	Fuel	Tires	Oil	Maintenance and repairs	Depreciation	Subtotal operating costs	Time	Comfort and convenience	Total costs
40	0–3	2.00	0.28	0.18	1.20	1.50	5.16	3.88	0	9.04
	3–5	2.10	0.33	0.18	1.20	1.50	5.31	3.88	0	9.19
	5–7	2.22	0.42	0.18	1.20	1.50	5.52	3.88	0	9.40
	7–9	2.53	0.57	0.18	1.20	1.50	5.98	3.88	0	9.86
44	0–3	2.09	0.34	0.21	1.20	1.50	5.34	3.52	0	8.86
	3–5	2.22	0.39	0.21	1.20	1.50	5.52	3.52	0	9.04
	5–7	2.35	0.51	0.21	1.20	1.50	5.77	3.52	0	9.29
	7–9	2.71	0.68	0.21	1.20	1.50	6.30	3.52	0	9.82
48	0–3	2.21	0.41	0.24	1.20	1.50	5.56	3.23	0	8.79
	3–5	2.35	0.47	0.24	1.20	1.50	5.76	3.23	0	8.99
	5–7	2.53	0.61	0.24	1.20	1.50	6.03	3.23	0	9.31
	7–9	2.95	0.81	0.24	1.20	1.50	6.70	3.23	0	9.93
52	0–3	2.34	0.47	0.29	1.20	1.50	5.80	2.98	0	8.78
	3–5	2.50	0.54	0.29	1.20	1.50	6.03	2.98	0	9.01
	5–7	2.72	0.71	0.29	1.20	1.50	6.42	2.98	0	9.40
	7–9	3.21	0.95	0.29	1.20	1.50	7.15	2.98	0	10.13
56	0–3	2.51	0.54	0.37	1.20	1.50	6.12	2.77	0	8.89
	3–5	2.71	0.62	0.37	1.20	1.50	6.40	2.77	0	9.17
	5–7	2.99	0.80	0.37	1.20	1.50	6.86	2.77	0	9.63
	7–9	3.58	1.07	0.37	1.20	1.50	7.72	2.77	0	10.49
60	0–3	2.73	0.56	0.52	1.20	1.50	6.51	2.58	0	9.09
	3–5	2.97	0.64	0.52	1.20	1.50	6.83	2.58	0	9.41

Restricted operation

Running speed, mi/h	Gradient class, percent	Fuel	Tires	Oil	Maintenance and repairs	Depreciation	Subtotal operating costs	Time	Comfort and convenience	Total costs
28	0–3	1.81	0.24	0.15	1.20	1.50	4.90	5.54	1.00	11.44
	3–5	1.89	0.27	0.15	1.20	1.50	5.01	5.54	1.00	11.55
	5–7	1.98	0.30	0.15	1.20	1.50	5.19	5.54	1.00	11.73
	7–9	2.22	0.48	0.15	1.20	1.50	5.55	5.54	1.00	12.09
32	0–3	1.85	0.28	0.15	1.20	1.50	4.98	4.84	1.00	10.82
	3–5	1.93	0.33	0.15	1.20	1.50	5.11	4.84	1.00	10.95
	5–7	2.02	0.43	0.15	1.20	1.50	5.30	4.84	1.00	11.14
	7–0	2.25	0.57	0.15	1.20	1.50	5.67	4.84	1.00	11.51
36	0–3	1.94	0.33	0.16	1.20	1.50	5.13	4.31	1.00	10.44
	3–5	2.03	0.38	0.16	1.20	1.50	5.27	4.31	1.00	10.58
	5–7	2.14	0.50	0.16	1.20	1.50	5.50	4.31	1.00	10.81
	7–9	2.38	0.67	0.16	1.20	1.50	5.91	4.31	1.00	11.22
40	0–3	2.08	0.39	0.18	1.20	1.50	5.35	3.88	1.00	10.23
	3–5	2.18	0.45	0.18	1.20	1.50	5.51	3.88	1.00	10.39
	5–7	2.32	0.59	0.18	1.20	1.50	5.79	3.88	1.00	10.67
	7–9	2.61	0.79	0.18	1.20	1.50	6.28	3.88	1.00	11.16
44	0–3	2.30	0.47	0.21	1.20	1.50	5.68	3.52	1.00	10.20
	3–5	2.41	0.55	0.21	1.20	1.50	5.87	3.52	1.00	10.39
	5–7	2.58	0.71	0.21	1.20	1.50	6.20	3.52	1.00	10.72
	7–9	2.96	0.95	0.21	1.20	1.50	6.82	3.52	1.00	11.34

Normal operation

32	0–3	1.85	0.23	0.15	1.20	1.50	4.93	4.84	0.50	10.27
	3–5	1.92	0.26	0.15	1.20	1.50	5.03	4.84	0.50	10.37
	5–7	2.01	0.34	0.15	1.20	1.50	5.20	4.84	0.50	10.54
	7–9	2.23	0.46	0.15	1.20	1.50	5.54	4.84	0.50	10.88
36	0–3	1.91	0.27	0.16	1.20	1.50	5.04	4.31	0.50	9.85
	3–5	2.00	0.31	0.16	1.20	1.50	5.17	4.31	0.50	9.98
	5–7	2.10	0.40	0.16	1.20	1.50	5.30	4.31	0.50	10.17
	7–9	2.34	0.53	0.16	1.20	1.50	5.73	4.31	0.50	10.54
40	0–3	2.00	0.32	0.18	1.20	1.50	5.20	3.88	0.50	9.58
	3–5	2.10	0.36	0.18	1.20	1.50	5.34	3.88	0.50	9.72
	5–7	2.22	0.47	0.18	1.20	1.50	5.57	3.88	0.50	9.95
	7–9	2.53	0.63	0.18	1.20	1.50	6.04	3.88	0.50	10.42
44	0–3	2.11	0.38	0.21	1.20	1.50	5.40	3.52	0.50	9.42
	3–5	2.23	0.44	0.21	1.20	1.50	5.58	3.52	0.50	9.60
	5–7	2.39	0.57	0.21	1.20	1.50	5.87	3.52	0.50	9.89
	7–9	2.75	0.76	0.21	1.20	1.50	6.42	3.52	0.50	10.44
48	0–3	2.27	0.45	0.24	1.20	1.50	5.66	3.23	0.50	9.39
	3–5	2.42	0.52	0.24	1.20	1.50	5.88	3.23	0.50	9.61
	5–7	2.62	0.67	0.24	1.20	1.50	6.23	3.23	0.50	9.96
	7–9	3.11	0.90	0.24	1.20	1.50	6.95	3.23	0.50	10.68
52	0–3	2.51	0.53	0.28	1.20	1.50	6.02	2.98	0.50	9.50
	3–5	2.71	0.60	0.28	1.20	1.50	6.29	2.98	0.50	9.77
	5–7	2.98	0.79	0.28	1.20	1.50	6.75	2.98	0.50	10.23
56	0–3	2.89	0.60	0.37	1.20	1.50	6.56	2.77	0.50	9.83
	3–5	3.18	0.68	0.37	1.20	1.50	6.93	2.77	0.50	10.20

† American Association of State Highway Officials (1960, pp. 129–131).

Table 9-11 Road User Costs for Passenger Cars in Rural Areas on Tangent Two-Lane Highways with Pavement in Good Condition in 1959†

User costs, cents/vehicle-mi for

Free operation

Running speed, mi/h	Gradient class, percent	Fuel	Tires	Oil	Maintenance and repairs	Depreciation	Subtotal operating costs	Time	Comfort and convenience	Total costs
32	0–3	1.85	0.21	0.15	1.20	1.50	4.91	4.84	0	9.75
	3–5	1.92	0.24	0.15	1.20	1.50	5.01	4.84	0	9.85
	5–7	2.01	0.31	0.15	1.20	1.50	5.17	4.84	0	10.01
	7–9	2.23	0.42	0.15	1.20	1.50	5.50	4.84	0	10.34
36	0–3	1.91	0.26	0.15	1.20	1.50	5.02	4.31	0	9.33
	3–5	2.00	0.30	0.15	1.20	1.50	5.15	4.31	0	9.46
	5–7	2.10	0.39	0.15	1.20	1.50	5.34	4.31	0	9.65
	7–9	2.34	0.52	0.15	1.20	1.50	5.71	4.31	0	10.02
40	0–3	2.00	0.32	0.18	1.20	1.50	5.20	3.88	0	9.02
	3–5	2.10	0.37	0.18	1.20	1.50	5.35	3.88	0	9.23
	5–7	2.22	0.48	0.18	1.20	1.50	5.58	3.88	0	9.46
	7–9	2.53	0.64	0.18	1.20	1.50	6.05	3.88	0	9.93
44	0–3	2.11	0.40	0.21	1.20	1.50	5.42	3.52	0	8.94
	3–5	2.23	0.46	0.21	1.20	1.50	5.60	3.52	0	9.12
	5–7	2.39	0.60	0.21	1.20	1.50	5.90	3.52	0	9.42
	7–9	2.75	0.80	0.21	1.20	1.50	6.46	3.52	0	9.98
48	0–3	2.27	0.50	0.24	1.20	1.50	5.71	3.23	0	8.94
	3–5	0.58	0.58	0.24	1.20	1.50	5.94	3.23	0	9.17
	5–7	2.62	0.75	0.24	1.20	1.50	6.31	3.23	0	9.54
	7–9	3.11	1.00	0.24	1.20	1.50	7.05	3.23	0	10.28
52	0–3	2.51	0.63	0.28	1.20	1.50	6.12	2.98	0	9.10
	3–5	2.71	0.72	0.28	1.20	1.50	6.41	2.98	0	9.39
	5–7	2.98	0.95	0.28	1.20	1.50	6.91	2.98	0	9.89
	7–9	3.65	1.26	0.28	1.20	1.50	7.89	2.98	0	10.87

Normal operation

Running speed, mi/h	Gradient class, percent	Fuel	Tires	Oil	Maintenance and repairs	Depreciation	Subtotal operating costs	Time	Comfort and convenience	Total costs
28	0–3	1.81	0.19	0.15	1.20	1.50	4.85	5.54	0.50	10.89
	3–5	1.89	0.22	0.15	1.20	1.50	4.96	5.54	0.50	11.00
	5–7	1.98	0.28	0.15	1.20	1.50	5.11	5.54	0.50	11.15
	7–9	2.22	0.38	0.15	1.20	1.50	5.45	5.54	0.50	11.49
32	0–3	1.85	0.23	0.15	1.20	1.50	4.93	4.84	0.50	10.27
	3–5	1.93	0.26	0.15	1.20	1.50	5.04	4.84	0.50	10.38
	5–7	2.02	0.34	0.15	1.20	1.50	5.21	4.84	0.50	10.55
	7–9	2.25	0.46	0.15	1.20	1.50	5.56	4.84	0.50	10.90
36	0–3	1.94	0.29	0.16	1.20	1.50	5.09	4.31	0.50	9.90
	3–5	2.03	0.33	0.16	1.20	1.50	5.22	4.31	0.50	10.03
	5–7	2.14	0.43	0.16	1.20	1.50	5.43	4.31	0.50	10.24
	7–9	2.38	0.58	0.16	1.20	1.50	5.82	4.31	0.50	10.63
40	0–3	2.08	0.36	0.18	1.20	1.50	5.32	3.88	0.50	9.70
	3–5	2.18	0.41	0.18	1.20	1.50	5.47	3.88	0.50	9.85
	5–7	2.32	0.54	0.18	1.20	1.50	5.74	3.88	0.50	10.12
	7–9	2.61	0.72	0.18	1.20	1.50	6.24	3.88	0.50	10.59
44	0–3	2.30	0.45	0.21	1.20	1.50	5.66	3.52	0.50	9.68
	3–5	2.41	0.52	0.21	1.20	1.50	5.84	3.52	0.50	9.86
	5–7	2.58	0.67	0.21	1.20	1.50	6.16	3.52	0.50	10.18
	7–9	2.96	0.90	0.21	1.20	1.50	6.77	3.52	0.50	10.79
48	0–3	2.71	0.56	0.24	1.20	1.50	6.21	3.23	0.50	9.94
	3–5	2.85	0.64	0.24	1.20	1.50	6.43	3.23	0.50	10.16
	5–7	3.13	0.84	0.24	1.20	1.50	6.91	3.23	0.50	10.64
	7–9	3.45	1.12	0.24	1.20	1.50	7.51	3.23	0.50	11.24

56	0–3	2.91	0.75	0.37	1.20	1.50	6.73	2.77	0	9.50
	3–5	3.18	0.86	0.37	1.20	1.50	7.11	2.77	0	9.88
	5–7	3.60	1.12	0.37	1.20	1.50	7.79	2.77	0	10.56
60	0–3	3.64	0.84	0.52	1.20	1.50	7.71	2.58	0	10.29
	3–5	4.05	0.97	0.52	1.20	1.50	8.24	2.58	0	10.82

Restricted operation

20	0–3	1.83	0.18	0.14	1.20	1.50	4.85	7.75	1.00	13.60
	3–5	1.91	0.21	0.14	1.20	1.50	4.96	7.75	1.00	13.71
	5–7	2.06	0.27	0.14	1.20	1.50	5.17	7.75	1.00	13.92
	7–9	2.34	0.36	0.14	1.20	1.50	5.54	7.75	1.00	14.29
24	0–3	1.81	0.21	0.14	1.20	1.50	4.86	6.46	1.00	12.32
	3–5	1.87	0.24	0.14	1.20	1.50	4.95	6.46	1.00	12.41
	5–7	1.99	0.31	0.14	1.20	1.50	5.14	6.46	1.00	12.60
	7–9	2.25	0.42	0.14	1.20	1.50	5.51	6.46	1.00	12.97
28	0–3	1.82	0.24	0.15	1.20	1.50	4.91	5.54	1.00	11.45
	3–5	1.89	0.28	0.15	1.20	1.50	5.02	5.54	1.00	11.56
	5–7	1.98	0.36	0.16	1.20	1.50	5.19	5.54	1.00	11.73
	7–9	2.23	0.48	0.15	1.20	1.50	5.56	5.54	1.00	12.10
32	0–3	1.87	0.29	0.15	1.20	1.50	5.01	4.84	1.00	10.85
	3–5	1.95	0.33	0.15	1.20	1.50	5.13	4.84	1.00	10.97
	5–7	2.03	0.43	0.15	1.20	1.50	5.31	4.84	1.00	11.15
	7–9	2.27	0.58	0.15	1.20	1.50	5.78	4.84	1.00	11.54
36	0–3	2.00	0.36	0.16	1.20	1.50	5.22	4.31	1.00	10.53
	3–5	2.09	0.41	0.16	1.20	1.50	5.36	4.31	1.00	10.67
	5–7	2.19	0.54	0.16	1.20	1.50	5.59	4.31	1.00	10.90
	7–9	2.45	0.72	0.16	1.20	1.50	6.03	4.31	1.00	11.34
40	0–3	2.27	0.45	0.18	1.20	1.50	5.60	3.88	1.00	10.48
	3–5	2.39	0.52	0.18	1.20	1.50	5.79	3.88	1.00	10.67
	5–7	2.53	0.67	0.18	1.20	1.50	6.08	3.88	1.00	10.96
	7–9	2.81	0.90	0.18	1.20	1.50	6.59	3.88	1.00	11.47

† American Association of State Highway Officials (1960, pp. 132–134).

mates of the costs of oil were developed as a function of these factors, assuming an oil price of 35 cents/quart. Also, empirical studies were undertaken to determine the cost of tire wear, which increases sharply with speed and with poorer surface conditions, as well as with increased propulsive force or braking force required as under conditions of congestion, steep grades, and improperly superelevated roads.

Maintenance and repair costs for automobiles were estimated to be 1.2 cents/mi on paved surfaces, 1.8 cents/mi on loose surfaces, and 2.4 cents/mi on unsurfaced roads.

One-half of the depreciation of passenger vehicles was assumed to be assigned to wear and tear of the vehicle due to distance traveled and one-half

Table 9-12 Road User Costs for Passenger Cars in Rural Areas on Tangent Loose Surface Highways in Good Condition in 1959†

		User costs, cents/vehicle-mi for								
Running speed, mi/h	Gradient class, percent	Fuel	Tires	Oil	Maintenance and repairs	Depreciation	Subtotal operating costs	Time	Comfort and convenience	Total costs
20	0–3	2.07	0.44	0.18	1.80	1.50	5.99	7.75	0.75	14.49
	3–5	2.17	0.50	0.18	1.80	1.50	6.15	7.75	0.75	14.65
	5–7	2.37	0.63	0.18	1.80	1.50	6.48	7.75	0.75	14.98
	7–9	2.74	0.83	0.18	1.80	1.50	7.05	7.75	0.75	15.55
24	0–3	2.03	0.50	0.18	1.80	1.50	6.01	6.46	0.75	13.22
	3–5	2.12	0.57	0.18	1.80	1.50	6.17	6.46	0.75	13.38
	5–7	2.27	0.73	0.18	1.80	1.50	6.48	6.46	0.75	13.69
	7–9	2.61	0.95	0.18	1.80	1.50	7.04	6.46	0.75	14.25
28	0–3	2.03	0.57	0.19	1.80	1.50	6.09	5.54	0.75	12.38
	3–5	2.14	0.65	0.19	1.80	1.50	6.28	5.54	0.75	12.57
	5–7	2.25	0.83	0.19	1.20	1.50	6.57	5.54	0.75	12.86
	7–9	2.57	1.09	0.19	1.80	1.50	7.15	5.54	0.75	13.44
32	0–3	2.10	0.65	0.21	1.80	1.50	6.26	4.84	0.75	11.85
	3–5	2.19	0.75	0.21	1.80	1.50	6.45	4.84	0.75	12.04
	5–7	2.32	0.96	0.21	1.80	1.50	6.79	4.84	0.75	12.38
	7–9	2.63	1.26	0.21	1.80	1.50	7.40	4.84	0.75	12.99
36	0–3	2.21	0.75	0.22	1.80	1.50	6.48	4.31	0.75	11.54
	3–5	2.33	0.86	0.22	1.80	1.50	6.71	4.31	0.75	11.77
	5–7	2.47	1.11	0.22	1.80	1.50	7.10	4.31	0.75	12.16
	7–9	2.80	1.46	0.22	1.80	1.50	7.78	4.31	0.75	12.84
40	0–3	2.39	0.87	0.24	1.80	1.50	6.80	3.88	0.75	11.43
	3–5	2.53	0.99	0.24	1.80	1.50	7.06	3.88	0.75	11.69
	5–7	2.71	1.28	0.24	1.80	1.50	7.53	3.88	0.75	12.16
	7–9	3.12	1.69	0.24	1.80	1.50	8.35	3.88	0.75	12.98
44	0–3	2.69	1.01	0.29	1.80	1.50	7.29	3.52	0.75	11.56
	3–5	2.83	1.15	0.29	1.80	1.50	7.57	3.52	0.75	11.84
	5–7	3.09	1.49	0.29	1.80	1.50	8.17	3.52	0.75	12.44
	7–9	3.63	1.97	0.29	1.80	1.50	9.19	3.52	0.75	13.46

† American Association of State Highway Officials (1960, p. 135).

due simply to the passage of time. The average undepreciated base cost of the vehicles was taken to be $3000, with an average life of 10 years and 100,000 mi. Thus, the average charge for depreciation is 1.5 cents/mi.

These costs are presented in Table 9-10 for divided highways and Table 9-11 for two-lane highways, both for pavements in good condition. As would be expected, operating costs increase with increasing speed of travel. Also, at any given speed, type of roadway, and gradient, the cost under restricted operation is greater than under normal operation, and similarly greater under normal than free operation, reflecting the added cost of fuel and other factors due to the speed variations. Table 9-12 presents similar information for passenger cars on loose-surfaced highways, and Table 9-13 presents it for unsurfaced roads.

To adjust these operating costs for curvature, the following formula is used. As was described in Chap. 4, the coefficient of side friction of motor

Table 9-13 Road User Costs for Passenger Cars in Rural Areas on Tangent Unsurfaced Roads in 1959†

		User costs, cents/vehicle-mi for								
Running speed, mi/h	Gradient class, percent	Fuel	Tires	Oil	Maintenance and repairs	Depreciation	Subtotal operating costs	Time	Comfort and convenience	Total costs
16	0–3	2.70	0.53	0.25	2.40	1.50	7.38	9.69	1.00	18.07
	3–5	2.88	0.60	0.25	2.40	1.50	7.63	9.69	1.00	18.32
	5–7	3.31	0.77	0.25	2.40	1.50	8.23	9.69	1.00	18.92
	7–9	4.23	1.01	0.25	2.40	1.50	9.39	9.69	1.00	20.08
20	0–3	2.56	0.61	0.26	2.40	1.50	7.33	7.75	1.00	16.08
	3–5	2.72	0.69	0.26	2.40	1.50	7.57	7.75	1.00	16.32
	5–7	3.03	0.88	0.26	2.40	1.50	8.07	7.75	1.00	16.82
	7–9	3.69	1.16	0.26	2.40	1.50	9.01	7.75	1.00	17.76
24	0–3	2.51	0.71	0.27	2.40	1.50	7.39	6.46	1.00	14.85
	3–5	2.05	0.81	0.27	2.40	1.50	7.68	6.46	1.00	15.09
	5–7	2.88	1.04	0.27	2.40	1.50	8.09	6.46	1.00	15.55
	7–9	3.45	1.38	0.27	2.40	1.50	9.00	6.46	1.00	16.46
28	0–3	2.51	0.82	0.28	2.40	1.50	7.51	5.54	1.00	14.05
	3–5	2.66	0.94	0.28	2.40	1.50	7.78	5.54	1.00	14.32
	5–7	2.86	1.20	0.28	2.40	1.50	8.24	5.54	1.00	14.78
	7–9	3.39	1.59	0.28	2.40	1.50	9.16	5.54	1.00	15.70
32	0–3	2.62	0.96	0.30	2.40	1.50	7.78	4.84	1.00	13.62
	3–5	2.77	1.10	0.30	2.40	1.50	8.07	4.84	1.00	13.91
	5–7	2.96	1.41	0.30	2.40	1.50	8.57	4.84	1.00	14.41
	7–9	3.49	1.87	0.30	2.40	1.50	9.56	4.84	1.00	15.40
36	0–3	2.79	1.07	0.32	2.40	1.50	8.08	4.31	1.00	13.39
	3–5	2.98	1.22	0.32	2.40	1.50	8.42	4.31	1.00	13.73
	5–7	3.20	1.58	0.32	2.40	1.50	9.00	4.31	1.00	14.31
	7–9	3.79	2.09	0.32	2.40	1.50	10.10	4.31	1.00	15.41

† American Association of State Highway Officials (1960, p. 136).

vehicles rounding a curve (that may be superelevated) is given by the following formula:

$$e + f = \frac{V^2 D}{85{,}950} \tag{9-18}$$

where e = rate of superelevation (dimensionless)
f = coefficient of side friction
D = degree of curve
V = vehicle speed, mi/h

To increase operating costs on curves, the coefficient of side friction is calculated from this formula or from the left portion of Fig. 9-8. Then the factor by which operating costs on a tangent road is multiplied to yield the cost on the curve is read from the right portion of the figure. Thus, if the road had a running speed of 38 mi/h with a curve of 8.0° and a superelevation of 0.06°, the calculated coefficient of side friction would be 0.0950 and the cost would be multiplied by 1.095.

In addition to these costs for which actual money payment is made, the value of time, discomfort, and inconvenience due to traveling on various road services is included in user costs. In part, however, the value of time may reflect the time saving which is directly translated into a money saving, in those cases where there is a travel time saving for persons who are being paid for their time, e.g., operators of trucks and buses and persons in passenger cars on business. Nonetheless, it is clear from persons' actual behavior that they do place a money value on time, in the sense that they will often select a toll road in order to save time in preference to an alternative free route, even when the direct money costs of traveling on the toll road may be higher. However, it is extremely difficult to place a single value on travel time to convert it into a monetary equivalent. On page 126 of its 1960 book, AASHO recommends a value of $1.55/vehicle-h as

representative of current opinion for a logical and practical value for passenger cars. The typical passenger vehicle has 1.8 persons in it, and a time value of 86 cents/person/h results in a vehicle total of $1.55/h.

Recent work has led to the adoption of values for time of from 25 to 75 percent of a person's wage rate for trips to and from work, 100 percent for business trips, and 25 to 40 percent for leisure trips (Heggie, 1972, p. 88).

At the time this document and the associated procedures were developed, it was a common practice to attempt to assign money values to all of the benefits of highway improvements, including time saving to travelers. Since then, it has become preferred to treat the time savings separately from any dollar savings to users, often simply in the form of retaining separately information on both the dollar saving and the time saving, making the assignment of a value to time unnecessary. Fortunately, the AASHO cost tables as presented in Tables 9-10 through 9-13 permit treatment of these separately.

In addition to the direct time value, it was recognized that there is a value associated with traveling on a smooth road surface, in comfort, free of the inconvenience of interruption in travel speed and the stress and strain which that may engender in the driver and possibly passengers. As a result, AASHO

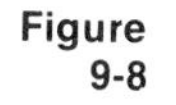

Figure 9-8 **Increase in passenger car operating cost due to curvature.** [***From American Association of State Highway Officials (1960, p. 24).***]

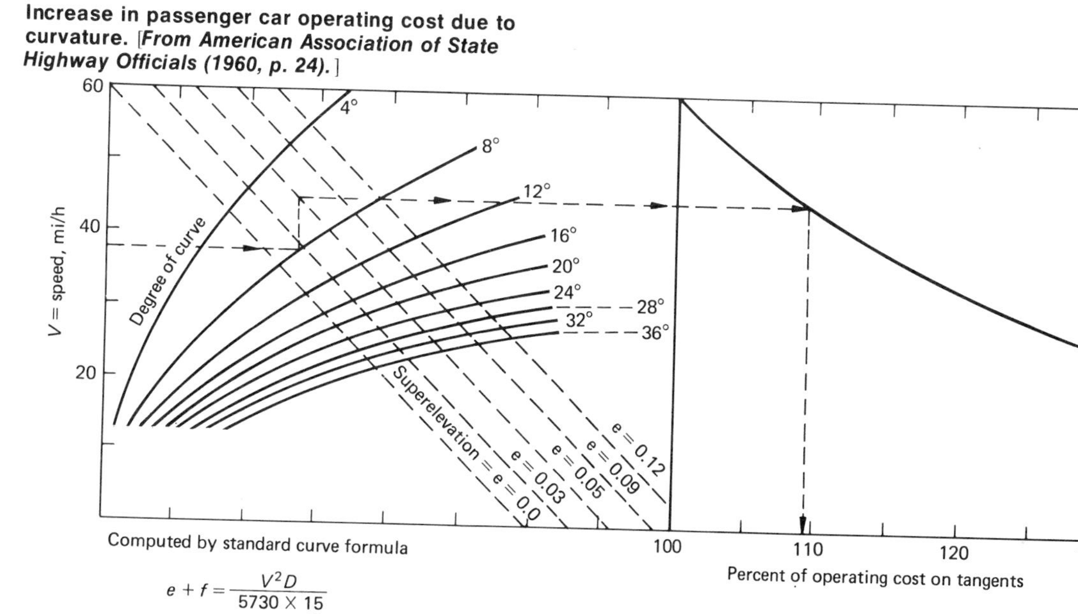

$$e + f = \frac{V^2 D}{5730 \times 15}$$

Example: Assume 38 mph operation on 8° curve, 0.06 superelevation. Follow arrows and read 109.50% as the factor to apply to correct the tangent road user cost values.

Table 9-14 Extra User Costs for Passenger Cars for a Vehicle Stop with No Standing Delay 1959†‡

Approach speed, mi/h	Additional cost per stop, cents				
	Fuel	Tires and brakes	Other operating costs	Time for acceleration and deceleration	Total
10	0.06	0.03	0.05	0.02	0.16
20	0.12	0.07	0.10	0.09	0.38
30	0.19	0.16	0.18	0.21	0.74
40	0.28	0.33	0.31	0.40	1.32
50	0.41	0.56	0.46	0.64	2.07
60	0.55	0.80	0.68	0.91	2.94

† Note: User cost of standing delay can be computed at the rate of $1.55/vehicle-h.
‡ American Association of State Highway Officials (1960, p. 137).

recommended that the operating cost per vehicle-mile be increased by 0 cents/vehicle-mi in free flow operation, 0.50 cent/vehicle-mi in normal flow, and 1.00 cent/vehicle-mi in restricted flow. For similar reasons, on unsurfaced roads, a value of 1.00 cent/vehicle-mi was added.

The added costs of vehicle stops are presented in Table 9-14 for passenger cars. As would be expected, this added cost depends upon the approach speed, which then determines the propulsive work required in acceleration, etc., and hence fuel and related costs. Also, the cost of time of the stop must be added, to be consistent with the other relationships at a rate of $1.55/vehicle-h.

While the AASHO red book does not give truck and bus costs directly, the AASHO recommends that costs of such vehicles be estimated as the following multiples of passenger car costs:

Single-unit trucks, 2 to 4

Combination trucks, 4 to 6

Buses, 2 to 4

Current use of these costs requires updating to current levels. The form in which these costs are given facilitates at least an approximation. In 1973, actual gasolene prices averaged about 43.7 cents/gal, and in 1974, 52.8 cents/gal (Motor Vehicles Manufacturers Association, 1975a, p. 4). Rather limited studies have indicated that the costs of operating motor vehicles in 1973 are 1.2 to 1.4 times the costs in 1959, so an approximation of the current costs would be to increase all costs by this amount, which is roughly the same proportional increase as the increase in fuel costs over this period. However, it should be borne in mind that these costs include the taxes collected on fuel and other items consumed in highway transportation, which are really charges on the users for the provision of roads and their maintenance, taxes which generally yield a substantial portion of the total cost of providing road facilities. Thus, if you wish to estimate properly the total cost of providing a particular road facility, it would be best to calculate the equivalent annual (or other time period) initial cost of the road, add to this the maintenance cost, and then add to this the operating cost excluding all user charges or taxes paid. Currently these taxes or user charges are 4 cents/gal for fuel, levied by the federal government, and typically from 7.5 to 10 cents/gal levied by states. Appropriate adjustments can be made using the information on fuel costs contained in these tables.

Traffic accidents are an extremely important cost of highway transportation. In 1973, for example, there were 55,800 deaths involving motor vehicles. However, it is important to point out that the death rate, measured in deaths per 100 million vehicle-miles, has been declining rather steadily since the 1920s, as has the death rate per 10,000 motor vehicles (National Safety Council, 1968). Lower speed limits, more widespread use of seat belts and other safety devices, as well as other safety-oriented programs are undoubtedly helping to reduce the accident rate, and preliminary indications are that the total number of deaths may be declining despite the increase in total number of vehicles and their usage in the United States at the present time. It is important to consider that on different types of roads there are different accident rates in the analysis of alternative designs. There appears to be no standard method for calculating accident costs, but one approach which is widely used was developed for the state of California (Smith and Tamburri, 1969). On the basis of data developed on accidents in Illinois, the cost of accidents per vehicle-mile on freeways and other facilities, in rural and urban areas, has been calculated. These values include the costs of reported accidents as well as an estimate of the number and cost of unreported accidents. On freeway facilities, the average total accident cost per vehicle-mile is estimated to be 0.18 cent in rural areas and 0.21 cent in urban areas. For facilities other than freeways the rural estimate is 0.47 cent and the urban estimate 0.69 cent. One of the advantages of diverting traffic to freeways is obvious from these calculations—the reduction in the total cost of accidents, which reflects a reduction in the number of all types of accidents except fatal accidents.

The use of this standardized cost model for road facilities can be illustrated by an example. We shall calculate the total daily transportation cost of 1 mi of an urban freeway which accommodates 38,400 vehicles/day in 1973 dollars, increasing AASHO user costs by 30 percent. The freeway cost \$2,500,000/mi to construct (including land) and is of the highest design standards.

At 10 percent interest and 30-year life (capital recovery factor of 0.106079) the daily cost is

$$\frac{\$2{,}500{,}000(0.106079)}{365} = \$726.57$$

The 30th highest hour volume is 2000 vehicles/h in one direction, and the peak hour factor is 0.91. From Fig. 5-11, it is seen that the maximum service volume at level of service C is 0.69 of the ultimate lane capacity of 2000 vehicles/h, or 2760 vehicles/h (on two lanes). Thus the volume-to-capacity ratio (as the term is used in the cost model) is 0.73. This corresponds to free operation. The speed, from Fig. 5-11, is seen to be about 50 mi/h, so the user costs will be based on interpolation between the 48 and the 52 mi/h speeds. Table 9-10 yields a total user cost of 8.785 cents/vehicle-mi. Since we will include the cost of road maintenance and construction, we delete the user tax, which equals (7.5/32.0)2.275 cents/vehicle-mi or 0.533. This results in a user cost in 1960 dollars of 8.252 cents/vehicle-mi. Multiplying this by 1.30 to reflect 1973 costs and adding accident costs at 1.2 cents/vehicle-mi on freeways, results in a total cost per vehicle-mile of 10.980. This yields a total daily user cost per mile of $4216.32. The maintenance cost per mile, from Table 9-9, is $4(4650)/365, or $50.96. Thus the total daily cost is

$$\$726.57 + \$4216.32 + \$50.96 = \$4993.85$$

It is interesting to note that about 84 percent of the total cost is user cost.

As a final note on road costs, we should mention that there are numerous studies updating the AASHO costs, but unfortunately they only treat portions of total user costs. However, for particular studies the costs found in Curry and Anderson (1972), Claffey (1971), and Winfrey and Zellner (1971) may be useful.

Choice of Technology and Cost-Output Relationships

As the example concluding the previous section illustrated, various choices which face engineers and planners can be facilitated by the development of appropriate cost information. Often there are situations where it is desired to select the means for performing a given transportation task at the least cost. This is essentially the classical problem in technological choice, which is treated abstractly and theoretically in many economics textbooks and which faces transportation engineers as well as other engineers every day.

The classical technological choice problem is very restricted: it involves selecting one from among many ways or technologies for producing precisely the same product. In this situation, the consumer of the product presumably is indifferent as to the way in which the product is produced and cares only for the characteristics of the product which do not change depending on the means selected for production. In this case, it is usually accepted that a reasonable goal is the selection of that technological means of production which is least costly to society. If we presume for a moment that the appropriate costs for each of the possible factors of production can be determined, then we are interested in finding that means of production of the given product which is least costly.

While economists typically treat a situation of an infinite number of possible means of producing the product, more typically there is a finite, usually a very small, number. In this case, the problem simply becomes one of identifying the possible technological means of production, estimating the costs associated with each, and selecting that technology involving the least cost.

In many transportation problems, the problem is often one of determining under what conditions different technologies should be used. In recent years, in which there has been considerable expansion of road systems and improvements in both developed and developing nations, a common type of question is under what conditions should roads be upgraded? The answer to this question has usually been sought in terms of total transportation costs, as we used this phrase above, including the sum of system, user, and accident costs. Often this question is asked in the context of roads on which the traffic is increasing, raising the question of what volume of traffic warrants an improvement over existing surfaces?

Figure 9-9 **Choice of technology based on minimum total cost: example of alternative road surface types.** *[From Odier, Millard, Santos, and Mehra (1971, p. 22).]*

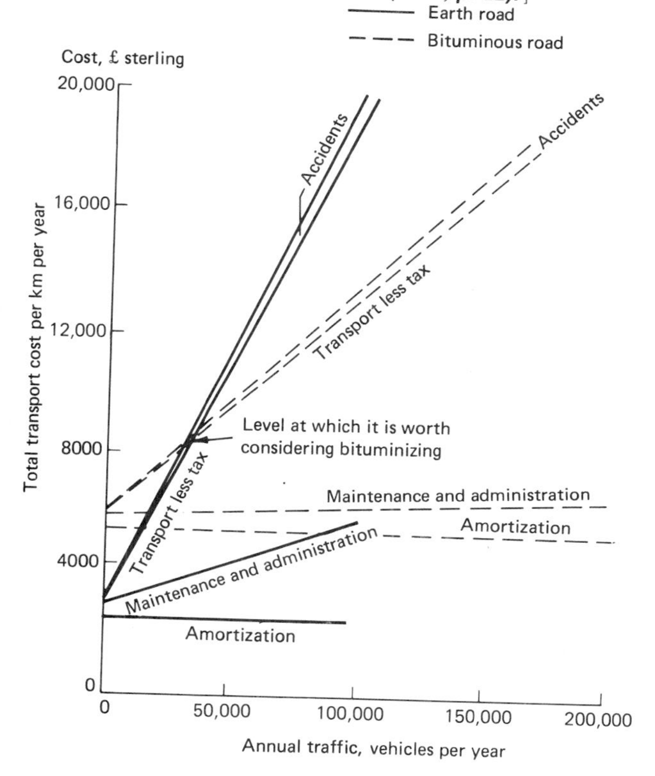

Typically in these contexts estimates of costs are made for the range of volumes experienced on the roads. A typical example is shown in Fig. 9-9, which gives the total transport cost per kilometer in pounds sterling versus annual traffic for earth and asphalt surface roads. In this figure the word amortization refers to payment annually of an amount equivalent to the initial investment, in other words, our capital recovery. Emphasizing the need to avoid double counting of costs, as we noted above, the cost of transport is referred to as "less tax" in this figure. As would be expected, the annual capital cost of earth roads is less than that of the improved roads, but the maintenance and user costs per vehicle-kilometer are greater. As a result, at volumes below approximately 33,000 vehicles/year, the earth road is cheaper, and above that the asphalted or improved road is cheaper. This type of information provides valuable input to nations' road departments in deciding which roads should be improved. Also, the savings on various roads can be compared, to decide where to allocate funds for road improvements first if the total funds available are insufficient to initiate all desirable projects.

One must be very careful, however, in using total transport costs as the sole basis for comparing different technological possibilities. The main problem which arises is that the alternatives may not be sufficiently identical in the transport level of service which they provide that users and others would be indifferent among them. In the case of upgrading a road, in many situations the improved road will encourage much more travel and possibly more rapid economic development of a region, important effects which must be considered in any overall evaluation. Also, users of the improved road would presumably benefit from the improvements, even if society as a whole saved money by maintaining the unimproved highway with its higher user costs. To take into account these differences, we must await the discussion of the demand for transportation and the effects of transportation on development, topics that will be covered in Chaps. 10 and 13.

Another example is provided by many of the alternatives at the forefront of urban transportation issues today. There have been numerous studies which have attempted to compare the costs of providing urban transportation service by private automobile on roads, by buses operating on public streets as well as on private bus roadways, by various forms of rail rapid transit, and, to a lesser extent, by some novel automated transport technologies. To make the quality of service among these as comparable as possible, transit systems are designed to operate at very low headways and to provide seats for everyone, to match the quality of service of the private automobile. Then on the basis of cost comparisons, attempts are made to identify the most appropriate circumstances for particular modes.

There are numerous dangers of this sort of generalization. An obvious one is that by forcing such a high quality of service on all public transit, the cost of that service is increased, even though many users of public transport may prefer a lower quality of service at a lower cost. Also, the use of only a single cost estimate for any particular mode does not account for variations in local conditions, which may make particular modes cheaper in one place and more expensive in others. Surely terrain, weather, availability of right-of-way, etc., all influence the cost of different types of transport facilities.

Equally misleading can be studies which compare different technologies solely on the basis of statements which have no basis in fact, but represent only the belief of the person making the comparison. This is often found in comparisons of level of service among modes, and even extends to costs, where quantitative analyses are quite readily performed! Sometimes reports of these studies take on more the air of promotion of one particular mode rather than professional engineering work. However, this is probably to a large degree a legacy of education which has tended to emphasize modal differences rather than similarities in principles of transportation engineering, and perhaps also the desire of many professionals to promote the particular mode with which they are affiliated, perhaps for obvious business reasons!

Freight Carrier Costs

Another important type of standardized cost model, for freight carriers, includes variables not found in the passenger models. It is developed and updated periodically by regulatory agencies to provide a uniform basis for estimating the costs of carrying any specific shipment of freight. These costs then can be used in deliberations regarding changes in freight rates, marginal costs being a lower bound (below which the carrier is worse off by carrying the freight) and average total cost being useful in determining whether or not a carrier's charges are excessive (including consideration of a reasonable profit).

This type of cost model is illustrated by ones developed by the Interstate Commerce Commission for railroads (1975) and trucks (1974), although other examples could have been chosen. The ICC models are in the form of tables broken down by the various factors which are taken into account, such as shipment weight, distance, etc. In the presentation below we shall use formulas which have been developed as close approximations to these tables; the differences are quite small. The ICC tables, or models, estimate two types of costs: variable and fully allocated. The variable costs are as we defined them, but with liberal allowances for many items, such as cars owned, to vary, so that they represent a relatively long period of adaptation. The fully allocated costs include allocations of all costs incurred to various units of traffic, with much arbitrary allocation of some costs, since it is unclear exactly how such costs would actually vary (if at all except in a very long period) with variations in traffic. An example would be the cost of owning the right-of-way of a railroad, a minimum width and length being necessary for a single track as long as any traffic were to be carried. Such fixed costs are a very small portion of the cost of truck service since the large fixed cost of providing the roads is paid for by the trucker and other users on the basis of use rather than as a fixed amount. In contrast, railroads pay the total cost of their tracks, etc., in the form of an essentially fixed cost, which makes their total costs appear quite different with respect to the fraction of all costs which are fixed. If someone else owned the rail lines and charged on the basis of use, then these costs would become variable.

The truck models estimate cost as a function of shipment weight, shipment density, and distance. As with many freight cost models, cost is estimated in cents per hundredweight, a hundredweight (abbreviation, cwt) being 100 lb. These models reflect the actual practice of consolidating small shipments, ones less than the weight and volume capacity of the truck being used, so that fuller utilization of the vehicle's capacity is obtained. The cost-estimating formulas for the northeastern part of the United States (precisely, the ICC's Eastern-Central Territory), for variable costs are, by shipment weight category:

$$\begin{aligned}
\mathrm{VC}_T &= (S/100)[924S^{-0.537} + (8.8286 + 0.16901L)(0.855 + 1.32^{-0.1447(D-2.5)}) \\
&\qquad + (0.29293S^{-0.736})L]100 && S < 100 \\
\mathrm{VC}_T &= (S/100)\{70.51 - 0.2907(S - 100) + (12.107 + 0.18098L)(0.855 \\
&\qquad + 0.68e^{-0.16176(D-7.5)}) + [0.00988 - 6.38 \\
&\qquad \times 10^{-5}(S - 100)]L\}100 && 100 \le S < 200 \\
\mathrm{VC}_T &= (S/100)\{41.44 - 0.1317(S - 200) + (6.75 + 0.17275L)(0.885 \\
&\qquad + 0.68e^{-0.16176(D-7.5)}) + [0.0035 - 2.99 \\
&\qquad \times 10^{-5}(S - 200)]L\}100 && 200 \le S < 300 \\
\mathrm{VC}_T &= (S/100)[28.27 - 0.0571(S - 300) + (2.706 + 0.1509L)(0.885 \\
&\qquad + 0.68e^{-0.16176(D-7.5)}) + 0.00051L]100 && 300 \le S < 437 \\
\mathrm{VC}_T &= (S/100)(23.153 + 0.13406L + 0.10261e^{-0.16176(D-7.5)}L)100 && S \ge 437
\end{aligned} \tag{9-19}$$

where VC_T = truck variable cost per shipment, \$

S = shipment weight, cwt

L = actual length of haul, mi

D = density of cargo, lb/ft^3

To obtain the fully allocated costs, the resulting values are simply multiplied by 1.11:

$$FAC_T = 1.11(VC_T) \tag{9-20}$$

where FAC_T = fully allocated cost per shipment, $

Railroad carload costs are of a similar form, but do not include a density term. The reason is that for a single shipment in a car, there is no consolidation of many shipments (except by the shipper, who would simply pay for the movement as one shipment), and it is assumed that the total weight of the shipment would be checked against the weight and cubic volume capacity of the car to ensure that it could be accommodated. Different cost relationships are developed for each different type of car. Those for a standard boxcar (in ICC jargon, boxcar, general service, unequipped, the latter term meaning not equipped with special load-restraining devices to reduce damage in transit and make loading easier), moving in the Official Territory (the northeast and midwest states) are given below:

$$VC_R = [116 + 0.00036S + L(0.31735 + 0.0001482S)] + (36.54I - 0.06669L) + (12.04Y - 0.06017L) \tag{9-21}$$

$$FAC_R = [116 + 0.02446S + L(0.31735 + 0.0002861S)] + (36.54I - 0.06669L) + (12.04Y - 0.06017L) \tag{9-22}$$

where VC_R = rail variable cost per shipment, $
FAC_R = rail fully allocated cost per shipment, $
I = number of railroad to railroad interchanges
Y = number of intermediate (not two end points) switchings in yards

In these equations, the three parts contained in brackets and parentheses are, respectively, train or line-haul cost, railroad interchange costs, and other classification yard costs. These costs include only movement in the rail car, not trucking to and from a rail line, if any (this being uncommon for most commodities).

Trailer on flat-car service costs models have also been developed. These include the cost of moving the truck trailer between the rail car yard terminal and the shipper or consignee (receiver), a cost which is often very substantial and which varies considerably from one area to another. In the official territory, for example, costs in 1973 to railroads for moving an empty trailer to a shipper's dock and then the loaded trailer to the rail yard averaged $42.17 in nonurban areas but were $99.37/trailer in New York and $66.33 in Chicago. For a movement between New York and Chicago the costs for one trailer shipment are

$$VC_{TOFC} = 193 + 0.00118S + (0.206 + 0.0001367S)L + (30.79I - 0.04266L) + (10.14Y - 0.033610L) \tag{9-23}$$

$$FAC_{TOFC} = 193 + 0.02528S + (2.206 + 0.0002622S)L + (30.79I - 0.04266L) + (10.14Y - 0.033610L) \tag{9-24}$$

All the terms are as defined earlier. For any other city pair, the first term can be reduced or increased by the difference in the local drayage costs.

The use of these cost models is very straightforward. Comparisons of the costs of transporting different sized shipments and the same shipment over different distances, by the same mode, and comparisons among these modes, are the subject of three of the problems. These illustrate general relationships.

TYPICAL CURRENT COSTS

Before leaving the technology of transportation and cost-estimating methods in particular, it is appropriate to present some information regarding the cost characteristics of different transportation technologies. There are dangers in doing this since cost varies widely among different situations even for the same transport technology, reflecting not only fixed local conditions such as terrain, but also management skills and the market conditions in the industries which supply transportation services and facilities (such as the construction industry). However, some familiarity with the approximate range of typical costs is important for the engineer.

Since most attention was devoted to urban transportation examples, the first generalized cost relationships will be presented for urban public transport carriers. Typical costs as presented in *The Characteristics of Urban Transportation Systems* handbook and elsewhere are presented in Table 9-15. This includes a recapitulation of rail rapid transit costs as they were used in the problem example, to which have been added the costs of subway and elevated line construction. The reason that surface line construction is so popular for modern rapid transit lines is obvious from this cost information. Also presented in this table are typical corresponding costs for bus transit, including costs for exclusive bus roadways—termed busways—which allow

Table 9-15 Typical Unit Costs of Urban Public Transportation in the United States†

Item	Cost, $	Notes
Rail rapid transit		
Guideway		
Surface	4.0–7.4 million/mi	Costs in reference reduced to delete station costs and exclude land cost
Elevated	11.2–16.6 million/mi	
Subway	25.0–50.0 million/mi	
Stations		
Subway	10.0–17.0 million/station	
Other	2.0–5.0 million/station	
Right-of-way	0.64–2.22 million/mi	
Cars	350,000/car	
Operating costs	1.01–2.79/car-mi	Includes track maintenance
Bus transit		
Busway	1.65–3.30 million/mi	On surface, excluding land
Right-of-way	0.85–2.95 million/mi	On surface
Stations	200–300 thousand/station	On surface, with pull-out lanes
Large bus	46,000/bus	51–53 seats, 12-year life
Operating costs	0.75–1.30/bus-mi	Note limitations on this form of cost, as described in text
Subway busway	27.8–46.6 million/mi	Estimated, with 2 stations/mi; none have been built
Administration	10 percent of operating costs	

† Sanders et al. (1974, chaps. 2 and 3, 1973 prices).

buses to achieve the high speeds typically associated with rail rapid transit lines. It should be noted that the costs of constructing busway tunnels are estimated on the basis of constructing other types of roadway tunnels and may not reflect costs which would actually be incurred. The other costs for bus transit, including surface busways, are based upon the costs of actual constructed systems.

Cost of regular public road facilities and automobile user costs were covered in detail in the previous section, and hence will not be repeated here.

There have been relatively few studies of the costs of intercity carriers, the most comprehensive of these having been completed in 1959 by Meyer et al. (1959). Although these costs are somewhat outdated, they do provide information on the relative cost of various forms of intercity transportation. In one of the pioneering applications of statistical cost-estimating procedures Meyer et al. developed long-run marginal costs for both passenger and freight intercity carriers in the United States. Also, estimates of fixed costs related to the size of the plant of various carriers, such as miles of railroad track and numbers of cars, were developed, but the presentation of these would be extremely cumbersome.

Table 9-16 presents the long-run marginal cost for intercity passenger carriers. These are presented in terms of cents per seat-mile and cents per passenger-mile for 1955, these being based upon the Meyer et al. study. The two measures are important because of the differences among modes in the actual utilization of seats available. The range in each case represents typical differences found due to differences in the types of vehicles operated and the operating conditions. For comparison purposes, the average total cost in cents per passenger-mile in 1974 is also presented. This average total cost, of

Table 9-16 Typical Costs of Intercity Passenger Service in the United States

	Long-run marginal costs in 1955		Average total costs in 1974, cents/passenger-mi
	Cents/seat-mi[a]	Cents/passenger-mi[b]	
Airplane	1.8–3.2	3.6–6.4	7.3[c]
Automobile	0.63–1.33	1.9–4.0	7.7[d]
Bus	1.25	2.5	3.9[e]
Railroad			
Day coach	1.3–1.4	2.6–2.8	
Overnight coach	2.2–2.3	4.4–4.6	
Parlor	2.7–3.0	5.4–6.0	
Pullman	5.0–6.0	10.0–12.0	
All service			13.1[f]

[a] Meyer et al. (1959, p. 158). Air costs are for jet aircraft and trips over 1000 mi. Auto costs are for a six-seat auto.
[b] Based on Meyer et al., using average load factors of 50 percent, except for auto and rail accommodations, 33.3 percent.
[c] Air Transport Association of America (1975).
[d] Motor Vehicle Manufacturers Association (1975), 1969 cost increased by 30 percent.
[e] National Association of Motor Bus Owners (1976, p. 23).
[f] Association of American Railroads (1975, p. 62), for Amtrak, including operating costs (11.9 cents/passenger-mi) plus 10 percent to reflect normal investment level in contrast to the higher level currently experienced.

course, includes fixed costs of the various carriers, although in most cases these fixed costs are very small over the long run, since vehicles, terminal buildings, guideways, etc., all must be replaced in the long run. As can be seen, relative costs remain about the same, with the automobile and bus being the least costly carriers, air somewhat more expensive, and rail typically the most expensive carrier. It should be noted that in all cases where these carriers use shared facilities they may not be paying exactly the full cost for their use, although these estimates include all identifiable costs including subsidies from various levels of government, which reach as high as over half the cost of operating Amtrak (National Railroad Passenger Corp.) railway passenger service, for example.

Table 9-17 presents cost information for intercity freight carriers in the United States. The first column again is taken from the 1955 study of Meyer et al. since this is the most recent general comparison of costs of all modes Within each mode, there is a considerable range of cost, reflecting variation in the average load per vehicle, the degree to which the vehicle must be returned empty, and variation in cost with distance, these costs being for distances between 200 and 1600 mi, with the exception of air, which is for the longer distance. There is also a distinction between the cost for bulk and merchandise traffic, bulk goods typically being much denser and hence yielding lower costs per ton-mile. For the movement of bulk commodities, water carriage is

Table 9-17 Typical Costs of Intercity Freight Carriers in the United States

Carrier and commodity type	Long-run marginal costs 1952–1955, cents/ton-mi*	Average or typical revenue in 1973, cents/ton-mi
Air—merchandise	u†	23.31‡
Inland water		
Bulk	0.105–0.332	0.25–0.80§
Merchandise	0.55–1.85	u
All commodities	u	0.378‡
Pipeline—oil	0.513–0.581	0.290‡
Railroad		
Bulk	0.390–0.810	0.41–1.60§
Merchandise	0.722–1.511	u
All commodities	u	1.62‡
Trailer on flat car	0.875–1.83	u
Truck		
Merchandise (TL)	1.82–4.90	1.93–3.02¶
All commodities (TL and LTL)	u	8.24‡

* Meyer et al. (1959, pp. 147–157). Ranges are for 0 to 100 percent empty returns, 200 to 1600 mi, and for rail operations include 30 percent added to line-haul costs to reflect yard and local freight costs.

† Indicates unavailable.

‡ Transportation Association of America (1975, pp. 7 and A-6). Complete current data on costs are not available; hence revenue is used as a surrogate for 1973. Revenue averages reflect differences in commodity types, shipment sizes, length of haul, and that the data for rail and truck (all commodities) are for Class I carriers only, for water are for ICC-regulated lines only, and air data are for scheduled domestic carriers only.

§ Mutschler, Evans, and Larwood [1973, p. 13 (rail coal movements) and p. 26 (barge coal movements)], all revenues for 1970.

¶ Costs of independent owner-operator truckers carrying primarily merchandise (Wyckoff and Maister, 1975, p. 115).

extremely cheap, pipeline and rail movement are typically somewhat more expensive. Water carriage is also among the cheapest for merchandise traffic, although under favorable conditions rail movement can be as inexpensive. However, these values underestimate the true cost of water transport because water carriers use government-provided waterways free of charge. The cost of moving freight by truck varies considerably, primarily because much truck traffic consists of small packages which, due to the extra handling involved and a typical low density, results in a high cost per ton-mile. Trucking costs per ton-mile also increase as shipment size decreases and length of haul decreases. In these estimates of rail costs, 30 percent has been added to the line-haul cost to reflect yard and local freight cost, this being typical of the values found in the study. Trailer on flat car (TOFC) or "piggyback" service appears quite attractive according to these estimates; it achieves costs almost as low as regular rail carload for merchandise movement and also has the ability of the truck to pick up and deliver freight to destinations removed from rail lines.

To bring these costs up to date, average revenue per ton mile in 1973 for these carriers is shown in the last column. This is, of course, an overestimate of cost, the amount of the overestimate being equal to profit and general taxes (as opposed to taxes for the use of facilities such as highways). However, profits in the transportation sector, as well as general taxes, are not so great that this significantly distorts the relative costs of the carriers. In interpreting these averages, though, one must recognize that differences in the length of haul, size of shipment, and the types of commodities shipped can make it appear as though there are more substantial differences between modes than actually exist. In particular it is likely that the long haul and relative high density of many commodities moving by rail in contrast to truck, in addition to the use of truck for a substantial amount of small parcel freight, magnifies the difference between rail and truck costs considerably. Also, the same differences probably make comparisons between water and pipeline on one hand, and railroad on the other, somewhat misleading.

General Relationships

In discussing the cost characteristics of various modes of transportation, the reader has undoubtedly noticed a number of relations between cost and characteristics of the transportation service provided which are common to all modes. While there are always exceptions to generalizations, many seem to hold rather widely, so it is useful to summarize them here.

One important generalization is that average total cost on any given transport system tends to decrease with increasing volume, holding level of service constant, up to a fairly high volume. This reduction in average cost is termed an economy of scale in economics. It usually results from the presence of some fixed costs spread over more units of output with increasing volume, and also from an ability to use resources more efficiently with more traffic. However, at some point diseconomies set in, such as might be due to problems of congestion or difficulties in managing effectively a large enterprise, resulting in an increase in average costs. Of course, the exact form of the cost curves, and the point where diseconomies set in, depends on the technology of transport, the network, characteristics of the traffic, the effectiveness of the management, and many other factors. The general form of the cost curve is illustrated in Fig. 9-1.

Figure 9-10 Typical relationship between cost and traffic volume. (*a*) Total cost vs. volume. (*b*) Average total cost vs. volume. Note: Level of service is held constant.

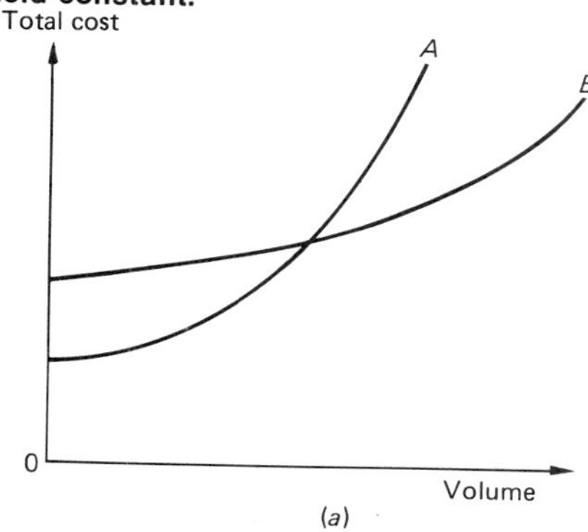

(*a*)

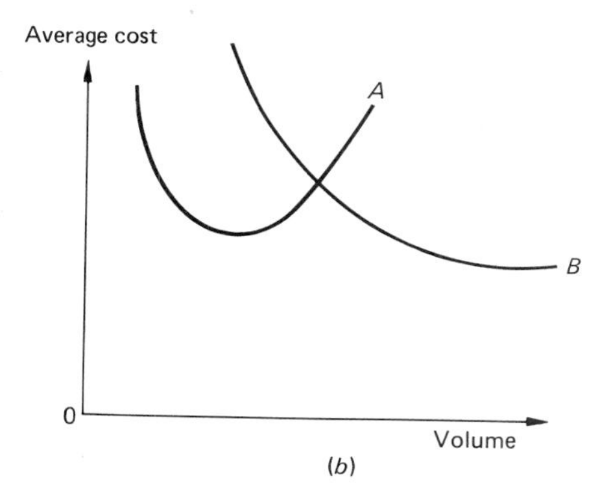

(*b*)

Often scale economies can be increased by altering the technology used at high volumes. This is usually accomplished by investments which increase fixed costs but decrease variable costs. This is illustrated in Fig. 9-9 by a road example, but the same principle applies to all modes. Figure 9-10 portrays this for a more general case in which diseconomies are also portrayed, for two different levels of investment, system *A* having a lower investment than system *B*. Sometimes *A* and *B* are interpreted as different technologies, such as truck *(A)* and rail *(B)*, but one must guard against excessive generalization, particularly between such different technologies.

Also, usually there is an increase in cost with improved levels of service, such as reduced travel time or increased reliability. In the case of speed, this is due to the need for a guideway designed for higher speeds, more powerful vehicles, greater fuel consumption, a more sophisticated control system, and so on. However, it is important to note that at some point, decreases in the level of service will result in increases in costs. This usually results from such factors as low utilization of equipment and labor resulting from low speeds, and the consequent need for more resources to handle any given amount of traffic. Thus the general relation between cost and level of service appears as in Fig. 9-11. Only average total cost is shown in this figure, since total cost exhibits exactly the same relation because the total volume is fixed. Of

Figure 9-11 **Typical relationship between average total cost and level of service for a transport system. Note: Traffic volume is held constant.**

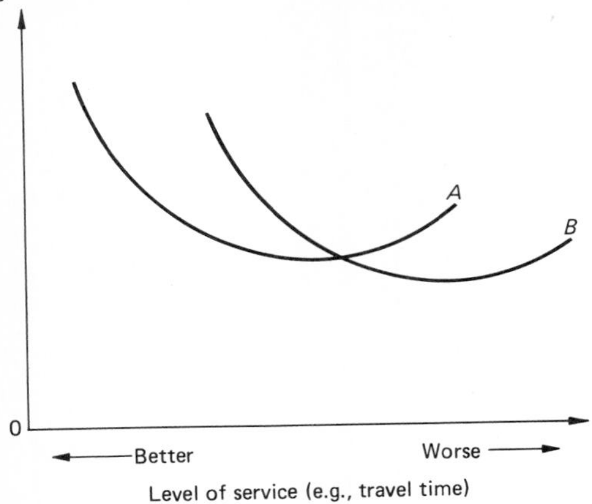

course, no management would normally operate a system in the range of increasing average cost with decreasing level of service, since other combinations of service level and cost—for example, that corresponding to minimum average cost—would surely be preferred by the user and the system owner (or operator) alike. Different technologies of transport or design variations within a technology (such as links with different speeds) might appear as shown by the curves *A* and *B* on this figure. The choice of technology or design would thus depend on the desired level of service.

Another useful generalization is that the average cost of transport per unit carried decreases with increasing size of shipments or number of persons in a party traveling together. In the case of persons, an obvious example is the reduction in the cost per person of automobile travel resulting from more persons in the car. The cost increase due to more passengers is so small it is hardly noticed, resulting in average costs essentially inversely proportioned to the number in the car. For large groups, a bus can be chartered, usually with further savings. In freight transport, some of the costs are fixed per shipment regardless of its size, resulting in reduced cost per ton-mile with increasing size, holding length of haul and other factors constant. And when the shipment is sufficiently large to fill an entire vehicle, terminal handling is reduced or eliminated, further reducing average costs. Thus the general relationship portrayed in Fig. 9-12 holds for most transport situations. Different technologies possess different characteristics in this regard, as illustrated by curves *A* and *B* in the figure. Curve *A* might reflect auto travel, while *B* might reflect charter bus travel.

A final generalization is one made in Chap. 2: the transport cost for any shipment [or person(s) traveling together] consists of a component which is invariant with distance and another which increases with increasing distance. This is illustrated in Fig. 9-13. Total cost thus increases with distance, but average total cost (e.g., per ton-mile or passenger-mile) will decrease with

Figure 9-12 **Typical relationship between average cost and freight shipment size or number of persons traveling together.**

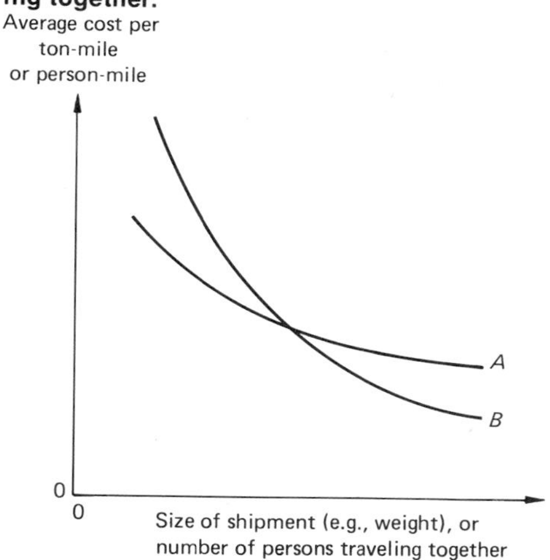

distance. Again the relationship will vary by mode, size of shipments, etc. Often curves such as *A* and *B* will be used to describe difference between modes, curve *A* representing a system with low non-distance-related costs and relatively high distance-related costs, compared with a system described by curve *B*. For example, curve *A* might represent truck movement and curve *B* "piggyback" movement, the extra non-distance-related cost reflecting the added terminal cost of drayage to and from the rail terminal and the loading and unloading process, and the lower distance-related cost reflecting rail movement in trains vs. single trucks.

Figure 9-13 **Typical relationship between cost and distance for a given shipment or party of travelers. (*a*) Total cost vs. distance. (*b*) Average total cost vs. distance.**

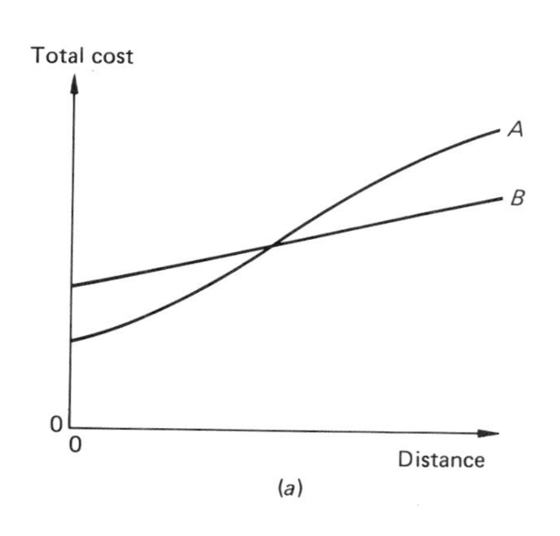

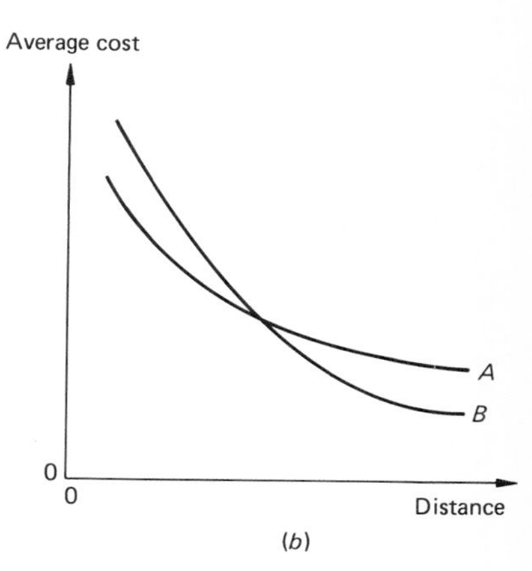

These generalizations should prove helpful in thinking about transport costs. However, as has been noted in the preceding discussion, one must be very careful in using them. Transport costs are determined by many factors—such as network, technology, operations plan, level of service, and traffic volume—all of which must be specified before costs can be estimated.

SUMMARY

1. Costs are extremely important for the engineer and planner, as one of a number of criteria upon which alternative plans and designs must be evaluated. Therefore, an understanding of the basic principles of economics as they relate to cost concepts and cost estimation is essential.
2. Different portions of the total cost to society of providing transport services are borne by different groups. Therefore, it is essential when discussing transport costs to specify which costs are being considered and to whom these accrue.
3. The costs for providing transportation services vary over time, as the facility or system passes through the stages of planning, construction, operation, rehabilitation, and possibly eventual abandonment. Costs of capital items, such as fixed plant and vehicles, can be converted to an equivalent annual amount, which depends upon the life of the item and the interest rate. Also, future costs can be discounted to an equivalent value in any year, the most common being present value (for present year), the present value depending upon the discount rate and the number of years removed from actual expenditure.
4. Important concepts of cost are total cost, fixed cost, marginal cost, and average cost. Whether a cost is fixed or variable depends upon the situation and the time period allowed for alteration of means of production; in the long run almost all costs are variable.
5. There are two basic types of cost-estimating models: statistical cost-estimating relationships and engineering unit cost models.
6. Both types of cost models typically divide costs into categories such as costs of owning fixed facilities, owning vehicles and containers, operating and maintenance costs, and administrative costs. Care must be taken in using these models to ensure that costs vary with the factors and characteristics which actually result in variation in costs.
7. Standardized cost models have been developed for almost all modes and contexts in transportation, and although these are not universal, their use is often warranted where the assumptions underlying the models are met.
8. While costs are an extremely important component of the evaluation of alternative plans and technologies in general, they are not the only consideration and must be supplemented with information on the benefits associated with the alternatives, as will be discussed in Chaps. 14 through 17.

PROBLEMS

9-1

Calculate the total cost of the rail rapid transit line used in the example in this chapter, with a daily required capacity of 32,000 passengers/day, if the trains operated in the peak were of 3 cars each operated at a headway of 2.25 min, and in the other periods were 2 cars each operated at a headway of 5.33 min. Compare the costs with those for the longer, less frequent trains used in the text example. What is the percentage increase in total cost? If the operating agency were responsible for only the continuing costs—maintenance, operation, and administration—what percentage increase in costs would it face?

9-2

How much would the costs of this transit line decrease if the number of stations were reduced to 10 (from 30) and train speeds increased to result in a round trip time of 0.9 h? What would be the advantages and disadvantages for attracting riders?

9-3

A pipeline with an 18-in diameter will cost $550,000/mi to construct and $0.005/ton-mi to operate. A 22-in-diameter pipe would cost $750,000/mi, but would cost only $0.004 to operate. If the pipeline lasted 25 years, with 10 percent interest, over what volume range should each be employed? What if the interest rate were reduced to 6 percent? Construction requires 1 year.

9-4

Compare the fully allocated cost of transporting a shipment of merchandise weighing 45 tons, with a weight density of 14 lb/ft^3, from New York to Chicago via rail, TOFC, and truck. The rail distance is 980 mi, truck 940 mi. Rail car capacity is 100 tons and 4900 ft^3; truck trailer capacity 29 tons and 2450 ft^3. TOFC movement is via one railroad, box car via two. Both movements have four intermediate yardings. What are the travel times, using Fig. 8-12? Plot cost vs. time for all three modes.

9-5

Plot the fully allocated cost per ton-mile of moving a carload (rail) shipment of freight varying from 20 to 70 tons over a distance of 1000 mi. Also plot the cost for distances of 500 and 250 mi. What is the pattern that emerges? This is typical of all means of freight transport.

9-6

Estimate the cost of providing the same 32,000 passengers/day capacity of the rail line example in the text with a bus service on an exclusive busway (at grade). You may devise a service with three zones and express operations if you wish. State assumptions and costs selected from the ranges given in the text. How do the costs compare to those for the rail line? How would level of service compare?

9-7

Consideration is being given to upgrading a section of rural highway 15,000 ft long which is expected to carry 3000 vehicles/day in the future year for which costs are to be compared. The facility now is two lanes, with a 9-ft lane width. The thirtieth highest hour volume is 410 vehicles/h, the capacity is 900 vehicles/h, and the design speed is 50 mi/h. Virtually all the traffic is automobile. The new road will be tangent and have a 60 mi/h design speed and a capacity of 1500 vehicles/h. The existing road has the profile given in Fig. 9-7, while the new road will be of constant grade. One-quarter of the present road consists of 6° curves, with a superelevation of 0.03. If the new road costs $1,000,000 and will cost $3000/mi to maintain per annum (compared to $5300 for the present road), it is justified on grounds of reducing the total transport costs? User costs may be assumed equal to 1.75 of those developed by AASHO in 1959. Use a road life of 30 years and a 10 percent interest rate.

9-8
A city is faced with a choice of two different types of transit systems. Type A has an initial cost of $500 million and a continuing cost per passenger-mile of $0.05. Type B has an initial cost of $700 million, but a continuing cost of $0.035. Each system will take 3 years to build, with equal costs in each year. Initially traffic will average 200,000 passenger-mi/day (average for 365 days/year). If this grows at a rate of 3 percent per annum for 25 years, which system would be least costly as measured by present value of all costs? If growth were at 6 percent? Use discount rates of 8, 10, and 15 percent per annum.

9-9
Calculate the cost of moving 50 tons of merchandise freight by truck over a distance of 300 mi. Consider five different shipment sizes into which this freight is divided: 1, 2.5, 12.5, 25, and 50 tons. Use Eq. (9-19). Plot the total cost versus shipment size, and the average cost per ton-mile versus shipment size. The pattern is a general one for freight: As shipment weight increases the cost per unit weight decreases. The cargo density is 12 lb/ft^3.

REFERENCES

Air Transport Association of America (1975), "Air Transport 1974," Washington, DC.

American Association of State Highway Officials (1960), "Road User Benefit Analyses for Highway Improvements," Washington, DC.

Association of American Railroads (1976), "Yearbook of Railroad Facts," Washington, DC.

Bolusky, Eric B., Edward K. Morlok, Robert M. Prowda, Mary Ann Semerano, and Peter E. Ward (1973), "Short-Haul Intercity Passenger Carriers: Their Cost, Capacity and Service Characteristics," Research Report. The Northwestern University Transportation Center, Evanston, IL.

Boyd, J. Hayden, Norman J. Asher, and Elliot S. Wetzler (1973), "Evaluation of Rail Rapid Transit and Express Bus Service in the Urban Commuter Market," Report No. DOT P 6520.1, U.S. Dept. of Transportation, Washington, DC.

Claffey, Paul J. (1971), "Running Costs of Motor Vehicles as Affected by Road Design and Traffic," National Cooperative Highway Research Program Report No. 111, Highway Research Board, Washington, DC.

Committee on Highway Capacity (1965), "Highway Capacity Manual," Special Report No. 87, Highway Research Board, Washington, D.C.

Curry, David A., and Dudley G. Anderson (1972), "Procedures for Estimating Highway User Costs, Air Pollution and Noise Effects," National Cooperative Highway Research Program Report No. 133, Highway Research Board, Washington, DC.

Federal Highway Administration, Construction and Maintenance Division (1976), "Price Trends for Federal-Aid Highway Construction," Washington, DC.

Heggie, Ian G. (1972), "Transport Engineering Economics," McGraw-Hill, London.

Interstate Commerce Commission (1974), "Cost of Transporting Freight by Class I and II Motor Common Carriers of General Commodities, 1973," Statement No. 2C1-73, Washington, DC.

Interstate Commerce Commission (1975), "Rail Carload Cost Scales, 1973," Statement No. 1C1-73, Washington, DC.

Kane, Anthony R., and Edward K. Morlok (1970), Road Capital Cost Functions: Theory and Estimation, *Transportation Research,* **4**(4): 325–337.

Lee, Robert R., and E. L. Grant (1965), Inflation and Highway Economy Studies, *Highway Research Record* No. 100, pp. 20–37.

Meyer, John R., John F. Kain, and Martin Wohl (1965), "The Urban Transportation Problem," Harvard, Cambridge, MA.

Meyer, John R., Merton J. Peck, John Stenason, and Charles Zwick (1959), "The Economics of Competition in the Transportation Industries," Harvard, Cambridge, MA, copyright © 1959 by the President and Fellows of Harvard College.

Morlok, Edward K. (1967), "An Analysis of Transport Technology and Network Structure," Northwestern University Transportation Center, Evanston, IL.

Morlok, Edward K. (1965), "Rail Transit Costs," Research Report, Northwestern University Transportation Center, Evanston, IL.

Morlok, Edward K., and John A. Warner (1978), Approximating Equations to I.C.C. Costs of Rail, TOFC, and Truck Intercity Freight Systems: Development and Application, *Transportation Research Record* No. 637.

Motor Vehicle Manufacturers Association (1976a), "1975 Automobile Facts and Figures," Detroit, MI.

Motor Vehicle Manufacturers Association (1976b), "1975 Motor Truck Facts," Detroit, MI.

Mutschler, P. H., R. J. Evans, and G. M. Larwood (1973), "Comparative Transportation Costs of Supplying Low-Sulfur Fuels to Midwestern and Eastern Domestic Energy Markets," Information Circular 8614, U.S. Department of Interior, Bureau of Mines, Washington, DC.

National Association of Motor Bus Owners (1976), "One Half-Century of Service," Washington, DC.

National Safety Council (1968), "Accident Facts," Washington, DC.

Odier, L., R. S. Millard, Pimentel dos Santos, and S. R. Mehra (1971), "Low Cost Roads: Design, Construction and Maintenance," Butterworths, London, © Unesco 1971. Figure reproduced by permission of Unesco.

Pignataro, Louis J. (1973), "Traffic Engineering: Theory and Practice," Prentice-Hall, Englewood Cliffs, NJ.

Poole, Ernest C. (1962), "Costs—A Tool for Railroad Management," Simmons-Boardman, New York.

Riggs, James L. (1977), "Engineering Economics," McGraw-Hill, New York.

Samuelson, Paul A. (1976), "Economics," McGraw-Hill, New York.

Sanders, D. B., T. A. Reynen, and K. Bhatt (1974), "Characteristics of Urban Transportation Systems—A Handbook for Transportation Planners," U.S. Dept. of Transportation, Urban Mass Transportation Administration, Washington, DC (updated regularly).

Smith, R. N., and T. N. Tamburri (1969), Direct Costs of California State Highway Accidents, *Highway Research Record* No. 255.

Transportation Association of America (1975), "Transportation Facts and Trends," Washington, DC.

U.S. Department of Commerce, Bureau of the Census (1976). "Statistical Abstract of the United States," 97th Edition, Washington, DC.

U.S. Department of Transportation, Federal Highway Administration (1973), "Highway Statistics," No. 5001-00088, U.S. Government Printing Office, Washington, DC.

Winfrey, Robley (1969), "Economic Analysis for Highways," International Textbook, Scranton, PA.

Winfrey, Robley, and Carl Zellner (1971), "Summary and Evaluation of Economic Consequences of Highway Improvements," National Highway Cooperative Highway Research Program Report No. 122, Highway Research Board, Washington, DC.

Wohl, Martin, and Brian V. Martin (1967), "Traffic Systems Analysis for Engineers and Planners," McGraw-Hill, New York.

Wyckoff, D. Daryl (1974), "Organizational Formality and Performance in the Motor Carrier Industry." Lexington, Lexington, MA.

Wyckoff, D. Daryl, and David H. Maister (1975), "The Owner-Operator: Independent Trucker," Lexington, Lexington, MA.

Transportation Demand

10.

It is essential that transportation engineers and planners understand the principles and techniques for estimating the demand for transportation. It is basically a reflection of the requirements for transport by users of the system, whether they be travelers or shippers of freight, and hence the demand for transport provides an important basis for the evaluation of transportation plans and facility designs. Without an understanding of transport demand, the efforts of engineers and planners may be misdirected and result in a system mismatched to the demand for transport and a waste of scarce resources.

THEORY

The theory of transport demand is derived largely from the economic theory of consumer choice. However, it is important to realize that persons from many professions other than economics have made significant contributions to the development of theoretically sound and practical means for estimating the demand for transportation in the context of actual planning and design studies. The current theory and techniques draw not only from economics but also from the fields of sociology, psychology, and marketing. In this section we shall first present the basic theory of demand for transport, illustrated with a few examples, and then in the following section shall present some models which are currently used for estimating transport demand.

The Derived Demand for Transport

The transportation of persons or things is generally not undertaken for its own sake, that is, people do not travel or convey goods because these activities are inherently desirable; transport is undertaken to achieve some other purpose. The demand for transportation is therefore referred to as a *derived demand,* stemming from the demand for some other commodity or service. Basically, it is derived from (1) the need for persons to travel to another location in order to partake in an activity (e.g., work, shop) there and (2) the need to transport goods to make them available where they can be used or consumed. The significance of the derived nature of transport demand for understanding transport demand, and for the development of practical methods for predicting demand, is substantial.

In the case of person travel, the derived character of the demand is reflected in the fact that travel is undertaken in order to achieve particular objectives or purposes, such as getting to work, purchasing food at a supermarket, swimming at a beach, etc. An important factor influencing the number of trips made to a particular place is thus the kinds of activities that can be engaged in there or, in other words, the degree to which specific trip purposes can be satisfied there. The cost of reaching that possible destination from the potential traveler's home or other origin of the trip is also important. The characteristics of the available means of transport from the person's origin to destination will largely determine the mode and route to be used. Furthermore, if we wish to determine or estimate the total number of trips which might be made from one place to another (such as between two towns), we will have to take into account the number of persons at the place of origin and perhaps such characteristics as their incomes and preferences for various kinds of activities. Thus the derived demand nature of travel makes it fairly complex to understand and hence to predict.

Similar considerations apply for the movement of goods. Freight is moved usually to make goods available at a location where there is a basic demand and where it would otherwise not be available. At the destination the goods could be consumed or could be used in the manufacture of some other product. Thus, again, it is necessary to take into account not only characteristics of the transportation system connecting two places, but also the demand for the good at the destination as well as the availability of that good at the origin of the movement. Because the movement of freight depends in part upon the demand for the good being moved (in addition to characteristics of the transport system such as the price for the movement), the demand for freight movement is also a derived demand.

Demand Functions or Schedules

The general economic theory of demand for commodities relates the quantity of a particular commodity which will be consumed (or purchased) to the price. In the usual presentation, the commodity is homogeneous, meaning that all of the units sold or exchanged in the market are identical. In general, as the price decreases, the quantity purchased by consumers increases (Fig. 10-1). It is usually hypothesized that above a certain price none of the commodity would be sold and there is a maximum quantity that would be sold corresponding to a zero price, although of course observations are rarely made at these two extreme levels of price. This type of demand curve is often referred to as "downward-sloping," which means that as price drops the quantity consumed or quantity demanded increases.

The reason for the downward slope of most demand curves is that, of course, as the price drops, additional persons will be able to afford the good and will purchase it. Also, some who had perhaps purchased a limited

Figure 10-1 **Typical form of demand function showing the concept of elasticity.**

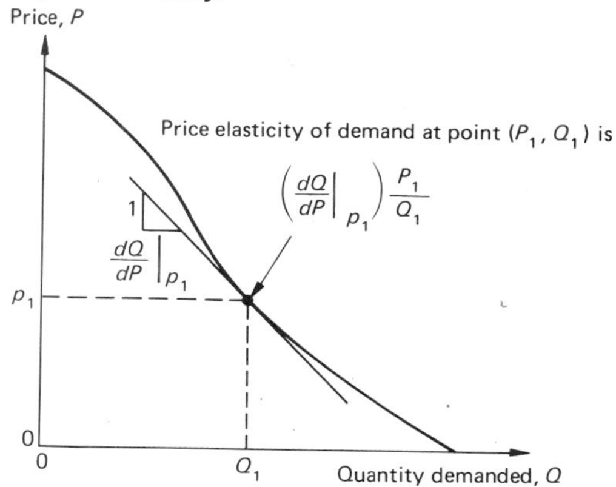

amount of this commodity at a higher price may now consume more. Depending upon the nature of the commodity and the various uses to which it might be put, consumers' total expenditures on the commodity may or may not increase as the result of a price reduction. It might be noted here that there are some commodities for which an increase in price has been observed to result in an increase in consumption, but this is usually due to a "snob appeal" of the commodity which makes it more highly valued by potential buyers as a result of the higher price. In effect, then, the nature of the commodity changes as the price varies. But this increase in quantity demanded with price is not typical of most commodities and certainly is not typical of transportation.

The demand curve, or demand schedule, as it is often called, as shown in Fig. 10-1, may take on a variety of shapes. Economists have defined the concept of price elasticity of demand to express a very important characteristic of the shape of the demand curve. Price elasticity is a measure of the rate of change of quantity demanded relative to the rate of change in price. In general, its value at one point will differ from its value at another point. A demand function or model relating the quantity demanded to price may be expressed as follows (the Roman style D indicating that it is a function defined on the variable P contained in the parentheses):

$$Q = \mathrm{D}(P) \tag{10-1}$$

and the point price elasticity of demand is defined by

$$\epsilon_P \equiv \frac{dQ}{dP}\frac{P}{Q} \tag{10-2}$$

where ϵ_P = point price elasticity of demand
P = price
Q = quantity demanded
$\mathrm{D}(P)$ = demand function

Precisely speaking, elasticity is a derivative defined at a point, as in Eq. (10-2). However, it is often defined in economics literature as the percentage change in quantity demanded resulting from a 1 percent change in price. Such a definition is usually given in order to make the concept of elasticity understandable to those who do not have a knowledge of calculus. As you may recall from the definition of a derivative, the derivative of y with respect to x (y and x being two variables related by a function) is defined as the ratio of a change in the value of y relative to a change in the value of x as the change in x approaches zero in the limit. The use of 1 percent in the approximate definition of elasticity is analogous to the use of a very small change in x, rather than the limit approaching zero, in the definition of the derivative.

In many situations, the elasticity of demand for a commodity with respect to its price has been found to be essentially constant. This situation arises with a demand schedule corresponding to the following mathematical model:

$$Q = \alpha P^{\beta} \tag{10-3}$$

where $\alpha, \beta =$ constant parameters of the demand function

That this corresponds to constant elasticity can be seen by differentiating this demand model with respect to price:

$$\frac{dQ}{dP} = \alpha\beta P^{\beta-1}$$

Substitution into the definition of elasticity, Eq. (10-2), yields

$$\epsilon_P = \alpha\beta P^{\beta-1} P Q^{-1}$$

With substitution of the original expression of demand we have

$$\epsilon_P = \alpha\beta P^{\beta-1} \left(\frac{P}{\alpha P^{\beta}}\right) = \beta \tag{10-4}$$

Thus β, the exponent or power of the price of the commodity P, is the price elasticity.

The concept of elasticity is used not only by economists but also by those concerned with the practical problems of industry, including transportation. For example in the public transportation industry in the United States it has been accepted as a general rule of thumb that a 1 percent increase in transit fares will result in a 3 percent decrease in revenue from passengers (Curtin, 1968; deNeufville and Freidlander, 1968). This is, of course, a statement regarding the elasticity of demand, in the form of the more common approximate definition, based on percentage changes, rather than the derivative definition. The elasticity of transit demand with respect to transit price is approximately -3, meaning that the number of passengers decreases (due to the minus sign) at a rate three times the increase in fare. The use of this in practice can be seen from a simple example. Assume a transit line carries 10,000 passengers/day at its present fare of 40 cents/ride. The transit authority is considering raising this fare to 45 cents and wishes to use this rule to estimate the number of passengers who would use the line with the increased fare. As derived above, the expression for the demand in terms of fare is given by Eq. (10-3), and substitution of the known quantity demanded, price, and elasticity of demand yields an estimate of the single unknown parameter α as follows:

$$10{,}000 = \alpha(40)^{-3.00}$$

Hence $\alpha = 6.40 \times 10^8$

and $Q = 6.40 \times 10^8 P^{-3.00}$ (10-5)

where Q = passengers/day
P = cents/trip

An increase in the fare to 45 cents will then yield for quantity demanded:

$$Q = 6.40 \times 10^8(45)^{-3.00} = 7023 \text{ passengers/day}$$

Thus the increase in fare from 40 to 45 cents, a 12.5 percent increase, is estimated to reduce the ridership on this line from 10,000 passengers/day to 7023 passengers/day, a 29.8 percent decrease.

It should be noted here that the definition of elasticity in terms of the effect of a 1 percent change in price often leads persons to use the elasticity incorrectly. Specifically, one might reason that if a 1 percent increase in price results in a 3 percent reduction in passengers, then a 12.5 percent increase in price should yield a proportionate reduction in passengers, specifically, a 37.5 percent reduction. This is obviously incorrect, as comparison with the calculations above reveals. In this case this erroneous use of elasticity would result in a prediction of a reduction of 3750 passengers, with a resulting ridership of 6250 passengers/day, a substantial underestimate of the expected ridership.

As mentioned above, elasticity provides information on the shape of the demand curve. This is illustrated in Fig. 10-2. Figure 10-2*a* illustrates the general form of a constant elasticity demand function, identical in form to that embodied in the transit managers' rule of thumb. A straight-line demand curve as shown in Fig. 10-2*b* yields a continuously varying elasticity which reaches infinity at the price intercept and zero at the quantity demanded intercept. A demand curve convex to the origin, as shown in Fig. 10-2*c*, also exhibits a range of elasticities, although there is very little change in elasticity at low quantities. Finally, Fig. 10-2*d* illustrates completely inelastic demand, in which the quantity demanded is the same regardless of price. Although this obviously could not hold over the entire range of possible prices (for at some point people simply would be unable to pay the price), it is a condition which holds under certain circumstances and within a certain range of prices, as for example when people are essentially insensitive to the price, perhaps because it is small.

Thus the range of values of elasticity of demand for a particular commodity with respect to its price for downward-sloping demand curves ranges from minus infinity to, but not including, zero. One of the reasons the concept of elasticity is so important is that it is closely related to the change in total revenue resulting from a change in price. At an elasticity of exactly -1, termed unitary elasticity, the total revenue remains the same regardless of price. However, if the demand is elastic, meaning that the elasticity is less than -1, then revenue will increase as a result of a decrease in price (i.e., the demand responds relatively elastically to price changes). And finally, inelastic

Figure 10-2 **Alternative forms of demand function and the associated elasticities. *(a)* Constant elasticity. *(b)* Linear demand. *(c)* Convex (to origin) demand function. *(d)* Completely inelastic demand.**

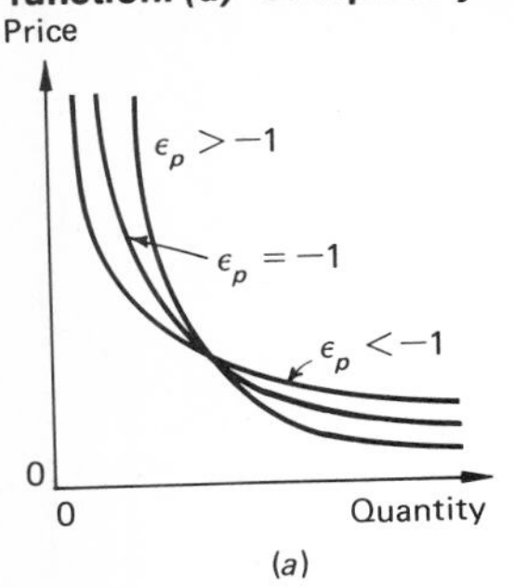

(a)

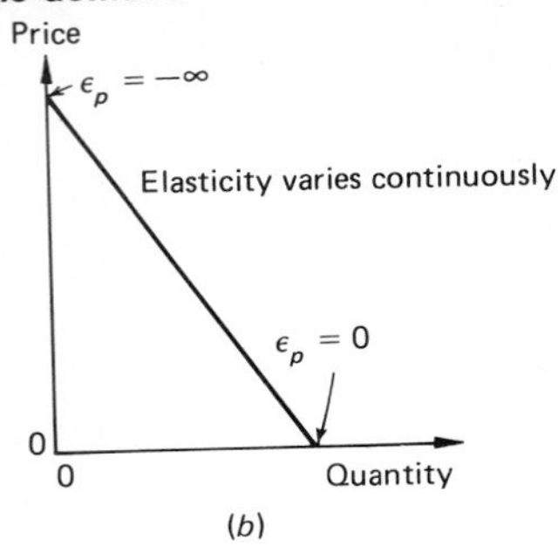

(b)

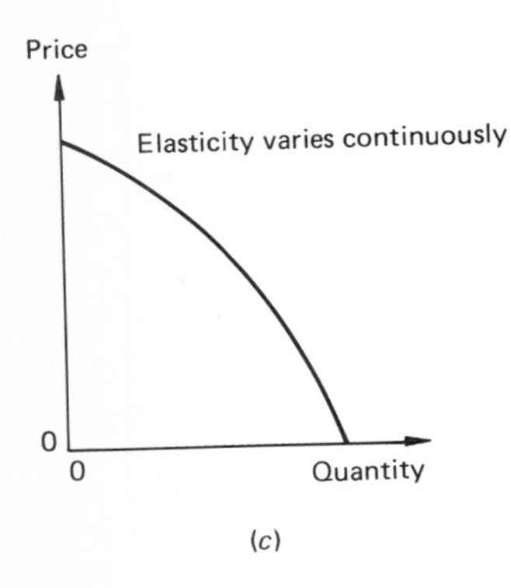

(c)

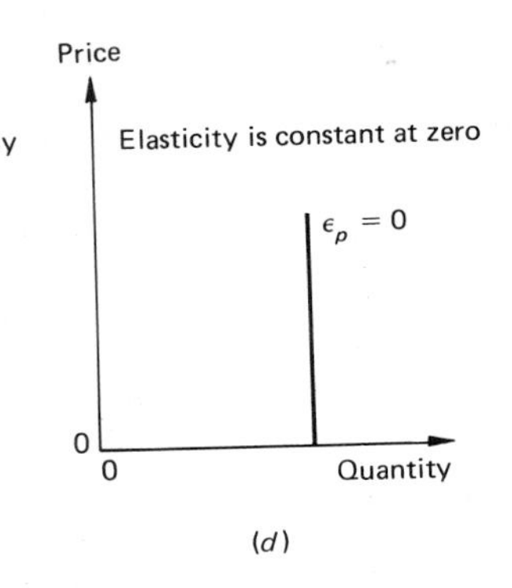

(d)

demand means that the elasticity lies between -1 and 0, and revenue will decrease as a result of a decrease in price. These conclusions regarding the effect on revenue can be verified easily by the reader.

Demand Models

At a number of points in the discussion we have mentioned the term demand models, in fact Eq. (10-5) is an example of a demand model, and the specific parameter values for α and β used in the transit example yielded a specific, quantitative, operational model for that problem. Most transportation demand models, however, are much more complex than this equation. Reasons for the complexity are the derived demand nature of transportation and the complexity of characteristics of transportation which are usually offered to a potential traveler or shipper, among others. A typical transportation demand model is of the following form:

$$d_{ij}^{pm} = D^{pm}(S_i, S_j; c_{ij}^m, c_{ij}^n, \ldots ; S_k; c_{ik}^m, c_{ik}^n, \ldots ; \ldots) \tag{10-6}$$

where d_{ij}^{pm} = quantity demanded of travel from city i to city j for purpose p via mode m

D^{pm} = function for estimating demand

S_i = socioeconomic characteristics of city i
c_{ij}^m = price and level of service characteristics of mode m from city i to city j
k = alternative destination city where trip purpose p may be satisfied
n = alternative to mode m

The reason for including transportation characteristics associated with the mode for which the model predicts the demand as well as for other possible competing modes is that the price and level of service characteristics of all of these modes will influence the usage of the mode of interest. Presumably as the price of the competing mode is reduced, or its level of service improved, the amount of travel on that mode will increase, partly as a result of drawing some traffic away from its competitor. The reason for including both prices and levels of service of the various modes is that the potential traveler responds to both of these. In the classical economic sense the traveler responds to prices. However, when making decisions regarding travel, the potential traveler will consider other factors such as total travel time, possible variation in that travel time due to unreliability of a particular mode, fatigue, perhaps due to the requirement of driving a motor vehicle or standing while traveling, discomfort due to a rough-riding vehicle or uncomfortable seat, etc.

The reason for the inclusion of socioeconomic characteristics of the persons making the trips and of possible destinations is embodied in the derived demand nature of travel. The purpose of the trip must be specified in order to evaluate the degree to which various possible destinations, or more precisely, the activity possibilities at those destinations, meet the objectives for making the trip. Similarly, the number of persons and their characteristics and preferences at the origin will determine the inherent demand or desire for making trips for various purposes as well as their ability to expend money and perhaps other resources in order to do so, and hence these characteristics are also important. The characteristics associated with the trip origin and possible destinations are basically socioeconomic in nature and are so termed.

Thus the demand for travel between two zones is a function of a number of characteristics of these two areas, the travel connections between them, as well as the characteristics of and connections to alternative destinations which might satisfy the same trip purposes. All this makes for an exceedingly complex model of the demand for transportation. The demand is thus seen to be a function of a number of variables, including but not limited to price. The need to include all of these factors in the estimation of travel demand has long been recognized by economists and engineers and others engaged in travel demand estimation. It is interesting that in recent years the need for including characteristics of commodities being consumed has been recognized by economists concerned with advancing the theory of consumer choice. The basic approach, that of predicting travel demand on the basis of characteristics of that travel rather than labels on modes and only their associated prices, is embodied in the recently developed extensions of the theory of demand. As has been stated by Lancaster (1966, p. 133),

The chief technical novelty lies in breaking away from the traditional approach that goods are the direct objects of utility and, instead, supposing that it is properties or characteristics of the goods from which utility is de-

rived. . . . Utility or preference orderings are assumed to rank collections of characteristics and only rank collections of goods indirectly through the characteristics that they possess.

While this may seem rather obvious from our discussions of transport demand, it does represent a very significant departure from the traditional, rather limited, theory of demand.

AN EXAMPLE TRAVEL DEMAND MODEL

Most of the concepts which we have discussed above can be illustrated by examining a model developed to help evaluate various alternative proposals for improving intercity travel in the heavily urbanized Washington-New York-Boston corridor. In the mid-1960s it was recognized that intercity travel in that corridor was becoming increasingly difficult and that the situation would be aggravated by increases in population as well as increases in per capita travel. As a result the Northeast Corridor Transportation Project was formed to determine future requirements for transportation in the corridor and to evaluate alternative improvements such as high-speed rail service, vertical take-off and landing aircraft, short take-off and landing aircraft, additional freeways and toll roads, etc. Included in this evaluation was an estimation of the amount of travel which would occur on various portions of the transportation system, including any new or improved services to be offered in the corridor. To do this it was necessary to have a travel demand model; one that was developed was the so-called Kraft-SARC model, so named because it was developed by Gerald Kraft of the Systems Analysis and Research Corporation. Although many other models could be used to illustrate the concepts we have discussed above, this one serves the purpose well.

The model is designed to estimate the number of travelers per year on each of four modes of travel: air, automobile, bus, and railroad. It is designed to do so under conditions where the characteristics of these modes might change, due to various improvements to the transport system and perhaps new pricing policies. It therefore includes two characteristic variables for each mode: travel time and price. In both cases the values are door-to-door, rather than terminal-to-terminal of the intercity carrier, partly because travelers presumably do react to total travel time and total travel cost rather than to only one component of the time or cost and also because improvements in intercity travel may include improvements in intra-urban travel, making access to terminals easier. Other variables describing the modes might have been included, such as measures of comfort, but available data and resources did not permit this. Also because the estimate would have to be made of travel sometime in the future, after improvements might be implemented, it was necessary for the model to reflect increases in population, employment, income, and perhaps other changes. Thus these socioeconomic variables had to be included in the model. Since the motivation for business travel is different from other travel, it was decided to develop two distinct models, one for business travel and one for personal travel. Again, in many contexts it would be desirable to distinguish further between travel types, such as between travel to visit friends and relatives and travel to recreational sites, but this was not done in this case.

Although the form of the model may appear at first to be very complex, in reality it is a rather simple extension of the constant elasticity model of Eq. (10-3). Given below is the form of the model for estimating travel on one mode, by purpose, between one pair of cities (Systems Analysis and Research Corp, 1963, pp. V-44,V-45). As in prior chapters where complex formulas have been presented on topics wherein the reader may wish to consult the original report, the notation of that report is retained.

$$D_{ij}^{mB} = e^{K_{mB}}(E_i E_j)^{e_m} \left[\prod_q (P_{ij}^{qB})^{p_{mB}^q}\right]\left[\prod_q (T_{ij}^{qB})^{t_{mB}^q}\right] (Y_i)^{y_{mB}}(A_j)^{a_{mB}} \tag{10-7}$$

$$D_{ij}^{mP} = e^{K_{mP}}(N_i N_j)^{n_m} \left[\prod_q (P_{ij}^{qP})^{p_{mP}^q}\right]\left[\prod_q (T_{ij}^{qP})^{t_{mP}^q}\right] (Y_i)^{y_{mP}}(A_j)^{a_{mP}} \tag{10-8}$$

where D_{ij}^{mp} = the number of round trips originating at city i going to city j via mode m for purpose p

K_{mp} = a constant for the demand model for mode m for purpose p

e_m = the elasticity of demand for trips via mode m for business with respect to the weighted employment product

E_k = the employment in city k weighted by trips per employee

n_m = the elasticity of demand for personal trips via mode m with respect to population product

N_k = the total population of city k

p_{mp}^q = the elasticity of demand for trips via mode m for purpose p with respect to the price of trips via mode q for purpose p

P_{ij}^{qp} = the one-way cost of travel between cities i and j via mode q for purpose p

t_{mp}^q = the elasticity of demand for trips via mode m for purpose p with respect to the travel time of trips via mode q for purpose p

T_{ij}^{qp} = the one-way travel time between cities i and j via mode q for purpose p

y_{mp} = the elasticity of demand for trips via mode m for purpose p with respect to per capita personal income

Y_i = per capita personal income in city i

a_{mp} = the elasticity of demand for trips via mode m for purpose p with respect to destination city attractiveness

A_j = the attractiveness of city j

Modes are R = rail, B = bus, A = air, C = auto and purposes are B = business, P = personal. $\prod_q(\cdot)$ indicates the product of all the terms $(\cdot)$ for the values of q, which in this case are the four modes $q = R, B, A,$ and C.

As can be seen, this model is basically similar to that in Eq. (10-3), with the addition of more variables, each adding an additional factor consisting of the variable raised to a power. In this model, the socioeconomic and transportation characteristics can change, to correspond to the cities i and j, the year, and the nature of the transportation alternatives being considered. These are all raised to appropriate powers, these powers, or elasticities, and the coefficient being the fixed parameters of the model which, once determined, remain fixed in application. This is analogous to the parameters α and β of the transit ridership model described above, in which these two parameters remained unchanged and the variable price reflected changed conditions.

One may naturally wonder how one goes about estimating the proper values of all these parameters. The process is conceptually very similar to the manner in which the parameter of the intercity bus cost model presented in Chap. 9 was estimated. The parameters are estimated on the basis of data on actual travel volumes, fares, etc. In the case of the bus cost model, which involved only two variables, cost and vehicle-miles, the process could be performed graphically, as illustrated in Fig. 9-5. For the travel demand model, actual data on the number of trips by the various modes of travel were gathered, along with figures of the population, employment, etc. in the Boston-Washington region. Then rather elaborate statistical regression techniques were used to estimate those values of the parameters which would minimize the difference between the travel demand by mode and city pair estimated by the model and the actual observed travel demand under those conditions. It is these parameter values which are then used for future situations to relate changed transportation characteristics or socioeconomic characteristics (such as population growth) to future travel demand.

The parameter values estimated in this study are presented in Table 10-1. In some cases, the parameter value was zero, indicating that that variable seemed to have no effect on the usage of that particular mode. In other cases, for theoretical reasons, certain variables were omitted from the estimating relationships for a particular mode. To see how these parameter values are used in practice, below we shall develop the equation for estimating rail travel. This will be done assuming a future-year population of city A and city B, the two cities between which travel is being estimated, of 1,500,000 and 5,000,000. In that same year, these cities will have mean income levels of \$7000 and \$6000, respectively. The constant $e^{K_{mp}}$ is 4.27×10^{-5}. Rail cost and time are \$25.00 and 354 min, respectively. Bus cost is \$17.74 and air time is 220 min. Auto time is 300 min. The attractiveness measure of city B is 12.2. Thus the general demand equation is

$$D_{AB}^{RP} = (4.27 \times 10^{-5})(7.5 \times 10^{12})^{0.854}(6 \times 10^{3})^{0.465} 12.2^{1.601} (17.74^{3.15} 25^{-3.003} 354^{-2.636} 220^{0.052} 300^{0.056})$$

Therefore the total number of personal trips by rail from city A to B and return, is

$$D_{AB}^{RP} = 2.508 \times 10^{3} \text{ trips/day}$$

URBAN TRAVEL-FORECASTING MODELS

In the last two decades there has been considerable research and development of methods for forecasting future travel within urban areas. In fact, this aspect of travel demand has probably received the most attention, due to transportation problems created by the rapid increase in the population and travel within urban areas since World War II. Methods of forecasting future needs for and usage of new urban transportation facilities have been developed largely by engineers and planners, although in most cases economic concepts of demand theory have been employed.

Sequential Demand Models

The standard methods for forecasting urban travel demand differ from the more common demand model consisting of a single equation, as illustrated

Table 10-1 Parameter Values of the Kraft-SARC Intercity Passenger Travel Demand Model†

Results of Constrained Regression Estimation for Business Trip Demand Models

	Elasticities of Business Trip Demand with Respect to:											
Mode	Employment product e_q	Rail cost p^R_{mB}	Bus cost p^B_{mB}	Air cost p^A_{mB}	Auto cost p^C_{mB}	Rail time t^R_{mB}	Bus time t^B_{mB}	Air time t^A_{mB}	Auto time t^C_{mB}	Income y_{qB}	Attractiveness a_{qB}	r^2
Rail	0.893	−0.354	2.283	0	0	−4.376	0	.361	0	‡	‡	0.91
Bus	0.802	0	−0.740	‡	0	0	−1.700	‡	0	‡	‡	0.73
Air	0.929	0	‡	−0.891	0	0.973	‡	−2.103	1.078	1.418	0.836	0.92
Auto	1.067	1.127	0	0	−0.358	0.844	0	0	−3.410	‡	0.333	0.71

Results of Constrained Regression Estimation for Personal Trip Demand Models

	Elasticities of Personal Trip Demand with Respect to:											
Mode	Population product n_q	Rail cost p^R_{mP}	Bus cost p^B_{mP}	Air cost p^A_{mP}	Auto cost p^C_{mP}	Rail time t^R_{mP}	Bus time t^B_{mP}	Air time t^A_{mP}	Auto time t^C_{mP}	Income y_{qP}	Attractiveness a_{qP}	r^2
Rail	0.854	−3.003	3.150	0	0	−2.636	0	0.052	0.056	0.465	1.601	0.89
Bus	0.673	0	−0.689	‡	0	0	−1.589	‡	0	2.542	1.869	0.78
Air	0.911	0	‡	−0.914	0.095	0.857	‡	−2.213	1.120	1.905	1.020	0.91
Auto	0.794	0.185	0	0.489	−0.929	0.458	0.074	0	−1.364	1.523	1.574	0.90

† Systems Analysis and Research Corp. (1963, p. V–92).
‡ This variable was not in the model.

by the Kraft-SARC intercity travel demand model. The urban travel forecasting demand model consists of a number of separate models used sequentially. The need for a number of distinct models which acting together predict travel demand undoubtedly reflects the complexity of urban travel, where consideration must be given to many possible destinations which might satisfy the same trip purpose, and to the many routes available to travelers within each mode.

The five steps or models that make up the urban forecasting procedure are portrayed in Fig. 10-3. The first step is to predict the land use pattern for the future year in which trips are to be predicted. The land use pattern portrays the spatial arrangement of human activities, described by the amount of each activity in each of a number of small areas, called zones, into which the region has been divided. Typically a zone includes 0.25 to 2 mi^2, and often the zones are delineated as squares, although sometimes they follow natural boundaries such as rivers. The zonal measures of activity include such items as the number of persons living there, the number of employees by types of industry, the number and sizes of shopping centers, etc. On this basis, the number of trips originating and terminating within each area is estimated, this being termed trip generation analysis. Then the trip origins are assigned to the various possible destinations, yielding the spatial distribution of trips in the process termed trip distribution. Once the origin and destination of the trip is known, then the various alternative modes can be compared to determine the likelihood of using each, which is done in the mode choice phase. Finally, once the mode of travel is determined, then the particular route to be used can be selected, termed traffic assignment. There are numerous variations of this process, although the order presented is the most common.

Figure 10-3 **The urban travel forecasting process.**

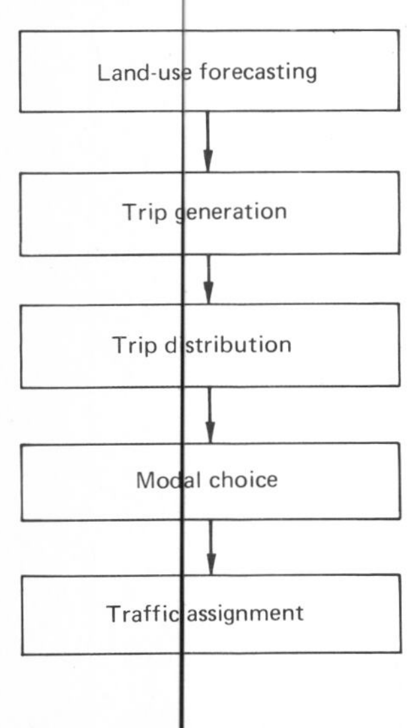

Land Use Forecasting

Ideally, transportation planners and engineers would like to be given estimates of the future pattern of human activities within urban areas for the future year (or years) with which they are concerned. Typically these data would be sought from city and regional planners, who are most concerned with the spatial pattern of activities and with the details of land use such as building codes and zoning. However, transportation professionals often found that such future land use projections are either unavailable or lack sufficient detail. Therefore, much attention is given in transportation planning efforts to predicting future land use patterns.

Many methods are in use, but one developed by the Chicago Area Transportation Study (1959) (acronym, CATS) is widely followed. The procedure is basically nonmathematical, relying heavily on the judgment of those involved in the forecasting. In the Chicago application of this method, the predictions were made by a team consisting not only of transportation engineers and planners, but also of real estate developers, community leaders, and others knowledgeable in expected land use development.

The procedure is based upon application of the following three rules: (1) land development decreases in intensity with increasing distance from the central business district; (2) available land that is already in use declines in density (activity per unit area) with increasing distance from the central business district; (3) proportional amounts of land devoted to different land uses are stable.

The method requires that future population and other measures of overall activity in the region be predicted, such predictions often being available from national and state planning organizations. Following this, based upon examining existing development plans, land use zoning restrictions and the rules mentioned above, an estimate is made of the capacity of each zone for accepting residences and other types of activities. On the basis of these considerations, the projected population growth is allocated to various zones, as are the amounts of projected growth of residential-serving activities, such as shopping centers and commercial and industrial development. The result of this process performed by such a team is a specific projection of the amount of future activity expected in various land use classes in each of the possible zones.

Trip Generation Models

Trip generation models are used to estimate the total number of trips originating in each zone, and the total number of trips which will terminate in each zone, by trip purpose. As will be revealed in the discussion of other aspects of travel demand, consideration of the purpose of the trip is important not only in determining the factors which will influence the number of trips to be made, but it will also influence the choice of mode, a very important consideration in the design of future transportation systems.

Trip generation models typically estimate the number of trips by purpose on the basis of land use or socioeconomic characteristics of each zone. Usually no explicit consideration is given to transportation system characteristics, even though according to the theory of travel demand, the cost and level of service of transport will influence the number of trips made. Perhaps the reason that transportation system variables do not enter trip generation relationships is that in most urban areas in developed countries where these models are most widely used, the cost of travel is relatively small compared to total income and the quality is fairly high (although there is obviously considerable pressure to improve the quality of service) so that transportation system price and service variations are not significant enough to influence the total number of trips being made. However, it should be noted that in estimating the number of trips generated in the extreme outer fringes of very large metropolitan areas, some professionals have found it necessary to revise downward the total number of trips, reflecting a tendency of people in such areas to combine in one trip the satisfaction of many trip purposes (e.g., linking trips for shopping with social or work trips), this presumably due to the relatively long distances which must be traveled in those fringe areas in order to satisfy the various trip purposes.

The mathematical forms of most trip generation models are presented below. The first form yields the total number of trips per zone, while the second gives trips per household. The latter is used to estimate trips per zone by multiplication by the number of households in the zone.

$$o_i^p = O^p(S_{i1}, S_{i2}, \ldots, S_{ij}, \ldots) \tag{10-9}$$

$$\frac{o_i^p}{H_i} = O^{p\prime}(S_{i1}, S_{i2}, \ldots, S_{ij}, \ldots) \tag{10-10}$$

where o_i^p = number of trips of purpose p originated in zone i

H_i = number of households in zone i

O^p, $O^{p\prime}$ = mathematical functions

S_{ij} = socioeconomic measures of activity j in zone i

For estimating the number of trips terminated or attracted to zone i, the models are of the same form, but an a and A replace o and O, respectively.

At this juncture it is perhaps appropriate to mention some definitions which are important to the understanding of current trip generation methods. Specifically, there are two categories of purposes of trips: home-based trips, for which either the origin or the destination is the home, and non-home-based trips, which include all other trips. Secondly, there are the concepts of trip origin and trip destination. Finally, there is the concept of attraction and production zones, the production zone being the zone in which the home is located for all home-based trips, regardless of whether that is the origin or the destination, while the production zone is the origin zone for non-home-based trips. Similarly, for home-based trips the attraction zone is the non-home zone, while for non-home-based trips, the destination zone is the attraction zone. These concepts and their definitions are important because of the manner in which the number of home-based and non-home-based trips are estimated. The production of home-based trips is based on characteristics of the home end of the trip, while the production of non-home-based trips is based upon characteristics of the origin zone. The production of non-home-based trips at zones away from the home is based on characteristics of the origin zone, while the attraction for non-home-based trips is determined by the characteristics of the destination zone.

There are many methods in use to estimate trip generation. First we will consider two methods for residential or home-based trip generation, and then turn to non-home-based or nonresidential trip generation. The two methods most widely used for residential trip generation are cross-classification analysis, often called category analysis in Europe, and regression analysis.

In cross-classification analysis, the first step is to decide what factors will be used as a basis for estimation of the number of trips made per household. The range of each of these characteristics is then divided into a set of classes, which may reflect either a natural classification for some variables, such as the number of cars owned or family size, or an arbitrary classification, such as one dividing income into a few ranges. Once this is done, then data on the actual number of trips made by each household (or in reality a sample of households in the region) are gathered, along with all of the other information on those households required to estimate trip generation. Each household is then included in the appropriate classification cell, and the mean trip generation rate measured by number of trips per household is used to estimate the number of trips made by households which in the future fall into that cell.

This abstract discussion is best made concrete by an example. Consider a particular trip purpose for which the number of persons in the family and the number of automobiles owned will be used as the basis for classifying households for estimating trip generation rates. Each of these variables takes on integer values naturally, so the classification can be by the number in the family, with 5 or more representing the upper classification, and the number of cars owned being 0, 1, or 2 or more. Actual data on households in the region have been gathered, and the number of households within each cell together with the number of trips made by such households are shown in Table 10-2*a*. For example, 1934 households, each consisting of exactly 2 persons and owning 1 automobile, made 6129 trips/day. The mean number of trips per household is then calculated from these data, as shown in Table 10-2*b*. Continuing with the same example, the households consisting of 2 persons and 1 automobile made an average of 3.16 trips/day for this purpose. To predict the future number of trips from a zone, the number of families within each of these cells is estimated. For example, in a particular zone, the number of families within each cell might be estimated as shown in Table 10-2*c*. For the 51 families consisting of 2 persons and 1 automobile, the total number of trips made per day would be 161 (161.2 rounded to the nearest integer). The total number of trips from the zone would be simply the sum of all the trip rates multiplied by the corresponding number of households.

It is important to note at this point that to use this model in future predictions, it is necessary to estimate all of the variables—in this case the number of households in each zone by family size and the number of automobiles owned—which are used to explain the number of trips made. Only if such characteristics can be predicted with reasonable accuracy can the model be expected to yield reasonably good results. Based on overall economic and population growth trends, such factors as family size and number of automobiles owned can be predicted reasonably well. However, one might criticize the use of number of automobiles in this case because presumably families could be influenced to purchase fewer automobiles if transit service were improved, although the extent to which auto ownership is actually influenced by transit usage is somewhat unclear. Also, this model assumes that a household would make the same number of trips per day now as it would in

Table 10-2 Example of Cross-Classification Approach to Home-Based Trip Generation Model Development†
***a* Cross-Classification of Households by Auto Ownership and Family Size, Giving the Total Number of Households and Total Number of Trips Made by Them in Each Class.**

	Automobile Ownership					
	0		1		2 or more	
Family size	No. of households	No. of trips	No. of households	No. of trips	No. of households	No. of trips
1	925	1,098	1,872	4,821	121	206
2	1,471	2,105	1,934	6,129	692	1,501
3	1,268	1,850	3,071	13,989	4,178	19,782
4 or more	745	1,509	4,181	18,411	4,967	25,106

***b* Trip Generation Rates, Trips per Household per Day**

	Automobile Ownership		
Family size	0	1	2 or more
1	1.19	2.57	1.70
2	1.43	3.16	2.17
3	1.45	4.55	4.74
4 or more	2.02	4.40	5.06

***c* Forecasted Number of Households in One Zone in Each Class**

	Automobile Ownership		
Family size	0	1	2 or more
1	24	42	8
2	10	51	107
3	11	31	158
4 or more	3	17	309

the future, even though preferences and tastes might change with the passage of time, which would also influence the number of trips.

The other common method for residential or home-based trip production estimation is regression analysis, a method that, as we explained earlier, is used to estimate the best values for parameters of a given mathematical relationship between two or more variables. Again, a major problem is determining the variables to be used to predict the number of trips. A typical form of regression model for home-based work trips originating in a zone is as given below:

$$o_i^1 = 10.2 + 1.68H_i + 2.09C_i + 1.98W_i \qquad (10\text{-}11)$$

where o_i^p = total number of trips of purpose p (work, $p = 1$) generated in zone i
H_i = number of households in zone i
C_i = number of cars owned by all households in zone i
W_i = number of workers residing in zone i

This is an aggregate totals model, as exemplified by Eq. (10-9). It is linear, in the sense that the number of trips is a linear combination of the values of the predicting (or in regression jargon, independent) variables. In this model, the variable estimated is the total number of trips for a particular purpose originating or terminating in a zone, rather than the number of trips per household, although the predicted variable (or dependent in regression jargon, dependent in the sense that its value depends upon the variables) may be total trips per zone or per household.

To aid in developing these models and to indicate the quality of these models in fitting the data on actual travel on which they are based, some of the same measures of goodness of fit described in the discussion of the linear regression models in the cost chapter are also used in this context.

The application of regression models is conceptually identical to that of the cross-classification models. Basically, all of the variables on which the total number of trips depend (termed the independent variables in this type of model), must be predicted first, then entering these values into the equation results in a specific estimate for the total number of trips in a zone or per household in the zone. In the latter case the number of trips per household is then multiplied by the total number of households expected in the zone, yielding total trips for the zone. Thus, if for zone 97, $H_{97} = 220$ households, $C_{97} = 341$ autos, and $W_{97} = 267$ workers, the number of work trips per day o_{97}^1 equals 1621.

Nonresidential trip generation models are conceptually very similar to residential models, although they are not so well developed. These models can be based upon cross-classification or regression methods, although the latter are much more common. Again, the number of trips is divided by purpose, and different models are developed for each trip purpose. Usually the number of trips generated and attracted are based upon the land use types, distinctions usually being made between offices, industry, commerce, shops, education, public buildings, open space, transportation and utilities, and vacant land.

Within each activity classification and trip purpose, a common procedure is to estimate an aggregate trip rate based on dividing the observed number of trips for that purpose by some measure of total activity within that land use type. Typically the measure of activity is a measure of the size of the facility, such as thousands of square feet or number of employees. Alternatively, regression models may be used; some typical examples are presented in Table 10-3.

It was mentioned above that trip generation models are extremely useful not only for long-range transportation planning, the context in which we are describing them as part of an overall travel-forecasting procedure, but also for use alone. Specifically, trip generation models are often used to estimate the total number of trips which will originate and terminate in major developments, such as new shopping centers, airports, and industrial parks. At the time such facilities are being constructed or are subject to major

Table 10-3 Selected Examples of Nonresidential Trip Generation Models
***a* Home-Based Trips to Shopping Centers†**

Equation	r^2	Standard error as percentage of $\bar{Y}$
$Y_1 = 3875 + 5.35X_2 + 291.9X_3 - 578.5X_4 - 0.65X_6 - 22.31X_9$	0.920	21
$Y_2 = 2841 + 3.23X_2 + 241.4X_3 - 410.8X_4 - 0.34X_6 - 10.45X_7 + 4.32X_8 - 25.70X_9$	0.892	25
$Y_3 = 801 + 0.06X_1 + 0.90X_2 + 31.2X_3 - 108.0X_4 + 35.7X_5 - 0.18X_6 - 1.47X_7 - 2.36X_8 + 1.67X_9$	0.985	14

Variable type	Symbol	Definition
Dependent	Y_1	All auto driver trips
	Y_2	Auto driver trips to shop
	Y_3	Other auto driver trips
Independent	X_1	Number of parking spaces
	X_2	Total person work trips
	X_3	Distance from major competition, 0.1 mi
	X_4	Age of study data, years
	X_5	Age of center at time of study, years
	X_6	Reported travel speed of tripmakers, mi/h
	X_7	Floor space for convenience goods, 1000 ft^2
	X_8	Floor space for shopping goods, 1000 ft^2
	X_9	Floor space for other uses, 1000 ft^2

† Keefer (1966, pp. 108–109).

extension, it is often necessary to consider the adequacy of transportation facilities (especially roads) in the vicinity of those facilities. By estimating the total number of trips into and out of such areas, the ability of nearby road transport facilities to accommodate that traffic can be assessed, and appropriate plans made for any necessary changes. Very often guesses will be made as to the direction in which traffic enters or leaves such areas, or historical information may be used to estimate expansion; these are found to be quite adequate for practical engineering design purposes. This use will be illustrated in one of the problems.

Trip Distribution

The primary objective of trip distribution is to distribute or allocate the total number of trips originating in each zone and among all possible destination zones. This phase of travel forecasting builds directly upon the output of the trip generation phase. There are many approaches to trip distribution, as it is one of the least understood parts of travel demand forecasting. We will describe what is probably the most commonly used method, the method which uses the so-called gravity model.

The gravity model approach can be derived by considering certain aspects of the problem of trip distribution. First, the basic problem is to determine in some manner the total number of trips from zone i to zone j such that the following two conditions hold: The sum of all trips into zone i must equal the

Table 10-3 *(Continued)*

b All Trips to Manufacturing Plants‡

Equation	r^2	Standard error as percentage of $\bar{Y}$
$Y_1 = 1449 - 3.02X_4 + 1.34X_5 + 1.18X_6 - 9.46X_7 - 0.97X_8 - 1.58X_9 - 0.79X_{10} - 0.62X_{11} - 0.64X_{12} - 1.22X_{13} + 2.32X_{14} - 0.01X_{15}$	0.98	13
$Y_2 = -287 + 0.78X_9 + 0.43X_{10}$	0.82	64

Variable type	Symbol	Definition
Dependent	Y_1	Auto driver trips
	Y_2	Transit passenger trips
Independent	X_1	Population within 5-mi radius, 1000s
	X_2	Automobiles within 5-mi radius, 1000s
	X_3	Residential land within 5-mi radius, 0.1 acre
	X_4	Net residential density in plant zone, persons/acre
	X_5	Net manufacturing density in plant zone, persons/acre
	X_6	Plant site area, acres
	X_7	Prime shift percentage, three highest morning hours
	X_8	Employees from car-owning households
	X_9	Employees not licensed to drive
	X_{10}	White-collar employees
	X_{11}	Male employees
	X_{12}	CBD–plant distance/CBD–cordon line distance
	X_{13}	Average distance, home to work, 0.01 mi
	X_{14}	Total work trips to plant
	X_{15}	Total manufacturing work trips, plant zone

‡ Keefer (1966, pp. 109–110).

total number of trips predicted to be originated (by the trip generation model) in that zone, by purpose:

$$\sum_{j=1}^{n} d_{ij}^{p} = o_{i}^{p} \tag{10-12}$$

Similarly, the total number of trips terminating in each zone must equal the number predicted in the trip generation phase:

$$\sum_{i=1}^{n} d_{ij}^{p} = a_{j}^{p} \tag{10-13}$$

where d_{ij}^{p} = number of trips of purpose p from zone i to zone j

Economic theory of demand suggests two general relationships which should apply to the values of the d_{ij}^{p}'s. First, for any given trip purpose, the number of trips from one zone to two other equally attractive zones for satisfying that purpose should be greater to the zone which is the least costly to reach (including travel time and other factors). Secondly, for any given trip purpose, the number of trips to each of two zones equally costly to reach should be greater to the zone which is more attractive for satisfying that purpose.

Table 10-3 *(Continued)*

c *Trips to and from Airports*§

Equation	r^2	Standard error as percentage of $\bar{Y}$
$Y_1 = -63 + 0.17X_3 + 0.78X_5 - 0.06X_7 + 0.76X_8 + 0.19X_9 - 0.43X_{12} - 0.62X_{13}$	0.86	60
$Y_2 = -2977 - 3.36X_1 + 2.50X_3 + 16.43X_5 - 2.27X_7 + 12.32X_8 + 13.06X_9 - 2.09X_{10} - 3.09X_{12}$	0.92	73
$Y_3 = -944 - 0.06X_2 + 0.69X_3 + 1.49X_4 + 3.01X_5 + 0.58X_6 - 0.48X_7 + 0.95X_9 - 0.41X_{10} + 0.33X_{11} + 0.30X_{12} + 1.51X_{13}$	0.35	83

Variable type	Symbol	Definition
Dependent	Y_1	Total aircraft departures performed in scheduled service, calendar year 1960
	Y_2	On-line revenue passenger originations in scheduled service, calendar year 1960
	Y_3	Y_2/100,000 Standard Metropolitan Statistical Area (SMSA) population
Independent	X_1	Population, 1000s
	X_2	Population, per mi²
	X_3	Nonwhite population, 0.1 percent
	X_4	Median age, 0.1 year
	X_5	Families with incomes over $10,000, 0.1 percent
	X_6	High school graduates, 0.1 percent
	X_7	Manufacturing employment, 100s
	X_8	Transportation employment, 100s
	X_9	Trade employment, 100s
	X_{10}	Institutional employment, 100s
	X_{11}	Manufacturing establishments with at least 100 employees
	X_{12}	Service receipts, $1,000,000 per year
	X_{13}	Unemployment rate

§ Keefer (1966, pp. 106–107).

These two conditions or relationships are met in a mathematical model which has a form similar to the Newtonian law of gravity:

$$F_{ij} = \delta \frac{m_i m_j}{s_{ij}^2} \tag{10-14}$$

where F_{ij} = force of attraction between bodies i and j

m_i, m_j = mass of body i and body j, respectively

s_{ij} = distance between the centers of mass of bodies i and j

δ = a constant of proportionality

If we think of the mass of body i as the total trips originating (for a particular purpose) in zone i, the mass of body j as the total number of trips attracted to zone j (for that purpose), and the distance as an overall measure of the cost of travel between these two zones, it can be seen that the force is analogous to the total number of trips from zone i to zone j. As the total number of trips generated increases, or as the total number of trips attracted increases, the total flow increases, with travel cost being held constant. Similarly, if the total trip productions and attractions are held constant and the travel cost is increased, the number of trips would decrease. Thus this gravity form seems to yield the desired relationships in terms of variations in the number of trips resulting from changes in its variables.

However, the model must be modified slightly so that the desired conditions on total production equal the estimated trips generated in a zone and total terminations equal the number of trips attracted. These modifications are derived below. This derivation and the resulting form of the gravity model was developed by a transportation engineer, Alan M. Voorhees (1955), as a result of the need to develop some practical means for trip distribution in urban highway planning studies. Since then the model has been modified somewhat, but it remains one of the most widely used models.

The most widely used form of the gravity model can be derived by first substituting the number of trips originated in zone i, o_i^p, the number of terminated in zone j, a_j^p, the number of trips from i to j, d_{ij}^p, all for purpose p, and the travel distance c_{ij} raised to the power b, into the gravity model of Eq. (10-14), to yield

$$d_{ij}^p = \delta \frac{o_i^p a_j^p}{(c_{ij})^b} \tag{10-15}$$

The factor δ must be evaluated. This is done in such a way that the constraint on trip originations given by Eq. (10-12) holds:

$$\sum_{j=1}^{n} d_{ij}^p = \sum_{j=1}^{n} \delta \frac{o_i^p a_j^p}{(c_{ij})^b} = o_i^p \tag{10-16}$$

Solving for δ, we obtain

$$\delta = \frac{1}{\displaystyle\sum_{j=1}^{n} \frac{a_j^p}{(c_{ij})^b}} \tag{10-17}$$

and hence, substituting the index k in the summation to avoid confusion, we get

$$d_{ij}^p = o_i^p \frac{\dfrac{a_j^p}{(c_{ij})^b}}{\displaystyle\sum_{k=1}^{n} \frac{a_k^p}{(c_{ik})^b}} \tag{10-18}$$

where d_{ij}^p = trips per unit time with purpose p from zone i to zone j
o_i^p = trips per unit time with purpose p originated in zone i
a_i^p = attractiveness of zone j for trips of purpose p

c_{ij} = travel cost (e.g., time) from zone i to zone j
b = cost or distance exponent
n = number of zones

It is important to note two characteristics of this original form of the gravity model as used in transport planning. When the model was originally developed, it was common to estimate trip attractions in terms of general characteristics of a zone, such as total area for shopping, rather than a precise number of trips terminating in the zone; hence Eq. (10-13) was not defined at that time and did not have to be met. These general characteristics are given in Table 10-4. Also, at that time, with distance (over the road) being the measure of travel cost, it was found that the exponent was different for different trip purposes, reflecting different degrees of willingness to travel long distances, depending upon trip purpose. Values for various trip purposes are given in Table 10-4. As can be seen, persons traveling to work are more willing to go long distances than others, as would be expected. An example computation of trip distribution by this form of the gravity model is presented in Fig. 10-4.

Two important changes have been made in the model since then. First, trips are estimated, giving rise to the need to satisfy Eq. (10-13). The second change is that the exponent of distance term, $(c_{ij})^b$, has been replaced by a more robust mathematical form. This is specified as $F(t_{ij})$, termed a friction factor or impedance function, where t_{ij} is usually the travel time. $F(t_{ij})$ is a decreasing function of t_{ij}, as was the exponent form (relative to c_{ij}). A typical form for this function is shown in Fig. 10-5. The form of this friction factor or impedance function is determined individually in each metropolitan area, on the basis of observed patterns of travel, in particular the relative frequency distribution of trips of various lengths. Along with this change, for some zone pairs the resulting estimated number of trips is multiplied by a so-called socioeconomic adjustment factor K_{ij} which is used very selectively to account for unusually close ($K_{ij} > 1.0$) or weak ($K_{ij} < 1.0$) ties between zones. It is useful where barriers, such as rivers with only a few crossings, have separated zones and reduced trips, so there are fewer than would be expected on travel time considerations alone, for example. For almost all zone pairs, it is taken as unity, yielding no adjustment.

Table 10-4 Attraction Measures and Distance Exponents of Original Voorhees Gravity Model†

Purpose of trip (p)	Measure of attractiveness (a_j^p)	Exponent of distance (b)
Work	Number of workers employed	0.5
Social	Dwelling units	3
Shopping		
Shopping goods	Commercial floor area	2
Convenience goods	Commercial floor area	3
Business	Floor area	2
Recreation	Floor area	2
Other	Floor area	2

† Voorhees (1955).

Figure 10-4 **Example of gravity model trip distribution estimation.**

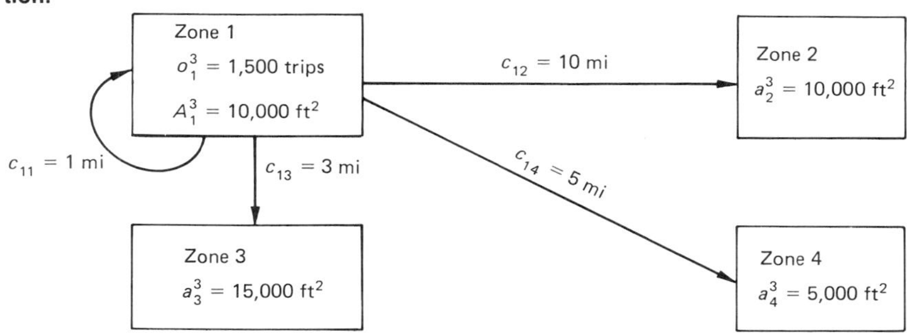

Trip purpose 3: Major Shopping
Distance exponent: 2

$$d_{ij}^3 = o_1^3 \frac{\dfrac{a_j^3}{(c_{ij})^2}}{\dfrac{a_1^2}{(c_{11})^2} + \dfrac{a_2^3}{(c_{12})^2} + \dfrac{a_3^3}{(c_{13})^2} + \dfrac{a_4^3}{(c_{14})^2}}$$

$$d_{11} = 1500 \frac{\dfrac{10{,}000}{1}}{\dfrac{10{,}000}{1} + \dfrac{10{,}000}{100} + \dfrac{15{,}000}{9} + \dfrac{5000}{25}} = 1500 \frac{10{,}000}{10{,}000 + 100 + 1667 + 200}$$

$$= 1500 \frac{10{,}000}{11{,}967} = 1253.5 \text{ trips/day}$$

$$d_{12} = 1500 \frac{\dfrac{10{,}000}{100}}{11{,}967} = 12.5 \text{ trips/day}$$

$$d_{13} = 1500 \frac{\dfrac{15{,}000}{9}}{11{,}967} = 208.9 \text{ trips/day}$$

$$d_{14} = 1500 \frac{\dfrac{5000}{25}}{11{,}967} = 25.1 \text{ trips/day}$$

The resulting form of the gravity model now used is

$$d_{ij}^p = o_i^p \frac{a_j^p F^p(t_{ij}) K_{ij}}{\sum_{k=1}^{n} a_k^p F^p(t_i) K_i} \tag{10-19}$$

where $F^p(t_{ij})$ = friction factor function for trip purpose p
K_{ij} = socioeconomic adjustment factor for zones i and j (applied selectively)

The exact means of estimating the friction factor and the socioeconomic adjustment factors *and* the iterations of the models required to satisfy Eq.

Figure 10-5 **Examples of gravity distribution model friction factor functions.** [*From Hutchinson (1974, p. 92).*]

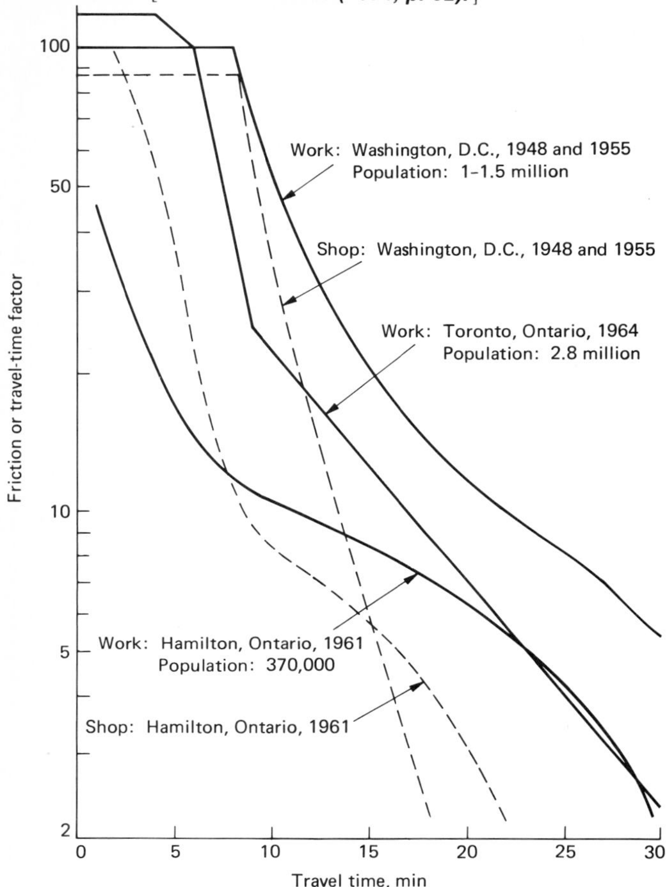

(10-13), all extremely complex and computationally cumbersome, are covered in detailed expositions of travel demand models in Stopher and Meyburg (1975, pp. 140–158) and Dickey et al. (1975, pp. 199–209). These texts also provide descriptions of other approaches to trip distribution.

The use of the gravity model is difficult in practice because of its computational complexity, but is not hard to grasp conceptually. Basically, what is required is the output of the trip generation analysis, an estimate of the travel time or travel cost measure between each pair of zones, and the value of the impedance function. With these, the estimation of the number of trips from each zone to all other zones, by purpose, is simply a matter of substituting these values into the equation, to yield the desired estimate of interzonal trips.

One may wonder how accurately models such as the gravity model estimate the future distribution of trips. There has been relatively little testing of these models, because a great amount of data is required and only recently has enough time elapsed between the time when the models were first developed and the present to make the test very meaningful. However, when the gravity model and two other commonly used distribution models were tested in Washington, DC, in their ability to predict over a 7-year period, it was found that the correspondence of the gravity model results and actual patterns was reasonably good, flows across major boundaries and along major corridors typically being within ±15 percent of the actual values (Heanue and Pyers, 1966). The data for travel in that metropolitan area in 1947 permitted estimation of the parameters of the gravity models so that they could be used in estimating the distribution in 1955, for which the actual trip distribution had been determined by a sample survey of all households. While this may seem like a large error range, in comparison to expectations in much engineering design, the errors are sufficiently small to enable rational planning of transport improvements. Also, it must be remembered that here the prediction is of peoples' behavior, not machines', and presumably many more factors influence human behavior than could ever be put into a mathematical model.

Modal Choice

Once the total number of trips from each origin to each destination has been estimated for each trip purpose, the next step is to estimate the number of travelers who will use each of the available modes. Typically in United States urban areas there are two choices, the automobile and one form of public transport (usually bus), although in many larger cities there may be two or more forms of public transport, possibly with different fares and levels of service. Mode choice has been hypothesized to depend upon characteristics of the mode which reflect the generalized cost of using that mode to the traveler, in much the same manner that these generalized costs of travel influence intercity mode choice as in the Kraft-SARC model. Important factors undoubtedly include the overall travel time by each of the mode alternatives from origin to destination, the total dollar cost from origin to destination, the comfort associated with each of the alternatives, and safety as it might be perceived by potential travelers. Each of these components can be divided further into a number of elements, travel time, for example, consisting of time consumed walking to and from the terminal (transit stop or auto parking facilities), time spent waiting for the vehicle (waiting for a bus or train or for the car to be delivered), and time spent riding in the vehicle, plus any additional time required, as for waiting at places where transfer from one transit vehicle to another must be made. Each of these times may influence the choice of travel differently. For instance, perhaps the greater output of energy and possible discomfort associated with time spent walking and waiting, in contrast to time spent seated in a vehicle, will result in a greater influence for that "out of vehicle" time than the "in vehicle" time. Also, there may be other factors influencing mode choice, such as the degree of advertising or other information affecting the public's awareness of transit services, the quality of transit stations and parking facilities (cleanliness, extent of air conditioning, etc.), and the perception of safety in such facilities and on transit vehicles. Also, attitudinal factors may enter, as has been observed in some areas where it is simply not "socially desirable" to be seen waiting for a transit vehicle or riding transit, while in other areas it seems fashionable to ride a commuter train to work.

The models developed to date deal particularly with travel time and cost. They divide time into components which reflect the different levels of comfort associated with each. However, there has been little success in dealing with some of the more subtle factors such as perceived safety and advertising.

Mode choice also presumably depends upon the purpose of the trip. Some trips (such as work trips) tend to lend themselves more easily than others to traveling via public transit, which often involves more walking and waiting than traveling by automobile. On the other hand, shopping trips and other trips where a large amount of "freight" must be carried are not so amenable to public transit. Also, often the cost of automobile travel decreases with increasing size of the party, the essentially fixed dollar outlays associated with running the vehicle, including tolls and parking charges, being divided among the number of travelers. This characteristic is, of course, automatically taken into account in the price of the trip. Groups of travelers, such as families, are very likely to travel via automobile for this and other reasons such as privacy and ease of carrying luggage, toys for children, etc.

The typical form for these models is as shown in Fig. 10-6. This particular set of models was developed for estimating transit and automobile use in the Washington, DC, metropolitan area, curves shown being based upon data from not only Washington but also Philadelphia and Toronto, as indicated (Deen et al., 1963). The models estimate the transit share of trips as a function of (1) the ratio of travel time via the best transit route to the travel time of the best automobile route (the travel time ratio in the figure), (2) ratio of cost to the traveler via transit to that via automobile (actual dollar outlays only, including only the additional cost of operating an automobile), (3) the travel service ratio, which is the ratio of the time spent walking, waiting, and/or transferring on the transit route to that for the automobile, (4) the economic status or income of the traveler, and (5) the purpose of the trip, differentiating between work and nonwork trips.

Each graph relates the transit share of trips to the travel time ratio via a particular modal split relationship or curve. A different graph is developed for each economic class, each cost ratio class, and transit to automobile level of service ratios, these being defined at the bottom of the figure. The solid portion of each curve represents the range of actual observations in Washington, DC. Also shown are curves based on data from Philadelphia and Toronto, for comparison and as an aid in extrapolation. The dashed lines for Washington indicate extensions of the observed relationships to cover the range of expected time ratios and other characteristics in the future plans. As can be seen from this figure, the curves based on observed behavior had to be extended considerably to include low travel time ratios which might occur with a modern, fast transit system, since none of the cities has an extensive amount of zone pairs between which total time via transit was much less than via automobile. This indicates one of the difficulties in predicting future use of various types of transport facilities, namely, that often existing facilities do not exhibit the same service characteristics as proposed facilities and hence experience with them must be used with considerable judgment in order to estimate use of future improved facilities. However, this problem is mitigated in this case by the natural upper bound of 100 percent of all trips being via transit; as can be seen, the curves were generally bent upward toward 100 percent in the lower ranges of travel time ratios.

To use these curves or models, it is necessary to know the number of travelers between each pair of zones who fall within each of the five income categories and also to estimate the times and costs for the best transit route and for the best automobile route between that pair of zones. Let us assume that between two zones there are 700 work trips and 500 nonwork trips. The transit routes include a time within the vehicle of 30 min and an excess time, including walking and waiting at the origin end and walking at the destination end, of 15 min, for a total time of 45 min; and a fare of 75 cents. Assume also that an automobile trip requires 35 min, with a total of 12 min walking time (5 min at the origin, 7 min at the destination) and a total cost, including the parking charge, of 90 cents. The traveler is in income category 5. The travel time ratio is 1.07, the service ratio is 1.25, and the cost ratio is 0.83, corresponding to category 2. From the appropriate curves, the percent of work trips via transit is estimated to be approximately 70, meaning that of the 700 total trips, 490 would be by transit and 210 by automobile. Of the 500 nonwork trips, with these times and costs, 53 percent would be via transit.

This model illustrates a number of assumptions that must be made in the prediction of modal split as well as with virtually any predictive model, assumptions which introduce considerable uncertainty. In particular, the use of this type of model assumes that persons' tastes and preferences for modes will remain in the future as they are now, at least within the same income category, although there is certainly considerable evidence to suggest that people have expected greater levels of comfort and other conveniences with the passage of time. Also, attitudes towards transport modes as well as other commodities may change in the future. The problems in ensuring adequate supplies of motor vehicle fuels have probably influenced some people to use public transport in order to reduce fuel consumption even if such travel is more time consuming or otherwise less desirable than automobile travel to them. Secondly, as noted above, that the range of historical data may not extend into the range in which proposed systems will fall may require the engineer or planner to extrapolate from past data into areas where actual behavior cannot be observed and hence models cannot be tested.

These uncertainties have led many people to criticize mode split models used in transportation planning, often with claims that they are inappropriate in prediction and often biased in their estimates. While there are certainly many limitations of the models and the results must be used with due recognition of the uncertainty of prediction, their use does have the advantage of making the assumptions regarding human behavior very explicit, so that weakness in these assumptions from various viewpoints can be easily identified and effects of alternative assumptions, such as shifting the mode split curves in favor or against a particular mode, can be made. Unfortunately, there has been no systematic evaluation of the accuracy of the mode split predictions against actual traffic on transit and highway routes, although this is often impossible because the particular system planned often has not yet been fully implemented, making the current system and the system for which the predictions were made not quite comparable. Very limited comparisons suggest that there has probably been a slight overestimate in general of total transit usage, with an underestimate of the fraction of trips which will use the

Figure 10-6 **The TRC mode choice model for Washington, DC.** ***(a)*** **Work trip relationships. [*From Deen, Irwin, and Mertz* (1963, pp. 118 and 119).]** ***(b)*** **Nonwork trip relationships. [*From Deen, Irwin, and Mertz* (1963, pp. 120 and 121).]** ***(c)*** **Stratification levels for cost ratio (CR), traveler economic status (EC), and service ratio (L). [*From Deen, Irwin, and Mertz* (1963, p. 99).]**

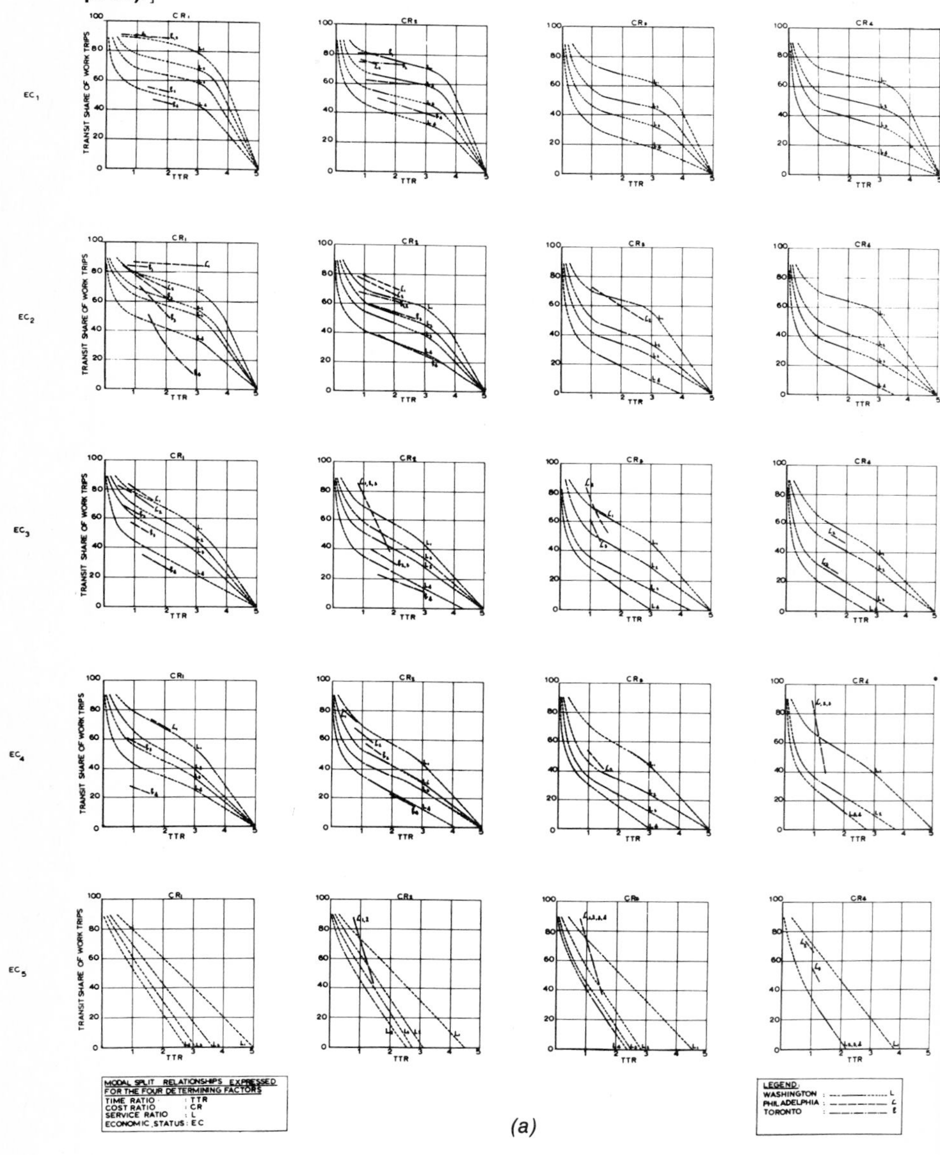

(a)

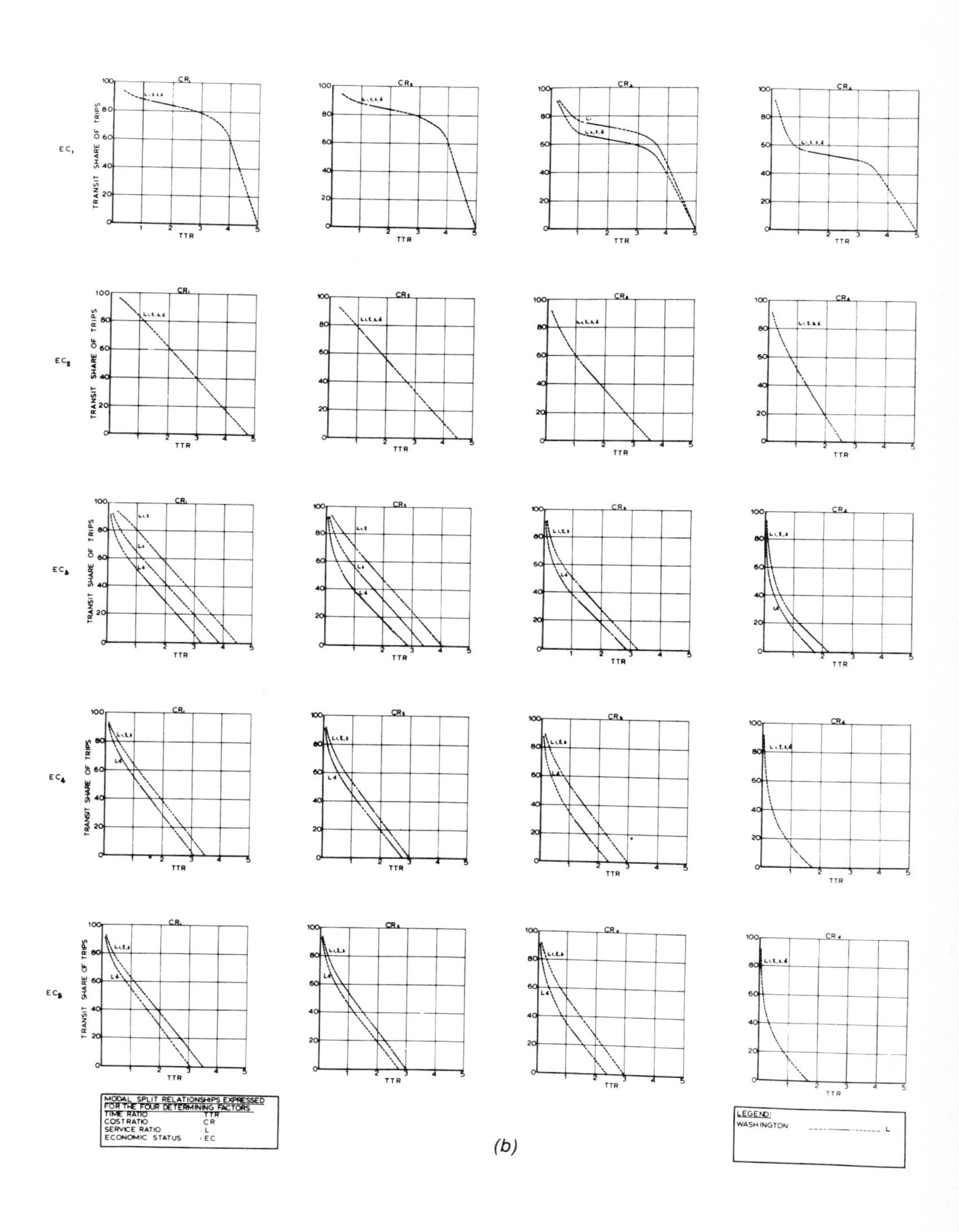
CR1
CR2
CR3
CR4
EC1
EC2
EC3
EC4
EC5
TRANSIT SHARE OF TRIPS
TTR
MODAL SPLIT RELATIONSHIPS EXPRESSED FOR THE FOUR DETERMINING FACTORS
TIME RATIO : TTR
COST RATIO : CR
SERVICE RATIO : L
ECONOMIC STATUS : EC
LEGEND:
WASHINGTON L

(b)

Fig. 10-6. ***Continued.***

CR_1 = 0.0 to 0.5
CR_2 = 0.5 to 1.0
CR_3 = 1.0 to 1.5
CR_4 = 1.5 and over

EC_1 = \$0 to \$3100 per annum
EC_2 = \$3100 to \$4700 per annum
EC_3 = \$4700 to \$6200 per annum
EC_4 = \$6200 to \$7500 per annum
EC_5 = \$7500 per annum and over

L_1 = 0 to 1.5
L_2 = 1.5 to 3.5
L_3 = 3.5 to 5.5
L_4 = 5.5 and over

TTR = transit time/auto time, origin to destination

(c)

system during peak periods, resulting in some unforeseen peak period congestion problems and greater than anticipated operating deficits (revenue being less and costs greater than anticipated, the latter due to the greater peaking as we noted in Chap. 9).

There are other approaches to mode choice. Some of these are simply different models that operate in essentially the same way as the models described above, differing in functional form (often a specific mathematical relationship between the percentage of travelers via transit and various service and other characteristics used) and in the types of variables used. Another approach which estimates the trips via transit originating or terminating in a zone, typically gives considerable weight to socioeconomic characteristics such as income and car ownership and includes only a rather generalized variable describing the overall level of service of the transit system relative to the automobile travel, rather than considering specific service characteristics of the individual origin to destination connection. This approach is probably more appropriate for occasional trips, where general characteristics of the transport system are more likely to influence one's choice of mode than specific characteristics for a particular zone pair. These alternative approaches are included in Fertal et al. (1966).

Trip Assignment

The final stage in travel demand estimation is the assignment of the trips to be made between each pair of zones, by mode, to particular routes through the network. This is primarily a problem within the highway mode, where there usually are numerous routes which a traveler could follow. In transit networks, typically the number of alternative routes is less; in many cases only one path connects two zones, and in others, even if there is more than one path, one route may offer such an overwhelmingly higher quality of service than the others that it is the obvious choice (e.g., a subway route connecting two distant zones in contrast to a bus or streetcar line).

The usual assumption made in traffic assignment is that travelers will choose the minimum time path, as was mentioned earlier in Chap. 8 in the discussion of transportation network characteristics. This assumption seems to hold rather well for motorists selecting a particular route when the only difference between these routes is travel time, in particular provided there is no cost difference between routes as created by a toll on one path. The use of the assumption that road travelers will take the minimum time path is almost universally accepted among transportation engineers, partly because it has been fairly widely tested and found to be a reasonable approximation to the behavior of most persons. However, detailed studies have revealed that under certain circumstances particular persons may choose routes which are not the minimum time path, for instance for some to avoid high-speed driving on freeways or for others to go out of the way to travel on such facilities (Michaels, 1965).

The major problem associated with the use of the assumption that travelers take the minimum time path in road travel is estimating the travel time on the various links which make up possible paths. The problem arises because, as we saw in Chap. 5, the travel time on any given road is dependent upon the volume of traffic on it; yet in analyzing future transportation systems, it is these demand models that are used to estimate the future volume, while at the same time the selection of route for any particular traveler depends on the travel time on the various links and hence on those volumes which are to be predicted. A routing of traffic is sought in which everyone takes the minimum time path from origin to destination and which also satisfies the condition that the travel times on all links (on which those minimum time paths are based) is consistent with the volume of traffic on those links as these two are related by the speed-volume function. This problem is quite difficult to solve in practice, and we shall discuss it in Chap. 12. In the context of various mathematical procedures for solving this problem, sometimes traffic is assigned to routes which in the mathematical model are not minimum time paths but somewhat inferior, this being done mainly to make the calculation process more efficient, although usually this procedure is accepted because it yields an approximation to the routing of traffic via minimum time paths.

The problem is somewhat different in the case where certain routes have tolls and other routes are free. In these situations, the approach to assignment is similar to that embodied in the modal split model discussed above, the curves giving the percentage of traffic via the toll route. These curves are called diversion curves, since they were originally developed to estimate diversion from free roads to toll roads.

In public transit networks, fares tend to be the same regardless of the route used, although there are exceptions, particularly in large metropolitan areas with more extensive ranges of public transit services. Usually the assumption is made that the traveler selects the minimum time path, where again the relevant time is the total time from origin to destination, including time walking and waiting for transit vehicles. In the application it is usually assumed that the traveler responds to the mean waiting time, which as discussed in Chap. 7, is equal to one-half the headway for either a constant arrival of passengers or random arrival at a constant average rate. Thus alternative routes through the transit network are compared on the basis of walking times at both ends of the trip plus one-half the headway of any route to be boarded, plus any time spent walking between routes if intermediate transfers are made, plus the time spent on the vehicles. In some situations, transit planners weight the waiting time and walking time more heavily or may assign a direct penalty of a certain number of minutes to a transfer to compensate for the added discomfort of these portions of the trip in comparison to the ride on the vehicles.

There have been a few studies of actual traveler route choice on transit systems where alternatives exist, but the results do not allow great generality. One early study in Toronto dealt with the choice between traveling on a surface street car for an entire journey or transferring to the subway at an intermediate point, this often leading to a lower total travel time than continuing on the street car to the destination (von Cube et al., 1958). Curves similar to the mode split curves presented above were developed which indicated that some travelers seemed to prefer remaining in the same vehicle (the street car) even if change reduced total travel time while others changed to the subway even if their total time was increased. A recent study of choice between express bus and subway–elevated train service in Chicago revealed that many travelers selected the bus even though its travel time to the downtown was generally greater than the rail service, indicating that even though the train would be faster, fear of crime on the subway-elevated line, particularly during the middle of the day and at night when traffic is lower than during the peak work-day hours, was the primary reason for their choice (Ferrari and Trentacoste, 1973). Travelers in New York City are now offered a choice between express bus and subway service between many points, the bus service being operated as a premium service with higher fares, with a policy of seating most passengers. Many are choosing the more expensive luxury bus service, which is proving profitable and expanding in traffic and routes. In summary, the use of minimum paths by transit passengers is probably a reasonable general rule, although it is clear that it must be modified in particular situations to reflect local characteristics.

Peaking of Urban Traffic

One of the most salient features of urban traffic is the substantial variation in volume by time of day and day of week. Within a weekday, travel is heavily peaked during the morning and evening periods commonly used for traveling between home and work. Figure 10-7 reveals this peaking in terms of the number of trips in Chicago in 1956 terminating in each of the hours of a weekday, by purpose of the trip. The availability of transport capacity undoubtedly influences to some extent the selection of travel time; thus this may not precisely reflect the most desired time pattern of travel, but it is probably true that the time most trips begin or end (particularly those to and from work) is not greatly influenced by congestion and hence probably indicates with reasonable accuracy the desired time pattern of travel. This does not mean that congestion is not a serious problem at certain hours of the day, as any peak hour traveler in a large metropolitan area knows, but simply indicates that congestion does not seem to be a significant deterrent to work trip travel during such peak periods.

Figure 10-8 presents the variation of traffic by month of year for selected

Figure 10-7 **Daily peaking of urban travel: example of Chicago metropolitan area.** [***From Chicago Area Transportation Study (1959, p. 35).***]

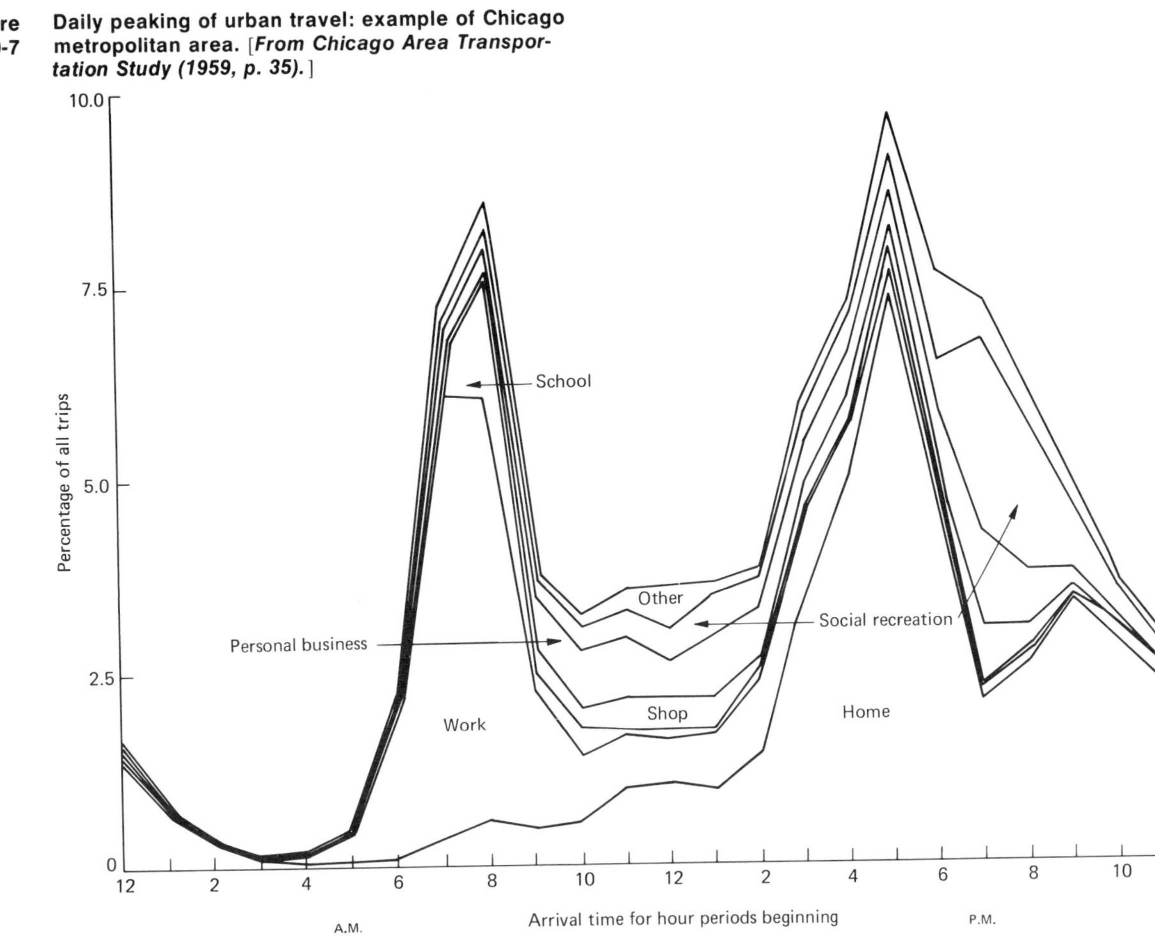

Figure 10-8 **Variation in travel by month of year on Wisconsin urban and rural roads.** [*From Martin, Memmott, and Bone (1963, p. 143).*]

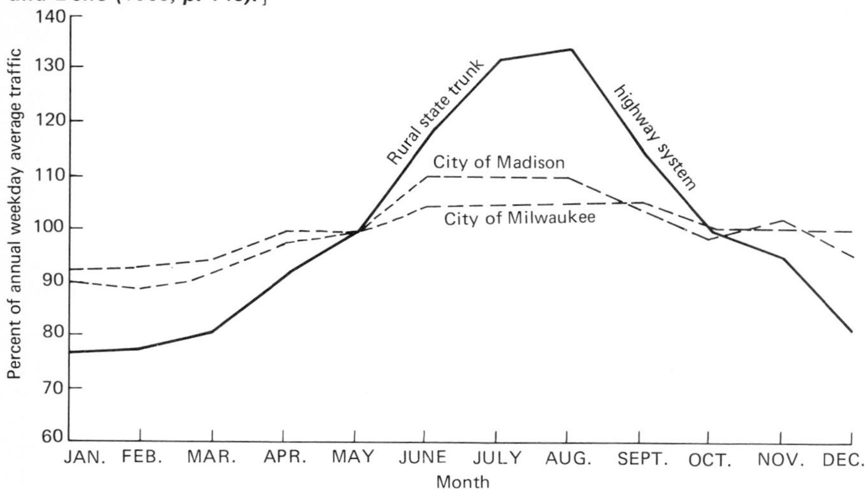

areas in Wisconsin. Urban area travel, as revealed by the curves for Madison and Milwaukee, varies about ±10 percent around the mean. (While the urban data are for road traffic only—this represents the majority of all trips—the pattern is undoubtedly similar for public transport.)

The primary methods used for dealing with the temporal aspects of movement, including variations by month and day of week, as well as time of day, are methods that allocate average annual daily traffic to various periods. As a result, these methods are used after the actual quantity of travel demanded between various origins and destinations is determined. As has been mentioned above, this, the result of the interaction of both the demand for transportation and the supply of transportation service, forms an equilibrium pattern of travel. Since we must first discuss the supply of transport and then methods for determining this equilibrium before the quantity of travel demanded between each pair of zones is determined, it is necessary to postpone the discussion of temporal aspects of travel to that point (Chap. 12).

Conclusion

This concludes a rather brief discussion of the methods now in widespread use for estimating the demand for urban travel. Although this five-step procedure may seem rather complex, particularly in comparison to the straightforward demand equation illustrated by the intercity model, the procedure is surely somewhat of a simplification of the actual process underlying the demand for urban travel. For instance, the number of trips generated is estimated with no consideration of the price and overall transport level of service, which surely would influence the number of trips made. Also, trip distribution is performed using methods which rely upon a single measure of the cost of traveling between zones, even though travelers perceive many different costs, including the time by road, the price paid for that travel, the time by public transit and its fare, etc. These shortcomings are widely recognized in the profession, and a great deal of research is being devoted to the development of better procedures. In fact, it is probably true that more research is being conducted on the demand for urban transportation than on any other aspect of urban transportation, as it is felt that accurate estimates of future travel are absolutely essential to the design of systems which will be responsive to future travel needs.

THE DEMAND FOR FREIGHT TRANSPORTATION

In contrast to the attention devoted to passenger transport demand, particularly in urban areas, little research has been done on the demand for either intercity or intra-urban freight movement. This is partly due to the fact that only recently has there been any recognition of problems associated with urban goods movement, virtually all of the attention of urban transport planning studies in the past having been directed toward developing facilities for person movement. Freight movement was viewed as simply an adjunct to person movement, primarily because the attention was directed toward new road and public transit facilities, with trucks accounting for perhaps 10 to 20 percent of total vehicles on the roads and because no freight is usually carried on transit. Similarly, there has been very little research on intercity freight transport demand, although in the past few years a number of specific research projects have been undertaken, partly in response to various freight carriers trying to better understand their markets, and partly in response to governmental concerns for the financial well-being of various freight carriers, most notably railroads. However, we shall try to cover a few aspects of freight demand which have particular relevance to engineering and planning.

Urban Freight Demand

A primary aspect of urban freight movement that has been studied and yielded results potentially useful in engineering and planning studies is the number of truck trips produced and attracted to various industries. The approach is essentially identical to that used in person trip generation modeling, i.e., the number of truck trips is related to characteristics of the activities being undertaken at the site of generation.

In a study of truck movements at about 240 manufacturing industries in metropolitan Toronto, Kardosh and Hutchinson (Hutchinson, 1974, pp. 415–416) developed the following equations for truck trip productions and truck trip attractions:

$$y_i^o = 11.4 + 1.53t_i \quad r^2 = 0.807$$
$$y_i^d = 12.5 - 0.86t_i \quad r^2 = 0.532$$

where y_i^o = truck trips originated per day at firm i
y_i^d = truck trips terminated per day at firm i
t_i = number of private trucks owned by firm i

Even though they attempted to relate the number of truck trips to characteristics of industries other than the number of privately owned trucks, it was found that the most generalizable relationships included private truck ownership as the primary independent or predictive variable. Despite the introduction of other variables and analyses conducted within individual industry classifications rather than to all industries, the number of privately owned trucks seemed to remain one of the primary determinants of truck trips. Private truck ownership seems to be very similar to automobile ownership in explaining trip making, it probably being an indicator of the degree to which the firm feels as though a large amount of transport will occur. It should be noted, however, that the Toronto study included a number of industries which typically own large private truck fleets for the movement of their own freight, such as food and beverage companies, newspapers, and ready-mix concrete plants.

To use these truck trip generation equations obviously requires an estimate of the number of privately owned trucks, which may prove extremely difficult except in studies of particular industries where those firms can be queried for their plans. However, such relationships probably would be quite useful in estimating the number of truck trips to and from such industries and thus would be very helpful in determining the adequacy of roads which provide access to those industries. These also would be particularly useful in assessing the adequacy of roads in the vicinity of new industrial parks and other major centers of industrial roads.

There has been much discussion of the use of models similar to those used for forecasting person travel to predict goods movements flows within urban areas in the future. However, none of these procedures has as yet been implemented on a regional basis.

It is interesting to note the temporal pattern of truck movements. This is illustrated in Fig. 10-9, which is for all truck trip origins in the Tri-State Region centered on New York City. As can be seen, the bulk of truck trips occurs

Figure 10-9 **Temporal distribution of truck trips in the New York City metropolitan area in 1963.** [***From Chappell and Smith (1971, p. 169).***]

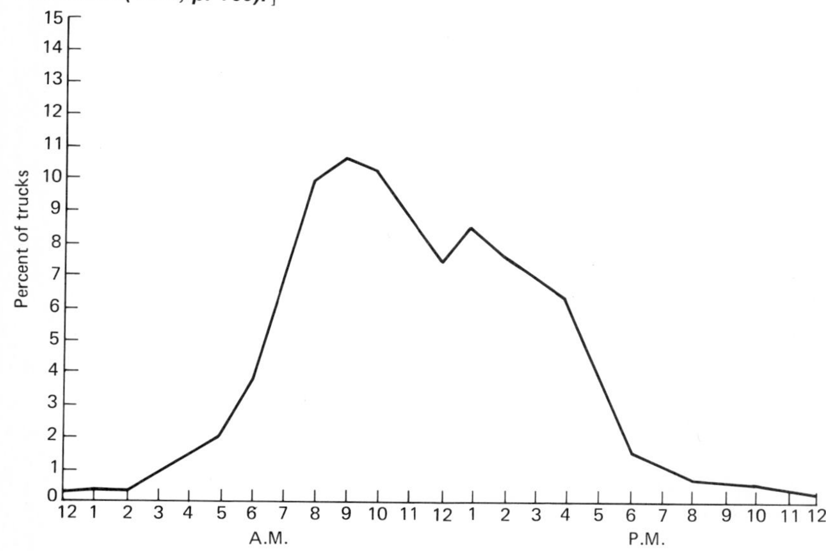

between about 6 A.M., and 6 P.M., the trips being much more uniformly distributed in this period than person trips. Such patterns are typical for urban truck movement although with respect to particular types of industries there is often more substantial peaking.

Intercity Freight Demand Models

There has similarly been relatively little work in the area of intercity freight demand until recently, and many methods for predicting intercity commodity flows and mode splits are being explored. There are methods of regional science and economics, mainly input-output analysis, which will yield interregional commodity flows. There have also been applications of models based on the concepts of urban travel forecasting models to intercity freight transport. The quantities and types of commodities generated in a region or zone are related to characteristics of the industries located there and to the population in the case of attractions of goods for final consumption. These flows can be distributed by gravity-type models, using freight transport rates and levels of service.

Modal choices of shippers have been researched extensively, revealing that they consider the following factors:

Travel time

Reliability (variation in travel time)

Price

Probability of loss and damage

Special packaging requirements

Convenience (e.g., need to take shipment to carrier terminal versus door-to-door service)

Availability of any required special services (e.g., refrigeration or water and feed for livestock)

Many other factors could undoubtedly be added to the list.

The relative importance of these factors varies considerably with the commodity being shipped and with other factors such as the policies of the receiver of the freight on inventorying a large extra supply of the commodity. The time a shipment is in transit represents lost money, in the sense that payment for the good (or receipt of it for use) is delayed. The value of this is equal to the interest which an amount of money equal to the value of the shipment would have earned in that time. Thus for a shipment valued at $100,000, a 20-day movement represents a cost, at 10 percent interest per annum, of $547.95. If the movement could be performed in half the time, $273.98 would be saved, some of which could be used to cover a higher charge of the faster carrier. Similar but more complex relationships apply to variations in transit time: the greater the variations, the greater the inventory of items which must be kept on hand to prevent running out before the next shipment arrives. In general, the greater the value of the freight, the more these factors favor fast and reliable carriers even if they charge more. The influence of the other factors listed is fairly straightforward.

PROJECTION TECHNIQUES

Probably the simplest of all means of predicting future traffic on transportation facilities is to project current realized demand or quantity demanded into the future. This is an extremely simple procedure, but one which implies a number of assumptions which tend to be true only in very restricted situations. Nevertheless, these situations do occur, and hence it is appropriate that we discuss projection techniques very briefly.

Basically the approach is to find a relationship between the demand, or more precisely the quantity demanded, and the passage of time. Such a relationship is usually found through examination of a graph of the quantity demanded versus time as shown in Fig. 10-10. A relationship is sought which tends to fit the historical data. Once such a relationship is found, it is used as a basis for estimating quantity demanded in future years. Typical mathematical forms for such relationships are shown in this figure. Often the relationships are fitted manually to plotted data with parameters estimated by trial and error procedures. Alternatively, some type of formal curve-fitting techniques such as linear regression can be used if desired.

The main limitation of these methods is that they assume that the same underlying growth (or possibly decline) process which has occurred in the years for which historical data are obtained will continue as far into the future as the projection will be made. This usually corresponds to assuming a world in which the changes in consumer preferences for transport (or shipper preferences and requirements for transport), available transport technologies, pricing policies, other service-determining policies, and regulatory agencies, etc., either will not change in the future or will change then at the same rate as now. Thus projection techniques are typically used to estimate future

Figure 10-10 Forecasting future demand by projection techniques. Possible function forms for curve *A* (nonlinear):
$d = \alpha(t - \beta)^{\gamma}$
$d = e^{\gamma(t-\beta)}$
Possible linear functional form for curve *B*:
$d = \alpha + \beta t$
In all cases, α, β, and γ are empirically estimated.

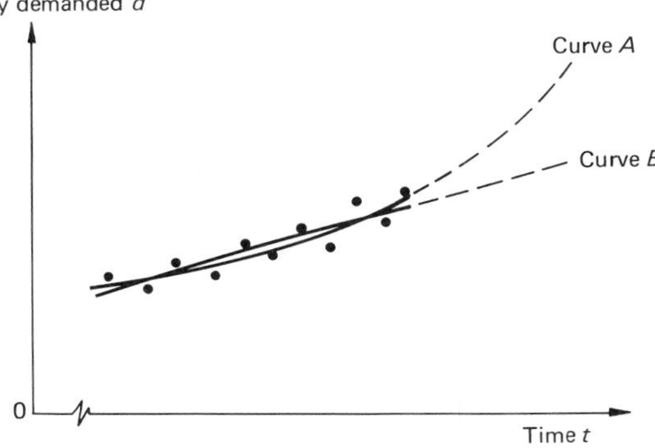

travel or freight shipments for periods of perhaps months or sometimes a few years into the future, but are rarely used except as very approximate indicators of future transport beyond a few years.

To realize the limitations on these methods, one only has to look back at changes in transport over the past two or three decades. In urban transport, there has been very substantial growth in the total number of trips, but the spatial pattern of these trips has changed greatly, reflecting the growth of urban areas in suburban rings around the center cities, and in some cases actual net migration of persons and businesses outward from center cities to these areas of rapid growth. Also during this period, in most urban areas the fraction of trips via public transport as well as the absolute number has declined, with the automobile becoming the dominant mode of urban travel in all metropolitan areas of the United States, and even in all but a few central cities. Without explicitly considering the changes in the pattern of land use—the location of residences, industries, shopping centers, etc.—and the linkages between residences and various places of possible trip destinations, these changes probably could not have been foreseen.

However, projection methods do have some role to play. They have been particularly successfully used in estimating the total traffic through major facilities serving fairly large regions where changes in the spatial organization of activities and the pattern of travel would not influence significantly the amount of traffic through these facilities. Examples include seaports, airports, and other intercity terminals serving entire urban areas. The total number of intercity trips through such facilities seems to be essentially independent of the location of residences, businesses, and industries. Also, projection methods have been successfully used to estimate such things as the total number of trips that might be made by all modes within a region or the total amount of freight to be shipped. Thus projection methods do have a role, but it is extremely limited, and a great deal of care must be exercised in accepting the results.

CONCLUSION

Thus many tools are available to the planner and engineer for the estimation of future traffic on transportation systems. As we shall see in later chapters, this is invaluable not only for estimating the capacity requirements of proposed transportation facilities, number of vehicles required, etc., but will provide much information useful in the evaluation of those proposed changes in the system.

It is interesting to speculate on the future trends in travel and the demand for transportation. One factor which will undoubtedly become very important in influencing these trends in the future is the development of new methods of communication, involving the transmission of not only data (words, numbers, etc.) and voices, but also increasingly images (Lathy 1975). There are experimental installations of videophones at different offices of the same firm, which enable conferences to be held regardless of the distance between participants, eliminating the need for travel to a single location. Also, it has been observed that many white-collar workers, especially in so-called "knowledge and information industries" (such as executives, consultants, and accountants) at offices remote from their firm's main production center communicate via telephone and sometimes via more sophisticated means, thus eliminating the need for much urban travel (Henneman and Krzyczkowski, 1974). Many stores now order items from warehouses by code using the keys of "touch phones," a procedure that might in the future be used by anyone wishing to shop, eliminating the need for a personal trip. Of course, this requires a truck trip to bring the goods to the home, but one trip might serve many customers, reducing total vehicle-miles of travel. Many observers believe that the combination of (1) improved communications technology, with a wider range of media and reduced costs, and (2) the seemingly increasing true cost of travel, due to more person trips congesting limited facilities (which often cannot be expanded substantially due to environmental considerations, as in the case of roads, or only at great expense, as in the case of subways) and the increased costs of energy to operate transport vehicles, all will result in a substantial substitution of communications for transportation in the future. On the other hand, perhaps small electric automobiles and radically new transport technologies will provide for all travel, with the ease and speed of the auto for short trips and the speed of the aircraft for long trips. If these have the low resource costs associated with high-volume mass transit carriers, then transport would remain very inexpensive. The efficacy of both of these hypothetical futures will be influenced by population and land use patterns, which now (1977) appear to be in the direction of low aggregate population growth and a migration toward smaller urban areas, away from large cities and their attendent problems. In smaller centers, activities can be fairly dispersed but trips are typically short, often within walking distance, which results in low transport expenditures. The future is obviously uncertain, but it is important to ponder possible futures and to be prepared to modify them to meet human (and environmental!) needs.

SUMMARY

1 The demand for transportation is derived from the desire on the part of individuals to partake of activities removed from their residences and, in the case of goods movements, from the desire to transport things from places where they are extracted or manufactured to places where they are consumed.

2 In the estimation of travel demand between two places, it is essential to consider the purpose of the trip, the characteristics of the origin area which will influence the total number of trips generated, the characteristics of the destination which will influence the total number of trips attracted for that purpose, and the overall price and level of service of the transport connection between these two places. Also, it is important to consider alternative destinations where that trip purpose might be satisfied, as well as the attractiveness of those destinations and the quality of the transport connections to them. Similar considerations apply in principle to goods movement.

3 There are basically two types of travel demand models, direct demand models and sequential demand models. Direct demand models consist of single equations embodying the factors identified above. Sequential demand models treat portions of the total demand-estimating problem individually and are typified by the models in widespread use for urban travel forecasting, which include estimation of land use patterns, trip generation, trip distribution, mode split, and route choice.

4 There has been relatively little development of practical demand models for freight transport, although some truck trip generation relationships have been developed for urban areas which will assist in planning of road facilities in the vicinity of major truck trip generators such as industrial parks, and there has been a limited amount of work on intercity freight mode choice.

PROBLEMS

10-1
Graph the demand curves for rail travel from city A to city B, with respect to rail price. Use the Kraft-SARC demand model, Eqs. (10-8) and (10-9), with modal elasticities given in Table 10-1. It is known that the total rail business volume is 5000 trips/day and nonbusiness is 3500 trips/day at the modal price and time values given below. From these and the total demand, the value of the remaining factor in the model (the first terms related to population, employment, etc.) can be calculated.

	Time, min	Cost, $
Bus	350	20
Rail	300	25
Air	200	30
Auto	285	21

10-2
Using the data of Prob. 10-1, construct a demand curve for total rail travel, as a function of rail travel time, for a range of total time of from 200 to 300 min. Plot demand versus line-haul time, which is 65 min less than the total travel time.

10-3
Using the data of Prob. 10-1 and a rail price of $20.00, what is the effect on rail business and nonbusiness travel of a reduction of the air fare by 30 percent?

10-4
Estimate the number of trips generated by homes in the zone for which the future number of households in each auto ownership and family size category are given in Table 10-2*c*. What is the total number of such trips for the entire zone?

10-5
Plot the number of trips per household (vertical axis) as a function of family size (horizontal axis) using the trip rate data presented in Table 10-2*b*. Con-

nect the points corresponding to the same auto ownership level. What pattern emerges?

10-6

Plot a demand curve for travel between two zones using the original form of the gravity model but with distance replaced by time. Use the zones and generation and attraction values of Fig. 10-4, but replace the distances by times at 20 mi/h. Then vary the time from zone 1 to zone 4 from 10 to 30 min, and plot the resulting demand curve.

10-7

Two zones in an urban area are connected by a highway and a local bus line. The average total travel time, excess time, cost, and excess cost by auto are, respectively, 10 min, 5 min, $0.30, and $0.05; and the same for transit are 17 min, 10 min, $0.35, and $0.00, respectively. What percentage of work and nonwork trips (of people whose income is in EC_2) will use transit, using the mode split model of Fig. 10-6?

If the total work trips are 500/day and nonwork are 700/day, what would the transit volume be? If there is an average of 1.4 persons/auto for work trips and 1.8 for other trips, what would be the auto volume per day?

10-8

Zones in a corridor are connected by road and a bus transit line. Various means of improving the latter are contemplated, ranging from a rail rapid transit line to a busway. You are to estimate the effect of various transit improvements on transit patronage between two zones which are 6 mi apart and between which 1000 work trips and 1500 shopping trips (each way) are made per day. Compare the current transit percentage and number of trips to those with alternative A, a rapid rail line, and B, a busway with zonal express service. All work trips are in the peak period. Assume auto characteristics do not change as a result of the transit changes and associated mode shifts.

		Current auto	Current bus	Alternative A	Alternative B
Total time, min	—peak	22	35	27	29
	—other	18	25	27	27
Excess time, min		7	11	15	11
Total cost, $	—peak	0.50	0.30	0.30	0.30
	—other	0.40	0.30	0.30	0.30
Excess cost, $	—peak	0.50	0	0	0
	—other	0.05	0	0	0

10-9

A shipper is trying to decide between rail and truck movement of a shipment valued at $15,000 and weighing 50,000 lb. He will make his decision based on the sum of inventory cost, the carrier charge, and his packaging and loading costs. The railroad will charge $0.018/cwt and the truck line $0.029/cwt. The rail time is estimated to be 8 days, and the truck, 2 days. The rail movement requires slightly more elaborate packaging, resulting in a total packaging and loading cost of $150, in contrast to $120 for the truck. Which mode is cheapest for the shipper, considering all these factors? He values capital at a 15 percent per annum interest rate.

REFERENCES

Chappell, C. W., and M. T. Smith (1971), Review of Urban Goods Movement Studies, *Highway Research Board,* Special Report No. 120, pp. 163–181.

Chicago Area Transportation Study (1959), *Study Findings,* vol. 1, Chicago, IL.

Curtin, John F. (1968), Effect of Fares on Transit Riding, *Highway Research Record,* No. 313, pp. 8–18.

Deen, Thomas B., William L. Mertz, and Neal A. Irwin (1963), Application of a Modal Split Model to Travel Estimates for the Washington Area, *Highway Research Record,* No. 38, pp. 97–123.

deNeufville, Richard, and Alex Freidlander (1968), Effect of Operating Policies of Urban Mass Transportation on Demand, *TRF Proceedings-Supplemental Papers,* **IX**: 81–92.

Dickey, John W., Robert C. Stuart, Richard D. Walker, Michael C. Cunningham, Alan G. Winslow, Walter J. Diewald, and G. Day Ding (1975), "Metropolitan Transportation Planning," McGraw-Hill, New York.

Ferrari, Neil D., and Michael F. Trentacoste (1973), Personal Security on Public Transit, *Transportation Research Forum Proceedings,* **XV**(1): 214–223.

Fertal, Martin J., Edward Weiner, Arthur J. Balek, and Ali F. Sevin (1966), "Modal Split," U.S. Government Printing Office, Washington, DC.

Heanue, Kevin E., and Clyde E. Pyers (1966), A Comparative Evaluation of Trip Distribution Procedures, *Highway Research Record,* No. 114, pp. 20–36.

Henneman, Suzanne S., and Roman Krzyczkowski (1974), "Reducing the Need for Travel," Report No. PB-234 665, National Technical Information Service, Springfield, VA.

Hill, Donald M., and Hans G. von Cube (1963), Development of a Model for Forecasting Travel Mode Choice in Urban Areas, *Highway Research Record,* No. 38, pp. 78–96.

Hutchinson, Bruce G. (1974), "Principles of Urban Transport Systems Planning," McGraw-Hill, New York.

Keefer, Louis J. (1966), "Urban Travel Patterns for Airports, Shopping Centers, and Industrial Plants," National Cooperative Highway Research Project Report No. 24. Highway Research Board, Washington, DC.

Lancaster, Kelvin J. (1966), A New Approach to Consumer Theory, *Journal of Political Economy,* **LXXIV:** pp. 132–157. Material reprinted by permission of the University of Chicago Press, publishers.

Lathy, Charles E. (1975), "Telecommunications Substitutability for Travel: An Energy Conservation Potential," U.S. Department of Commerce Office of Telecommunications Report No. OT 75-58, Washington, DC.

Martin, Brian V., Frederick W. Memmott, and Alexander J. Bone (1963), "Principles and Techniques for Predicting Future Demand for Urban Area Transportation," MIT Press, Cambridge, MA.

Michaels, Richard M. (1965), Attitudes of Drivers Determine Choice Between Alternative Highways, *Public Roads,* **33**(11): 225–236.

Stopher, Peter R., and Arnim H. Meyburg (1975), "Urban Transportation Modeling and Planning," Lexington, Lexington, MA.

Systems Analysis and Research Corporation (1963), "Demand for Intercity Passenger Travel in the Washington-Boston Corridor," Report No. PB 166 884, National Technical Information Service, Springfield, VA.

U.S. Department of Transportation, Federal Highway Administration, Urban Planning Division (1975), "Trip Generation Analysis," Washington, DC.

von Cube, Hans, R. J. des Jardins, and Norman Dodd (1958), Assignment of Passengers to Transit Systems, *Traffic Engineering,* **28**(11): 12–14, 50.

Voorhees, Alan M. (1955), A General Theory of Traffic Movement, *Proceedings of the Institute of Traffic Engineers,* pp. 46–56.

Wohl, Martin, and Brian V. Martin (1969), "Traffic System Analysis for Engineers and Planners," McGraw-Hill, New York.

Supply of Transportation

11.

Following the economic paradigm of transport network analysis, we now, appropriately, turn to the description and analysis of the supply of transportation services. In Chap. 12, we will bring together the supply of transport service and the demand for transportation to achieve something somewhat analogous to the equilibrium in the usual economic analysis of marketplaces. This is done in order to predict the total amount of transport which will occur under specified conditions and the associated prices and levels of service. But first, an understanding of the supply of transport services must be developed.

THEORY OF TRANSPORT SUPPLY

In describing the supply of transport services, we begin with the definition of supply in economic theory. The supply schedule (or supply function or supply curve) specifies the relationship between the market price for a commodity and the amount of this commodity that producers are willing to produce and sell. As in the case of the basic definition of the demand curve, this definition of the supply curve refers to a particular commodity, which is homogeneous in the sense that regardless of the number of items produced, the particular item retains the same characteristics and would be valued similarly by a potential buyer.

Economic Theory

A typical form for such a supply curve is shown in Fig. 11-1. The basic form is of a price increase resulting in an increase in the quantity produced and offered for sale (Samuelson, 1958, pp. 378–391). The reason for this increase in price with increasing quantity is simply that in order to induce a firm to

Figure 11-1 **Typical supply schedule or function.**

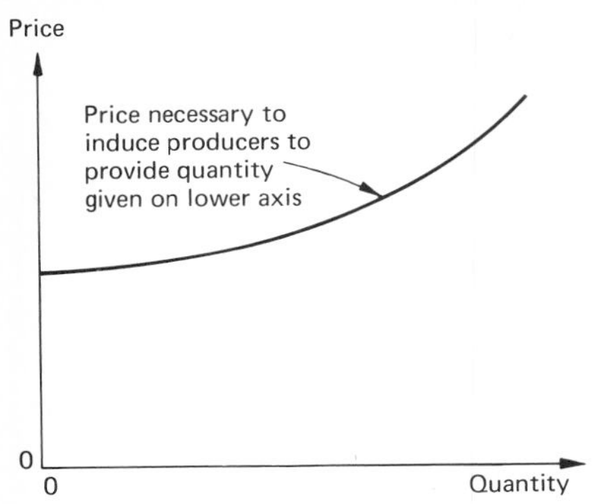

produce more of a particular product, a higher price must be obtained for that product. In equation form:

$$P = S(Q) \tag{11-1}$$

where P = price
Q = quantity
S = the supply function relationship

The reason for this hypothesized behavior and certain characteristics of the supply curve related to the period of time of the analysis bear additional discussion.

To make the discussion specific to transport, we will consider an example of truck freight service, between city *A* and city *B,* in which presumably many independent truckers, perhaps each owning one or a few trucks, provide service to a large number of shippers and receivers of freight within the service area. For any given level of expected future freight traffic, these truckers taken all together will have purchased a certain number of vehicles, invested in particular sizes of terminals, etc., the choices presumably having been made in order to operate the most efficient truck service possible. The resulting costs for this service might appear as portrayed in Fig. 11-2, the price P_1 corresponding to the quantity of freight, perhaps measured in ton-miles, of Q_1. If there were an increase in the amount of freight to be hauled, these truckers might be willing to accommodate it, but most likely they would insist upon receiving a somewhat higher rate. The reasons for this are that in order to operate more truck-miles to accommodate the additional freight, they would probably have to pay their operators overtime, and might also incur higher costs of maintenance of the equipment resulting from the abnormally high usage and reduced total time per vehicle available for maintenance. Also, their terminal facilities might become congested (remember the discussion of congestion and queuing phenomena in terminal facilities in Chap. 7), which would result in the need to hire additional terminal handlers, probably proportionally more than the increase in traffic. All this would result in an increase in the total cost of providing transport more than proportionate to the increase in total traffic. Stated in economic terminology (first presented

Figure 11-2 **Example supply schedules for truck service between two cities.**

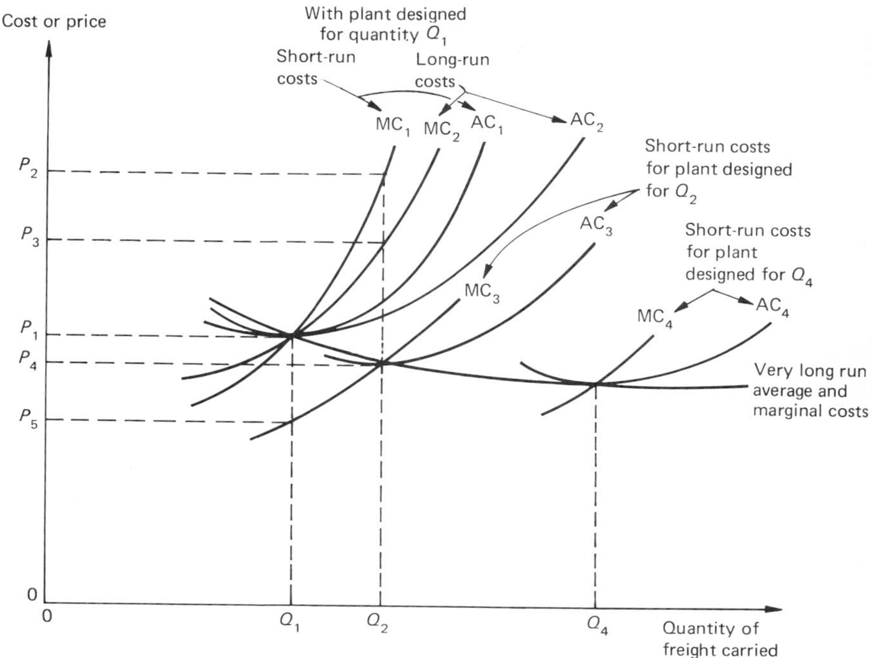

in Chap. 9), the marginal cost of additional traffic is greater than the average cost, resulting in the need for a higher price (equal to the marginal cost) per unit of traffic to persuade these trucking firms to provide more service and capacity. In Fig. 11-2, the short-run average cost curve AC_1 and marginal cost curve MC_1 are portrayed, with P_2 exemplifying the short-run price required to induce the truckers to provide a capacity sufficient to accommodate Q_2.

However, the firms would undoubtedly seek more efficient ways to accommodate increased traffic if this condition appeared likely to continue. Specifically, to provide for increased capacity at a somewhat lower cost, they would probably make arrangements to rent vehicles and perhaps hire additional temporary drivers. These measures would result in a lower cost, as indicated by P_3 on Fig. 11-2. If the increase in traffic appeared to be permanent, they would probably substitute purchased trucks for those rented, hire additional drivers as full-time permanent employees, and expand their terminal facilities to provide the most efficient service possible at traffic level Q_2. These adjustments and changes would take time to implement, perhaps months or even a year or two. Once these changes had been effectuated, new short-run average cost and marginal cost curves, shown by AC_3 and MC_3, would result. For this new level of ownership of vehicle and terminal facilities, etc., an entirely new set of cost curves apply. Assuming the vehicles and terminals, etc., are designed to be most efficient, at Q_2, Q_2 would be the quantity of least average cost, as shown. Similarly, additional curves could be drawn, each curve representing a plant—in this case trucks, terminals, etc.—designed on the basis of a certain expected volume of traffic. Such additional curves are also shown in Fig. 11-2.

In the long run—a period of time in which the firms can adjust their physical plants to meet the change of traffic conditions—the average price for this transportation service always reflects an optimum adjustment of the plant to the traffic to be carried. This corresponds to an envelope connecting the lowest points on these various supply curves, as shown by the very long run supply curve in the figure. In this particular example, the curve shown indicates a decreasing cost with increasing volume, although in any particular situation the curve may decrease or increase depending upon the cost of production, effects of the quantity of factors of production consumed in this sector—transportation—on the overall price in the market, and other factors.

One might ask, what would happen if the transportation plant were optimized for a particular expected level of traffic which failed to materialize, so that a lesser amount of traffic was transported. Refer again to Fig. 11-2, and consider a transport system with a plant size or capacity designed for traffic level Q_2. Once this plant is in place, if less than Q_2 traffic were to be transported, each of the firms involved would only have to cover its marginal cost in order to remain in business. Therefore, in the short run, the price of transport service would decrease along the marginal cost curve for traffic volumes less than Q_2, for example at Q_1, a price of P_5. Such lower prices could be maintained as long as the items represented by the fixed cost of these firms did not require replacement, i.e., as long as these costs were fixed, rather than variable, costs.

Once such fixed costs, terminal facilities or trucks or trailers, had to be replaced, a low price such as P_5 would not produce sufficient revenue to permit that purchase. At that price these firms would have to cease operation. In this situation, the price would have to rise in order to cover the cost of replacement of such equipment, and unless technology or prices of factors of production had changed, in the long run, the price would rise to the level indicated by the very long run curve.

As we shall discuss in greater detail below, the possible fluctuations in the prices, the quality, and other characteristics of transport services resulting from competition in a completely free marketplace is one of the reasons for much regulation of transport service. In particular, regulation is often directed towards transport situations where substantial price variations (as noted above) might result in very inadequate transport service, perhaps resulting in substantial injury to a region or economy. Also, in many situations it is not so easy to enter the market as it is in trucking. As a result normal competitive conditions may not result in prices at the level of marginal cost; the few firms engaged in the service may perhaps act in concert to charge abnormally high rates and make excessive profits.

An Extended View of Transport Supply

As we noted in the demand chapter, it is inadequate to describe the demand for transportation solely by the relationship between the price charged for a particular movement and the quantity of that movement which would occur. Rather, it is necessary to describe additional characteristics of the movement, in particular, the level of service, including such factors as travel time, reliability of service, comfort for passengers, or probability of damage of freight. The supply of transport must be described in a similar manner, including level of service as well as price, since it must mesh with the description of the demand of transport in order for us to use them jointly to determine the flow that will occur on a transport system (as we shall do in Chap. 12).

The level of service of transport is closely related to the volume, in a manner somewhat similar to that for pricing. This can be illustrated by continuing with the truck service example used in the discussion of pricing, but focusing on the time required from origin to destination (termed transit time for freight), one important level-of-service characteristic. Again, we will begin with a trucking service provided by many independent operators, who provide service based on a total plant capacity, including number and types of trucks, sizes of terminal facilities, etc., predicated on an expected traffic volume of Q_1. If suddenly an increase in traffic were to occur, as to Q_2, then likely the average time required for a shipment from A to B would increase from T_1 to T_2, as shown in Fig. 11-3. This curve is termed the level-of-service–volume curve.

The average travel time probably would increase because of increased congestion in the terminals, and probably some of the traffic would have to be accommodated on extra truck runs to be made in the evening or at other times which entail abnormally long waits for vehicle trips. Also, truckers would probably attempt to make greater use of available capacity on existing runs (increasing load factors), which would result in increased probabilities of delays for individual shipments waiting for available space on truck trips. As in the case of pricing behavior, if the increase in traffic were expected to continue for an extended period, then temporarily hiring trucks and drivers would probably reduce some of these delays. This would result in an average time T_3. Similarly, if the increase in traffic were expected to continue for a very long period, then additional trucks would actually be purchased, additional drivers hired as permanent employees, terminal facilities and maintenance

Figure 11-3 **Example level of service versus volume curves for a truck line between two cities.**

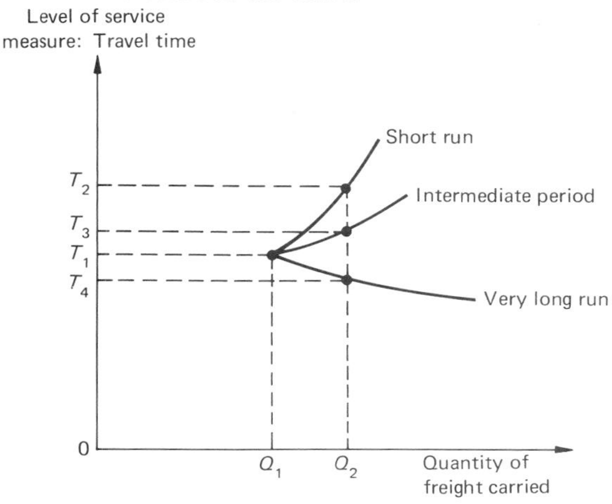

facilities expanded, etc., so that the total trip time would no longer reflect abnormal delays. In fact, as a result of the increased volume in traffic, there would probably be some reduction in average trip time due to the greater frequency of vehicle trips. This is as shown by T_4 in Fig. 11-3.

From this example, it is apparent that in the short run, congestion effects dominate the change in level of service resulting from an unexpected increase in traffic. They in general will cause a degradation of the quality of service, as illustrated by the average travel time example above. However, in the intermediate run or long run, if the transport system can be modified to accommodate the increased traffic in a more efficient manner, then the same or an even better level of service might result. A better level of service is especially likely where there is some waiting for vehicle trips, because with the increased volume, the extent to which the level of service will be affected by variations in volume will depend on the particular case. Figure 11-3 is simply meant to represent a common situation rather than to present a definitive form of the level-of-service–volume curve. In particular, if plant capacity cannot be expanded in an optimal way, for environmental and other reasons—it is becoming increasingly difficult to expand many transport facilities to accommodate increased traffic—a permanent degradation of level of service with increased volume may result.

Transport Technology, Production Functions, and Management Supply Behavior

The manner in which transportation service is provided and its effect on the level of service bears closer scrutiny, since it is extremely important for the description of all types of transport networks. In fact, in certain cases, such as automobile transport, the level of service is more important than price in determining amount of usage and satisfaction of users with the service, at least at price levels typically found in developed nations. In this section we shall attempt to clarify a few key concepts which will be used in the description of the supply of transport services.

In the trucking example above, the transit time of freight shipments from one city to another decreased with increasing volume because the managements of the various truck lines decided to provide the extra capacity required by increasing the frequency of truck trips. In general, the manner in which increased traffic volume can be accommodated is limited by technological possibilities, although as in this case there are many options from which a particular means must be selected. In this discussion, we shall focus on the so-called line-haul service, the movement from one terminal to another. However, similar considerations will apply to the choice of terminal design and operation and in fact to any other part of the system.

In deciding how to accommodate any given volume of freight, the truck line management must decide on the number of truck trips to operate per unit time (the frequency or volume of truck trips) and the carrying capacity of the trucks used. The choice is limited by certain technological relationships. Specifically, we will assume all of the trucks operated by one firm are of the same size or capacity. (This assumption can be relaxed, but it makes the analysis much more difficult, and the simpler situation illustrates the points well.) With this assumption, the capacity provided in each direction will be

$$q_c = fQ \tag{11-2}$$

where f = truck volume or frequency in one direction
Q = cargo capacity of each truck
q_c = total cargo flow capacity in one direction

This flow capacity must be greater than or equal to the volume of freight carried:

$$q_c \geqslant q$$

where q = actual flow of cargo

Assuming a constant load factor l,

$$q = lg_c = lfQ \tag{11-3}$$

The number of vehicle-miles operated per unit time also depends on the frequency:

$$m = 2Lf \tag{11-4}$$

where m = truck-miles operated per unit time
L = distance from one terminal to the other

Finally, the number of trucks required to operate the service is given by the following relationship, assuming that the round trip time for all trips is identical:

$$n = t_r f_r \tag{11-5}$$

where t_r = round trip or cycle time

In Fig. 11-4 these three relationships, Eqs. (11-3) through (11-5), are portrayed as they apply to a situation of carrying a constant volume of freight. To illustrate the effect of increased volume of cargo, the relationships for two different volumes are shown. Thus any given volume q_1 could be accommodated with f_1 vehicle trips made by trucks of capacity Q_1 or with f_2 trips with truck capacity Q_2. In these two cases the numbers of vehicles required would be n_1 and n_2, respectively, and the numbers of vehicle-miles operated would be m_1 and m_2, respectively.

These relationships specify the combinations of various inputs or resources needed to produce various outputs, various volumes of freight moved between the two terminals of this truck line. In the language of the economist, these form the production function (Samuelson, 1958, pp. 501–503). They are based on the technology of the productive processes involved and simply present the alternative ways in which a given quantity of output can be produced.

It remains for the management of the truck line to decide what particular combination of these inputs to use to produce the desired output, which would be represented by a point on the appropriate output quantity curve. For example, a truck line may decide to use vehicles of only one size regardless of traffic volume or quantity. In general, the choice of combination of inputs to use will depend on the output level and on the relative prices of the various inputs, which are really just factors of production. In principle, a firm would choose that combination which is the least costly or as little costly as

Figure 11-4 **A production function for a truck line serving two cities.**

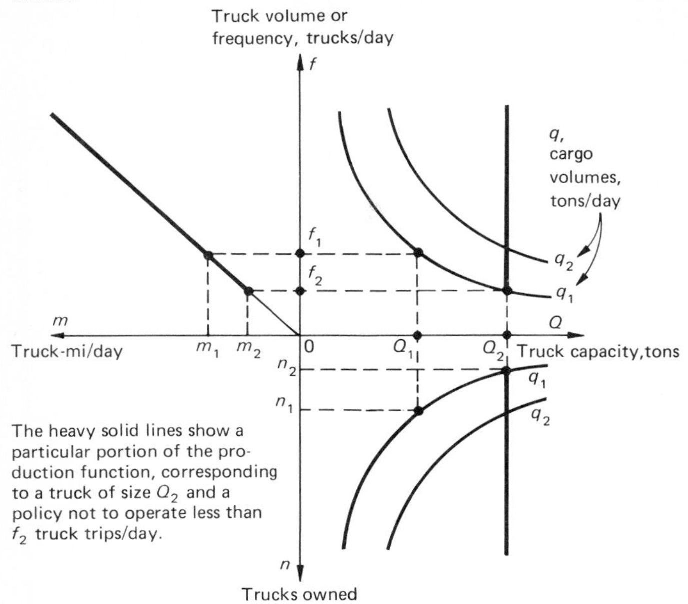

available information and firm policies permit. Of course, all inputs must be considered, including those of the terminals, administration, etc., and here we have just presented those related to the line-haul portion of the operation.

Truck lines often select the largest vehicle which can legally be operated; this option is also portrayed in Fig. 11-4, as a truck of capacity Q_2. The corresponding frequency options are indicated by the heavy lines, which are shown only for a range of frequencies above a certain minimum, in this case, f_2. The reason is that often a transport carrier will operate at least a certain minimum frequency of vehicle trips on a route even if the traffic volume does not fill those vehicles. This may be done in order to provide a reasonable frequency of service, perhaps to meet a requirement imposed by a regulatory agency that such a minimum service be provided. In Fig. 11-4, this frequency is shown to result in an effective capacity (maximum truck capacity multiplied by the load factor) equal to the cargo volume of q_1. This is interpreted as meaning that for any cargo volume of q_1 or less, the management will operate trucks as if the cargo volume were q_1. Alternatively, we could have measured the output as capacity q_c instead of cargo carried, and replaced the q_1 curve with a q_1/l curve, this curve being the least level of effective cargo volume capacity the carrier would provide. The only problem with this is that a more detailed specification of inputs may require knowledge of the cargo carried, as it would in this case for the input of fuel consumed, since this depends on the weight of the vehicle and cargo.

Given any such management decision, the level-of-service–volume relationship follows directly. This is because the average transit or travel time of the cargo from one origin in one city to destination in the next is equal to the sum of pickup time, terminal processing time (weighing, computation of charges, etc.), plus waiting for the next outbound intercity truck, truck travel time to the other city, and delivery. The greater the frequency of truck trips, the lower the waiting time at the terminal. (As may be recalled from Chap. 7, if the trucks operate on a uniform time headway, when the volume is doubled, the waiting time is halved.) Thus the travel-time–volume relationship appears in the long run as shown in Fig. 11-3.

As we mentioned in the previous chapter, on demand, the travel time of freight and passengers represents a valuable resource being consumed in transportation. For freight, the monetary value can be estimated by considering the value of the cargo, the interest rate that the owner of the freight pays on borrowed funds (or maximum rate that could be obtained on alternative uses of the money tied up in goods in transit), and the transit time, as discussed in Chap. 10. Thus one can conceive of a total cost to a user of the system consisting of the price paid to the carrier for the transportation service, a monetary value associated with the time the goods are in transit, and any other costs (such as packaging) involved in the movement. This forms a relationship with volume that might appear as in Fig. 11-5*a*, termed the user

Figure 11-5 Example of user cost–volume relationships for truck service. *(a)* Long-run relationship. *(b)* Short-run relationship.

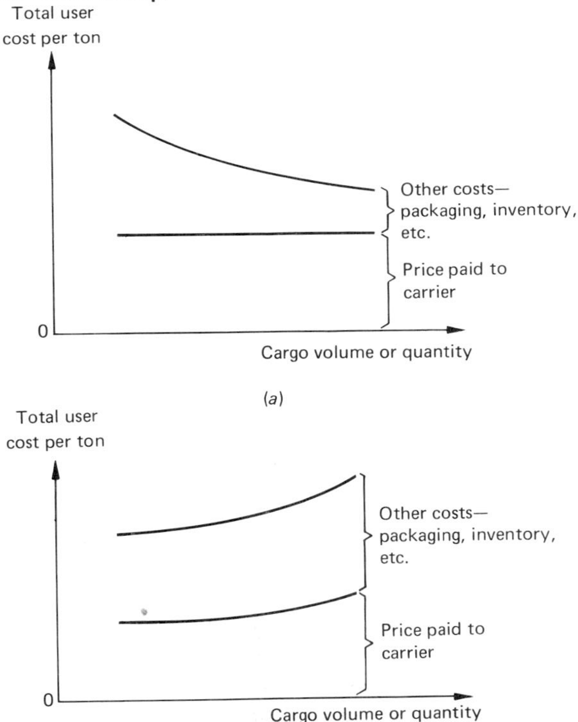

cost–volume curve (Meyer and Straszheim, 1970, p. 46; Wohl and Martin, 1967, p. 118). It is drawn with a constant price charged by the carrier, reflecting typical transport charges for an industry such as trucking where long-run marginal costs appear to be roughly constant as a function of volume, although often in other transport modes carrier costs vary considerably with volume. In contrast, the short-run user cost curve would probably appear as in Fig. 11-5*b*, increasing with increasing volume due to the combination of congestion effects at terminals, etc., increasing travel time, and the increase in carrier charges following the short-run marginal cost curve (as described in Figs. 11-2 and 11-3).

In fact, it has been argued by many that the travel time and other level-of-service properties of transport really are an integral part of the production function. The reasoning is that the travel time and other costs of transport as perceived by the user are really resources or inputs to the transport productive process. Obviously, without the consumption of the time of the cargo, its movement from place to place could not occur, and the same applies to travelers. Time and other user-perceived costs do have value, even if in some cases (such as passenger transport) we cannot put a precise monetary value on them. If this argument is accepted, then the level-of-service or user cost relationships become parts of the production-possibility curves and of the production function. However, again it is appropriate to note that the carrier usually does not incur these user costs and hence may not consider them in deciding on the particular production method to follow. Fortunately, whether user costs are part of the production function or merely follow from it is not critical to the analysis.

To summarize, then, the production function provides information on the resources or inputs required to produce any given level of output. Since various combinations of inputs can in general by used to produce the same output, it portrays options available to the management, designers, planners, or others who choose the particular combination of resources (i.e., the particular technology) to be used. The range of options in any given situation may be limited by a number of factors, such as franchise restrictions or other regulations which a carrier must follow, environmental or safety regulations, design standards for facilities and vehicles, and limitations on the funds available. Given any such constraints and the goals of the managment group or other decision-making unit, a particular choice of the method of production is made, usually considering such factors as the expected traffic volume, revenues, etc. Once this choice is made, the relation between the volume of traffic and user costs is determined. The relationships among these factors—the production function, management choices, and the user cost versus volume relation (including traditional supply functions)—are portrayed in Fig. 11-6.

SUPPLY CHARACTERISTICS OF TRANSPORT FACILITIES

One important class of transport supply functions and related user cost–volume functions is for transportation facilities, such as roads (both "free" and toll), airports, ports, parking facilities. The distinguishing feature of these is that a facility is provided for a "user," who provides a vehicle, whether it be a personal automobile, a truck, an aircraft, etc. The provision of a facility is

Figure 11-6 **Determinants of supply relationships.**

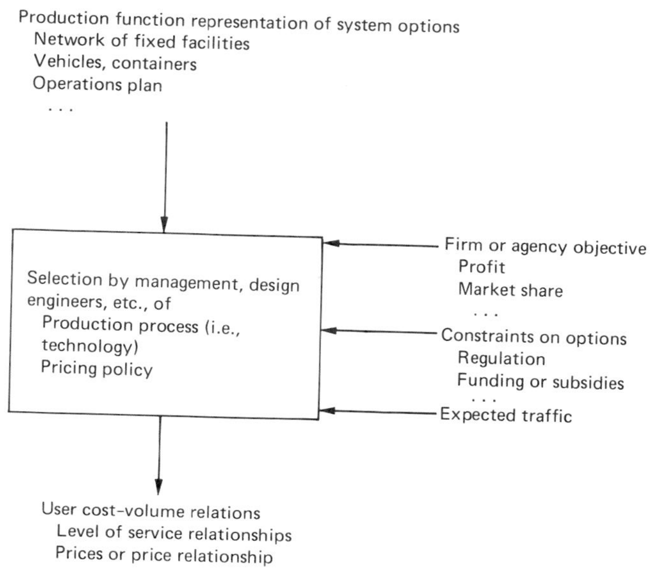

distinguished from the service provided by a carrier, who provides the vehicle (and everything else) required to move the cargo or passenger.

Pricing

The pricing of transport facilities varies considerably. At one extreme is the pricing of such facilities as sidewalks and hallways of buildings for which the user pays no direct fee. In these instances the cost is borne by another party—general taxes in the case of sidewalks and the owner of the building (and through rents, the tenants) in the case of hallways, stairs, etc. It is reasoned that the cost of these facilities is small and it would be extremely difficult technologically and very costly to charge users directly. So the facility is financed by a fee on those who benefit from the facility, i.e., those whose property is made accessible by the facility. While use and benefits from the facility may not be very well reflected in the fee schedule, the amount is so small and a more use-related charging scheme so difficult to implement that such financing is usually acceptable.

At the other extreme is a facility which must finance itself out of the revenues it receives from users. Examples include toll roads, bridges, some ports and airports, and portions of the "free" public road system. The basic principle is that the average price should equal average cost plus profit, the latter usually consisting solely of the interest on bonds since these facilities usually are owned by "nonprofit" governmental agencies. Typically the costs of these facilities are largely fixed, with only a small variation due to operating and maintenance costs that vary with use. Hence the price charged tends to follow a rapidly decreasing average cost curve, as shown in Fig. 11-7.

However, in practice there may be deviations from such a pricing policy. Firstly, at very low volumes the facility may be subsidized by local and other

Figure 11-7 Typical total and average cost curves for a transportation facility. *(a)* Total cost vs. number of users per unit time. *(b)* Average cost vs. number of users per unit time.

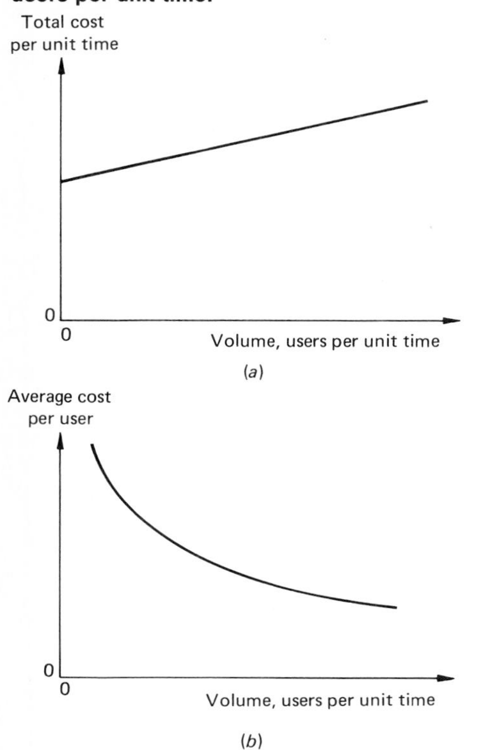

governments, which results in a less sharply changing price-volume curve. Also, there may be political impediments to changing the price once it has been set; deficits are made up by subsidies from general tax funds, and profits are used to finance or subsidize any operations of the agency which lose money. Thus it is common for "profitable" toll road facilities to subsidize deficit-producing public transit lines or airports or to provide the capital necessary for other projects such as road, airport, or port expansion.

Also, there are often many different types of users of a single facility, each of which imposes different requirements on the facility and results in different costs being incurred. In these situations a pricing scheme which reflects these differences and which in total recovers the entire expenditure is desired. This raises serious problems, for by definition fixed costs cannot be interpreted as being caused or incurred because of the use of the facility by any particular class of user. Any allocation of the fixed costs is necessarily arbitrary! This problem is particularly troublesome in road transport, for the amount of money involved is very large indeed and there are issues of fairness in the treatment of all modes of transport which are also involved.

An alternative form of pricing would be to price at marginal cost. Since in principle, at least, the marginal cost of each user can be identified, this should be workable. But there are many problems with this scheme. First, in most cases the marginal cost of providing and maintaining the facility is well below the average costs, which means that revenues do not equal costs and the facility would have to be subsidized. Second, it is difficult to identify true marginal costs, for their magnitude depends upon the time period involved. Third, the value often depends upon the management policy on operations and maintenance. For example, one management may decide to perform major maintenance every decade, at great expense then, but with very little expenditure between such heavy maintenance periods. Is the true marginal cost almost zero in those intermediate years and then much greater when the time for scheduled major maintenance arrives? In contrast, another management may decide to perform maintenance more frequently, which would result in a more nearly uniform marginal cost of keeping the facility open for use. This raises the question of which cost to use in any analysis. While one can dismiss these questions by observing which is the more efficient management policy, a workable means to do this is not always available.

Another problem with marginal cost pricing is that it may be institutionally very difficult to effect throughout the economy, particularly to services not publicly owned. How would you subsidize a privately owned transport carrier and ensure that the incentive to remain efficient continued? What would be a fair subsidy? Who would pay the taxes to support the subsidy?

For all these reasons, marginal cost pricing of facilities has not been used on a large scale. However, as we shall see when we discuss the combined marginal costs of providing the facility plus the costs of the users, there may be a practical role for these concepts which does not have all the disadvantages identified above.

Example of Road Pricing

We next consider the means used to determine the most appropriate level of prices for the use of roads. While a number of methods have been proposed, most of them are arbitrary. The method presented below seems to be reasonably consistent with pricing that would occur in the private sector.

The basic approach is to divide the users of the facility into classes, each class causing a different level of expenditure per unit of use (e.g., vehicle-mile or axle-mile). Let us assume that the users of a facility have been divided into N classes, 1 entailing the least cost per unit of use, 2 the next least cost, etc. The amount of use by each of these has been calculated to be U_1 for class 1, U_2 for class 2, U_3 for class 3, etc. The cost of providing the facility for all users, classes 1 through N, assuming all of them modified their vehicles so as to fall in class 1, is calculated, and termed C_1. Then, the additional cost of providing the facility for all users in class 2 and above, assuming they modified their vehicles to meet the limits of class 2, is calculated as C_2. This process is continued, finally to yield the added cost of providing the facility for users of class N, for users actually in that class, C_N. Note that the actual total cost is thus $C_1 + C_2 + C_3 + \cdots + C_N$. Then a set of partial prices is calculated, the first being for all users and based on the cost C_1 and the level of use $U_1 + U_2 + U_3 + \cdots + U_N$, this being $p'_1 = c_1/(U_1 + U_2 + U_3 + \cdots + U_N)$. Then a partial price for class 2 is calculated, C_2 divided by all users in this class and above (more costly), $p'_2 = C_2/(U_2 + U_3 + \cdots + U_N)$. This continues to class N, where $p'_N = C_N/U_N$. Then the price per unit of use for each class is calculated as the sum of all the partial prices up to and including that class. Thus, in equation form:

$$p_1 = p'_1 = C_1/(U_1 + U_2 + U_3 + \cdots + U_N) = C \Big/ \sum_{i=1}^{N} U_i$$

$$p_2 = p'_1 + p'_2 = p'_1 + C_2/(U_2 + U_3 + \cdots + U_N) = p'_1 + C_2 \Big/ \sum_{i=2}^{N} U_i \qquad (11\text{-}6)$$

$$\cdots$$

$$p_N = p'_1 + p'_2 + p'_3 + \cdots + p'_N = p'_1 + p'_2 + p'_3 + \cdots + C_N/U_N$$

That this results in a total revenue, price multiplied by respective level of use, summed over all user classes, exactly equal to total cost, is revealed by the following substitution:

$$p_1U_1 + (p_1 + p_2)U_2 + (p_1 + p_2 + p_3)U_3 + \cdots + (p_1 + p_2 + p_3 + \cdots + p_N)U_N$$

$$= p_1 \sum_{i=1}^{N} U_i + p_2 \sum_{i=2}^{N} U_i + p_3 \sum_{i=3}^{N} U_i + \cdots + p_N U_N$$

$$= \frac{C_1}{\sum_{i=1}^{N} U_i} \sum_{i=1}^{N} U_i + \frac{C_2}{\sum_{i=2}^{N} U_i} \sum_{i=2}^{N} U_i + \frac{C_3}{\sum_{i=3}^{N} U_i} \sum_{i=3}^{N} U_i + \cdots + \frac{C_N}{U_N} U_N$$

$$= C_1 + C_2 + C_3 + \cdots + C_N$$

Now we shall turn to an actual application, taken from Meyer et al. (1959, pp. 64–85), which although dated, is one of the most recent in widely available literature. It illustrates the method very well. The example is taken from actual studies of appropriate road user charges in Louisiana (Ross, 1955). It should be borne in mind that the costs are only those to be borne by the state, and that a small amount has been subtracted from them to reflect property tax contributions for local land access of about 6 percent of total road expenditures. Also, costs are calculated separately for each of four different types of roads, although the method used in each is identical.

The first step is to calculate the cost of providing the roadway for each different type of facility (high, medium, and low) for those vehicles accommodated at the least cost per vehicle-mile, those with weights per axle less than or equal to 6000 lb. The cost includes the cost of right-of-way, construction of the facility, maintenance of the facility, and administration of user charges, which in this case are primarily fuel and taxes. In the case of the Louisiana study, it was estimated that in 1955 it would cost \$42.3 million/year to provide a so-called high-type pavement for vehicles with this range of axle loads. This information is shown in Table 11-1. Also, it was estimated that all vehicles would operate approximately 17.7 billion axle-mi over each mile of such roadway. Dividing the total annual cost by the annual axle-miles results in a total cost per axle-mile of \$0.002386, also indicated in the table.

The next step is to determine the added cost of providing or designing those same facilities to accommodate vehicles which have axle weights ranging from 6001 to 10,000 lb. The additional cost for such vehicles was esti-

Table 11-1 **Example of Incremental Cost Procedure for Cost Allocation: Roads in Louisiana†**

Size-weight increment, thousands of lb	Cumulative axle-miles, thousands	Increments of cost, $ thousand									Total cost per axle-mile, mills§	Cumulative cost per axle-mile, mills§
		Right-of-way	Grading and drainage	Surface	Structures	Maintenance	Administration	Debt service	Fuel and oil tax administrations‡	Total		
High-type pavement												
0–6	17,735,095	3835	5904	11,711	11,066	2615	2150	4782	253	42,316	2.38600	2.38600
6–10	1,403,097	...	610	1,241	528	638	...	...	...	3,017	2.15024	4.53624
10–14	1,035,988	...	305	2,633	2,541	953	...	...	...	6,432	6.20857	10.74481
14–18	812,237	...	...	2,168	360	742	...	...	...	3,270	4.02592	14.77073
Medium-type pavement												
0–6	2,840,755	290	2050	2,751	1,444	935	568	...	41	8,079	2.84396	2.84396
6–10	272,287	...	151	1,720	98	671	...	...	...	2,640	9.69565	12.53961
10–14	208,949	...	402	1,433	120	1006	...	...	...	2,961	14.17092	26.71053
Low-type pavement												
0–6	1,275,662	114	1062	2,240	542	1390	211	576	19	6,154	4.82416	4.82416
6–10	94,554	...	...	1,600	55	1543	...	...	...	3,198	33.82199	38.64615
Gravel												
0–6	809,940	36	256	554	113	1882	36	...	12	2,889	3.56693	3.56693

† Meyer et al. (1959, p. 80), based on Ross (1955).
‡ Distributed between surface types on the basis of the distribution of traffic.
§ One mill is one-thousandth of a dollar

mated to be \$3.02 million/year, and it was estimated that there would be 1.4 billion axle-mi/year by vehicles of axle weights above 6000 lb. Dividing the added cost by the axle-miles operated with weights of over 6000 lb yielded an additional cost per axle-mile (partial price) of \$0.00215. Vehicles which weighed between 6001 and 10,000 lb/axle would then be charged the sum of the basic cost to provide the minimal roadway plus the added cost per axle-mile of providing a road adequate to accommodate the additional traffic, the value in this case being \$0.004536.

Similarly, as can be seen from the table, the incremental or additional cost of providing pavements for heavier axle loads was calculated, divided by the total number of axle-miles with that weight or greater, and the resulting incremental cost per axle-mile was added to the preceding cost to yield the total cumulative cost per axle-mile assignable to each vehicle class. Such calculations were done for each of the different types of pavements provided, as can be seen from Table 11-1.

Accepting the premises of this study, we would next try to devise a pricing system which would generate revenues roughly equivalent to these costs. As can be seen from Table 11-1, the cost per axle-mile for a particular type of vehicle varies with the different types of pavement, particularly for the heavier vehicles, which results in a very substantial total cost for lower-design-standard pavements. In order to arrive at an average cost per axle-mile throughout the entire statewide system, the fraction of miles which vehicles of each type would spend on each of the different types of pavement is calculated and used to weight the various costs. The result is then an average cost per axle-mile for each different axle weight class. Then the problem would be to devise a taxing scheme for fuels used by the vehicles which would reflect the different costs. In practice, there is a basic fuel tax per gallon of fuel and similar taxes per unit on various other items consumed in motor vehicle operation. In order to set these tax rates, the average miles operated per gallon is calculated. From it the desired level of tax per gallon of fuel is determined. In practice, it appears as though a heavy vehicle should pay more per gallon than light vehicles, and this is achieved in some states to a rather limited extent by imposing an extra tax on fuel or other items consumed by large vehicles.

Any taxing scheme or pricing scheme is, of course, subject to much criticism. Many criticisms derive from the fact that a tax on fuel or other factors of production very imperfectly compensates for the cost of providing highway facilities, because the amount of damage to highways or the nature of the facility required is only very indirectly related to the amount of fuel or other factors of production consumed. The only way to overcome this would be to apply some sort of toll scheme, which, of course, would be very expensive if applied generally unless some of the electronic tax collection schemes now being developed could be applied inexpensively. Also, the prices so developed are somewhat arbitrary, in effect charging the fixed cost of road facilities to the users rather than only the marginal cost. However, with the exception of right-of-way, the road pavement must be renewed regularly—if at long intervals—and this pricing scheme approximates the long-run marginal cost.

Also, it is often pointed out that this sort of pricing of highways does not include payment to communities in lieu of property taxes which a private owner of a roadway or other transport facility, such as a railway or pipeline, would have to pay. The issue of whether or not money should be paid to municipalities and other units of government in lieu of various forms of taxes on road facilities is an extremely complex one, for there are obviously many differences in road facilities and their uses: some are used not only by motor vehicles but also by pedestrians, bicyclists, etc. (privately owned transport facilities usually do not have these joint uses). Nevertheless, underpricing one aspect of transport relative to another would induce distortion in modal usage and even in the total amount of transport. Unfortunately, although the pricing of road facilities is an area in which it is not too difficult for thoughtful persons to identify most of the major issues—and we have only mentioned a few of the more important ones above—it requires a substantial amount of new research for most of the questions raised to be answered. Until that research is undertaken, the pricing of road facilities will remain more a matter of political expediency and bargaining than a result of careful economic and engineering logic.

Turning from theory to practice, a recent study (Bhatt et al., 1977) of the revenues from users and the expenditures for roads in the United States in the period 1956 through 1975 sheds considerable light on the actual relation between user charges and expenditures. Taken as a whole, over this period all classes of vehicles contributed user charges (through fuel, tire and other excise taxes, and licenses) of $277.3 billion, while direct costs for construction, maintenance, and administration totaled $306 billion. However, the revenues from urban roads exceeded expenditures by about 20 percent, while on rural roads expenses exceeded revenues by about 49 percent. Furthermore, heavy vehicles were found to contribute revenues only slightly in excess of the additional cost associated with accommodating such vehicles, while other vehicle classes contributed far more than their marginal costs. These results, plus the more conceptual issues discussed in preceding paragraphs, indicate that there is room for much research and careful policy formulation in the area of road pricing.

Average Total Transport Cost

In Chap. 9 the concept of total transportation cost was presented. This cost is meant to represent all of the scarce resources used in the production of transport service, including not only the expenditure for right-of-way, construction, user costs, and road maintenance, but also a valuation of the time spent by persons and shipments while traveling, the cost of accidents, and so on. The concept has many theoretical and practical shortcomings, as mentioned in that discussion, mainly with respect to attempting to place a monetary value on scarce resources such as travel time, the value of which undoubtedly varies widely among different road users. Still it is interesting to examine average total cost functions for typical road facilities. This concept of total transport cost undoubtedly has considerable validity in situations wherein almost all of the costs of transport can be assigned a monetary value, such as where the cargo and commercial vehicles form the bulk of movement over the road facility, so that travel time, accidents, damaged goods, and so forth, can all be evaluated monetarily without a great deal of difficulty. The primary use of road facilities for goods movement often occurs in developing countries.

Figure 11-8 **Average total transportation cost-volume relationship: example based on road costs in a developing nation. Note: Based on Fig. 9-9.**

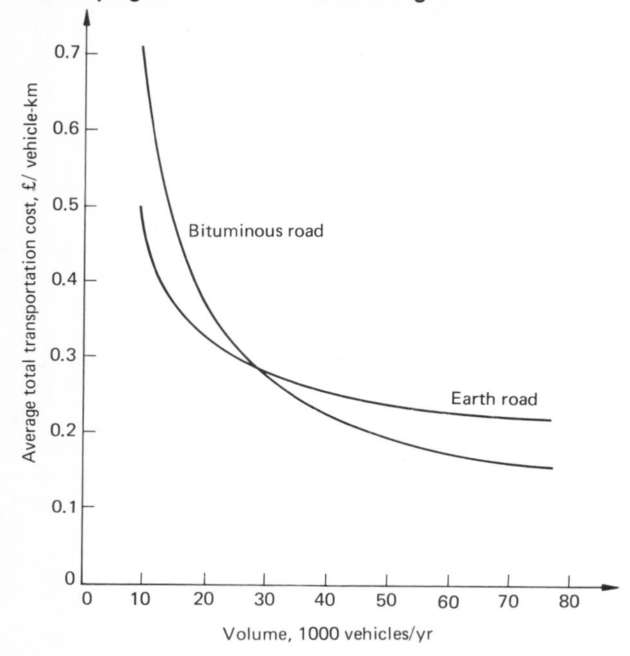

By way of example, in Chap. 9, Fig. 9-9 presented two typical transport cost relationships for a road facility which could remain an unimproved earth road or be improved as a bituminous road, for various volumes of traffic. As indicated there, total transport cost can be used as the criterion for deciding at what volume a road facility should be improved. The total transport cost presented in that figure can be divided by the average annual traffic to yield an average cost per vehicle-mile. The resulting average total transport cost per vehicle is presented in Fig. 11-8. As would be expected, the average total cost decreases rather markedly with increases in volume, and a further decrease caused by the substitution of an improved bituminous road for an earth road occurs above a volume of 29,000 vehicles/year. It should be noted that all of these volumes are sufficiently low that very little if any congestion is likely to arise, and hence even with the basic two-lane road used for all these computations the cost per vehicle continues to decrease with increase in volume. Of course, at some point the volume becomes so great that congestion becomes a factor and without an increase in the number of lanes or other improvements, the average cost increases.

Costs Perceived by the User

In the case of the use of transport facilities, direct dollar outlays are not the only important cost, since there are also the user travel time and other costs related to level of service. In fact, in making choices among road routes it has been found that travelers respond more to travel time than to any other cost item. It is instructive to examine the cost of travel as it would be perceived by users of a facility such as a highway. Since the travel time is the major cost perceived by travelers, we shall pay particular attention to it. The relationships revealed will also apply to other types of transport facilities.

The total cost perceived by a motorist traveling over 1 mi of highway consists of the time consumed, any unusual discomfort or stress due to difficult traffic flow conditions or poor road design (e.g., very sharp curves and steep grades), some of the operating and maintenance costs of the vehicle, and, of course, tolls, if any. People undoubtedly vary in their perception of the money costs of motor vehicle travel, some probably being aware of the added costs of fuel, oil, and maintenance per mile, while others probably ignore most of them since the actual expenditures are made only intermittently. As we discussed in Chap. 10, studies have shown that in making decisions between alternative routes, motorists usually choose the route with the least travel time, although deviations from this rule are observed due to differences in tolls, scenery, or other neighborhood characteristics. Thus it is not clear exactly what costs users actually perceive and to what extent they are aware of their true magnitudes.

However, we might assume that users perceive travel time, and focus on that cost. In Chap. 5, we explored the relationship between average speed of traffic on a road and the volume of traffic. The relationship differed by type of road, number of lanes, etc. But given a particular design, traffic volume, and mix of vehicles, we can estimate the travel speed and hence estimate the time per mile, which is just the inverse of the speed. We shall do this for a freeway in rolling terrain, with three lanes in each direction, 8 percent trucks and 15 percent buses, and a design adjustment factor of 0.95. Using Eq. (5-11), which is repeated below, we obtain a capacity of 3537 vehicles/h.

$$c = 2000NWT_cB_c \tag{5-11}$$

$$T_c = \frac{100}{100 - 8 + 4(8)} = 0.807$$

$$B_c = \frac{100}{100 - 15 + 3(15)} = 0.769$$

$$c = 2000(3)(0.95)(0.807)(0.769) = 3537 \text{ vehicles/h}$$

Figure 5-11 gives the relationship between average traffic speed and volume-to-capacity ratio. Hence we can determine the speed for various volumes, take the inverse to get travel time per mile, and plot this versus volume as in Fig. 11-9. (Incidentally, often the travel time per unit distance is termed slowness.) Here we are ignoring the backward-bending part of the speed and travel time curves, since roads do not often operate in that regime, although it could be plotted if desired. As can be seen, travel time per mile increases with increasing volume. This is the same relationship that was observed in Chap. 5 between time delay at intersections and volume, and in fact it is the general relationship between time and volume in all types of transport facilities, including terminals, as we noted in Chap. 7! In short, user time costs either

Figure 11-9 Typical road travel time versus volume relationship: freeway.

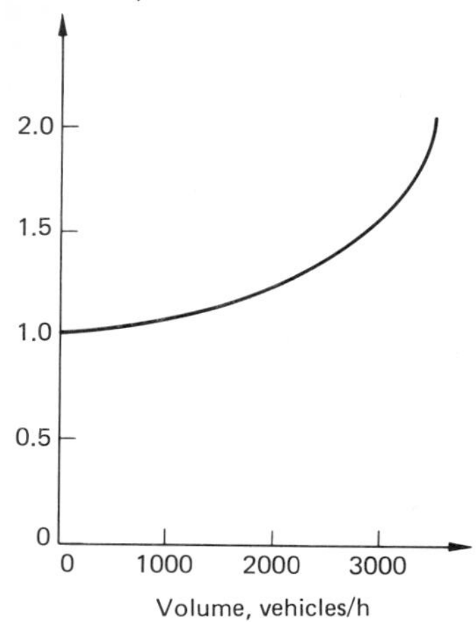

remain the same or increase with increasing volume, the latter applying to most typical volume ranges.

Before leaving the discussion of road level of service, we point out that many aspects of road level of service are directly under the control of the individual road user. For example, owners and operators of their own private automobiles can select from among a wide variety of vehicle types and styles the combination of various comfort attributes most nearly ideal for them. Similarly, operators of buses and trucks and other commercial vehicles can select from among vehicles offering different level-of-service attributes (such as seating configuration and rest rooms on buses) the ones which conform most nearly to the wishes of the users. Also, if traffic is flowing under extreme free-flow conditions, deliberate choices can be made regarding the speed at which a vehicle is operated, presumably up to but not very much in excess of any legal maximum (or below minimum) speed limits.

SUPPLY CHARACTERISTICS OF CARRIERS

Carriers are distinguished from providers of transport facilities in that the former provide the vehicle (or possibly container) in which the goods or passengers are to be transported. Thus they include bus, rail, taxi, and air passenger services, and the freight services of railroads, steamship lines, truck lines, etc. The supply characteristics of these are different from those of facilities as viewed by potential users and follow the general pattern described in the initial discussion of supply characteristics of a truck line.

Little research on the supply characteristics of freight and passenger carriers has actually been undertaken. However, to illustrate the principles involved, we shall focus on the provision of urban bus transit service, after a brief description of pricing policies and regulation and some other institutional aspects of transport.

Price and Level-of-Service Principles

The truck line example used to introduce many of the economic concepts of supply is typical of the pricing and service quality characteristics of any carrier which is operating in a reasonably competitive market where entry of new firms is easy and where prices are not fixed by either a regulatory agency or by collusion among the carriers. Many freight transport markets are probably of this type, including not only trucking wherever it is not subject to price and entry regulation, but also some air freight and water services. Also, some passenger transport markets are similar, such as taxicab and even some scheduled services (such as small bus lines) where entry is not limited. However, in practice these conditions are often not met, and it is therefore important to discuss the principles governing pricing and service quality in these other situations.

In practice, carriers are guided by many principles in the setting of prices. One of these has already been discussed, namely, the charging of marginal costs in a perfectly competitive market. In fact, the principle of marginal cost pricing goes beyond a single type of market. A widely accepted economic theory is that the setting of prices equal to marginal costs will result in the most efficient allocation of resources in an economy and hence that all goods and services should be so priced. This raises the question of the time period for the marginal cost. Most seem to feel that long-run marginal costs are appropriate, since otherwise substantial fluctuations in prices might result, which would be very disruptive to the economy.

The implications of this marginal cost policy are numerous. It means that the price is greater during peak traffic periods than at other times, since peak periods require many additional vehicles and containers and greater facility capacity. Also, prices are not in general identical for movement between the same two points in both directions since the direction with unused capacity has a very low price, reflecting the ability to carry things in the empty vehicles. As indicated, these principles are now followed to some extent in such areas as urban transportation, where different fares depending on direction and time of day are often found.

A closely related principle is to charge at what is in effect an average cost (rather than marginal cost), but with differentiation in prices to reflect the differences in the marginal costs of different users. The reason for charging at an "average" cost level is to ensure that total revenue equals total cost, which of course it would not with marginal cost pricing if marginal costs are below average costs. This avoids the problem of subsidizing any transport firms or agencies which have marginal costs below average costs. Usually the difference in the price to different users is set so as to equal approximately the difference in their marginal costs. An example of this type of pricing is the pricing of road facilities.

Another principle is value-of-service pricing. Under this scheme carriers charge an amount roughly proportional to the value of the commodity being carried. The reason is that the more valuable a cargo, the more purchasers (or sellers) are able to pay for transport. This was the basis for much railroad pricing before the advent of trucks and inland waterways as viable economic alternatives for long-distance movement. The rates were set so that total revenue equaled total cost (plus profit). But once other carriers were available, the railroads found themselves losing the high-value freight on which the high prices applied (such prices often being well above trucking costs, for example) and being left with freight, which, because of its low value, was often priced well below average and even marginal costs! Since many different modes of transport now have somewhat similar costs, the feasibility of this type of pricing has diminished considerably.

Another principle of pricing is the maintenance of specific transport price differentials between pairs of cities for particular commodities. The concept is basically that if in the past the transportation of a particular commodity from *A* to *B* has been *Y* dollars more expensive than the shipment of the same amount of that commodity from *C* to *B*, then that difference should be maintained in the future. The reason for this is that the locations of industries in these towns, the amounts shipped, etc., are based upon this difference and that this pattern should not be disrupted. Maintenance of existing differences has played an important part in the pricing of monopolies and of rigidly regulated transport carriers, but obviously it is not workable if a part of the system is not regulated. Thus the advent of trucks and roads, and the ease with which a shipper might operate a private truck fleet if charged an excessive amount by a carrier, has made rigid adherence to this principle impossible.

A final principle is that of equity in prices. Equity can be defined in many ways, but in general it involves prices being the same in some respect. At one extreme, all prices might be made identical, regardless of the distance traveled, the direction of movement, time of day, etc. While this may seem to be an unusual scheme, it is in fact used in many urban transit systems throughout the world as well as in many postal and small-package services. Another extreme would be to charge the same rate per unit distance per unit weight. Alternatively, it may simply mean that the same price applies for the same quantity and distance, a rule often approximated in passenger carrier prices.

All of these principles are or have been used to some extent in transportation. Since so much of transport is regulated, we turn to a discussion of regulations.

Some Institutional Characteristics: Types of Carriers, Regulation, and Subsidization

A common carrier is a transportation firm which will carry any freight or any passenger between points on its routes at an established fare applied to all identical movements without discrimination (Grossman, 1959, pp. 187–212). The carrier has an obligation to transport the freight or passengers to their destinations reasonably rapidly and without damage or loss in the case of freight or injury in the case of persons. These general responsibilities are of course subject to some qualifications. In particular, some carriers may be equipped to carry only passengers; and others, only freight of particular types. Also, they may serve only certain places and would be under no obligation to serve others. They may also refuse to transport something when no capacity is available, but presumably would have to accommodate that person or shipment as soon as capacity did become available. And finally, common carriers are usually expected to charge "reasonable" prices, presumably ones which do not yield "excessive" profits.

These concepts of the common carrier date back to the medieval period. In fact, some aspects have been traced back to Roman times. In exchange for charging transport carriers with these responsibilities, governments have also provided them with certain benefits. In particular, in many cases the number of firms which can engage in common carriage is limited, in exceptional cases, to one carrier in a particular area (often with regard to a particular commodity), but more often to the number which can reasonably be expected to earn a fair profit. Also, governments have paid much attention to providing public ways in which common carriers can operate their vehicles and to keep these free of thieves, unjust taxes which might be imposed by communities through which carriers operate, etc., aspects which were of extreme importance in medieval times when the concept of common carriage was first being developed (or perhaps redeveloped), as noted in Chap. 2. Also, governments may set up mechanisms to determine the character of the services provided by common carriers and the rates which they charge, considering both the public or societal need for service and the need of the carrier to cover costs and earn a fair profit.

In recent decades the technology of transport has changed, and of course the volume of passenger and freight traffic to be moved has increased very substantially. As a result, common carriers no longer provide the sole or necessarily the dominant form of transportation. In the realm of freight movement, many firms ship such large volumes of freight that they can afford to operate their own truck fleets, and in a few cases, their own aircraft and boats. This is known as private carriage, carriage in which the owner of the freight transports it in his or her own vehicle. Also, there is an intermediate form of carrier, called a contract carrier. Such a carrier obtains from appropriate regulatory bodies the right to engage in transporting particular commodities for particular firms, entering into contracts with those firms and in no sense claiming to be a common carrier. Also, there are many so-called indirect carriers of freight who take responsibility for the movement, collect the charges, etc., but do not actually perform the movement from origin to destination. Examples are the postal services and many package express companies, whose line-haul movement is often on another carrier (e.g., railroad or airline). A very important class of such indirect carriers are freight forwarders, who consolidate many movements between origin and destination regions after performing their own pickup and delivery service and then ship the entire collection of freight via another carrier. The freight forwarder thereby obtains the advantage of a low price from the carrier, because of the large size of the shipment; part of this gain can be passed on to the shipper, and often the local freight forwarder can perform a more efficient specialized local service than a large intercity carrier could. Finally, there are many areas of transport in which any person can enter the business and offer to transport freight at any price, with only a minimum of regulations designed to ensure safety standards and legitimate business practices. An example is the movement of agricultural commodities in the United States by truckers whose rates and quality of service are completely unregulated.

The dependence of travelers on common carriers has also been diminished by technological developments. The automobile provides for movements over land: as we noted in Chap. 2 most travel in the United States—about 80 to 90 percent, depending on the type of travel being considered—is via auto. Not only can one purchase a car, but one can easily rent one for short periods. Small aircraft and boats can also be owned or rented by individuals, but their use is limited. The range of common carriers is substantial, including bus, air, rail, taxi, and, to a limited extent, water transport services. Often two or more firms provide service between the same points, providing some competition.

At the present time, regulation of transport in the United States and in most nations includes control over prices charged, entry or exit from the business, quantity of service (or overall capacity), quality of service, and limitations on the commodities to be carried and the locations served. As an example, the extent to which these various types of regulations are applied to various modes of freight transport in the United States is indicated in Table 11-2. The full range of degree of regulation can be seen, with railroad freight service rather extensively regulated, while other forms of freight service are not so extensively regulated. Also, within each mode, there are various degrees of regulation. Almost all intercity and local passenger service is regulated as to entry, price, and often level of service.

The significance for regulation in the modeling of transport supply is that the regulations will influence the prices and the quality of service provided. The decisions of the regulatory bodies in particular cases as well as general regulations must be taken into account in any attempt to portray the supply of transport service. This regulation may achieve prices which are very close to those that would be charged by the carriers acting on their own. However, it also may be true that regulators force carriers to charge abnormally high prices, perhaps in order to protect the interest of another carrier or firm which

Table 11-2 Intercity Freight Transport Regulation in the United States[a]

Type of regulation	Modes			
	Rail	Highway	Water	Pipeline
Rates	Full	Partial[b]	Partial[c]	Partial[d]
Entry or exit through carrier expansion or contraction	Full	Full	Full	None
Entry through merger or acquisition, same mode	Full	Full	Full	None
Entry through merger or acquisition, another mode	Partial	None[d]	None[e]	None[e]
Overall quantity of service	None	None	None	None
Quality of service, including route	None	Full	None	None
Limitations on commodity and direction of traffic	None	Full	None	None

[a] Capron (1971, p. 167).

[b] The degree of rate regulation is largely a function of the form of carrier organization. Common carrier rates are regulated most; agriculturally exempt and "private" carrier rates (or internal prices) are not regulated. Contract carriers fall in between.

[c] This is similar to highway regulation, except that exemptions are granted for other reasons besides the agricultural commodities provision.

[d] Rates are regulated on a rate-of-return basis; hence individual rates are the object of very little regulatory concern.

[e] There is no explicit prohibition. Entry is subject to ultimate interpretation by the Interstate Commerce Commission and the courts.

is inherently more costly, as has been charged in the case of many railroad rates being held abnormally high relative to truck rates. On the other hand, regulatory agencies may force carriers' rates below what would otherwise be charged. This may result in short-run benefits to the communities so served, but may present serious consequences in the long run, due to the carriers' inability to develop and adopt technological innovations or to maintain the quality of service at a satisfactory level and their need to charge abnormally high rates elsewhere (Gellman, 1975). This is not to call into question the general principle of regulation, but rather merely to point out that it can result in distortions of the normal market mechanism and result in what may be viewed as gains in the short run but possibly unfortunate consequences in the long run. The conditions of inherent national monopolies of single common carriers which characterized much transport in the last century simply do not generally hold today, due in large measure to the advances in and diversity of transport technology.

Another related characteristic of transport service which tends to differentiate it from many other sectors of the economy is that many transport services are subsidized, some very heavily. The immediate result of this is, of course, that the prices charged for transport are somewhat below those that

Table 11-3 Magnitudes of Typical Transport Subsidies in the United States in 1974† Net Federal Subsidy‡ as a Percentage of the Net Federal Plus User Expenditure, per Unit of Transportation Service (Units: freight—ton-mi, passengers—passenger-mi)

	Percent
Urbanized area passenger travel	
Private auto	1.9
Taxi	0.2
Bus	29.2
Rapid rail	58.5
Rail commuter	23.5
Other domestic passenger travel	
Private auto	nil
Bus	nil
Rail	23.0
Air carrier	5.0
General aviation	13.0
Domestic freight	
Air	2.1
Highway	0.9
Rail	0.7
Marine§	40.0–52.2

† U.S. Department of Transportation (1975, p. 53).

‡ Net federal subsidy is defined as total federal expenditures in 1974 minus all user charges received by the federal government in 1974.

§ Depends on allocation of (*a*) marine safety expenditures between passengers and freight, (*b*) marine water pollution expenditures between shore and waterborne sources, and (*c*) search and rescue expenditures between rescue associated with aviation and marine, and within the marine category between domestic marine freight haulage and other marine activity (for example, foreign ships, fishing vessels, recreational boating etc.).

would otherwise be charged according to normal economic criteria. In the long run, charging prices set below costs results in an increase in the amount of transport consumed and a tendency of industry and residences to be more widely spread throughout the land.

Some idea of the magnitude of transport subsidies in the United States can be gained from Table 11-3. While this table is developed on the basis of rather incomplete information in many cases and includes only federal government subsidies, it does reveal that surface carrier passenger transport in particular seems to be very heavily subsidized, especially rail passenger transport and urban public transport of all sorts. It appears as though road transport as a whole is fairly heavily subsidized, although as we discussed earlier, it has not been firmly established how different road users (e.g., autos versus trucks) share this subsidy. Also, one might argue that there are differences in the costs that different carriers must bear, and these may result in some implicit subsidies or excess charges to certain carriers. In particular (freight) railroads and pipelines as private carriers must pay property taxes on all facilities and pay corporate income taxes. In contrast, public roads and airports and most port facilities and waterways, provided by governments and governmental authorities, are generally free of property taxes and income taxes and make few payments to general government funds, if any, in lieu of these. As a result, the costs of these forms of transport may be based upon total costs different from those for rail and pipeline service, and this condition may give unfair advantage in the form of an abnormally low cost to these carriers. This exceedingly complex issue cannot be adequately treated here for it involves many ramifications for the taxing of publicly owned and private facilities as well as the principles of taxation, issues which extend far beyond transport.

Some Actual Prices

Freight rates—or tariffs as they are sometimes called—are not usually the result of the application of a formula to characteristics of the shipment such as its weight, volume, and distance, but rather are the result of complex and detailed cost estimation, considerations of prices of alternative carriers (including private trucks), and the effects of numerous historical rate increases (and decreases!) approved by regulatory commissions in the case of regulated carriers. This results in freight rates displaying uneven variations with distance, weight, etc., and difficulty in generalizing about rates. However, a recent study (Samuelson et al., 1976) has been somewhat successful in developing formulas which predict prices based upon a number of characteristics of the shipment. Since these formulas illustrate the influence of many of the factors discussed above, we include examples here. The formulas for rail, truck (truckload and less than truckload), TOFC, and barge are presented in Table 11-4. The influence of factors such as distance and weight are clearly evident. However, the type of cargo and its value also seems to influence rail, truck, and barge charges, indicating that value-of-service principles are being used. It should be noted that some of these formulas are based on samples of movements in particular areas of the United States and do not apply everywhere. The truck data are for regulated carriers in the mideastern states (Middle Atlantic Region); barge data are for Mississippi River flows; and rail data are for the entire United States.

Table 11-4 Models for Estimating Typical Intercity Freight Carrier Prices†

Rail carload

$LR = 9.311 + 0.565\ LM + 0.137 \times 10^{-3} M - 0.799\ LW + 0.466 \times 10^{-5} W + 0.153\ LV + 0.305\ G \qquad r^2 = 0.89 \qquad 10{,}000 \text{ lb} \leq W \leq 200{,}000 \text{ lb}$

TOFC

$LR = 10.889 + 0.478\ LM + 0.291 \times 10^{-3} M - 0.929\ LW + 0.886 \times 10^{-5} W - 0.325 \times 10^{-1}\ LD - 0.112\ L \qquad r^2 = 0.81 \qquad 10{,}000 \text{ lb} \leq W \leq 80{,}000 \text{ lb}$

Truckload

$LR = 13.56 + 0.155\ LM + 0.106 \times 10^{-2} M - 1.022\ LW + 0.160 \times 10^{-4} W + 0.835 \times 10^{-1}\ LV - 0.832 \times 10^{-1}\ LD \qquad r^2 = 0.84 \qquad W \leq 60{,}000 \text{ lb} \qquad M \leq 1{,}000 \text{ mi}$

Less than truckload

$LR = 5.66 + 0.256\ LM + 0.228 \times 10^{-3} M - 0.156\ LW + 0.247 \times 10^{-1}\ LV - 0.1681\ LD$
$r^2 = 0.71$ minimum charge about \$6 $\quad M \leq 1000$ mi

Barge

$LR = 0.574 + 0.414\ LM + 0.189 \times 10^{-3} M + 0.796 \times 10^{-1}\ LV - 0.261\ P \qquad r^2 = 0.79 \qquad W \leq 300{,}000 \text{ lb}$

† Samuelson et al. (1976, pp. 6–19).
LR = natural log of rate, rate in cents/cwt
M = distance via carrier used, mi
LM = natural log of M
V = value of shipment (at 1972 wholesale prices), \$
LV = natural log of V
D = weight density of shipment, lb/ft^3
LD = natural log of D
W = weight of shipment, lb
LW = natural log of W
G = artificial variable which takes on a value of 1 if shipment is a gas, 0 otherwise
P = artificial variable which takes on a value of 1 if shipment is of particles, 0 otherwise
L = artificial variable which takes on a value of 1 if shipment is a liquid, 0 otherwise

The use of these formulas is straightforward, but care must be exercised in units and in the use of logarithms. For example, suppose a rail shipment had the following characteristics: 100 mi, 50 tons of merchandise, valued at \$0.25/lb. The independent variables would be

$$M = 1000 \text{ mi}$$
$$LM = 6.908$$
$$W = 50 \text{ tons} \times 2000 \text{ lb/ton} = 100{,}000 \text{ lb}$$
$$LW = 11.513$$
$$V = \$0.25/\text{lb} \times 100{,}000 \text{ lb} = \$25{,}000$$
$$LV = 10.127$$
$$G = 0$$

Hence
$$\begin{aligned} LR &= 9.311 + 0.565(6.908) + 0.137 \times 10^{-3}(1000) - 0.799(11.513) \\ &\quad + 0.466 \times 10^{-5}(100{,}000) + 0.153(10.127) + 0.305(0) \\ &= 6.167 \\ R &= 476.75 \text{ cents/cwt} \end{aligned}$$

The total charge of this shipment is therefore

$$(476.75 \text{ cents/cwt})(1000 \text{ cwt})(\$1/100 \text{ cents}) = \$4767.50$$

Table 11-5 Models for Estimating Typical Intercity Passenger Carrier Prices†

Air—first class‡
$P = 27.38 + 0.095M$
$r^2 = 0.974$

Air—tourist class§
$P = 10.81 + 0.06626M$
$r^2 = 0.983$

Air—commuter (local) service‡
$P = 11.53 + 0.112M$
$r^2 = 0.955$

Bus§
$P = 3.827 + 0.03996M$
$r^2 = 0.969$

Rail—coach§
$P = 3.095 + 0.04818M$
$r^2 = 0.981$

Rail—Pullman‡
$P = 16.09 + 0.078M$
$r^2 = 0.986$

† Where P = price, \$/person, one way
M = distance via carrier, mi
‡ Sincoff and Dajani (1975, p. 91).
§ Miller (1975, p. 155).

Table 11-5 presents similar models of price for intercity passenger carriers. Most fares seem to be set using a formula which includes a constant plus an amount proportional to distance. Of course, if a carrier's route is very circuitous relative to a competitor, a lower fare than that given by the formula might be used for competitive reasons. Also, there are many special fare reductions on specific routes, usually designed to encourage travel in low-volume periods. Also, a carrier will occasionally initiate a reduced fare to help induce business, and competitors will often offer a similar reduction. However, the passenger fare structure is basically much simpler than that for freight.

SUPPLY RELATIONSHIPS FOR AN URBAN TRANSIT LINE

To illustrate the principles of supply of services by a carrier we shall employ an example of urban bus transit services. This example is drawn from one of the relatively few analyses of supply characteristics of carriers (most of the research having been done on facilities, particularly roads), and it is based upon Morlok (1976).

There are many different aspects of the supply of transit services which might be treated. One is the price, which usually is a key variable in supply analysis. However, in the case of most urban common carrier transit services, prices are not determined by the transit management in response to market forces, but rather are determined within the political arena, prices usually being less than average costs, and subsidies from taxes are used to make up the deficit. Hence we shall not focus on prices. It is interesting to note, however, that the total revenue—fares, advertising, etc., plus subsidies—expected by a transit operator has been found to influence decisions regarding the total amount of service planned to be operated over the entire system, service in effect being expanded or contracted to consume all of the revenue expected (Gaudry, 1975).

Therefore, our focus will be on level of service rather than on price. In particular, the service on a single route will be modeled, assuming that the route to be followed (i.e., the points to be served) is fixed and also that the transport technology is fixed (in this case, buses operating on public streets or exclusive busways). In the very long run, even the technology is not fixed, although in practice usually the substitution of new technology requires a major planning decision, governmental funding of any new guideways, etc., and thus the decision includes others besides the transit management. Also, there is the question of the time period of analysis. Here we shall consider both short-run and a longer-run period for adjustment.

Short-Run Supply Relationships

The short run will be defined as a period in which the transit management cannot adjust its schedules or the number of buses and drivers assigned to a particular route. Thus the management will have decided on a particular frequency of bus trips and assigned an appropriate number of vehicles and drivers to the route.

The effect of variations in the number of passengers on the travel time of buses and hence on the travel time of individual users can be illustrated by an example. Consider the operation in the weekday morning peak period of a route 11 mi in length. Buses are operated on a 10-min time headway from one end of the route to the other, stopping at any terminals (bus stops) at which a passenger wishes to board or alight. If a bus makes no stops between the end terminals (i.e., no passengers wish to board or alight), then the running time will be 50.6 min (this odd value being chosen so the example will match assumptions made in a later example).

If intermediate stops are made for passengers, the running time will increase. From the acceleration and deceleration characteristics of the bus, using the methods discussed in Chap. 4, it has been determined that each stop will add 12 s to the running time, due to the deceleration to rest and the subsequent acceleration. The dwell time at each stop can be calculated from the equation for bus dwell times (in the A.M. peak) given in Table 7-4:

$$Y = 0.5 + 1.3A + 2.2B - 0.1AB$$

where Y = dwell time, s
A = number of passengers alighting
B = number of passengers boarding

The running time for a bus will therefore depend upon the number of stops made and the number of passengers boarding and alighting at each stop. Assuming 1 passenger boards the bus at each of 10 different stops, the added time due to these 10 stops is

$$10(12) + 10[0.5 + 1.3(0) + 2.2(1) - 0.1(0)(1)] = 147 \text{ s}$$

And if these 10 persons alight at 4 CBD stops, the added time will be

$$4(12) + 4[0.5 + 1.3(2.5) + 2.2(0) - 0.9(2.5)(0)] = 63 \text{ s}$$

Thus the total running time will be increased by 3.5 min to 54.1 min. Having 10 passengers/bus, at 6 buses/h, corresponds to a passenger volume past the peak load point on the route of 60 passengers/h. Similarly, with 20 passengers/bus, boarding at 20 stops and alighting at 4 stops in the CBD results in an added time of 4.90 min for boarding and 1.27 min for alighting, yielding a total running time of 56.8 min. These values are plotted in Fig. 11-10, along with other values, up to a maximum of 75 passengers/bus, the normal nominal capacity of the typical 51- to 53-seat urban transit bus. For all values of passengers per bus from 20 to 75, it was assumed that the bus made 20 stops (an average of 2/mi) outside the CBD and 4 in the CBD. The pattern of increasing travel time with increasing volume is readily apparent. If travelers between intermediate points outside the CBD were included, the relationship

Figure 11-10 **Short-run relationship between travel time and passenger volume on an example bus line. Note: Buses are operated on a headway of 6 min and have a seating plus standee capacity of 75 persons.**

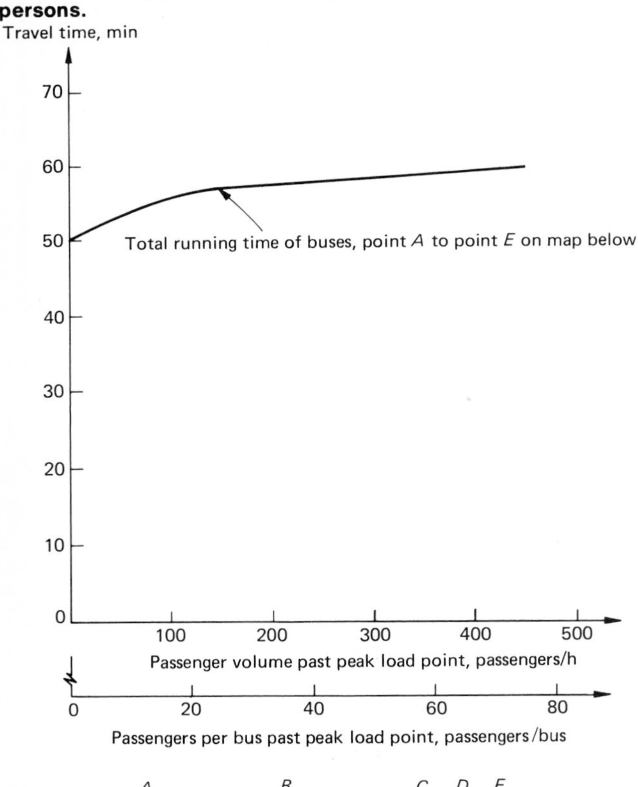

would be the same but shifted upward to reflect even larger increases in dwell times and perhaps in the number of stops.

A reasonable question is, what would happen above 75 passengers/bus? If such a sudden unexpected overload were to occur, drivers might allow somewhat more than 75 passengers to board each bus. It would probably result in considerably increased dwell times due to aisle congestion. Still, passengers might arrive at a faster rate than buses could carry them away, leading to queues that would last until the arrival rate of passengers dropped (as shown on the cumulative flow chart of Fig. 7-7, for example). Another problem with the increased dwell times is that the round trip times of buses might increase to the point where buses arriving at a terminal would arrive after their scheduled departures, which would result in canceled trips and further increases in the passenger delays. [The maximum volume that these vehicles could accommodate in any one direction can be calculated from Eq. (8-11).]

However, such unforeseen increases in traffic are rare. Also, transit systems are quite flexible, so when traffic on one route is unexpectedly high, often vehicles and drivers can be pressed into service to accommodate the loads. In a recent example, traffic on a new extension of a rail rapid transit line was about twice that planned for, which volume had required four trains originally. The high volume of riders was observed early in the morning peak hour period, and the four trains initially assigned to this line were supplemented in less than an hour with four additional trains and crews called from a pool, and the traffic was accommodated without great strain. While such unexpected increases could not be accommodated simultaneously on the entire system, they can be on a single or even a few routes.

Intermediate-Run Supply Relationships

The intermediate run is used here to describe the period of time required for transit management to adjust schedules, vehicle assignments, etc., to the expected volume of traffic on each route. It is distinguished from the long run in that the latter would be used to describe a period in which even the transport technology could be modified. As noted earlier, most transit services in North America are subsidized by public authorities, and the cost of these services usually exceeds by a considerable amount the revenues derived from transit services. Strong pressures exist to minimize the deficit, which induces a very strong cost minimization orientation on the part of transit operators. As we noticed in Chaps. 8 and 9, the cost is largely determined by the weekday peak period operation, because that determines both the number of vehicles and the number of drivers or other operating personnel required. Thus much of management's attention will be directed toward the peak period. Regulatory control of the services provided by transit authorities is minimal, although sometimes a minimum service frequency or a requirement to operate a particular unremunerative route is imposed. Thus management usually has a great deal of discretion over the service provided.

Transit management usually has relatively little information on the demand curve which it faces for its services. Rather, it has information on the actual flow on its various routes, in the form of counts of passengers past the peak load points on each route during the peak period and sometimes during other periods. Usually no information is available on the actual origin and destination pattern of traffic or on other demand characteristics such as the trip length distribution.

Discussions with those responsible for the scheduling of transit routes in large cities revealed that they attempt to schedule bus routes in the following manner. During the weekday peak period, an attempt is made to operate a number of buses just sufficient to accommodate the traffic. Usually this nominal capacity is 75 passengers/bus, as we used in the short-run analysis above. However, a smaller value is used in cases where the typical trip is long, to provide seats for most or all travelers, and where there is a large amount of simultaneous boarding and alighting at intermediate stops, which requires relatively free movement in the aisles. Also influencing the scheduling is a requirement to operate vehicles at headways no greater than a specified maximum, often of the order of 10 to 15 min during the day and 30 min in the early morning. Outside of the peak periods the attempt to utilize fully the capacity of each vehicle was far less strong, the minimal frequency requirement usually resulting in route capacity greater than actual traffic.

In developing an intermediate-run model of the supply of transit service, it is useful to use the simple route employed above, that of a single 11-mi-long bus route in which all of the buses operate in local service from one end of the route to the other. In this case it will be assumed that transit management has no control over the running time of buses from one end of the route to the other, this time being determined by the transit vehicle's acceleration and speed capabilities and the prevailing speed of traffic on the roads comprising the route.

In this context, the only decision variable open to management is how frequently to operate buses. The general discussion of managerial behavior above suggests that the management would follow the following two rules: First, buses would be operated with a frequency at least equal to the minimum frequency considered acceptable for the transit service:

$$f \geqslant F = \frac{1}{H} \tag{11-7}$$

where f = frequency of bus departures in one direction, buses/h
F = minimum acceptable frequency, buses/h
H = maximum acceptable headway, h/bus

Since the volume of passengers could exceed that which can be accommodated in buses operated at the minimum frequency, the frequency also has to be greater than or equal to that which is required to accommodate the passenger flow during any particular period:

$$f \geqslant \frac{p}{Q} \tag{11-8}$$

where p = passenger flow past peak load point on route, passengers/h
Q = capacity of bus, passengers/bus

Assuming that management operates the minimum number of trips in order to meet these two service-related criteria, the rule which management should follow in determining the frequency of buses would be the following:

$$f = \max\left(F, \frac{p}{Q}\right) \tag{11-9}$$

Following such a rule would probably minimize the cost of operating the service, since the number of buses, operators, and vehicle-miles would be minimized for the given route conditions.

A further complication is that the rate of passenger flow may vary within any period. This would lead to a nonuniform headway if more than the minimum frequency service is required, i.e., passenger flow is greater than FQ, and if management policy permitted nonuniform headways. If a constant headway were required, then headway would be adjusted for the peak passenger flow within each schedule period. This would lead to an average load less than vehicle capacity and could be incorporated in the model by appropriate selection of the value of Q. It also should be noted that the number of bus trips made over the entire day must be an integer, thereby possibly requiring a slight adjustment in the frequency of operation in each period, but the adjustment is small in comparison to typical total frequency values and therefore is ignored here. Similarly, variations in the passenger flow during any one period will be ignored here, in effect we presume that periods are selected to correspond to roughly constant flow rates.

Given the managerial behavior described in Eq. (11-9) above, it is possible to estimate the supply of transit service on a route as it would be viewed by a traveler. One important aspect of the supply of service is the travel time from origin to destination, for any particular traveler, including the waiting time as well as on-vehicle time. If we assume uniform or random passenger arrivals at the origin stop and constant time headways between buses, the average waiting time would be one-half the headway, as was derived in Chap. 7:

$$w = \frac{h_t}{2} = \frac{60}{2f} \tag{11-10}$$

where h_t = bus headway, min/bus

As can be seen from this expression, as the passenger traffic increases above that amount required to fill the minimum frequency of buses, travelers' waiting time would decrease. Assuming that the volume of traffic is sufficient to fill all the vehicles at the minimum frequency, the average number of passengers per vehicle past the peak load point will be independent of the traffic volume. Assuming the origin-destination pattern of traffic does not vary with volume, the same number of passengers will board each vehicle regardless of volume. Therefore under these conditions, the number of stops and the dwell time would be independent of volume and hence the bus running time will be independent of volume. (Specifically, it would be 60 min from A to E in the earlier short-run example, since that corresponds to 75 passengers/bus.) In this case travel time between any points i and j would equal

$$t'_{ij} = V'_{ij} + \tfrac{1}{2}h_t \tag{11-11}$$

where V'_{ij} = vehicle running time between stops i and j, plus one-half the dwell time required for alighting at j, min
t'_{ij} = total travel time between stops i and j, min

The V'_{ij} term must include one-half the time required for the alighting of passengers at j, since this is part of their time on the system. However, a comparable inclusion of boarding time at i is not required, since this is already accounted for in the waiting time.

Two points regarding the V'_{ij} term should be made. First, in the case where the passenger traffic is less than sufficient to fill the minimum number of vehicle trips operated (the minimum frequency), then the travel time presumably will be slightly less, as we discussed in the short-run section. Of course, this assumes a time table adjustment by the management which may not be made. Such an adjustment will not be considered further here. The portion of V'_{ij} due to the time required for unloading at stop j is also likely to be very small, particularly since most urban buses have two or more doors, both of which are usually used for unloading whenever the number of alighting passengers is large. The maximum time which would be included would then be the time required to unload 75 passengers through two doors, about 40 s using the equation for the A.M. peak described earlier. Since most values would be less than this, in further discussion this unloading time will be ignored, and V'_{ij} replaced with V_{ij}, the running time of the bus from departure at i to arrival at j.

Virtually all studies of modal choice have revealed that passengers value waiting time considerably more than on-board vehicle time. Thus the perceived average travel time for any traveler between stops i and j will be the following:

$$t_{ij} = V_{ij} + \frac{60W}{2f} = V_{ij} + \frac{30W}{\max(F, p/Q)} \qquad (11\text{-}12)$$

where t_{ij} = total perceived travel time between i and j, equivalent min
W = relative weight of waiting to on-board vehicle time
V_{ij} = vehicle running time from i to j, min

Most studies indicate that the value of W is 2.5 to 3.0 (e.g., Pratt, 1970).

An understanding or appreciation of how total travel time as perceived by a traveler would vary with volume on an all-stop local service bus route can be obtained from an examination of relationships for a particular hypothetical route. Assuming the same route as before, which is 11 mi long, for 10 mi of which the bus can operate at a speed of 12 mi/h (including stops) and 1 mi of which is in the central business district, where average speed is 6 mi/h, and there is a minimum service frequency of 6 buses/h (equivalent to a maximum headway of 10 min), the relationship between travel time and volume is as shown in Fig. 11-11. Passengers are assumed to originate with a uniform distribution in the 10-mi portion and be destined with a uniform distribution along the 1-mi portion within the central business district.

Figure 11-11 shows the relationship between average total travel time measured in equivalent minutes and the volume of passenger traffic on the route. The average of all passengers follows the line for the travel time between points B and D. The travel time remains constant until the capacity provided by the minimum frequency of vehicle trips is fully utilized, after which the increasing frequency required will result in a reduction in waiting time. This reduction can be fairly substantial. In this example, for volumes up to 450 passengers/h, the average total equivalent travel time is 45 min; a traffic increase to 1500 passengers/h reduces the time to 34.5 equivalent min, a 23.3 percent reduction. Of course, because the reduction in travel time with increasing volume is the same regardless of the length of the trip, the percentage decrease is much more significant for shorter trips. This example

Figure 11-11 **User travel time versus volume for an example all-stop bus route.** [*From Morlok (1976), p. 361.*]

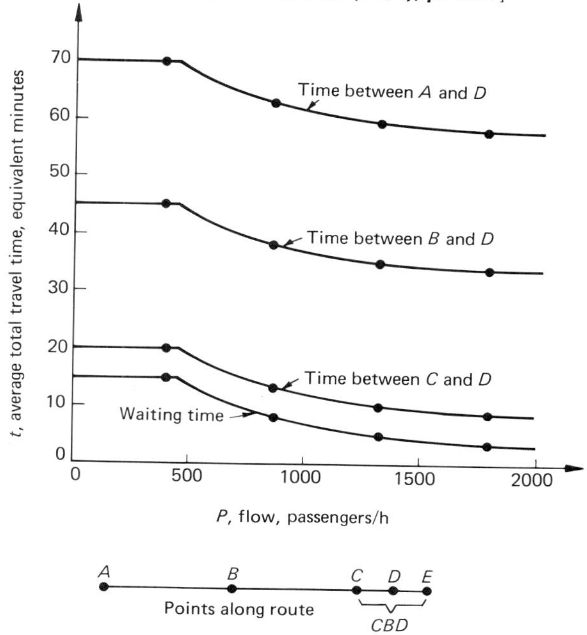

serves to illustrate the substantial influence that increasing frequency will have on reducing waiting time and hence perceived or equivalent passenger travel time.

We turn for a moment to the cost to the operator of providing the service. In Chaps. 8 and 9, we determined that there are basically three determinants of the cost of operating any given bus route: the number of buses, the driver-hours, and the vehicle-miles of operation. Costs of this service have been estimated using cost parameter values for 1970, but the same pattern would emerge using more recent costs. Also, total transport cost can be calculated, using a value of traveler's time of $3.00/h ($9.00/h for waiting time).

The results for the hypothetical route are shown in Fig. 11-12. As can be seen, the operator's cost per passenger-mile begins at low volumes at a fairly high value of 20 cents, but drops as traffic increases, up to 450 passengers/h, because more passengers are using the system (passenger-miles generated) while costs are remaining constant (the number of buses operated remains constant). Beyond 450 passengers/h the operator's cost remains approximately 10 cents/passenger-mi, with minor variations due to the requirement of an integer number of vehicles and vehicle trips. Total transportation costs continually decrease with volume, due to the reduced waiting time.

Variations in Intermediate-Supply Relationships Due to Operating Policies

In operating a bus service along a particular route, transit management can choose not only the frequency of service to be provided with a given stopping policy, but also select the particular pattern of stops which buses will make.

Figure 11-12 **Carrier cost and total transportation cost versus volume for example all-stop bus route.** [***From Morlok (1976), p. 362.***]

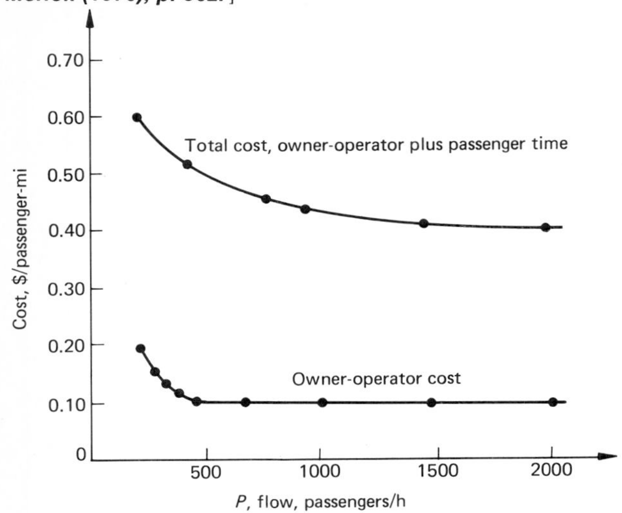

Many bus routes in North America are divided into two or more sections during heavy-volume periods. Buses may be operated between the central business district and the closer of the two sections in local service, making all stops (as was shown in Fig. 8-10). In order to provide service to the outer section, some buses may operate from the CBD, passing through the inner zone without stopping (operating as an express), then making all stops in the outer zone. A third route may cover the entire distance as a local.

In this manner, the running time of buses to the outer section is substantially reduced. In general, this benefits the travelers between the outer section and the central business district. Also, by increasing the average speed of buses, more revenue trips can be obtained from each vehicle in any particular time period, and vehicle and driver costs are reduced. However, there are disadvantages to zone service since the frequency in any one zone usually will be less than that which would prevail with an all-stop route.

If we assume that the example route is divided at a point such that one-half of the traffic to the central business district originates in one zone and one-half in the other, then the previously applied rules regarding the frequency with which vehicle trips will be operated become the following:

$$f_e = f_l = \max\left(F, \frac{p}{2Q}\right) \tag{11-13}$$

where e and l refer to the zone express and zone local services, respectively. This formula, analogous to Eq. (11-9), assumes that the same service frequency and bus capacity standards will be applied.

The total travel time of travelers using the zone services will be the following from either zone.

$$t_{ij}^{z} = V_{ij}^{z} + \frac{30W}{f^{z}} = V_{ij}^{z} + \frac{30W}{\max(F, p/2Q)} \tag{11-14}$$

where t_{ij}^{z} = total perceived travel time between i and j on zone service z, min
V_{ij}^{z} = running time of vehicles between i and j on zone service z, min
f^{z} = frequency of service in each of the zones (here all assumed equal), $f^{z} = f_e = f_l$, buses/h

Provided the traffic is greater than that required to just fill the minimum frequency service, all travelers will face a reduced frequency of service in comparison to that which would occur with the all-local service. Although this results in an increased waiting time, the longer-distance travelers will experience a reduced on-board vehicle running time. In general, the average speed which vehicles can maintain in the inner-zone local service will be somewhat less than what they could maintain in the nonzone local service, because more passengers would be boarding in a shorter distance than in the nonzone case and hence the number of stops and dwell times would be longer. However, the gains for the longer-distance traveler can be very substantial.

The same numerical example as used earlier can be used to illustrate the effects of the introduction of zone services. These computations are per-

Figure 11-13 **User travel time versus volume for example bus route operated with all-stop service and with zone express service.** [*From Morlok (1976), p. 363.*]

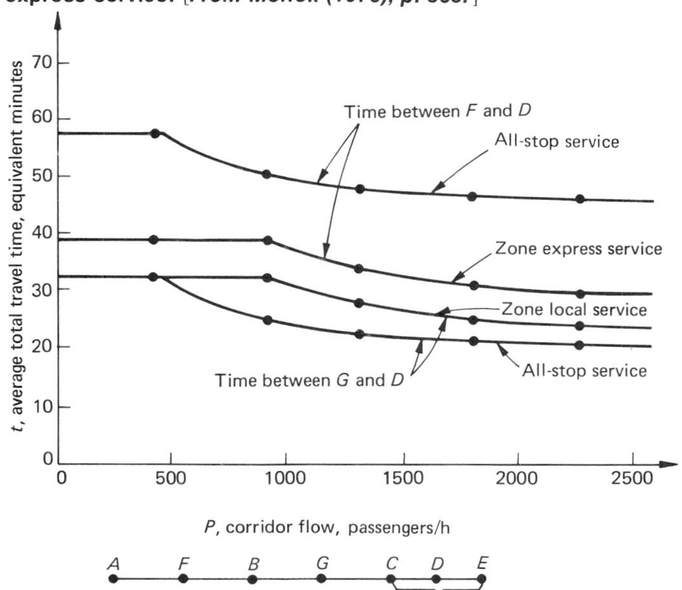

Zone service:

Express service: Stops A–B, Nonstop B–C, stops C–E

Local service: Stops B–E

All-stop service: Stops A–E

formed for the same numerical assumptions as in the nonzone all-stop service case, with the exception that it is assumed that the bus operating in the purely express mode could average 45 mi/h, in conformity with a freeway-type operation. Figure 11-13 shows the average travel time for persons traveling between each of the two zones and the central business district. These times are also shown for the all-stop local bus service and for the zone express or the zone local service. The average time for travelers between the outer zone (which extends between points *A* and *B* and the central business district) is equal to the time between points *F* and *D*. As can be seen from the figure, the reduction in average travel time for these due to the zone service is very substantial, reductions of one-third being common. The increases in time for the travelers in the near zone (which extends between points *B* and *C*) are significant, although the increase is the least at large volumes. The average travel time between the inner zone and the central business district is identical to that between points *G* and *D*. At 1000 passengers/h, the average times on nonzone and zone services are 24 and 31 min, respectively, while at 2000 passengers/h, the values are 21 and 24 min, respectively.

The cost to the carrier in providing the zone service has been calculated using the same 1970 parameter values as used for the nonzone all-stop service. Figure 11-14 contains the results of these computations. As would be expected, in the lower-volume range the all-stop nonzone service is less expensive, reflecting the greater penalty of the minimum frequency requirements on the zone service. However, once the volume exceeds 700 passengers/h, the zone service becomes substantially less costly than the all-stop service, the difference increasing with volume. Thus, it appears that a cost-minimizing transit management would attempt to introduce some type

Figure 11-14 Carrier cost and total transportation cost for example bus route operated with all-stop service and with zone express service. [***From Morlok (1976), p. 364.***]

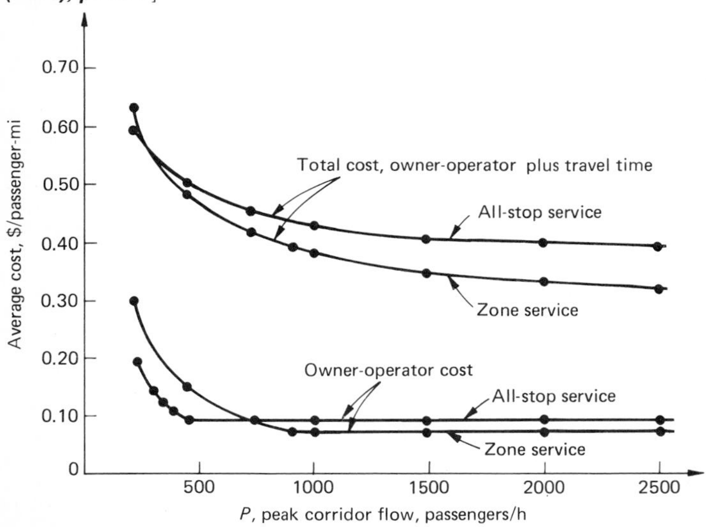

of zone local and express service on high-volume routes. It is important to note that the traffic need not reach that point required to fill completely buses operated at the minimum frequency on the two-zone service for that zone service to be cheaper than the all-stop service. This is because at some lower volume the increase in vehicle utilization made possible by the higher average speed of the zone express service will more than outweigh the increased number of trips required for that service. Thus the zone service may be less expensive than all-stop service even if the former requires the operation of partly empty vehicles. Also shown in Fig. 11-14 is the total transportation cost. Again is revealed the substantial overall gain possible with this type of operation as well as the general pattern of declining cost with volume.

SUMMARY

1 The quantity, quality, and price charged for transportation service or the use of a transportation facility is determined by the management of that service or facility. The management's choices are limited by the network of fixed facilities, the associated technology of transportation used, the vehicles and containers available, and in many cases regulatory constraints.

2 The representation of the supply of transportation follows the paradigm of economic equilibrium analysis, in which the supply function relates the quantity of a good that firms will offer for sale as a function of the price obtained. In transportation, it is generally necessary to extend the considerations of supply to include characteristics of the service offered, which often will change as a result of the amount of traffic on a system.

3 Also, it is important to realize that the consumer of transportation usually incurs additional costs beyond the price paid the transportation carrier or facility operator: travel time and discomforts of person travel, cost of money tied up by inventories in transit, loss and damage of goods, etc. All of the attributes that the user perceives as costs or undesirable features of transport are divided into two categories: prices or direct monetary costs and level-of-service characteristics. The entire set, a vector, is often referred to as the user cost vector or simply user costs. However, if all are converted to a monetary scale and summed, the sum is referred to as the total user cost or simply user cost.

4 In general, the price paid to the owner of a facility decreases or remains constant with increasing volume. However, other user costs such as travel time ultimately increase with volume due to congestion, as we noted in Chaps. 5 and 7. Hence typically the total user cost increases with volume, on any particular facility, in the absence of expansion of facility capacity and of price changes.

5 The costs to users of a carrier, on the other hand, typically increase with volume in the short run due to congestion and perhaps to increases in prices charged. But in an intermediate period sufficient for the carrier to assign more vehicles to a route, the level of service typically improves, due to the more frequent service and perhaps to changes in the operations plan which tend to favor more nonstop services with increasing volume. Although research on carrier level of service has been limited, such reductions in total user cost have been observed in both urban public transit and airline passenger service. Over such intermediate periods, prices usually remain the same or decrease, due to a combination of competitive conditions, regulation, and some economies of higher volumes. In the long run, these decreases in user cost and possibly in price are even more pronounced, because the technology of transport can be changed to match more ideally the higher volumes.

6 There are many principles or guidelines for pricing of transportation services, including value of service, marginal costs, average costs (including allocation of common costs), and equity. All find some application in practice, particularly in freight transport, where prices seem to follow a very irregular pattern.

PROBLEMS

11-1

Calculate the automobile user cost versus volume for travel over a 1-mi-long section of freeway, which consists of three lanes in each direction, has 10 percent trucks and 10 percent buses, is in flat terrain, has a design speed of 60 mi/h, and has an adjustment factor of 0.95. The user cost should be in dollars per vehicle-mile, equaling the costs estimated by AASHO in 1959 and presented in Table 9-10 for normal operation. Plot the results.

11-2

A transit authority in a major city has been found to schedule buses according to the following formulas, where f is the frequency in buses/h and p is the passenger volume past the peak load point in passengers/h. If buses have running speed of s mi/h as given and operate at uniform headways, plot the user travel time (wait plus riding) versus volume curves for a range of 100 to 2000 passengers/h. If buses seat 53 persons, plot the fraction of passengers seated past the peak load point.

Weekday peak periods (7–9 A.M., 4–6 P.M.): $f = 4.10 + 0.013p$, $s = 12$ mi/h

Weekday base period (9 A.M.–4 P.M.): $f = 4.36 + 0.016p$, $s = 14$ mi/h

Weekday evening period (7 P.M.–midnight): $f = 4.34 + 0.012p$, $s = 15$ mi/h

11-3

A rail transit line is 10 mi long. Trains stop at all stations and operate by shuttling from one end of the line to the other. The running one-way time is given by the following empirical formula:

$(30 + 0.02p + 0.00002\,p^2)$ min

where p is the total number of passengers boarding each train. (a) Plot one-way train running time versus passengers/train for 0 to 2250 passengers/train. (b) Then calculate the number of round trips that each train could make in 2 h, assuming there are 10 times as many passengers in the peak direction as in the other direction. (c) Finally, plot peak direction passenger volume in passengers/h (trains/h multiplied by passengers/train) versus passengers/train assuming that the line has 15 trains and all trains are in continuous operation. (d) What can you conclude about transit line capacity?

11-4
Plot the price and travel time of truck and railroad movement of a merchandise shipment of 23 tons, with a volume of 2800 ft^3, and a value of $50,000. Use Table 11-4 to estimate the prices and Fig. 8-12 to estimate the times. The distances are 980 mi by truck and 1025 mi by rail. What factors would you consider in deciding whether to ship by rail or truck?

11-5
Using the data given in Fig. 11-9, what dollar value of toll should be charged a user at a volume of 3000 vehicles/h so that she would perceive the effect of her using the road on other users? Assume a value of time of $4.00/vehicle-h. This is called congestion pricing. What might be its benefits?

11-6
"Public transit service in many cities is operated by a single agency, and the government will not allow anyone to operate a competing bus service. Since it is a monopoly, it must be regulated as to its prices and quality of service." Discuss this statement in light of current travel habits and transport services typically offered in a large city in the United States.

11-7
Management decisions regarding the supply of service on rail rapid transit lines in a large city have been found to be replicated by the following model relating peak period train frequency f in trains per hour, and passenger volume past the peak load point p in passengers per hour.

$$f = 9.94 + 0.00058p$$

Also train running time between stations (start to start) can be estimated by the equation below relating time t in minutes to distance m in miles:

$$t = 0.8015 + 1.4217\,m, \quad m \geqslant 0.5 \text{ mi}$$

Assuming uniform train headways, random passenger arrivals, and stations spaced 1.0 mi apart, plot average travel time, station to station, including waiting time, versus volume from 300 passengers/h to 7000 passengers/h for a 6-mi trip. If the management institutes *A* and *B* skip-stop service at 1000 passengers/h, what would the user time versus volume curve be? Plot this on the same graph. Assume one-half the passengers use the *A* train, that the *A* train station spacings are all 2 mi for this trip, and that the frequency and running time models apply to *A* and *B* service also.

REFERENCES

Bhatt, Kiran, Michael Beesley, and Kevin Neels (1977), "An Analysis of Road Expenditures and Payments by Vehicle Class," The Urban Institute, Washington, DC.

Capron, William, ed. (1971), "Technological Change in the Regulated Industries," Brookings Institution, Washington, DC.

Gaudry, Marc (1974), A Note on the Economic Interpretation of Delay Functions in Assignment Problems, in Michael A. Florian, ed., "Traffic Equilibrium Methods," Lecture Notes in Economics and Mathematical Systems 118, Springer-Verlag, New York, pp. 368–381.

Gaudry, Marc (1975), An Aggregate Time-Series Analysis of Urban Transit Demand: The Montreal Case, *Transportation Research,* **9** 4: 249–258.

Gellman, Aaron J. (1975), The Impact of Economic Regulation on Technical Change and Innovation in Surface Transportation, in Allen R. Ferguson and Leonard L. Lane, eds., "Transportation Policy Options: The Political Economy of Regulatory Reform," Public Interest Economics Foundation, Washington, DC.

Grossman, William L. (1959), "Fundamentals of Transportation," Simmons-Boardman, New York.

Jessiman, William A., John Basile, Donald E. Ward, and Daniel P. Maxfield (1970), "Intercity Transportation Effectiveness Model," 2 vols., Report Nos. PB-200 469 and PB-200 470, National Technical Information Service, Springfield, VA.

Meyer, John R., Merton J. Peck, John Stenason, and Charles Zwick (1959), "The Economics of Competition in the Transportation Industries," Harvard, Cambridge, MA.

Meyer, John R., and Mahlon R. Straszheim (1970), "Techniques of Transport Planning," vol. 1, Pricing and Project Evaluation, The Brookings Institution, Washington, DC.

Miller, James C. (1975), An Economic Policy Analysis of the Amtrak Program, in James C. Miller, ed., "Perspectives on Federal Transportation Policy," American Enterprise Institute for Policy Research, Washington, DC.

Morlok, Edward K. (1976), Supply Functions for Public Transport: Initial Concepts and Models, in Michael A. Florian, ed., "Traffic Equilibrium Methods," Lecture Notes in Economics and Mathematical Systems 118, Springer-Verlag, New York, pp. 322–367.

Pratt, Richard H. (1970), A Utilitarian Theory of Travel Mode Choice, *Highway Research Record* No. 322, pp. 40–53.

Ross, William B. (1955), "Financing Highway Improvements in Louisiana," Louisiana Highway Dept., Baton Rouge, LA.

Samuelson, Paul A. (1958), "Economics," McGraw-Hill, New York.

Samuelson, Ralph D., Steven R. Lerman, Paul O. Roberts, and James T. Kneafsey (1976), "Models for Freight Tariff Estimation," Center for Transportation Studies Report No. 76-7, MIT, Cambridge, MA.

Sincoff, Michael Z., and Jarir S. Dajani, eds. (1975), "General Aviation and Community Development," NASA Langley Research Center and Mechanical Engineering Dept., Old Dominion University, Norfolk, VA.

U.S. Department of Transportation, Office of the Secretary (1975), "A Statement of National Transportation Policy," Washington, DC.

Wohl, Martin, and Brian V. Martin (1967), "Traffic System Analysis for Engineers and Planners," McGraw-Hill, New York.

Transportation Network Flows

12.

In order to estimate the magnitude of flows that will actually occur on a transportation system it is necessary to merge the demand relationships discussed in Chap. 10 with the relationships for the supply or availability of transport service described in Chap. 11. It is only from merging the two types of relationships that the actual amount of usage of the system can be predicted and the prices and user costs estimated. Our approach is analogous to that used in the equilibrium analysis of economic markets, but differs in that we must be concerned not only with the price paid for various transport services but also with the various other costs perceived by transportation users, such as travel time, reliability of service, comfort of passengers, and damage to freight. The discussion will begin with the economic analogy and then generalize to the multidimensional analysis required in transportation systems.

THEORY

Economic Market Equilibrium

The basic economic theory of market equilibrium applies to a situation in which the price of a homogeneous commodity bought and sold in a market is determined in such a manner that the total amount produced equals the total amount purchased. This can be illustrated by returning to the example treated in the previous chapter of the provision of intercity truck service by a number of independent firms, none of which was large enough to dominate or control the market. In that situation, in the short run, it was concluded that it was possible to increase the quantity of traffic accommodated only by increasing the price charged for freight movement, such that the increase was sufficient to provide the additional capacity. This resulted in a supply curve which increased with increasing price, such as that presented in Fig. 12-1. As

Figure 12-1 Equilibrium of supply and demand of a homogeneous good in a market.

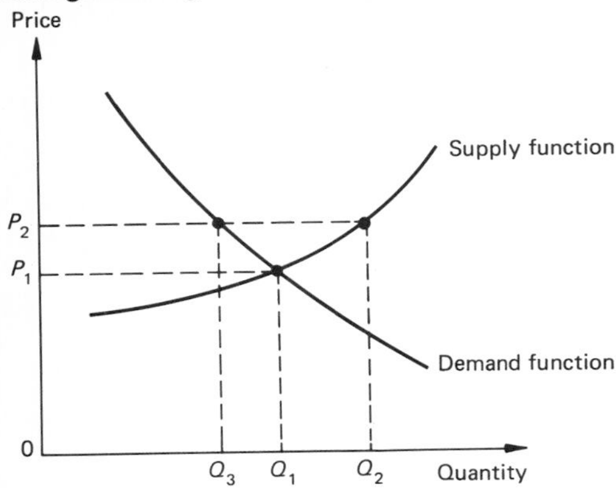

we noted in Chap. 10, the demand for transportation decreases with increasing price, resulting in a typical demand function as shown in Fig. 12-1.

Given these supply and demand curves, the only stable situation is one in which the quantity of freight Q_1 is carried at a price of P_1. This can be readily seen by considering alternative prices and quantities being produced. For example, the quantity Q_2 might be the capacity provided by the truck lines, this capacity being the effective capacity, equal to the quantity of cargo they planned to carry, obtained by deducting the capacity not used due to temporal variations in traffic and other factors from the maximum traffic that could be carried if all trucks were fully loaded. The price expected would be P_2, but at this price only Q_3 would be offered for carriage by those who demanded this service. Thus the total revenue derived by the truck lines would equal P_2Q_3, far less than the P_2Q_2 which they require in order to sustain the quantity Q_2. As a result, they will tend to contract the capacity provided, and in an effort to obtain more business, will begin to cut the price. This process will continue until the equilibrium price of P_1 and quantity of Q_1 is reached, at the intersection of the demand and supply curves. A similar analysis can be performed for other values of price and quantity to see how the price and quantities produced and demanded will change in order to reach the equilibrium values.

Extension to Include Level of Service

As we noted in the discussions of both the demand for transportation and the supply of transportation service, it is generally necessary to include variations in the quality of transportation service when treating the demand or supply of transportation. The reason is that the amount of traffic on either a transportation facility or on a transport carrier affects the quality of service. Furthermore, this variation is usually of significance to the potential user of that facility or service, and degradation in the quality of service tends to decrease the quantity demanded. There are basically two approaches to considering these variations in level of service.

The simpler approach is to relate the quantity of transportation demanded to a measure of total user cost, which will include not only the price paid for the service but any level-of-service factors which vary with the quantity of transportation. This approach is particularly widely used in the analysis of vehicle flow on roads and of other types of transportation facilities, mainly because in those contexts it can often be assumed that the primary or perhaps the only cost to which the user responds is the travel time. For example, a motorist on a nontoll (or "free") road typically would be rather insensitive to the cost of operating a motor vehicle (which, through gasoline and other taxes, includes the price paid for use of the road) because these costs usually vary to only a small degree among different road types and because the exact cost per mile is hidden due to the rather intermittent purchase of gasoline and other items. However, the user is quite aware of the amount of time taken for traveling. For this reason, travel time is often the single measure of user costs.

Even if it is not reasonable to assume that users respond to only one cost (such as travel time), the concept of a single measure of total user cost may be employed. In principle at least, it should be possible to ascertain the amount of money which is exactly equivalent to each of the nonmonetary user costs. The equivalent monetary values of all the level of service factors which users respond to and monetary costs which users perceive are then summed to yield the total user cost, which is precisely what was done in the development of the total cost of operating an automobile in Chap. 9. The resulting total user cost per vehicle-mile (as presented in Tables 9-10 through 9-12) includes not only the monetary costs of fuel, maintenance, depreciation, and so forth, but also amounts reflecting the travel time (which depends on speed) and discomfort of driving (which depends on the quality of traffic flow and the condition of the road surface). However, the development of such a measure of total cost requires assumptions regarding the value of travel time, discomfort, etc., and the assumption that users are aware of all the monetary expenditures for automobile operation—assumptions which are not valid in all situations. Therefore we clearly separated the various components of user cost so that the different types of cost could be treated in different ways depending on the problem situation. Given the difficulties in establishing a value of time and of other user costs, conversion of all user costs to monetary units is often impossible. But, as noted earlier, it is often acceptable as an approximation to consider only one of the user costs, such as travel time, and thereby eliminate the problem of conversion of many distinct costs into monetary units.

It is possible to draw a demand function for use of a particular road which relates the quantity of travel (such as vehicles per hour or per day) to the user cost, where perhaps this user cost includes only time or may include a few other factors which have been translated into a single dimension. Similarly, the relationship between this user cost and the volume of travel on the roads can readily be estimated using the methods we summarized in Chap. 11 and described earlier in Chaps. 5 and 9. These demand and user cost curves appear as shown in Fig. 12-2. By reasoning analogous to that of an economic market, there is one equilibrium corresponding to the intersection of the demand curve and the user cost–volume curve. Thus, in this case the equilibrium volume is q_1 and the equilibrium user cost, t_1.

The fundamental similarity between the equilibrium volume of traffic on a

Figure 12-2 Equilibrium flow and user cost level on a transportation facility.

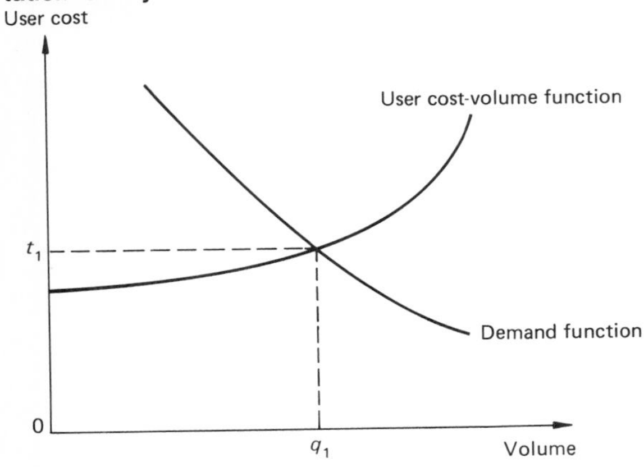

road as portrayed on Fig. 12-2 and the economic market equilibrium as shown in Fig. 12-1 is striking. It has resulted in the borrowing of the term "equilibrium" to describe the volume of traffic. However, it is important to note that the demand curve for use of the facility is different from the traditional economic demand curve in that the latter is related to the price of the commodity while in the road traffic flow case the quantity of flow demanded is related to a combination of many different costs perceived by the users, which costs may or may not include the price paid for use of the road facility. Also, while the road user cost–volume relationship bears a striking resemblance to the economic market supply function, they are not identical. The supply function portrays the quantity of a homogeneous good that various producers are willing to offer at various prices, while the user cost–volume relationship primarily portrays a fundamental technological characteristic of vehicle flow on the road and portrays only to a very limited extent, if at all, the price which the owner of a road facility charges users.

An alternative means for relating the demand for transportation to the characteristics and prices of the services offered is required in those situations where both price and quality of service must be considered distinctly. This is of course the more general case, and it applies to situations where transport carriers are being considered and often to situations where use of facilities is being considered. The means for doing this is portrayed in Fig. 12-3. Here there are three axes, representing the quantity of transport, such as flow on a particular facility or carrier between two points, the price charged, and the level of service. Often more than one level of service dimension will be required; one is used here simply for ease of graphical portrayal. It might be noted that the user cost used in the first approach (above) includes both the price and all level of service measures represented on these separate axes in this alternative approach.

In this space of three (or more) dimensions, the demand function can be drawn as a surface as shown in Fig. 12-3. The quantity demanded decreases with increasing price, as would be expected, and also with increasing value on the level of service axis, indicating that larger values have increasing disutil-

Figure 12-3 **Equilibrium flow treating price, volume, and level of service as distinct variables**

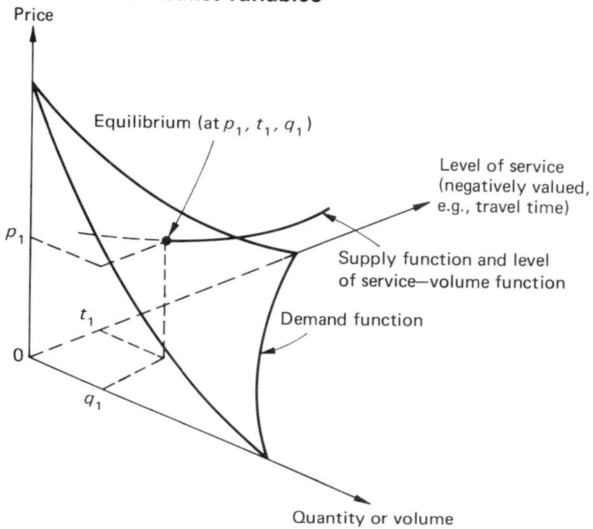

ity. This would be the case for travel time, number of transfers, and percentage of passengers who are standees, for example. However, a positively valued measure could be used also, in which case quantity demanded would increase with increasing numerical value.

Also shown in the figure is the relationship among quantity of transport, the price which the carrier or facility operator will charge, and the level of service which will result from the manner in which the system management decides to operate the service as a function of volume. As we noted in Chap. 11, the underlying choices made by the management of the facility or service which result in this function are quite complex (they were portrayed in Fig. 11-8), but in general, given a particular technology or production function, they will result in a unique price and unique level of service for each volume of traffic. Thus these are portrayed as a single line in this three-dimensional space.

The equilibrium quantity of travel is that given by the intersection of this line representing the supply of transport and the demand function. This is shown at the values p_1, q_1, and t_1 in the drawing. If any price other than the value corresponding to this equilibrium is charged, the quantity demanded will be inconsistent with the supplied quantity and there will be changes in the price.

Again, it must be emphasized that the relationship among price, quantity, and level of service representing the availability of transport service changes with the time period of the analysis. In the very short run, it may be impractical to consider a change in prices, in which case the relationship would lie in a plane parallel to the level of service and quantity axes. This relationship would presumably reflect the change in level of service resulting from different volumes. Thus, in this very short run case, the level of service would become worse as volume increases on facilities or carriers. However, if the time period for adjustment were sufficiently long to permit a carrier to adjust the number of vehicles and trips operated, then the degradation of level of service with increasing volume might not be so severe, and the price charged probably would vary unless there were regulatory or other constraints. In the long run, a carrier often is able to adjust capacity to readily accommodate any volume, often with an improvement in level of service with increasing volume and perhaps even a reduction in price resulting from scale economies. Thus the form of this function, and hence the equilibrium quantity, price, and level of service, will depend upon the time period allowed for adjustment of the system to the traffic demand.

Now that the theory of the treatment of transport with its multifaceted user costs has been presented briefly, we turn to the methods used for particular types of facilities and services. Also, it is necessary to extend the analysis from that of a single market—one origin to one destination—to a network in which traffic flows between many different points.

NETWORK EQUILIBRIUM

Most of the research in transportation network analysis has resulted from the need in recent years to expand the capacity and extent of urban transportation networks, due to the rapid increase in population and spatial growth of urban areas. As a result, most of the research has focused upon road network planning, and to a much lesser extent until very recently, the planning of public transportation systems. However, here we will attempt to present the underlying theory and methods in a manner which generalizes to other modes and situations. Since it has been found that urban travelers, particularly on roads, seem to react primarily to travel time and only to a limited extent to other user costs (including the price for use of roads), these methods have dealt primarily with a single measure of user cost, as portrayed in Fig. 12-2, rather than a vector consisting of level of service and price.

Concepts

Although the concepts of networks have been introduced previously (mainly in Chap. 3), it is useful to review them before using them again. The demand for transportation is described in terms of the quantity demanded between two areas, termed zones. The network or system connecting the zones consists of nodes and links, links being either undirected, allowing for flow in both directions, or directed (also termed arcs), allowing flow in only one direction. Each zone has a node termed a centroid, at which the traffic (passengers and freight) is assumed to begin and end for purposes of movement over the network. In the case of road networks, travel time and other user costs are associated only with the links. However, links can be used, of course, to represent different paths through intersections as we discussed in Chap. 11. Each link or arc is described by a user cost–volume relationship of the type portrayed in Fig. 12-2.

Single Road

The simplest situation involves a single road connecting two points. These two points are the zone centroids at which traffic originates and terminates, and the single road connecting them is a link. The demand function for traffic from one zone to the other appears as was shown in Fig. 12-2, which also includes the user cost–volume relationship for flow in that direction. The equilibrium volume and user cost are as shown in that figure. The analysis is quite straightforward. It should be noted that it can be performed for flow in one direction, as explained here, or for flow in both directions when total flow in one direction is equal to flow in the other direction, as it often is in transportation networks when the period of analysis is fairly long relative to any directional peaking of traffic. For example, the one-way analysis would have to be used for a peak period in an urban area while a combined two-way analysis could be used for an entire day.

A change in the road facility quite simply results in a new equilibrium corresponding to the intersection of the demand function with the user cost–volume function for the new road, as shown in Fig. 12-4*a*. Similarly, a change in demand, perhaps resulting from new developments in the region or an increase in familys' disposable incomes, results in a shift in the demand curve and a new equilibrium. This is portrayed in Fig. 12-4*b*.

Figure 12-4 **Shifts in equilibrium user costs and volumes resulting from shifts in the user cost–volume function and the demand function. *(a)* Effect of a change in the road user cost-volume function. *(b)* Effect of a change in the demand function.**

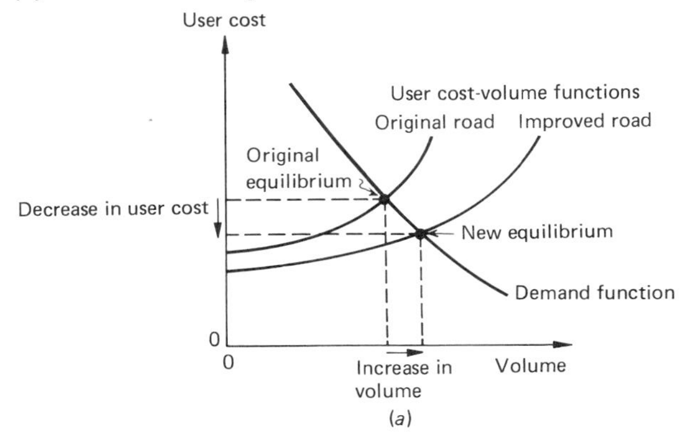

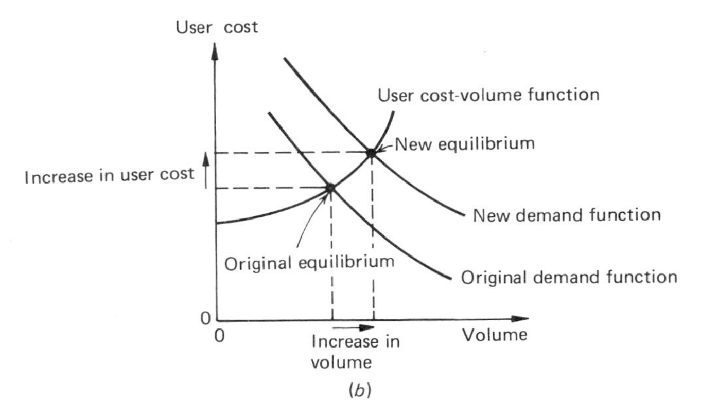

Figure 12-5 **Equilibrium flows with two arcs in series connecting one origin to one destination. *(a)* The network. *(b)* The sum of arc user costs yields path user cost. *(c)* Equilibrium of path user cost-volume and demand functions.**

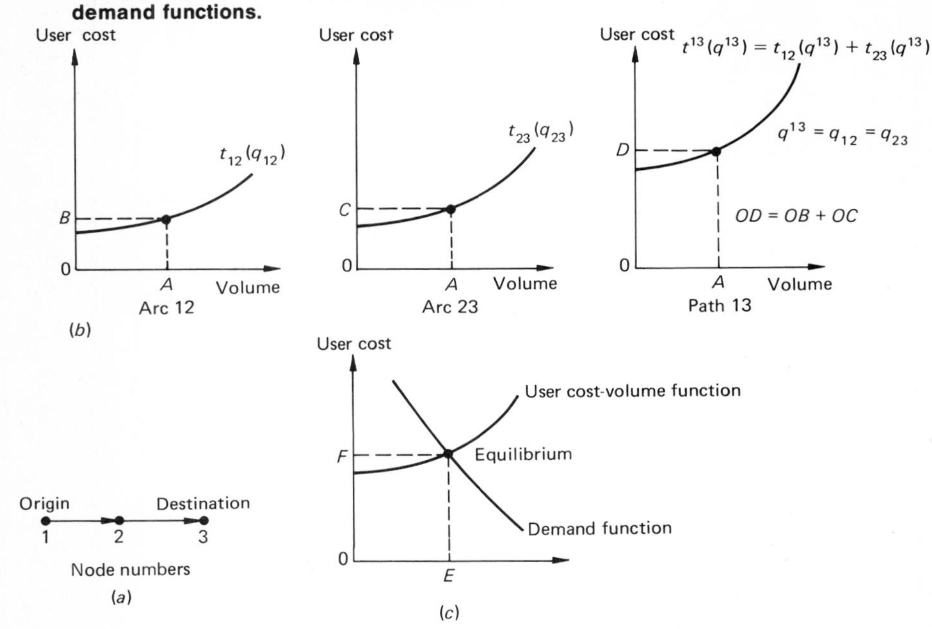

Two Roads Connecting Two Points

A simple extension of this analysis is the situation where two or more roads join together to form a chain of links (or arcs) or path between two traffic generating zones, as shown in Fig. 12-5*a*. The demand curve presents no problems, but the means for developing a relationship between the user cost from one zone to another over the various road links must be specified. The user cost–volume functions for the individual links making up the path are as shown in Fig. 12-5*b*. In this particular situation, the volume or quantity of traffic in any one direction on one link is exactly equal to the quantity of traffic on the other links moving in the same direction, since there is no possibility of traffic entering or leaving the system at the intermediate nodes. Therefore, the total user cost in going from one traffic zone to the other is simply the sum of the user costs on the individual links making up the path from one zone to the other. Since it is desired to have this user cost as a function of the volume or quantity of traffic, it is necessary to sum the user costs on the various directed links at each volume and to plot the total user costs versus that volume, as shown in Fig. 12-5*b*. Mathematically, the operation is as follows:

$$t^{ij}(q^{ij}) = \sum_{rs \in P_{ij}} t_{rs}(q^{ij}) \qquad (12\text{-}1)$$

where $t^{ij}(q^{ij})$ = user cost from node i to node j as a function of volume q^{ij}, with the superscripts indicating origin to destination nodes

$rs \in P_{ij}$ means rs is in the set of arcs comprising the path P_{ij}
$t_{rs}(q^{ij})$ = user cost on arc rs at volume q^{ij} ($q^{ij} = q_{rs}$), with the subscripts indicating arc or link nodes

The resulting user cost–volume relationship for travel from the origin zone 1 to the destination zone 3 is then introduced into the graph containing the demand function, as shown in Fig. 12-5*c*, and the equilibrium value is obtained as before. The equilibrium volume is *OE* and the user cost *OF*.

Two Alternative Routes Between Two Traffic Zones

A somewhat more complex situation exists when there are alternative routes between any two zones. This is in fact the most common situation in road and other types of transportation networks. This case is more complex than the others, because not only must users determine whether or not to travel, but they must also decide which of the two routes to use if they do travel. The former decision is really specified by the demand curve as we have been using it. However, in using that demand curve it is necessary to specify what travel time (or other user cost) is used as the argument (price and/or level of service variables) of the demand curve. In this case, behavioral studies of motorists have revealed (as noted earlier) that motorists tend to choose the route with minimum travel time. If we assume that all users do in fact select the shortest time path, then it is presumably that shortest time which should be used as the argument of the demand function. If users do choose to travel only on the shortest of the two routes, it is clear that they will use only one route unless the travel time on the two routes happens to be the same. This latter condition is in fact common and quite possible on road networks since the travel time increases with increasing volume and as the presence of traffic increases times on the various links there is some chance that the travel times on two or more routes might be equal.

It is instructive to consider the situations under which one route or two routes might be used, and therefore we shall consider a situation of one origin zone connected to a destination zone by two different arcs, as illustrated in Fig. 12-6*a*. Let us assume that the two routes have user cost–volume relationships as shown in Fig. 12-6*b*. As can be seen from these two curves, route *a* has a lower travel time than route *b* (even with the latter having zero volume) until route *a* has a volume of at least *OB* vehicles/h, the travel time on route *a* at a volume of *OB* being *OA* and that on route *b* at a volume of *O* being *OA*. Thus at volumes below *OB*, all users will use route *a*. However, if traffic is increased slightly above *OB*, some vehicles will use route *b* and others will use route *a*. They would divide themselves between the two routes in such a manner that the travel times of both routes were identical. For example, if the resulting travel time were *OC*, then *OD* travelers would use route *a* and *OE* travelers, route *b*. If this same total number of users were to use both routes but more would be using, say, route *b*, some of those using route *b* would find it advantageous to shift to route *a*, reducing their travel time and slightly increasing the travel time of others already on route *a*. Thus the only stable situation with any given total volume of traffic is for the division or assignment of traffic among these two routes to be such that travel time on all routes used is identical and less than (or conceivably equal to) that on unused routes.

Mathematically, for parallel links or arcs connecting node *i* and *j*:

Figure 12-6 **Equilibrium flows with two parallel arcs connecting one origin to one destination.** ***(a)*** **The network.** ***(b)*** **The sum of volumes at each user cost level yields user cost-volume function from origin to destination.** ***(c)*** **Equilibrium of origin to destination demand and user cost-volume functions.**

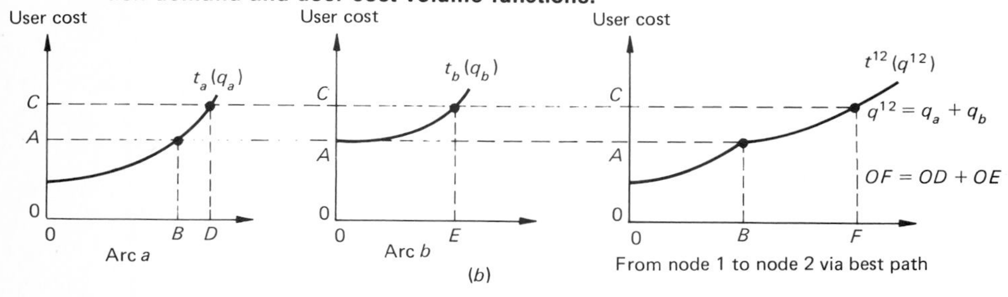

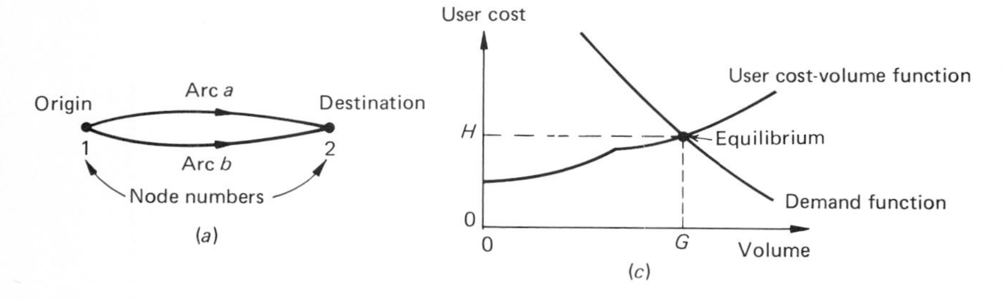

$$t^{ij}(q^{ij}) = \{q^{ij}(t^{ij})\}^{-1}$$

$$q^{ij}(t^{ij}) = \sum_{k \in P_{ij}} q_k(t^{ij})$$

where $\{\cdot\}^{-1}$ = the inverse function of
$q^{ij}(t_{ij})$ = the volume–user-cost function from origin i to destination j
$q_k(t^{ij})$ = the volume–user-cost function for the arc or path k with user cost $t^{ij}(t^{ij} = t_k)$
$k \in P_{ij}$ = path k in the set of paths P_{ij} from i to j

With the resulting user cost–volume relationships for the total flow and the possibility of using both routes where appropriate, it is then again possible to merge this with the demand curve to obtain an equilibrium flow (Fig. 12-6*c*). As can be seen, the total volume is *OG* and the user cost experienced by all users, regardless of the arc used, is *OH*. To find the volume on each of the two individual roads, we simply go back to each of the individual road user cost–volume curves of Fig. 12-6*b* to obtain the volume on each road corresponding to this travel time. Because of the manner in which we constructed the user cost–volume curve of Fig. 12-6*c*, the sum of the volumes on the individual routes will correspond to that on the equilibrium curve.

Multiple Demands and Shared Routes—A Simple Case

The situation of travel demand between more than two zones, involving some links shared by two or more paths connecting different origin-destination

Figure 12-7 **Network for example problem illustrating simultaneous solution of user cost–volume and demand equations.**

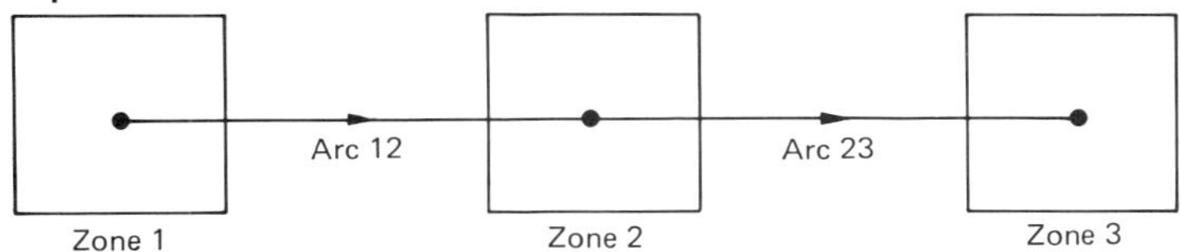

zone pairs is of course a more general network case. To solve this graphically is usually impossible. The complexity of solving this problem can be illustrated by considering a numerical example and attempting to solve for the equilibrium volumes by simultaneous solution of the appropriate equations.

The network of this very simplified problem is shown in Fig. 12-7. The demands and flows are all in one direction. To make the analysis as simple as possible, linear relationships are used for both demand and user cost–volume functions. The demand functions are

$$d_1^2 = 2000 - 15t_1^2$$
$$d_1^3 = 2500 - 10t_1^3$$
$$d_2^3 = 2000 - 20t_1^3$$

where t_r^s = travel time from origin node r to destination node s, min
d_r^s = volume of travel demanded from origin node r to destination node s, vehicles/h

The supply or user cost–volume relationships for the two arcs are

$$t_{12} = 20 + 0.04q_{23}$$

$$t_{23} = 25 + 0.05q_{23}$$

where t_{ij} = arc travel time, min
q_{ij} = arc volume, vehicles/h

Here again linear relationships are used to simplify the calculations; any functional form could be used.

From the network portrayed in Fig. 12-7, it is clear that two of the demand flows will make use of each of the directed links. Hence consideration must be given to the fact that the total flow on arc (2, 3) will equal that originated at node 2 destined for node 3, plus that entering node 2 via arc (1, 2) but less that terminating at node 2. Such summing of flows into and out of a node is performed by what are termed node conservation of flow relationships. These relationships treat traffic for each destination independently, a separate relationship being written at each node for each destination. While it will not be proven in this text, it might be rather obvious and certainly is true that in general to ensure that all of the flow from one node to another actually reaches its destination it is necessary only to keep track of the destination of the flow and is not necessary to retain information on the origin. The reason that this is true is basically the same reason that in routing a letter to its destination it is necessary to have only the address of the destination and not that of the origin or the reason why when asking directions via telephone on how to reach a particular destination one need not specify the origin but only provide information on where one is at that moment. These node conservation of flow relationships specify that the total traffic entering a node for a particular destination must equal the total traffic leaving that node for that destination, including consideration of traffic actually being generated at that node and traffic leaving the system at that node because it has reached its destination. Thus at node 2 the conservation of flow relationship for destination node 3 is

$$q_{12}^{3} + d_{2}^{3} = q_{23}^{3}$$

where q_{ij}^{s} = volume on arc (i, j) with destination s

And for destination 2 at node 2 it is

$$q_{12}^{2} = d_{1}^{2}$$

These expressions are simply specific examples of the general conservation of flow relationships which can be expressed in network terminology as below. As may be recalled from the discussion of networks in Chap. 3, a directed link or an arc is defined by two nodes, an A-node and a B-node, and it is directed from the A-node to the B-node. The general specification of a node conservation of flow relationship is

$$\sum_{i \in A_j} q_{ij}^{s} + d_{j}^{s} = \sum_{k \in B_j} q_{jk}^{s} \qquad (12\text{-}2)$$

where d_j^s = volume of traffic generated at node j destined for s, provided $j \neq s$

$-d_j^s$ = volume of traffic terminated at node j destined for s, provided $j = s$

$i \in A_j$ means node i is in the set of A-nodes of arcs whose B-nodes are j

$k \in B_j$ means node k is in the set of B-nodes of arcs whose A-nodes are j

The complete set of conservation of flow equations for this problem is as given below:

$$d_{1}^{2} = q_{12}^{2}$$
$$d_{1}^{3} = q_{12}^{3}$$
$$q_{12}^{2} = d_{2}^{2}$$
$$q_{12}^{3} + d_{2}^{3} = q_{23}^{3}$$
$$q_{23}^{3} = d_{3}^{3}$$

One of these equations is redundant in each destination set. This can be seen by comparing the first and third equation directly, and deduced by comparing the other three for destination 3. In general, the node conservation of flow relationship needs to be specified for all nodes but one for each destination, and in general the equation dropped is that for the destination node (i.e., for $j = s$).

The combination of the three conservation of flow relationships, three demand functions, and two user cost–volume functions completely specify this problem. It will be noted that there are eight unknown variables (three levels of demand, two user costs, and three arc volumes by destination) and eight equations. Although the total volume on each arc may be thought of as an additional class of unknown variable, once the arc volume by destination is specified, the total volume is known as simply the sum of these destination-specific volumes, i.e.,

$$q_{ij} = \sum_s q_{ij}^s \tag{12-3}$$

Thus there is a single unique solution to this problem. The solution to the problem, not difficult but rather tedious, is reproduced below:

The conservation of flow relationships reveal that

$$q_{12} = d_1^2 + d_1^3$$
$$q_{23} = d_1^3 + d_2^3$$

Substitution into the user cost–volume functions yields

$$t_{12} = 20 + 0.04(d_1^2 + d_1^3)$$
$$t_{23} = 25 + 0.05(d_2^3 + d_1^3)$$

Substitution of these into the demand equations yields

$$d_1^2 = \begin{cases} 2000 - 15[20 + 0.04(d_1^2 + d_1^3)] \\ 2000 - 300 - 0.6d_1^2 - 0.6d_1^3 \end{cases}$$
$$1.6d_1^2 = 1700 - 0.6d_1^3$$
$$d_1^3 = 2500 - 10[20 + 0.04(d_1^2 + d_1^3) + 25 + 0.05(d_2^3 + d_1^3)]$$
$$1.9d_1^3 = 2050 - 0.4d_1^2 - 0.5d_2^3$$
$$d_2^3 = 2000 - 20[25 + 0.05(d_2^3 + d_1^3)]$$
$$2d_2^3 = 1500 - d_1^3$$
$$0.5d_2^3 = 375 - 0.25d_1^3$$

Subtracting the last two yields

$$1.65d_1^3 = 1675 - 0.4d_1^2$$

Multiplying this by 4 and then subtracting from the first yields

$$0 = -5000 + 6.0d_1^3$$
$$d_1^3 = 833.33 \text{ vehicles/h}$$

Further substitution yields the total volume of traffic demanded between each of the three centroid node pairs:

$$d_1^3 = 833.33 \text{ vehicles/h}$$
$$d_1^2 = 750.00 \text{ vehicles/h}$$
$$d_2^3 = 333.33 \text{ vehicles/h}$$

Also, the volume and travel times on the various arcs are as given below:

$$q_{12} = 1583.33 \text{ vehicles/h}$$
$$q_{23} = 1166.66 \text{ vehicles/h}$$
$$t_{12} = 83.33 \text{ min}$$
$$t_{13} = 83.33 \text{ min}$$

The reader can verify that these travel times, when appropriately summed, yield the predicted demands.

The complexity of solving this problem has undoubtedly revealed that the solutions of network equilibrium problems in general are quite difficult. As we will explain in a later section, the process is in fact so difficult that it is necessary to have elaborate computer programs to solve even fairly small practical problems. It is important to remember that the problems we are solving here are made more simple than many because we are treating road systems on which users are assumed to respond to travel time only. The inclusion of prices and other variables would complicate the analysis considerably.

A Carrier Example

All of the examples up to this point have been of roads. We now consider one or two examples for carriers, since they are equally important. The procedure is identical to that for roads, provided that there is a single user cost to which travelers or shippers are responding. Such a single-dimension user cost is appropriate in the analysis of many public transportation networks where the price charged is largely determined externally to the management of the system and in general does not vary with the amount of traffic using particular routes. As we derived in Chap. 11, the relationship between user travel time and volume for a transit route over a period such that service frequency can be altered to suit the traffic is as shown in Fig. 12-8. This form of the

Figure 12-8 Demand and user cost–volume equilibria on a public transportation route. *(a)* Example with a single equilibrium. *(b)* Example with multiple equilibria.

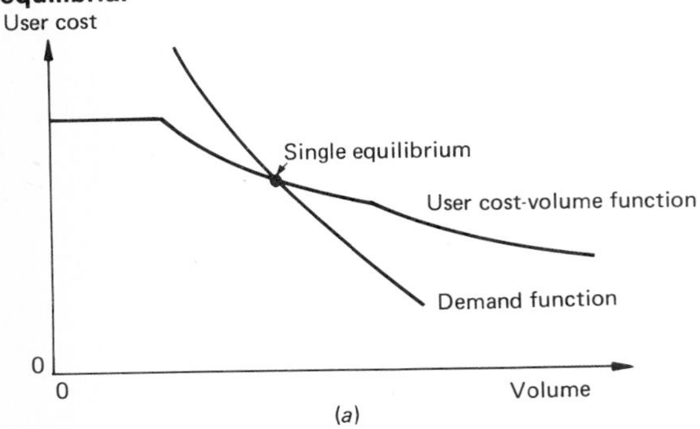

(a)

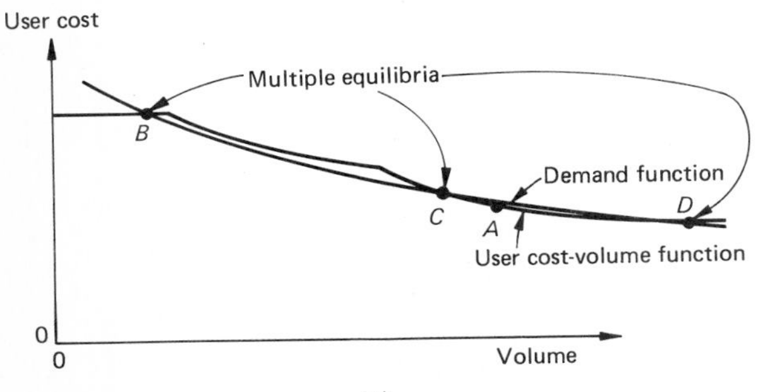

(b)

relationship applies to both total travel time and equivalent (or perceived) travel time (in which waiting and other time outside of the major mode vehicle is weighted more heavily), the curve simply having a steeper negative slope in the latter case. The demand curve for transportation would indicate a decrease in the quantity demanded with increasing user cost, as shown. The equilibrium of the demand and user cost–volume relationships would be as shown in Fig. 12-8. In part *a* of that figure, the situation of a single equilibrium is indicated. However, since both the demand and user cost–volume curves slope downward with increasing volume, it is possible that there could be more than one equilibrium value, as shown in part *b* of that figure.

It is important to note that, in situations of multiple equilibria such as portrayed in Fig. 12-8*b*, the direction of progression from a nonequilibrium point toward one of the equilibrium points depends upon the relative slopes of the user cost–volume and demand curves. Starting at a point such as *A*, the movement toward equilibrium from this unstable point would not be to the closest equilibrium point (measured along the volume axis) *C*, but rather to the more distant one, *D*. In general, if the quantity demanded at a particular user cost level is greater than that which would lead the carrier to provide service at that user cost (as it is in this figure at point *A*), then the carrier will decrease the user cost by expanding the capacity or traffic planned for. In this instance, this process will lead to an equilibrium at point *D*. Where the demand curve is below the user cost–volume curve, as to the right of point *D* and between *B* and *C* in this figure, the movement to equilibrium is in the other direction (toward *D* or *B*, respectively).

While these carrier curves have been discussed without reference to the underlying network or zone pattern, they apply to analyses of one or many origin-destination zone pairs. Thus they apply to a transit route with a number of stations along it, each corresponding to a zone in which traffic is generated. The relevant measure of volume of passenger traffic is the flow past the peak load point on the route. This volume typically is the sum of all boarding passengers on a route which carries travelers to a central business district, for example. Of course, some travelers might travel to outlying points and get off before passing this peak load point just outside the CBD, but their presence would not affect the volume past this point and hence would not enter into the analysis.

A Two-Mode Example

An illustration of the manner in which two different modes interact can be provided by a fairly simple example portraying the interaction of motor vehicles on a road facility and a public transport line. The transit line operates on its own right-of-way, and hence is unimpeded by general road congestion. For this example, we shall assume that people choose that mode which has the least travel time, and that the total quantity of travel between the two zones served by the road and the transit line is fixed. As we indicated in the demand chapter, neither of these two assumptions is precisely correct, but more typical assumptions would make the analysis almost impossible to perform graphically. The user cost–volume curves for these two paths are assumed as shown in Fig. 12-9*b*.

The analysis will be essentially identical to that used earlier in the case of two alternative roads connecting the same points. However, in this case the

Figure 12-9 **Simplified two-mode equilibria for connections between two zones.** ***(a)*** **The network.** ***(b)*** **Modal user cost–volume functions.** ***(c)*** **Equilibria for the two parallel paths.**

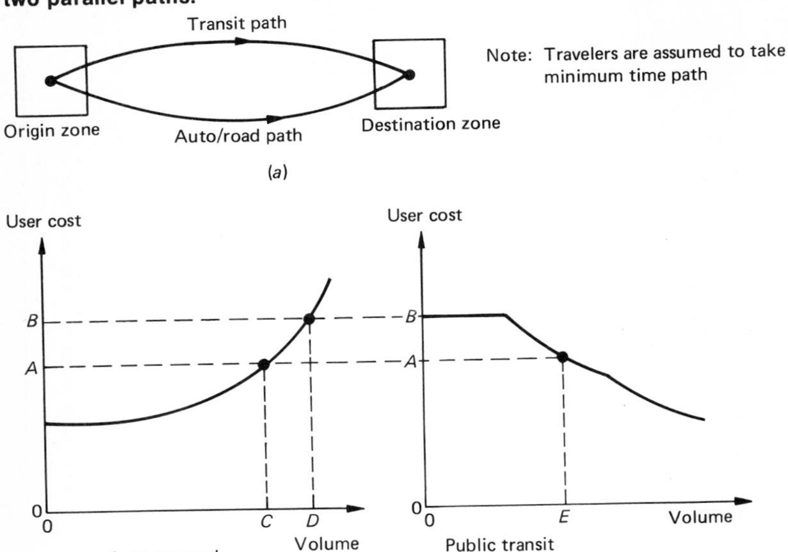

shape of the user cost versus total volume (sum of volume on the two routes) curve will appear quite different. The procedure for combining the two curves is again that of summing the volumes on each of the two routes corresponding to the same travel time. This is performed in Fig. 12-9*c*. A few selected points have been labeled on these curves to illustrate the precise relationship between them. In particular, note that the road volume decreases with increasing total volume in that range of volume where the transit line is used, in that both the road and the transit users have identical travel times. Thus the slope of the user cost curve is less (negatively steeper) than it is on the transit curve alone. The surprising result is that the user cost–volume relationship entails increasing cost with increasing volume in the lower volume ranges, up to a maximum, and then a reduction with further increases in volume, which corresponds to the improvement of the quality of transit service resulting from increased traffic. Of course, at some point the transit line will become saturated and travel time will increase, unless the analysis is for such a long period that additional fixed plant can be provided. Notice that the maximum user cost within this more reasonable range is provided by the maximum travel time on the transit route, corresponding to low-volume use of that path.

Superimposed on this resulting user cost–total volume curve in Fig. 12-9*c* is a demand function. It is deliberately drawn to provide two equilibria. Depending upon which equilibrium is chosen, the transit line might not be used at all or might be used very extensively to the virtual exclusion of highway travel. Of course, in principle if transit travel time can be brought below that on the road even at very low volumes, then the highway would be completely unused. These curves and these two equilibria provide a great deal of insight into the rationale behind many of the transit experiments undertaken recently, which are directed to shifting the equilibrium flow of traffic in such a manner as to both reduce highway congestion and increase the volume on transit to the point where its level of service can be maintained economically at higher levels.

Full-Scale Network Equilibrium

Now that we have considered a number of examples of specific equilibrium flow problems, we turn to the procedures used in practice for large, complex multimodal networks. The most fully developed methods have been designed for urban transportation planning. Hence they make use of the sequential series of models for estimating travel demand, which we explained briefly in Chap. 10. The operation of these is as portrayed in Fig. 12-10 (Wohl and Martin, 1967, pp. 129–131). The entire procedure depends upon a forecast of future land use activities, as we discussed in Chap. 10. Then, a specific user cost or level of service is assumed for each of the links on the highway and transit networks. On the basis of these assumptions, trips are generated, distributed, and then the mode split is estimated. Also, the routes which would be followed by travelers in each of the modes, generally assuming that they will take their minimum time paths, are determined. This results in the assignment of travelers from each origin to each destination by each available mode to a particular sequence of links on that mode. With this information, the total amount of traffic on each link can be estimated, and the expected user level of service with that volume of traffic can be estimated from the user cost–volume (or volume-to-capacity) relationships for that mode. These values estimated on the basis of predicted volume can then be compared to the values assumed initially in the estimates of the flows. If they are sufficiently in conformance, then the equilibrium flows presumably have been found. However, if they are not, then adjustments in the level of service values must be made and the process repeated until this conformance is achieved. In this manner, the sequential models for estimating travel demand can be

Figure 12-10 **Network equilibrium flow prediction procedure used in urban transportation planning.** [***From Wohl and Martin (1967, p. 130).***]

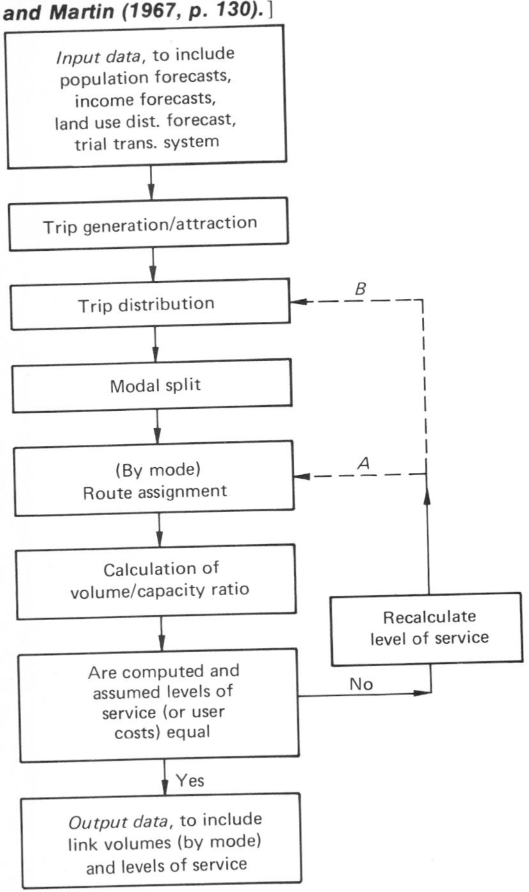

used in a manner that provides for an approximation of network equilibrium to be achieved.

The simple examples used above indicate that this process is extremely complex and computationally cumbersome. In general even simple problems cannot be solved without the aid of elaborate computer programs, such as those developed by the U.S. Federal Highway Administration (1973), termed the Urban Planning Package, and more recently by the U.S. Urban Mass Transportation Administration, termed the Urban Transportation Planning System. Both, extremely complex sets of programs, will not be covered here. These programs are designed to model entire urban areas, with as many as 2000 zones and perhaps 5000 (undirected) links.

TRAFFIC ASSIGNMENT

Traffic assignment is that part of the process of predicting equilibrium flow on urban transportation networks which deals with the steps following the distribution and modal split of the traffic. Thus, the origin to destination trip table for trips via the particular mode in question is known at this stage. In this sense, it may be considered that the total demand for transport is fixed. However, the main focus of the method is on the alternative routes which travelers might choose within a mode between any particular origin-destination pair. To the extent that these are choices made by the users, the demand is not completely fixed. There are various approaches to traffic assignment, but only three fairly widely used methods will be discussed here.

All-or-Nothing Assignment

All-or-nothing assignment is basically an extension of the finding of minimum paths through a network. It is called all-or-nothing because every path from an origin zone to a destination zone has either all of the traffic (if it is the minimum path) or none of the traffic. Also, with this type of assignment it is assumed that there is a known travel time on each of the links comprising the network and that this time will not change with variations in the volume. Thus to determine the pattern of traffic routing through the network, all that is required is that the minimum path algorithm discussed in Chap. 3 be applied to determine the minimum path from each origin node to each destination node, it being assumed that all travelers will use that minimum path. Once all of the minimum paths have been determined, the flow between each origin-destination pair is associated with that path. It is then possible to sum all of the flows on each of the arcs to obtain the total flow. The result of the process, termed the traffic assignment, is the total flow on each of the various arcs or links.

More formally, the steps are

1 Find the minimum path tree from each of the zone centroid nodes to all other nodes.

2 Assign the flow from each origin to each destination node, obtained from the trip table, to the arcs comprising the minimum path for that movement.

3 Sum the volume on each arc to obtain the total arc volume. If (undirected) link volume is desired, sum the flows on the two arcs that represent any bidirectional link.

This procedure can be illustrated by a small example. The network is shown in Fig. 12-11*a*. The minimum path trees are developed using the algorithm presented in Chap. 3, and the resulting trees for all home nodes are shown in part *b* of that figure. The origin-to-destination flow corresponding to each cell of the origin-destination trip table (Fig. 12-11*c*) is assigned to or associated with the arcs which make up the minimum path. These flows are shown in Fig. 12-11*d*. Then the volumes on each arc are summed to yield the total arc volumes, presented in Fig. 12-11*e*.

The resulting traffic assignment can be presented in many different ways. One of the most common is to associate the volume with the arc or link, as in the example above. Another is to portray the volumes graphically, the width of a line along each arc or link indicating the magnitude. This is shown in Fig. 12-12. Another is to present the information in the form of the ratio of volume to capacity for each arc or link. Of course, it is necessary to specify the level of service at which the capacity is calculated.

Figure 12-11 **Example of all-or-nothing traffic assignment.** ***(a)*** **The network.** ***(b)*** **The minimum path trees.** ***(c)*** **The origin-destination trip table.** ***(d)*** **The assignment of trips to minimum path trees.** ***(e)*** **The assigned traffic volumes.**

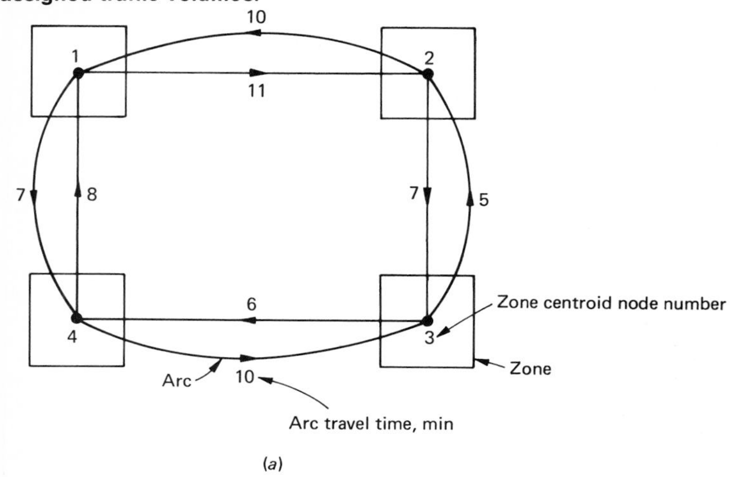

(a)

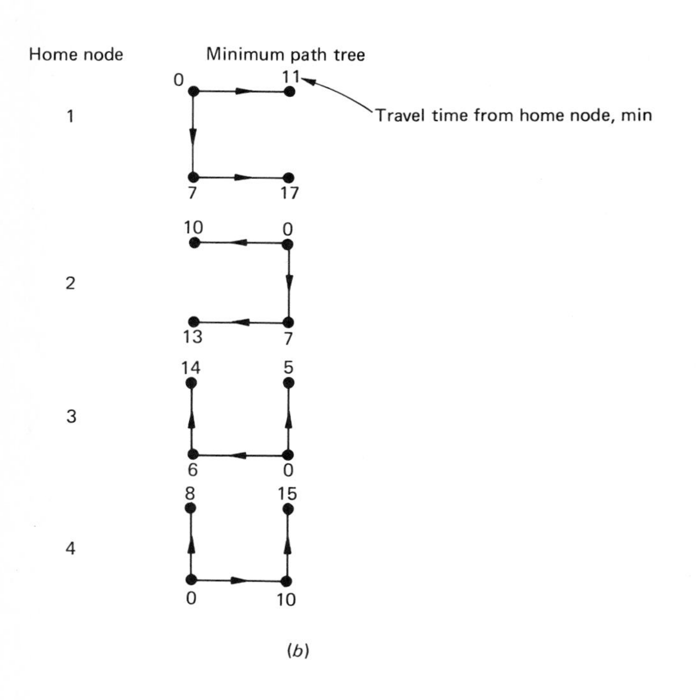

(b)

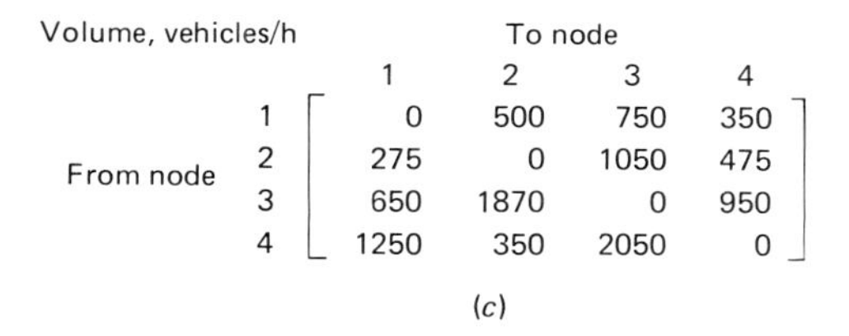

Volume, vehicles/h

From node \ To node	1	2	3	4
1	0	500	750	350
2	275	0	1050	475
3	650	1870	0	950
4	1250	350	2050	0

(*c*)

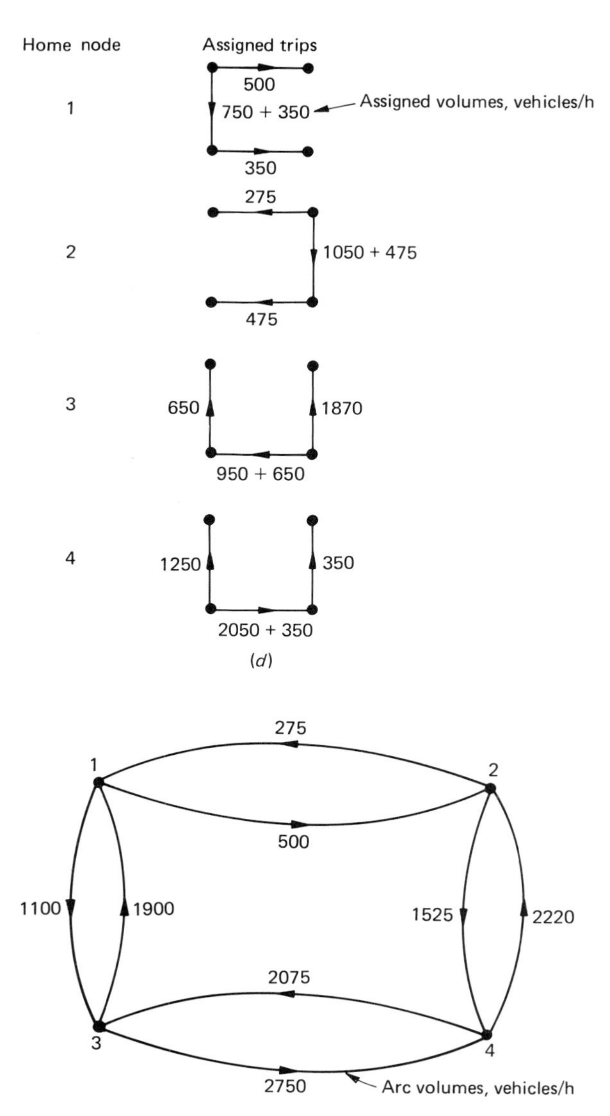

(*d*)

(*e*)

Figure 12-12 Example of a computer-prepared network band plot showing assigned traffic volumes. [*From U.S. Department of Transportation, Federal Highway Administration (1973, p. 183).*]

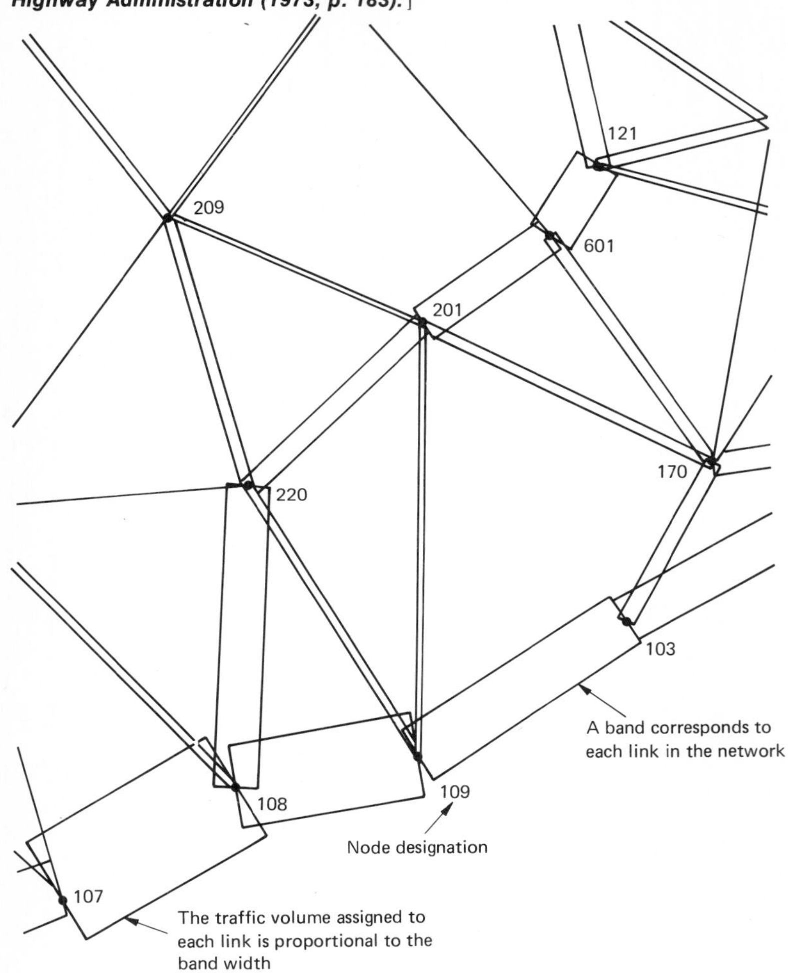

Capacity-Restrained Assignment

The obvious shortcoming of the all-or-nothing assignment is that it completely ignores the effect of the volume of traffic on a link or arc on user cost or level of service. While in some cases it may be possible to guess at the level of service likely to correspond to the resulting volume of traffic on an arc (particularly if the volumes are quite low), there is often little basis for such a guess when modeling future conditions on an urban network. Both the system itself and the travel demands may be very different from present conditions. Thus a means to account for the effect of volume on level of service is necessary.

Figure 12-13 Road travel time–volume relationships used in the 1967 metropolitan Toronto urban transportation planning study. [*From Hutchinson (1974, p. 92).*]

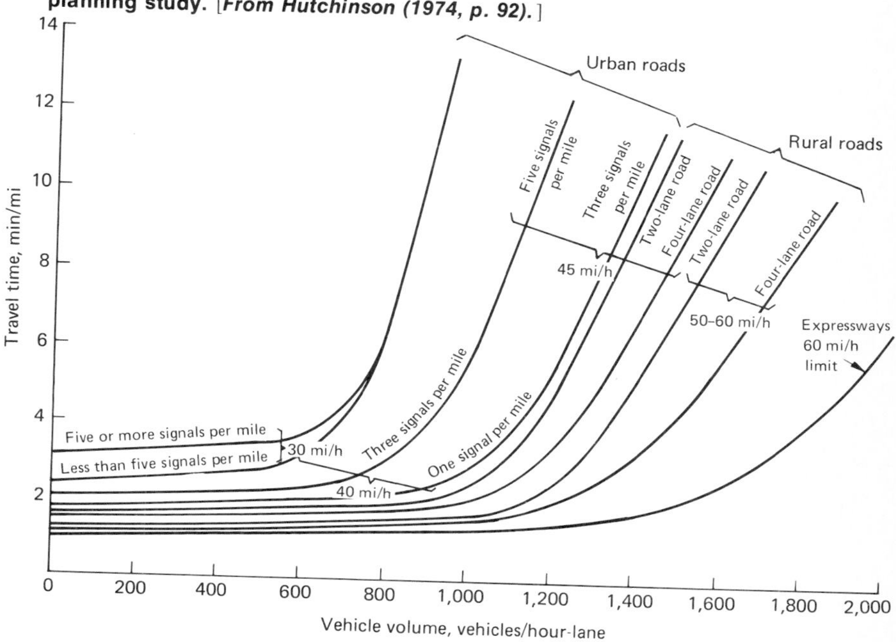

Usually traffic assignments are made for a period of many hours, most often for an entire day. As a result the user cost–volume relationship will differ from that for a short period (such as an hour) because of the variation in traffic throughout the day and the use of an average time or user cost for the entire period. Much attention has been devoted to the development of user cost–volume relationships applicable to traffic assignment problems. Typical relationships are shown in Fig. 12-13, which were developed in 1967 for application to the Toronto area. The most noteworthy difference between these and those developed in Chap. 5 on the basis of consideration of uniform flow (or nearly uniform flow, as with the peak hour factors near unity for hourly capacity estimates) over a short period (typically 1 h) is that the curves do not bend backward. The reason for this is simply that even if very high volumes are accommodated during one or two peak hours, which would result in high travel times and sometimes operation in the backward-bending part of the travel time–volume curves for a few minutes, these conditions are quite temporary and add only a small amount to the average travel time measured for a much longer period such as an hour or an entire day.

Somewhat similar functions are widely used in almost all assignment models. In the computer programs for traffic assignment developed by the U.S. Federal Highway Administration, for example, the following relationship between travel time and volume is employed (U.S. Department of Transportation, 1973, p. III-15).

$$t = t_0[1 + 0.15(v/c)^4] \tag{12-4}$$

where t = travel time on the arc
t_0 = free-flow travel time (the inverse of mean free speed, the speed at zero volume)
v = volume of traffic on the arc
c = practical capacity of the arc

Practical capacity is a measure similar to the capacity at an acceptable service level, the service volume described in Chap. 5. The term was used prior to the introduction of the concept of level of service. For most purposes, the practical capacity can be taken as the service volume at the lowest acceptable level of service, as indicated in Fig. 5-10.

The process of capacity-restrained assignment is similar to the all-or-nothing assignment in that it begins with an assumed travel time on each of the links, usually the free-flow travel time. However, once that assumed travel time has been used to obtain minimum time paths and the traffic assigned, by the all-or-nothing process described earlier, the resulting volumes on each link are used to estimate again the travel time on each of the links. This travel time is compared with the travel time assumed earlier, and if there are any substantial discrepancies, the entire process is repeated. In this repetition, termed an iteration, the level of service or travel time is adjusted upward or downward to be closer to that which would occur with the previously assigned volume of traffic on that link. This process of iteration is continued until the resulting volumes have associated with them travel times which are essentially identical to the travel times assumed in the calculation of minimum time paths which led to those volumes.

The actual process is computationally rather complex, usually requiring computer analysis. In the U.S. Federal Highway Administration programs, the assignment iteration process is performed in the following manner.

1 Using the free-flow arc times, calculate minimum time path trees. Designate these arc times by a superscript (1), i.e., $t^{(1)}$.

2 Assign the traffic as in all-or-nothing assignment.

3 Calculate the arc travel times using Eq. (12-4) and the assigned volumes. This time for each arc is labeled with a superscript (2), i.e., $t^{(2)}$

4 Calculate a new minimum path trees based on arc times given by

$$t^{(3)} = 0.75t^{(1)} + 0.25t^{(2)}$$

5 Return to step 1, using the $t^{(3)}$ arc times to develop the minimum paths. Then follow steps 2 through 4, increasing the index of all superscripts by 3 units. Current practice is to continue this process until four separate assignments have been made, then go to step 6.

6 The volume assigned to each arc is taken as the average of the volumes of the four assignments.

Clearly, this is an approximation to the desired result. It has been found to yield useful assignments, ones that are usually sufficiently accurate for many planning purposes. However, there are always doubts as to the validity of the results, and as a result recently much research has been and continues to be devoted to computational methods that will yield the desired equilibrium assignments (e.g., see LeBlanc et al., 1976). Also, research has been devoted to rules for assignment which allow for travelers' selection of other than the best path, as might occur due to imperfect information or pure whim, an example being a widely accepted assignment rule that spreads traffic probabilistically among various routes (Dial, 1971). In fact, some of this research is directed toward the entire traffic forecasting process, not just the assignment phase (e.g., see Evans, 1976). It is likely that in the next few years engineers and planners will have at their disposal computerized methods which will yield estimates of arc flows which are truly consistent with both the demand and the user cost–volume functions.

User- and System-Optimal Assignments

In our analyses of road traffic above, it has been assumed that users always select their own routes. As we observed in Chap. 11, a user selects a route on the basis of the travel time which she or he incurs and ignores the extra travel time which her or his presence on the road forces on every other user of that road. Because of this difference between the travel time to which the user is reacting and the total additional travel time which she or he causes all users to incur, users often select routes which are best from their own individual standpoints but are not best from the standpoint of keeping the total time expended by all users as low as possible. In this section we shall present a simple example of how this occurs (adapted from Beckmann et al., 1956, pp. 83–87) and discuss its significance.

Let us consider two zones which are connected by two roads and focus on the traffic flow in one direction only, the total flow being fixed. The total volume of travel to be accommodated on these two roads is indicated by the length of the horizontal axis *AB* in Fig. 12-14. The user cost–volume curves for each of these two roads is shown, each beginning at the extreme ends of the horizontal axis. Curve *CD* is for road 2, and curve *FG* is for road 1. With users taking their minimum time paths, both roads will be used. The equilibrium user cost (time) will be *XJ*, and the equilibrium volume on road 2 will be *AX* and on road 1 will be *BX*.

Also shown on this figure is the marginal cost of an additional user on each road, by curve *CE* for road 1, curve *FH* for road 2. This marginal cost is estimated according to Eq. (9-4) and hence it includes the additional user costs to all users resulting from the addition of one more unit of volume of traffic on each of the roads. At the equilibrium achieved on the basis of users choosing their own paths at *X*, it can be seen that the marginal cost on road 2 is far greater than that on road 1. The significance of this is that if one user were to switch from road 2 to road 1, the savings in total travel time to all users on road 2 would more than offset the added total travel time to all users on road 1. Thus the totality of all users would be better off, in the sense that total vehicle-hours would be reduced, although the cost for the user on road 1 would increase and would decrease on road 2. Similarly, additional units of traffic could be profitably taken from road 2 and shifted to road 1 to result in a lower total vehicle-hours of travel.

This gain in total vehicle time resulting from shifts from road 2 to road 1 will occur until the marginal costs are equal, which occurs at volumes on road 2

Figure 12-14 **User-optimal and system-optimal assignments of traffic to two parallel arcs.**

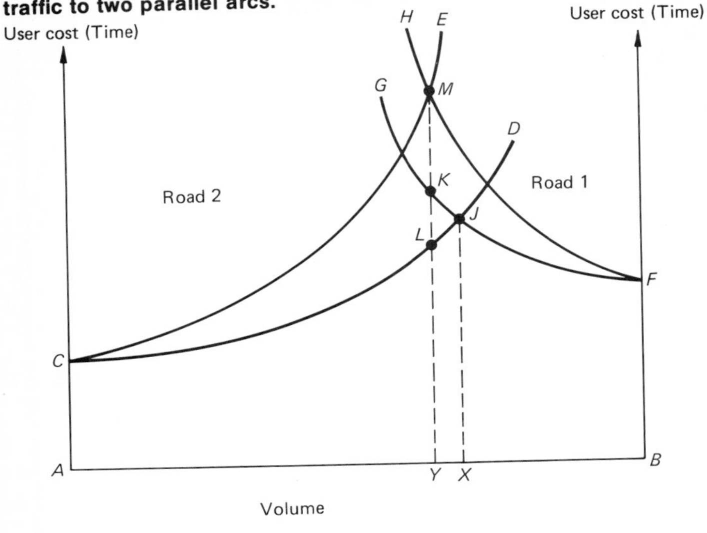

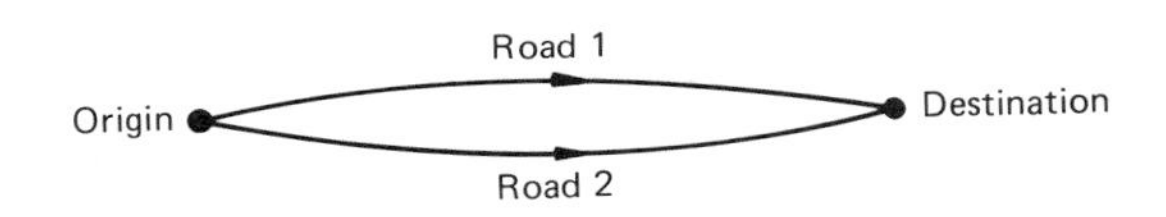

of *AY* and on road 1 of *BY*. At that point, by definition, no change can be made which would produce greater gains on one road than the losses on the other, and hence no change would result in reduction of total user travel time. Thus, the total travel time experienced by all users with an assignment of traffic between the two roads based on an equality of marginal user costs is less than the total travel time based on users experiencing equal average travel cost. This assignment, which achieves the least total (system) travel time, is termed the system-optimal assignment, in contrast to a user-optimal assignment, in which users select paths to minimize their own individual travel times.

The significance of this result is the following. On the basis of users selecting their own individual minimum time paths, total road travel time consumed by all users in general will be greater than that for a situation in which road users are assigned to routes on the basis of minimizing total system travel time. In the second case, users could be allowed to choose their own paths, but they would have to select their paths on the basis of their marginal cost to the system rather than their average cost. At the system-optimal assignment in Fig. 12-14, the users would have to perceive their marginal cost of *YM*, rather than their average cost (i.e., the travel time each actually experiences) of *YL* on road 2 and *YK* on road 1. Thus users must perceive a cost in addition to their travel times, the additional cost (measured in units of time) being equal to *LM* for road 2 users and *KM* for road 1 users. This amount of time, or its money equivalent, is termed a congestion toll.

Thus the congestion toll is a means whereby traffic can be reallocated in order to reduce the total vehicle time. However, in this reallocation some travelers would be made better off in the actual travel time that they experience (in this example, road 2 users) and others worse off (road 1 users), requiring compensation of those who are worse off. But the toll exactly compensates for this difference, making them both experience a cost (time plus money) equal to the marginal costs. This concept will be discussed in greater detail in Chap. 17.

CONCLUSIONS

Although the methods of network equilibrium have been developed largely out of the practical need for means to estimate traffic on road systems, they provide a general framework for the estimation of traffic on any transportation system. The complexities of analysis involving the interaction among many modes and the consideration of both price and level of service variables require the use of the computer for solving even small network flow problems; methods for doing this are still in the development or early prototype stage. However, in many situations, focusing attention on a single price or level of service measure can yield very useful results, as it does in road and many public transport situations. For these problems, computer methods are available which are practical to use and give results which are reasonably widely accepted. Nevertheless, many subproblems of the general network equilibrium problem are amenable to manual computation, as are the traffic assignment problems involving small transit and road networks and demand-supply equilibrium problems involving single equation functions and known routing.

SUMMARY

1 The concept of network equilibrium merges the demand for transportation as described in Chap. 10 and the availability of transportation services as described by supply functions or the user cost–volume relationship in Chap. 11 to yield information on the actual quantity of transport which will occur and the associated prices and levels of service.

2 Network equilibrium is much more complex than the usual economic market equilibrium because it requires considering both price and level of service, the latter often being in many dimensions. Also, the interaction is over a network rather than in a single market, which requires keeping track of commodities which represent flows between different origins and destinations via different modes and paths.

3 Most methods for computing network equilibrium flows apply to situations where attention can be realistically focused on a single measure of user cost (such as travel time), as seems to be the case with road networks and public transport systems, where prices are fixed (or often not perceived by travelers). Problems involving very small networks can be solved graphically or analytically. However, for the solution of large-scale problems, computer methods are mandatory.

4 Even if a problem cannot be solved completely in its ideal quantitative form according to the concepts of network equilibrium, the overall method provides a framework within which the possible effects of different types of changes can be identified and traced. Often this is extremely helpful in practical problem solving.

PROBLEMS

12-1

Three directed road links, designated by their A-nodes and B-nodes as (1, 2), (2, 3), and (3, 4), are arranged in series. If their individual travel time–volume relationships are as given below, what is the travel time–volume relationship for the entire distance from node 1 to node 4, assuming each link experiences the same volume of traffic? The units are t, min; v, vehicles/h.

$$t_{12} = 10\,[1 + 0.15(v/2000)^4]$$
$$t_{23} = 15\,[1 + 0.15(v/2300)^4]$$
$$t_{34} = 8\,[1 + 0.15(v/2500)^4]$$

12-2

In the network shown below, the demand functions for travel are given, along with the travel time–volume relationships for the three directed links. Nodes 1, 3, and 4 are traffic generators, while node 2 is merely an intersection. Solve these equations for the equilibrium volumes and travel times.

Demand:

$d^{13} = 2000 - 10t^{13}$ $\qquad$ d^{ij} = quantity demanded from i to j, vehicles/h

$d^{43} = 4000 - 15t^{43}$ $\qquad$ t^{ij} = travel time from i to j, min

User cost–volume:

$t_{12} = 25 + 0.05\,q_{12}$

$t_{23} = 30 + 0.06\,q_{23}$ $\qquad$ t_{rs} = arc travel time, min

$t_{42} = 30 + 0.08\,q_{42}$ $\qquad$ q_{rs} = arc volume, vehicles/h

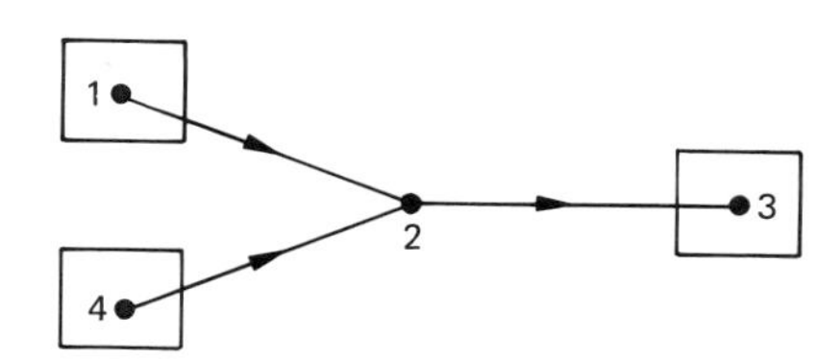

12-3

In Prob. 12-2, assume that the demand curve from node 1 to node 3 has shifted upward as a result of new land development and is given by

$$d^{13} = 4000 - 20t^{13}$$

What are the resulting equilibrium flows and travel times in the network?

Compare with the results for Prob. 12-2. Why does a shift in the demand curve from node 1 to node 3 affect the volume from node 4 to 3?

12-4

In Prob. 12-2, the link (2, 3) has been improved, resulting in a new user cost–volume relationship of

$$t_{23} = 20 + 0.04q_{23}$$

What effect does this have on the equilibrium flows and travel time?

12-5

In Prob. 12-2, arc (1, 2) is improved, resulting in a travel time–volume curve of

$$t_{12} = 15 + 0.04q_{12}$$

What is the resulting equilibrium flow? Why does this change, which directly affects only travel from node 1 to node 3, alter the equilibrium flow from 4 to 3?

12-6

a In the network shown below, there are 6 vehicles/h which wish to travel from node o to node r. The travel time in minutes (t) is given as a function of volume in vehicles/h (x). Find the equilibrium flow in which users choose their own minimum time paths.

Show that this equilibrium flow is individual-user-optimal by transferring 1 vehicle from the route via q to that via p, resulting in greater travel time via p than q.

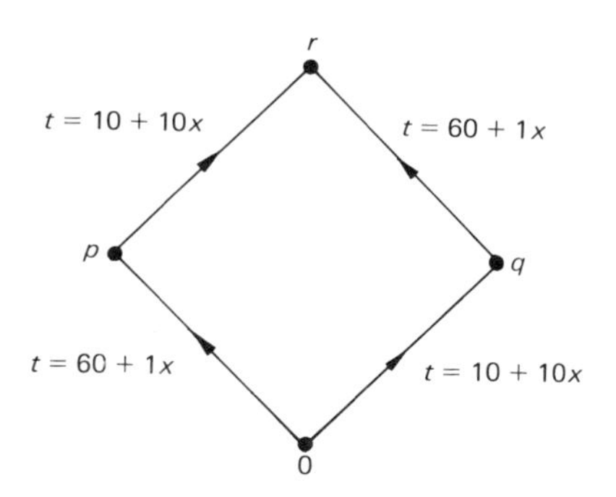

b A directed link is now added to the network, from q to p, as part of a general road improvement program, as shown on page 528. Starting with an assignment in which 3 vehicles/h traveled via nodes o, p, r and the other 3 vehicles/h via o, q, r, show that a vehicle (1 vehicle/h) traveling via o, q, r can switch to o, q, p, r and be better off (i.e., reduce its travel time). What are the times from o to r via the three routes? Now show that a vehicle that had been traveling via o, p, r can also switch to o, q, p, r and be better off.

c As a result of these two switches, we have the new equilibrium. Show that the times via the three routes are equal. Compare this time with that on the network of part **a**, in which there is no arc (q, p).

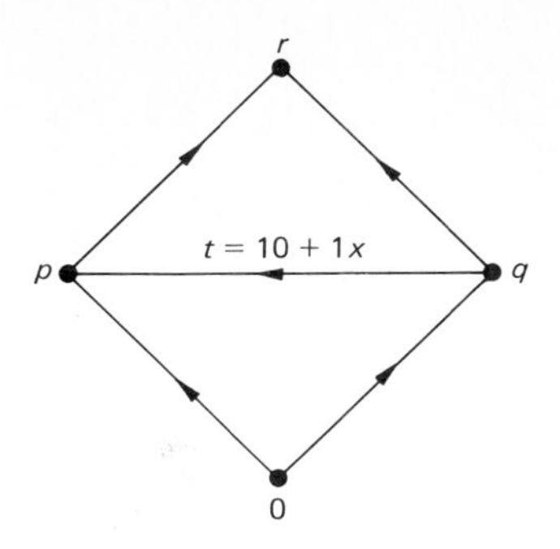

Comment: It is true that a link can be added to a road network with the result that *all* travelers are made worse off! This is known as Braess' paradox of traffic flow. Transport network flows are indeed complex and undoubtedly contain many counterintuitive phenomena. [For a more complete discussion of this, see Murchland (1970), from which this example was taken, with slight modification.]

12-7

Assign the vehicle trips shown in the origin–destination trip table to the network, using the all-or-nothing assignment technique. Travel times, in minutes, are given on the network. Make a list of the links in the network, as directed arcs, and indicate the volume assigned to each. Also calculate the total vehicle-minutes of travel. Show the minimum path tree and assigned traffic for each of the five home nodes.

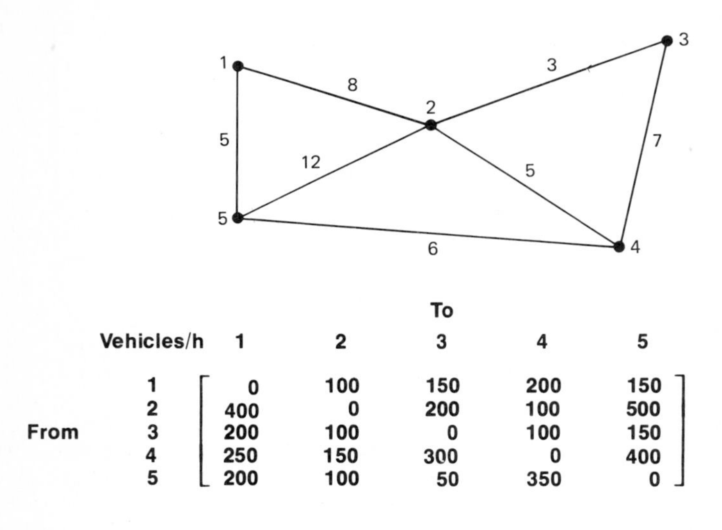

	Vehicles/h	To 1	2	3	4	5
From	1	0	100	150	200	150
	2	400	0	200	100	500
	3	200	100	0	100	150
	4	250	150	300	0	400
	5	200	100	50	350	0

12-8

A rail rapid transit line connects the three zones as shown on page 529, operating all-stop trains with running times (start at one station to start at next) as shown. Below this is the morning peak period trip table for travel among these three zones. Find the equilibrium amount of passengers on the transit line and the corresponding train frequency (trains per hour) in the peak flow direction, assuming the transit management selects train frequency according to the function given. Also, assume that road travel time is not affected by diversion to the transit line (implying that most road users are traveling to and/or from points not served by the transit line). Use the TRC (Washington) modal split model of Chap. 10 for work trips and economic class 5.

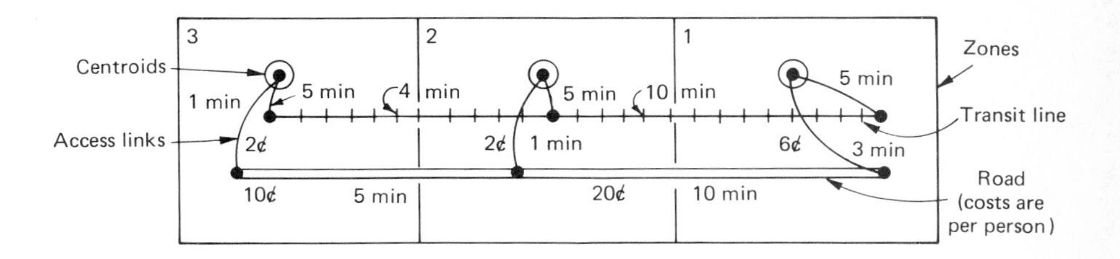

Rail fare: 50 cents

Parking charge in zone 1 only: $1.50/day (average of 50 cents/person-trip considering both directions and auto occupancy)

Trains/h = 6.0 + 0.0035 (passengers/h) in peak load direction

		To		
Person trips/h		**1**	**2**	**3**
From	**1**	**0**	**1000**	**500**
	2	**3000**	**0**	**400**
	3	**2000**	**1200**	**0**

Hint: Plot a graph of total transit passengers past the peak load point, arc (2, 1), versus average waiting time and plot the train frequency function in the same graph.

12-9
In performing a full analysis of a new transit line such as that in Prob. 12-8, what other aspects of travel demand in addition to modal choice should be considered?

12-10

a Two parallel roads take traffic from node 1 to node 2. If the demand function (d in vehicles/h, t in min) is

$$d_1^2 = 5000 - 105t_1^2$$

what is the equilibrium flow and time on each road assuming users select their own minimum time paths?

The user cost–volume relationships are as given in Eq. (12-4), with the parameters

Arc	t_0, min	c, vehicles/h
a	10	2000
b	15	2000

b What would be the flows if users were charged and reacted to a travel time equal to the marginal time by a system of tolls? If users valued time at $4.00/vehicle-h, what would the dollar toll be on each road? What would the actual travel times be? Compare the two equilibrium solutions.

12-11
Assign the trips contained in the origin–destination trip table of Fig. 12-11*c* to the network below, using an all-or-nothing assignment procedure. Arc times are in minutes. Show each step.

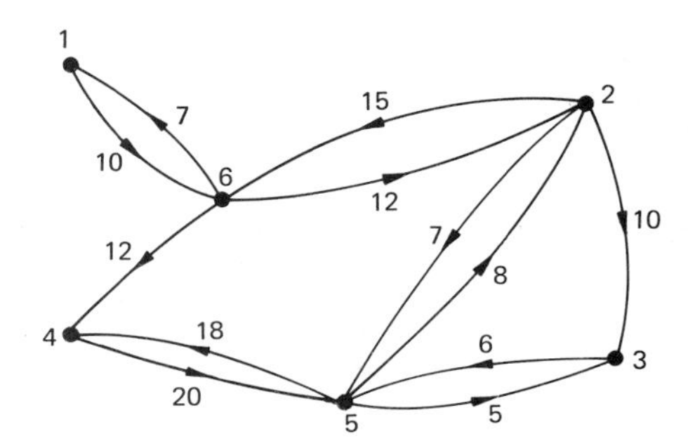

REFERENCES

Beckmann, Martin, C. B. McGuire, and Christopher B. Winsten (1956), "Studies in the Economics of Transportation," Yale, New Haven, CT.

Creighton, Roger, L. (1970), "Urban Transportation Planning," University of Illinois, Urbana.

Dial, Robert B. (1971), A Probabilistic Multipath Traffic Assignment Model Which Obviates Path Enumeration, *Transportation Research,* **5**(2): 83–112.

Evans, Suzanne P. (1976), Derivation and Analysis of Some Models for Combining Trip Distribution and Assignment, *Transportation Research,* **10**(1): 37–57.

Hutchinson, Bruce D. (1974), "Principles of Urban Transport Systems Planning," McGraw-Hill, New York.

LeBlanc, Larry J., Edward K. Morlok, and William P. Pierskalla (1976), An Efficient Approach to Solving the Road Network Equilibrium Traffic Assignment Problem, *Transportation Research,* **9**(5): 309–318.

Murchland, John D. (1970). Braess' Paradox of Traffic Flow, *Transportation Research,* **4**(4): 391–394.

Potts, Renfrey B., and Robert M. Oliver (1972), "Flows in Transportation Networks," Academic, New York.

U.S. Department of Transportation, Federal Highway Administration (1973), "Traffic Assignment Manual," Report No. 5001-00060, Washington, DC.

U.S. Department of Transportation, Federal Highway Administration (1972), "Urban Transportation Planning," Washington, DC.

Wohl, Martin, and Brian V. Martin (1967), "Traffic System Analysis for Engineers and Planners," McGraw-Hill, New York.

Environmental Impacts

13.

Up to this point we have focused on the workings of the transportation system itself, its technology and its economic and service characteristics and the demand for transportation. But the provision of transportation, like the provision of almost any other material good or service, carries with it many side effects. Many of these effects are desirable, including broadening the range of goods available for people to purchase and consume and increasing the range of options with respect to styles of living and living standards (as we discussed in Chap. 2). But, in addition, transportation carries with it many unintentional side effects, such as the creation of air pollution by vehicles and the severance of established communities by new freeways and transit facilities. It is the purpose of this chapter to discuss these various effects of transport system changes which occur outside of the system itself, effects which occur in the environment of the system. Although once almost totally ignored, these environmental effects, or environmental impacts, as they are often termed in the planning and engineering profession, have assumed an increasingly important role in transportation decision making, and in some situations are the overriding considerations in deciding what types of changes to make in transportation systems and services. Hence, it is essential that these impacts be understood and taken into account in any transportation decision-making context.

A CONCEPTUAL FRAMEWORK

Before discussing in detail the various types of impacts of transportation system changes, we place these into the perspective of the transportation system and the services it provides. Such a perspective is provided by an input-output diagram of the transportation system, in which the transportation system is placed in the context of its environment. This is done in Fig.

Figure 13-1 **The transportation system and its environment.** [*Adapted from Thomas et al. (1970, p. 20).*]

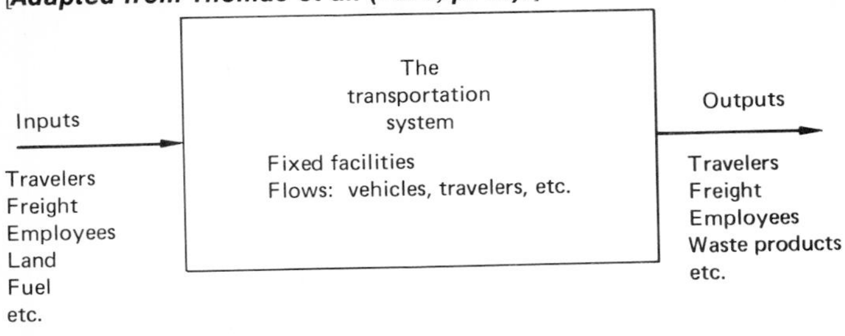

13-1, where the box represents the transportation system, and all that lies outside the box is the environment of that system. Thus the box includes such items as the way links, vehicles, terminals, and other functional components of the transportation system, as well as the travelers and shipments of freight being carried on that system at any time. The environment includes the physical space outside of the system, the various human activities taking place at residences, manufacturing plants, parks, etc. Crossing the line which represents the boundary between the system and its environment are the various items which enter and leave the transportation system. Included among these are the travelers and freight shipments, as well as the various physical resources which may be consumed in producing transport, including such items as land, vehicles, fuel, and human labor to run the system. Some of these leave the system in largely the same form in which they entered, such as the employees, who return home after a day's work, and many items are discarded regularly by the system, such as vehicles which have exceeded their useful life, parts, etc. Until now, we have been concerned with basically two types of items—the objects being transported by the system (passengers and freight) and the various resources required by the system to operate.

There are, however, many other inputs and outputs of the system. For example, pollutants are emitted from the system into its environment as a result of the burning of fuel and other processes attendant to providing the transportation service. Similarly, there is noise. Also, there are many inputs to the system which are not represented as resources to be paid for in any market transaction but nevertheless are resources consumed or altered by their use in providing transport. Fresh air, taken in to mix with fuel in order to provide the propulsive forces, is an example. Similarly, often water and other essentially free resources are consumed or altered in this manner. Even though these items—inputs and outputs—are not represented in economic markets, they are real resources that must be taken into account in a complete analysis of transport costs in their most general sense.

Furthermore, the prices paid for certain inputs and outputs may not reflect their true value, as for example in the case of land which, when converted to a particular transport use, might create an undesirable barrier across a wilderness area. Again, the treatment of these inputs and outputs in this framework explicitly accounts for such resources.

Also, transportation systems have been observed to influence their environments in ways not directly represented by the various inputs and outputs of the system and the transformations which might occur on particular items. As we observed in Chap. 2, the provision of cheaper transport may radically influence the spatial pattern of human activities such as the location of houses, factories, stores, etc. Such locational effects are often referred to as reorganization impacts or development impacts (Garrison et al., 1959). In the framework of Fig. 13-1, these changes in the location of activities occur in the environment of the transportation system, although of course the associated changes in travel patterns (and freight-movement patterns) are reflected in changes in the flows in the transport system itself. Since the location and amount of various human activities are important factors in the overall standard of material well-being and quality of life in any society, it is important that these developmental impacts be understood and taken into account in any analyses of the consequences of changes in the transport system.

Thus, by use of an input-output diagram, we have identified two types of impacts of transportation system changes. One is inputs and outputs of the system which are not represented by the costs or movement of persons and freight with which we have been concerned, inputs and outputs which are normally referred to as impacts on the natural environment. The second type is changes in the manmade environment, typified by reorganization of land uses. We shall first discuss the physical environmental impacts and then turn to the reorganization or developmental impacts.

IMPACTS ON THE NATURAL ENVIRONMENT

There are four types of physical environmental impacts of transportation facilities which seem to be of particular importance at the present time: noise, air pollution, ground and water pollution, and vibration. Each of these will be discussed in turn. In addition, the very presence of a transportation facility often is deemed undesirable, perhaps because of its size or its outward appearance. Since this type of effect can be treated in much the same manner as other land use impacts, it will also be discussed in the following section. It should be noted that the list of negative and possibly positive impacts is subject to change, as human preferences or tastes change, and perhaps more importantly as our concern for our environment and the natural environment changes. Thus this list cannot be treated as fixed.

Noise Impacts

Noise is unwanted sound. Most sound from transportation systems is unwanted, primarily because it has the potential for disturbing human or other activities, and in a few cases it may in fact be injurious to persons or other living organisms. However, there are at least a few instances in which a certain type of noise may be desirable, as in the case of the bell and other noises of San Francisco cable cars adding to the environmental quality of restored areas in that city, and the beneficial effect that noise of vehicles may have on safety at places where two or more transport streams cross, like the whistle of trains crossing highways. However, with the exception of such rather specialized situations, in general noise from transport systems is unwanted.

Figure 13-2 **Noise transmission from emitter to receiver.**

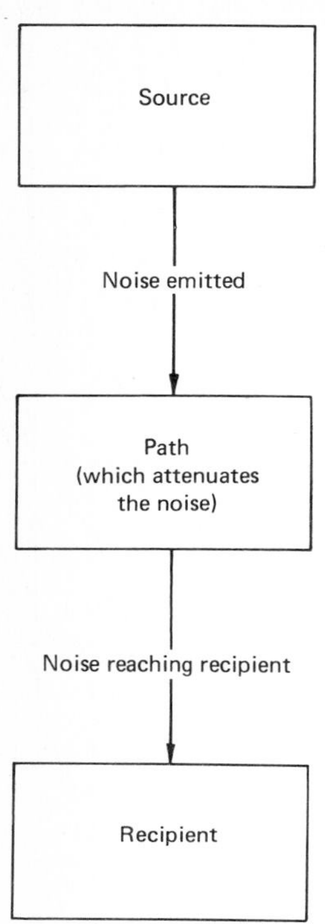

Consideration of transport noise requires explicit reference to three important elements. One is the noise emitted. Second is the recipient of the noise, a person or activity which is particularly bothered by the noise. Finally is the path over which the noise is transmitted from the source to the receiver, the characteristics of which can determine the loudness of the noise or the extent to which the noise remains disturbing or injurious by the time it reaches the receiver. These are portrayed in Fig. 13-2.

Any quantitative discussion of noise must include specific quantitative measures of its loudness. Two scales are in common usage for treating noise from transportation sources. Noise is commonly measured as a pressure, which is the ratio (multiplied by 20) between a particular noise pressure and a standard low pressure which represents the approximate limit of human audibility (0.0002 dyne/cm^2). This measure is called the sound pressure level and is usually measured in decibels (dB). By itself the sound pressure level does not indicate the human response to noise, since human annoyance with sound varies with the frequency or pitch of the noise as well as with the intensity, higher frequencies being more annoying than lower frequencies. Through considerable experimentation and data analysis, accousticians have found ways for combining the sound pressure levels at various frequencies representing any particular noise to reflect approximate human annoyance or response. One criterion thus developed is the A scale, the A indicating a particular weighting of intensity at various frequencies. Provided the noise has no overwhelming single pitch component and is not a sharp impact noise, this A scale is useful for measuring noise loudness. It is the most commonly used scale for ground transportation noise. The dBA is the symbol for the measure of loudness in decibels which uses this A scale for combining the sound pressure levels at different frequencies. Values associated with various types of common noises are indicated in Fig. 13-3.

Since air transport noise often has particular pitches which are overwhelmingly intense and may also be characterized by sharp impacts, another weighting scheme and hence another scale is used for noise from that source. The aircraft noise is evaluated using the measure of the effective perceived noise level, EPNL or EPNdB. The values on this scale of various aircraft noises are also shown in Fig. 13-3. Thus the effective perceived noise level measure is similar in concept to the decibel but since it is on a different numerical scale, distinction must be made between them.

Noise from land transport is primarily due to road vehicles, although there are instances where noise from other sources is troublesome, such as elevated urban transit railways. Because the annoyance of noise from new freeways and other roads with high volumes of traffic has been extremely severe, attempts have been made to develop standards for the maximum noise to which various human activities should be subjected. One collection of such standards is shown in Table 13-1. It is important to note that there remains considerable disagreement as to what constitutes acceptable noise levels, different agencies recommending different values, and considerable variation existing among the standards of various nations. Also, it is important to recognize that the degree of annoyance associated with a particular noise level depends very heavily on the person and the types of activities engaged in. Thus it may be unrealistic to attempt to follow a single standard everywhere, even for residences. Presumably those who live in high-density developments with a great deal of surrounding traffic are willing to experience higher noise levels in exchange for low rents or other benefits of such living, in contrast to those who, if their finances permit, choose to expend large portions of their income on large tracts of land removed from other persons and noise-creating activities, such as transport, in order to enjoy a quiet environment.

The noise level created by a transport facility in the vicinity of a noise-sensitive activity can be estimated approximately without a great deal of difficulty. In the case of highways, equations have been developed to estimate the noise levels at various distances from the highway. The level of noise depends upon the volume of traffic, the speed at which that traffic is moving, and the mix of vehicles (specifically the percentage of trucks). The noise generated by road traffic flowing at approximately a constant speed with a volume such that there is almost a continuous stream of vehicles is given by the following equation (Galloway et al., 1969):

$$\bar{L} = 10 \log_{10} q - 10 \log_{10} d + 20 \log_{10} u + 20 \tag{13-1}$$

where $\bar{L}$ = mean noise level at receiver located a distance d from the source, dBA

d = distance between receiver and pseudolane at the center of the traffic lanes

q = traffic volume, vehicles/h

u = mean speed of traffic, mi/h

Figure 13-3 **Typical noise levels in dBA and EPNdB of some common noises. Note: Approximate relation between EPNdB scale and dBA scale is shown in this comparison of various noise sources. [*From U.S. Department of Transportation (1972, p. 3).*]**

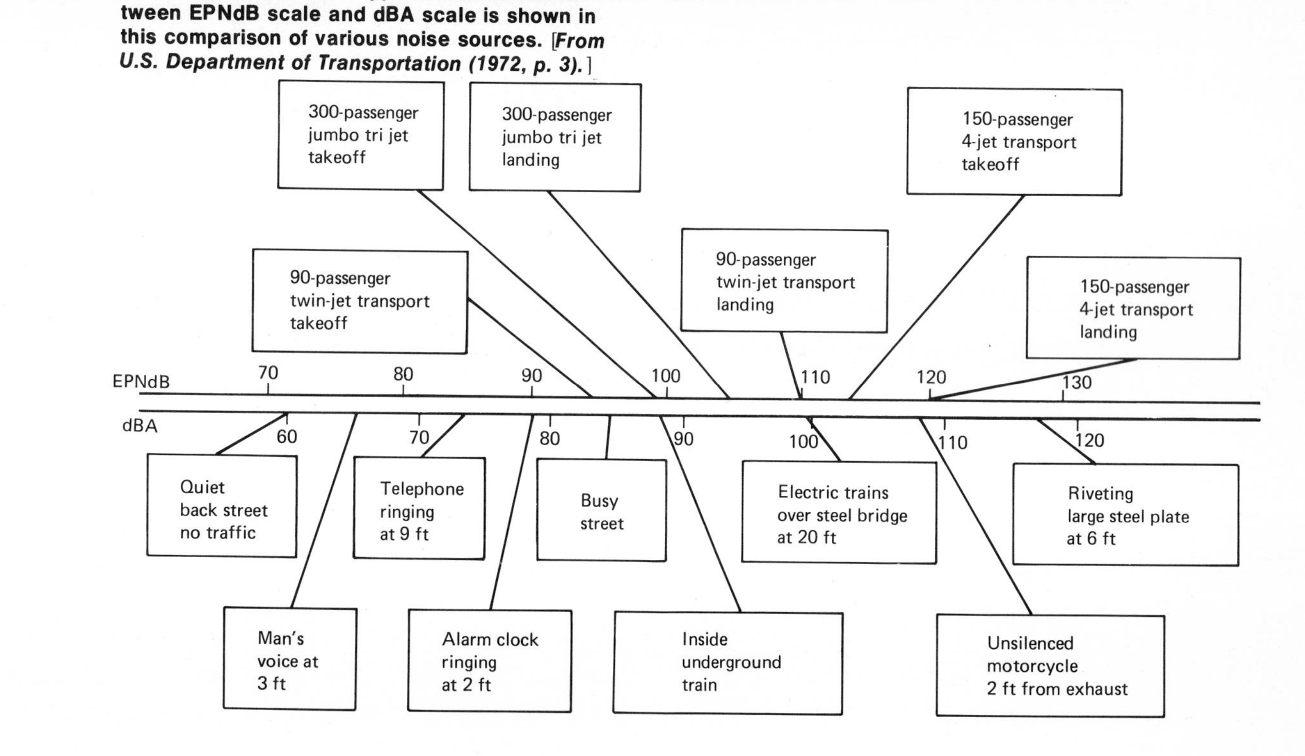

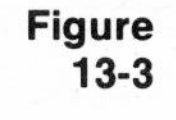

Table 13-1 Residential Noise Level Guidelines Recommended by the U.S. Department of Housing and Urban Development†

General external exposures, dBA
Unacceptable
Exceeds 80 dBA 60 min in 24 h
Exceeds 75 dBA 8 h in 24 h
Discretionary—normally unacceptable
Exceeds 65 dBA 8 h in 24 h
Loud repetitive sounds on site
Discretionary—normally acceptable
Does not exceed 65 dBA more than 8 h in 24 h
Acceptable
Does not exceed 45 dBA more than 30 min in 24 h

† U.S. Department of Transportation (1972, p. 26).

This equation is valid for volumes above about 1000 vehicles/h. It assumes that there is no obstruction (such as a building or high wall) between the roadway and the point at which the noise level is being predicted, other than perhaps a few trees or shrubs.

The location of the pseudolane is based upon the approximate location of a single line source of noise which would produce the same characteristic level of noise as many lanes of traffic. If there is only a single lane, then the pseudolane is the center line of that lane. If there are many lanes, the pseudolane lies between the center line of the nearest lane and the center line of the farthest lane, but is closer to the nearest lane. The distance from the center line of the nearest lane to the pseudolane center line is simply the geometric mean (square root) of the distance between the center line of the nearest lane and the center line of the farthest lane. For example, on a freeway in which the distance between the two outer-lane center lines is 96 ft, the distance from the center line of the nearest lane to the pseudolane is 9.8 ft. Thus, if the receiver were located 100 ft from the center line of the nearest lane, the distance d in the above equation would be 109.8 ft.

Many factors in addition to volume, speed, and distance influence the level of noise in the vicinity of highways; for instance, on very rough pavement, such as gravel or rough concrete, the noise level is approximately 5 dBA greater than that predicted by this equation. Also, the greater the percentage of trucks in the traffic stream, the greater the noise. Although the noise emission characteristics of trucks vary widely, apparently due to differences in maintenance and sound-deadening design features, an approximation is that one additional dBA be added for each 2.5 percent trucks in the traffic stream. Furthermore, on grades of up to 5 percent, about 2 dBA should be added to reflect the additional noise due to higher power requirements from engines. Unfortunately, little definitive information on the change in noise levels due to the presence of motorcycles and sports cars is available.

Little information is available on the production of noise in stop and go traffic. However, fitting an approximate equation to rather scanty data yields the following relationship:

$$\bar{L} = 65 - 10 \log_{10} d \tag{13-2}$$

Figure 13-4 **Effects of variations in highway design on noise transmitted into environment.**
A_1 = flat section at grade
B_1 = 20-ft elevated section on structure, or on narrow shouldered fill
C_1 = 20-ft elevated section on fill with 36-ft shoulders
D_1 = 20-ft depressed section
A_2 = flat section with 11-ft solid noise barrier
B_2, C_2, and D_2 are same as B_1, C_1, and D_1, but with 8-ft solid noise barrier at edge of highway (or, for D_1, at ground level) [*From Beaton and Bourget (1968, p. 5).*]

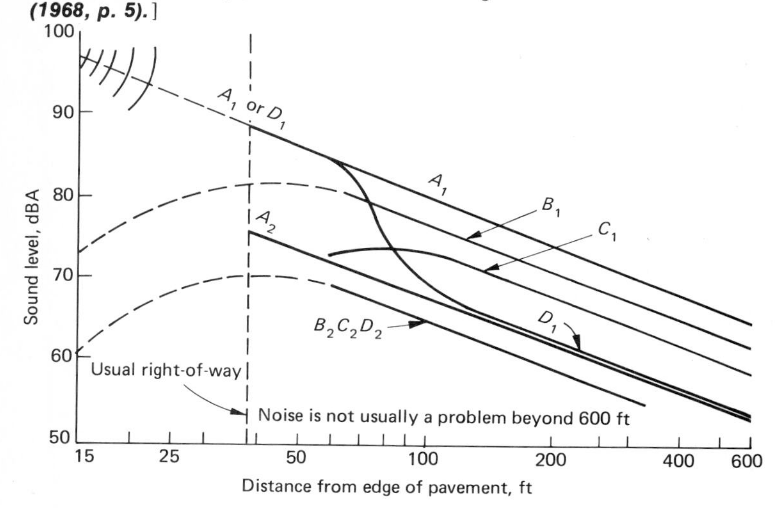

Unfortunately, no statements regarding the effects of speeds or mix of vehicles can be made, since information on these aspects is too meager. However, it can be assumed that this equation applies to relatively low speed operation typical of major arterial streets, probably in the vicinity of 20 to 30 mi/h.

The level of noise reaching a recipient at some distance from the highway can be influenced significantly by design features of the road and appurtenances which affect the paths of noise transmission. This is revealed in Fig. 13-4, where the effect of such design variations as elevating or depressing the highway is revealed. Elevating roads substantially reduces noise in the immediate vicinity, and depressing roads reduces noise in the more distant areas. Also, barriers paralleling the road can substantially reduce noise, although their unsightliness will undoubtedly limit their use unless they can be made attractive through landscaping or other treatment. In addition trees and shrubbery can reduce noise by approximately 2 dBA. Interior noise levels are generally much lower than exterior levels, noise levels typically being about 10 dBA less inside a building with open windows and about 20 dBA less with closed windows (Bolt et al., 1967, pp. 4–5).

As mentioned earlier, noise from rapid transit lines and rail lines in cities is a problem in situations where the lines are in close proximity to residences or other noise-sensitive activities. However, this noise can be reduced and the path modified, probably to the point where the noise represents no particular difficulty in most situations. The use of welded rail as opposed to individual rail sections, appropriate rail and wheel maintenance to maintain smooth rolling surfaces, and the layout of routes with superelevation (often termed "banking" in casual conversation) on curves so that unbalanced forces are no longer present (as will be discussed in detail in Chap. 16), reduce transit noise substantially, up to about 11 dBA according to some estimates. Also, noise barriers can be easily installed along the right-of-way, reducing noise at 50 ft by 10 to 13 dBA and by 7 to 8 dBA at 500 ft. Bridges or other structures still present a problem, noise typically increasing by about 3 dBA on concrete structures and 10 dBA on steel structures with decks. Typical levels of wayside noise for rapid transit trains of various lengths, with modern design features but without noise shields, operating on the surface of the ground, are given in Fig. 13-5.

Objectionable noise from aircraft is being reduced by new federal aviation regulations regarding noise emissions from aircraft and by altering flight paths and controlling development in the vicinity of airports to reduce noise intrusion into noise-sensitive activities. The Federal Aviation Administration has recently promulgated regulations which limit the effective perceived noise level of aircraft as a function of the gross take-off weight of the aircraft and the distance from take-off or landing. Specifically, for very large aircraft (above about 600,000 lb gross weight), the maximum effective perceived noise level in take-off, measured at 3.5 nmi from brake release along the axis of the runway, is limited to 108 EPNdB. Similarly, the approach noise 1 nmi from the threshold for those aircraft is limited to 108 EPNdB. Sideline noise is limited to the same amount, measured 0.35 nmi from the axis of the path for four-engine aircraft.

A very useful chart for describing the exposure to noise of various places in the vicinity of the transportation facility is contained in Fig. 13-6. Lines

Figure 13-5 **Typical noise emissions from rail rapid transit line. Note: Trains are traveling at 40 mi/h. [*From U.S. Department of Transportation (1972, p. 17).*]**

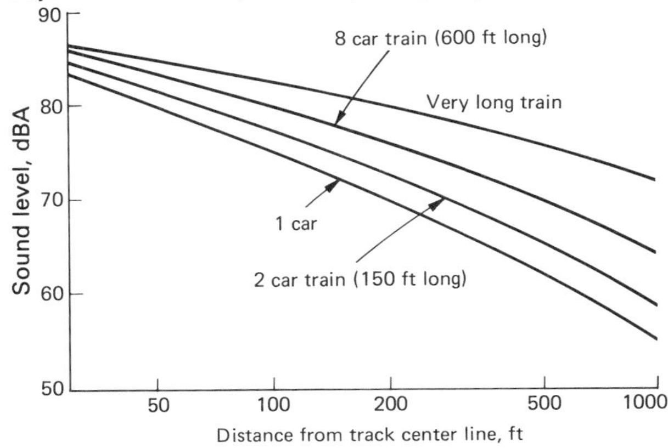

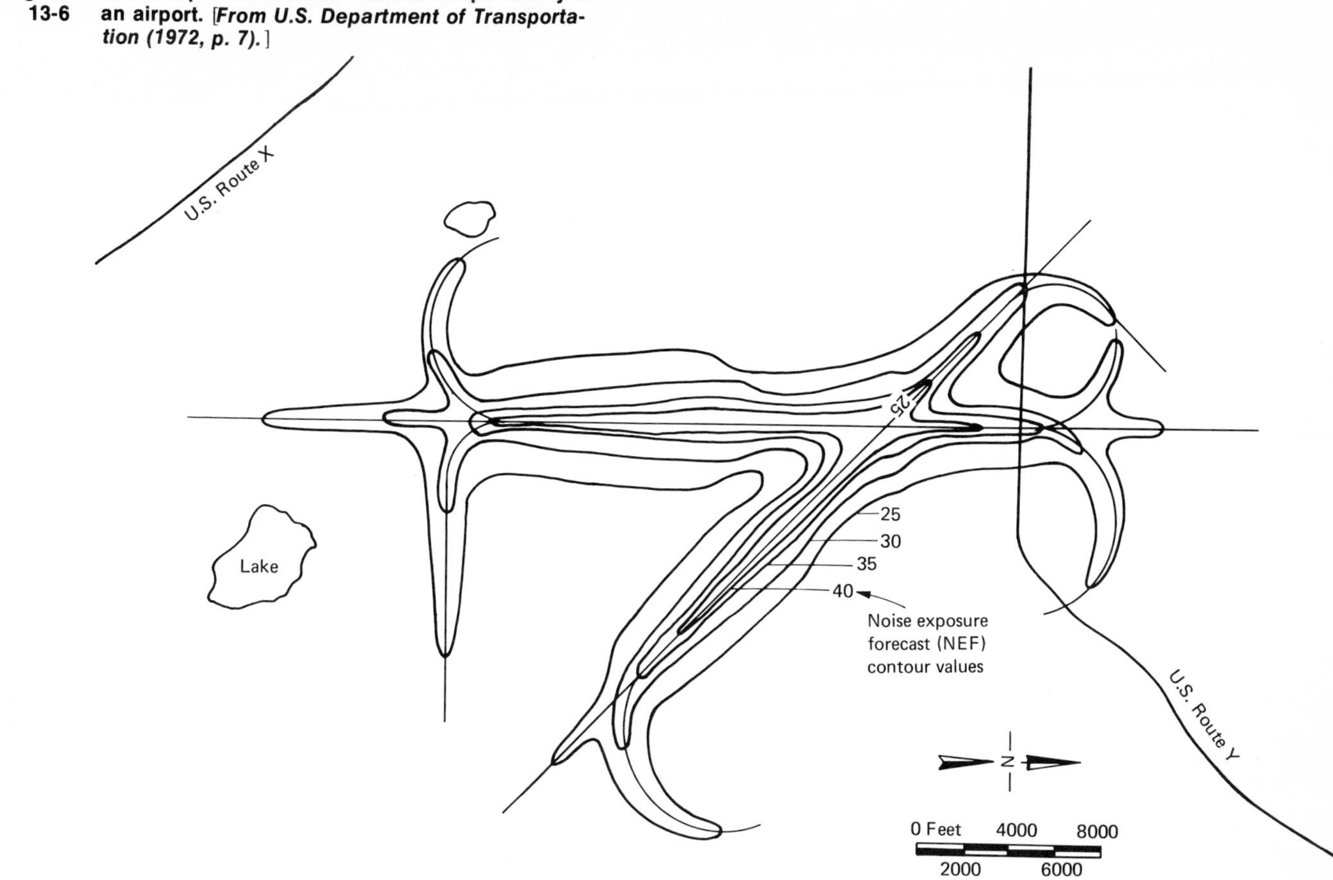

Figure 13-6 **Noise exposure forecast contour map: vicinity of an airport.** [*From U.S. Department of Transportation (1972, p. 7).*]

portraying the locations at which equal levels of noise are experienced can be drawn to reveal the contour of noise exposure in a manner similar to that revealed by contour lines on a map indicating the terrain characteristics. An example of this is shown in Fig. 13-6, for noise exposure in the vicinity of an airport. The particular values used in that figure are the values of so-called noise exposure, which is a weighted combination of the noise experienced from various aircraft in the course of a day. Similar contours can be drawn using decibel measures.

Before leaving the discussion of noise, it is important to point out the effect of two or more sources of sound. It is not correct to simply add the decibels from two sources to yield the total noise level, because the scale is not linear. Instead noise must be added logarithmically. Specifically, if the two sounds are equal in decibels, the resulting noise is only 3 dBA greater than either alone. And if the difference between the noise levels of the two sources is 10 dBA or more, the sum is approximately equal to the louder of the two. If the higher level exceeds the smaller by 0 to 1 dBA, the sum exceeds the higher level by 3 dBA; if by 2 to 4 dBA, the sum exceeds the higher by 2 dBA; and if by 5 to 9 dBA, the sum exceeds the higher by 1 dBA.

An Example Noise Problem

In the course of a preliminary location study for a freeway, the highway engineers considered a location which passed near a quiet residential area with modern, air-conditioned homes. They were considering the placement of a full interchange at the intersection with a state highway, as shown in Fig. 13-7. The first design provided for the freeway at grade, with the state highway crossing on a bridge. Would this be acceptable, and if not, what alternative would you suggest?

Because the area near the freeway interchange is residential, analyses must be performed for noise levels during the day and night, for inside and outside houses. Hence peak, off-peak, and very low volume (night) traffic conditions are relevant.

First, standards of acceptability must be set. These must be somewhat subjectively set, because of the state of the art. It seems reasonable to select 45 dBA for interior levels, based upon Table 13-1. During the daytime peak periods, outside noise levels probably should not generally exceed 65 dBA, so as to meet the normally acceptable standard in that table. If these prove very costly to meet, trade-off analyses of costs against noise standards should be made. Also, other benefits of noise-reducing devices, such as improved aesthetics of a depressed roadway, should be considered.

During the daytime peak periods the noise level at the residential property boundary and away from the interchange, using Eq. (13-1) is

$$q = 4000 \text{ vehicles/h} \qquad d = 200 + \sqrt{72} = 208.5 \text{ ft} \qquad u = 50 \text{ mi/h}$$
$$\bar{L} = 10 \log_{10}(4000) - 10 \log_{10}(208.5) + 20 \log_{10}(50) + 20$$
$$\bar{L} = 69 \text{ dBA if autos only}$$

Adding 2 dBA for 5 percent trucks yields 71 dBA at the property line. This is unacceptable in relation to the 65 dBA standard.

At this point, alternative designs must be considered. One is to elevate the freeway, but this probably would be unacceptable aesthetically. Another is to depress it, which would enhance the view of both the freeway and the state

Figure 13-7 **Map of area for example highway noise problem.**

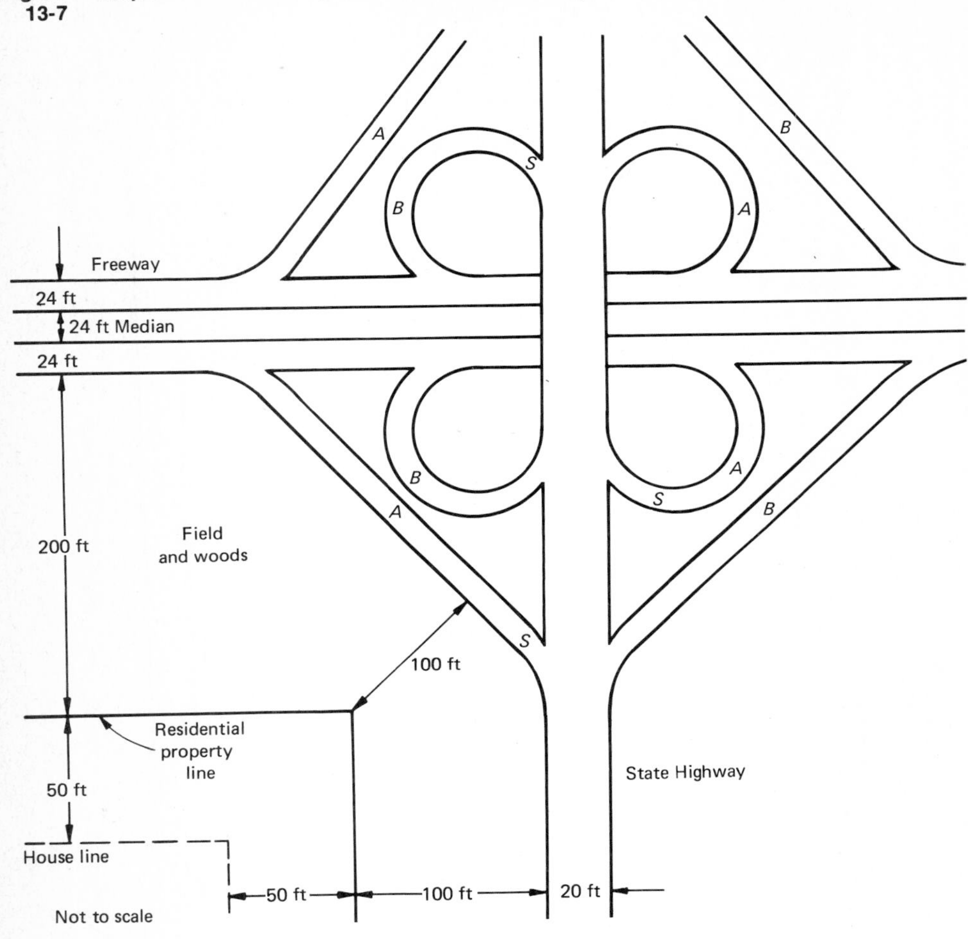

Design year traffic predictions

Segment	Peak: Volume, vehicles/h	Peak: Percent trucks	Peak: Speed, mi/h	Nonpeak: Volume, vehicles/h	Nonpeak: Percent trucks	Nonpeak: Speed, mi/h
Freeway (all lanes)	4000	5	50	1000	3	65
State highway	1000	5	35	300	3	50
Ramps (*A* and *B* together as pair)	200	3	25	50	3	25

highway, since the bridge could now be at grade. Only this alternative will be analyzed, although in an actual study others might be.

From Fig. 13-4, it can be seen that at a distance of about 200 ft, a depression with sloping walls reduces noise by about 7 dBA. This yields a noise level of 65 dBA, which just meets the standard. If the 200 ft were heavily wooded, the noise would be further reduced by perhaps 1 to 2 dBA, to about 63 dBA, making it quite acceptable.

Interior levels of noise must also be checked. The distance from the freeway to house interiors is 50 feet greater than that to the property lines, so using Eq. (13-1) and the modifications for trucks, depression, and woods, the noise at house walls is

$$\bar{L} = 10 \log_{10}(4000) - 10 \log_{10}(258.5) + 20 \log_{10}(50) + 20 + 2 - 7 - 2$$
$$\bar{L} = 62 \text{ dBA}$$

With windows open, the interior noise would be reduced by about 10 dBA, yielding 52 dBA, about 7 dBA too high. If windows facing the freeway were closed, the level would be reduced by about 20 dBA to 42 dBA, quite acceptable. Hence the depressed freeway seems acceptable in the daytime even during peak periods, since these homes are all air-conditioned. This design should be checked for other periods also, as is called for in the problems, before it can be considered acceptable.

Air Pollution

The emission of various gases and particles from transportation activities into the atmosphere also creates some rather severe problems of environmental degradation. The importance of various transportation sources relative to all sources of air pollution of particular types was indicated in Table 2-2. This table is based upon the U.S. Environmental Protection Agency's 1971 ambient air quality standards, which established tolerable levels of air pollution concentration for each type of pollutant in accordance with its relative toxicity. This table attributes overall air pollution to its sources in proportion to the contribution of that source to the level of the EPA standard for each type of pollutant. Transportation is seen to be a major pollution source, but its responsibility for 16.7 percent of all pollution, according to this calculation, indicates that it is somewhat less important as a pollution source than stationary power plants and industrial processes. Yet, by the same measures, transportation is responsible for almost 70 percent of the pollution in Los Angeles and San Diego, and over 50 percent on the average in many urban areas throughout the United States. Thus it is essential to consider air pollution from transportation sources and attempt to reduce it.

Air pollution is fundamentally different from noise, in the sense that once emitted into the atmosphere, pollutants remain there for an extended period and can be carried by air currents to other locations. Thus the diffusion of air pollutants into the atmosphere may lead to spreading of the pollution, which results in diminution of concentrations and hence of any danger, or it may result in a further concentration to potentially dangerous levels of pollutants from many different sources. In contrast, noise exists only momentarily, not accumulating over time.

It has been found possible to estimate the emissions of air pollution from transportation and other sources without difficulty. However, it has been extremely difficult to predict likely levels of concentration of pollutants, and hence it is difficult to estimate air quality as a result of particular travel patterns and amounts of emissions. Therefore we shall present estimates of the emissions, presuming that in most cases that if emissions are reduced, air quality is not made worse.

Table 13-2 Present and Expected Air Pollution Emissions from Road Traffic in the United States†‡

	Type of pollutant (1972), grams/mi			Type of pollutant (1990), grams/mi		
Speed, mi/h	Carbon monoxide	Hydro-carbons	Oxides of nitrogen	Carbon monoxide	Hydro-carbons	Oxides of nitrogen
			Pollutants emitted on freeways			
60.0	30.89	6.08	7.39	7.23	0.95	1.86
55.0	31.66	6.31	6.85	7.23	0.95	1.86
50.0	33.75	6.67	6.35	7.71	1.00	1.72
45.0	34.93	6.80	6.35	7.94	1.09	1.63
40.0	37.20	7.03	5.90	8.39	1.18	1.50
35.0	41.91	7.71	5.67	9.53	1.22	1.45
30.0	47.17	8.16	5.44	10.43	1.36	1.36
25.0	56.25	8.66	5.22	12.70	1.54	1.32
			Pollutants emitted on arterial streets			
30.0	46.95	8.16	5.44	10.89	1.36	1.36
25.0	55.79	8.85	4.90	12.70	1.59	1.32
20.0	68.95	9.53	4.54	15.88	1.81	1.22
15.0	84.10	10.75	4.54	19.96	1.91	1.22

† Data based on vehicle mix of 83.04 percent passenger cars, 6.81 percent 2-ton trucks, 3.26 percent 6-ton trucks, 3.29 percent 20-ton trucks, and 3.60 percent 25-ton trucks. (Trucks are diesel powered.) Data for 1972 and 1990 are based on 12 years preceding the calendar year of interest; the data reflect Environmental Protection Agency best estimates of air pollution emission factors for each model.
‡ Sanders et al. (1974, pp. 4-8, 4-9), based on Turner (1973).

The amounts of three important pollutants—carbon monoxide, hydrocarbons, and oxides of nitrogen—emitted by motor vehicles on various types of highways at various speeds have been estimated. These are presented in Table 13-2, based upon a typical mix of vehicle types, as indicated in the table, including vehicles constructed in the 12 years preceding the calendar year of interest. The emission rates per vehicle-mile are given for 1972 and 1990. The values for 1990 represent expectations regarding improvements in vehicle propulsion system design which will yield the substantial reductions in pollution emissions indicated. However, much progress has already been made, current (1977) emissions being far less than those for the 1972 vehicle mix, particularly for carbon monoxide and hydrocarbons. As can be seen from these tables, increased vehicle speed results in lower emissions of carbon monoxide and hydrocarbons per vehicle-mile, while the emission of oxides of nitrogen increases per vehicle-mile with increasing speed. Because all three of these pollutants are undesirable, there is no general rule regarding the best direction of speed change from the standpoint of air quality.

The primary means for reducing air pollution from transportation sources is thus to attempt to reduce the pollution emissions at the source. This is an excellent example of an attempt to put a so-called "technological fix" on an environmental problem. Assuming that future standards for pollution emissions can be met through technological advances and that adherence to those standards will be enforced, pollution from road transportation sources should diminish very substantially in the future.

Air pollution from transportation sources other than road vehicles is a problem, but generally much less so. Air, rail, and water systems emit much of their pollutants in rural areas and other places where the adverse effect is minimal. Urban public transportation systems do of course generate pollutants, but systems powered by electricity supplied from a central power source tend to emit them in areas which often are not particularly sensitive, and internal-combustion-powered transit vehicles tend to be only a small fraction of the total vehicles in any area.

Vibration

Vibration from transportation sources appears to be a limited problem. It undoubtedly occurs in the vicinity of major surface transportation arteries where heavy vehicles are operated in close proximity to structures containing human activities that are extremely sensitive to vibration. Probably the most notable among these is rail subway lines, where vibration has been observed to be a problem in structures which adjoin the facility, particularly where the foundations of the building may be directly connected to the rail tunnels. However, it has been also observed that at distances above about 50 ft separating the structure and the tunnel, vibration is usually not sufficient to be annoying or disruptive of most human activities. Although road vehicles are somewhat more cushioned by their pneumatic rubber tires than solid-wheel rail vehicles, vibration also can be a problem in the vicinity of major roads on which vehicles are subject to bouncing. While problems of vibration have only occasionally forced relocation of planned transport routes or relocation of vibration-sensitive activities (e.g., laboratories), vibration must be considered a potential source of problems in any new transport facility planning.

Ground Water Pollution

There is considerable opportunity for excessive pollution of the ground resulting from certain types of emissions from transportation systems. For example, salt deposited on highways to reduce ice and snow problems and oil dropped from railroad trains can be carried by water through the ground and collect in particular locations to the point where the ground would be incapable of supporting certain types of plant life. Also, the substances used to kill weeds along surface transport rights-of-way, most notably rail lines, could also be toxic to various wildlife and shrubbery. Although published information on this source of pollution seems rather limited, perhaps indicating that the problem is not perceived by many transport agencies or environmental groups, what little has been published suggests that the problems are in fact rather minimal, although a great concentration of transport activities which would emit unusually high amounts of pollutants could pose a threat to ground water purity.

Environmental Capacities of Streets

It is clear from the above discussion of various types of impacts of transportation activities on the natural environment that in general an increase in traffic results in an increase in negative environmental impacts, assuming all other conditions remain unchanged. For example, examination of Eq. (13-1) reveals that an increase in road traffic, assuming speed does not change, will result in an increase in the noise level at any given distance from the road. Similarly, an increase in traffic volume, *ceteris paribus,* will result in an increase in air pollution emissions and therefore presumably lead to a degradation in air quality. Similar considerations apply to other impacts on the natural environment. Thus in order to maintain any given level of environmental quality, traffic must not exceed a certain volume.

This limitation on traffic volume may be considered an environmental capacity, in the sense that environmental impact considerations result in the specification of the capacity, in contrast to the more usual considerations of traffic speed or level of service as the basis of capacity of road facilities, as described in Chap. 5. The two concepts, basically similar, differ in that in one case users are the basis for the capacity determination and in the other case nonusers are the basis. The concept of environmental capacity probably originated with the Buchanan report (Buchanan et al., 1963, pp. 71–76), which was concerned with the accommodation of the automobile in towns in Great Britain but since then has been increasingly adopted as a basis for rational road planning in other areas as well.

There are many criteria on which the environmental capacities of roads can be based. The most prominent include effects on the natural environment such as noise, odors, and air pollution in general, vibration, disturbances to television and radio reception, dirt and litter, damage to vegetation and wildlife and also some effects on humans such as pedestrian safety and conflicts with various movement patterns (Sharpe and Maxman, 1972, p. 35). Thus the establishment of the environmental capacity of a road depends on a number of factors. First of all are the characteristics of the road itself, its design speed, number of lanes and their widths, number of intersections, etc. Also influencing the environmental capacity will be the composition of the traffic using the street, particularly the number of trucks and other vehicles that might emit unusually large amounts of pollutants. Finally, there are the characteristics of the land uses and activities on the road, which will determine the maximum acceptable levels of various types of undesirable pollutants or other negative impacts of the road.

The usual approach to the establishment of environmental capacities is to survey persons who live, work, etc., in the vicinity of a road. They are queried regarding the extent to which they find the vicinity acceptable with respect to the various possible impacts of traffic noted above. Surveys are conducted along many roads such that the range of values of any one pollutant, for example, noise, is very large, and in particular, such that the range of responses with respect to any particular impact includes some roads which are found acceptable for any given characteristics of the road, traffic volume and composition, and nearby human activities. A fraction or percentage of the people affected who consider the impact acceptable is selected as a criterion level for the determination of the environmental capacity. Specifically, the environmental capacity of a road with these characteristics is the maximum volume of traffic such that at least the criterion fractions of persons affected feel that the impact is acceptable. This then establishes the environmental capacity of a particular type of road, mix of traffic, and type of nearby human activity for each particular type of impact.

The environmental capacity of any particular road can be limited by any one of the possible impacts. Therefore, the environmental capacity is defined as the lowest volume for all of the various impacts. Thus, for example, if the environmental capacity of a particular type of road and type of human activity was 10,000 vehicles/day based upon air pollution, 7000 vehicles/day based upon pedestrian safety, and 5000 vehicles/day based on noise, the environmental capacity would be 5000 vehicles/day.

The environmental capacity of roads clearly may vary considerably among different cities and nations, primarily because people may consider different types of impacts differently in those different situations. Until more research on environmental capacity is undertaken, it will not be possible to generalize across cities and cultures easily. However, as an example of the types of results obtained from these studies, Table 13-3 is presented. This table presents the environmental capacities developed in Louisville, Kentucky in 1971. Under the column labeled prototype characteristics is a description of the road facility and the types of activity along it. Also presented is the current traffic volume and then the environmental capacity along with the controlling environmental factor. It is interesting to note that many of the facilities examined in this survey have actual daily traffic volumes which exceed the environmental capacities.

Environmental Impact Statements

The concern for the possible harmful effect of transportation projects on the environment, especially the natural environment, has led to the practice of including a statement of expected environmental effects in reports on proposed new facilities. These statements usually include a prediction of the impact from the project and consideration of alternative designs which might reduce the negative effects. In the United States, as well as in some other nations, such reports on environmental impacts are required before a project using federal funds can be undertaken.

IMPACTS ON LAND USE AND VALUE

In addition to the impacts of transportation systems on the natural environment, there is also an impact on the uses of land and on land values. Perhaps the most obvious of these impacts is the taking of land for rights-of-way for transportation facilities, changing the land use to a transportation usage. Also, changes in the transportation level of service (and prices) in an area might induce a certain type of land use which would not have occurred otherwise, a consequence that was described briefly in Chap. 2. This has the potential effect of changing not only the spatial pattern of land uses but also through such changes, the overall quality of life in an area and, closely related, the values of land, which result from altered demands for land in various portions of a region.

Land Taking and Relocation Impacts

Of all the impacts associated with the construction of new transportation facilities, the taking of land seems to raise the most difficult problems and controversies. In principle, the taking of land for a transportation facility is like the purchase of land for any other economic activity which will replace a prior use. Thus perhaps in principle it should be no more difficult than the purchasing of a plot or plots of land for construction of a new apartment building, a

Table 13-3 Examples of Environmental Capacities of Roads†

Prototype characteristics	Current volume, ADT‡	Environmental capacity, ADT‡	Controlling environmental factor
Commercial and institutional, two lanes each way at grade	7,500	14,100	Air pollution
Commercial, institutional, and industrial mixed, two lanes each way at grade	12,500	35,700	Noise
Commercial, institutional, and residential, two lanes each way at grade	17,500	14,100	Air pollution
Commercial, institutional, and recreational, two lanes each way at grade	25,000	14,100	Air pollution
Commercial and institutional, two lanes each way, some at grade and some elevated	25,000	35,700	Noise
Commercial, institutional, and industrial mixed, two lanes each way, some at grade and some elevated	12,500	35,700	Noise
Commercial and institutional, some streets two lanes each way and some two lanes one way, at grade	12,500	14,100	Air pollution
Commercial, institutional, and industrial, some streets two lanes each way and some two lanes one way, at grade	25,000	14,100	Air pollution
Commercial and institutional, some streets two lanes each way and some three lanes one way, at grade	2,500	35,700	Noise
Commercial, institutional, and industrial, some streets two lanes each way and some three lanes one way, at grade	7,500	35,700	Noise
Commercial, institutional, and residential, some streets two lanes each way and some three lanes one way, at grade	12,500	35,700	Noise
Commercial, institutional, and recreational, some streets two lanes each way and some three lanes one way, at grade	17,500	14,100	Air pollution
Commercial and institutional, some streets two lanes each way and some four lanes one way, at grade	7,500	14,100	Air pollution
Commercial, institutional, and industrial, some streets two lanes each way and some four lanes one way, at grade	12,500	14,100	Air pollution
Commercial, institutional, and residential, some streets two lanes each way and some four lanes one way, at grade	17,500	35,700	Noise
Predominantly residential with some commercial and institutional, the streets two lanes each way at grade	15,000	13,300	Noise
Predominantly residential with some commercial and institutional (60 percent or more residential), with streets two lanes each way at grade	15,000	13,300	Noise
Residential with some industrial, the streets two lanes each way at grade	17,500	13,300	Noise

Table 13-3 Examples of Environmental Capacities of Roads *(Continued)*

Prototype characteristics	Current volume, ADT‡	Environmental capacity, ADT‡	Controlling environmental factor
Residential, some streets two lanes each way at grade and some two lanes each way elevated	17,500	14,100	Noise
Residential with commercial and institutional, some streets two lanes each way and some two lanes one way, at grade	12,500	15,500	Public safety
Residential with industrial, some streets two lanes each way and some two lanes one way, at grade	17,500	15,500	Public safety
Residential with some streets two lanes each way and some two lanes one way, at grade	25,000	15,500	Public safety
Residential with commercial and institutional, some streets two lanes each way and some three lanes one way, at grade	10,000	21,300	Public safety
Residential with industrial, some streets two lanes each way and some three lanes one way, at grade	23,200	21,300	Public safety
Residential with some streets two lanes each way and some three lanes one way, at grade	17,500	19,400	Noise
Residential and recreational, some streets two lanes each way and some three lanes one way, at grade	25,000	21,300	Public safety
Residential with most streets two lanes each way and some four lanes one way, at grade	12,500	19,400	Noise
Residential with commercial and institutional, some streets two lanes each way and some one lane each way, at grade	2,500	14,100	Noise
Residential and industrial, with some streets two lanes each way and some one lane each way	7,500	14,100	Noise
Residential with most streets two lanes each way and some one lane each way	25,000	14,100	Noise
Residential with streets two lanes and three lanes each way, at grade	12,500	13,300	Noise
Residential with some commercial and institutional, the streets two lanes each way at grade and three lanes each way elevated	12,500	21,300	Public safety
Residential with industrial, the streets two lanes each way at grade and three lanes each way elevated	17,500	21,300	Public safety

† Sharpe and Maxman (1972, p. 38).
‡ Average daily traffic, vehicles/day (sum of both directions if bidirectional).

factory, etc. However, since a transportation facility must be placed on contiguous strips of land which will route that facility in the areas where it is needed, particular parcels must be taken, rather than any parcel within a general area. In the past it has been found that many persons whose parcels must be purchased in order for a facility to be constructed in the desired location will refuse to sell their property in the open market. As a result governments in general have given transportation firms and authorities the right to purchase property at a fair market value regardless of the willingness of the seller to sell the property, in effect requiring the occupier of that land to move in exchange for payment of a fair market price. This is termed the right of eminent domain, and it has existed in most nations since the development of canals and railways in the early 19th century. Forcing people to leave whether they want to or not naturally creates a great deal of ill will and controversy. Moreover, beyond the difficulty of determining the fair market value of the land, the value of a property surely varies considerably with the personal preferences of the user. It is a further complication that many persons who are forced to relocate would not normally move even if offered substantial sums for their property; the personal costs and problems associated with relocation may far exceed the monetary compensation for the loss of their home.

Some of the most common specific problems raised by new transport facility construction in urban areas are expressed well in a recent article by Christensen and Jackson (1969, p. 1):

When expressways run through a major city, large numbers of people and many businesses are displaced. Unfortunately, the highways are frequently routed through the least desirable sections of the city, and those who are displaced are the poor, the aged, and those who are least able to take care of themselves, and there is little likelihood that many of them will use the expressway that displaces them . . . .

In theory, relocation assistance is simple. In practice, it is difficult, complicated, and time-consuming. Frequently, successful relocation depends upon solving personal problems, both financial and social, in addition to finding replacement property.

Another problem associated with transportation land taking is that the new land use, a transportation facility, has a number of characteristics which often make it an undesirable neighbor. For instance, the facility may sever many local streets and sidewalks, dividing the neighborhood into two sections. Also, the new facility may be very unsightly and carry with it increased levels of pollutants and other undesirable features, as we noted in the previous section. Finally, most transportation facilities do not pay property taxes, unlike most private land uses which may have existed on the sites before the construction of transportation facilities. Thus the town and perhaps other levels of local government face a loss of property tax revenues from the land on which the transportation facility is located. Of course, if the value of land nearby the facility increases sufficiently—the subject of later discussion—then this effect may be mitigated.

In order to overcome many of the problems associated with land taking and the relocation of existing land uses, fairly liberal provisions have recently been included in laws specifying the manner in which land can be taken for various publicly owned transportation facilities, particularly road and transit facilities. It is often possible to pay more than the fair market value for the site and any structures on it, if it is found that equivalent property is more costly than the property being taken. Also, there is almost invariably payment for the entire cost of moving and also considerable assistance in locating equivalent property and making the social adjustment to the new neighborhood. These provisions have undoubtedly gone a long way toward helping those who must be relocated for new transportation facilities to be at least as well off after the move as they were before.

There have been numerous studies of the success with which land taking and relocation have been carried out, particularly from the standpoint of those who must be relocated. Numerous problems seem to remain. In particular, it has been found that in many large cities sufficient housing is not available for those who are relocated, particularly when those being relocated are among the poor and minority groups, whose housing options are often quite limited. Also, major transportation projects may have the effect of eliminating a sizable fraction of housing of a particular type, quality, and price level, resulting in a reduction of the supply of that type of housing with no corresponding reduction in demand, which increases housing costs. Business relocation can also lead to problems, particularly for marginal and small businesses, such as family grocery stores and barber stores, that depend heavily upon a particular immediate neighborhood for their clientele. When these are relocated, they often experience considerable economic loss starting up anew elsewhere. Finally, psychological problems also abound among relocated families, particularly elderly families which find it extremely difficult to adjust to new neighborhoods and who may lose contact with old friends due to increased spatial separation and lack of mobility. With these problems, it is hardly surprising that there is considerable opposition to major new transportation facilities which require displacement of families and businesses in large numbers. However, changes in financial compensation rules to consider individual problems and relocation assistance will tend to ameliorate these difficulties.

Land Values

It is plausible that improvements in transport service in a particular area will result in an increase in the value of land in that area, provided all other conditions remain unchanged. Usually persons and businesses assume that the ease of transport to other places—often termed accessibility—of a particular parcel of land is increased as a result of the transportation system improvement, so presumably the price which potential buyers might pay for a parcel of land will be increased.

Before we examine the empirical studies of changing land values resulting from transportation system improvements it might be useful to consider for a moment more carefully the process by which land values might increase as a result of an improvement in transport. In order to do this we shall consider a highly simplified model of what is in reality a very complex process. However, this simplified representation will elucidate a number of important characteristics. We will focus upon land in use for residences, in an area in which all residential land lies essentially along a single transport line radiating out from a central business district, where all persons in the area work. Furthermore, we shall assume there is a fixed total number of households and that these households have a fixed total amount of income (per household) that they can spend on housing and on transportation to and from work. Thus it is assumed that all other expenses, such as for food, clothing, and recreation, will be constant regardless of variations in the cost or characteristics of housing or work transportation.

In Fig. 13-8, these people live along the line *OA*, shown as the horizontal axis, while the costs of transportation and housing are shown as the vertical axis. The original transportation system before the improvement has a cost for traveling any distance between the CBD and a location in this residential line indicated by the line *BC*. If the cost of maintaining a house is given by the amount *DE*, then the maximum distance persons could afford to live from the CBD is at point *F*. At any point closer to the central area, such as point *G*, the sum of household maintenance plus transportation expenditures is less than the total amount persons are willing to pay, and as a result the price of those dwelling units will be bid upward, to the level indicated by *HD*, yielding the so-called location rent paid for the privilege of residing in that particular location.

Although this is a very simplified model, it reveals in essence why land values tend to fall with increasing distance from the major center of work or business activity within an area. Of course, in practice, the value of land and homes is influenced also by variations in the quality of housing and in the quality of the environment of that housing, so in reality the rent for land does not follow precisely this reduction with increasing distance from the central area. But with these assumptions, the total area available for housing the population extends from *O* to *F*.

Now assume that a transportation improvement has been made, reducing the cost of travel to follow the line *IJ*. As a result, persons can now live

Figure 13-8 **A simplified model of the effect of transport cost reductions on the location and price of residences.**

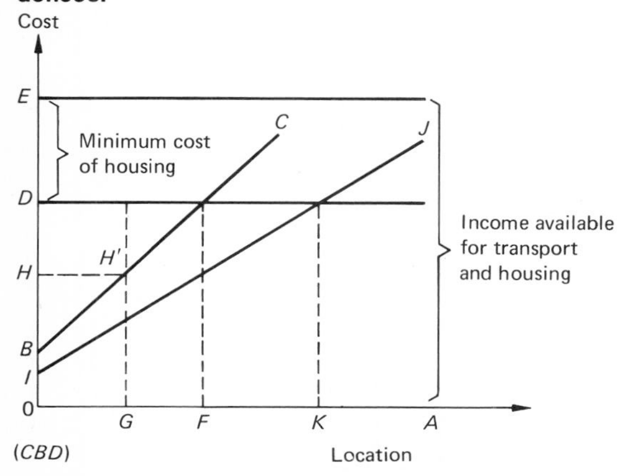

anywhere up to the distance *OK* from the center, which increases the total land available for residences. It would be expected that this would reduce the average density of residences (i.e., households per square mile), since more land is available for the fixed number of households in this case. Also, notice that the location rent which can be paid for land anywhere along this line increases, since transportation costs are reduced everywhere. However, if sufficiently more land is made available for residences, then the reduction in households per unit area in the region from *O* to *F* may result in a reduction in the price per unit area paid for land there, even though the price per household might increase. However, in the area from *F* to *K*, the area newly available for residences, the location rent paid will definitely increase, since in this idealized residence-only land use situation there was no possible use and hence no location rent paid for land beyond *F* with the original, unimproved, transport system.

This highly simplified example has revealed two important characteristics of transportation improvements. First, a reduction in the cost of transportation makes income available for other uses, which may include increased expenditures on housing. Secondly, a reduction in the cost of transportation in general will bring more land into the area which is usable for residences or for other economic activities, with the likely result that the average density of uses will decrease. Thirdly, it appears as though the value of some land will increase as a result of a transportation improvement, while the value of other land not positively affected by the transportation improvement may decrease. This may occur even if the improvement reduces transportation cost or increases accessibility for all parcels of land, because some parcels may be more positively affected than others. While a more elaborate and realistic model would elucidate these and other points more fully, this simplified example has served to illustrate some of the major effects which transport improvements typically have on land value.

There have been a number of empirical studies of changes in land values associated with transportation improvements. In particular, there have been numerous studies of the effect of new urban and rural freeways on the value of land near those facilities. Horwood et al. (1965) have summarized the effects of land-value impact studies in Dallas, Houston, Atlanta, and San Antonio. The results of this effort are summarized in Table 13-4. The changes in land value were observed for varying distances from the new freeways, band A in the table referring to the area closest to the freeway and band D being so far away as to experience almost no effect. Thus changes in band D are assumed equal to land value changes which would have occurred in A in the absence of the freeway. A distinction was made between completely unimproved or vacant land, i.e., land on which no improvement or structure exists, and land with improvements. The change in value of improved land was calculated in two ways: for the land alone (without the improvements) and for the land with the improvements. As can be seen from inspection of the table, the increase in land value in the band or area immediately adjacent to the freeway generally is many times larger than the increase in land value in areas removed from the freeway. This lends considerable plausibility to the argument that transport improvements increase the value of land and thus can provide benefits to society in this way in addition to the immediate transportation benefits.

Table 13-4 Examples of Changes in Land Value for Different Distances from New Freeways in the United States†‡

Location	Average increase over base year, percent											
	Land with improvements				Land without improvements				Unimproved (vacant) land			
	Band A	Band B	Band C	Band D	Band A	Band B	Band C	Band D	Band A	Band B	Band C	Band D
Dallas												
1941 annexation	431	100	139	106	623	123	185	130	518	383	291	166
1946 annexation	127	26	(22)	31	1027	538	...	104	1179	766	...	136
Houston												
Unadjusted	250	130	50	90	282	150	38	76				
Adjusted	245	125	(15)	44	190	96	(70)	(12)				
Atlanta												
West	99	4	11	102	...	...	...	...	197	12	53	148
East	40	18	(35)	102	...	...	...	...	247	35	(58)	148
San Antonio	251	181	71	(2)	377	264	127	30				
Overall average	206	83	28	68	500	234	70	66	535	299	95	149
Overall range	40–251	4–181	35–139	2–106	100–1027	96–538	70–185	12–130	197–1179	12–766	58–291	136–166

† Horwood et al. (1965, p. 19).
‡ Note: Band A is along the freeway, B further away, etc.

However, care must be exercised in interpreting these results because it certainly is possible that the increase in value of land near the transportation improvement may really be a transfer of value from land far removed from these improvements—land which decreases in value as a result of these improvements. It is extremely difficult to measure these potentially negative effects on land far removed from the facility, primarily because if the negative value is spread throughout an entire area except for that area served by a new facility, then the negative effect on any individual parcel would be extremely small and therefore difficult to detect. Also, it is possible that the increase in land values is simply a translation of the savings in transport cost resulting from the new facility and thus these increases in value are really other ways of measuring exactly the same beneficial effect as the reduction in travel time and other transportation cost. The extent to which the increase in land value is a transfer of reduced value from elsewhere and the extent to which it simply mirrors changes in transport costs of those persons located on land near the new facility, are unanswered questions which are being addressed by current research. They remain extremely difficult questions to answer; it may be many years before reasonably definitive answers are obtained from research on the subject.

A Transit Impact Model

A relatively simple, yet interesting, model of the impacts of a new rapid transit line on travel costs and thereby on property values has been developed by Boyce and Allen (1973). This model considers travel to and from a central business district in a single radial corridor. The residences are located throughout the corridor, and travel by automobile is modeled as direct (i.e., via a straight line) to the CBD, while travel on the new transit line is via a station along the single radial line, the station chosen presumably the one that results in the least total travel cost. The travel cost via either route—all auto or auto plus transit—is expressed in a very generalized way and can represent direct dollar expenditures, travel time, a combination of these, or a combination of these and other characteristics.

The model can be derived with reference to Fig. 13-9*a*. The cost of driving to the CBD from any particular point such as i is

$$c''_{oi} = Ad_{oi} + P_o + T_{oi} \tag{13-3}$$

where c''_{oi} = total cost of auto trip from point i to point o
A = auto operating cost per unit distance
d_{oi} = distance (straight line) from point o to point i
P_o = parking charge at o (one-half for each trip)
T_{oi} = any road toll charged from o to i

The cost of the automobile per mile, A, can be composed of a vehicle operating cost only, a travel time, or other components. For example, time and cost can be combined by using an appropriate value of time, as done in this study, where a value of time of \$2.40/h was used.

Similarly, the user cost associated with using the auto plus rapid transit route is

$$c'_{oi} = Ad_{pi} + F_{op} \tag{13-4}$$

Figure 13-9 **A model of travel cost savings from a new transit line. *(a)* The model. Rapid transit route follows x axis; x and *y* are coordinates. *(b)* Loci of equi cost savings for users of station *p*.**

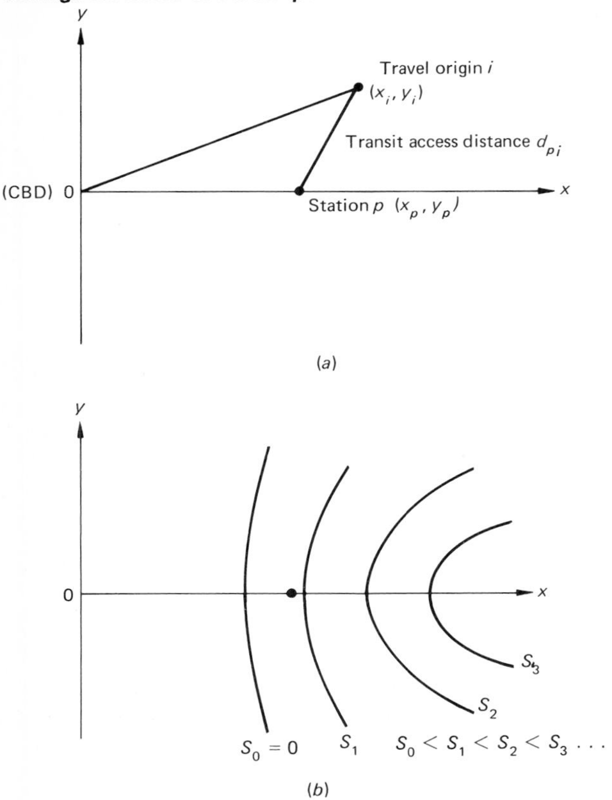

where c'_{oi} = total cost of auto and transit trip from point *i* to point *o*
d_{pi} = distance (straight line) from point *i* to station *p*
F_{op} = cost from station *p* to point *o*

Between any particular point and the CBD it is thus possible to compare the costs for using either of the two major modes and to determine which mode might be used.

The difference between the cost via automobile and the cost of rapid transit can be expressed as follows:

$$S_{oi} = Ad_{oi} + P + T - Ad_{pi} - F_{op} \tag{13-5}$$

where S_{oi} = savings from using transit route

This expression of the savings from using the new rapid transit line thus expresses the potential saving in travel costs for those who live in the area where the saving would be greater than zero. Lines of constant saving value can be determined by solving this equation for the combination of distance variables which yield a constant saving, as follows:

$$d_{oi} - d_{pi} = (S_{oi} + F_{op} - P - T)/A \tag{13-6}$$

Examples of such lines are drawn in Fig. 13-9*b*. Since the right-hand side of this equation is a constant for a given level of savings and the left-hand side is the difference between two distances, this equation results in an hyperbola form for the equal travel savings or isosavings lines. These lines bend around each station and away from the CBD, as illustrated in the figure.

By substituting in the value of savings equal to zero, we obtain the line which divides areas in which persons choosing their minimum cost mode of travel will use the rapid transit line from those areas in which they will continue to use the automobile. Again, this line is shown in Fig. 13-9*b*. The fact that with the introduction of the new rapid transit line certain areas do not experience an improvement in the best mode of travel further underscores the point made above that some areas will not benefit from a transportation improvement and therefore may experience a decline in property values if the people move from these areas to other areas.

This model was used to estimate the travel savings resulting from the introduction of a new high-speed rail rapid transit line in a corridor emanating from Philadelphia. The major conclusions of this study are (Mudge, 1974):

1 The Lindenwold line did have a positive impact on nearby residential property values in the suburban area served. While the increases in value on individual properties are small in comparison with total value of those properties (typically increases of a few hundred dollars on property valued between $15,000 and $35,000), the summation of these gains results in a very sizable increase in value.

2 The impact seems to be closely related to the travel savings.

3 In addition to spatial variation expressed by the travel savings, the impact varies with socioeconomic groups, being more apparent in lower- and middle-income areas than in higher-income areas.

4 Some evidence was found that at least a portion of the impact in the corridor with the new transit line is a transfer from nearby corridors.

Spatial Reorganization

The theoretical discussions above indicate that not only will the value of land change as the result of transportation systems, but also the spatial location of various activities will change too. The discussions above have been primarily oriented toward transportation improvements which extend the area available for use by various activities. This in general would be expected to increase the dispersion of activities and to increase the amount or distance of travel.

However, transportation improvements need not necessarily be directed toward reducing the cost of travel throughout an entire region or reducing the cost of long-distance travel. Rather, some improvements may be directed toward improving shorter-distance travel or travel within a particular area. For example, instead of radial freeways or high-speed transit lines designed to reduce travel time for long-distance commuters, improvements may be directed toward the arterial street system or toward building a dense network of local transit routes in a more limited area. Such improvements would be expected to make that area more attractive and result in at least some shift of growth from other areas toward that particular central area. Such a policy of improving central city transit might, for example, result in a diminution of the tendency toward urban sprawl, which seems to be the result of more typical road and transit improvement policies.

Also, transport improvements might be carefully selected to encourage development within particular areas within a region. For example, new highways or transit facilities might be constructed with access only at particular locations, thereby encouraging development around those points of access and, at least in relative terms, discouraging location further away. This is particularly true in the case of transit facilities designed to encourage walking access rather than auto access, because the distance over which people can walk is, of course, quite limited. It seems as though auto access is generally so rapid and inexpensive that a household many miles from a transit station with parking can reap almost the same benefits from the new facility that a household within walking distance of the facility can. Of course, if parking charges were substantially increased or the cost of automobile operation were to increase substantially as a result of higher user taxes, then location within walking distance of a transit line would be encouraged.

Lest one be left with the impression that transportation system changes are in general a very powerful tool for shaping urban form, it is well to point out that there are many other factors which influence the spatial pattern of development of a metropolitan area. Obviously the attractiveness of various sites for residential and other uses, the quality of schools, the adequacy of sewers and electric power and other utilities, and zoning restrictions, and the difficulties in getting them changed, will all affect where new development occurs. Travel throughout most metropolitan areas in developed nations is fairly easy at the present time, speeds typically being fairly high in all but congested inner cities and costs apparently being very low relative to typical incomes. As a result, minor differences in locations due to differences in travel costs or time probably will have relatively little effect on the relative attractiveness of locations for new development.

In fact, a recently completed study of the land development effect of the $1.7 billion San Francisco Bay Area Rapid Transit System has concluded that so far it has had very little effect on the location of residences and other activities, although since the line has been open only a few years it may be far too early to detect such effects (Webber, 1976). However, similar observations for new rapid transit lines and new freeways cutting through already developed areas do lend caution to the expectation that such new facilities and the associated changes in travel costs will substantially influence land development patterns.

On the other hand, it may be true that changes in the transportation system—perhaps both in new facilities which reduce travel time and in prices which create additional accessibility differences—coupled with other metropolitan area policies which influence land development, such as zoning and land taxation, have an important impact on land development. Much current research on these issues suggests that transportation may be a powerful control on the spatial form of metropolitan development (e.g., Putman, 1976).

Regional Development Effects

In the preceding section we were primarily concerned with the reorganization of land uses within an urban area, that is, with the redistribution of a fixed population or fixed total of other activities within an area. In addition to that reorganization effect, improved transport might increase the total level of economic activity within a region.

In theory, reducing the true cost of transport will result in an increase in the total amount of goods and services available in a society. The reason for this is savings in transportation costs can then be used to produce more of other goods and services. Alternatively, and more likely, consumers would choose to purchase some additional transport and some additional amounts of other goods and services, the change in these amounts being chosen such that the maximum value to that society is achieved. In any case, clearly consumers would choose to change the amount of transport consumed only if they felt they were better off in terms of the total collection of goods and services available for consumption, and therefore it is likely that with a reduction in the cost of transport, the total real income of any society will increase.

Numerous case studies of transport improvements in developing nations support the conjectures above regarding the beneficial effects of transport improvements. Most of these relate to situations where the new transportation facilities or services permitted the sale of a natural resource to other nations, thereby increasing the monetary income of that nation and enabling it to purchase items which it previously could not. Also, the new transport can make available raw materials and resources which can be processed directly within the country and be made available to its own citizens, thereby increasing the range and amount of goods available for consumption.

Tools for estimating the likely effects of different transportation improvements within these contexts are only now being developed. These tend primarily to be models of national economies, with only one relatively small but nevertheless important component being the transportation system. Small reductions in transport costs make available monies for investment in other sectors of the economy, which possibly lead to long-term growth. Also, extensions of transport into hitherto undeveloped areas can make available natural resources which can be used by that society to improve its economy, again with long-term and possibly very substantial effects on economic growth and income. Because of the importance of transport in economic development, it is likely that models for estimating the effects of transportation improvements within national contexts will be developed in the near future.

SUMMARY

1 Transportation systems, like most other productive processes and sectors of the economy, produce many unintended by-products in addition to the primary product, which is transportation. Some of these unintended products or effects are beneficial, while others are undesirable, and in many cases the attitude toward the effect depends upon the viewpoint taken.

2 One important class of these effects is impacts on the natural environment. These include the emission of noise from transportation facilities, pollution

emissions from vehicles, pollutants such as oil and grease which contaminate ground water, and vibration. The first two are particularly troublesome, but they are reasonably well understood, and changes in the facilities, vehicles, or operations plan (such as reductions in speed) can be effected to reduce these negative environmental impacts. Also troublesome for the natural environment—especially wild animals—may be the barrier effect of new land transportation facilities.

3 Impacts on the manmade environment include the taking of land for facilities, the associated removal of human activities formerly occurring on those facilities, and possible negative or positive effects on the quality of life and property values in locations near transportation facilities. The increase in pollution near a facility degrades the environment, while any increase in ease of travel or accessibility resulting from proximity to the facility improves the quality of that area. The net effect depends upon the strength of these two tendencies and the activity engaged in.

4 Numerous studies have been undertaken on the effect of new transportation facilities on the value of nearby land, and by implication therefore on the overall change in quality of that environment for various human activities. It is difficult to generalize from these studies, although most tend to reveal an increase in the value of land near new transport facilities such as freeways and rapid transit lines. However, it must be borne in mind that this increase in value may be largely due to attractiveness of that land for new types of uses, and the activities previously on those sites may be negatively affected by close proximity to the new facility.

PROBLEMS

13-1
Estimate the noise levels at the residences near the freeway interchange in Fig. 13-7 considering daytime nonpeak freeway traffic.

13-2
Do the same (as in Prob. 13-1) for daytime peak period traffic on the interchange ramps, considering it like stop-and-go city traffic. Also calculate the noise emitted from vehicles on the state highway.

13-3

a Graph the noise from vehicles on a freeway as a function of the volume of traffic on the freeway. Calculate the noise considering the effect of increasing volume on average speeds, using the speed-volume relationships of Chap. 5. The freeway consists of four lanes (two in each direction) with a 60 mi/h design speed, 6 percent trucks, 2 percent buses, flat terrain, a design adjustment factor of 0.93, a PHF of 0.83, and equal volumes in both directions. Estimate the noise 200 ft from the centerline. The pavement's lanes are 12 ft wide, with a 40-ft-wide median area between the edges of the traffic lane pavements in the two directions. Consider volumes up to a volume-to-capacity ratio of unity.

b What does the resulting curve suggest about ways to control noise from freeways?

13-4

Consider Fig. 13-9 as a map of a freeway rather than a rapid transit line, with the access points at the station locations.

a Compare the total pollution emissions per vehicle trip for each of the three pollutants via the freeway and via arterial streets (straight-line route) for vehicles traveling between the CBD and a point located 3 mi out in the x direction (along the freeway) and 2 mi in the y direction. Calculate this using the data on average vehicle emissions for 1972 in Table 13-2. Assume a freeway speed of 50 mi/h and an arterial speed of 20 mi/h. The freeway on-ramp is located 3 mi from the CBD.

b Compare the results for 1972 and 1990 vehicles. What is the percentage reduction in each type of pollutant?

13-5

Using the travel cost model for trips to or from the CBD presented in Eq. (13-4), determine the boundary between use of two stations on a rapid transit line (or interchanges on a freeway). The stations are located 3 and 6 mi from the CBD along the x axis. The values of the model parameters are

A = \$0.10/vehicle-mi
F = \$2.00 near station
F = \$2.20 far station

Draw the isosavings lines and the boundary.

13-6

Compare the savings from building a rapid transit line with stations every 3 mi with those for a line with stations every mile. The average speed of the former is 45 mi/h, the speed of the latter is 25 mi/h. Other characteristics are a value of time of \$2.50/h on the transit line, \$3.00/h driving, an auto operating cost of 5 cents/mi, an average auto occupancy of 1.5, a transit fare of \$0.25, a parking charge of \$2.00/vehicle-day, and a toll (each way) of 50 cents. Draw the isosavings lines for the area extending from the CBD for 5 mi along the (straight) transit line and for 3 mi on either side of it. What areas benefit the most from these designs?

REFERENCES

Beaton, John L., and Lewis Bourget (1968), Can Noise Radiation from Highways Be Reduced by Design?, *Highway Research Record,* No. 232, pp. 1–8.

Bolt, Beranek, and Newman (1967). "Noise Environment of Urban and Suburban Areas," Report to U.S. Department of Housing and Urban Development. Los Angeles, CA.

Boyce, David E., and W. Bruce Allen (1973), "Impact of Rapid Transit on Suburban Residential Development," Research Report, Regional Science Department, University of Pennsylvania, Philadelphia, PA.

Boyce, D. E., W. B. Allen, and F. Tang (1976), Impact of Rapid Transit on Residential-Property Sales Prices, in M. Chatterji, ed. *Space, Location and Regional Development,* Pion, London, pp. 145–153.

Buchanan, Colin et al. (1963), "Traffic in Towns," Penguin, Harmondsworth, England.

Christensen, A. G., and A. N. Jackson (1969), Problems of Relocation in a Major City: Activities and Achievements in Baltimore, Maryland, *Highway Research Record,* No. 277, pp. 1–8.

Galloway, William J., Weldon D. Clark, and Jean S. Kerrick (1969), "Highway Noise: Measurement, Simulation, and Mixed Reactions," National Cooperative Highway Research Program Report No. 78, Highway Research Board, Washington, DC.

Garrison, William L., Brian J. L. Berry, Duane F. Marble, John D. Nystuen, and Richard L. Morrill (1959), "Studies of Highway Development and Geographic Change," University of Washington Press, Seattle, WA.

Horwood, Edgar M. et al. (1965), "Community Consequences of Highway Improvements," National Cooperative Highway Research Program Report No. 18, Highway Research Board, Washington, DC.

Mudge, Richard R. (1974), "The Impact of Transportation Savings on Suburban Residential Values," Paper No. P-5259, Rand, New York.

Putman, Stephen H. (1976), Preliminary Results from an Integrated Transportation and Land Use Models Package, *Transportation,* **3**(3): 193–224.

Sanders, D. B., T. A. Reynan, and K. Bhatt (1974), "Characteristics of Urban Transportation Systems," U.S. Department of Transportation, Washington, DC.

Sharpe, Carl T., and Robert J. Maxman (1972), A Methodology for Computation of the Environmental Capacity of Roadway Networks, *Highway Research Record,* No. 394, pp. 33–40.

Thomas, Edwin N., and Joseph L. Schofer (1970), "Strategies for the Evaluation of Alternative Transportation Plans," National Cooperative Highway Research Program Report No. 96, Highway Research Board, Washington, DC.

Turner, Roy E. (1973), "TRANS Technical Notes: Air Pollution Amounts," U.S. Department of Transportation, Federal Highway Administration, Washington, DC.

U.S. Department of Transportation, Federal Highway Administration (1974), "Social and Economic Effects of Highways," Washington, DC.

U.S. Department of Transportation, Office of the Secretary (1972), "Transportation Noise and Its Control," Washington, DC.

Webber, Melvin M. (1976), The BART Experience: What Have We Learned? *The Public Interest,* **45:** 79–108.

Winfrey, Robley (1969), "Economic Analysis for Highways," International Textbook, Scranton, PA.

Wingo, Lowdon, Jr. (1961), "Transportation and Urban Land," Resources for the Future, Washington, DC.

Engineering Design, Planning, and Management Applications

4

Decision Making in Transportation Management and Planning

14.

With this chapter we shall begin the discussion of the application of the principles and methods presented in earlier chapters to the various types of problems facing transportation engineers and planners—problems in long- and short-range planning and in management and operations. In order to do this it is essential that we have an understanding of how transportation systems are evaluated, that is, how measures of value or worth are placed on the various alternatives. This permits the evaluation of existing systems for possible problems and the evaluation of various alternative solutions for those problems. Also, it is essential that the processes by which decisions in transportation are reached be understood, decisions involving the choice of the best alternative action, whether it be a plan for the future, a design, or an operational change in the existing system. Usually transportation engineers and planners work in firms or public agencies in which the ultimate responsibility for decision making lies with other individuals, generally elected public officials or the officers of private firms. This raises questions as to the proper role of engineers and planners in this decision-making context and also questions as to how the analysis tools presented in preceding chapters and the evaluation procedures to be presented in this chapter will be used.

Preceding chapters have described characteristics of the transportation system, its major components, their characteristics, and how the system functions as a whole. Also covered was the relationship between the transportation system and its environment, one of satisfying the need for transport and also of producing various environmental impacts and consuming valuable resources. These characteristics of transportation systems and their environments provide an essential basis for the analysis of the various types of transportation problems, but before we attempt to present methods for evaluation and decision making, we summarize certain characteristics of the transportation system and its relationships to its environment which are of great importance in determining appropriate approaches to evaluation and decision making.

TRANSPORTATION DECISION MAKING

Transportation in modern society is different in many ways from most other productive activities. As a result of these differences, governmental concern for transportation is substantial, leading in some cases to direct regulation of transportation services and prices, and in other cases to government ownership and often operation of portions of the transportation system. We review some of the characteristics which have led to such substantial governmental involvement and also discuss certain peculiarities of governmental decision making as opposed to private decision making.

Characteristics of Transportation Problems

One of the most important characteristics of transportation activities which tends to set it apart from other economic activities is that the interrelationship between the transportation system and the entire socioeconomic system which it serves is very strong. On one hand, transportation is an integral part of many social and economic activities, and as a result it is essential that transportation service of adequate quality and capacity be provided. The quality and price of transportation service can substantially affect the spatial pattern of human activities, and by affecting this pattern as well as the quantity and quality of various economic goods to be produced and consumed, can radically influence the material quality of life in a society. Similar effects occur with respect to social phenomena, also. As a result of these considerations, the effects of impacts of changes in the transportation system can be very widespread, and even if each individual effect is rather small, the total combination can be quite large, especially over an extended period of time. Because of this potentially profound influence of transportation on the quality of life in a society, governments act in various ways to ensure that the transportation service provided is not only adequate to meet immediate needs but will also help to guide the development of that society along desirable lines in the future.

A corollary of these considerations is that objectives for transportation may appear to be quite far removed from transportation itself. For example, in developing nations, the choice of a particular transportation plan may depend more on its effect on the rate of growth of gross national product than on transportation level of service alone. And in developed nations considerations of environmental impacts and social effects are very important, often being the primary consideration in selecting among alternative plans. Often these considerations lead to adoption of plans which would not have been selected on transportation considerations alone.

Another important characteristic of the transportation system is that the range of alternative ways of changing that system is extremely great. On one hand, there are many alternatives with respect to fixed plant improvements, including location, type of fixed plant (e.g., mode or transport technology), and design capacity and level of service. There also is a wide variety of options with respect to how the system is operated, including such aspects as schedules, pricing, and routing of traffic. Also, there are many options with respect to ownership of the system and with respect to the allocation of authority for making operating and pricing decisions, the range including full government ownership and operation, ownership and operation by private carriers, combinations of the two, and varying degrees of regulation of private carriers by the government. The richness of the range of alternatives leads to substantial complexity in the analysis of various solutions to problems, because it usually means that a very large number of alternatives must be considered before it can be reasonably presumed that the best alternative (or one close to the best) has been identified.

Also, certain of these alternatives have particular characteristics which make for further difficulties in decision making. Fixed plant improvements tend to take very long times to implement: it typically takes a decade or more to make any major addition to the system. This requires the planners to look ahead for a decade and further, in order to evaluate a particular fixed plant option. Also, many of the alternatives seem to be very inflexible, in that once implemented, they are very difficult to modify so as to provide a transportation service different from what was originally intended, even though a change in the service may be needed to meet altered travel demands. Perhaps the most obvious example of this is the inability to move an element of the fixed plant if traffic should make this desirable. Instead, an entirely new facility must be built. Fortunately, in some instances an innovative plan of operations can be instituted in order to adapt the fixed plant to the changed usage.

As mentioned above, much of the transportation system is owned and operated within the public sector, either directly by governments or by agencies which are wholly owned arms of various levels of government (such as port authorities and transit authorities). In this context it is clear that it is appropriate for such organizations to attempt to consider all impacts of any change in the transportation system, not just those effects which would be directly felt by that particular governmental body or agency. This would include impacts on not only the system owner and operator, but also effects on users and nonusers of the system, as well as on the entire region which the system serves. The consideration of impacts must also extend not only to benefits or gains from the project but also to the full range of costs, those borne directly through monetary expenditures and also environmental and so-called social costs which might not appear in any market transactions. Although comprehensiveness in the consideration of effects of a particular action obviously cannot be achieved completely, it is nevertheless a principle which must guide public decision making.

Another often difficult characteristic of public decision making is that there are alternative uses of government funds besides the investment in one or another transportation project. This means that the benefits and costs of the various transportation projects must be identified in such a manner that makes it possible to compare them with the benefits and costs from other government investments of the same funds, such as in schools, social programs, or conceivably even in reduction of taxes. Of course, in many situations taxes or user charge revenues from particular facilities (such as highway user charges) are specially earmarked for particular types of transportation projects, and the competition of transportation expenditures with other types of expenditures does not arise directly. Nevertheless, in principle such a comparison must be made at the time decisions are made to implement or to restrict expenditure of the funds to particular categories of projects, or otherwise public monies may be expended on projects which do not generate the maximum public benefit.

Despite heavy government involvement in transportation, a very large portion of transportation facilities and services is provided within the private sector. In North America, carriers are privately owned, as are most of the fixed facilities with the exception of roads, ports, and airports. Even in nations where all major facilities and common carriers are owned by the government, much of the actual movement is carried out by private firms or individuals, primarily in private trucks and automobiles. In the private sector, the posture of the firm (or individual) with respect to the range of impacts of transportation decisions which should be considered is not so clear-cut. In some cases, the firms are left to choose their own criteria. In such cases decision making is probably based on a combination of profit, some regard for general societal impact, considerations of balanced growth of the firm, and perhaps some idiosyncratic characteristics of the firm and its management. Many firms do consider the external (to the firm) effects of management decisions, including impacts on employment, the physical environment, and other impacts, which are only imperfectly if at all connected with the firm's profit-making potential. In addition, much of the private sector of the transportation system is regulated with respect to some or all of the following characteristics by government agencies: safety, starting or terminating the business, location of points served and connections between them, level of service (e.g., frequency of trips), and prices. Ideally, the regulatory agencies would take into account the broad range of impacts resulting from any significant change in these characteristics of the carriers.

These characteristics of transportation problems lead to three basic properties of transportation evaluation and decision making which summarize the sources of the tremendous practical difficulties for any methodology in this area. One characteristic is that the problems and analysis of solutions tend to be extremely complex, primarily because of the wide range of alternatives and the large number of effects—extending far into the future—which must be considered. Secondly, there is often no single criterion by which the best alternative might be measured, simply because the alternatives usually have numerous effects, some of which are perceived as beneficial by some groups and as undesirable by others. These considerations raise questions as to how to balance the gains and the costs of various groups in order to reach a decision. Finally, since changes in the system may take many years to implement, the evaluation must be made with due consideration of future gains and costs and the preferences of persons at that time, which may be quite different than those held now.

Structure of Transportation Decision Making

Perhaps the discussion above has indicated the reason why so much attention was paid to presenting the overall structure of the problem-solving process in Chap. 1. This process is quite useful in structuring the approach to decision making in transportation planning, design, and management. The size and complexity of the problems typically dealt with requires an explicit structure in order that the various elements of decision making be identified so that no

Figure 14-1 **The problem-solving process showing the uses of models of technology, demand, and impacts covered in Parts 2 and 3.**

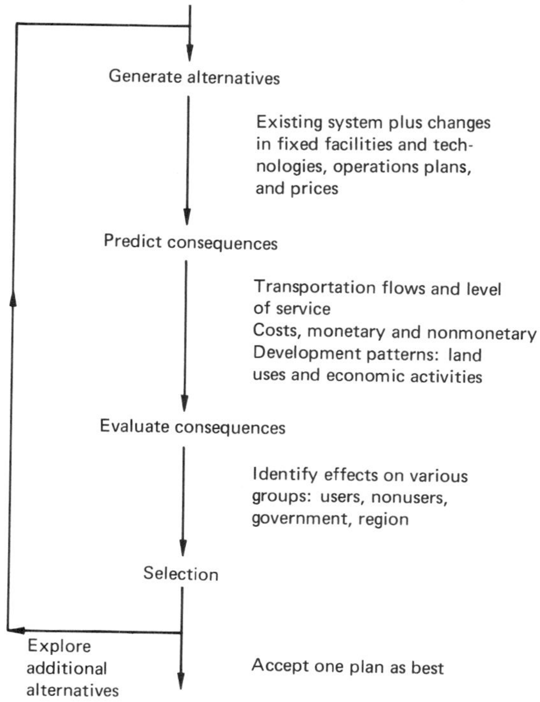

elements which are important are forgotten. Although that process was covered in some detail in Chap. 1, we review it briefly and identify the roles that the various principles and methods discussed in preceding chapters might have in this process.

Figure 14-1 presents the problem-solving process with notes at the various stages indicating the typical types of information which might be contained in a rather large-scale systems planning problem. As can be inferred from this figure, Part 2 of this book, concerned primarily with the technology of transportation, forms the basis for two parts of the process: It provides one very important class of alternatives for improving the transportation system and also provides the basis for predicting some of the effects of potential changes in the system, particularly effects with respect to the quality of transportation service and the costs of providing that service. Part 3, which was concerned with the relationships between the transportation system and its environment, including both the demand for transportation and the impacts of various transportation systems changes on the natural and human environment, also provides information in these two areas. One common means of dealing with transportation problems is to attempt to rearrange the pattern of land uses to change the demand for transportation. Also, of course, an important effect of any change in the transportation system is the effect on demand and hence on equilibrium flows and on the long-term effects on land use patterns.

However, as can be seen from this figure, we have yet to describe methods by which the existing transportation system and its relationships to its environment can be evaluated or methods by which future changes in the system might be evaluated. Nor have we discussed specific decision-making processes which lead to the selection of a particular alternative, whether the decision making be in the context of long-range planning, in that of designing facilities and vehicles, or in the sphere of operations and management. This is the subject of the next two sections.

There are many dozen different ways for evaluating alternative public projects. Most of these have been developed in response to the need to have methods in the public sector which would give explicit recognition of the many different aspects of public projects which must be considered and the different effects which these projects can have on different groups. However, after discussing certain requirements for any evaluation procedure, we shall describe a general type of evaluation procedure representative of many currently used in transportation. Also, we shall present one particular evaluation procedure which is based upon economic theory and which is commonly used in Western Europe and Great Britain to evaluate transportation and other public projects and which is often used as one component of evaluation in North America.

Desired Characteristics of Evaluation Methods

Many different specific requirements are imposed upon evaluation methods by the characteristics of transportation projects and the contexts in which they are evaluated. One of the most important and desirable characteristics of an evaluation method is that it permit presentation of information on the differential impacts of any transportation change on various groups. Also, it is important that a method to be able to accommodate many different measures of impact, since the scales or dimensions on which various impacts are measured will generally be quite different. For example, the effects of an improvement in user costs will probably be measured in time and monetary units, while the effect on air pollution would typically be in pounds of particles emitted per day or perhaps the number of days in a year in which various air pollution standards are exceeded. Also, some of the impacts undoubtedly are not quantifiable, and the method should be able to accommodate nonquantitative information. The latter might consist of photographs or sketches for aesthetic considerations and verbal descriptions of impacts on social patterns, etc.

Also, the method should be adaptable to different situations, allowing consideration of different impacts and measures in different applications. The impacts which are important for one application might be quite different from those which are important in another, and the method should reflect this. In fact, in one application some dimensions of impact which are considered in another problem context might be completely ignored. Also, the degree to which various impacts might be described in detail will vary from problem to problem and perhaps even within different stages of analysis for the same problem.

MULTIPLE-OBJECTIVE EVALUATION AND SELECTION METHODS

In the past decade or so a number of methods for evaluation have been developed which largely meet the above requirements. Although the methods differ, they share a common basic approach, which will be presented below.

A Generalized Method

The basic approach is to present information on the impacts of the various alternatives in a manner which conveys the extent of each impact identified in the analysis. The most straightforward manner in which to do this is to present the information in the form of a matrix. The columns of this matrix represent the various alternatives, and the rows of the matrix, the various impacts. Each cell, corresponding to a particular impact and particular alternative, is the location of information on the extent of that impact for that particular alternative. The various methods seem to differ in the manner in which they organize information on the impact and the particular types of classifications used.

Figure 14-2 presents one example of this type of matrix. Here the impacts are organized according to particular groups, such as users, nonusers, the region, etc., and presumably are for a single future year. Within each group, the various types of impacts on that group are identified, and then within

Figure 14-2 Hypothetical example of an evaluation matrix.

	Alternatives			
Affected Groups and Impacts	1 Extensive Freeway System (Plan A)	2 Extensive Rapid Transit System	3 Extensive Rapid Transit and Freeway System	4 Extensive Freeway System (Plan B)
Users travel time, person-h $\times 10^6$/day				
Suburbanites				
General	10.5	13.7	9.9	12.5
To CBD	2.6	2.4	2.4	3.2
City residents				
General	4.2	3.8	3.6	4.8
Transit "captives"	1.8	1.2	1.2	1.8
Goods distributors, vehicle-h	1.7	2.0	2.0	2.1
Nonusers				
Land consumed, acres	1895	27	1897	2275
Homes, no.	2200	4	2200	2350
Others, no.	376	12	379	425
Air quality (days/yr below standard)				
CBD	16	2	4	17
Elsewhere	0	0	0	0
System owner-operator				
Road capital costs, $\$10^6$	1950	25	1975	2560
Transit capital costs, $\$10^6$	0	860	860	0
etc.	etc.	etc.	etc.	etc.

each of these impact types individual subgroups are identified in order to reveal variations in the impact within each group. For example, within the category of transportation system users, user costs such as travel time and dollar expenditures are identified. The total travel time experienced by various subgroups within the region are then identified under that category, and similarly the costs experienced by various subgroups are presented. As can be seen from this matrix, this method of presentation provides information on the time of various groups of travelers consumed by their making trips on each of the various alternative proposed transportation systems.

The advantage of this type of presentation is that it shows the extent to which various groups will benefit from various alternatives and where the costs are being borne. Alternative 3, for example, generally seems to have characteristics at least as good and in many cases better than the other alternatives with respect to the average travel times of travelers for various purposes. However, it requires much more expensive facility construction, resulting in the loss of more homes and land than other alternatives, and it also results in somewhat more travel. As a result, the total resources expended by society on transportation will be greater, both in dollar terms and also in terms of degradation of air quality and other amenities. Alternative 4, with a generally higher level of prices charged for transportation services throughout the region, experiences less travel in total, but seems to maintain this high quality. Also, the lessened amount of travel permits some reduction in resources used and certainly improves environmental quality. It also is less of a burden on general taxes to construct and maintain the system. However, deciding between alternative 3 and alternative 4 obviously raises a number of important social questions regarding the extent to which travel should be encouraged by making transport inexpensive (in its broadest sense) to the consumer, in contrast to discouraging travel in order to maintain a high level of environmental quality and to minimize energy consumption.

This example has revealed an important characteristic of this general type of evaluation procedure, namely, the procedures merely present evaluative information on the various alternatives, but they do not in general lead to the specification of a particular alternative as best. Instead, the selection of a best alternative is necessarily left as another state in the process. The selection of the alternative plan or action to be implemented must consider the consequences for all affected groups, balancing the interests of all of them. This is essentially a societal or political choice in all but the most limited cases.

It should be noted that there are many ways of organizing this information. Another common approach is to identify and use a number of specific objectives which transportation system improvements should attempt to achieve. For example, one objective might be to minimize users' travel time, another to minimize users' travel costs, and the third to minimize land consumed for transport facilities. Another approach would be to organize the information around particular subregions of the area in which the improvements are being considered, such subregions perhaps being municipalities or other natural or political units. All of these different approaches lead to some type of matrix presentation of information. The most common varieties of matrix organization are presented in Fig. 14-3.

Another important aspect of the evaluation of transportation systems is the change in the nature of consequences with the passage of time. Such varia-

Figure 14-3 **Alternative forms of organization of evaluation matrices.**

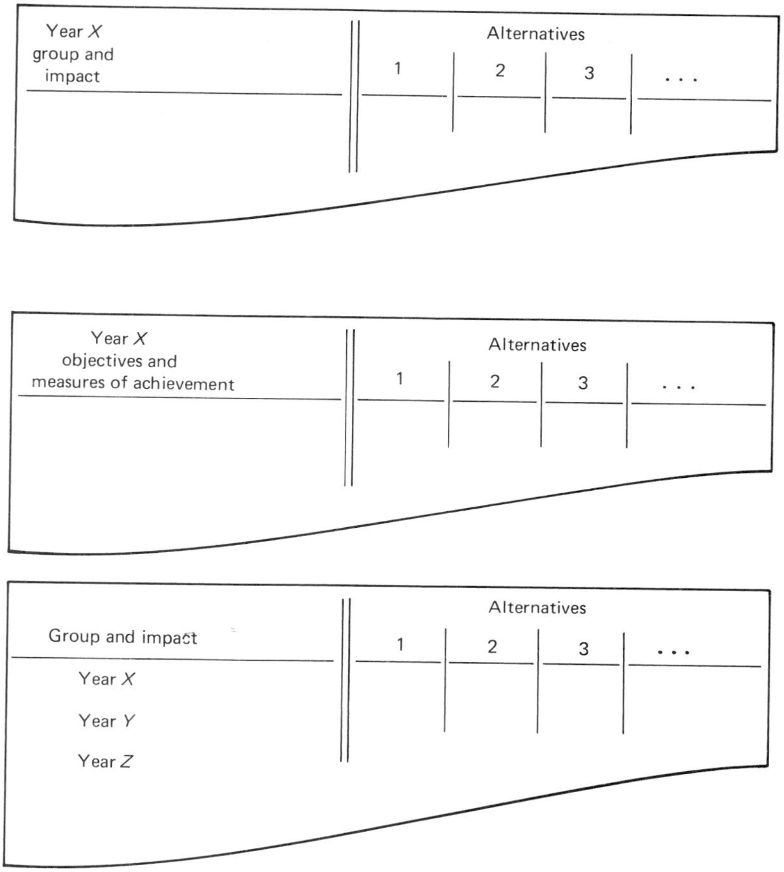

Year *X* group and impact	Alternatives			
	1	2	3	. . .

Year *X* objectives and measures of achievement	Alternatives			
	1	2	3	. . .

Group and impact	Alternatives			
	1	2	3	. . .
Year *X*				
Year *Y*				
Year *Z*				

tions in impacts are all too often ignored, but they can be very significant and important. The effects of a transportation project can be quite negative during construction, when system performance may be degraded and when negative environmental impact due to noise and dirt can be very substantial. But after the facility is opened, that same area may be very much enhanced. Treatment of changes over time are easily accomplished in this matrix presentation. Different matrices can be developed for different years, or a single matrix might contain information on the impacts in various years throughout the life of a plan or facility.

Some idea of the wide variety of types of impacts to be considered in any evaluation of large-scale system changes is presented in Fig. 14-4. This matrix contains a list of objectives for transportation improvements developed for Chicago for evaluating both system plans and individual projects in that metropolitan region. As can be seen from this list, considerable concern is expressed not only for user impacts or objectives and cost-related items, but

Figure 14-4 **Objectives and measures used to evaluate alternative transportation plans in Chicago†**

Objective	Unit of Measure	Existing System
1 Support the concept of high accessibility corridors	no. of HAC	4.8
2 Maintain the relative accessibility of the Chicago CBD so that it is preserved as the focus of the region	no. of relative accessibility	6.22
3 Encourage the development of regional economic centers within the development corridors and high-intensity development areas served by transportation networks	no. of economic centers served	44.2
4 Decrease open space of the 1995 Regional Plan that would be consumed by transportation	no. of cells traversed	0
5 Support development corridors and high intensity development areas	no. of relative accessibility in dev. corridors	56.66
6 Reduce door-to-door travel time	min	38.0
7 Reduce fatalities of transportation system	no. of fatalities	2.18
8 Reduce nonfatal accidents on transportation system	no. of accidents	1,674.2
9 Increase flexibility in choice of mode	no. of people	2,687,852
10 Increase accessibility of work opportunities for residents of economically deprived areas	no. of manufacturing jobs accessible	9.75
11 Increase the opportunity to utilize the transportation system	no. of people	8,405,000
12 Decrease the monetary user costs of trips	no. of $/trip	0.799
13 Decrease public costs of transportation	no. of $	727,614,839
14 Decrease carbon monoxide	tons/day	481
15 Decrease hydrocarbons	tons/day	67
16 Decrease nitrogen oxides	tons/day	200
17 Decrease sulphur dioxide	tons/day	12
18 Decrease particulates	tons/day	5
19 Decrease energy resources consumed for transportation	BTUs	$15{,}510 \times 10^8$
20 Decrease dislocation of residences	no. of cells traversed	0
21 Decrease dislocation of businesses and services	no. of cells traversed	0

† Hoffman and Goldsmith (1974, p. 414).
‡ Reference point is set at 10 percent above the maximum performance achieved by any network for each objective.

also nonuser impacts which would affect the quality of life in particular areas. Also shown in this matrix are the specific measures used to evaluate the degree of achievement of each of the objectives, and the corresponding numerical values for six possible transportation systems in 1995—the "do-nothing" alternative (the existing system) and five alternative plans.

The level of achievement of objectives by an alternative is often termed its effectiveness. Thus alternative *B* in Fig. 14-4 is the most effective alternative in meeting goal 1, supporting corridors of high accessibility, or ease of travel to important activity centers. On the other hand, this alternative is not particularly effective in achieving some of the other objectives, such as

Interim Plan	Alternative A	Alternative B	Alternative C	Alternative D	Reference Point‡
8.4	7.5	10.7	5	8.5	11.8
6.00	5.59	5.00	4.36	5.53	6.84
42.5	57.4	60.9	67.9	61.9	74.7
25,307	10,080	17,493	3,787	12,953	27,837
56.70	54.10	52.14	53.26	53.36	62.37
29.9	33.6	31.9	36.5	33.1	42.8
2.11	2.16	2.14	2.18	2.16	2.40
1,485.2	1,579.2	1,550.9	1,637.7	1,638.5	1,841.6
2,970,652	3,973,843	3,128,803	3,583,809	3,429,928	3,942,190
12.35	11.7	11.28	13.18	14.50	15.95
8,452,808	8,488,874	8,551,229	8,762,775	8,788,895	9,667,985
0.809	0.799	0.799	0.796	0.793	0.890
1,310,540,603	984,064,494	1,154,482,857	885,127,207	974,693,207	1,441,594,663
479	469	466	540	467	594
67	65	65	66	63	74
210	188	204	201	200	231
9	6	5	5	4	13
4	4	3	3	3	6
15,730 $\times 10^8$	15,540 $\times 10^8$	15,430 $\times 10^8$	15,350 $\times 10^8$	15,360 $\times 10^8$	17,303 $\times 10^8$
12,693	4,347	9,280	1,387	4,080	13,963
8,380	3,627	6,213	773	2,853	9,152

cost (13), where it is the second most expensive, and dislocation of residences and businesses for new highway and transit facilities (20 and 21).

Distinct from the concept of effectiveness is the concept of efficiency, which expresses the value of the return from an action relative to the resources required to implement that action and hence obtain that return. Measures of efficiency are therefore usually expressed as a ratio, typically a ratio of value or benefit to resources or costs. Such ratios could be constructed from the information presented in Fig. 14-4—for example, the ratio of high accessibility corridors supported to the number of cells (zones) of dislocated residences.

However, such efficiency ratios must be used with care. While one may tend to select the alternative with the largest ratio, one must remember that any such ratio considers only two of the many objectives which are to be considered in the problem. Furthermore, the alternative with the largest ratio may not be the most desirable. For example, by the efficiency measure of number of high accessibility corridors supported to the number of residential cells traversed by new facilities, the existing system has the largest value and hence is the preferred alternative. But alternative *B* supports over twice as many such corridors, and the additional "cost" in terms of the residential land taken may be justified by the gain. Because of these problems of interpretation, ratio measures of efficiency are most widely used in situations where the returns and the costs are all-inclusive measures, as when all returns and all costs can be expressed in monetary units and hence total returns and total costs can be used in the ratio. These efficiency measures in monetary units will be described in a later section.

Another important and perhaps surprising characteristic of many public sector evaluations of transportation improvements is that often what are normally thought of as costs and hence as negatively valued characteristics to be minimized, are in reality thought of neutrally or even as benefits. Specifically, many transportation projects in the jurisdiction of one level of government (e.g., city or state) are funded heavily by higher levels of government, with as much as 90 percent of the funds coming from that higher level provided the remainder is generated at the local level. The expenditure of large sums within a particular area will obviously create a large number of jobs, and then those workers in spending that money create additional expenditures and jobs throughout the economy of that area. This expenditure of money which comes from the higher-level government will thereby help stimulate the economy of the local area. As a result, cities and regions often deliberately try to arrange transportation projects in such a manner that they fully expend the available funds from other levels of government so that the beneficial effects on the local economy of these expenditures is maximized. While this is certainly a perversion of the usual function of costs in a society, it nevertheless is a condition given by the earmarking of large amounts of money for specific types of projects—whether needed or not—and has become a very important consideration in local area decisions. It seems to be a very unfortunate byproduct of a system of allocating government funds according to particular project categories without specific regard to the need for expenditures in those areas relative to expenditures in other areas and will undoubtedly lead to overexpenditures in some areas and neglect in others.

A final observation is that all possible impacts or objectives certainly do not apply to all decision situations. In many cases the range and types of alternatives being considered will simply not have all of the impacts which transportation projects in general might have. As a result, the list of objectives or impacts to be considered must be tailored to suit the particular project alternatives being considered. In fact, it has been argued by some recently that a very attractive evaluation strategy would be to generate the list of impacts to be considered only after the alternatives have been identified, to tailor the impacts specifically to provide information on the differences between the alternatives and not provide any information in areas where they are identical. Since it is only the differences among alternatives which will influence the decision between them, this approach seems to have considerable merit, particularly in view of the general difficulty of identifying a priori all of the conceivable impacts which changes in the transportation system might create.

Selection Processes

The process by which the best alternative is selected is one of the least understood processes in transportation decision making, whether it be in the public or private sector. The reason is that the selection process is internal to the decision maker or decision makers. However, since the analysis and evaluation processes are all undertaken in order to provide information to support the selection of the best action, the professional must have some understanding of this process to provide the proper types of information.

One part of the selection process is undoubtedly the search for dominance among alternatives. Dominance refers to a condition in which one alternative is said to be better than another alternative if it is at least as good in all characteristics and better in at least one characteristic. This condition applies only to quantitatively measured impacts, of course. Mathematically, dominance can be expressed by the following relationship. Suppose the consequences of each alternative are measured on scales such that the lower the value, the more it is preferred. If the value of the impact of alternative *a* in category *i* is C_{ai}, then in this category alternative *a* is preferred to alternative *b* provided

$$C_{ai} < C_{bi}$$

and they are considered equally good, provided

$$C_{ai} = C_{bi}$$

This dominance of alternative *a* over alternative *b* requires the following relationship holds for every impact measure

$$C_{ai} \leq C_{bi}$$

and that at least one impact (say, *k*) be such that

$$C_{ak} < C_{bk}$$

In vector notation, this is expressed as

$$\begin{bmatrix} C_{a1} \\ C_{a2} \\ \cdot \\ \cdot \\ \cdot \\ C_{ai} \\ \cdot \\ \cdot \\ \cdot \\ C_{an} \end{bmatrix} \leq \begin{bmatrix} C_{b1} \\ C_{b2} \\ \cdot \\ \cdot \\ \cdot \\ C_{bi} \\ \cdot \\ \cdot \\ \cdot \\ C_{bn} \end{bmatrix} \tag{14-1}$$

and

$C_{ai} < C_{bi}$

for at least one measure i. In the example presented in Fig. 14-2, alternative 1 dominates alternative 4 because it is better or at least as good in all dimensions and better in at least one.

In practical decision problems dominance rarely occurs among so many pairs of alternatives that one alternative ultimately is shown to dominate all others. However, it often can eliminate a few alternatives, and any reduction in the number of alternatives makes the selection process easier.

Also, conditions of near-dominance can be sought, in which case if the nearness to dominance is sufficiently close, the nearly dominated alternative can be eliminated from any further consideration. The definition of nearness is of course up to the decision maker, but presumably would be indicated in any array such as that in Fig. 14-2 by numerical values which are extremely close to one another relative to the possible range of values represented by the alternatives considered. In effect, the elimination of a nearly dominated alternative means that the difference between the two alternatives in those dimensions where the alternative to be eliminated is not inferior is so small as to be insignificant. Thus alternative 3 really dominates alternative 2 in this figure.

Once the dominated alternatives have been eliminated, then the process becomes very difficult. Usually an attempt is made to find an alternative which will benefit all groups, more or less in the same degree, except in cases where there is particular need for improvement of the condition of a particular group. This consideration is something like a principle of equity for choosing among alternatives. It usually involves comparing an alternative with the existing situation from the standpoint of a particular group. It also might involve comparing an alternative from the viewpoint of a particular group with an alternative which is known to be highly favored by that group and which therefore might be considered to represent the expectations of that group regarding the extent to which the transportation system would be improved in the future. It is essential that the alternative selected not be one which is very disadvantageous to any particular group, unless some compensation is made which would result in a net gain for that group.

A concept which is often very helpful in making a selection from among various alternatives is tradeoff analysis. The concept can be illustrated by an example which involves only two measures of impact and two alternatives. To be specific, we shall consider objectives 6 and 13 and the existing system and alternative *B* from the impact matrix of Fig. 14-4. As can be seen from that matrix, the existing system has an average travel time per trip of 38.0 min, while alternative *B* has a time of 31.9 min, 6.1 min less, making alternative *B* the preferred option considering only the travel-time objective. But in terms of the other objective of minimizing the public cost of the transportation system, alternative *A* will cost $1.154 billion, while the existing system will cost only $727 million, or $427 million less, making the existing system the preferred alternative. Suppose that the decision-making entity decided that alternative *B* is the preferred alternative, considering only these two objectives. This decision would imply that the reduction in user travel time by 6.1 min per trip was worth more than the additional expense of

$427 million. The added expense of $427 million is in effect traded off for the reduction in user travel time of 6.1 min per trip. By comparing the gain with the loss or cost, the decision about whether or not the tradeoff is desirable is facilitated. Thus the basic concept of tradeoff analysis is to identify the gains and the losses resulting from choosing one alternative over another, so that these gains and losses can be compared and the desirability of the switch can be assessed. The method does not identify the best alternative but rather simply helps to organize information so that reaching a decision is facilitated.

Sometimes it is useful to express the tradeoff in terms of a ratio of gains to losses (much like the efficiency measures described earlier), or vice versa. In this example, the ratio of total additional cost expenditure necessary to the gain in total user travel time is a useful ratio. Assuming that there are 15 million trips per day regardless of which alternative plan is implemented, we find that the savings in travel time would be 557 million person-hours per year. The ratio of cost to time savings would be $0.0767 per hour, a relatively small expense per unit of time savings. By expressing the tradeoff between time savings and additional costs in the ratio form, it is possible to compare the value with other information on values of time and thereby be better able to access whether or not the gain is worth the expense.

The preceding example has been simplified by the consideration of only two measures of impact, as if these were the only ones of interest. While there are some situations where that would be the case, in general there are many impacts of interest, as the list of objectives in Fig. 14-4 indicates, for example. The concept of tradeoff analysis can be extended to such situations. The approach is still to compare two alternatives at a time, considering one as a base and determining whether or not the other is preferred to it and hence should replace it as the base for further comparisons. However, rather than focusing on only two measures of impact, all are considered. This is accomplished by grouping together all the impacts for which there is a gain, with the associated measures of the gains, and grouping those for which there is a loss in another group. Such a presentation of information facilitates the subjective comparison of gains and losses, so that the preferred alternative can be selected.

The selection process still remains one involving subjective judgments, but such analyses of dominance and tradeoffs aid those making decisions considerably. The selection can often be aided further by concentrating on decisions between only two alternatives at a time, often called pairwise comparison. At each step, one alternative is considered the base—that alternative being the best found so far and against which all alternatives not previously considered will be compared. Usually that alternative which corresponds to no action (e.g., the existing system of Fig. 14-4) is taken as the first base, this being replaced as the basis for comparison once a preferred alternative has been identified. The comparison actually can proceed through all the alternatives in any order, but it is important that all the alternatives be considered.

Also significantly influencing the selection of a best alternative are any constraints on the characteristics of an acceptable alternative. Undoubtedly the most common restriction is a budgetary one, most often a limitation on the total budget available from local funds or some other level of government either for construction or operating subsidy. Since these limitations sometimes change from year to year, often they cannot be taken into account in the analysis and planning process but rather must be dealt with at the time a decision among alternatives is to be made. In contrast, more technical restrictions, such as those on air pollution and noise, can usually be taken into account in the technical planning process because the standards tend to be more enduring.

The process by which particular alternative courses of action are selected within the public sector and then ratified or approved at the various levels of government seems to be changing rather rapidly. In most cases the problems are perceived at the local level and first dealt with at that level, resulting in a particular plan of action being selected. Then this plan is transmitted to various higher levels of government for approval, and at these levels modifications of the plan might be required. At these various levels of review conformance with requirements for environmental impacts and with specified technical standards for proposed designs are determined. Once the project has been approved at all the relevant levels, then detailed design and construction plans for implementing the plan are developed.

Transportation (and other) plans or projects involving the use of federal money in the United States are required to be discussed in public hearings. Such hearings are held in areas which might be affected by those projects. The intent of these public hearings is to enable the public, organizations, and any interested groups to comment on proposed plans and to recommend changes. After a rather long history of planning transportation projects largely to meet user requirements and minimize costs, it was discovered that there are other very substantial impacts which need to be considered and that these largely fall upon nonusers in the immediate vicinity of new facilities. Thus these hearings were originally intended to enable nonusers as well as individuals and groups in other roles to provide input to the planning process.

While there seem to be very few specific operational guidelines for determining how to balance all of these potentially conflicting viewpoints, the public hearing process has undoubtedly sensitized transportation engineers and planners to many of the local problems created by transport improvement projects. In fact, the hearing process has been one of many factors which has led to a very diminished level of construction of new transportation facilities within urban areas. Although apprehension of possible problems of new facilities may be somewhat unfounded in many cases, it is important that the nonuser viewpoint be given serious consideration. One problem with the public hearing process which is very difficult to overcome is that it is devoted entirely to citizens and groups in the area of potential impact yet gives no attention to the future citizens who are not in the area now but who will experience the benefits and costs of the proposed system change. However, undoubtedly through experience with the citizen participation process, many of these weaknesses of the existing procedures will be overcome.

AN ALTERNATIVE APPROACH: ECONOMIC EVALUATION METHODS

The traditional engineering economic approach to evaluation is similar to evaluation in the private sector. The basic approach is to attempt to estimate the change in benefits resulting from a proposed action, relative to the "do nothing" alternative (like the change in revenue from an action in the private sector), and the corresponding change in costs. The differences between these—the so-called net benefits—are calculated for each proposed alternative.

That alternative which exhibits the largest net benefit, like the largest profit (provided that the net benefit was greater than zero), would be selected. If none were greater than zero, then the null or do-nothing alternative would be selected. In addition to the method of selection based on the criterion of maximum net benefits, there are other approaches, such as one based on the ratios of benefits to costs, although they all tend to yield the same alternative as best. Since that based on net benefit maximization seems to be most popular, we shall focus most attention on it. But first it is essential to clarify what is meant by the terms benefits and costs.

Benefits and Costs

The important question of the definition of the costs and benefits to be included in these calculations differs widely among different application areas, even within the transportation sphere. In many instances, the definition of benefits is largely up to the analyst, and it is not uncommon for all conceivable benefits to be evaluated in monetary units and included in the sum of benefits. For example, in a project which will reduce the travel time to a particular portion of an urban area, part of the benefit may be this reduction in travel time for users and another part an expected increase in land values in that area resulting from improvement in accessibility. Obviously, the increase in land value is largely a result of the reduced transportation costs to persons desiring to go to this area and hence is likely to be largely if not entirely an example of double counting of the same benefit. Also, the magnitude of the benefits can be significantly influenced by the magnitude of any transformation values which convert gains measured in nonmonetary units into monetary units, such as the value of travelers time or the value of a reduction in air pollution. As we noted in previous chapters, the methods for estimating these values are extremely imprecise and lend themselves to the selection of transformation values which will yield any desired magnitude of benefits, such that the decision can be influenced either positively or negatively. These are but a few of the reasons why these methods are largely being replaced by the more vigorous and less quantitatively demanding evaluation methods discussed earlier.

However, for the economists who developed these methods there is a correct means of measuring the benefits. This is to measure the benefits using the demand curve for the particular product which is being evaluated. Mohring and Harwitz (1962), in a discussion of the benefits of highway improvements, identify four types of benefits which can be rephrased as general transportation improvement benefits: (1) benefits to previously existing transportation that continues after improvement, (2) benefits arising from the increased use of transportation entailed in substituting transport for other goods or services, (3) benefits arising out of increase in the productive capacity of the economy associated with the transportation improvement, and (4) benefits resulting from the triggering of investment and further economic or social development.

Figure 14-5 **The consumer's surplus measure of the benefits of a transportation improvement.** [***Adapted from Mohring and Harwitz (1962, p. 23).***]

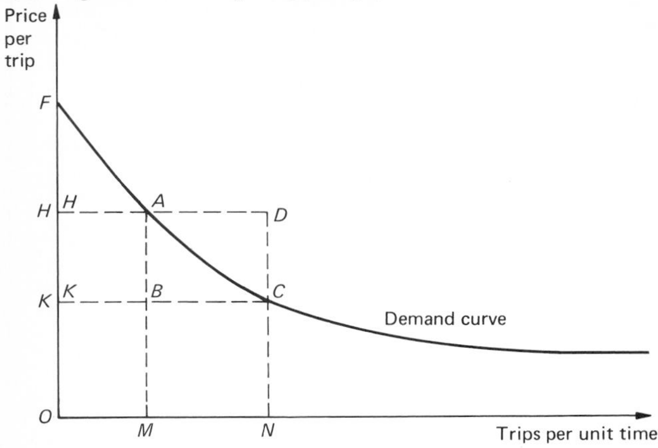

The first two classes of benefits are very clearly measured directly by the demand curve for transportation. Specifically, this can be seen by reference to Fig. 14-5. The total value to be derived by a particular traveler or shipper by shipping freight is expressed by this demand curve. The reason for this can be seen by considering two specific points on the demand curve. First, consider a situation in which the price is very high, such that no person travels (i.e., a price greater than *OF*). As the price is reduced, no one travels until the price reaches *OF,* at which point one trip is made. The person who takes that trip presumably perceives a benefit from taking that trip equal to *OF,* assuming that this person is rational in the sense that the trip will be taken only if the benefit is at least as great as the cost (or price). Since the trip is made only with the price equal to or less than *OF, OF* must be the benefit perceived by this traveler. As the price is reduced further, more trips are made, presumably because the price then is equal to or less than the benefits of these trips. For example, as the price is reduced from *OH* to *OK,* the number of trips taken increases from *OM* to *ON*. These additional trips are ones that have benefits less than *OH* but greater than or equal to *OK.* The conclusion from this is that the benefits perceived by travelers resulting from any given quantity of trips is equal to the area under the demand curve between the origin and that quantity. Thus, in Fig. 14-5, if the quantity of travel were *ON* (corresponding to a price of *OK*), the total benefits perceived by the travelers would be equal to the area bounded by the lines connecting the points *O, M, N, C, A, F, H, K,* and *O* or, in typical geometry notation, the area *ONCF.* Furthermore, the net benefit to each traveler is equal to the benefit minus the price paid. For example, if the price were *OK,* the trip corresponding to the point *M* on the quantity axis would have a benefit of *MA,* a price of *MB,* and hence a net benefit of *BA.* Thus the total net benefit of all trips taken at a price *OK* would equal the area *ONCF* minus the area *ONCK,* or the area *KCF.* This area is also termed the consumers' surplus in economics. Similarly, if the price were *OH,* the total benefit would be *OMAF* and the net benefit would be *HAF.*

Now suppose the price of transport were reduced from *OH* to *OK* due to an improvement in the transport system. The increase in the net benefits to previous trips would be equal to the area *KBAH,* since the total net benefits under this new condition to those trips would be *KBAF.* However, there is also an increase in the total number of trips made, which results in a further increase in net benefits of *BCA.* Thus the total increase in net benefits resulting from this system improvement equals the area *KCAH.* Mohring and Harwitz demonstrate in their book that this area *KCAH* is a very close approximation to the total benefits accruing in benefits classes (1) and (2) above. The substitution benefits, those appearing in the area *BCA,* represent the total amount of those benefits regardless of whether these are transferred by their initial recipients to other members of the population. It represents these benefits regardless of whether they arrived out of the substitution of transportation of one type for another form of transportation, the substitution of transportation-intensive activities for other types of activities, or the substitution of transportation-intensive means of production for other means of production. It should be noted that much of the increase in land value associated with transportation projects really arises out of this substitution.

The third type of investment benefit, benefits entailed as a result of an increase in productive capacity of the economy, is potentially quite significant. By reducing the resources required to transport things from place to place, the total productive capacity in the economy is increased. However, transportation investments are by no means unique in their ability to increase the productive capacity of the economy in this way. To the contrary, any productive investment will provide precisely the same sort of benefit. In fact, presumably no one would willingly undertake an investment unless it held promise of reducing the cost of producing a good or a service or radically improving the quality of a particular good or service. The value of these resources released for alternative employment would be precisely represented by the change in price on the price axis in Fig. 14-5. Thus again this type of benefit is already included in this net benefit estimation. The fourth type of benefit, also a very important class of benefit, has been observed frequently as a result of transportation improvements. However, such transportation improvements would result in additional benefits only to the extent that three conditions are fulfilled: (1) the investment would not have been made in the absence of the transportation improvement, (2) the improvement utilizes resources that otherwise would have remained unemployed, and (3) it does not displace economic activity that otherwise would have taken place. As we observed in the preceding chapter, many of the triggering benefits from transportation improvements seem to involve a shifting of an activity from one place to another because of the transportation improvement, and thus there are no significant additional benefits. For these reasons, the generally used measure of net benefit is the change in consumer surplus occurring under the demand curve.

The change in net benefit, total benefits minus total costs, is equal to the area *KCAH* provided the prices paid by users reflected the true costs or resources used in providing the service. This is often not the case in transportation, due to subsidies and imperfect pricing, as we noted in Chap. 11. In many cases prices reflect especially high taxes on transport, while in others they reflect large subsidies. Thus the difference between true resource costs and prices paid must be incorporated into the analysis. In general, this means that any resource costs not included in the price (either before or after the improvement) must be added to that price and any taxes or other surplus subtracted from it. Otherwise the method remains the same.

This is widely used in the United Kingdom (McIntosh and Quarmby, 1970) and a number of other countries as an estimate of the total net benefit resulting from transportation improvements (as well as changes in other sectors). While it is often very difficult to estimate precisely the demand curve, the methods which we discussed in Chap. 10 can be used for this purpose. In most applications the estimation of the entire demand curve between the old price and the new price would be very time-consuming and difficult. In practice, it is common to estimate the user cost or price and quantity demanded for the two alternatives being compared and then approximate the demand curve between these two points as a straight line. Thus in Fig. 14-5 the change in consumers' surplus would be approximated by a trapezoid connecting the points *HACK*. In the usual economic notation,

$$CS_{12} = \tfrac{1}{2}(P_1 - P_2)(Q_2 - Q_1) \tag{14-2}$$

where CS_{12} = change in consumer surplus resulting from a decrease in price from P_1 to P_2
P_i = price i
Q_i = quantity demanded at price P_i

In Fig. 14-5, point *A* corresponds to index 1 in this equation and point *C* to index 2.

However, the main deficiency of this method, which seems to have limited its use in North America, is that it provides very little information on the distribution of gains and costs throughout the society. Nevertheless, it is a valid procedure for estimating benefits. Hence the economic procedures for evaluation and selection do have application when a society wishes to select its public investments according to the total magnitude of benefits (and costs) rather than the distribution of these throughout the society.

The Net Benefits Selection Method

In the case of major transportation projects and other investments with long lives and periods of use, the calculation of costs and benefits must include the period of life of the plan or project in order for the evaluation to be complete. The costs and benefits in various future years would be translated into their equivalent value at the present time, their present value, by discounting, as we described in Chap. 9. Discounting is similar to compounding interest but obviously works in the opposite direction, in that it gives a present equivalent value of a future gain or cost, according to the following formula:

$$PV = \frac{1}{(1 + I)t} FV_t \tag{9-6}$$

where PV = present value
FV_t = future value t years hence
I = discount rate

All of these present values are summed to yield the total present value of the net benefit of each particular alternative.

In precise mathematical terms, then, this method is based upon the calculation of the present value of net benefit for each alternative as follows:

$$V_j = B_j - C_j = \sum_{t=0}^{N_j} \frac{1}{(1+I)^t} (B_{j,t} - C_{j,t}) \tag{14-3}$$

where V_j = present value of net benefits for alternative j
I = discount rate
$B_{j,t}$ = benefits or gains of alternative j in time period t (t time periods hence)
$C_{j,t}$ = costs of alternative j in time period t
N_j = life of project j

Any project for which V_j is less than zero is rejected. Of those remaining, the one with the largest value of V_j is the best (designated j^*):

$$j^* = (k \mid V_k \geq V_j, j = 1, 2, \ldots, M) \tag{14-4}$$

where M = number of alternatives
j^* = optimal alternative

This technique, the net benefit of present value approach, along with a number of other approaches (cost-benefit ratio, rate of return method) which use identical information and generally result in the same best alternative are widely used and form an important part of any evaluation. Computer programs for the laborious calculations have been developed.

Other Engineering Economic Evaluation Methods

There are numerous other methods of economic evaluation. In the public sector, the benefit-cost ratio method (more precisely termed the incremental benefit-cost ratio method) has been in widespread use. In the private sector, two commonly used methods are the internal rate of return and the payback period methods. These will be described very briefly.

The incremental benefit-cost ratio technique is similar to the present worth method, and it results in the same selection of the best project. Results of this method answer, for each alternative, the question, is the rate at which we accrue benefits greater than the rate at which we spend money? Each alternative is analyzed relative to another alternative.

First, the projects are labeled in terms of increasing costs, as project 1, 2, . . . , j, . . . , M; costs $C_1, C_2, \ldots, C_j, \ldots, C_M$; and benefits $B_1, B_2, \ldots, B_j, \ldots, B_M$; respectively,

where $C_j = \sum_{t=0}^{N_j} \frac{1}{(1+I)^t} C_{j,t}$

$$B_j = \sum_{t=0}^{N_j} \frac{1}{(1+I)^t} B_{j,t}$$

Then the incremental benefit-cost ratio for each successive pair of projects is calculated to determine if it is equal to or greater than unity. The first project is usually the null or do-nothing alternative, which is of course usually the cheapest project. If the first ratio is greater than or equal to unity, then project 2 has benefits at least equal to its cost, and it is termed acceptable. This alternative becomes the base for the next ratio comparison with unity, and so on. If a ratio is less than unity, the project whose incremental benefits of costs are being compared to unity is deemed unacceptable, and the ratio is calculated for the next most costly project using the same base (the most costly acceptable project up to that point in the analysis).

An example may assist in understanding. Consider a case of four alternatives ($M = 4$).

Step 1 $\quad \dfrac{B_1}{C_1} \overset{?}{\geq} 1 \quad$ If yes, project 1 is better than doing nothing. Go to step 2. If no, project 1 is unacceptable. Substitute B_2 and C_2 for B_1 and C_1 and test.

Step 2 $\quad \dfrac{B_2 - B_1}{C_2 - C_1} \overset{?}{\geq} 1 \quad$ If yes, project 2 is better than project 1. Continue to step 3. If no, project 2 is unacceptable. Substitute B_3 and C_3 for B_2 and C_2.

Step 3 $\quad \dfrac{B_3 - B_2}{C_3 - C_2} \overset{?}{\geq} 1 \quad$ If yes, accept project 3 and go to step 4. If no, project 3 is unacceptable. Substitute B_4 and C_4 for B_3 and C_3.

Step 4 $\quad \dfrac{B_4 - B_3}{C_4 - C_3} \overset{?}{\geq} 1 \quad$ If yes, project 4 is the optimum alternative. If no, project 3 is the optimum.

The best project is the most expensive project for which the incremental benefit-cost ratio is greater than unity and for which all the more expensive alternatives have benefit-cost ratios less than unity. Even though the ratios of alternatives considered earlier in the analysis may be greater than that of the final choice, they will not be the best.

In precise mathematical terms, this method can be expressed as follows. First, order all alternatives from least costly to most costly, $i = 1, 2, \ldots, M$. Then calculate

$$R_{i+1} = \frac{B_{i+1} - B_k}{C_{i+1} - C_k} \overset{?}{\geq} 1 \tag{14-5}$$

Start with $i = k = 1$, and if $R_2 \geq 1$, project 2 is acceptable and becomes the new base k. If $R_2 < 1$, project 1 remains the base k. Increase i by 1, to get $i + 1 = 3$. Perform the same test. If $R_3 \geq 1$, the new base is project 3; if not, the base remains the same. Continue to test successive projects until all are exhausted. The most costly of the acceptable projects is the best.

The internal rate of return method is quite similar to the benefit-cost method in that it involves a pairwise comparison of alternatives and the most costly accepted alternative is optimal. The rate of return method is widely used in private industry, the idea being to find the maximum discount rate which the incremental project can sustain and remain beneficial to the investor. All increments having a rate of return equal to r greater than the interest rate i would be acceptable.

The projects are ranked from least costly to most costly. Beginning with the first alternative having $r \geq i$, comparisons are made by solving the present worth equation for the incremental benefits and costs of projects:

$$\sum_{t=0}^{N} (\text{PWF}_{r,t})(C_{j,t} - C_{k,t}) = \sum_{t=0}^{N} (\text{PWF}_{r,t})(B_{j,t} - B_{k,t}) \tag{14-6}$$

where r = internal rate of return

$$\text{PWF}_{r,t} = \frac{1}{(1+r)^t}$$

N = project life (all assumed equal)

If $r < i$, j is discarded and r is computed for the increment between k and the next project. If $r \geq i$, the increment between j and the next project is taken. The optimal project is the most expensive project having an incremental $r \geq i$. Since the internal rate of return is easily compared to the cost of borrowing funds, this method is widely used in private industry. However it is somewhat unpopular because there may be multiple roots r to Eq. (14-6), leading to ambiguous results.

The payback period analysis asks the question, how long must a project continue in operation, with the costs and benefits predicted for it, to pay back the initial costs? The optimal solution is the one with the shortest payback period. The reason is that future benefits or revenues are uncertain and that the uncertainty of a positive net return is minimized by minimizing the period required to reach a positive net discounted return.

Mathematically, the method is to select that project whose value of P, the years required for payback, is the least. The payback period P is defined as the shortest period such that

$$\sum_{t=1}^{P} \frac{1}{(1+I)^t} B_{j,t} \geq \sum_{t=1}^{P} \frac{1}{(1+I)^t} C_{j,t} \tag{14-7}$$

This method also has particular appeal in the private sector.

An Example

This example illustrates the use of the economic evaluation methods and the concept of consumers' surplus as a measure of user benefits. It is typical of benefit-cost analyses of transport projects, and is drawn from Dickey et al. (1975, pp. 321–323). The problem concerns a road relocation project in which the proposed alignment is much shorter than the present one. The question of whether or not the relocation should be undertaken will be conducted using the benefit-cost ratio method. To simplify the calculations, it will be performed for an average year in the future, using capital recovery factors to estimate annual costs equivalent to initial costs (as described in Chap. 9).

Present traffic on the existing route is 1500 vehicles/day and it is estimated that the average traffic for the next 30 years, the analysis period, will approximate 2500 vehicles/day on the new route and 2000 vehicles/day on the old. This traffic is composed of passenger cars only. Assignment of type of operation is made on the basis of the ratio of the 30th highest hourly traffic volume to the volume at level of service D. The proposed facility will have a pavement width of 20 ft with improved alignment and grades, yielding free operation, while the 18-ft pavement of the existing facility provides for lower service level normal operation. Table 14-1 shows the data used to arrive at this differentiation.

The curvature proposed on the new facility is minimal. It is to be superelevated (banked), so a correction for curvature can be ignored. However, the existing highway has a large number of curves approaching the maximum curvature for a 50 mi/h design speed. It is estimated that 50 percent of its length will have an average curvature of about 4°, with superelevation negligible. From Fig. 9-8 a correction of about 8 percent is read for these speeds. Since only one-half of the roadway is curved, the correction is halved to 4 percent and applied for the whole of the route to the values from Table 9-11. The calculated unit cost is shown in Table 14-1. The average annual road user benefits measured by the change in consumer surplus are, using Eq. (14-2):

$$CS_{12} = \tfrac{1}{2}[(2.0)(0.1024) + 0.4(0.1061) - 1.5(0.0901)](2500 + 2000)(365) = \$92{,}000$$

The average annual unit maintenance cost on the existing route is estimated to be $1100/mi or a total of 2.4($1100) = $2640, and that of the proposed route to be $880/mi or a total of 1.5($880) = $1320.

Table 14-1 Costs and Benefits for Road Improvement Evaluation Problem†

		Existing sections	Proposed sections
1	Future volume, vehicles/day	2000	2500
2	30th highest hourly volume, vehicles/day	375	450
3	Service volume, level *D*, vehicles/day	450	715
4	Ratio	0.83	0.63
5	Type of operation	normal	free
6	Design speed, mi/h	50	60
7	Running speed, mi/h	37	42
8	Lanes, no.	2	2
9	Length, mi	2.0	1.5
10	Grade class, percent	0–3	0–3
11	Surface and condition	paved–good	paved–good
12	Curvature	50 percent–4°	negligible
13	Unit cost, cents/vehicle-mi	10.24	9.01
14	Estimated pavement life, years	...	20.00
15	Estimated right-of-way life, years	...	60.00
16	Estimated life, other, years	...	40.00
17	Pavement costs, $	...	66,000
18	Right-of-way cost, $	...	33,000
19	Grading and other cost, $	...	451,000
20	Annual pavement cost, $	...	5290
21	Annual right-of-way cost, $	...	1740
22	Annual other cost, $	...	26,290
23	Total annual capital cost, $	...	33,320
24	Maintenance cost, $/mi-year	1100	880

† Dickey et al. (1975, p. 322).

The total estimated cost of the proposed improvement is \$550,000 and the interest rate is 5 percent. This interest rate is applied with reference to the cost and life expectancy of the individual items of improvement to compute the annual cost. Capital recovery factors, calculated using Eq. (9-12), are 0.0802, 0.0528, and 0.0583 for the pavement, the right-of-way, and the construction (other) capital items, respectively.

The change in costs resulting from the new facility is

$$(0.0802)(\$66{,}000) + (0.0528)(\$33{,}000) + (0.0583)(\$451{,}000) + \$1320 - \$2640 = \$32{,}008.90$$

Therefore the the ratio of change in benefits to change in costs is

$$\frac{\$92{,}000}{\$32{,}008.90} = 2.87$$

This clearly is a worthwhile project by this criterion, with benefits almost three times the costs.

Problems with Above Approaches

A number of very serious conceptual problems arose with respect to the use of single monetary estimate of benefits minus costs for public projects. These arose primarily because of concerns of distribution of benefits and costs among the population and the difficulties in assessing the exact nature and monetary value of many of these benefits and costs.

An obvious requirement of the above method is that benefits and costs be estimated in monetary units. While this is certainly possible for any item being transacted in the marketplace and might be possible for other items which are clearly traded off against dollars, such as travel time in the choice among modes, there are other impacts for which dollar estimates are virtually impossible to obtain. An example is the loss of a park in a residential neighborhood, or the loss of transit service in a community due to the provision of more road and parking capacity and resultant decline in transit usage. Clearly there is no marketplace transaction for these items which would yield reliable information on the monetary value of the effects, but such impacts are not uncommon as a result of major public investments.

Another difficulty associated with this method is that even if a monetary value can be associated with each impact upon each person or group affected, there remains a problem of combining the value of the impacts in some manner to obtain a total. The simplest means to do this would be to add the net benefits for each affected individual or group. Yet this implies that \$1 of net benefits to one person or group is worth the same as \$1 of net benefits to another. It implies that a time saving of 1 h to a laborer who values his time at, say, \$1.00/h, is exactly equivalent to a time saving of 6 min to a businessman who values his time at \$10.00/h. Clearly such a value judgment is a monumental one and often it would be disagreed with, because it implies that an impact on any particular individual or group is to be measured according to that person's willingness to pay, which is in turn a direct function of income.

A further difficulty with this approach is that it mixes together the positive and negative effects not only upon individuals but also among many individuals and groups. Thus it implies indifference between, for example, doing nothing and undertaking an action which results in a loss to one group of $1000 and a gain to another of $1000. This, again, is too monumental an assumption to be generally accepted.

It is for these reasons that these so-called engineering economic methods are primarily used in the context of broader evaluation and selection methods, of the types described earlier. While these engineering economic methods are useful as an overall indicator of the magnitudes of certain types of benefits and costs, their limitations usually result in the requirement of much additional evaluative information for presentation to decision makers, particularly in the public sector.

ROLE OF THE ENGINEER AND PLANNER

The traditional role of the engineer and planner in transportation system evaluation and decision making included not only providing technical information—identifying the alternatives and the gains and costs of each—but also selecting among the alternatives the best project, the project that will be implemented. This was particularly true up to about 20 years ago, when engineering economic evaluation methods such as those presented above were used, especially for highway investment and other government facility projects. In those cases the technical tools provided a basis not only for evaluation but also for ranking and selection of projects. Hence treating both of these aspects as part of the technical tasks of professionals was possible.

With the recognition of the diversity of the types of impacts of transportation improvements and the different patterns of benefits and costs which these have, the role of the professional has gradually changed to one of providing information on which others would base their decisions. The reason for this change is of course in part that there is no technical procedure which will in general lead to the selection of a single alternative as best when using one of these more robust multiobjective or multi-impact evaluation methods. Furthermore, given the recognition that different groups will benefit to differing degrees from various public investments, it seems appropriate that the choice among alternatives be made by those responsible to the citizens for their actions, elected public officials. However, such decision makers often seek the advice of professionals for the best course of action, but the responsibility for the decision still lies with the elected officials.

In recent years it has become increasingly apparent that there is another important role for engineers and planners in addition to that of simply providing information for decision makers. This second role is identifying the major issues involved in the selection of particular alternatives and clarifying those issues so that the information on which an intelligent decision might be based is available. Issues arise whenever there is a potential conflict regarding the attractiveness of a particular alternative from the standpoint of different interests or affected groups. Professionals should try to provide information on the extent to which various groups would benefit or suffer as a result of particular alternatives and attempt to devise means by which groups which would otherwise suffer from an alternative could benefit from it, by changing certain characteristics of that alternative or providing some type of compensation for negative effects.

A logical extension of this dual role of professionals is an objective for any planning or design process "to achieve substantial, effective community agreement on a course of action which is feasible, equitable, and desirable" (Manheim and Suhrbier, 1971). Such a goal undoubtedly will be very difficult to achieve in many problem contexts, but it seems to be a goal worthy of pursuit by all transportation professionals in the public sector.

The primary result of pursuit of this goal seems to be that the engineer or planner must participate in the political decision process. This participation includes the provision of technical information and also requires much attention to revision of plans or designs to achieve the "substantial and effective agreement" sought, and in many cases undoubtedly leads to creation of entirely new alternatives which attempt to combine the desirable features of others but not include the negative effects. Also, the alternatives undoubtedly include reference to specific provisions of monetary (or other) compensation for negative effects that cannot be eliminated.

It is not clear how the goal presented above and the broadened role of professionals applies to private sector decisions. Obviously the issue is not whether or not impacts external to the private firm should be considered in its decision making, for such firms now consider many of the economic, environmental, and social effects of their actions. This is required by law or regulation to some extent, and partly dictated simply by common business ethics. However, complete agreement by all affected parties on desirability of an action seems far too strong a goal, since this would imply agreement of the firm's competitors. Such a condition would surely inhibit change, including changes which would benefit the society as a whole. Thus it is unclear how this goal must be modified to apply to the private sector.

SUMMARY

1 Decision making can be structured by the general problem-solving process, the sequences of steps in this process being generation of alternatives, prediction of consequences, evaluation, and finally selection of the preferred alternative. This chapter focuses on evaluation and selection.

2 For many years evaluation and selection were combined, through the use of so-called economic evaluation methods, which measure benefits and costs only in monetary units and which contain methods for the selection of the best alternative. Because many important characteristics of alternatives cannot be converted into monetary values and because single values obscured the distribution of benefits and costs, these methods now typically play a lesser role than more general methods, which distinguish between evaluation and selection.

3 Evaluation is concerned with placing values on the various alternatives. In effect, this means developing descriptions of the alternatives which will be of use in selection, including information on who is affected, in what manner, and to what extent.

4 Because of the multiplicity of objectives to be achieved and the fact that the same alternative usually is not the best one for all the objectives, the best alternative usually cannot be identified by a rule for selection. Rather the selection involves much judgment and the weighing of the relative advantages and disadvantages of the alternatives. It is therefore a subjective process, properly done by the decision makers, not the professionals.

5 A consequence of this is that the role of engineers and planners primarily involves providing information on the range and consequences of alternatives. Furthermore, it is argued by many that professionals should take active roles in the evaluation and selection process, trying to identify who gains and loses from each alternative, articulating the resulting issues of distribution of costs and benefits, and attempting to find alternatives which are satisfactory to all persons or groups affected.

6 These principles and procedure will be illustrated in the following chapters where they are applied to particular planning, design, and operations management problems.

PROBLEMS

14-1
Select a transportation problem in your area. Structure an analysis of it, identifying major types of alternatives, prediction methods you would use, evaluation procedure, and the expected choice process. Who will be the primary decision makers?

14-2
Traffic on a road which might be reconstructed for higher speeds is growing at 10 percent per annum, and this growth is expected regardless of whether or not the road is rebuilt. The present 5.5-mi-long road costs $56,000/year to maintain, and on it user costs average 27 cents/vehicle-mi. The rebuilt road will cost $3,600,000 initially, the expenditure to be made in equal parts over 3 years. The maintenance cost will average $35,000/year, starting the fourth year, when traffic can use it. User costs on the 4.6-mi route are estimated to average 23 cents/vehicle-mi. Current traffic is 12,000 vehicles/day, including flow in both directions. With a 20-year life (from now), and a 10 percent discount rate, which alternative, reconstruction or not, is preferred on the basis of net present value?

14-3
What is the effect on the conclusion in Prob. 14-2 of a reduction in the discount rate to 6 percent? Of a reduction in the traffic growth rate to 8 percent? Of both?

14-4
What factors other than those included in the economic analysis might be relevant in the decision in Prob. 14-2? The road is in a rural area.

14-5
Compare the characteristics of the alternatives presented in Fig. 14-4. Do any alternatives dominate others? If not, are there examples of near-dominance?

On the basis of data presented in that figure and your subjective feeling as to the relative importance of various objectives, can you select a best alternative?

14-6
A controversy in an urban area centers around whether or not a new freeway should be built connecting an essentially industrial area with an intercity turnpike about 10 mi away. The route would pass through residential areas and would relieve considerable congestion on the present two-lane highway. Whom would you expect to be affected by this decision? List objectives and measures of achievement which you would use in an evaluation of these alternatives.

REFERENCES

Bronitsky, Leonard, and Joseph Misner (1975), "A Comparison of Methods for Evaluating Urban Transportation Alternatives," Report No. UMTA-MA-06-0053-74-1, Transportation Systems Center, U.S. Department of Transportation, Cambridge, MA.

Dickey, John W., Robert C. Stuart, Richard D. Walker, Michael C. Cunningham, Alan G. Winslow, Walter J. Diewald, and G. Day Ding (1975), "Metropolitan Transportation Planning," McGraw-Hill, New York.

Hill, Morris (1973), "Planning for Multiple Objectives: An Approach to the Evaluation of Transportation Plans," Monograph Series No. 5, Regional Science Research Institute, Philadelphia.

Hoffman, John, and Martin E. Goldsmith (1974), A Comprehensive Evaluation of Transportation Alternatives for a Large Urban Area, *Proceedings of the Transportation Research Forum,* **15**(1): 413–427.

Jordan, D., S. Arnstein, J. Gray, E. Metcalf, W. Torrey, and F. Mills (1976), "Effective Citizen Participation in Transportation Planning," 2 vols., U.S. Department of Transportation, Federal Highway Administration.

Kuhn, Tillo E (1962), "Public Enterprise Economics and Transport Problems," University of California Press, Berkeley.

McIntosh, P. T., and D. A. Quarmby (1970), "Generalized Costs, and the Estimation of Movement Costs and Benefits in Transport Planning," MAU Note 179, Mathematical Advisory Unit, Department of the Environment, Government of the United Kingdom, London.

Manheim, Marvin L., and John H. Suhrbier (1971), Community Values in Transport Project Planning, *Proceedings of the Transportation Research Forum,* **12**(1): 297–310.

Manheim, Marvin L., John H. Suhrbier, Elizabeth D. Bennett, Lane A. Neumann, Frank C. Colcord, and Arlee T. Reno, Jr. (1975), "Transportation Decision-Making: A Guide to Social and Environmental Considerations," National Cooperative Highway Research Program Report No. 156. Transportation Research Board, Washington, DC.

Mohring, Herbert D., and Mitchell Harwitz (1962), "Highway Benefits: An Analytical Framework," Northwestern University Press, Evanston, IL.

Thomas, Edwin N., and Joseph L. Schofer (1970), "Strategies for the Evaluation of Transportation Plans," National Cooperative Highway Research Program Report No. 96, Highway Research Board, Washington, DC.

Long-Range Transportation Planning

15.

Planning pervades almost all aspects of transportation decision making. However, it is important to distinguish between various types of planning, particularly between long-range and short-range planning and between planning for fixed facility improvements and planning for operations and pricing policies. The term "transportation planning" has traditionally been used to describe long-range planning. Such planning naturally focuses on major improvements to the fixed plant of the transportation system and policies which will influence transportation operations over an extended period. However, since many transportation planning agencies which had been limited to long-range planning are now broadening their scope to include short-range planning, it is best to qualify terms such as transportation planning with descriptive modifiers, such as long-range, operations, etc. In this chapter, long-range transportation planning will be discussed, and following chapters will cover the planning of individual facilities (often termed design) and operation planning or management.

The largest transportation planning activity in recent years has undoubtedly been urban transportation planning, where the main focus of attention was on planning future road and public transportation facilities. It is within the field of urban transportation planning that most of the research and development of new modeling tools has been undertaken and where most of the experience in long-range transportation planning has been developed. Since the basic principles of long-range transportation planning and the general character of the methods used seem to apply to all modes and contexts of transportation, not just urban transportation problems, we shall attempt to present principles and methods in a general way. The preponderance of urban examples and references simply underscores the fact that urban problems have received more attention than others. Recently, however, methods and models developed for urban transportation have been adapted and applied to such diverse planning problems as preparing long-range transportation plans for developing nations and restructuring the railroad system in the northeastern portion of the United States. Undoubtedly the principles and methods of transportation planning will find even wider areas of application in the future.

TYPES OF PLANNING

There are many different types of planning; it is useful to review them before delving into the specifics of transportation planning processes. One of the most important characteristics of almost all planning is the natural hierarchy which exists among the various components of the complete planning effort. The basic structure of a hierarchy is that there are many different levels of activity. Within each level, there usually are many different activities, which might be quite distinct. However, when bringing some or all of the activities of a particular level together, a higher-level activity is created. There may be many distinct activities at the higher level, some or all of which can be brought together to form an even higher level group, and so on.

One view of this applied to transportation planning is shown in Fig 15-1. The highest-level activity shown in the figure is national planning, concerned with general policies regarding economic and population development and the associated transportation infrastructure. The level below this corresponds to comprehensive regional planning, in which not only the transportation system but also land use patterns are the focus of attention. The results of actions at this level lead to specific requirements for transportation infrastructure, and this leads naturally to the next lower level in the hierarchy, regional transportation planning. Below this level is the planning for transportation in particular subareas, such as the central business district, a residential neighborhood, or a corridor. This leads to the planning of particular facilities, their construction and availability for use as part of the entire transportation system. Finally there are the plans for operation of those facilities, including transit schedules, road traffic regulations, etc.

This particular hierarchy reflects an essentially functional division of the transportation system and its relationship with its environment, in which the system is divided into smaller and smaller geographic or spatial units. There are many alternative ways in which the hierarchy could be constructed. For example, another hierarchy might be in terms of the governmental units responsible for planning at various levels. The highest level would still be national or regional long-range planning, which would then feed into state or province planning, then to county or metropolitan planning, and finally local city agency or private firm planning. Regardless of the particular form of hierarchy preferred, the main function of hierarchical concepts is to point out that the many elements of the transportation system naturally fit together as interrelated components of a much larger whole. Similarly, the transportation system may be viewed as simply one element of a very complex socioeconomic system.

Transportation planning is undertaken for a variety of reasons. One of the most important reasons is that a very long period of time is required to implement most major changes in the transportation system, particularly the construction of new facilities. Hence the reaching of a rational decision re-

Figure 15-1 The hierarchical structure of land use and transportation planning.

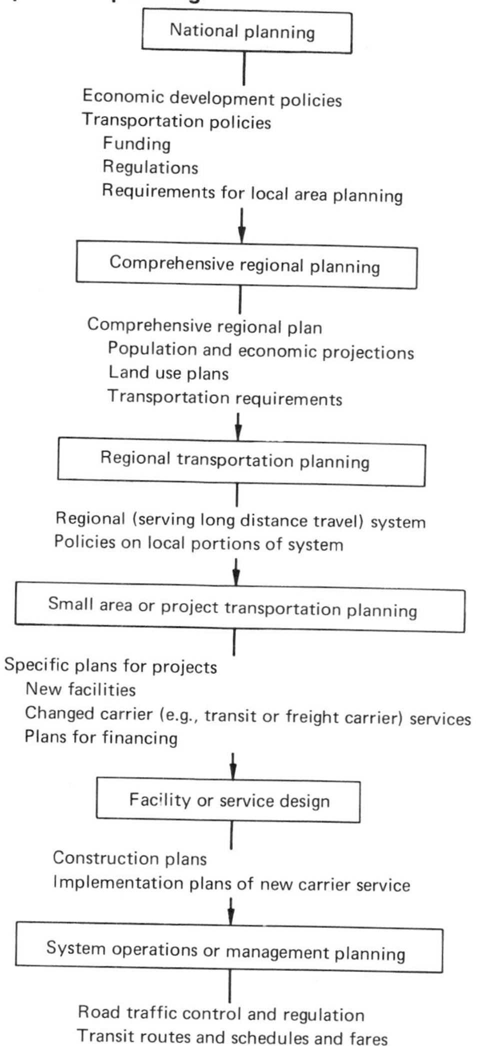

garding whether or not to construct a particular facility requires looking ahead into the future, at the period when it will be used and the benefits from its use will occur. Also, for the same reasons, it is essential to look into the future in order to identify potential problems sufficiently early that new facilities or other changes could be implemented to alleviate those problems before they become intolerable.

It is also important to plan in order to make sure that various changes in the system will work together in a compatible manner and that they yield maximum benefits to the region. If each new facility or other change were planned entirely independently of other changes, then a system which has many good parts but does not function well as a whole might emerge. Also, the process of specifying intended future changes in a plan that is made public in advance of implementation helps to identify any problems associated with the planned actions before funds and other resources are committed or partially expended. This opportunity for reaction is particularly helpful when plans are presented to the public so that various interested persons and groups can respond to them before a final commitment is made. The publicizing of future plans for such an important component of modern society as transportation enables private industry and other groups to take into account those plans in deciding their own future actions with regard to such decisions as where to locate new factories, etc., thus avoiding what otherwise might turn out to be costly mistakes. Planning also permits the identification of future financial resources needed to implement the plans, providing a basis for developing appropriate funding arrangements or modifying plans if funding cannot be obtained.

In addition to the above reasons for undertaking long-range planning, which apply to planning in any situation, either public or private, in many instances transportation agencies in government are required to develop long-range plans. Such requirements are very common when a higher level of government (or external agency such as an international financing agency) is providing part or all of the funding for transportation projects. The requirement is based on the premise that the funding agencies should be able to see what the entire transportation system will be like and how it will perform when the projects which they are being asked to fund are implemented. The plan thus serves as a test for the wisdom of pursuing these investments. Also, it is often required in the public sector that any transportation project be an element of a comprehensive long-range transportation plan for the region. This ensures that any change in the transportation system will be made in conjunction with other parts of the system and ensures that funds are not wasted on duplicative projects. Similar considerations apply to plans within the private sector, of course.

Although long-range plans have been discussed as if they remain fixed once decided on, they are usually quite dynamic. They are more often statements of future intent, subject to change with the passage of time. Although at any one moment a plan may specify all of the changes intended to be made in the transportation system within a period of, say, 20 years (thus specifying what the transportation system would be 20 years hence), it is unreasonable to assume that the system will actually achieve that particular condition in such a period. Such a plan for the distant future, often called a master plan, is really used as a guide for investments and other changes in the system that will be made in the near future, since it is only those near-term projects to which a commitment of resources must actually be made.

Elements of the plan to be implemented in the distant future need not actually be carried out, since no resources are devoted to them now, nor do contracts have to be signed for construction or other tasks of implementation. Also, it is not inconceivable that some future projects now planned will not be undertaken, as a result of future contrary decisions. In fact, with the passage of time, it is usually found that land use patterns and other factors affecting transportation change in directions not fully anticipated in the original master plan, and as a result that plan is reexamined and changes are made in it. This process of reexamining the plan and in effect developing a new plan is usually a continuous process, or nearly so, with the result that the master plan for the distant future is always in a state of flux. However, at any one moment the master plan in effect at that time is the major instrument by which particular projects are selected for implementation. Thus the master plan provides a long-range guide for the development of the system, but that guide is nevertheless modified as conditions change.

THE URBAN TRANSPORTATION PLANNING PROCESS

Urban transportation planning is undoubtedly the most common form of long-range transportation planning. The field has developed very rapidly, essentially since World War II, primarily as a result of rapid population growth and increased spatial extent of urban areas (often termed "sprawl"), especially in developed nations. In fact, as we noted in Chaps. 10 through 12, most of the methods for modeling entire transportation systems—including their relationships with the socioeconomic environment—have emerged as a result of the practical need for tools to assist in decisions regarding the best transportation investments to make in these rapidly growing urban areas. The field of urban transportation planning is still changing and maturing, and it is widely recognized that the methods and procedures which have been developed are not entirely adequate to deal with the very complex problems raised by rapid growth in urbanized areas and the demands for a higher quality of life. However, the general approach as well as many of the specific methods represent a significant advancement over earlier approaches to planning.

Overall Approach

Urban transportation planning has been institutionalized in the United States and many other nations as the process responsible for the planning of all publicly owned transportation facilities within a metropolitan area. Furthermore, transportation plans must be coordinated with land use and other plans for the region. This planning must be undertaken continuously, so that the long-range plans as well as the immediate action programs can be modified to meet changing needs. As a result of these requirements, the urban transportation planning process has become known as the 3C transportation planning process—continuing, comprehensive, coordinated transportation planning process.

The basic steps of a typical urban transportation planning process are shown in Fig. 15-2. Since planning must start with data that represent the region served by the transportation system, one of the first steps is the gathering of data, or development of the inventory, as it is often termed. These data provide the basis for identifying problems that might not have been identified previously by more casual observations and help sharpen the specification of known problems. Also, they provide the basis for the development of the various models used to forecast future land use, determine travel patterns, and model the performance of various proposed changes in the transportation system. These models have already been described in Chaps. 10 through 12. Analysis of the data by professionals, leading to problem identification, and involvement with community and political leaders, etc., leads to the

Figure 15-2 **The urban transportation planning process.** [*Adapted from Creighton (1970, p. xix).*]

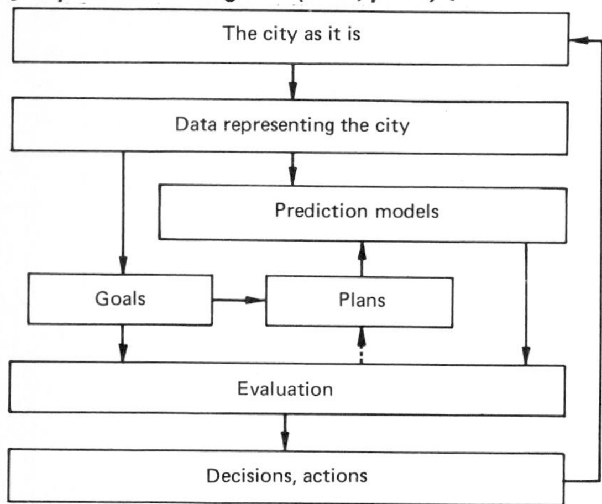

specification of goals or requirements for future transportation systems. From this statement, alternative plans are then formulated. These are tested using the various models of urban development and the transportation system, and it is the prediction of the consequences of any particular plan which provides the data for evaluation of each alternative plan in terms of the degree of achievement of these goals. As the dashed arrow between evaluation and plans indicates, usually the steps of formulating plans, predicting their consequences, and evaluating them are passed through many times, each time a new plan which might represent an improvement over the plans tried earlier is tried. Finally, an acceptable plan or set of plans is developed and this information is then passed on to various decision-making units. This selection process typically includes the interaction of professionals, community groups, and various levels of government, as we noted in Chap. 14.

This plan is then implemented, in the sense that those projects earmarked for early implementation are undertaken. Since the metropolitan area is constantly changing, the process is a continuous one. This is indicated by the arrow on the right side leading back to the changing metropolitan area and the iteration through the planning process again.

As we mentioned earlier, the transportation system is but one important element of the infrastructure of the entire socioeconomic system which makes up a metropolitan area. As a result, changes in the transportation system must be coordinated with other plans, particularly land use and the provision of other elements of public infrastructure such as water supply, gas lines, sewerage, etc. The need for this coordination is widely recognized, although existing institutions in many developed nations, particularly in North America, do not foster this type of coordinated planning to a large degree. Regional control over transportation decisions and changes is quite substantial, primarily because of the funding arrangements which give considerable power to various levels of government. A similar hierarchy usually does not exist for land use and other infrastructure planning, however, and in these cases often the smallest units of government make decisions, often largely independently of decisions or consequences in other areas. However, if such coordination is to be further developed and is to lead to desirable outcomes rather than simply to an extended bureaucracy, many conditions must be met. These are clearly stated by Creighton (1970, p. xxii):

The foregoing does not suggest that coordination is undesirable or permanently unworkable—quite the contrary. It does suggest we have to be realistic. If coordination of multiple programs is going to be obtained, and particularly if land use planning is to be coordinated with transportation planning, there must exist the ability to measure the important relationships (and they are not all important) between different functional activities, an agreed set of goals, and the ability to evaluate objectively the impact of alternative arrangements of human activities upon the quality of human life. Out of this there can come a sound national and regional land development policy and this can then be meaningfully served and supported by planned transportation systems.

In the next few sections we will discuss the steps in the transportation planning process in more detail. In particular, we will emphasize the steps dealing with data gathering, plan implementation or action, and generation of alternative plans, as these steps have not been covered in prior chapters. In fact, the generation of alternatives is so important that we will devote an entire section to it. But first the use of models will be reviewed.

The Use of Models

Returning to the transportation planning process, a more detailed view of the role of modeling is provided in Fig. 15-3. This figure stresses the aspects of the planning process associated with predicting patterns of land use and other activities, associated transportation demands, and performance of particular transportation systems in the context of those activity patterns. The roles of the various models of the transportation system and its relationship to its environment described earlier can be seen. Land use models are used to predict the future pattern of activities, from which estimates of the demand functions for transportation are developed. These transportation demands are then merged with the supply of transportation facilities and services (we use the term "supply" somewhat loosely in comparison to the precise economic meaning we noted in Chap. 11) to yield estimates of the actual amount of travel which will occur and the resulting quality of service, prices, etc. Since the pattern of transportation service will influence the direction of land use development, this effect or impact is accounted for by the feedback (dashed line) from the transportation system flow pattern to the land development model.

This set of models would be operated in the following manner. First, a projection of land use patterns at a time 2 to 5 years in the future is made, with the pattern of development being influenced by the current transportation system. Then at this point in the future the transportation system is modeled to yield actual flows and performance and then this is used as a basis for prediction of the pattern of land development over another increment of time. Of course, the pattern of land development is influenced by land use policies

Figure 15-3 **Transportation models as used in urban transportation planning.** [***Adapted from Manheim (1966, p. 52).***]

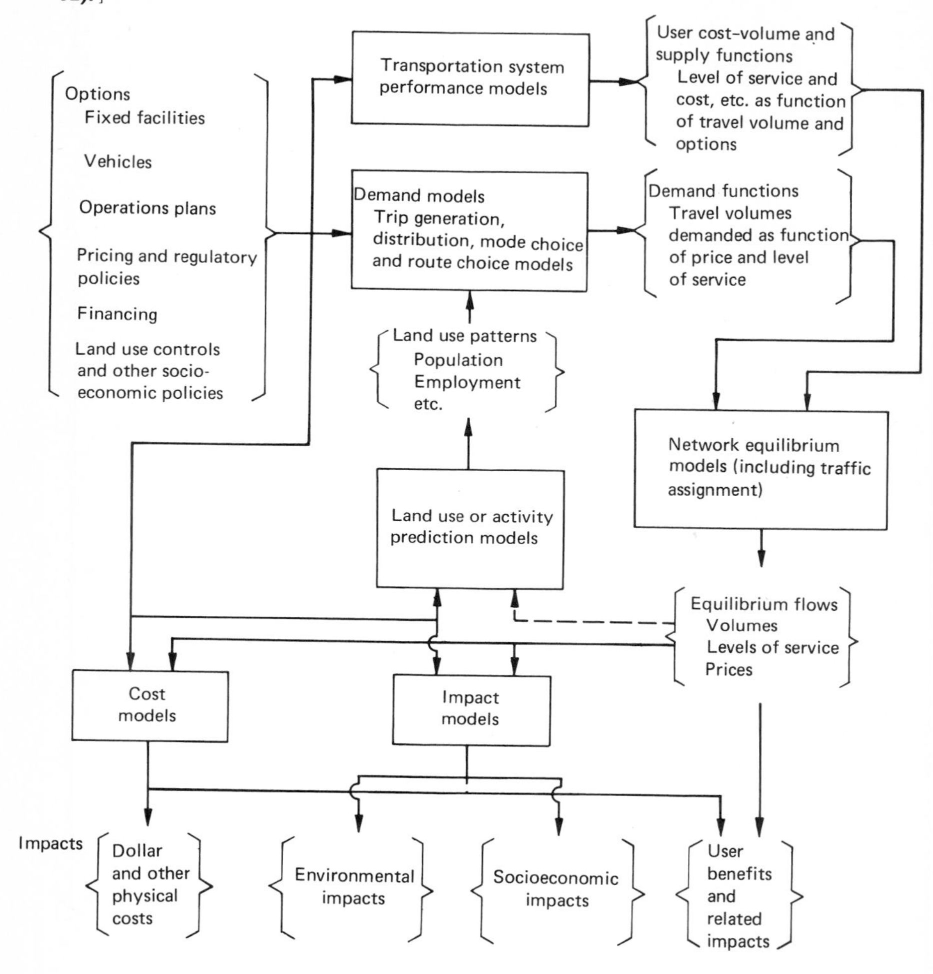

and other development policies, including such factors as zoning, the siting of major new centers of activity, etc. Then at this point in the future the transportation system is modeled again, and the process continues. This modeling by using the land use model to jump ahead over a period of years and then simulating the transportation system at that particular time in the future is performed as many times as is necessary to reach the end of the planning period, termed the planning horizon. This same process is followed every time an alternative plan for transport, or conceivably for land use or other policies, is to be evaluated.

In the context of urban transportation planning, it is most common to use the particular models or methods for land use planning and travel demand modeling which we described in Chap. 10. Specifically, the travel demands are typically modeled by a combination of a trip generation model, a trip distribution model, a mode choice model, and a traffic assignment model. However, in other planning contexts other types of models for these same processes may be used, as in the case of intercity transportation planning, where a model such as the Kraft-SARC model might be used. Thus this basic structure for long-range planning applies to any context in which formal mathematical models will be used for predicting consequences of the various plans.

Data Collection

In order to use any of the models of the transportation system or its environment which we described earlier, it is necessary to have estimates of the various parameters of those models. In transportation planning it has been discovered that parameters which apply to one city or metropolitan area often do not apply to another, and therefore every time a planning activity is undertaken in a particular area a basis for estimating the model parameters for that area must be developed. The primary means for doing this is to obtain survey sample data on the characteristics of that particular metropolitan area and its transportation system, and from these estimate those parameter values which best replicate the data.

Although the process of data collection may seem rather mundane, it is in fact one of the most difficult, time-consuming, and expensive portions of transportation planning. In many urban transportation planning studies, for example, up to one-half of the budget is spent on data collection and the development of parameters of various standardized models. Of course, the data collection is not undertaken solely to support the use of formal models. It also has a great deal of value in simply describing the current state of transportation in the metropolitan area for which the plan is being developed and in helping to identify present and potential problems for that area. While the exact methods for gathering data actually fall somewhat beyond the scope of this book and are most appropriately studied after one obtains a considerable background in statistics, it is useful to have some basic understanding of the data collection phase.

Three main types of data must be collected to support urban transportation planning: (1) inventories of travel, (2) inventories of land use and associated human activities, and (3) data on the transportation system or networks over which traffic is accommodated.

Travel surveys are designed to gather information on all of the trips made in the metropolitan area; these data are useful in assessing the adequacy of transportation facilities as well as in developing models of travel demand. For purposes of data gathering as well as other purposes, travel can be usefully categorized into a number of types. These are illustrated in Fig. 15-4. The area to be studied is bounded by what is termed the cordon line. Internal trips are those whose origin and destination are within the cordon line boundary. External trips are defined as having one end inside the survey area and one end outside the area. Finally, there are a few through trips, whose origin and destination both lie outside the survey area but which pass through that area and hence use facilities of interest in the planning. Sometimes these three types of trips are called internal-internal, internal-external, and external-external, respectively.

Figure 15-4 **Types of urban trips.** [***From Creighton (1970, p. 154).***]

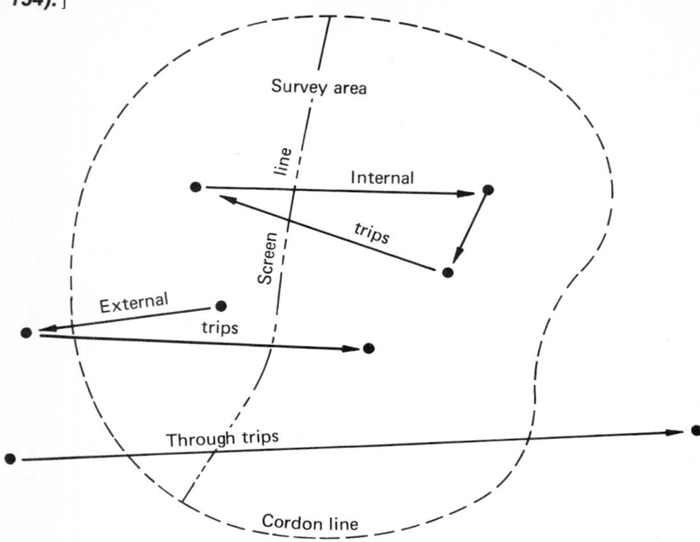

Also shown in this figure is a screen line, which is a line passing continuously from one point on the cordon line to another. Often one or more screen lines is set up in an area in order to assist in checking the extent to which the sample of trips actually yields a reasonable estimate of the total number of trips. It is very easy to measure the number of trips passing a line, this often being done by automatic vehicle counters on roads and through use of readily available information from transit lines. It is therefore possible to compare an estimate of the total number of trips which should cross the screen line made from the survey with the actual number of trips observed to cross that line. In this way errors in the surveys can be detected and corrected before the data are used. Screen lines are usually set up along natural boundaries, such as rivers or mountain ranges, through which there are relatively few transportation arteries.

There are many techniques for obtaining information on trips within an urban area. These include home interviews, telephone surveys, and mail surveys. Usually only a small fraction of all households or travelers is interviewed, simply because it takes a considerable amount of time to obtain complete information on all of the travel of an individual in the course of a day. Typically the sample size for home interviews is around 1 percent in the largest of metropolitan areas, increasing to 10 percent in very small cities. The data gathered include information on each and every trip made by a person in the course of the day, the origin and destination address for each trip, the trip purpose, the land uses at the origin and at the destination, the mode of travel, often the route of travel, the time of day, and, if by automobile, information on any other persons in the car. Similar surveys are usually conducted for trips by trucks and other commercial vehicles, again by sampling a small fraction of total trip makers.

For trips which originate outside the cordon line, the information on trips is obtained by surveying the traveler at a point where he or she crosses the cordon line. Usually this is done at the roadside for automobile and other road vehicle trips, with only a sample of all vehicles being subject to interview in this manner. Similarly, travelers on common carriers can be interviewed while on board their vehicles.

There also may be a number of specialized surveys which reflect particular trips of interest. For example, if improvements are planned for central business district distribution systems, there may be a special survey of persons walking in that area to determine characteristics of that travel. Whenever such special surveys are conducted, combining those data with data from home interviews and other sources must be done very carefully to avoid double counting of trips.

These data are then used as a basis for estimating the total population of trips within the metropolitan area. This is done by multiplying the number of trips within any category of interest observed in the survey by the inverse of the fraction of the population surveyed. For example, if 1 in every 100 households was interviewed, then trip data obtained from that survey would be multiplied by 100 to obtain the total population of trips by members of households. However, in practice the degree to which trips are made but not actually reported through interviews varies considerably with the type of trip, and as a result the multiplier used may vary somewhat from this inverse value. Often the values will be determined by checks of total trips made against screen line counts, which are usually highly reliable.

Also, it is essential that data on land uses and associated activities in the area be obtained. The basic data sought are on the type of land use within any travel zone, the area of land devoted to that activity, the location of the particular parcel of land (often by both street address and a geographic coordinate location), and often additional data such as parking facilities, street access, railroad siding availability, and zoning. Often many of these data are available from municipal records and records of various utility companies, but many nevertheless must be obtained by field surveys. These data are then combined with the travel data in order to provide estimates of the parameters of various demand models for transportation.

Although it may seem rather strange, there usually exists no single inventory of all the transportation facilities within an urban area. The reason for this is simply that responsibility for providing and maintaining the facilities is divided among many organizations. In addition, some may not have very precise information on the facilities for which they are responsible, particularly if those facilities were constructed many years ago. The network inventory thus is oriented toward identifying all of the fixed facilities in existence in an urban area, including their location, length, design features such as width, capacity, or number of tracks or lanes, special characteristics such as type of signals and design speeds, and the locations of intersections or points of interchange with other facilities. Also, in the case of facilities over which common carrier services are operated, the important level of service and price characteristics of those services must be obtained.

Closely related to this are surveys of actual use of existing transportation facilities. Included here are measures of the total vehicle-miles of travel on each of the facilities, usually with specification of variations by time of day and perhaps by vehicle type. These data are usually obtained by either manual count on roads or by automatic counters connected to tubes crossing the road that count axle passes rather than vehicles, with the former (manual) method having the advantage of being able to distinguish among vehicle types. Also, actual travel times on roads are often measured by driving in the traffic stream. To obtain vehicle-miles per unit time (e.g., hours), the measured volume is simply multiplied by the link length. This information on usage and level of service obviously is particularly useful in identifying places where congestion and overcrowding of facilities exists.

These data are then sufficient not only for the description of the existing status of the region with respect to land use and transportation characteristics, but also for supporting the development and application of the various models designed to predict the future state of the region and to assess the effectiveness of various transportation plans. [For more detailed information on surveys, see Creighton (1970, chap. 7), and National Committee on Urban Transportation, 1958.]

Evaluation of Plans

The evaluation of alternative plans is of critical importance to the success of the entire planning process. In recent years much research and development has been directed toward evaluation methods for transportation planning. The results of this effort were presented in the previous chapter in the discussion of multiple-objective evaluation methods and the presentation of examples. Since they were covered there, with an example drawn from urban transportation planning, there is no need to repeat the same material. However, a closely related aspect of transport planning, that of deciding when to implement each improvement, not yet covered, will be treated below.

Programming Improvements

An important element of any plan is a statement of when various changes in the transportation system should be implemented. This is essential in the case of long-range planning, for the period covered through the horizon year typically is 20 or more years, and implementation of improvements will be spread out over that period.

It is generally impossible to implement all the elements of a plan immediately, because of limitations of budgets and on total resources available within the construction industry, which would preclude massive sudden new construction even if funds were available. The specification of when various improvements will be instituted is termed programming, and the result of it is termed a programmed or staged plan.

The sequence of projects can be selected on the basis of a number of criteria. It is desirable to implement first those portions of the plan which will yield substantial benefits even without the presence of projects that will be implemented later in the period. The relevant benefits are those that would result provided that other projects identified for earlier implementation are already in place. Also, the order of projects and the total number of projects to be implemented in any one year should be consistent with the total amount of funds which will be available in that year. Care must be exercised in identifying the amount of funds which will be available to include any funds which may have become available earlier but remained unexpended, provided that

Figure 15-5 **A programmed long-range plan: example of freeways in the Chicago metropolitan area. *(a)* Map of existing and proposed freeways. *(b)* Planned period of investment in each segment.** [***Adapted from Eash, Haack and Civgin (1970).***]

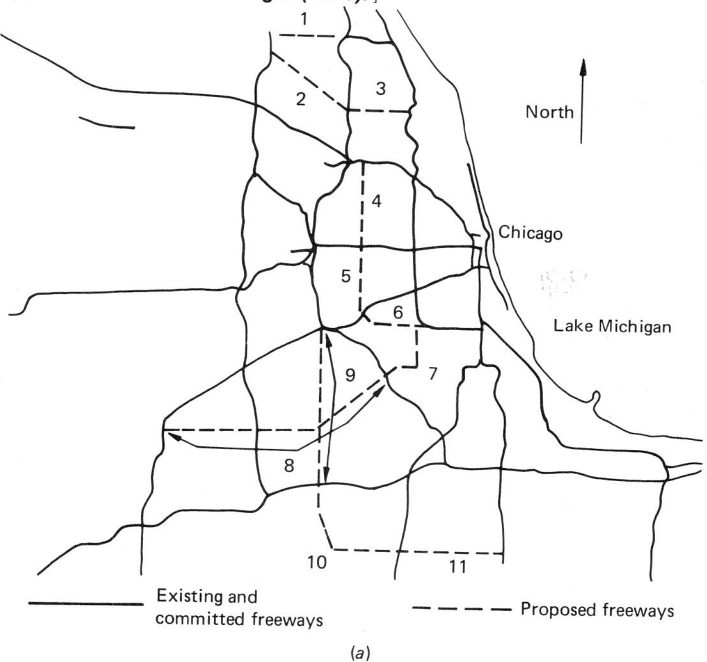

(*a*)

Segment	Year 1	2	3	4	5	6	7	8	9	10
1									Both	Both
2			Both	Both	Both	Both				
3							Both	Both	Both	Both
4	Both	Both								
5	Both	Both								
6					Both	Both	Both	Both		
7	Both	Both								
8	Purchase right of way	Purchase right of way	Construct	Construct						
9	Purchase right of way	Purchase right of way			Construct	Construct				
10	Purchase right of way	Purchase right of way			Construct	Construct				
11					Both	Both				

– – Purchase right of way —— Construct ▬ Both

(*b*)

those unexpended funds can be retained (and perhaps draw interest) until they are used. Finally, consideration should be given to the fact that projects identified for implementation in distant years may in fact never be implemented, because of changes in the plan with the passage of time, and therefore the system which emerges at any point in time (as a result of prior project implementation) should be one which is amenable to future additions and changes which may or may not have been in the original plan. Since it is impossible to predict the future with certainty, it may be found that the future land use and other conditions which made a particular plan attractive will not actually occur, and the actual pattern of activities may call for revisions in the long-range plan. Therefore it is important that the original plan be flexible enough to accommodate changes. These changes might be due to increases in travel on certain facilities, decreases in travel on some, or major changes in the spatial pattern and lengths of trips, so that a substantially different transport system is called for. (Many of these considerations are covered in more detail in problems at the end of the chapter.)

The actual process of staging projects may be carried out in a number of ways. Sometimes this is done by ad hoc procedures, by trying out different sequences of project implementation to see how they perform and selecting the best from among the alternatives tried, in much the same way that the overall planning process results in a final system plan. However, recently, analytical methods for staging projects have been developed using advanced mathematical techniques such as mathematical programming.

An example of one method applied to a problem of staging highway improvements in Chicago is contained in Eash et al. (1970). Their example indicates when various sections of a planned expressway network for Chicago are to be implemented (Fig. 15-5). As can be seen, many of the sections take 2 or 3 years to construct, and as a result funds are earmarked for expenditure over that period. Also indicated in this figure is the purchase of right of way before construction in order to take advantage of low land prices and to forestall development of the right of way.

ALTERNATIVES AND THEIR GENERATION

It goes without saying that one of the most important elements of transportation planning is the generation of alternatives. It is axiomatic that a transportation plan can be no better than the set of alternatives from which it was selected. Therefore, a great deal of attention is properly devoted to the process of developing transportation plan alternatives. There are many different approaches, all of which are helpful but none of which necessarily leads to the "best" plan. However, in any context it may be useful to utilize many of the alternative system concepts described below and the various strategies that have been developed to aid in the process of generating good alternatives.

Land Use and Travel Patterns

Although transportation planning is primarily concerned with only transportation system changes, as the name implies, it is important to remember that the effectiveness of any transportation system is very much influenced by the land use and travel patterns in the area it serves. Many transportation problems would be substantially alleviated if not eliminated simply by land use changes which resulted in changes in travel patterns. Thus it is important, in transportation planning, to be aware of the interrelationships with land use, and in certain cases it may be appropriate to consider alternative patterns of land use in addition to the more usual transportation facility and service options. In this section we shall try to highlight some of the many different forms which land use patterns can take, and the travel patterns and the transport systems associated with them.

Before discussing specific land use patterns and associated travel patterns, it may be useful to review some of the principles which have guided land use planning in the past. At a time when transport was relatively expensive—up to the early part of this century—the arrangement of various activities within urban areas was very much constrained by the need to provide most of the activities in which people engage within walking distance of their households. Thus factories were often located next to workers' housing, and within residential areas one typically found schools, recreation and social facilities, stores, and most other land uses which support households.

This mixing of various land uses had numerous disadvantages, particularly in an era when the burning of fossil fuel was the source of most energy for manufacturing and other activities. As a result, when transport became cheaper relative to typical family income, planners attempted to separate the various land uses which tended to conflict with one another, and at the same time the public seemed to be gravitating toward such preferences also. This often led to substantial distance between residences and places of work, a tendency which has been further exaggerated by the suburban location of much population growth in recent years. Undoubtedly the construction of freeways and radial rapid transit lines (both usually with low user costs and large subsidies) have helped induce this tremendous separation of various activities in which people engage. As large shopping centers have replaced neighborhood shopping areas, travel by automobile or public transit has become necessary for shopping. Similarly, regional schools have replaced neighborhood schools (and the single-room country schoolhouse!), with a resulting increased need for student travel. However, in recent years the character of much workplace activity has changed markedly, with the result that it is quite possible to provide housing near workplaces without negative side effects of unsightliness, noise, or pollution. This is due in part to the technological changes in many industries: cleaner fuels have replaced coal, and less noisy production facilities are used. Also it is due to the increasing fraction of all employment in service industries, such as offices of many types, which can be good neighbors to residential areas. The effect of these changes can be seen in some of the redeveloped inner cores of cities, where very attractive housing is provided on the periphery of the central business district, and also in some new planned communities (often called new towns). Thus, it seems that the technology of housing, shopping, some types of industry, etc., permit the various activities to be located fairly close to one another without major conflict if that should prove desirable for other reasons, such as reducing the need for mechanized travel.

Given the apparent wide range of options with respect to the relative spatial locations of various activities, it is useful to consider the various types of spatial arrangements which might occur in the associated travel patterns.

Figure 15-6 **Idealized patterns of land use and travel in a circular metropolitan area. *(a)* A single core center. *(b)* Multinucleated centers. *(c)* Diffused hierarchical centers.**

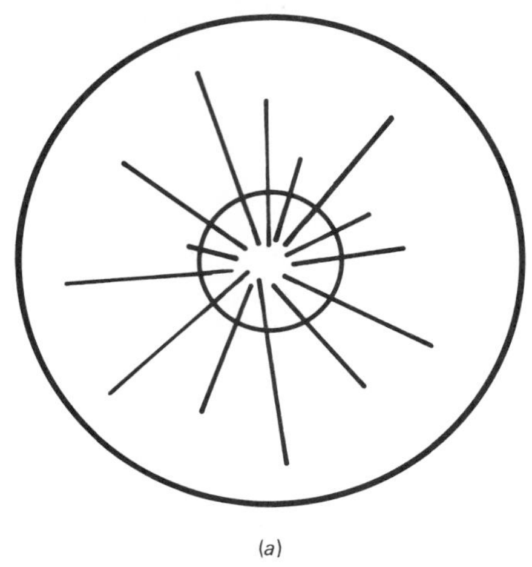

(*a*)

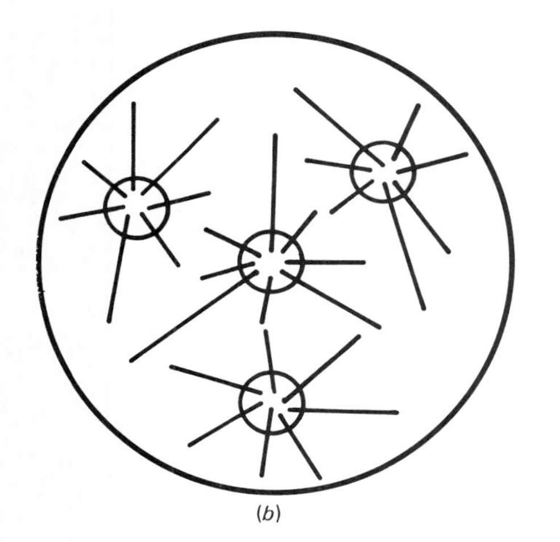

(*b*)

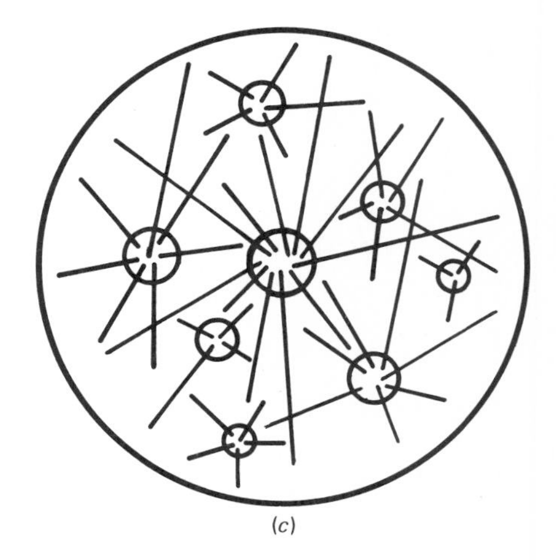

(*c*)

○ a major activity center (e.g., industrial area, shopping center) with diameter proportional to amount of activity

— a trip

Figure 15-6 presents various idealized forms of land uses and the associated travel pattern. One idealized form would focus all travel on a central core, resulting in relatively long trips if the area is large, as in Fig. 15-6*a*. This is in fact the typical form of travel in smaller municipalities where there is only one central business district, which provides the major place of employment as well as the focus of shopping, recreation, and education. Larger urban areas typically have many centers, on each of which a great deal of travel is centered. Obviously the replacement of a single center with many centers, each of which is equally attractive, tends to reduce considerably the length of trips, as in Fig. 15-6*b*. Large urban areas typically have many nuclei, of different character, such that certain trips can be satisfied by traveling short distances and more specialized trip purposes must be satisfied by traveling longer distances to one of a few centers which would cater to that particular need. This tends to lead to a rather diffused pattern of travel which includes both short and long trips. In many areas centers for various types of activities are diffused throughout the region, leading to a very diffuse pattern of travel, as in Fig. 15-6*c*.

In addition, variations in overall shape of metropolitan areas or regions have strong influences on travel patterns. Although most urban areas tend to be roughly circular in shape, with the exceptions due to natural boundaries, it is interesting to consider other shapes in the context of planning new towns. Three idealized shapes are shown in Fig. 15-7, including a line, a star, and a circle. The primary advantages of the star pattern from a land use standpoint are that it provides for much open space in the vicinity of residences. From

Figure 15-7 **Idealized shapes of metropolitan areas and associated patterns of travel. *(a)* Circular form (see also Fig. 15-6). *(b)* Star form. *(c)* Linear form.**

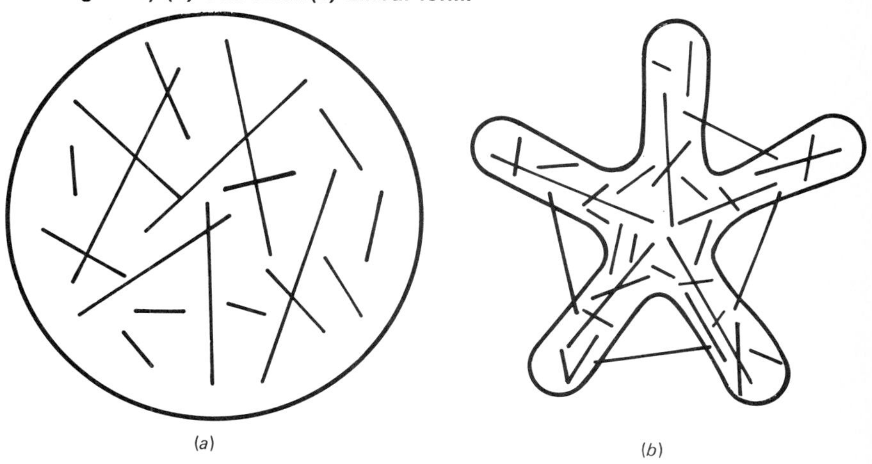

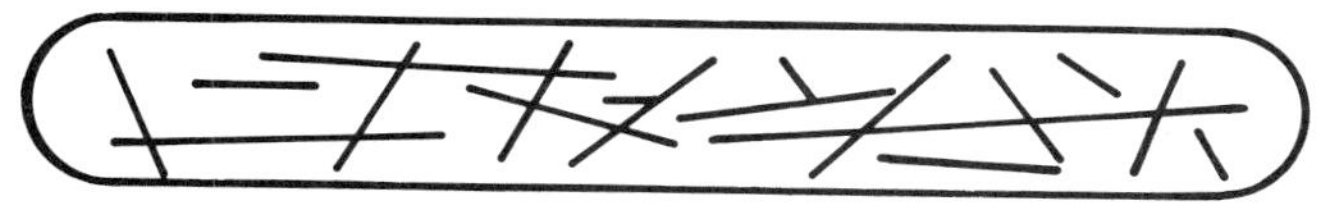

(*c*)

the transportation standpoint, the efficacy of each of these forms depends very much upon the location within these patterns of origins and destinations of travel. For example, the linear form could be easily served by a single linear transportation route of appropriate capacity, but the concentration of all activities at the midpoint of this linear strip might create severe capacity problems in that area and lead to substantial underutilization of transport capacity at the extremities.

With the advent of much motor vehicle traffic in towns and cities a new concept of coordinated road planning and land use planning has emerged. This concept is that of an environmental area, in which the road traffic is kept to a volume such that it does not interfere with other activities. This concept has been popularized by a major study of urban transport in Great Britain, entitled *Traffic in Towns* (Buchanan et al., 1964). It is described as follows:

This is the only principle on which to contemplate the accommodation of motor traffic in towns and cities, whether it is a design for a new town on an open site or the adaptation of an existing town. There must be areas of good environment-urban rooms—where people can live, work, shop, look about, and move around on foot in reasonable freedom from the hazards of motor traffic, and there must be a complementary network of roads—urban corridors—for effecting the primary distribution of traffic to environmental areas. These areas are not free of traffic—they cannot be if they are to function—but the design would ensure that their traffic is related in character and volume to the environmental conditions being sought. If this concept is pursued it can easily be seen that it results in the whole of the town taking on a cellular structure consisting of environmental areas set within an interlacing network of distributory highways. It is a simple concept, but without it the whole subject of urban traffic remains confused, vague, and without comprehensive objectives. . . .

Applied to a whole town, it would produce a series of areas within which considerations of environment would predominate. These areas would be tied together by the interlacing network of distributory roads on to which all longer movements would be canalized without choice. The relationship between the network and the environmental areas would therefore be essentially one of service: the function of the network would be to serve the environmental areas and not vice versa.

This concept is illustrated in Fig. 15-8, where freeways and arterials are designed to take through traffic away from the streets of the environmental areas, thereby reducing noise, pollution, and the barrier effect of a heavy-volume street.

Another important characteristic of land uses and their associated travel patterns is the temporal pattern of travel demand. Activities (land uses) may be so widely separated that the high-capacity transport facilities which are necessary to meet the peak period demands are used to full capacity only for a small fraction of the day, due to the sharply curtailed demand in the off-peak periods. This situation seems to apply to many central business districts in larger North American cities, where the primary activity is weekday work, which requires a substantial peak period capacity, and where the activities occurring at other times of day, such as shopping and recreation, do not fully

Figure 15-8 **The environmental area concept.** [*From Hutchinson (1974, p. 236).*]

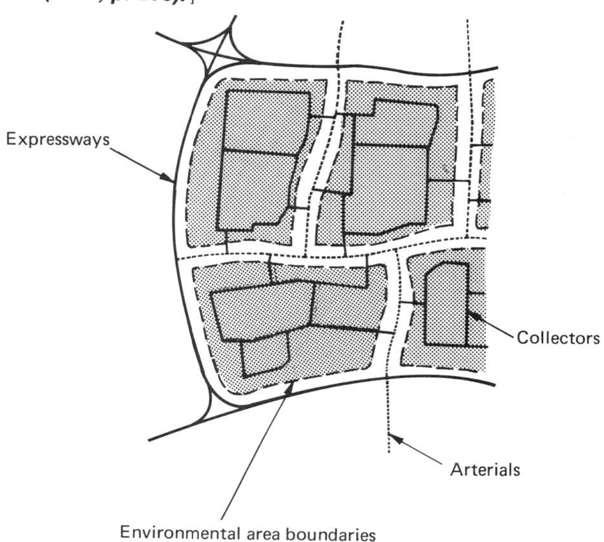

utilize the capacity of transport facilities. If land uses are mixed such that the transport facilities needed to provide capacity for travelers for one type of activity during one portion of the day can be used to satisfy the capacity needs of travelers for another activity at another time of the day, then that transportation capacity is better utilized and the total cost of facilities is reduced. Similarly, facilities built for weekday travel may also be useful for major recreational flows on weekends if the location of activities is appropriately arranged.

Probably a great deal can be done to encourage use of a facility for many types of travel, also. While it may be unacceptable to most people to ride a rather spartan public transport service for social or recreational trips, even though those facilities are quite acceptable for work travel, a transport facility which can provide a higher level of service during periods of recreational or social travel might become a much more useful facility at very low additional cost. The allocation of some lanes of major arterial roads and expressways to buses and perhaps other high-occupancy vehicles only during work trip peak periods, while allowing those same lanes to be used by all classes of vehicles at other periods is one example of a change to reflect the different requirements of different trip purposes.

Transportation Networks

Closely related to the spatial pattern of travel is the form of the transportation network appropriate for meeting the requirements of that travel. Setting aside for the moment the question of what technologies or modes are used on the network, various idealized forms of transportation networks can be conceived. Figure 15-9 presents a number of these. Perhaps the most common form of transportation network within metropolitan areas is the grid network,

Figure 15-9 **Idealized types of transportation networks. *(a)* Grid network. *(b)* Radial network. *(c)* Ring-radial network. *(d)* Spine network. *(e)* Hexagonal network. *(f)* Delta network.**

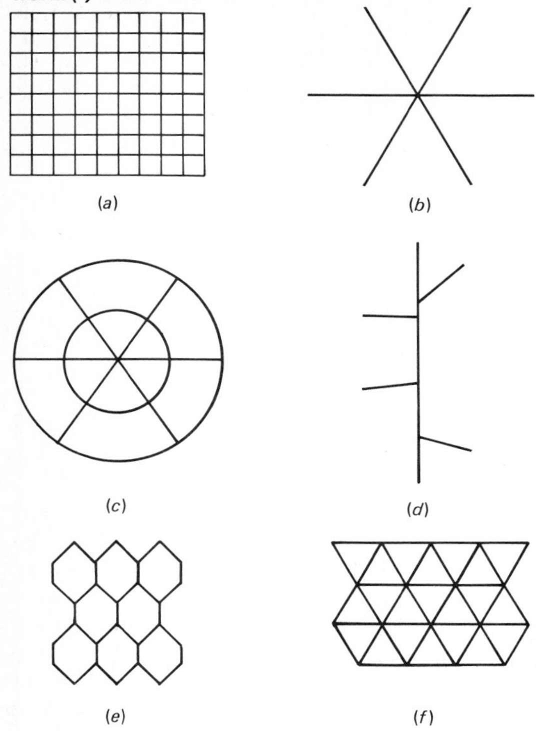

which is the form of the street network for most cities which have planned road networks. The grid network is particularly suitable for a situation where travel patterns are highly dispersed, as is typical for most large metropolitan areas at the present time, and also where an equal quality of transport service to all areas is desired. Another standardized type of network is a radial network, which focuses on a particular core area such as the central business district. In many cities a network of radial arterial streets and perhaps expressways is imposed on the basic grid network, expressing the importance of the central business district over various other centers of activities in the region. Another popular type of network, particularly for major arterial roads and expressways, is a combination of radial and ring facilities, which not only provides good access to the core area but also is suitable for traffic to and from outlying centers and for routing of traffic passing through the metropolitan area around the congested center. There is also the spinal network, typical of intercity transportation networks in many of the highly developed urbanized corridors such as in the northeastern portion of the United States. Finally, there are other abstract forms which are possible but do not seem to be used, such as the hexagonal network, which has the advantage of intersections with merges and diverges but no direct crossings, and the delta network.

In the context of urban transportation planning, a number of rules of thumb have been developed to assist in the designing of networks (Creighton, 1970, pp. 228–244). While these rules of thumb are by no means universally valid, they do seem to be widely accepted in the profession, and hence probably have considerable validity. With respect to freeways, a general rule is to avoid facilities that have stub ends, which in effect would turn freeway traffic onto local and arterial streets. This rule seems to presume that the freeway would carry a high volume of traffic, a likely assumption in urban areas but one which may not apply in rural areas where the freeway might be built to provide for very high speed travel rather than for a high level of capacity. Also, a general rule is to try to avoid intersections of more than four freeway approaches, because with more than this number the interchange becomes very complex for the driver and requires far too much land. Another rule is the number of lanes coming into an intersection should equal the number going out, so bottlenecks will not occur.

Also, much attention has been devoted to the general shape of freeway networks for urbanized areas in the United States. In this context, where the bulk of travel is very dispersed, with a rather modest focus on the central business district and other major activity centers, it has been found that a pattern which is essentially a combination of the grid and ring-radial patterns is most appropriate. In this hybrid pattern, the distances between parallel or nearly parallel facilities is small in the vicinity of the central business district and other major activity centers and then gets larger with increasing distance from those areas of concentration of traffic. A purely ring-radial system focuses far too much traffic on the congested core area links, yet it is desired to have a closer spacing of freeways there because of the higher traffic volume. Hence the compromise pattern, typified by that shown for Chicago in Fig. 15-5, seems best.

There seems to be less agreement over general principles for public transit networks. In many cities, as a matter of policy, a basic local transit service is provided on a grid pattern, so as to provide mobility via public transport throughout the entire region. However, the bulk of transit travel in the United States today is to and from central business districts, which the networks reflect by forming radial route patterns particularly for rapid transit and other high-quality types of services. In many cities a grid of bus routes serves local trips in all directions, and CBD trips are served by radial rapid transit lines, combining forms *a* and *c* of Fig. 15-9.

It is generally felt that a route with one end in an outlying location and the other end in the central business district is less desirable than a route which extends from an outlying point through the central business district to another outlying point because the through service requires fewer vehicle-miles of travel to provide the same effective capacity, since two separate routes would completely overlap within the central business district. Also, if a separate right-of-way is provided, the fewer vehicle trips resulting from the paired routes can reduce capacity requirements and hence initial construction and fixed plant maintenance costs. Of course, for paired routes to be efficient the amount of traffic on both should be roughly the same, and to gain maximum efficiency the routes should be paired such that a vehicle serving peak flows in one direction on one route reaches the end of the other route at a time that permits it to serve peak period flows on that route also. On the negative side, long routes tend to result in less reliable service compared to short routes simply because disturbances to schedule adherence have a longer period of time to take effect before vehicles reach a point at which there is a layover and hence time available to make up for the delay (Welding, 1957). Another problem with through routes is that usually vehicles are run from one end of the route to another, even though the capacity required may be such that termination of the trip at the central business district would be most appropriate. For this reason many commuter railroads and suburban bus lines actually run many vehicle trips into the central business district in the morning, store the vehicles at the terminal or at a nearby location during the workday, and commence the outbound trip in the evening. Obviously in this case through running would simply increase vehicle-miles and operating costs substantially.

Another important consideration in the design of public transit networks is the attempt to match the capacity actually provided to the traffic occurring on the route at each point. If travel patterns are such that there is a focus of travel on a single point, as is typically the case with the central business district, the amount of traffic on a route will typically decline with increasing distance from the central business district. Reductions in capacity corresponding to the reduction in traffic can be achieved by various means. One is to divide a route into zones, as has been discussed in Chap. 8. Also, certain vehicle trips can be terminated at intermediate points along a route even when all vehicles provide all-stop service, although this creates inconvenience and confusion for travelers.

In both highway transportation and public transit, the increasing diversity of technologies available for providing movement seems to be creating a hierarchical structure in both systems within urban areas. In the case of a highway system, the hierarchy includes freeways which are designed for high-speed, high-volume, relatively long distance travel; arterial streets, which are somewhat lower in level of service and capacity; collector streets, which feed the arterials; and finally local streets, which provide for land access. While short trips may involve movement over only local streets or local and collector streets, longer-distance trips involve traveling on the higher-type facilities and possibly some circuity of travel in terms of distance in order to realize the gains in travel time and perhaps operating cost resulting from the other facilities. A similar pattern seems to be emerging in public transport, where commuter rail systems and rapid transit lines with long distances between stations provide for movement over long distances at high speeds within urban areas. Shorter-distance travel can be provided by streetcar or bus transit or the more conventional rail rapid transit with fairly short station spacings (typically of the order of 0.5 mi), these catering to shorter trips and also providing access to and from the regional system. Similarly, local bus and streetcar service, service by jitneys and dial-a-buses, and high-activity-center distribution systems, which are similar to horizontal elevators in concept, can provide access to and from any of these systems, as can the private automobile in conjunction with appropriately located parking lots. Examples of these two types of hierarchies are presented in Fig. 15-10.

Figure 15-10 **Typical forms of hierarchical networks in urban areas.** ***(a)*** **Road network.** ***(b)*** **Public transit network.**

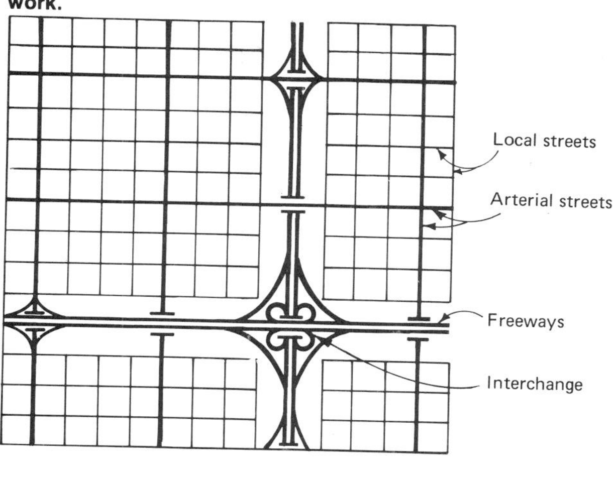

(a)

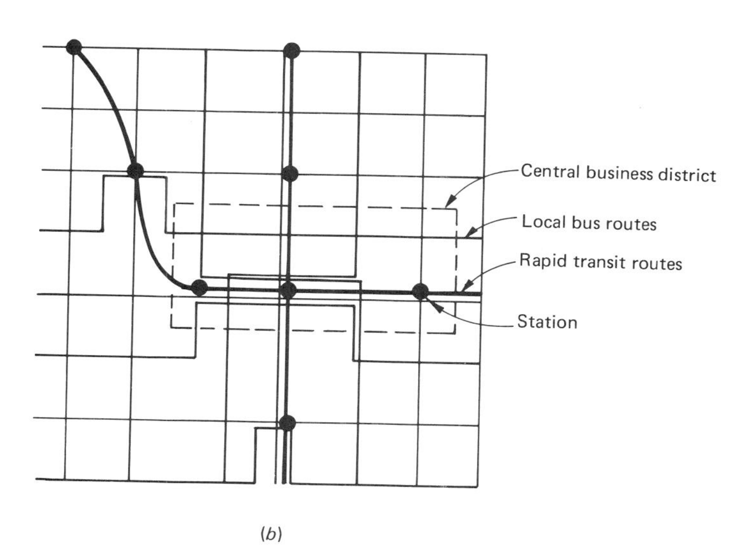

(b)

Creating Alternatives

One of the most difficult tasks in any planning is the process of generating or recreating alternatives. While the preceding discussions have identified a wide range of types of transportation activities for urban areas, the planner must conceive of the specific alternatives to be evaluated in any particular planning context. A number of methods have been devised which seem to provide useful results in creating alternatives.

One of these is termed brainstorming. This is basically a very unstructured process in which anyone somewhat knowledgeable in the subject is asked to join a group which will attempt to conceive of as large a number of alternatives as possible. It seems to be most successful when little or no attempt is made to evaluate an alternative as it is generated, so that the participants feel as free as possible to invent new alternatives.

Another particularly useful approach is to attempt to create alternatives which are particularly well suited to individual goals for the area and then combine their best features. This starts with the attempt to create one or more transportation systems which conform as closely as possible to a particular goal. Then attention is addressed to another goal, and again the process is directed toward identifying system alternatives which conform most closely to this goal. This step is repeated until all of the goals have been treated. Then an attempt is made to combine the best features of all of these alternatives in such a manner that one or more alternatives are created which are attractive from the standpoint of all the goals. This process seems to work best when there is relatively little conflict among the goals, for obvious reasons.

Probably the most common problem in transportation planning is to meet major requirements for capacity and quality of service. The capacity requirement is approached by developing a map of future travel patterns. This map consists of bands which represent the major travel flows between zones, the beginning and end points of the band being the origin zone and destination zone, and the width of the band being proportional to the total volume. It is usually impractical to represent all origin-destination flows in this manner, and one of two means of reducing the number of flows is used. One is to simply portray very large flows, meaning that many of the origin-destination zone pairs would not be represented by desire lines. This is illustrated in Fig. 15-11*a*. Alternatively, many desire lines may be combined provided that they are close to and approximately parallel to one another and hence might be accommodated by the same facility. This is usually done by assigning all trips to a very complete network, covering the entire area, usually a grid or a network of a grid plus both diagonals (Fig. 15-11*b*). These maps of desire lines can be generated by computer with automatic plotting, which facilitates their employment in transport planning, since the sample survey data and the models for predicting future trips are both invariably computerized.

Information on the length distribution of trips is particularly revealing in that it tends to indicate the relative importance of short trips which might be carried on local facilities and of long trips, which by their nature tend to require facilities with high levels of service. An example of such a trip length distribution for all automobile trips in the metropolitan area of Chicago is shown in Fig. 15-12. Perhaps the most striking feature of this is that even though this metropolitan area is some 60 mi from end to end, travel is dominated by short trips, roughly half of all vehicle trips being less than 3 mi. The importance of short trips is also illustrated by the related measure shown in the figure, which portrays the fraction of all person miles or vehicle miles of travel attributed to trips of various lengths.

A very useful form of portrayal of transport network service characteristics is the so-called isochron map of transport times or cost. Such a map presents the travel time (or cost) from a selected point to all other locations in the study area, the times given by lines of equal time, termed isochrons, as shown in Fig. 15-13*a*. These are usually prepared by computer, with times calculated

Figure 15-11 **Portrayal of travel patterns using desire lines and spider networks.** ***(a)*** **Desire line map. Note: Only volumes greater than 10,000 are shown.** ***(b)*** **Trips assigned to a spider network.**

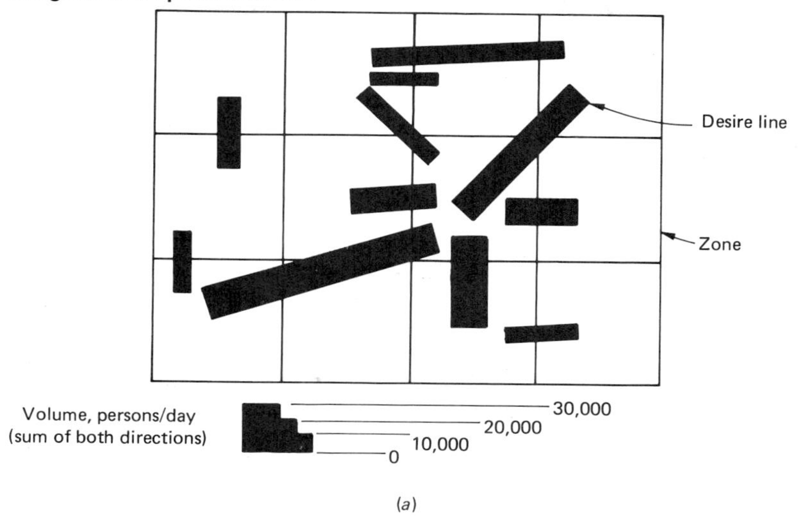

(a)

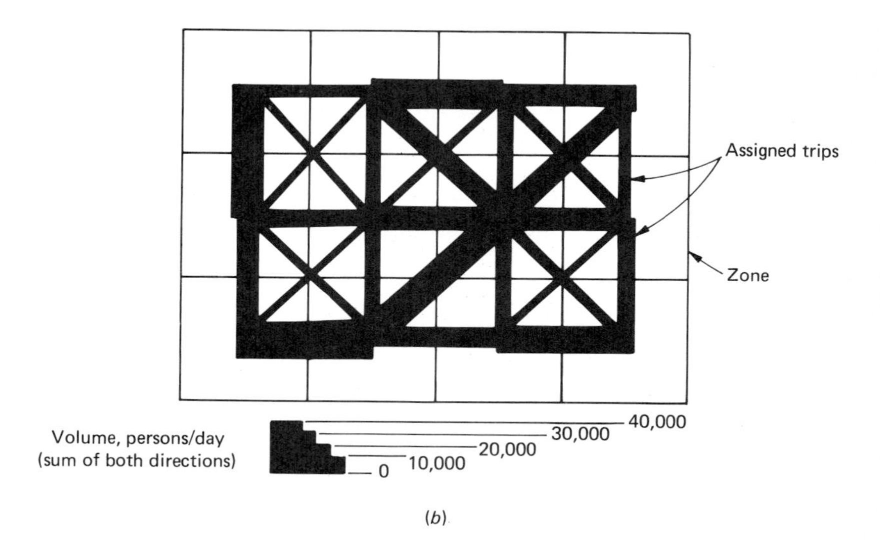

(b)

via minimum time paths on whatever mode or modes are specified. A different map is generated for each point of focus of the travel. These are particularly useful in identifying what areas are well served and what areas are poorly served.

Conceptually very similar to this are graphs showing the relationship between travel time (or cost) and distance between points. A single graph can include times and distances between all pairs of points (or zones) in an area,

Figure 15-12 **Typical distribution of trip lengths in a large urban area: example of Chicago automobile trips in 1956. *(a)* Trip length frequency distribution. [*From Creighton (1970, p. 31).*] *(b)* Vehicle-miles of travel frequency distribution by trip length. [*From Creighton (1970, p. 32).*]**

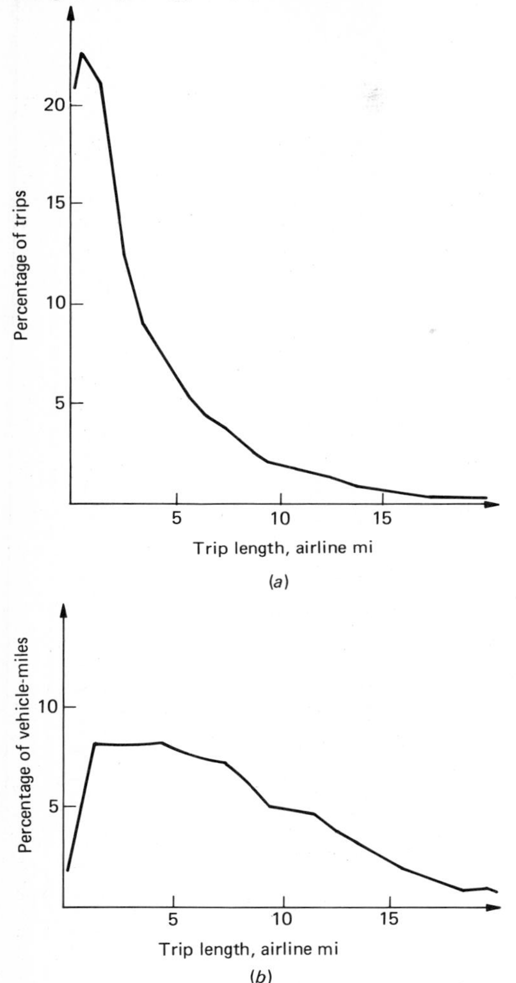

not just those to a single point (or zone) as in the case of a map. A target or standard for the time-distance relationship can be established (e.g., by regression) and all points not meeting the standard identified as not having adequate service. Attention can then be directed to those pairs of points so identified. This approach is illustrated in Fig. 15-13*b*.

Closely related to these portrayals of network level of service is a map that shows where a particular type of transport service is available for use. Specifically, it shows the portions of a region that are served by a carrier of by

Figure 15-13 **Portrayal of network level of service by isochron maps and time-distance plots. *(a)* Isochron lines of travel time. *(b)* Origin to destination time-distance plot.**

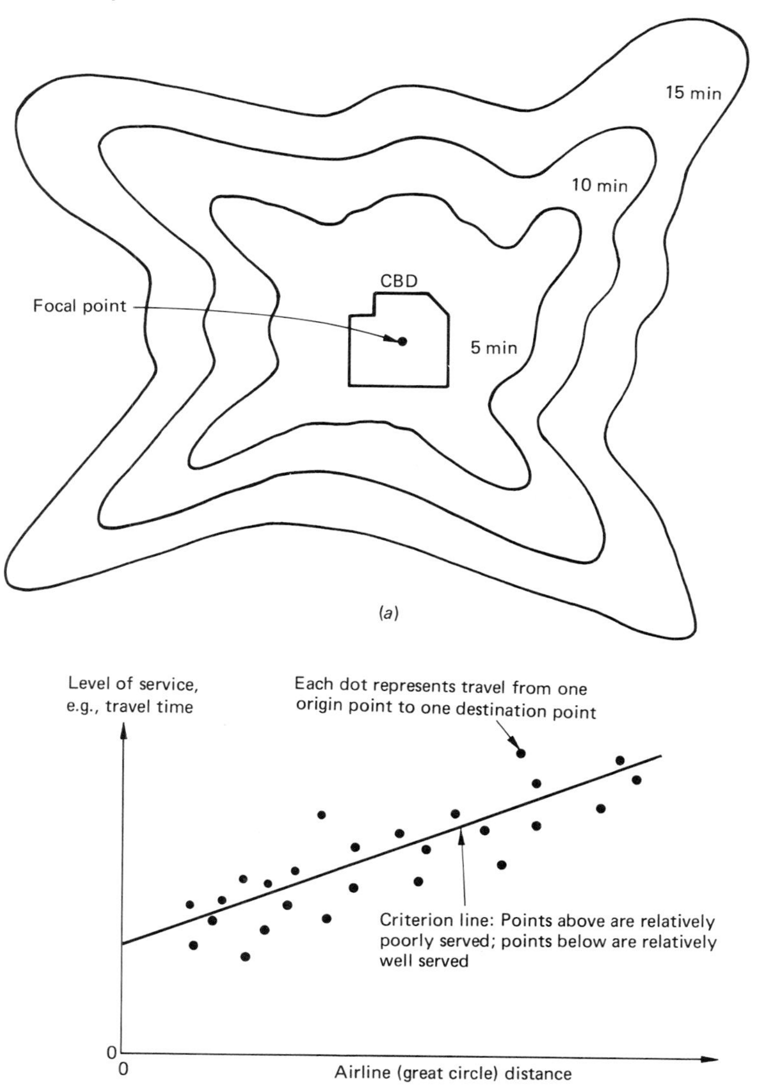

facilities of a particular type, usually those which meet certain service standards. Probably the most commonly used form of this is used to show the area served by public transit routes, as illustrated in Fig. 15-14. In this example the area served is delimited as that which is within walking distance of transit stops, the maximum walking distance assumed to be 5 min in the city

Figure 15-14 **Typical map of area coverage of a transit system: example of Dortmund, West Germany. Note: Area served is defined as within 5 min walk of a transit route in the central business district, 10 min elsewhere. [*From Leibbrand (1970, p. 197).*]**

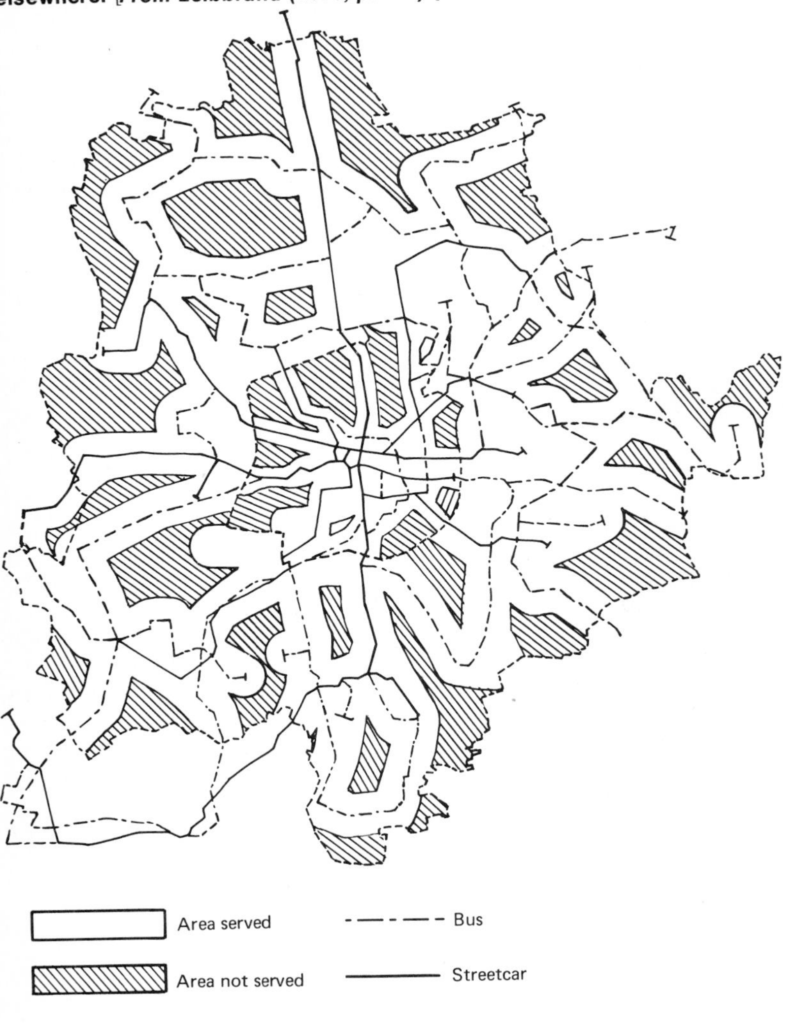

center and 10 min outside that area. Such a map shows clearly the area coverage, and suggests where additional routes might be placed. The fraction of an entire region within the service area is a measure of the overall coverage of the service. The concept can be modified easily to suit other types of transport, such as freeways. Care must be exercised in the choice of maximum access distances.

All of these methods can be used to help to identify problems and to create alternatives that will help alleviate them. Recently these methods have been augmented to computerized methods for creating alternatives which will achieve various goals. These methods rely heavily on mathematical programming, and hence will not be covered in this text, but the interested reader can learn more about these in papers such as that by Hay et al. (1967).

Technological Alternatives

In addition to the overall pattern of the transportation network, there are alternatives resulting from different designs of the facilities and services offered on those networks. Many of these are variations in design that make little difference in the more important characteristics of the system such as travel time or speed of travel on a particular facility, the capacity of a facility or carrier service, the cost to construct and then operate the system, and the environmental and land development impacts. An example of a design option which would make little difference is that between operating a commuter railroad with diesel-electric locomotives or with electric propulsion (usually with each car being powered) with the power obtained from an overhead wire, although even here there are some differences, such as the unsightliness of the overhead wire versus the noise of the diesel locomotives. On the other hand, obviously a commuter rail service is different, in many ways that are important in planning, from a commuter bus service operated on local streets parallel to the railroad. Although we shall delve into design options in considerable detail in the next chapter and into operational options (including transit schedules) in detail in the following chapter, it is important to cover some of the more significantly different transportation technologies and their operational characteristics in this chapter. We shall focus on urban transportation options by way of illustration, but the corresponding alternatives for other planning contexts should be apparent.

Turning first to the road system, the technological options are primarily freeways and arterial streets, local access streets presumably being provided when the land is first developed. Examples of such facilities are shown in Fig. 15-15*d* and *e*. The differences between freeways and arterial streets with respect to capacity, speed (the speed-volume curve), initial cost, user costs (including accidents), land required, and environmental impacts were all covered in earlier chapters and will not be repeated. In general, freeways are more efficient carriers of any given large volume of traffic, in terms of all of these criteria, but new freeways in already built-up areas create serious problems of neighborhood disruption.

An important class of road system options involves the manner in which the facilities are used, alternatives including one-way arterials, exclusive lanes for certain users (such as transit buses illustrated in Fig. 15-15*a*), and the banning of certain types of vehicles from certain streets (as autos from downtown streets to create a pedestrian mall, as in Figure 15-15*b*). Parking policies are also important, including the extent to which off-street parking is provided, restrictions on street parking, the prices of parking, and who provides parking facilities (e.g., private firms, public authorities, or both). Although few urban road facilities now have tolls, the imposition of tolls to pay for roads in addition to the usual reliance on the fuel tax (a very crude user price mechanism), is also an important option. These road options are summarized in Table 15-1.

Public transit is also characterized by numerous options, in fact, in a sense even more than the road system. This is because all transit systems need not

(a)

(b)

(c)

(d)

(e)

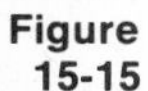

Figure 15-15

Road system options in urban areas. ***(a)*** **Transit bus-only lane on city street. [*Courtesy of Chicago Transit Authority.*]** ***(b)*** **Pedestrian mall with transit service created from a central business district. [*Courtesy of U.S. Dept. of Trans., Federal Highway Administration.*]** ***(c)*** **Off-street parking garage.** ***(d)*** **High-capacity freeway: middle lanes (center) are used in direction of peak flow only. [*Courtesy of U.S. Dept. of Trans., Federal Highway Administration.*]** ***(e)*** **Central business district arterial street beautified by trees and flowers in area formerly used for parking. [*Courtesy of U.S. Dept. of Trans., Federal Highway Administration.*]**

Table 15-1 Road System Options

- **Fixed facilities**
  - Freeways (full access control—access only at interchanges)
    - Location
    - Design speed
    - Number of lanes
    - Interchange location
  - Arterial streets
    - Location
    - Design speed
    - Number of lanes
    - Lane control (i.e., to alter direction)
  - Local streets
    - (Same options as for arterial streets)
  - Intersections
    - At grade versus grade separated
    - Design capacity and speeds
    - Control (e.g., signalization)
  - Parking facilities
    - Location
    - Size
- **Operations plan**
  - Routing
    - One-way versus two-way streets
    - Turn prohibitions at intersections
    - Prohibition of certain classes of vehicles (e.g., no trucks)
    - Vehicle-free zones
  - Allocation of capacity
    - Assignment of lanes to traffic directions and turns
    - Prohibition of parking
    - Exclusive lanes (e.g., for buses, high-occupancy autos)
    - Dynamic adjustments of traffic signal timings
  - Pricing
    - Tolls versus "free" roads
    - Parking charges
    - Congestion tolls
- **Institutional**
  - Financing of initial and continuing costs
    - From current revenues versus bonds
    - From users entirely versus others (e.g., general taxes) or combination
    - Taxing power of road agencies
  - Compensation for negative impacts
    - Extent of compensation for damages (e.g., only for right-of-way versus for noise damage also)
    - Form of payment (single or continuing)
    - Payment of property taxes
  - Ownership
    - Role of public authorities or private firms (e.g., for toll roads)

have technologies that are compatible with one another, unlike the road system, which must be designed for standard types of vehicles throughout since it is fully interconnected. Here we will only sketch the major options, detailed design variants being covered in the following chapters. One major distinction among different types of transit carrier services is whether or not

(a)

(b)

(c)

(d)

(e)

(f)

Figure 15-16
Illustrations of various types of private right-of-way urban public transit alternatives. *(a)* Rail rapid transit line on ground surface with street crossings at grade to reduce construction costs. *(b)* Rail rapid transit line in freeway median. [*Courtesy of Chicago Transit Authority.*] *(c)* Rail rapid transit in modern subway station. [*Courtesy of Washington Metropolitan Area Transit Authority.*] *(d)* Bus rapid transit showing exclusive busway in freeway median. [*Courtesy of Northern Virginia Transportation Commission and Mattox Photography.*] *(e)* Light rail transit: streetcar on private right-of-way. [*Courtesy of Boeing Vertol Co.*] *(f)* Rail rapid transit terminal with extensive parking, auto passenger drop-off area (kiss 'n' ride), and feeder bus stop. [*Courtesy of Washington Metropolitan Area Transit Authority.*]

the right-of-way (ROW) vehicles operate on is on the ground surface, in a subway, or elevated, as illustrated in Fig. 15-16*a* through *c*. Another important distinction is between a private ROW, meaning it is used exclusively by transit vehicles, and a shared ROW, such as public streets used by buses and others. A private ROW allows faster vehicle speeds than mixed traffic (although free-flow freeways allow speeds equal to most transit lines), permits the use of specially designed vehicles which can be very large (e.g., a subway train carrying 2000 passengers), and also allows for automation in vehicle operation and fare collection, which may reduce costs. Such systems have a wide variety of variations which do not necessarily affect characteristics important in planning, such as rail subways with rubber tires (e.g., Paris and Montreal) versus steel wheels, although obviously such variations can affect characteristics such as noise and acceleration capability. Usually it is assumed that details such as these will be addressed in the design phase, and that the design engineers will make choices which will meet the performance requirements specified in the planning.

The use of streets and other existing roads for transit reduces costs; and through the reduced fixed cost of a route, a more complete coverage of an area is possible, often with routes only 0.5 mi apart. Nevertheless, for short trips, street transit is probably preferable to many people, because of shorter distances to stops, which are at street level and are usually only a couple of blocks apart.

Finally, there are ways of combining the two forms of transit, as is appropriate in situations where trips have one end located in a large, low-density area, such as a suburban residential area, and the other end in a central city business district. Here street transit may be used for collection at the residential end to bring travelers to a line-haul facility designed for high-speed, high-capacity movement to the CBD. This can be accomplished by local buses coordinated with a separate rapid transit line-haul route or with buses operating on streets and then on a bus-only busway as in Fig. 15-16*d*, or with streetcars operating on streets and then on their own private ROW into the CBD, as in Fig. 15-16*e*. Alternatively the residential collection can be by auto, with parking at the rapid transit (bus or rail) stations, as on many suburban transit lines and commuter railroads now (Fig. 15-16*f*).

It is important to recognize that any of the various public transit technologies can be operated in many different ways so as to tailor them to a particular task. For example, it may be desirable to have a subway line serve as a distributor of trips within the central business district. Stations might be only 0.5 mi apart, and trains run every 2 or 3 min (even if this means short trains), to provide a service like an elevator in a building. Alternatively, buses could be used for the same purpose, perhaps on streets from which all other traffic is banned. In some new towns, buses operating on private roadways provide for all transit services within the community, in an attempt to reduce automobile usage to a minimum. Essentially the same technologies—conventional bus or rail—can be used for regional transit service, with stations many miles apart, although of course the design of the vehicles and the ratio of seats to places for standees would be different. If auto traffic is banned from an area (this plan has been adopted in a few CBDs and is proposed for many more), the parking facilities would be coordinated with such distributor transit services to provide for ease of access to the area.

(Currently a number of automated transit technologies are being developed for this distributor function, and these will be discussed in Chap. 18.)

The third major distinction among types of transit services is that between fixed route schedule carriers and so-called demand-responsive carriers, which go where travelers are or wish to go on request. Conventional bus and rail services fit the former category. Taxis are an example of the latter. Typically taxi service is much more expensive than bus or rail service, but it can serve low-population-density areas with dispersed trips well, as its rapid growth attests. In an effort to make taxilike service cheaper, the dial-a-bus concept has been conceived. In this scheme, small vehicles (seating perhaps 5 to 10 people) are routed on demand to where passengers want to go, like a taxi, but the routing is arranged to combine as many passenger trips in the vehicle as possible, to spread the cost. The travel time is greater than for a taxi, often with waits at the origin of 10 to 20 min, but the fares can be much less. This concept enables relatively low cost transit service to be provided to areas where it could not be provided economically by other means.

In addition to options related to the types of transportation facilities and carrier services offered, there are other important options. One class of options is centered around the financing of the system. At the present time both the road system and the public transit system in most cities are owned and operated by governmental agencies, with the prices charged being somewhat less than the average cost of providing the system, which results in the need for subsidies from general tax funds. An alternative would obviously be to charge prices equal to average costs, with higher transport costs for users, but this would put transport on an equal footing with most other economic activities.

Another class of options is centered on the question of who should provide transport services, private firms, governmental agencies, or a mixture of both. Currently there is much debate as to the effectiveness of governmental agencies which in effect have a monopoly on such services as public transit, with allegations that such agencies are unresponsive to traveler needs, since there is no incentive for management to be responsive. Also, there are charges of gross inefficiencies in publicly owned carriers for which deficits are made up by subsidies. In many of the larger United States cities there are illegal transit services supplementing conventional bus and rail service provided by public agencies, these often with lower fares and yet still making a profit. On the other hand, the public agencies often must provide uneconomical services as part of their common carrier obligations, and it is alleged that they usually have safer vehicles and more complete insurance coverage than their competitors. Without taking sides on this complex issue, we see that a strategic decision is the degree to which there will be freedom of entry into local transit services and the degree to which fares, hours of service, routes, and schedules are to be regulated. Perhaps local transit services within a community could be better provided by small firms or public agencies controlled by persons in the area served. The answers to these questions undoubtedly depend upon local conditions. All of these transit options are summarized in Tables 15-2 and 15-3.

Another important class of options relates to freight movement. Currently there is almost complete reliance on trucks for urban goods movement, and while new technologies are being explored (as will be covered in Chap. 18),

Table 15-2 Options for Private Right-of-Way Public Transit†

- Fixed facilities
  - Way links and intersections
    - Location
    - Design speed
    - Capacity
    - Number of tracks or lanes in each direction
    - Degree of intrusion control (e.g., fully fenced ROW or not, at grade intersections with roads or fully grade-separated)
  - Terminals
    - Location and spacing
    - Capacity (number of berths)
    - Separation from running lines
    - Fare collection (at terminal versus on vehicle)
    - Provision for parking
- Control systems
  - Design (in conjunction with fixed facility and vehicle design and operations plans)
    - Control aids (e.g., railroad block signal system)
    - Degree of automation
    - Monitoring to detect problems as they occur
- Vehicles
  - Design
    - Size and capacity (including options of coupling to form trains)
    - Accommodations (e.g., all seated, places for standees)
    - Source of power (e.g., diesel engine, electricity from overhead wires or third rail)
- Operations plan
  - Use of ROW
    - Exclusively transit
    - Shared (e.g., with freight service or intercity passenger service)
  - Scheduled, fixed route service
    - Schedule plan (e.g., all-stop, skip-stop, zone express)
    - Departure frequency
    - Passenger capacity
    - Coordination with other services
  - Fares and charges
    - Parking fees
    - Constant versus distance-variable fare
    - Self-supporting versus subsidized

† For example, rail rapid transit in its various forms and busways.

much thinking is being devoted to ways of reducing costs with existing technology. One idea is to ban all auto traffic from congested areas, thereby freeing trucks of road delays and parking congestion, and the idea seems to have worked well in a number of cities. Also, truck services may be coordinated, with one truck rather than the trucks of many firms serving an area which would reduce the number of separate truck movements and hence congestion and costs. However, this idea is opposed by many shippers (or receivers) of freight as well as truck lines, because it means that they will no longer have control over the shipment from origin to destination and quality of service may deteriorate with a single firm or government agency having a monopoly on the service. These options are summarized in Table 15-4.

While this discussion by no means exhausts the range of urban transporta-

Table 15-3 Options for On-Street Public Transit

- Fixed facilities
  - Way links and intersections
    - Must be compatible with road limitations
    - May operate on pavement (buses) or on track in pavement (streetcars)
  - Terminals
    - On-street loading (most common) versus off-street or pull-out berth
    - Design (e.g., simple corner stop with sign versus shelter)
- Vehicles
  - Design
    - Size and capacity (e.g., 10-seat jitney, 53-seat bus)
    - Accommodations
    - Source of power (e.g., diesel engine, electricity from overhead wires)
- Operations plan
  - Scheduled fixed route service (e.g., bus, streetcar)
    - Routes
    - Schedule plan (e.g., all-stop, skip-stop, zone express)
    - Departure frequency
    - Passenger capacity
    - Coordination with other services
  - Demand-responsive service (e.g., dial-a-bus, taxi)
    - Area served
    - Level-of-service standards (e.g., target waiting times, directness of travel)
    - Passenger capacity
    - Provision for user with special requirements (e.g., handicapped)
  - Fares and charges
    - Transit service
    - Parking fees

Table 15-4 Freight Transport Options in Urban Areas

- Fixed facilities
  - Roads
    - Maximum vehicle size and weight designed for
    - Special truck roads (e.g., as for deliveries in congested shopping areas or truckways)
  - Terminals (ports, truck, TOFC, etc.)
    - Location
    - Capacity
    - Range of commodities
  - Other specialized systems
    - Use of rail movement
    - Extent of pipelines for gas, water, etc.
- Operations plans
  - Coordination
    - Union terminals
    - Coordinated pickup and delivery services
  - Traffic regulations
    - Restrictions on truck operations on streets
    - Special parking areas on streets for trucks
  - Economic regulation
    - Extent
    - Type (e.g., entry, pricing, level of service)

tion alternatives, it hopefully has indicated their breadth. More detail on many of them will be presented in the next two chapters.

CONCLUSIONS

The discussion of transportation planning has largely been drawn from methods developed in the context of urban transportation planning in the period since World War II. However, the principles have hopefully been presented in a sufficiently general form that their application to other areas of transportation planning is clear. The field is still growing and changing, and undoubtedly many of the methods and procedures used today will be modified in the future, resulting in a more responsive and useful transportation planning process.

SUMMARY

1 Planning is undertaken in both the private and the public sector in order to provide a rational basis for current and future actions.

2 Transportation planning exists with a hierarchy of other types of planning or decision making, bounded on the one hand by considerations of transportation within the context of national and regional social and economic development and on the other by attention to the design and operation of particular portions of the transportation system.

3 The urban transportation planning process has been developed to be comprehensive in its coverage of all modes, coordinated with land use planning, and continuing so that the plan can be appropriately updated as conditions change. Urban transportation planning focuses on development of long-range plans primarily for major fixed facility improvements to the transportation system. Important elements of the planning process are the gathering of data on the existing state of the region and its transportation system, the development of models and methods for predicting future requirements on the transportation system, the generation of alternative future plans, the evaluation of those plans, and the selection of a particular set of plans for transmission to appropriate decision-making units. Also important is the programming of the components of the plan so that particular changes can be implemented at the most opportune times.

4 Regional transportation planning is characterized by a very large number of options, options not only with respect to the transportation system itself, but also with respect to land use patterns and the associated travel demands. Regions tend to be unique in their land use patterns and travel demands, and hence the best form of transportation network for one region may be quite different from the best form for another.

PROBLEMS

15-1

A rapid transit line is being proposed in a well-developed corridor. To illustrate the approach and methods involved, consider two zones, each 1 mi^2, 5 mi apart. One is a high-income residential area, the other the central business district. A total of 10,000 persons/day travel between them (each way). Currently there is no transit service, only a 60-mi/h design speed freeway with six lanes (three in each direction), on which a total of 60,000 vehicle trips/day are made (average of 1.5 persons/auto), with 10 percent of the vehicles being trucks. The freeway interchange is an average of 0.75 mi away from the actual origin or destination of trips, access being via local roads at an average speed of 10 mi/h, with a 2-min walk at the residential end and a 5-min walk at the CBD end, on the average. Parking in the CBD costs $1.50/vehicle-day or fraction thereof. The proposed transit line has one station in each zone, and the average walking distance to them is expected to be 0.5 mi, with an average walking speed of 3 mi/h. Trains of sufficient size will be operated at a 6-min headway for 18 h each day, with cars seating 50 and standing 25 (with standees in the peak hours only). Forty percent of all trips are during the peak 4 h, with an 80 percent–20 percent directional split. Nonpeak traffic is uniformly distributed over 14 h with a 50 percent–50 percent directional split. What patronage would be expected on the transit line, assuming only travelers between the two zones used it? The line will be on an elevated guideway, and a fare of 50 cents will be charged in the peak period, 25 cents at other times. Will the line make a profit? What will be the reduction in road congestion due to the line during peak periods and other periods? What will be the reduction in noise from the road, at 200 ft from the center line? Use the TRC modal split curves of Chap. 10 for the highest income group and assume that all peak trips are work trips, all others being nonwork trips. For transit costs, use Eq. (9-16), and assume an average station to station speed of 50 mi/h, with a 5-min dwell time at each terminal.

15-2

The prices, level of service, and usage of each of four modes connecting two metropolitan areas which are 200 mi apart are given below.

Mode	Access (Sum of both ends) Time, h	Access (Sum of both ends) Cost, $	Line-haul time, h	Cost, $	Usage passengers/day, one direction
Air	1.5	4.00	1.2	25.00	1000
Auto	0.8	1.00	4.0	10.00	3000
Bus	1.1	1.00	4.0	8.00	800
Rail	2.3	1.00	3.7	15.00	1580

Various improvements to the rail service are proposed. The current service is 7 trains/day, one every 2 h, but this could be increased to yield 1-h or 0.5-h headways. Also, running times could be reduced from 3.7 h to 3.0, 2.5, or 2.0 h. The costs of line improvements and operating costs (including car ownership and all maintenance) are given on page 637. Possible fares are $15, $20, and $25. Current rail demand is 850 business trips and 730 personal trips each way per day.

Running time, h	Line improvement costs, $ $\times 10^6$ †	Train operation, $/train-mi
3.7	0	$2.00 + 0.50n$ ‡
3.0	100	$2.25 + 0.60n$
2.5	200	$2.40 + 0.65n$
2.0	350	$2.90 + 0.90n$

† Life = 30 years.
‡ n = cars/train.

With a 10 percent interest rate, what is the maximum profit level of improvement and price, assuming the demand curve does not shift? Assume a 50 percent average load factor and that rail cars seat 75 passengers. Use the Kraft-SARC demand model presented in Chap. 10. Access times include one-half the headway as waiting time. [Note: Each student can examine one option, with the overall comparison in class. *Hint:* Simplify the demand model of Eqs. (10-7) and (10-8), to include as variables only the rail costs and times; then the data presented are sufficient for calibration.]

15-3
Draw an isochron map for the transit and automobile modes of a metropolitan area for which you can obtain a map of the transit and road system. Try to use public transit time tables if you can't make reasonable assumptions on average speeds, such as 12 mi/h for local buses, 25 mi/h for rail rapid transit, 15 and 25 mi/h for arterial streets in the peak and base periods, respectively, and 30 and 50 mi/h for freeways, peak and base, respectively. Assume a 3 mi/h walk speed. Draw the isochrons in 10- or 15-min intervals for a single major point. What types of improvements are suggested? Where? (Note: As a class project, maps for a number of points of focus can be drawn.)

15-4
From a transit map for a metropolitan area, identify corridors with large flows. Could a busway or rail transit line be placed in these corridors? Are existing rights-of-way (e.g., freight rail lines) available?

15-5
Using the same map as in Prob. 15-4, draw a map showing the area coverage of the region by the transit system. Assume that people will walk up to 0.5 mi to and from transit stops. On street transit routes, you can assume that stops are so frequent that all land within ± 0.5 mi of the route is within walking distance.

15-6
What do you think would be the long-term effect of charging all motorists and transit users the full cost (i.e., average cost) of the facilities and services they use? In larger metropolitan areas, would this lead to a return of population and industry to the central cities? Would there be less traffic congestion?

15-7
Do you think there is a need to develop new transit technologies? If so, what types are needed?

15-8
What do you think would be the effect of a no-fare policy on public transit in your city?

15-9
A metropolitan area planning commission has concluded that a freeway (or a rapid transit line, in an open cut) should be built through an already developed residential area. What types of neighborhood impacts should be considered and how might those adversely affected be compensated? The route requires the taking of homes, stores, and park land and will lead to an increase in traffic on many streets which lead to the interchanges (or stations).

15-10
What do you think the effect of advancements in communications, particularly videophones which might be as inexpensive as telephones, will be on the future demand for urban travel?

REFERENCES

Baerwald, John E., Matthew J. Huber, and Louis E. Keefer, eds. (1976), "Transportation and Traffic Engineering Handbook," Institute of Traffic Engineers, Prentice-Hall, Englewood Cliffs, NJ.

Buchanan, Colin et al. (1964), "Traffic in Towns," Her Majesty's Stationary Office, London.

Cleveland, Donald E., ed. (1964), "Manual of Traffic Engineering Studies," Institute of Traffic Engineers, Washington, DC.

Creighton, Roger L. (1970), "Urban Transportation Planning," University of Illinois Press, Urbana, IL.

Dickey, John W., Robert C. Stuart, Richard D. Walker, Michael C. Cunningham, Alan G. Winslow, Walter J. Diewald, and G. Day Ding (1975). "Metropolitan Transportation Planning," McGraw-Hill, New York.

Eash, Ronald W., Harvey O. Haack, and Mehmet Civgin (1970), An Investment Approach Toward Developing Priorities in Transportation Planning, *Highway Research Record* No. 314, pp. 87–97.

Hay, George A., Edward K. Morlok, and Abraham Charnes (1967), Toward Optimal Planning of a Two-Mode Urban Transportation System: A Linear Programming Formulation, *Highway Research Record* No. 148, pp. 20–48.

Hutchinson, Bruce G. (1974), "Principles of Urban Transport Systems Planning," McGraw-Hill, New York.

Leibbrand, Kurt (1970), "Transportation and Town Planning," MIT Press, Cambridge, MA., trans by Nigel Seymer from Kurt Leibbrand (1964), "Verkehr und Städtebau," Birkhäuser Verlag, Basel, Switzerland.

Manheim, Marvin L. (1966), Transportation, Problem-Solving and the Effective Use of Computers, *Highway Research Record* No. 148, pp. 49–58.

Manheim, Marvin L. (1969), Search and Choice in Transport Systems Analysis, *Highway Research Record* No. 293, pp. 54–81.

National Committee on Urban Transportation (1958), "Better Transportation for Your City," Public Administration Service, Chicago, IL.

Welding, P. I. (1957), Instability of a Close-Interval Service, *Operational Research Quarterly,* **8**(3): 133–148.

16. Location and Design

The design of the physical elements of a transportation system—fixed facilities, vehicles, and other components (including the location of fixed facilities)—is one of the most important tasks of transportation engineers. The function of the location and design process is to translate general plans for new facilities and the associated operations plans into the specific, detailed form needed in order to commence actual construction. In practice most design work is performed by specialists concerned with a particular component, such as roads, bridges, airports, railway cars, etc. However, since the work of the transportation system engineer is usually a direct input into the engineering design process, it is important even for the generalist to understand the basic elements of the engineering design process. This will facilitate communication between component location and design engineers and transportation system planners and engineers and thereby improve the quality of the system which is their mutual concern. Also, many persons who have their first introduction to the transportation professions through a study of transportation system engineering and planning may wish to specialize in design, and this introduction will serve to expose them to the nature of design work.

As the discussion above indicates, this chapter must concentrate on the general principles and approach to engineering design of components of transportation systems. These will be illustrated by examples, mainly roads, airports, and, as integrated with operations planning in Chap. 17, urban transit. The main emphasis will be on those aspects of design which are based on the principles of transportation systems and their interaction with their environment, as described in previous chapters. It should be borne in mind that the design of a component usually involves the application of not only transportation system theory but also theory from one or more other fields. For example, the design of a road involves structural engineering, soil mechanics, and hydraulics; and the design of a vehicle involves structural engineering, thermodynamics, engineering mechanics, and if a passenger vehicle, human factors engineering. Thus a component design engineer will generally have studied not only transportation engineering but also one or more other engineering disciplines.

THE DESIGN PROCESS

The design process typically consists of a number of stages undertaken in sequence. This is perhaps best illustrated by considering design of a road or railroad link. The desirability of having a new route is identified in the planning process, and the primary characteristics of the link are determined. These include such items as the end points of the route, approximate locations of key intermediate intersections or terminals, and the capacity and level of service to be provided. The first step in the process is to determine the location of the facility. Typically it would be selected from among options lying within a corridor or strip of land which would meet the requirements for the facility. Once the approximate location has been decided, the next step is to design the facility for the site selected. This includes specifying the exact size and location of the facility, standards to be met in the construction, materials to be used, etc. Following this, the construction can be planned and undertaken. These steps are shown in Fig. 16-1.

Figure 16-1 The sequence of steps in planning and design.

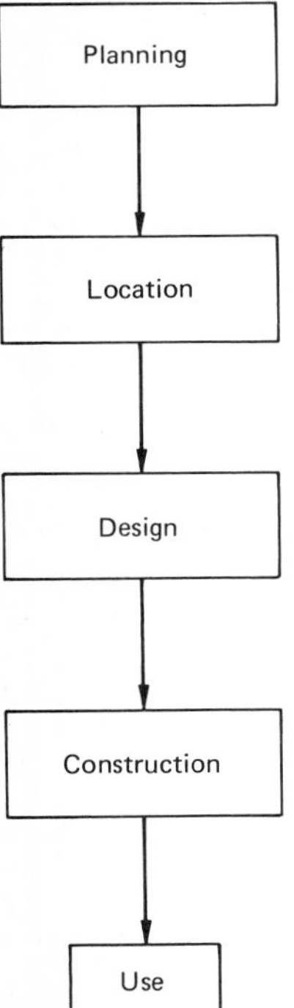

A similar process is followed in the design of other fixed facilities, such as terminals, ports, airports, and pipelines, and also in the design of vehicles.

Design is often thought of as a very routine process, consisting primarily of the straightforward application of standard methods. Some design is of this type. But much of it is not, but rather requires the use of considerable creativity in developing designs which will meet requirements unique to each particular problem at a reasonable cost (in market and nonmarket senses). With the passage of time, design problems which seemed very difficult when they first appeared become solved so many times that the solution becomes well-known and routine, so that design engineers can turn their creative talents to new problems. New problems are arising with incredible frequency, partly as a result of the concern for the environment—to make transport facilities "good neighbors" rather than detriments to their immediate environments—both natural and manmade. Also, there is increasing emphasis on efficiency in the use of resources, not only scarce environmental ones such as open land, but costs in general. This is made very difficult in a field where construction costs have risen very rapidly relative to other costs. For example, as noted in Table 9-5, road construction costs have doubled in a period when consumer and wholesale prices rose an average of about 40 percent.

Many if not most problems of design are subject to some regulations and standards. Some of these are based on environmental considerations, such as maximum noise and pollution levels and maximum height of structures. Also, standards are established in order to ensure compatibility among facilities and vehicles throughout the region to which they apply (often an entire nation or group of nations). These include minimum widths of channels for vehicles, minimum radii of curves, minimum design speeds for routes, etc.

In addition to these requirements there are many standard types of designs for particular components which are usually adopted but which can be deviated from in exceptional cases. Examples include standard curve radii and other characteristics of road intersections, standard sizes for road shoulders, and standard pavement markings and signs. These are adopted to provide uniformity in the road system so that all drivers will know what to expect in the way of curves, sight distances, merging and cross traffic, etc. The existence of such standards also helps the design engineer in making decisions which otherwise would have to be made on the basis of engineering judgment, and where the exercise of judgment alone could cause undue controversy. For example, without standard road widths, each engineer would have to decide how much wider than the maximum width vehicle to make each road lane. Lane widths might vary considerably from one section of a road to another, perhaps reflecting differences in judgment among different engineers as well as availability of funds at the time of construction. In periods of severe limitation on funds for construction, lanes much narrower than those recommended by engineers may be used to reduce costs. And later, after these roads are in use, if the accident rate on the narrower sections were greater than on the wider sections, the designers of the narrower sections, or the government agency which built and maintains them, might be subject to law suits claiming improper or negligent design. The existence of standard designs largely eliminates these problems, the standards presumably prescribing what is economical (in both initial cost and continuing cost), safe for the user, and compatible with the maintenance of environmental quality. Of course, such standard designs must be updated as public preferences and tastes change, as research uncovers new considerations, and as available technology changes.

Most design involves the analysis of trade-offs in satisfying stated design objectives within constraints imposed by available resources and the environment. It is not difficult to envision an "ideal" design for a transportation facility to satisfy common transportation objectives, e.g., a straight and level route or a terminal with direct connections. But the process of design usually involves balancing available resources and externally imposed constraints, on the one hand, against maximum allowable deviations from the "ideal" on the other.

Thus design is not a mindless application of a set of components and standards selected for a particular system. A good design cannot result from computation alone. Hence, there can be "computer-aided" design but not "computerized" design of transportation facilities. All transportation design of any significant complexity involves human judgment, creativity, and decision making.

Public Involvement

Most decisions involving the planning, location, or design of any major transportation facility (except for some decisions regarding privately owned facilities) receive close public scrutiny and are, in fact, usually subject to public approval. This is as it should be, given the fact that most such facilities are financed, at least in part, by public funds.

Much of this public scrutiny and approval is formalized and legislated. Public hearings are integral parts of urban transportation studies and airport master plan studies. In urban transportation planning there is usually a hierarchy of public hearings corresponding to different stages of the planning-location-design process. For example, there may be one public hearing for the regional plan, another for the general location and types (e.g., freeway, widened arterial, rapid transit line) of facilities being planned for a corridor, and a third hearing for the final facility location and design. Usually there are specific requirements that alternative plans be presented at each stage, so that the public can choose (at least in part) the location and type of facility to be built. While such requirements add to the work of the professional and often delay decisions, the prospect of public scrutiny has undoubtedly resulted in much more careful analysis at each stage of the process. In particular, there surely is much more attention given to analysis of a wide range of alternative ways of meeting the transportation objectives of a project and to consideration of community impacts.

LOCATION OF FIXED FACILITIES

A primary consideration in the provision of transportation facilities is their location, which must be selected on the basis of initial cost, continuing costs for maintenance and operation, and environmental impacts. First we shall present basic principles and factors, and then their use will be illustrated with a road example.

General Principles

The process of selecting a location for a new facility is usually quite complex and time-consuming, primarily because of the diversity of consequences of a decision. Usually the final selection is not ideal from any one viewpoint but represents a compromise among many, often conflicting, objectives. The process of selecting a location for a major facility is usually quite long, often taking a number of years, because of the many groups whose interests must be satisfied with the final selection.

A list of the various criteria which typically are considered in selecting a location is presented in Table 16-1. Opposite each criterion category are listed the more important factors which will influence the consequences or impacts of the facility as viewed within that criterion category. Obviously a major factor is the initial construction cost, which includes such items as purchasing the necessary land, preparing it—including demolishing any structures on it and relocating families and businesses whose property is being taken (often with special compensation for the disruption and difficulty of moving)—and construction of the facility itself. Special costs may be incurred during construction, such as compensation to businesses while access to their stores is denied or made difficult by construction equipment and activity and to property owners along the routes used by trucks, which lose some of their loads while traveling and thereby make the areas dirty. The initial cost may differ considerably for different routes, due to differences in gradients, curvature, and even length. Since usually more direct, lower-grade, routes can be built only at greater initial cost, this added cost must be weighed against other factors such as reduced operating costs.

Table 16-1 Criteria to Be Considered in Facility Location Decisions

Criteria	Influencing factors
Construction costs	Facility design type Topography and soil conditions Current use of land
User costs	Usage of facility (and alternative facilities) Facility design features (e.g., gradients on a road, ease of access to a terminal) Operations plan Ease of future expansion and change
Impacts on natural environment	Proximity to sensitive areas or activities Facility design features (e.g., noise barriers)
Impacts on human activities	Creation of boundary around or division of neighborhood Aesthetics of design Reinforcement of desired development patterns
Acceptance by relevant interest groups	Private firms and government agencies (owners or financial supporters) Governments of affected areas Public support

The extent of variation of user volume with location depends significantly upon the type of facility being considered. For some facilities, such as airports and rural freeways, total usage may not vary much with location, while in urban areas different locations of freeways or transit routes and variations in the access points can affect markedly the number of users. In the latter case, a proper evaluation must include the impact of each alternative on all of the system users who are affected by any of the various alternatives being considered. Some of the group of system users may actually use only certain of the alternatives, but they must be considered in evaluating each of the alternatives. Thus user costs must be treated carefully, so that all affected groups (in this case, all system users who might be affected by one of the alternatives being considered) are included in the analysis. This is simply following the guidelines for evaluation described in Chap. 14.

Environmental impacts have become increasingly important in recent years, partly because of the change in societal attitudes toward the environment and partly because of research which uncovered the very negative effects that transport facilities could have on the environment. Different locations often result in vastly different environmental effects, mainly through avoidance of sensitive areas, such as wildlife sanctuaries and parks. Also, different designs may produce vastly different impacts, as in the case of noise from a freeway at grade in contrast to one depressed and with a noise barrier. Certain environmental conditions may preclude a particular location of a facility, as in the case of a jet airport in an area with a large bird population, which would be very dangerous because of the likelihood of birds being drawn into the jet engines, causing the aircraft to lose power and crash.

The impacts on human activities, while often more subtle and difficult to predict quantitatively, are equally important. Particularly critical is the degree to which a new facility will contribute to the kind of land development desired or lead to undesirable development. A freeway, for example, can be used to create a barrier between an industrial district and a residential area, and this may be considered beneficial. In contrast, if it passes through the middle of a residential neighborhood, it may create a barrier to interaction, making the neighborhood less desirable and tending to lead to its decay. But the facility might be located so as to pass through areas of decayed, abandoned structures, and thereby be used as an instrument of urban renewal. By creating better access to jobs, recreation centers, etc., the new facility might also help the area it passes through. What the effects of a facility will be, and how the communities where it is proposed to be located will view it, depends upon local circumstances.

This leads to the final category of criteria listed in Table 16-1, that of acceptance of the proposed location by various affected groups. Prior to about 1965, location decisions in the United States could be made unilaterally by the agency constructing a road or airport. With the right of eminent domain, that agency could then obtain the needed property, with payment of a fair or market price to the owner (the value could be contested with court settlement, if necessary), and commence construction. Now public hearings are required, and while there are no firm rules to determine when the degree of acceptance is adequate or inadequate, the effect of this process has been to stop most projects when there is substantial citizen opposition. Thus locations now in effect require support of not only all of the government agencies involved and important local government bodies (town councils, etc.), which control the funds, but also of the general public.

Such widespread support is generally not required for the private development of transportation facilities. However, mainly port facilities, terminal buildings (as at airports), pipelines, and railways are still constructed without government funding. In most cases these are located in areas designated by a public, political process for that type of activity or the facilities are in rural areas. If there were significant public opposition, projects probably might be stopped or delayed on environmental or other grounds.

The actual processes used in reaching a decision on a route or terminal location are identical to those presented in Chap. 14. Usually one of the multiple-objective methods is used. This will be illustrated by a road example in the next section.

A Road Location Example

An example of the wide range of factors considered in a facility location problem is provided by a study of alternative routes for a rural highway in Maine. The study is of alternative ways of either upgrading the existing highway or constructing a new road which bypasses the existing facility in the area where congestion exists, and the data are taken from a general report on evaluative methods which used this as an example (Manheim et al., 1975, pp. 40–48), as this was adapted from the original report (Edwards and Kelcey, 1972). The area is shown in Fig. 16-2, with the present intercity routes, US 1A and US 1, converging at the main business district in this small (population about 500) community. The alternatives to retaining the existing road (the null or "do-nothing" alternative) considered were, using the numbers appearing on the figure:

1 Bypass to the south of the town, with two 12-ft lanes and 8-ft shoulders, crossing the salt marshes and river on a new bridge.

2 Upgrade the existing road with better alignment and geometric design on the present right-of-way and build a new bridge on US 1A over the river.

3 Partially build a new route, using an existing bridge and route through the town center.

4 Build a new bypass to the north of the town, using an existing bridge (route US 1 on map).

The evaluation centered around an attempt to identify all the groups likely to be affected by any of these alternatives. These groups included not only system users (motorists) and the provider of the facilities (the State Department of Transportation), but also residents of the area who are affected by the traffic and also those who live in homes displaced by a new route, town businesses, which could lose business due to the bypasses, the town government, which might lose tax revenue generated from land and buildings taken for a new road, and residents who enjoy or depend on the physical environment of the area. The latter are especially important here, because of the natural beauty of the area and the dependence of the local economy on lobster fishing.

Figure 16-2 Map showing alternative highway locations in Harrington, Maine. [*From Manheim et al. (1975, p. 44).*]

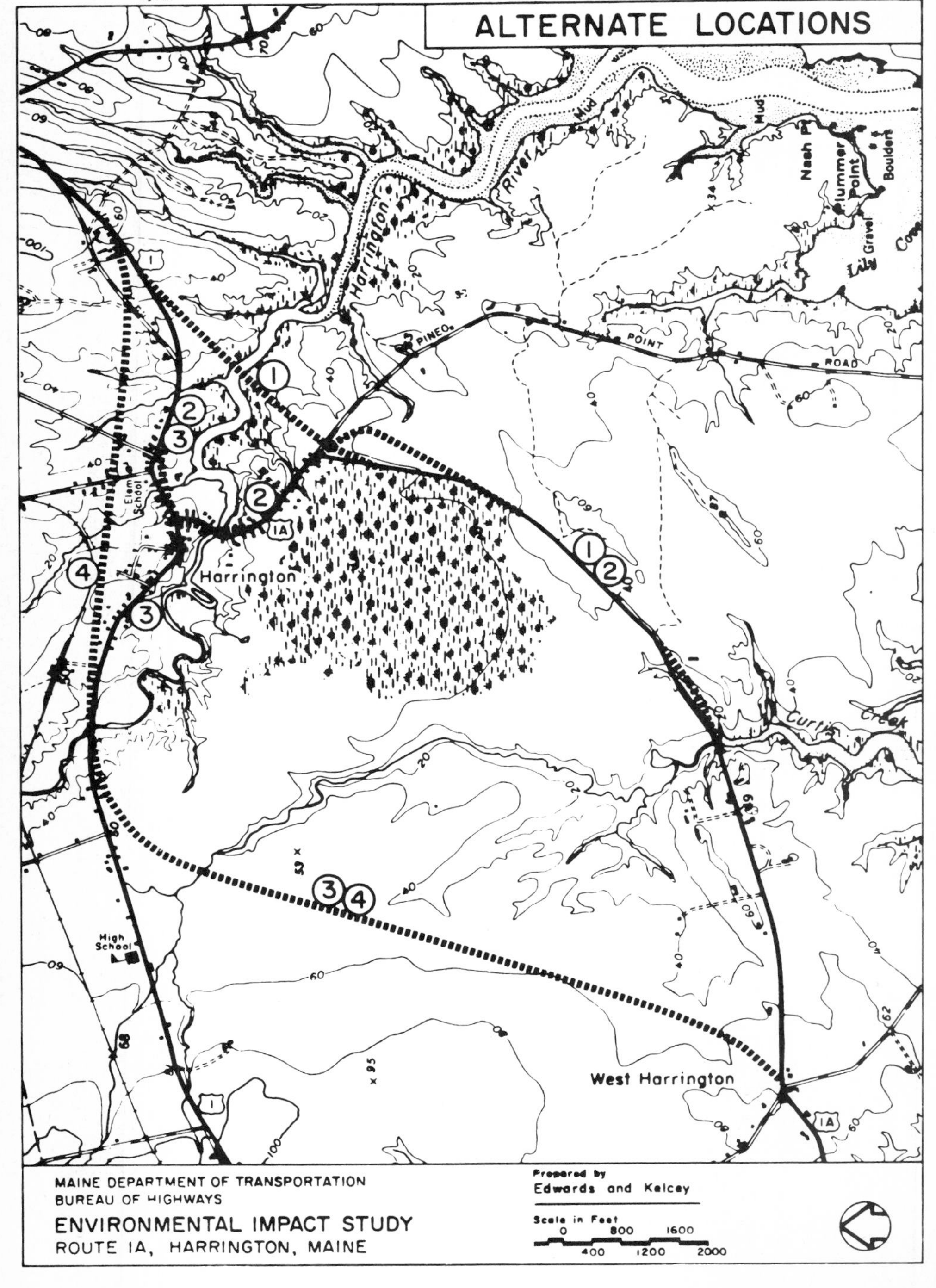

Then an attempt was made to estimate the magnitude of the effects. The results of these predictions are presented in Table 16-2. It is clear from this table that no one alternative dominates all others. Even if the transportation-user-related criteria (which tend to favor a new route) are ignored, some of the new routes have positive impacts, such as improving the upstream salt

Table 16-2 Predicted Impacts of Various Highway Alternatives: Example of Route through Harrington, Maine†

Interest/Impact type	Alternative 0	1	2	3	4
Through traffic					
Speed (avg), mi/h	25	55	30	30	55
Distance, mi	3.7	3.2	3.8	3.8	3.7
Accident rate factor‡	4	1.2	3.5	2.5	0.6
State Department of Transportation					
ROW cost ($1000)	0	22	205	129	122
Construction cost ($1000)	0	1447	1377	1050	1416
Total cost ($1000)	0	1499	1582	1179	1538
Town center businesses					
Displaced	None	None	Gas station, auto parts store	Same as 2	None
Loss of business from through traffic	None	Yes—to grocery store, lunch counter, gas station (about 10 percent of total retail sales)	None	None	Yes—to restaurant, grocery store, lunch counter, three gas stations (about 20 percent of total retail sales)
Town center residences					
Displaced	None	None	7 single family	3 single family	None
Relocation problems	None	None	Yes	Yes	None
Town center environment					
Through traffic volume per day					
1970:	2620	1400	2620	2520	1250
1990:	4350	2325	4350	4180	2075
Safety	Poor	Good	Poor	Fair	Good
Air quality (μg/m³ CO 1990)§	825	306	825	536	386
Noise (dBA*—1990)					
At 50-ft distance	73	70	73	73	70
Methodist church	70	70	70	74	70
Baptist church	73	70	73	73	70
Elementary school	58	54	58	58	65
Visual	None	Mixed	Mixed	Small	None
School safety impact	None	None	None	None	150′ from school
Town government—tax base loss	None	Slight	High	Moderate	Slight
Other areas					
Residential displacements	None	None	None	None	3 units
Air quality	None	None	None	None	None
Noise	None	Some	Slight	Some	Extensive
Visual					
View of Harrington River	None	Yes	None	None	None
Natural ecology					
Salt marsh displacement	None	1.5 acres	None	None	None
Upstream salt marsh	None	Improved	Improved	Some	Some
Downstream salt marsh	None	Some	Some	Considerable	Considerable
Increased run-off	None	Some	Some	Much	Much
Tree acreage	None	Slight	Slight	25 acres	28 acres
White-tail deer migration	None	None	None	Some disruption	Some disruption
Lowering of water table	None	Slight	Slight	Some	Some
Erosion from new slopes	None	Slight	Slight	Some	Some

† Source: Manheim et al. (1975, p. 45).
‡ Relative to statewide average for this type of facility.
§ Worst condition of several town center locations.
* dBA value exceeded only 10 percent of time.

marsh and reducing traffic noise in builtup areas. Thus the final selection must be made through a subjective balancing of the relative importance of all of these factors, a decision best left to those responsible to the citizens of the area and the state (including users of the roads) rather than to the project engineers.

If a new road is constructed, clearly it would be desirable to not only compensate the owners of land and buildings taken, as is commonly done now, but also to compensate the community for loss of property tax revenues, to require maintenance of road drains so as to minimize the harm of storm water run-off to the natural environment, and perhaps to replace natural lands lost with comparable forest land elsewhere. Such comprehensive compensation would probably make transport improvements much more acceptable than they are today.

GEOMETRIC DESIGN OF LINKS

The process for design of road links illustrates many of the characteristics of design processes in general. The approach to road design has been standardized to a large extent, for the reasons identified earlier: (1) similarity of requirements from situation to situation, (2) desirability of meeting minimal standards for uniformity and safety, and (3) provision of guidance for judgmental aspects of design. The process and standards for designs have evolved over many years and are promulgated in the publications of such organizations as the American Association of State Highway and Transportation Officials (AASHTO) and the Transportation Research Board (TRB).

The primary design features of a road are its location and its cross section. The location is partly specified by the horizontal alignment, its position in a horizontal plane relative to a coordinate system in which the positive y axis is north, the positive x axis east, and the positive z axis is elevation above mean sea level (msl). It is further specified by the profile, which as noted in Chap. 4 specifies the elevation at various locations along the route. Often a number of points, called stations, are identified along the route, at which the distance from some reference point and the elevation are given. These fully specify the location. The cross section of the road gives its width and other features such as pavement thickness, lane width, location of curbs, etc. We shall discuss all of these factors below, first concentrating on the cross section.

For purposes of design, the AASHTO classifies roads according to (1) traffic volume, (2) mix of vehicle types, (3) design speed, and (4) weight of vehicles. The first three are used in the design process in ways that are based on transportation relationships, while the last is related through structural and soils engineering, so we shall concentrate on the first three.

Design Volume and Cross Section

The traffic volume influences design by determining, in conjunction with the speed and vehicle mix, the number and width of lanes required. The specific volume used for this is the design hourly volume, designated DHV, which is usually the 30th-highest hourly volume occurring in a year. The DHV on any given type of road is usually a constant fraction of average daily traffic, ADT:

$$\text{DHV} = \frac{K}{100}\text{ADT} \tag{16-1}$$

where DHV = design hourly volume, two directions, vehicles/h
ADT = average daily traffic, two directions, vehicles/day
K = constant of proportionality, percent

For design purposes, the volume in the maximum flow direction for the design hour must be specified, usually as the directional split, termed *D,* the percentage of DHV in the maximum flow direction. Also important is the percentage of trucks in the traffic stream, designated as P_T. With the formulas of the 1965 *Highway Capacity Manual,* as presented in Chap. 5, it is also necessary to specify the fraction of buses P_B. Typical values for all of these factors are presented in Table 16-3, along with the definition of current and future traffic. It should be noted that it is now most common to predict future traffic by use of the demand methods presented in Chap. 9 rather than by use of a simple projection of future traffic as described in this table. The design speed, designated V, is selected on the basis of the function of the road and cost considerations, and is usually specified in the planning process.

These factors will determine the number of lanes (and their widths) required to accommodate the traffic provided the design type of the road is specified. A road with full access control is a freeway; one without is a highway or arterial street. The method used was presented in Chap. 5, where the capacity of roads was discussed, except that here the service volume is specified and the number of lanes is to be determined. Basically this involves trial and error solution of Eq. (5-13), substituting various values for N until a value yielding an SV greater than or equal to D(DHV)/100, and such that $N - 1$ lanes is insufficient for D(DHV)/100, is found. This is the required number of lanes. Note that if the number of lanes allocated to traffic in each direction is varied by time of day, this must be taken into account.

The use of these formulas is straightforward. Consider a rural freeway in rolling terrain on which the DHV is 6000 vehicles/h, with a D of 60 percent, a design speed V of 60 mi/h, a P_T of 8 percent, a P_B of 3 percent, and a design adjustment factor of 0.95. As noted in Chap. 5, it is common to assume level of service B on rural roads and C on urban roads for the DHV. The service volumes for flow in one direction on two, three, and four lanes are

$$\text{DHV}(D) = 6000(0.60) = 3600 \text{ vehicles/h}$$

$$\text{SV} = 2000\,N\,\frac{V}{C}\,WT_cB_c$$

$$T_c = \frac{100}{100 - P_T + E_TP_T} = \frac{100}{100 - 8 + 4(8)} = 0.806$$

$$B_c = \frac{100}{100 - P_B + E_BP_B} = \frac{100}{100 - 3 + 3(3)} = 0.943$$

$$\text{SV}_{\text{two lanes}} = 2000(2)(0.5)(0.95)(0.806)(0.943) = 1444 \text{ vehicles/h}$$
$$\text{SV}_{\text{three lanes}} = 2000(3)(0.58)(0.95)(0.806)(0.943) = 2513 \text{ vehicles/h}$$
$$\text{SV}_{\text{four lanes}} = 2000(4)(0.63)(0.95)(0.806)(0.943) = 3639 \text{ vehicles/h}$$

Hence four lanes (one direction) are required.

Table 16-3 Traffic Volume Concepts Used in the Design of Highways†

Traffic element	Explanation and nationwide percentage or factor
Average daily traffic (ADT)	Average 24-h volume for a given year; total for both directions of travel, unless otherwise specified. Directional or one-way ADT is an average 24-h volume in one direction of travel only.
Current traffic	ADT composed of existing trips, including attracted traffic, that would use the improvement if opened to traffic today (current year specified).
Future traffic	ADT that would use a highway in the future (future year specified). Future traffic may be obtained by adding generated traffic, normal traffic growth, and development traffic to current traffic, or by multiplying current traffic by the traffic projection factor.
Traffic projection factor	Future traffic divided by current traffic. General range, 1.5 to 2.5 for 20-year period.
Design hour volume (DHV)	Future hourly volume for use in design (two-way unless otherwise specified), usually the 30th-highest hourly volume of the design year (30HV) or equivalent, the approximate value of which can be obtained by the application of the following percentages to future traffic (ADT). The design hour volume, when expressed in terms of all types of vehicles, should be accompanied by factor P_T, the percentage of trucks during peak hours. Or, the design hour volume may be broken down to the number of passenger vehicles and the number of trucks.
Relation between DHV and ADT (K)	DHV expressed as a percentage of ADT, both two-way; normal range 11 to 20, 7 to 18 percent in urban areas. Or DHV, expressed as a percentage of ADT, both one-way, normal range, 16 to 24.
Directional distribution (D)	One-way volume in predominant direction of travel expressed as a percentage of two-way DHV. General range, 50 to 80. Average, 67. In urban areas this approaches a 50–50 ratio.
Composition of traffic (P_T)	Trucks (exclusive of light delivery trucks) expressed as a percentage of DHV. Average 7 to 9 on rural highways. Where weekend peaks govern, average may be 5 to 8.

† Adapted from American Association of State Highway Officials (1965, p. 70).

Obviously one of the most important features in the design of any transportation facility is the accommodation of vehicles in a safe and economical manner. Most facilities must accommodate vehicles of many different types, sizes, and speed performance levels. In order to design a road for a range of vehicle types, it is necessary to identify a vehicle (or more precisely, a set of characteristics) which will determine the size and configuration of the facility. This vehicle's characteristics must be selected so that when the facility is designed to accommodate it, that facility will also be able to accommodate all other vehicles which are to use it. The American Association of State Highway Officials (1965) defines the design vehicle as "a selected motor vehicle, the weight, dimensions, and operating characteristics of which are used to establish highway design controls. . . . The design vehicle should be one with dimensions and minimum turning radius larger than those of almost all vehicles in its class." With minor modification, namely, omission of the word "highway," the above definition applies for any mode of transportation.

For highway design, and the design of related facilities, such as parking garages, there are a number of standard design vehicles, passenger cars and trucks, as shown in Fig. 16-3. For most links and intersections which must accommodate all types of vehicles, the largest truck or bus governs the design.

One important characteristic of all roads is the pavement width. To accommodate the largest vehicles, the width of lanes are usually as indicated below:

Local (land access) streets: 9 ft, but often no lanes identified in a pavement of at least 27 ft

Collector streets: 9 ft, but again often no lanes in a pavement of at least 36 ft

Arterial streets: 11 to 13 ft

Freeways: 12 to 14 ft

Parking lanes are usually 6 to 8 ft wide where parallel parking is provided. If there is no parking, a lane or shoulder for breakdown is usually provided. High-speed high-volume roads are best designed as freeways, on which access is restricted to ramps (i.e., no driveways or intersecting streets) and traffic in one direction is separated from that in the other by a median strip 12 to 40 ft wide. Also, some major arterials are divided similarly.

Typical cross sections for an urban freeway and a rural two-lane highway are shown in Fig. 16-4. The freeway design is for a relatively restricted right-of-way width, necessitating the use of a narrow median with a barrier (rather than a much wider median) and retaining walls along the entrance and exit ramps. The rural road design is for an intermediate type road, designed for speeds of 50 to 60 mi/h and thirtieth highest hour volumes of over 200 vehicles/h.

Horizontal and Vertical Alignment

Another important feature of a road design is the horizontal alignment, and, in particular, curves. Sometimes these are laid out as simple curves, that is, curves of constant radius. Such a curve is illustrated in Fig. 16-5, along with most of the standard terminology of simple curves and the basic trigonometric relationships. Given the location of the two tangents to be connected by a curve and the radius (or degree) of the curve, the determination of the horizontal alignment is straightforward, using the formula presented in this figure. It should be noted that in these formulas, the degree of curvature is defined as the central angle subtended by an arc of 100 ft rather than the angle subtended by a chord of 100 ft as we defined it in Chap. 4. In highway engineering the arc definition is generally used, but in railroad engineering

Figure 16-3 Characteristics of highway design vehicles. *(a)* Design vehicle types, dimensions and turning radii. [*From American Association of State Highway Officials (1973, pp. 269–270).*] *(b)* Turning path of a WB-50 design vehicle. [*From American Association of State Highway Officials (1973, p. 275).*]

			Vehicle dimensions, ft					Turning dimensions, ft		
Design vehicle type	Symbol	Wheelbase	Front overhang	Rear overhang	Overall length	Overall width	Height	Minimum radius	Minimum inside radius	Corresponding outside radius
Passenger car	P	11	3	5	19	7	—	24	15.3	25.8
Single unit truck	SU	20	4	6	30	8.5	13.5	42	28.4	43.9
Single unit bus	BUS	25	7	8	40	8.5	13.5	42	20.3	47.1
Semitrailer combination, intermediate	WB-40	13 + 27 = 40	4	6	50	8.5	13.5	40	19.9	41.2
Semitrailer combination, large	WB-50	20 + 30 = 50	3	2	55	8.5	13.5	45	19.8	46.2
Semitrailer-fulltrailer, combination	WB-60	9.7 + 20.0 + 9.4* + 20.9 = 60	2	3	65	8.5	13.5	45	22.5	45.55

* Distance between rear wheels of front trailer and front wheels of rear trailer.

(a)

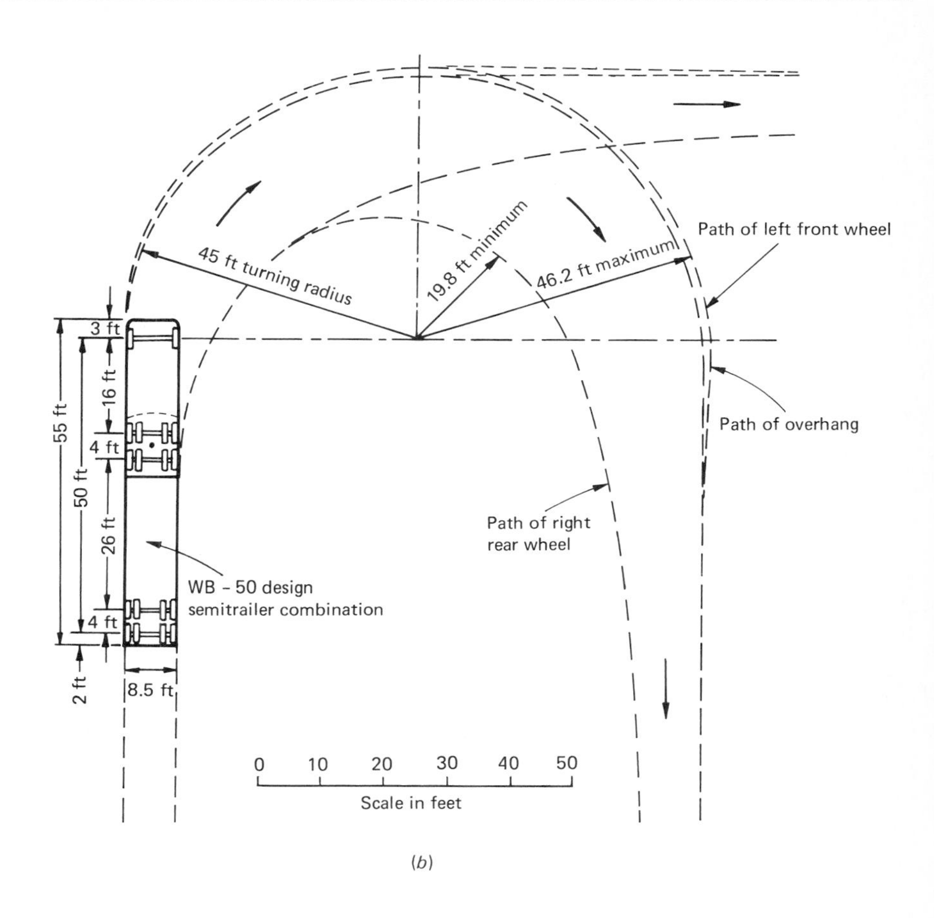

(*b*)

the chord definition is used. The radius or degree of curvature will depend on the speed of vehicles, the extent to which the road is superelevated, and the land available for the road right-of-way. The interaction among these is complex. The subject will be covered after presentation of the spiral curve, which is generally used in conjunction with simple curves on main highways.

In general, simple curves on main highways are less than 5° to 7°, except in mountainous terrain where 10° is the usual maximum. Pavements are generally widened on curves to allow for the fact that the rear of a vehicle, especially an articulated one, does not follow the path of the front. Present practice is for no widening if the curve is less than 10° and the pavement is at least 12 ft wide. Usually 2 to 4 ft is added to pavement width on sharper curves.

A problem with simple curves is the sudden transition from a straight to a curved path, with the associated change in driver behavior required and increase in lateral force on the vehicle and its contents. To ease the transition, the radius is often gradually decreased (from the infinite radius of the tangent) using an easement curve. Easement curves are typically used on high-volume or high-speed roads where the curvature is greater than 3°. The easement curve usually follows the form of the spiral, which is also used on railroads. The terminology of an easement curve is illustrated in Fig. 16-6.

Figure 16-4 **Typical highway cross sections.** ***(a)*** **Depressed urban freeway.** [***From American Association of State Highway Officials (1973, p. 436).***] ***(b)*** **Two-lane rural highway.** [***From American Association of State Highway Officials (1965, p. 263).***]

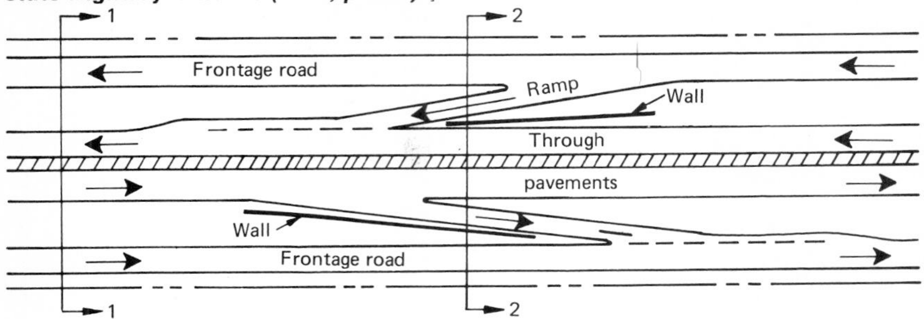

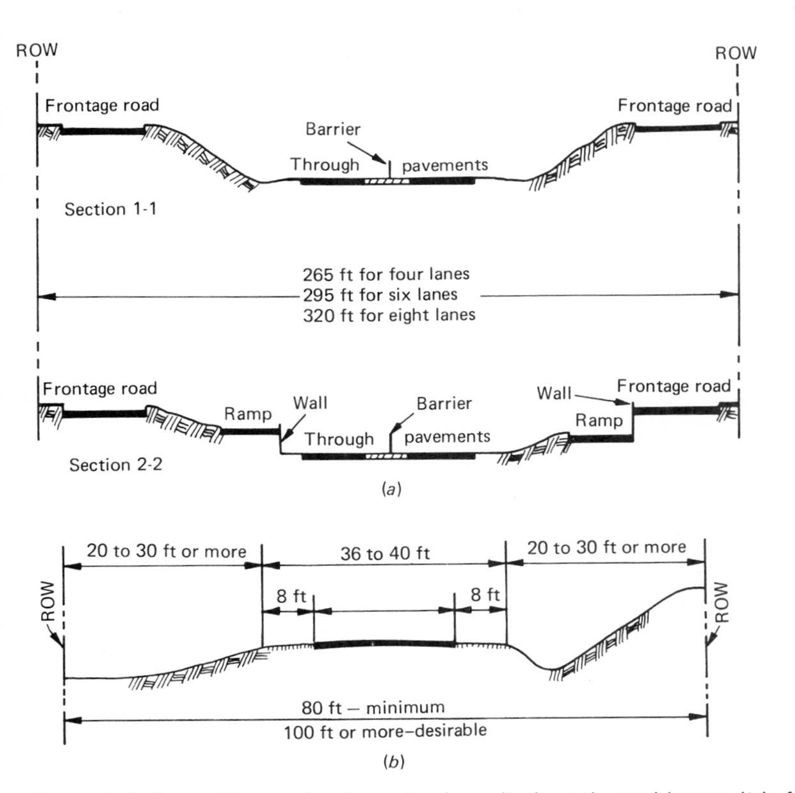

In the spiral, the radius varies linearly along its length, and hence it is fully specified by the two end radii and its length. The formulas for the x, y coordinates of any point a distance l from the end of a spiral can be derived easily. Using the notation of Fig. 16-6, where TS refers to the point of connection with a tangent and SC the point of connection with a simple curve of degree D_c, we can write for an infinitesimal circular segment with a central angle $d\theta$ and an arc length dl (as in Fig. 16-6*b*), by definition:

Figure 16-5 **A simple horizontal highway curve.** [*From Ritter and Paquette (1967, Fig. 5-10).*]

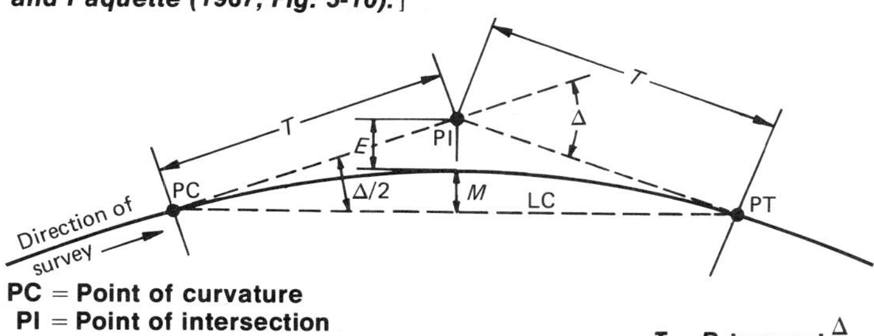

PC = Point of curvature
PI = Point of intersection
PT = Point of tangency
***E* = External distance**
***M* = Middle ordinate distance**
***R* = Length of radius of curve**
***T* = Length of tangent (PC to PI and PI to PT)**
***D* = Degree of curve (angle subtended at the center of curve by an arc of 100 feet)**
***L* = Length of curve, ft**
LC = Long chord
Δ = External angle, deg.

$$T = R \text{ tangent} \frac{\Delta}{2}$$

$$E = R \text{ exsecant} \frac{\Delta}{2}$$

$$L = 100 \frac{\Delta}{D}$$

$$M = R \text{ versine} \frac{\Delta}{2}$$

$$LC = 2R \sin\frac{\Delta}{2}$$

$$d\theta = \frac{dl}{R}$$

From the linearity assumption, one can write by direct proportion:

$$\frac{R}{R_c} = \frac{L_s}{l} \quad \text{or} \quad \frac{D}{D_c} = \frac{l}{L_s}$$

Hence $R = \dfrac{L_s R_c}{l}$

and $d\theta$ may be written

$$d\theta = \frac{l\,dl}{L_s R_c}$$

Integration yields

$$\theta = \frac{l^2}{2L_s R_c}$$

Now consider the infinitesimal right triangle bounded by dl, dx (and the included angle θ), and dy. We can write

$$dx = dl \cos \theta$$
$$dy = dl \sin \theta$$

It is convenient in this case to substitute the series expansions of $\cos \theta$ and $\sin \theta$, which yields

$$dx = dl \left(1 - \frac{\theta^2}{2!} + \frac{\theta^4}{4!} - \cdots \right)$$

Figure 16-6 **Horizontal highway curves with easements.** ***(a)*** **Layout and terminology. [*From Oglesby (1975, p. 297).*] *(b)* Notation used in derivation of equation of spiral easement curve.**

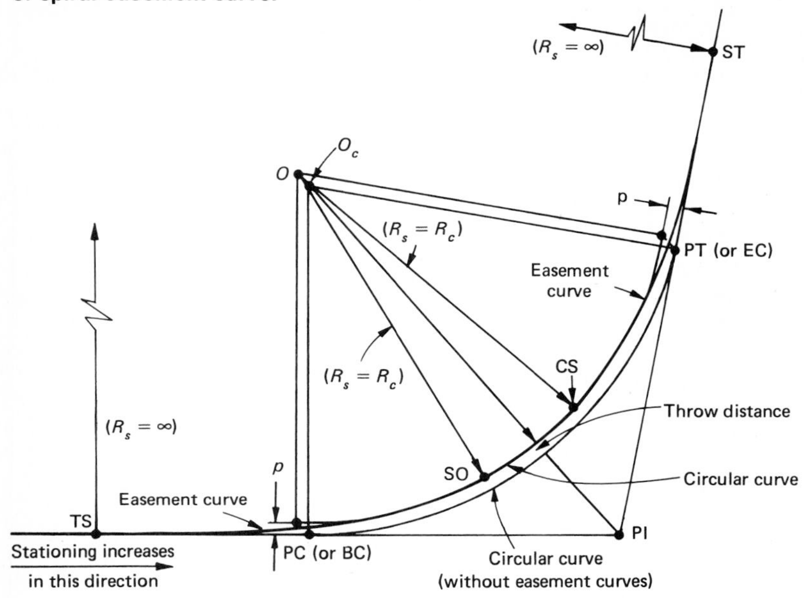

(*a*)

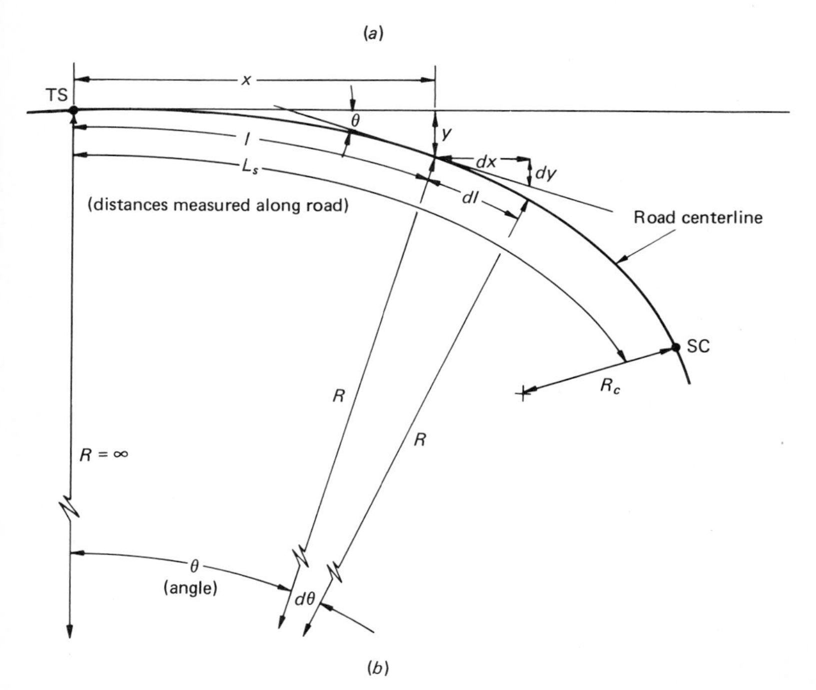

(*b*)

$$dy = dl\left(\theta - \frac{\theta^3}{3!} + \frac{\theta^5}{5!} - \cdots\right)$$

Substitution of $\frac{l^2}{2L_sR_c}$ for θ gives

$$dx = dl\left(1 - \frac{l^4}{8L_s^2R_c^2} + \cdots\right)$$

$$dy = dl\left(\frac{l^2}{2L_sR_c} - \frac{l^6}{48L_s^3R_s^3} + \cdots\right)$$

Integration of both sides of the above set of equations yields the desired result:

$$x = l - \frac{l^5}{40L_s^2R_c^2} + \epsilon$$

$$y = \frac{l^3}{6L_sR_c} - \frac{l^7}{336L_s^3R_c^3} + \epsilon \qquad (16\text{-}2)$$

The ϵ in the above equations represents terms of the expansion that converge very rapidly to zero and hence may be ignored. Equation (16-2) permits the computation of the coordinates (x, y) of a point anywhere along a spiral curve.

Standards have been set for the length of spirals from tangent to simple curve sections, to ensure a gradual transition. This minimum is given by the formula (Paquette et al., 1972, p. 381):

$$L_s = 1.6\,\frac{V^3}{R} \qquad (16\text{-}3)$$

where L_s = length of transition, ft
V = speed, mi/h
R = radius at end point, ft

Curves on roads are superelevated in order to reduce the centrifugal force on the vehicle and its contents, as we noted in Chap. 4. This not only increases comfort, but also reduces the side friction coefficient necessary between the tires and the pavement to prevent the vehicle from skidding sideways. There we derived the following approximate formula relating the necessary coefficient of side friction, the superelevation measured in elevation per unit horizontal distance (along the radius), the speed, and the radius or degree of curvature, for movement around the curve:

$$e + f = \frac{V^2}{15R} = \frac{0.067V^2}{R} = \frac{V^2D}{85{,}900} \qquad (16\text{-}4)$$

where e = rate of superelevation, dimensionless
f = coefficient of side friction
V = speed, mi/h
R = radius, ft
D = degree of curvature

Also, we noted that at very high speeds, the vehicle may overturn if the maximum coefficient of side friction is sufficiently large, as it might be on a dry pavement. This critical overturning speed is derived below, using the notation of Fig. 4-7. The moments of forces about the outer wheel yield (with clockwise being positive)

$$\frac{T}{g}\frac{V^2}{R}(\cos\theta)H - \frac{T}{g}\frac{V^2}{R}(\sin\theta)\frac{Y}{2} - T(\sin\theta)H - T(\cos\theta)\frac{Y}{2}$$

For overturning, this must be greater than zero. Hence, upon simplification, we have

$$V^2 \geq \frac{gR}{H - \frac{1}{2}Ye}(He + \tfrac{1}{2}Y) \qquad (16\text{-}5)$$

For overturning to take place the maximum realizable value of f must be at least that given by Eq. (16-4), for the velocity that satisfies Eq. (16-5). Obviously this speed should not be approached on any road.

Finally, there is the prospect of sliding off the road surface, where f is sufficiently small. Once the absolute value of f in Eq. (16-5) exceeds that attainable with the actual tire and road conditions, the vehicle will either slide outward when speed is high or inward when speed is low or zero. A large e reduces the possibility of the former, but it increases the possibility of the latter. Where snow and ice are common, f at zero speed can be as low as 0.10 (Table 4-4 contains typical values), and this possibility restricts maximum superelevations to $e = 0.08$. Elsewhere where only rain reduces f, a superelevation up to 0.12 is accepted.

Values for f vary widely with speed as well as road conditions. Current practices allow values up to 0.16 for vehicles moving at 30 mi/h, 0.15 at 40 mi/h, 0.14 at 50 mi/h, 0.13 at 60 and 65 mi/h, and 0.11 at 75 and 80 mi/h. With typical values of H for motor vehicles, this means that Eq. (16-4) will govern design speeds rather than Eq. (16-5). Hence (16-4) can be used, with the appropriate e and f values, to determine the speed limit on any road, or conversely to determine the minimum radius for any given speed.

Superelevation is usually attained in the transition or spiral section of a curve and in some of the prior tangent section, as necessary. It is standard practice to do this in such a manner that vehicles do not have to slow down, as the transition from a normal cross section to a superelevated one is very gradual. The AASHTO has specified limits on the rate of change in cross section (see American Association of State Highway Officials, 1965, for details).

Roads also contain vertical curves, as transitions from one gradient to another. Gradients are typically limited to the following values on main roads or highways:

Flat country: 6 percent at 30-mi/h design speed
3 percent at 80-mi/h design speed

Mountainous terrain: 9 percent at 30-mi/h design speed
5 percent at 70-mi/h design speed

Local streets often have grades up to 15 percent.

Vehicles differ greatly in their ability to surmount grades at speed. This was indicated in Chap. 4, where Fig. 4-30 specified the reduction in speed of trucks on various grades. Since trucks can be traveling at much lower speeds than passenger cars, they are often provided with special climbing lanes on high-speed roads—an extra lane only for slow-moving vehicles in the upgrade direction. Similarly on particularly long, steep downgrades, trucks are often restricted to lower speeds than automobiles due to their poorer braking characteristics and are given a special lane.

A curve in the vertical plane is required at the points of intersection of road sections of differing gradient. Usually the parabola is used, primarily because of its convenient mathematical properties in laying out the profile in the field. The parabolic vertical curve is most conveniently described by an equation in which the x axis is measured horizontally from the point at which the vertical curve begins and the y axis is the elevation relative to that point. Those axes and the point of beginning of the vertical curve (BVC) are shown in Fig. 16-7. Using that notation, the equation for the curve is derived as follows.

The general form of a parabola is usually expressed $y = ax^2 + bx + c$. It is desirable to define the coefficients a, b, and c of the above equation in terms of common vertical curve parameters and notation. Assume the origin of the x coordinate axis to be the BVC and the origin of the y coordinate axis to be msl. Then, if $x = 0$, y represents the elevation of the BVC. The first derivative of the parabolic equation with respect to x is

$$\frac{dy}{dx} = 2ax + b$$

Evaluation of the first derivative at $x = 0$ implies that $b = g_1$. Taking the second derivative with respect to x yields

$$\frac{d^2y}{dx^2} = 2a$$

Figure 16-7 **Notation used in derivation of parabolic vertical curve.**

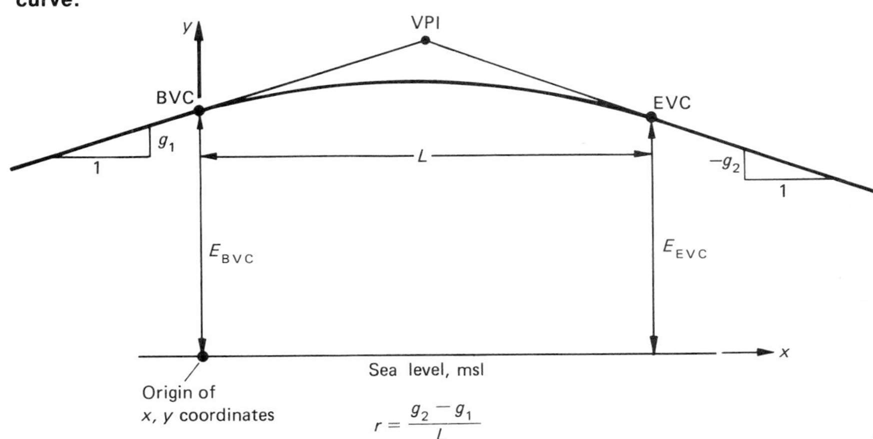

This is to say, a parabola has a constant rate of change of grade (slope) equal to $2a$. From Fig. 16-7, it is clear that the grade changes from g_1 to g_2 in length L parallel to the x axis. Therefore, r is the rate of change of grade and

$$a = \frac{r}{2} = \frac{g_2 - g_1}{2L}$$

The equation of a parabolic vertical curve can now be written

$$y = \frac{r}{2}x^2 + g_1x + E_{\mathrm{BVC}} \tag{16-6}$$

where y = elevation at point x
r = rate of change of grade
g_1 = grade at BVC
E_{BVC} = elevation at BVC

One must be very careful of the signs of g_1 and g_2 when using these formulas; upgrades, as seen from the BVC are positive, downgrades are negative.

On rare occasions, compound vertical curves are used to fit a particular desired profile. A compound vertical curve consists of two or more successive parabolic curves of different r values, without intervening tangents. These compound curves are used when there are special clearance or drainage problems or at certain intersections of transportation routes.

Thus the critical parameter in determining the parabolic vertical curve in any given situation with fixed values of g_1 and g_2 is the length of the curve L. There are two main restrictions on L, namely, (1) that the underside of vehicles not be struck by the road and (2) that the sight distance of drivers is adequate for safe operation at the contemplated speeds. In all but very slow-speed situations involving very long wheelbase or long overhang vehicles, the sight distance restriction governs.

The restriction related to the vehicle striking the road can be analyzed graphically. This requires very careful construction of the road profile and vehicle longitudinal cross section or side elevation, but is usually easier than solving the rather complex algebraic relationships.

The other important factor governing the vertical alignment of roads and influencing horizontal curvature in conjunction with objects which interfere with a view of the road ahead around curves, is the sight distance. The overall design must be such that the driver has a sufficiently long clear vision to avoid hitting unexpected obstacles. Sight distance is the distance a driver can see ahead. Stopping sight distance is the minimum distance required for the design vehicle to stop from the design speed, measured for the time of first sighting an obstacle in the vehicle's path. As we saw in Chap. 4, this distance depends upon vehicle type, initial speed, and gradient. (It may be useful to refer to Fig. 4-18 again.) The sight distance on a road should be at least the stopping sight distance, with the driver's eye assumed to be 3.75 ft above the pavement and the object to be seen only 0.5 ft high. The actual sight distance at any location can usually be determined from the profile and plan maps, as shown in Fig. 16-8, although equations for certain simple situations can be

Figure 16-8 **Measurement of stopping and passing sight distances. *(a)* Stopping sight distance. *(b)* Passing sight distance.** [***From Oglesby (1975, p. 287).***]

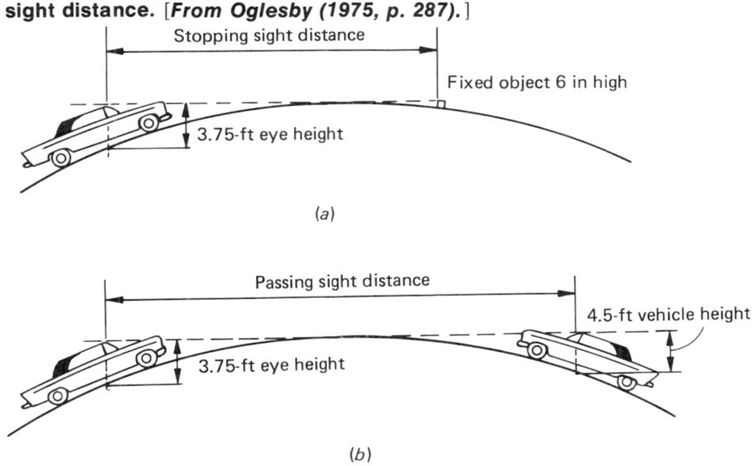

derived, as in Prob. 16-2. Based on these considerations, the AASHO has set minimum lengths for vertical curves, as shown in Fig. 16-9.

However, in the case of two- and three-lane roads where a lane is used for passing by vehicles in both directions the object to be seen is assumed to be 4.5 ft high. Based on considerations of vehicle speeds and acceleration rates, the AASHO has concluded that minimum sight distances for passing where the lane is used in both directions should be 800, 1300, 1700, 2000, and 2300 ft for design speeds of 30, 40, 50, 60, and 70 mi/h, respectively.

These values are based upon an analysis of passing maneuvers. Assumptions regarding the speed behavior of drivers is necessary, and the ones used by AASHO are as follows: The vehicle being passed travels at a uniform speed, and the passing vehicle is forced to travel at this same speed when it cannot safely pass. When a section allowing safe passing is reached, the driver requires a short period of time to reach a decision on whether or not it is safe to pass. Once she does decide to pass, she accelerates her vehicle and enters the passing lane, reaching and maintaining a speed 5 to 10 mi/h greater than the speed of the vehicle being passed. A vehicle appearing from the other direction at the instant the passing maneuver begins must be seen by the driver of the passing vehicle for a sufficient distance ahead that the oncoming vehicle will arrive alongside the passing vehicle only when the maneuver has been completed.

The movement of three vehicles involved in such passing is shown in Fig. 16-10. The distances identified there are

d_1 = distance traveled during perception and reaction time and during the initial acceleration to the point where the vehicle will turn into the opposite lane
d_2 = distance traveled while the passing vehicle occupies the left lane
d_3 = distance between the passing vehicle at the end of its maneuver and the opposing vehicle

Figure 16-9 **Design controls for vertical highway curves.** ***(a)*** **Crest vertical curves.** [***From American Association of State Highway Officials (1965, p. 205).***] ***(b)*** **Sag vertical curves.** [***From American Association of State Highway Officials (1965, p. 210).***]

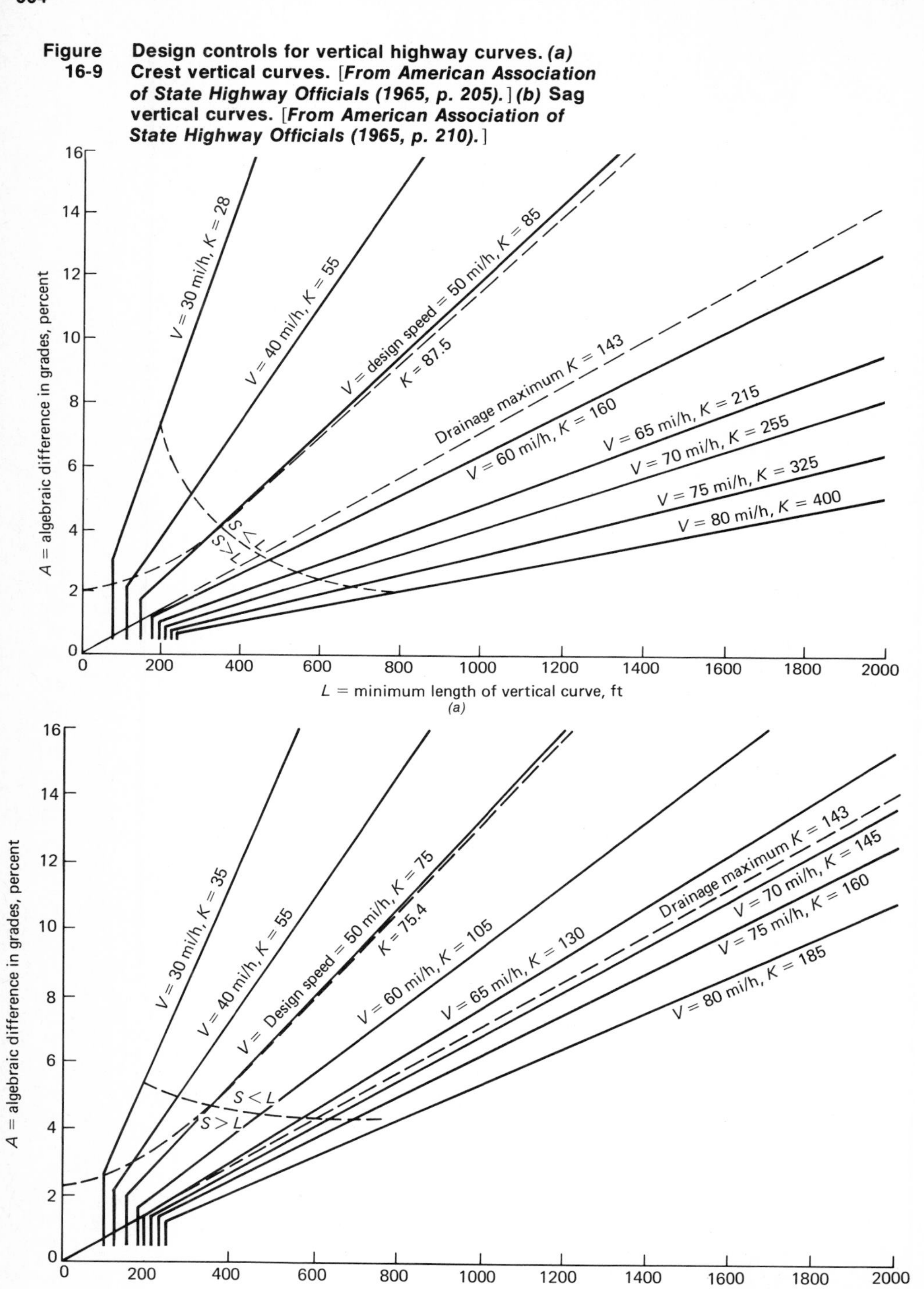

Figure 16-10 **Calculation of passing sight distances on two-lane highways.** [***From American Association of State Highway Officials (1965, p. 143).***]

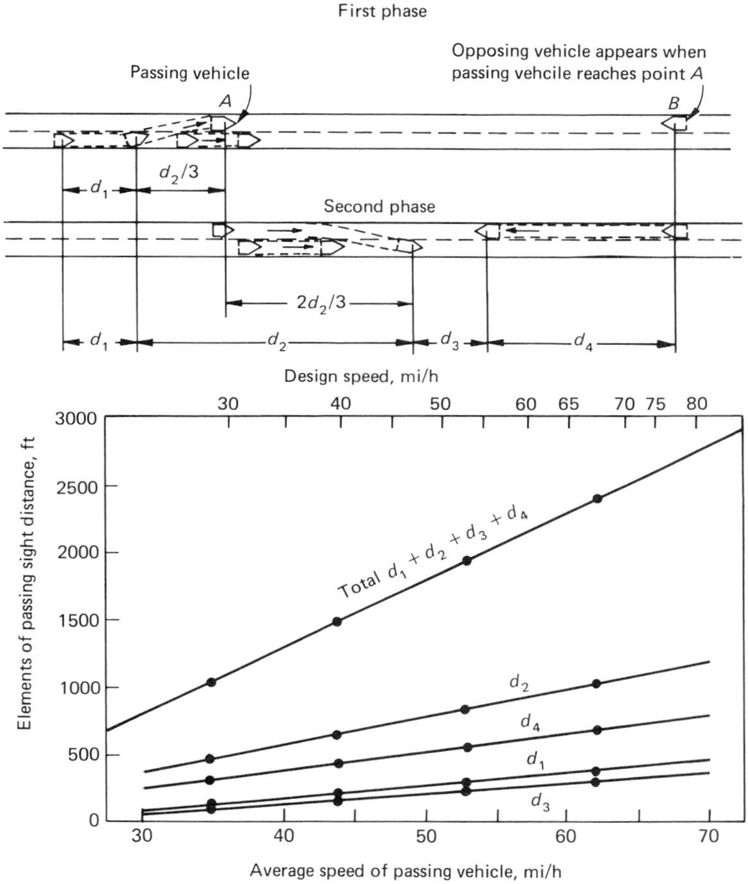

d_4 = distance traveled by the oncoming vehicle for two-thirds of the time the passing vehicle occupies the left lane

The sum of these four distances is the minimum sight distance for safe passing. Its computation is straightforward. Consider a road designed for 50 mi/h, on which the average vehicle speed is 42.6 mi/h and the average passing speed is 52.6 mi/h. Observations of actual passing maneuvers indicated that at this road design a vehicle intending to pass arrives behind a slower vehicle at a speed of 46.3 mi/h, from which it accelerates to a speed 5 to 10 mi/h greater (6.3 mi/h in this example) than the passed vehicle (over the distance d_1 in the figure), at an average acceleration rate of 1.47 mi/(h)(s), and that the passing requires 10.7 s in the left lane (covering the distance d_2) in the figure. The clearance distance d_3 is assumed to be 250 ft. By Eqs. (4-38) and (4-40), the time and distance, respectively, for these maneuvers can be calculated. For the first phase d_1 and the corresponding time t_1 are as below. [For these computations, the distance d_1 in Fig. 16-10 corresponds to $d_{i+1} - d_i$ in Eq. (4-50) and corresponds to $t_{i+1} - t_i$ in Eq. (4-48).]

$$t_1 = t_{i+1} - t_i = \frac{6.3 \text{ mi/h}}{1.47 \text{ mi/(h)(s)}} = 4.3 \text{ s}$$

$$d_1 = d_{i+1} - d_i = \left[\frac{46.3 \text{ mi/h}(6.3 \text{ mi/h})}{1.47 \text{ mi/(h)(s)}} + \frac{1}{2}\,\frac{(6.3 \text{ mi/h})^2}{1.47 \text{ mi/(h)(s)}}\right] \frac{5280 \text{ ft h}}{3600 \text{ s mi}} = 310.5 \text{ ft}$$

The second phase consumes 10.7 s, so

$$t_2 = 10.7 \text{ s}$$

$$d_2 = 52.6 \text{ mi/h}(10.7 \text{ s}) \frac{5280 \text{ ft h}}{3600 \text{ s mi}} = 825 \text{ ft}$$

The third phase, the clearance distance, is assumed by AASHO to be 250 ft at this speed:

$$d_3 = 250 \text{ ft}$$

Finally, a vehicle in the opposite direction could appear just as the passing vehicle reaches point A in Fig. 16-10, the point at which it is assumed the decision to pass cannot be reversed. This is assumed to be at a point one-third from the beginning of the occupancy of the left lane (which starts at d_1). Assuming the opposing vehicle is traveling at the same speed as the passing vehicle, then

$$d_4 = \tfrac{2}{3} d_2 = \tfrac{2}{3}(825 \text{ ft}) = 550 \text{ ft}$$

Hence the sight distance for safe passing on a road with vehicles of these

Figure 16-11 **Typical signs and pavement markings for no-passing zones on highways.** [***From Ritter and Paquette (1967, Fig. 5-20).***]

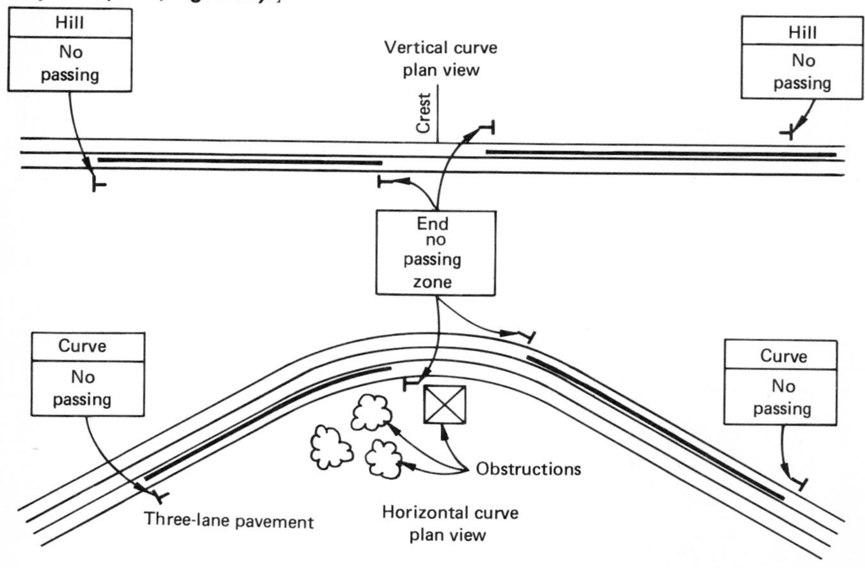

Table 16-4 Typical Design Standards for Streets in Residential Subdivisions†
Development Density, 2–6 dwelling units/acre

	Local streets			Collector streets		
Terrain, cross slope, percent	0–8	8–15	15	0–8	8–15	15
Right-of-way width, ft	60	60	60	70	70	70
Pavement width, ft	32	34	34	36	36	36
Sidewalk distance from curb face, ft	6	6	6	10	10	10
Minimum sight distance, ft	200	150	110	250	200	150
Maximum grade, percent	4	8	15	4	8	12
Design speed, mi/h	30	25	20	35	30	25
Minimum center line radius, ft	250	175	110	350	230	150
Maximum cul de sac length, ft	500	500	500			
Minimum cul de sac radius, right-of-way, ft	50	50	50			

† As adapted by Paquette et al. (1972, p. 413) from Box et al. (1967).

speeds, typically for a road with a design speed of 50 to 60 mi/h, is

$$d = d_1 + d_2 + d_3 + d_4 = 1935.8 \text{ ft}$$

On the basis of computations such as these, AASHO recommends sight distances for passing on roads with design speeds of 30, 40, 50, 60, and 70 mi/h of 1100, 1500, 1800, 2100, and 2500 ft, respectively.

It is important to introduce proper markings on a road to ensure that motorists know where it is safe to pass. Typical markings are illustrated in Fig. 16-11.

Before turning to intersections, it is appropriate to mention the typical design characteristics of residential subdivision streets. These roads are primarily designed for local land access, as we noted in Chap. 15. Hence their design is based primarily on considerations of simply physically accommodating each of the various types of vehicles that might use them, including parking vehicles, in contrast to a design also based on considerations of high volumes and speeds. Typical design standards from such streets are presented in Table 16-4. The cross slope is the average gradient of the land, measured in a vertical plane perpendicular to the axis of the road. Collector streets are wider simply because they accommodate more traffic and often have more parking vehicles, often being sites of small commercial areas.

GEOMETRIC DESIGN OF INTERSECTIONS

A primary characteristic of road transportation is that each driver is free to choose her or his own route through the network, and it is therefore essential to provide intersections to ensure safe and efficient passage from one road link to another. Road intersections fall into two major categories: at-grade intersections and grade-separated intersections (often called interchanges).

Figure 16-12 **Typical types of at-grade road intersections.** [***From American Association of State Highway Officials (1965, p. 388).***]

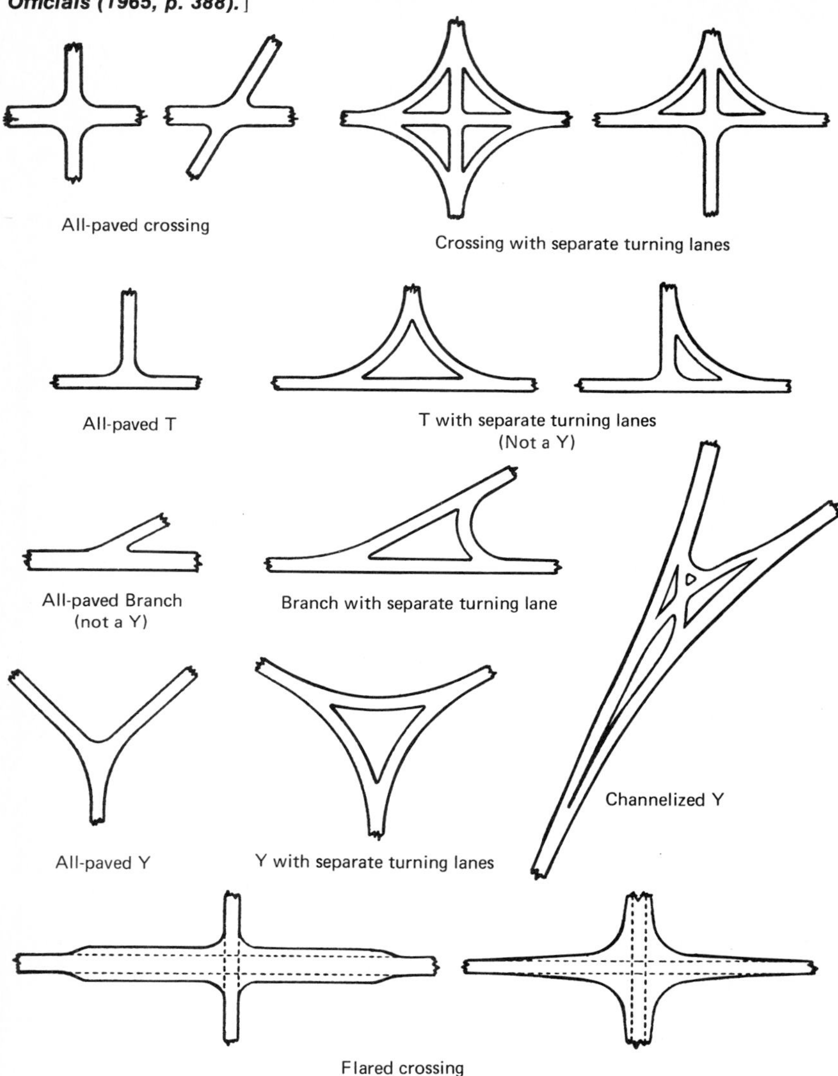

At-grade intersections are those in which the various roads or approaches to the intersection direct traffic onto paths which can conflict with one another, as is usually the case with intersections of city streets. In contrast, grade-separated intersections provide for the separation of traffic on different paths, such that the only intersection of paths of vehicles occurs where vehicles diverge from or converge onto the same path.

At-Grade Intersections

The many different types of at-grade intersections reflect the arrangement of the approach roads, the degrees of separation of certain conflicting movements, the volume of traffic to be accommodated, and the speed and the total amount of land devoted to the facility. Examples of many different types of at-grade intersections are presented in Fig. 16-12.

As we noted in Chap. 5, it is often advantageous to separate vehicles with different paths. Channelization is widely used at intersections where there is a high volume of traffic or where the size of the intersection is so great that without some channelization the particular path to be followed by a vehicle moving through the intersection would be unclear. Examples of channelization using islands are shown in Fig. 16-13. Islands are areas which are forbidden for vehicles to use, often consisting of raised sections identifying them as areas not to be traversed by vehicles. Also on occasion they are simply areas painted on the street surface, perhaps with a few signs to indicate their function. Their primary purpose is to identify paths which vehicles should follow through the intersection, and thereby to increase the safety of vehicle movement and the speed with which vehicles can pass through the intersection, and thereby to increase the capacity of that intersection. They can also be used as places where pedestrians can wait safely before crossing traffic lanes, these islands being termed pedestrian refuge islands.

Since intersections comprise in part curved sections of roadway, the physical principles that determine the performance of vehicles on curves also apply to intersections. Particularly important in intersection design are the radii of curves and the superelevation, if any. Based on an assumed design speed and likely values of side friction and superelevation, the minimum radius for a turn at an intersection can be calculated, using Eq. (15-4). Such calculations for typical design speeds and other characteristics are shown in Table 16-5*a*, along with the minimum radius of curves for that design speed recommended by the American Association of State Highway and Transportation Officials. For design speeds on turns at intersections of 40 mi/h and greater, curves should be designed in accordance with the methods for curves on main roadway sections described previously.

Also of importance in the design of intersection curves is the type of vehicle expected to use the curve. Minimum radii at the edge of pavement for various vehicle types and various angles of turns at low- and medium-speed intersections (up to 40 mi/h) are presented in Table 16-5*b*. It will be noted that many of the turns should be designed with three-center compound curves at the edge of the pavement, the curve consisting of two end sections of large radius and a middle section of much smaller radius. Also, these curves are to be offset, by the distances shown, as calculated using the definition of offset shown in Fig. 16-6.

Freeway Intersections

There are also many different types of freeway intersections, which are usually called interchanges. If the intersection involves only two freeways, then the entire intersection presumably would be grade-separated so that there would be no vehicle paths crossing other vehicle paths at grade. These also may be applied in intersections with arterial streets. Examples of such interchanges are shown in Fig. 16-14*a* through *c*. However, intersections between freeways and arterial streets can include some at-grade intersections. Particularly common among these are various diamond interchanges, which are widely used in urban areas because they require much less land than clover-

Figure 16-13 **Examples of road traffic islands.** [*From American Association of State Highway Officials (1965, pp. 370 and 462).*]

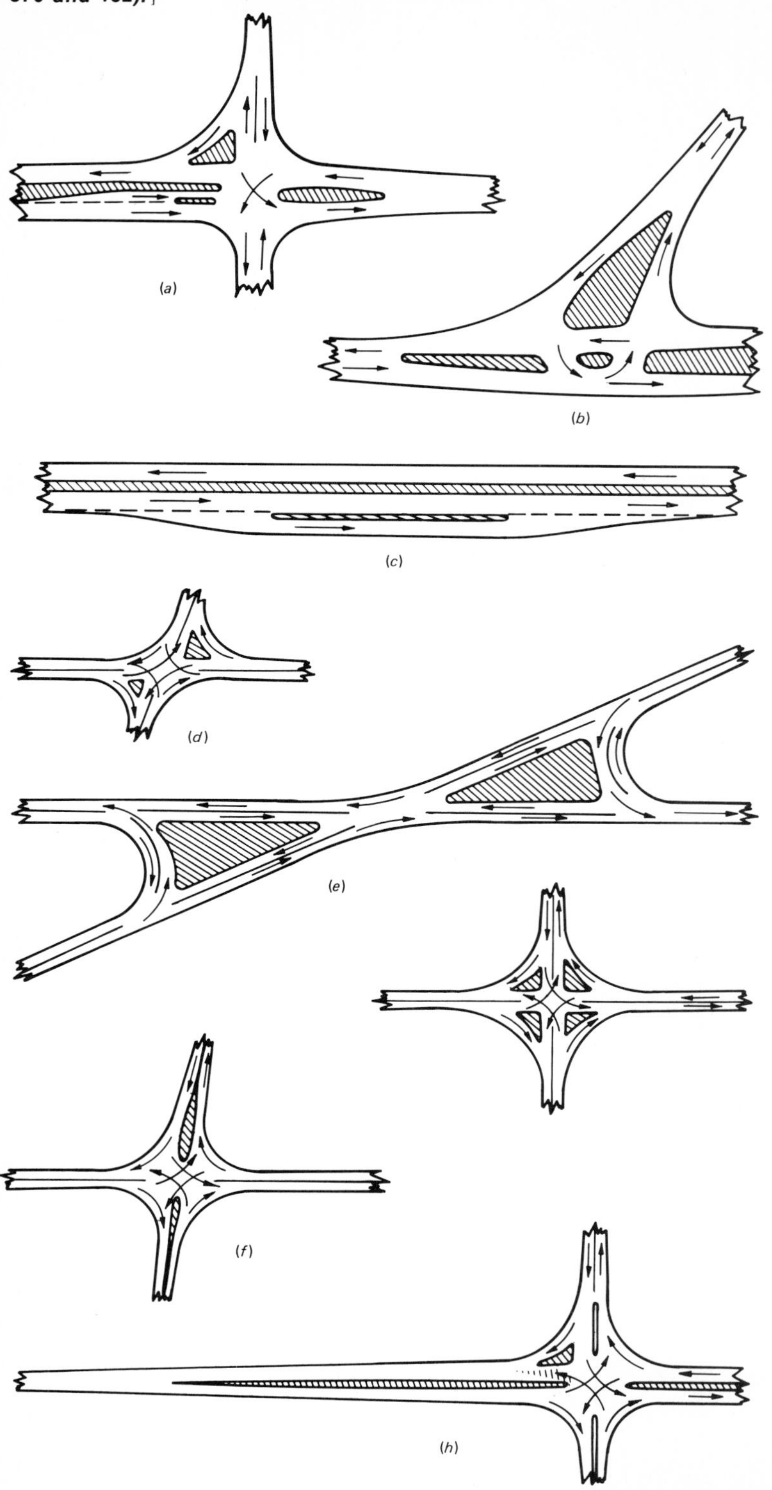

Table 16-5 **Design Characteristics of Intersections**

a **Minimum Radii for Intersection Curves†**

Design (turning) speed *(V)*, mi/h	15	20	25	30	35	40
Side friction factor *(f)*	0.32	0.27	0.23	0.20	0.18	0.16
Assumed min. superelev. *(e)*	0.00	0.02	0.04	0.06	0.08	0.09
Total $e + f$	0.32	0.29	0.27	0.26	0.26	0.25
Calculated min. radius *(R)*, ft	47	92	154	231	314	426
Suggested curvature for design						
Minimum radius, ft	50	90	150	230	310	430
Maximum degree of curve	...	64	38	25	18	13
Average running speed, mi/h	14	18	22	26	30	34

b **Minimum Edge of Pavement Designs for Intersection Turns‡**

Design vehicle	Angle of turn, deg	Simple curve radius, ft	Three-centered compound curve, symmetric: Radii, ft	Three-centered compound curve, symmetric: Offset, ft	Three-centered compound curve, asymmetric: Radii, ft	Three-centered compound curve, asymmetric: Offset, ft
P	30	60				
SU		100				
WB-40		150				
WB-50		200				
P	45	50				
SU		75				
WB-40		120				
WB-50		170	200–100–200	3.0		
P	60	40				
SU		60				
WB-40		90				
WB-50		...	200–75–200	5.5	200–75–275	2.0–6.0
P	75	35	100–25–100	2.0		
SU		55	120–45–120	2.0		
WB-40		85	120–45–120	5.0	120–45–200	2.0–6.5
WB-50		...	150–50–150	6.0	150–50–225	2.0–10.0
P	90	30	100–20–100	2.5		
SU		50	120–40–120	2.0		
WB-40			120–40–120	5.0	120–40–200	2.0–6.0
WB-50		...	180–60–180	6.0	120–40–200	2.0–10.0
P	105	...	100–20–100	2.5		
SU		...	100–35–100	3.0		
WB-40		...	100–35–100	5.0	100–35–200	2.0–8.0
WB-50		...	180–45–180	8.0	150–40–210	2.0–10.0
P	120	...	100–20–100	2.0		
SU		...	100–30–100	3.0		
WB-40		...	120–30–120	6.0	100–30–180	2.0–9.0
WB-50		...	180–40–180	8.5	150–35–220	2.0–12.0
P	135	...	100–20–100	1.5		
SU		...	100–30–100	4.0		
WB-40		...	120–30–120	6.5	100–25–180	3.0–13.0
WB-50		...	160–35–160	9.0	130–30–185	3.0–14.0
P	150	...	75–18–75	2.0		
SU		...	100–30–100	4.0		
WB-40		...	100–30–100	6.0	90–25–160	3.0–11.0
WB-50		...	160–35–160	7.0	120–30–180	3.0–14.0
P	180	...	50–15–50	0.5		
SU	U-turn	...	100–30–100	1.5		
WB-40		...	100–20–100	9.5	85–20–150	6.0–13.0
WB-50		...	130–25–130	9.5	100–25–180	6.0–13.0

† American Association of State Highway Officials (1965, p. 325).
‡ American Association of State Highway Officials (1965, p. 318).

Figure 16-14 Examples of freeway interchanges. *(a)* T or trumpet. *(b)* Cloverleaf. *(c)* Semidirect T junction. *(d)* Conventional diamond. *(e)* Diamonds with collector-distributor roads. *(f)* Diamond with left-hand ramps. [*Adapted from American Association of State Highway Officials (1965).*]

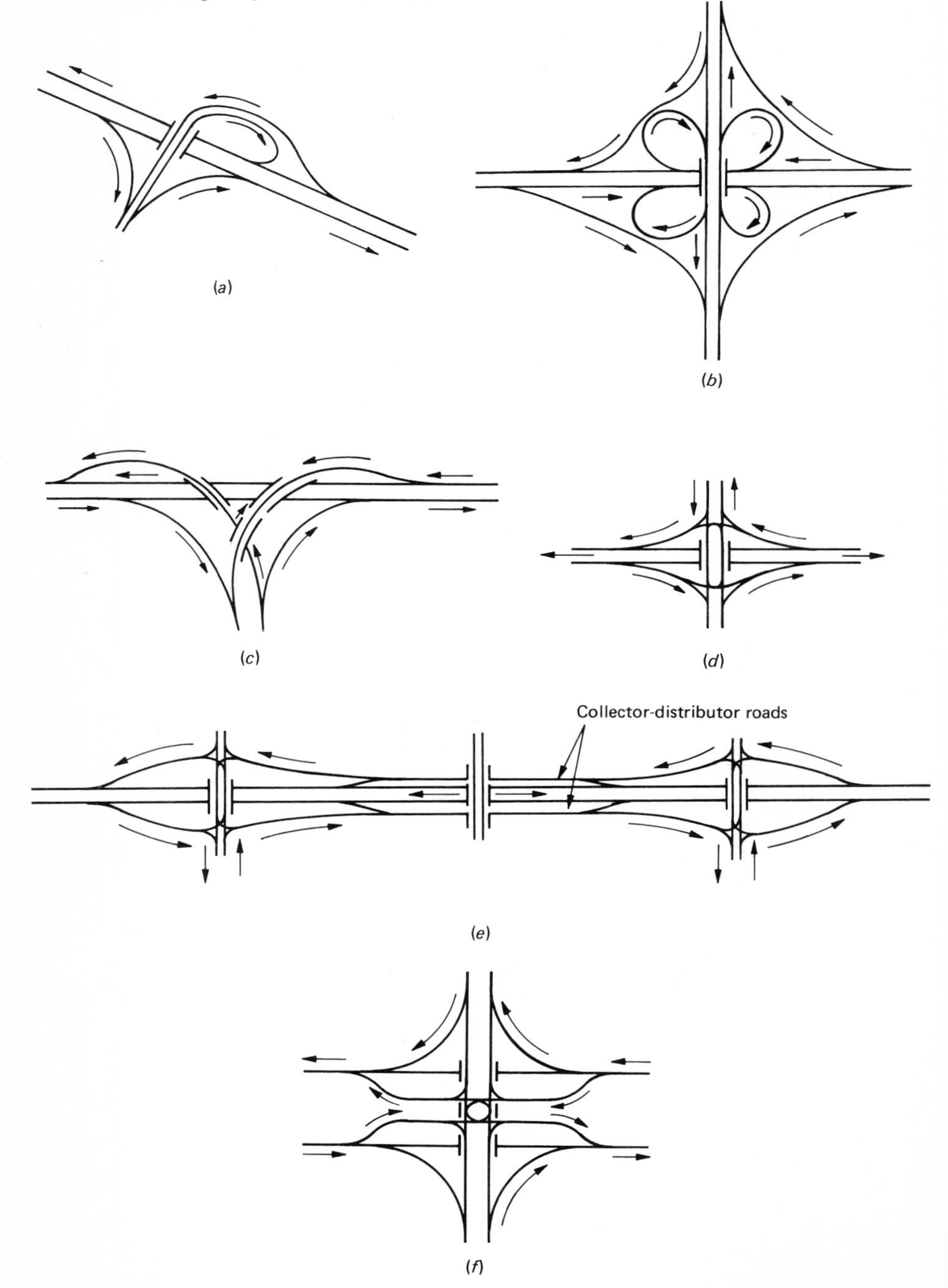

leaf interchanges. Examples of these are shown in Fig. 16-14*d* through *f*. The design of either of these types of interchanges involves the use of the same principles discussed above. In the case of ramps which involve large curve radii and high speeds, the procedures used for main line roads often apply. However, where the design speeds are low and at-grade intersections are involved, then the principles and practices for at-grade intersections discussed previously are used.

The advantages of fully grade-separated intersections are numerous. The absence of conflicting vehicle paths tends to reduce accident rates. Also, by providing for uninterrupted flow along any path through the intersection, the speeds at which vehicles travel through the intersection are increased. Similarly, capacity is increased because flow on any one path is never interrupted to allow for flow on other paths. However, in many cases fully grade-separated intersections cannot be used: They require a considerable amount of land, land which often is not available in urban areas. Also, they are usually more expensive than other types of intersections, although sophisticated traffic control devices and costly ramp construction can make diamond interchanges very expensive also.

OTHER FACTORS IN DESIGN

In addition to the considerations of geometric design which we have described above, there are many other aspects of the design of links and intersections. One is the provision of adequate drainage across the pavement and away from the roadway, which requires proper slopes perpendicular to and along the axis of the highway pavement as well as proper slopes and drainage troughs and culverts along the roadway. Also, design of the pavement itself and the subgrade support is important, the design depending upon the materials to be used, the nature of the soil and any particular form of subgrade to be employed, and the number, weight, and speed of vehicles to use the facility. As mentioned previously, all of these additional factors require a background in areas of engineering apart from transportation engineering and therefore will not be discussed here. The interested reader can consult texts such as Oglesby (1975).

GEOMETRIC DESIGN OF TERMINALS

As we discussed in Chap. 7, a wide variety of transportation terminals serve different modes and fulfill different functions. While it would be impossible to cover the detailed procedures and practices of design of all these different types of terminals, it is important that the basic steps applicable to the design of all terminals be covered. We will illustrate these with particular examples, drawn primarily from parking facilities and air terminals. The problems illustrate the application of these principles to other types of terminals.

Steps in the Process

There seems to be a similarity in the sequence of steps used in the design of all types of terminals. The steps are

1. Determine the functions to be performed by the terminal (e.g., transfer of passengers from one vehicle to another versus passenger transfer plus servicing of vehicles).

2. Estimate the volume of the various types of traffic to be accommodated by the terminal, including temporal variations.

3. Specify the minimum standards for level of service of the various components of the terminal, taking into account any standards prescribed by appropriate authorities (e.g., delay standards for airports, as discussed in Chap. 7).

4. Generate and evaluate alternative terminal designs. This step may include consideration of alternative locations.

In particular instances some of these steps may be performed in prior planning efforts, while in other cases all four stages will require extensive analysis in the course of terminal design. For example, previous regional or national planning efforts may determine the functions to be performed at a particular terminal and the volumes of traffic to be accommodated. Also, there may be considerable interaction between the analyses and results in the four steps. In some cases, alternative locations of a terminal will affect the total volume of traffic to be accommodated. In other cases, such as for metropolitan area airports, the total volume of traffic may not vary with different locations.

Geometric Design of Parking Facilities

Perhaps the most common type of a terminal is the parking facility. Although it may seem extremely simple at first glance, the design of parking facilities is reasonably complex and illustrates many of the principles and techniques involved in the design of terminals in any mode.

Parking facilities are classified according to three primary characteristics. The first is whether the parking is provided on-street or off-street. On-street parking usually is of a very simple design along the curb, while off-street parking facilities may be extremely complex. The second classification is according to whether or not the parking is by the driver of the vehicle, termed self-parking, or by attendants, termed attendant parking. Self-parking is generally preferred by most motorists, but with attendant parking it is often possible to have smaller parking spaces and aisles, resulting in a greater parking capacity in any given land area. Finally, a distinction is made between single-level parking facilities, which are usually provided at ground level, and multiple-story parking facilities, which must include ramps connecting the various levels (hence often called ramp parking facilities).

The design of curb parking facilities is extremely simple. Generally curb parking is provided with the parked vehicle parallel to the curb, this type being termed parallel parking. This takes less street space from running lanes than other designs which call for vehicles to pull into spaces at varying angles (up to 90°) to the curb. Also, experience has indicated that there are fewer accidents with this type of parking than with the angled parking. The primary design options for curb parking are thus the length and width of the space for each vehicle, and the angle relative to the curb. Consideration of the length and width of existing passenger automobiles, their turning radii, and their maneuverability in backing into and pulling out of such spaces has led to a standard space approximately 8 ft wide and 22 ft long. Also, it is generally recommended that a lane at least 10 ft wide be available for maneuvering the vehicle into and out of the parking space.

Parking lots and garages, in addition to providing parking spaces, provide aisles to reach the spaces. Also, if the lot or garage is one at which a charge is made, and that charge is determined on the basis of the time elapsed between the pickup of a ticket at entrance and presentation of the ticket to the collector upon exit, then space for obtaining the ticket at the entrance and for paying at the exit must be provided. Finally, if the facility is designed for attendant parking, space must be allocated for the arrival of vehicles, where drivers and passengers leave their vehicles, and attendants to pick them up to take them to the parking area, and similarly space must be provided for attendants to drop off vehicles and for passengers and drivers to pick them up at the exit. These areas must be designed particularly carefully because the arrival rates and discharge rates are subject to random variations, and hence queues or waiting lines can build up.

The design of the areas for parking spaces and aisles must be based upon a design vehicle. Generally the space allocated to each vehicle is approximately 8.2 to 8.5 ft wide, and 18 to 20 ft long. The necessary aisle width varies depending on the angle at which vehicles are parked with respect to the aisles. The smaller the angle between the access of the parked vehicle and that of the aisle, the less maneuvering space required and the narrower the aisle can be. Also, if the garage has supporting columns, it is necessary to design aisles and spaces so that cars can maneuver easily around them. Any ramps must be designed with vertical curves giving adequate clearance to vehicles. Standard designs for parking lots are illustrated in Fig. 16-15, which shows the variation in aisle width. The designs in this figure are for square areas 100 ft on a side; larger lots can be designed by combining portions of these modules to fill the lot in an efficient manner. It should be noted that in some designs there are dead spaces, spaces where a vehicle can be parked but where it is inaccessible whenever certain other stalls are occupied.

Another important characteristic of the geometric design of lots is the sizing of any entrance and exit areas, particularly where vehicles stop and a queue can form. Generally, such areas are designed with a number of servicing channels and a size of area for the waiting line such that the probability that any queue would spill over into the streets or the aisles is below a certain very small value. The methods used for the design of any facilities at a parking lot entrance or exit where vehicles are stopped and where queues may build up were discussed in Chap. 7 in the sections on simulation and queueing theory. In the case of parking lots, it is generally assumed that vehicles arrive at entrances or exits according to the Poisson distribution, which can be applied to any period in which the mean arrival rate is essentially constant. The service time at any server (such as a pay station or ticket machine) depends upon the nature of the facility and the transactions involved, but a reasonable assumption which we shall use in this example is that the service time is exponentially distributed. Under these circumstances, it was shown in the discussion of Table 7-9 that the probability $p(n)$ of any given number of vehicles n in the system (the serving facility plus the queue) was

Figure 16-15 Arrangements of stalls and aisles in parking lots. [*Adapted from U.S. Department of Commerce (1956, p. 126).*]

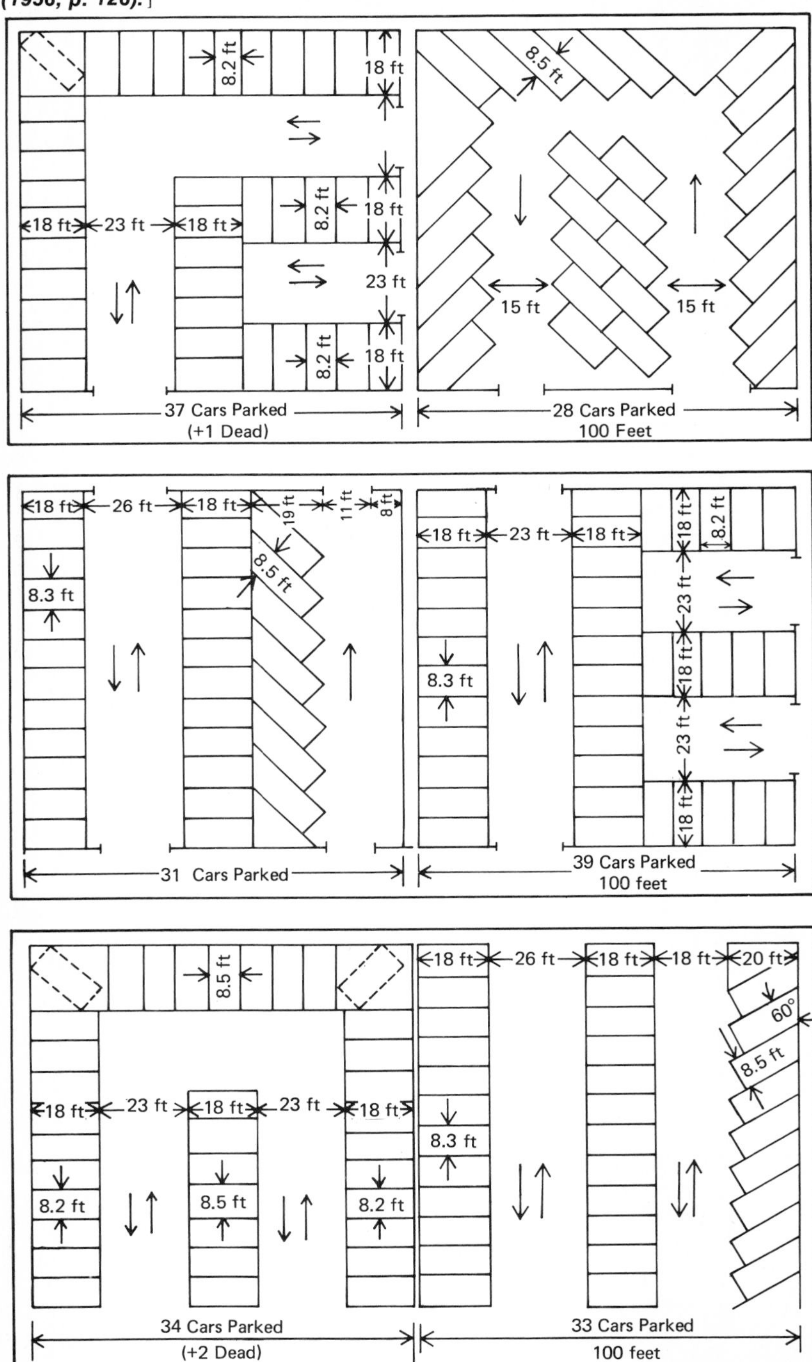

$$p(n) = \frac{1}{n!}\left(\frac{\lambda}{\mu}\right)^n p(0) \qquad 0 \leq n \leq k-1$$

$$\text{and } p(n) = \frac{1}{k!k^{n-k}}\left(\frac{\lambda}{\mu}\right)^n p(0) \qquad n \geq k$$

$$\text{with } p(0) = \frac{1}{\left[\sum_{n=0}^{k-1} \frac{1}{n!}\left(\frac{\lambda}{\mu}\right)^n\right] + \frac{1}{k!}\left(\frac{\lambda}{\mu}\right)^k \frac{k\mu}{k\mu - \lambda}}$$

where k = number of channels
μ = mean service rate of each channel
λ = mean arrival rate

Using these equations, there are $n - k$ vehicles in the waiting areas, provided $n \geq k$, and none otherwise. To design the area such that the probability that vehicles will overflow a facility which accommodates A vehicles is less than some value P simply requires that we determine the number of channels such that the probability that more than $A + k$ vehicles will be in the system is less than P. Mathematically, this requires the following relationship:

$$\sum_{n=A+k+1}^{\infty} p(n) \leq P$$

$$\text{or } \sum_{n=0}^{A+k} p(n) \geq 1 - P$$

In the case of a single-channel entrance to a parking lot, at which vehicles arrive at an average rate of 1 vehicle every 30 sec (λ = 2 vehicles/min) and where vehicles are served by the ticket machine and gate server in an average time of 20 sec (μ = 3 vehicles/min), the probability of any given number n of vehicles being in the system is (using the simplification of the equations for this case of $k = 1$):

$$p(n) = \left(\frac{\lambda}{\mu}\right)^n \left(1 - \frac{\lambda}{\mu}\right)$$

If it is desired that the probability of overflow of the waiting area be less than 0.02, then the number of waiting line spaces in the waiting area (A) must be 8, since:

$$\sum_{n=0}^{8+1} p(n) = 0.983$$

$$\text{and } \sum_{n=0}^{7+1} p(n) = 0.974$$

Note that any waiting area with more than 8 spaces would also satisfy the requirement.

Thus the sizing of entrance and exit facilities at parking facilities is a rather straightforward application of queuing theory. Attendant parking can also be treated as a multiple-channel queue, in which each attendant is considered to be a serving facility. The mean service time is the mean time required by the attendant from pickup of one vehicle to pickup of the next vehicle, including

Figure 16-16 **Layouts of ramp parking garages.**
(a) **Staggered floors with one-way aisles.**
(b) **Sloping floors with two-way aisles.**
(c) **Sloping floors with one-way aisles.**
(d) **Horizontal floors with two one-way helical ramps.**

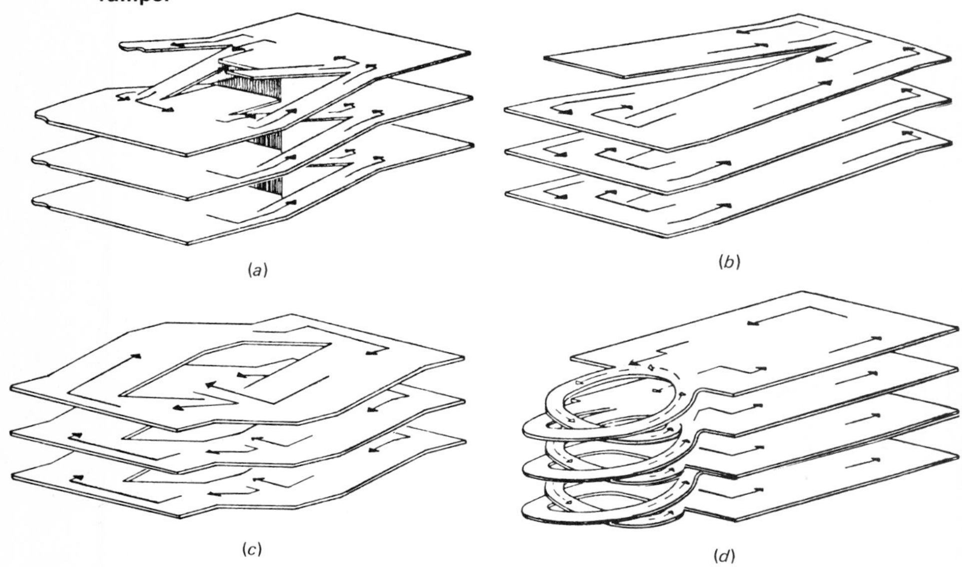

not only the parking but also the return to the area where vehicles are to be picked up. Again this is a rather straightforward application of queueing theory.

The other important aspect of parking facility design is the design of multistory facilities. The most common of the many alternative designs are presented in Fig. 16-16. In almost all of these an attempt has been made to separate the traffic flow in the inbound direction from that in the outbound direction, in order to minimize interference and delays as well as accidents. An important feature that tends to increase the capacity of many of the designs is the use of the parking floors as the ramps between the different levels. An interesting economic problem which may or may not fall within the purview of design engineers and architects is whether or not a parking facility should be a single-level lot or multistory garage, and if the latter, how many floors and spaces it should have. The economic principles underlying this decision are illustrated in one of the problems at the end of this chapter.

Air Terminals

Air terminals are among the largest terminal facilities and therefore they illustrate the options and issues faced in designing most types of transport terminals. In this discussion we will focus on the major options of terminal design, mainly those associated with the overall layout of the facility, and not delve into the details of design of individual components, since any treatment of that would consume a volume the size of this entire text. (For those interested in pursuing this subject, it is covered thoroughly in de Neufville, 1976, and Horonjeff, 1975.) We will treat two aspects of airport design: the overall layout of the entire facility and the layout of the passenger terminal building itself.

An airport consists of a number of components: (1) the runways for take-off and landing, (2) taxiways connecting the runways with other parts of the terminal, (3) the apron, where planes park for loading and unloading, (4) the terminal building itself, where passengers and baggage and possibly freight are processed, (5) the hangars for storage and maintenance of aircraft, and (6) the various control facilities including a control tower as well as various lighting systems and other guidance systems for assisting aircraft in landing and take-off. These components can be arranged in many different ways on the airport site. Some typical airport configurations are illustrated in Fig. 16-17, ranging from extremely simple single-runway airports to the multiple-runway designs of large airports. These drawings, schematic, indicate the relative location and orientation of terminals, runways, taxiways, etc. The

Figure 16-17 **Typical airport configurations.** [***From Horonjeff (1975, pp. 200–201).***]

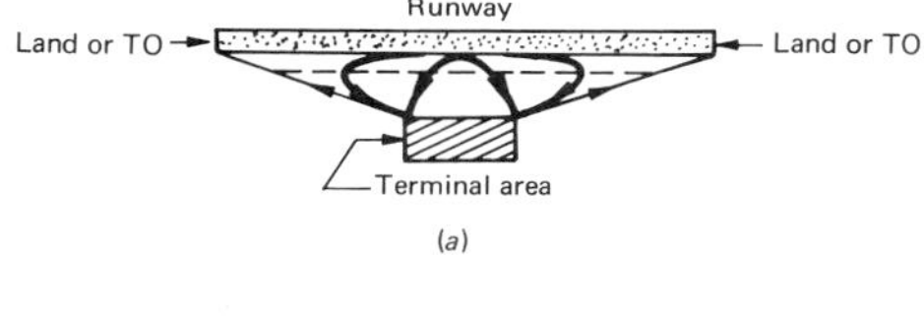

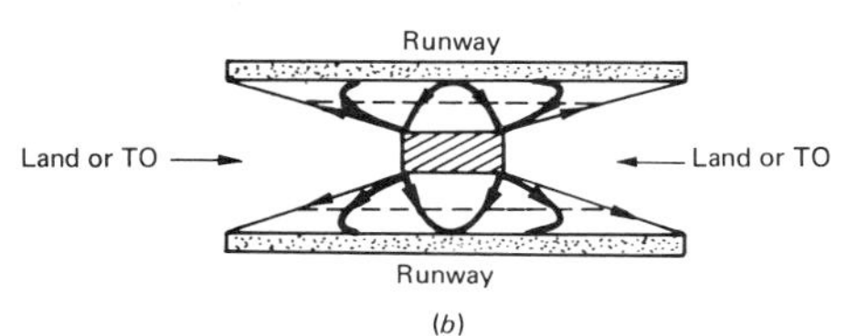

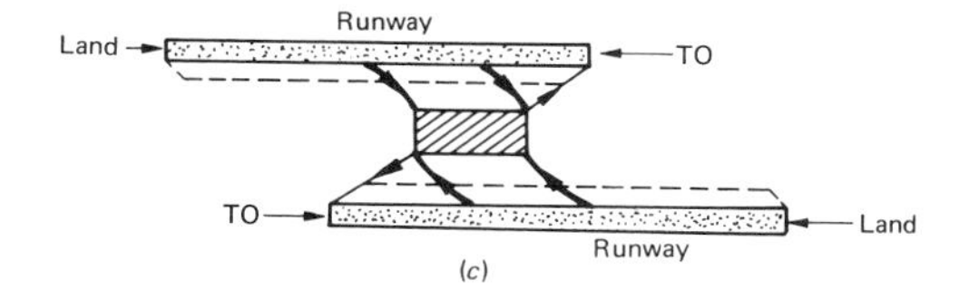

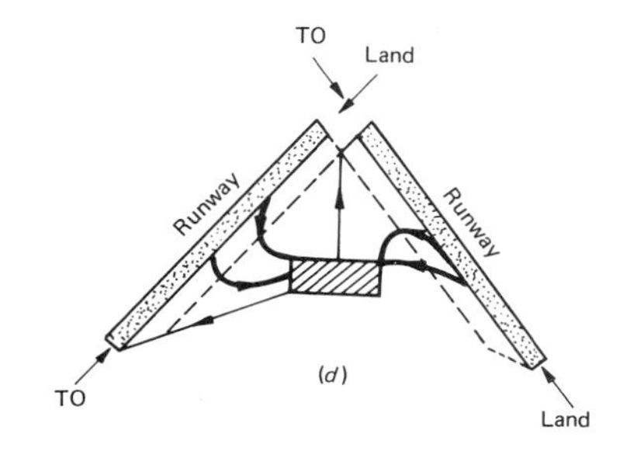

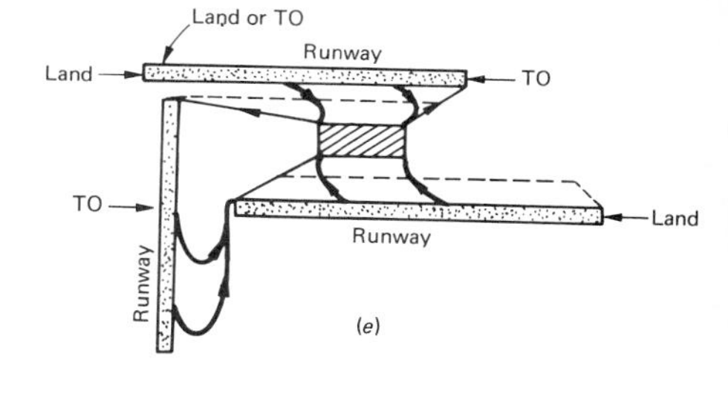

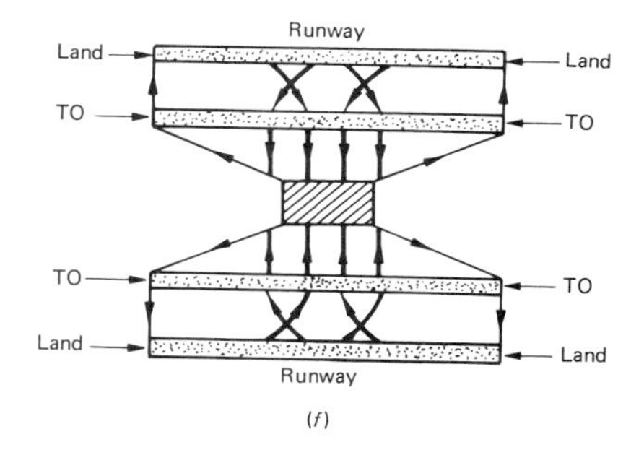

Taxiways for departing aircraft
Taxiways for arriving aircraft
Interconnecting parallel taxiways
Terminal area
TO means take-off

exact distances and sizes can vary considerably depending on the facility. The selection of a particular layout depends primarily on the shape of the land area available for the terminal facility, the size and number of particular facilities necessary to accommodate the expected traffic, the terrain, and the weather conditions, especially wind directions and velocities, which influence the orientation of runways.

In Chap. 7 we described the methods used to determine the capacity of airport runways; the capacity of common types and arrangements for various mixes of aircraft were presented in Table 7-5. Using this table, with estimates of the annual volume of flights to be accommodated and the hourly maximum volume, those runway configurations which will accommodate the traffic can be identified. These then become candidate alternatives for this particular airport. Also, knowledge of the actual annual volume and the capacity of any particular design permits estimation of the total aircraft delay experienced in the course of the year in aircraft minutes, from Fig. 7-11, from which the average delays per aircraft can be calculated.

For example, consider a situation in which it is desired to have an airport which will accommodate 300,000 flights/year during IFR (Instrument Flight Rules) weather conditions. From Table 7-5 it is determined that only two types of configurations have adequate capacity: (1) the design with two parallel runways separated by at least 5000 ft and (2) the four-runway design with two pairs of parallels, each pair being close but sufficiently separated to be independent. The average annual delay for each pair can be calculated from Fig. 7-11. The two-runway independent IFR arrivals and departures design has an annual capacity of 340,000 operations. Therefore with a volume of 300,000 operations, the total annual delay will be 395,000 aircraft-min. In contrast, the delay with the larger capacity (four-runway) airport will be only about 50,000 aircraft-min/year. Obviously the four-runway design will be more expensive to construct and probably more expensive to maintain, and hence it will be necessary in the design process to trade lower aircraft delays and greater future capacity for additional investment.

There are numerous alternatives with respect to the design of terminal facilities. Figure 16-18 presents the most common alternative designs for the relative positions of the aircraft for loading and unloading and of the terminal building, where passenger processing occurs. Originally almost all terminals were of type *a*, but this type of facility requires an extremely long terminal building to accommodate a large number of flights simultaneously. As a result, this design was augmented with fingers emanating from the terminal to increase the number of spaces for aircraft, as shown by types *b* and *c*. Even greater demands for accommodation of large numbers of aircraft are leading to the creation of terminals with the loading and unloading facilities located away from the terminal, to consolidate passenger processing and at the same time allow for an even larger number of aircraft to be accommodated. This leads to the design concept of the remote or satellite terminal type, where the flight interface may be connected with the terminal by passenger conveyor, a horizontal elevator system, or a bus or similar large conveyance.

The number of gates required at an airport can be determined rather easily from the number of flights expected during a peak hour. Knowing the average time a gate is occupied by an aircraft (including time required for arrival and

Figure 16-18 **Alternative layouts of passenger-handling facilities at airports.** ***(a)*** **Centralized gate arrival.** ***(b)*** **Pier finger.** ***(c)*** **Pier satellite.** ***(d)*** **Remote satellite.** ***(e)*** **Mobile conveyance.** [***From Horonjeff (1975, pp. 247–248).***]

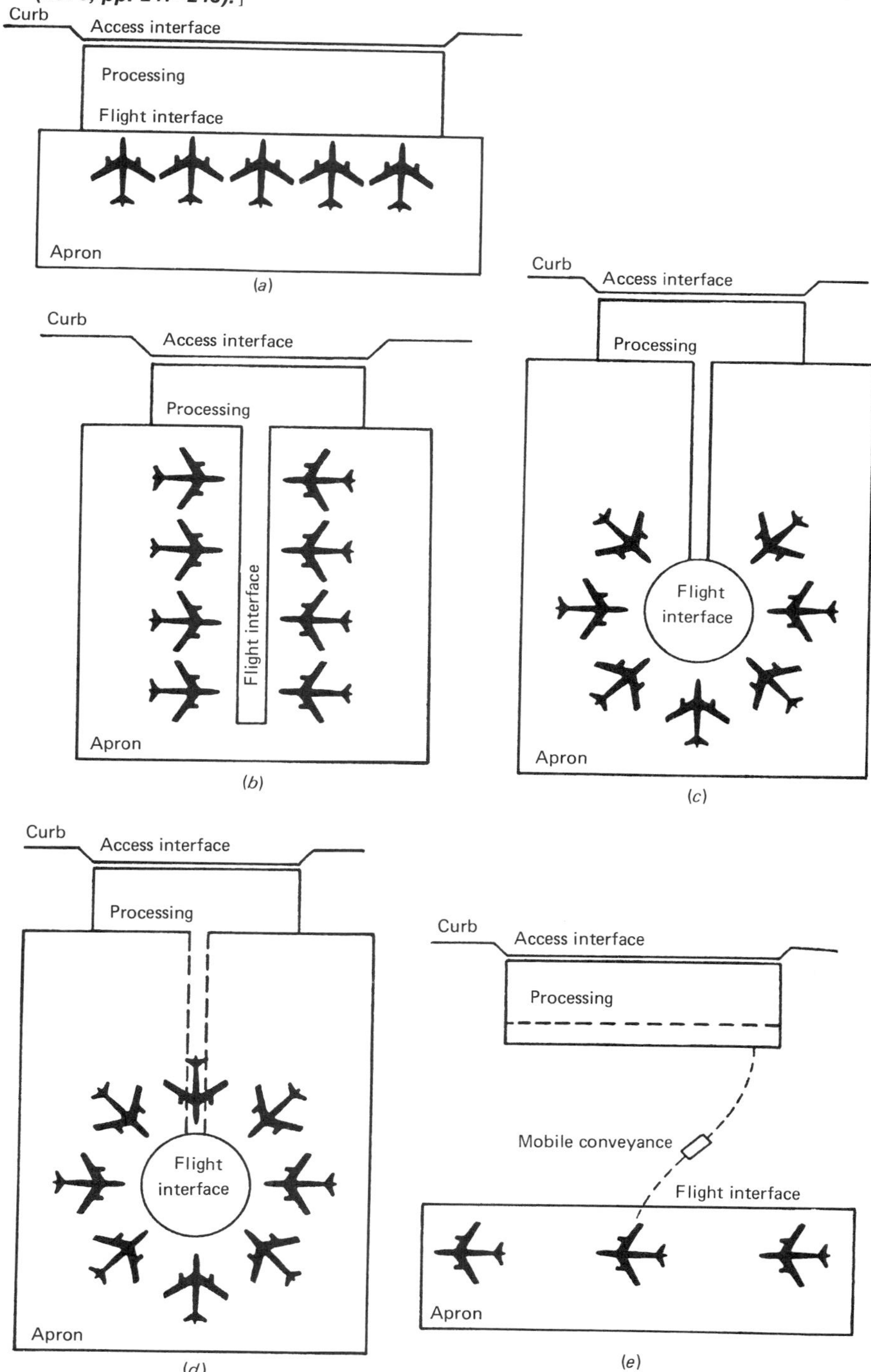

Table 16-6 Typical Space Requirements of Airport Terminal Buildings†

Facility	Space required, in 1000 ft^2 or 100 m^2, per 100 typical peak hour passengers
Ticket lobby	1.0
Baggage claim	1.0
Passenger waiting and assembly	2.0
Visitor waiting rooms	1.5
Immigration	1.0
Customs	3.0
Amenities (including eating facilities)	2.0
Airline operations	5.0
Total gross area (domestic)	25.0
Total gross area (international)	30.0

† Horonjeff (1975, p. 258).

departure), the number of gate positions can be calculated from the following equation:

$$G = \frac{q}{U} T \tag{16-7}$$

where q = design aircraft volume, vehicles/h
T = mean time of aircraft at gate, h/vehicle
U = design traffic utilization of each server, usually 0.5 to 0.8/gate

Typical average times for aircraft to remain at the gate are 30 min for domestic flights and 60 min for international flights.

The approximate size of a terminal building and the facilities housed therein can be estimated from guidelines regarding the average space required per passenger during the peak period. These typical terminal building space requirements are presented in Table 16-6 for the various functional components typically found in passenger terminals. Once the total space of the terminal building has been estimated, then alternative arrangements of the various functional elements, including multistory options, can be explored. In the case of large terminals, good practice dictates that alternative designs be examined with reference to the time required for passengers to be accommodated through the various facilities and to travel from one facility to another. This is best done using the processor model of terminal facilities presented in Chap. 7 and an associated tabulation of typical times under average and peak conditions for passengers to be processed at each point and also to travel from one facility to another.

Special Considerations

There are many special considerations that arise in the context of particular design problems. For example, in the design of many terminals it is important to consider revenue from shops, parking facilities, or offices. This is especially true in many downtown passenger terminals, as well as airports, where the profit from such activities can pay a substantial part of the total cost of the terminal. And of course the rental value for such purposes is enhanced by the accessibility the terminal provides. Environmental limitations, such as zoning and the need for the terminal to fit into its environment aesthetically, are also important considerations.

Undoubtedly one of the most important considerations in any facility (or passenger vehicle) is meeting the requirements of "barrier-free design." This concept refers to a design such that persons with physical limitations such as blindness or the need to use a wheelchair can use the facility without difficulty. Until recently, this design goal was largely ignored, but it can often be achieved by simple design modifications, such as wide doors, ramps parallel to steps, and elevators wherever changes in elevation are too great to use ramps. It is increasingly common for governments at all levels to require such barrier-free design features in all facilities and vehicles for which they contribute funds (as in urban public transit) and also may be required as a part of building codes. And of course there are real benefits to the transportation system by minimizing impediments to its use, not to mention the benefits for society as a whole.

CONCLUSIONS

Design of most transportation components is constrained by design standards dictated by considerations of compatibility with the rest of the transportation system, as well as the necessarily intuitive nature of many decisions which must be made in the design of components. However, the existence of these constraints and of standard designs does not relieve the engineer of the obligation to consider alternative designs and to seek that design which best balances quality of service to the user and capacity, cost of initial construction, compatibility with the natural environment and neighboring human activities, and flexibility for future growth.

This chapter has attempted to present some of the basic concepts and methods of engineering design of transportation systems, particularly focusing on the design of fixed facilities. The problems which follow are designed to illustrate the use of these principles in a wide variety of modal contexts.

SUMMARY

1 The process of designing any components of a transportation system usually involves the following steps: (1) identification of requirements to be met by the design, (2) development of alternative designs, (3) evaluation of those alternatives and selection of the best for implementation.

2 The design of fixed facilities, and often other components, if they are not purchased "off the shelf" as most vehicles and control systems are, occurs after the need for the facility has been determined in the planning process. The plan usually specifies approximately characteristics such as capacity, level of service, and expected initial and continuing costs, leaving much to be determined in the design process.

3 The range of options in design is usually considerable, and it is incumbent upon the design engineer to explore these so that the final design is the best that can be prepared under the time and financial constraints of the effort. Although the specific types of options vary depending upon the particular project, typically the following types are involved: overall functional layout or configuration (horizontal and vertical), sizes of the components, types of equipment to be installed, and flexibility for adaptation to changed requirements.

4 Usually a design must meet specific standards for that type of facility, such as in geometric design of elements used by vehicles, and environmental constraints.

5 Typically the evaluation of alternative designs will involve trade-offs between construction and operating costs, cost and environmental impacts, and user level of service and initial costs.

PROBLEMS

16-1
A highway design includes the intersection of a plus 4.8 percent grade with a minus 4.2 percent grade at station 1052 + 75 and elevation 851.5 ft above mean sea level. Calculate the center line elevations for every 50-ft station on a parabolic vertical curve of 600-ft length. What would the maximum safe speed (in 10 mi/h increments) be on this curve?

16-2
A two-lane rural highway has a curve of 8° extending for 650 ft along its center line. It is laid out with 12-ft-wide lanes, a 9-ft-wide shoulder on each side, and a corner of a building 10 ft from the edge of the shoulder. What is the maximum safe speed (choosing multiples of 10 mi/h) considering safe stopping distances? Measure the sight distance along the center line of the traffic lane on the inside (i.e., closest to the center) of the curve. If R is the radius of the curve and m is the distance between the object and the center line of the inside running lane, what is the formula for S, the sight distance along the inner lane?

16-3

a For the same curve as in Prob. 16-2, what would be the superelevation e on this curve such that the value of f, the coefficient of side friction, would be zero at the design speed? For this value of e, what would be the value of f for a vehicle stopped on the curve? Compare this with typical values of f under rain and ice conditions.

b What would be the values of e and f if e were selected such that f at design speed was equal (although opposite in sign) to f with the vehicle stopped?

c For the value of e calculated in Part b, what speed is required for a vehicle to overturn? This vehicle, a trailer truck, has a center of gravity of 7.5 ft above the road, and the outsides of the wheel treads are 8.0 ft apart. What value of f does this imply? Compare this value with those presented in Table 4-4.

16-4
A two-lane road has been tentatively designed with a 6 percent grade extending for 3.2 mi. The speed limit is 50 mi/h. Consideration is being given to

providing a climbing lane for trucks. Compare the cost of operation of an automobile, based on the methods presented in Chap. 9, up this grade at 48 mi/h and at 28 mi/h, as if following a heavy truck. Give the dollar costs, travel time, and imputed dollar value of time costs separately, multiplying the 1959 costs by 1.5 to yield current costs, and assuming normal operation. How many automobiles would have to be so delayed for the savings (including travel time) from a climbing lane to equal the estimated initial costs of $450,000 plus $4000/year added maintenance? Use a 10 percent interest rate and a 25-year road life.

16-5

A depressed urban freeway is to be designed so as to accommodate peak period traffic with the following characteristics: ADT = 78,000 vehicles/day (sum of both directions), $K = 10$ percent, $D = 55$ percent, $P_T = 5$ percent, $P_B = 10$ percent. The design speed will be 50 mi/h, and during the peak hour the flow should be at level of service *C*. Design characteristics yield a $W = 0.95$.

a How many lanes are required in one direction?

b What would the speed be in the reverse peak direction, with the number of lanes calculated in Part a?

c If each lane is 13 ft wide, the shoulders are 10 ft wide (both sides of traffic lanes in each direction), the median grass strip is 15 ft wide, and the side slopes are 20 ft wide, how wide is the needed right-of-way? Sketch the cross section, showing lanes and other items.

16-6

Lay out a Y (or, wye) interchange of the type depicted in Fig. 16-13*b*. All three approaches are two-lane roads, with 12-ft-wide lanes and 6-ft-wide shoulders on the right side only. The angle made by the two right-hand approaches is 80°, and the median on the right approach is 10 ft wide. The design should be for speeds of 20 mi/h on right turns and 15 mi/h on left turns. Assume there are no right-of-way restrictions on size.

16-7

Lay out parking spaces in an area 150 ft wide and 225 ft long. Try to maximize the number of accessible spaces (i.e., not including "dead" spaces), while meeting the lane and space limitations presented in Fig. 16-15.

16-8

A developer is considering alternative designs for a central business district self-parking facility. The land (cleared) costs $600,000, but in the analysis the corresponding daily cost should include only interest (15 percent per annum) since the land presumably would be sold (at this amount, present value) whenever the parking use is terminated. A ground level lot would cost $28,000 to construct on the land, and would have 96 spaces. A garage would cost $45,000 for the ground floor (75 spaces) plus $50,000 for each additional floor (85 spaces). Operating expenses for the lot (including fee collectors, maintenance, insurance, etc.) would be $21,000/year, and would increase by $2000/year for each additional floor. Potential usage has been estimated at up to 250 all-day parkers at $1.50, up to 100 parkers up to 4 h at $1.00, and up to 100 parkers less than 1 h at $0.50. Turnover rates (parkers per space per day) are 1, 2, and 6, respectively. Prices can be set so as to exclude any class of parkers. How many floors should be built, and what rates should be charged? Garages in this area are limited to four floors, including the ground floor. The garage will be open 300 days/year, and will have a life of 15 years.

16-9

Reconsider the parking entrance waiting area design problem treated in the text for the case of two parallel ticket servers rather than only one. There will be a single waiting line, so motorists will enter whichever serving channel is free first. With the same target of no greater than a 0.02 probability that the number of autos waiting will exceed the size of the waiting area, how many waiting spaces will be needed? If each server requires 240 ft^2 and each space requires 160 ft^2, which design—one or two channels—requires less space?

16-10

A typical terminal station on a rail rapid transit line is laid out as shown below. The problem is to determine whether or not this design is adequate for trains arriving (and departing) on a 90-s time headway, with a 60-s dwell time. The trains are 250 ft long and are limited to 15 mi/h in the vicinity of the station and turnouts (this design being called a "scissors" crossover). They stop 10 ft from the end of track and accelerate and decelerate between 0 and 15 mi/h in 7 s, covering 75 ft. The crossover is governed by signals at points labeled *A* through *L*, using the two-block three-aspect signal system (see Fig. 5-6) and the train operator must see a "proceed" indication at least 50 ft before passing a signal to avoid slowing down. Can this terminal accommodate trains on a 90-s headway? If not, what is the minimum time headway and capacity (trains/h)?

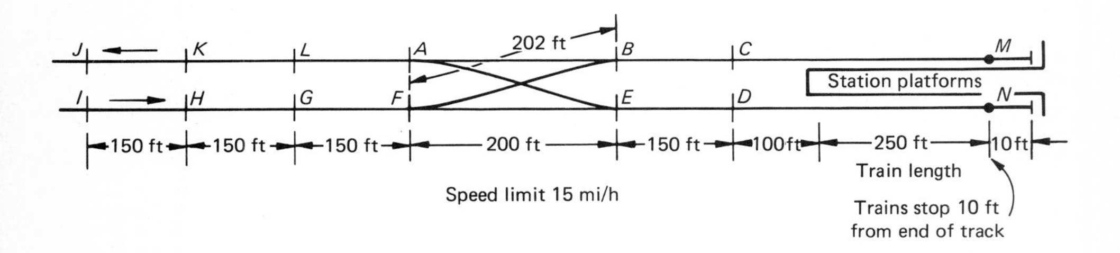

[Hint: The critical section of track is the crossover. Calculate the times trains enter and leave that section, as follows. For the first train, calculate the moments at which it arrives at *H*, clears *E*, arrives at *N*, leaves *N* (and hence needs "proceed" signal at *D*), and clears *A*. Do the same for the second train, arriving at *H*, clearing *B*, etc. Are these compatible?]

16-11

In Prob. 16-10, would the same station design work with a dwell time of 30 s? The problem can be alleviated by moving the crossover. Using a time-space diagram for this terminal (showing both tracks and the crossover on it), relocate the crossover so that this arrangement would work.

16-12

If in Prob. 16-10, the trains required very long dwell times, say 3 min (as commuter railroad trains often do at a CBD terminal), what would be the capacity of the two-track terminal? What would be the capacity of a three-track terminal as below. The order of usage of tracks by trains is 1, 2, 3.

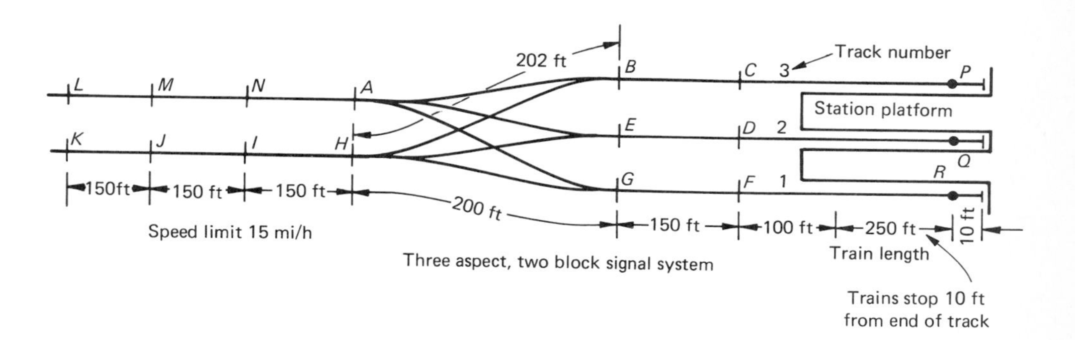

16-13

Compare the capacity of the two-track terminal in Prob. 16-10 with that of a loop terminal, as drawn below. Train dwell time and acceleration and deceleration are as indicated in the prior problem. Signals are all 150 ft apart at the locations marked, except as noted. Speed limit is 15 mi/h between signals *A* and *B*, 50 mi/h elsewhere.

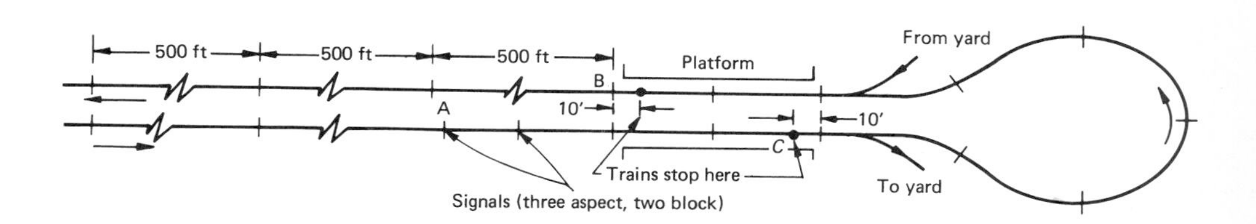

REFERENCES

American Association of State Highway Officials (1965), "A Policy on Geometric Design of Rural Highways," Chicago, IL.

American Association of State Highway Officials (1973), "A Policy on Design of Urban Highways and Arterial Streets," Washington, DC.

American Railway Engineering Association (1975), "Manual for Railway Engineering," Chicago, IL.

Box, Paul C., William Dean, Rod Engelen, Peter G. Koltnow, Harry Porter, Jr., Michael A. Powills, Jr., Martin A. Wallen, and Earl C. Williams (1967), Recommended Practices for Subdivision Streets, *Traffic Engineering,* **37**(4): 15–29.

de Neufville, Richard (1976), "Airport Systems Planning," MIT Press, Cambridge, MA.

Edwards and Kelcey, Inc. (1972), "US Route 1A, Harrington, Maine, Final Environmental Impact Statement," Prepared for Maine Department of Transportation, Newark, NJ.

Hay, William W. (1953), "Railroad Engineering," Wiley, New York.

Haynes, D. Oliphant (1957), "Materials Handling Equipment," Chilton, Philadelphia.

Horonjeff, Robert (1975), "Planning and Design of Airports," McGraw-Hill, New York.

Howard, George P., ed. (1974), "Airport Economic Planning," MIT Press, Cambridge, MA.

Leibbrand, Kurt, trans. by Nigel Seymer (1970), "Transportation and Town Planning," MIT Press, Cambridge, MA.

Manheim, Marvin L., John H. Suhrbier, Elizabeth D. Bennett, Lane A. Neumann, Frank C. Colcord, and Arlee T. Reno, Jr. (1975), "Transportation Decision-Making: A Guide to Social and Environmental Considerations," National Cooperative Highway Research Program Report No. 156. Transportation Research Board, Washington, DC.

Oglesby, Clarkson H. (1975), "Highway Engineering," 3d ed., Wiley, New York.

Paquette, Radnor J., Norman Ashford, and Paul H. Wright (1972), "Transportation Engineering," Ronald, New York.

Quinn, Alonzo deF. (1972), "Design and Construction of Ports and Marine Structures," McGraw-Hill, New York.

Ricker, E. R. (1957), "Traffic Design of Parking Garages," Eno Foundation, Saugatuck, CT.

Ritter, Leo Jr., and Radnor J. Paquette (1967), "Highway Engineering," 3d ed., The Ronald Press Co., New York.

Smith, Wilbur, and Associates (1965), "Parking in the City Center," Motor Vehicle Manufacturers Association, Detroit, MI.

U.S. Department of Commerce, Bureau of Public Roads (1956), "Parking Guide for Cities," Washington, DC.

System Operation and Management

17.

Transportation engineers and planners have long been concerned with many aspects of the operation or management of the transportation system. In many instances, this involvement in management was a natural outgrowth of previous involvement in the planning or design of systems and facilities. This is true in traffic engineering, for example, where it was found in the 1920s that in addition to designing roads suitable for motorized traffic, it was essential for efficient accommodation of traffic to manage the traffic flow by means of regulations and operational aids such as traffic signals and channelization. Transportation engineers have long been concerned with such typically managerial functions as monitoring the use of facilities, changing regulations governing traffic movement as needed, and maintaining the facilities to proper standards.

In most other fields of transportation, engineers are heavily involved in the maintenance of facilities simply because engineering provides the basis for determining when and what type of maintenance is required. In addition, engineers and planners have often become involved in all aspects of the management of transportation systems, both private and public, since so many management decisions involve engineering. Thus it is quite common to find persons with engineering backgrounds among the top officials of transportation companies and of government-owned transportation authorities, where they are concerned not only with decisions regarding the physical system but also questions of finance, labor relations, and marketing, to mention a few.

In recent years there has been renewed interest in educating persons to assume managerial responsibilities in transportation. This stems partly from a recognition that many of the methods used for planning and design are equally applicable to some management decisions. Also, there seems to be increasing recognition that many of the old distinctions between different portions of the transportation system which have led to completely independent managerial units (particularly in the public sector, such as between highways and transit), should be reduced to utilize the entire transportation system more effectively.

Obviously, much of the knowledge and skill required to manage properly transportation systems is based on the same principles used in the management of any other major enterprise, whether public or private. However, there are a number of specific types of problems, relationships, and methods which apply only to transportation systems. On these this chapter will focus. They naturally fall into two categories, those oriented toward the management of transportation facilities and those concerned with the management of transportation carriers.

MANAGEMENT OF FACILITY USE

There are many components of transportation systems which are managed primarily as facilities available for the use of any person (or organization) who has an appropriate vehicle and meets the licensing requirements for operating that vehicle. Examples include the public road network, airports, and ports, to name a few. Since the number and type of traffic units which would use any particular facility at any time are not determined by the facility management but by the decisions of many individual users, there often arise problems of controlling the flow of traffic in order to provide for efficient operation and a reasonable quality of service. The experience of many decades has led to the development of a wide range of techniques which assist in controlling traffic flow. Probably the greatest amount of work has been done in the area of road traffic engineering, where, because of the ubiquity of the highway network and the very large volume of traffic accommodated on it, the problems have become extremely difficult. However, even though most of the methods will be identified and illustrated with respect to this particular mode, many apply to any type of transportation facility.

Transportation System Management

Traditionally the responsibility for management of transportation facilities has been divided by mode, and also divided further within each mode, to create units of a manageable size. This seems to have created a fragmentation of responsibility and authority which can produce a lack of coordination among various interrelated portions of the transportation system. Thus traffic control on a turnpike or a bridge facility operated by a particular public authority may be uncoordinated with the traffic engineering decisions made on the street and arterial road network which feeds these facilities. Furthermore, responsibility for operating an urban public transit system is almost invariably in a different organization from that which has responsibility for controlling the traffic flow on streets, even though most (or all) transit vehicles operate on those streets. By coordinating decisions regarding the operation of the transit system as well as the streets, better traffic flow for both motorists and transit users might be achieved.

For these reasons, the concept of transportation system management has evolved. It has arisen primarily in the context of urban transportation, where strong interrelationships are evident among various parts for which there are now separate management entities. The basic idea is to bring together the managerial units concerned with the various parts of the urban transportation system and foster cooperation in much the same manner that comprehensive regional planning has already led to joint decisions regarding any major capital or new facility improvements anywhere within the system. The primary orientation is toward unifying the management of publicly owned portions of the transportation system, which usually consist of the road network, major facilities such as bridges, tunnels, and some parking garages, and the public transit system. However, through the control over use of roads, there is also some degree of control over private carriers such as truck lines and bus lines, and taxicab companies, whose operations also influence the overall quality of traffic flow.

The transportation system management concept involves a hierarchy of managerial units and responsibilities. At the highest level is the responsibility for major policies regarding traffic movement in the region. This includes decisions regarding the major regional roads and the regional public transit system, both of which cater primarily to relatively long-distance traffic. (Remember, even within the largest urban areas, the average trip length may be no more than 3 or 4 mi, so long trips might refer to all trips greater than 5 mi.) This unit would determine the basic guidelines under which decisions at lower levels in the management hierarchy would be made. The next level in the hierarchy is comprised of the managerial units concerned with particular major portions of the system, such as the regional transit system, a regional highway network, etc. At this level are probably the major existing organizations, such as the regional transit authority, a regional port, tunnel and bridge authority, and perhaps a turnpike authority. Within each of these organizations is a further breakdown of responsibility for a particular facility or portions of the system, such as the rail transit routes or particular bridges or tunnels. This level probably also includes the managements of more localized roads, such as those controlled by counties or specific towns. Here lies the major responsibility for traffic control, the setting of traffic signals, channelization, etc., on all but the major regional routes. This organization is probably divided further into units responsible for day-to-day maintenance and operation. These various levels are shown in Fig. 17-1.

Methods for managing the use of transportation facilities seem to have developed on an ad hoc basis. As indicated above, probably the largest amount of work and hence the richest source of managerial concepts has emerged from road traffic engineering, although other areas, particularly the control of traffic in and around airports, have also seen development of sophisticated traffic control methods. In the next few paragraphs, we shall attempt to present many of the more widely used ideas of traffic control in these various modes, although the list is by no means exhaustive (a more detailed treatment for urban traffic is found in Remak and Rosenbloom, 1976, for example), and in fact it is undoubtedly growing every day because of the pressing nature of many of the problems.

Diverting Traffic from Congested Areas

One of the oldest problems in traffic management is the high level of congestion in the central business districts of cities. This problem apparently arose throughout the world even before the advent of automobiles, although

Figure 17-1 **Example of hierarchy of responsibilities for transportation system management (TSM) in urban areas.**

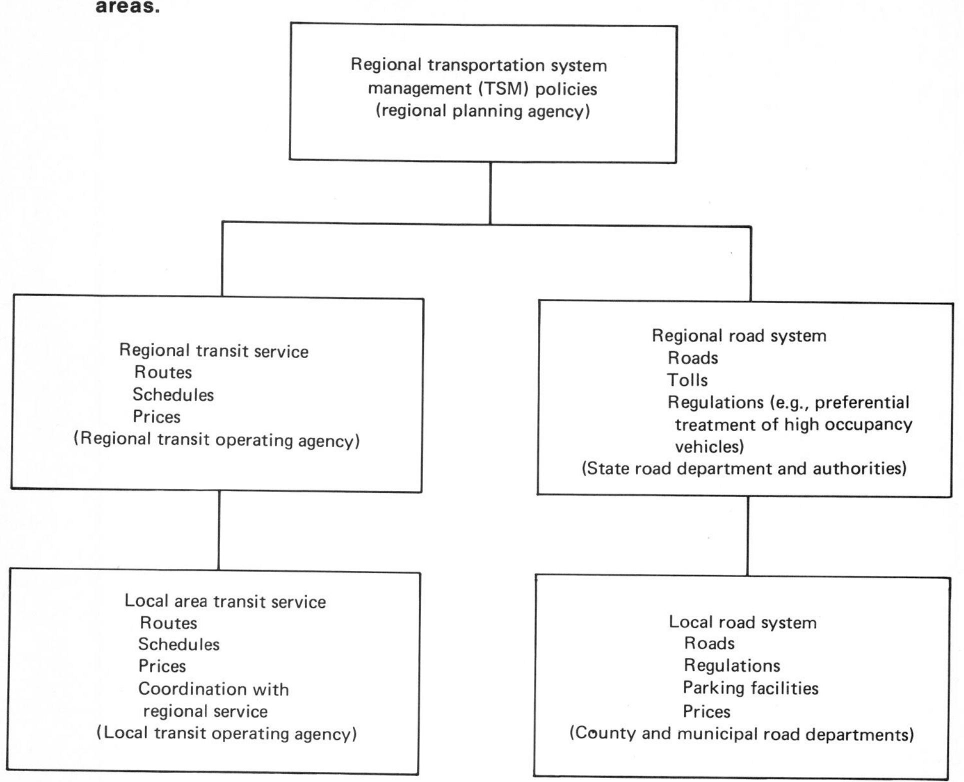

motorized individual transport has certainly aggravated it considerably. The problem is made especially troublesome in the central business area because of the conflict between vehicle flow and other activities which involve a considerable amount of pedestrian traffic.

Various means have been tried to remove traffic from central areas. In many cities it is observed that a great deal of the traffic moving directly in the central area is through traffic, traveling from an origin on one side of the central area to a destination on the other side, as shown in Fig. 17-2*a*. This observation, based on some of the earliest origin-destination surveys of traffic, led early planners to attempt to find ways of routing through traffic around the central business district. In some cities this can be done by traffic engineering means applied to routes which ring the central area. Often the removal of through traffic from the central area can be accomplished by simply making other routes which circumvent the central area more attractive than those passing through it, primarily by reducing the travel time on these alternate routes, as in Fig. 17-2*b*. Sometimes this can be achieved by coordination of traffic signals or otherwise giving traffic on the circumferential routes priority treatment, while in others it may require the building of entirely new freeways. By these measures, motorists would choose to use the ring

Figure 17-2 **Traffic management schemes for reducing CBD congestion. *(a)* Situation creating CBD congestion with both through traffic and traffic destined for the CBD. *(b)* Through traffic routed around CBD by providing more attractive routes. *(c)* Through traffic prohibited from CBD by use of traffic precincts.**

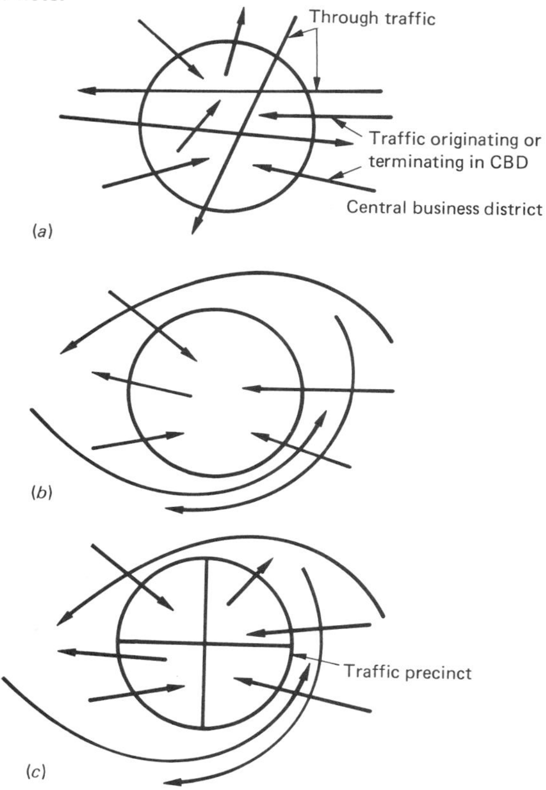

route rather than go through the central area, and the desired reduction in traffic would be achieved.

In some cities, however, it has not been found possible to improve the quality of service on ring routes sufficiently to induce motorists to do this of their own volition. In those cases, one successful approach is to ban traffic from passing through the central area but still to allow that traffic to enter the central area which involves business or other activities in that area. This is accomplished by dividing the area into a number of zones (or precincts, as they are called), such that one can drive into or out of the precinct, from a ring or circumferential road, but cannot pass from one zone directly into another. This is illustrated in Fig. 17-2*c*. This approach forces all through traffic to use the ring road, but still provides for very easy access by auto to the central area so as not to diminish its accessibility or attractiveness.

A closely related concept is that of creating pedestrian malls, in which all road traffic may be removed from certain streets, to turn these streets com-

Figure 17-3 **A pedestrian mall, with bus transit service, created from a city street.** [***Courtesy of U.S. Dept. of Transportation/Federal Highway Administration.***]

pletely over to pedestrians. One such mall is illustrated in Fig. 17-3. This, of course, forces the rerouting of traffic and eliminates traffic congestion on the mall streets. However, provision must be made to accommodate the diverted traffic. In some of the larger business districts a special transit service is provided to carry people around the mall or traffic-free area. This can be done in many different ways, such as by conventional or specially designed buses with wide doors for rapid loading and unloading and a minimum number of seats or by a special fully grade-separated transit route (e.g., subway or elevated). Many firms are now developing special transit systems designed for circulation within major activity centers. These systems are typically operated on a fixed guideway, much like a conventional railroad train, but with fully automated operation of small vehicles which can be operated singly or in trains, depending on traffic demand. These systems are often designated "people mover systems" or "major activity center distribution systems."

Priority Treatment of High-Occupancy Vehicles

Another type of potentially very effective change in road operations is the provision of priority treatment to transit vehicles and other high-occupancy vehicles (HOVs). High-occupancy vehicles are defined slightly differently in different areas, but all definitions prescribe a minimum number of persons who must be in the vehicle, usually three or four, sometimes not counting the driver if he or she would not make the trip except for being paid (e.g., a taxi or bus driver). The rationale is that the provision of priority treatment to HOVs tends to improve their level of service, which tends to increase their usage and hence decrease the usage of other vehicles.

As applied to public transportation vehicles, there are various forms which this priority treatment may take, but perhaps the most common and easiest to implement is reserving a lane for the exclusive use of transit vehicles (Levinson et al., 1973). Where transit vehicles are expected to make stops to pick up and discharge passengers, this reserved lane is often parallel to the curb. However, if special passenger loading platforms can be provided in the center of the street (as is common for streetcars), then the reserved lane is usually a center lane. The disadvantages of the curb lane are that it is often used by private vehicles for stops to pick up and discharge passengers, resulting in delays to transit service although such use of the lane by private vehicles can be prohibited. Also, since it is necessary for all vehicles making right turns to use the right-hand lane there are delays to transit vehicles at major intersections, especially if there are conflicts between pedestrians and right-turning vehicles. Also, the curb lane often is used for water drainage, and on a heavily crowned road it can be the least comfortable lane from the standpoint of ride quality. Where transit vehicles do not stop to pick up and discharge passengers or where they leave the reserved lane to do so, it is common to reserve the center lane to avoid these problems, although then left turns may be a problem.

Typical arrangements of reserved lanes for transit service are shown in Fig. 17-4. Part *a* of this figure shows a common form of bus service oriented

Figure 17-4 Example of network with exclusive bus lanes in area where roads are congested. *(a)* Original operation entirely in mixed traffic lanes. *(b)* Improved operation with exclusive bus lanes.

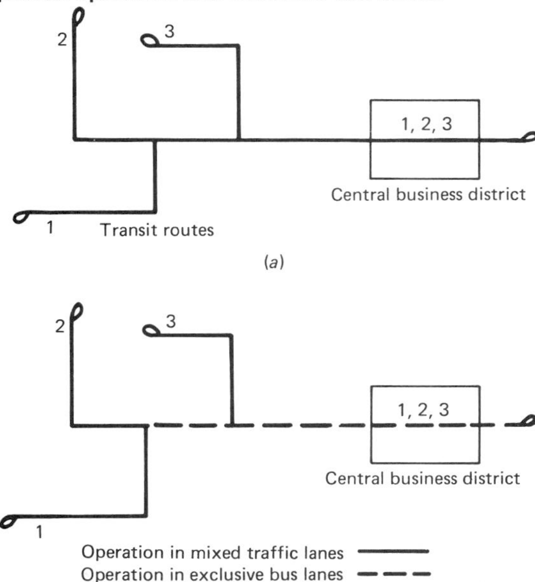

toward a central business district, in which a number of feeder lines join to form a major trunk route entering the central business district. The exclusive bus lane can be provided along the trunk line, benefiting all the bus routes. Particularly where buses operate nonstop between the CBD and outlying points, the provision of such reserved lanes on arterial streets or freeways is quite feasible and often very beneficial. On one-way streets, it has often been found advantageous to make the reserved bus lane operate in the direction opposing the other traffic flow, which discourages the use of that lane by unauthorized vehicles. Another attractive feature of lanes reserved for buses is that the street or freeway lane is available for use by other vehicles during periods when priority treatment of transit vehicles is either unnecessary or undesirable. Thus, the lane is available for use by other motorists on weekends, perhaps when the pattern of trip making is such that extensive transit use is not possible. In those periods, the number of other vehicles using the lane can be restricted so that transit speeds are not reduced.

Other forms of transit vehicle priority treatment include the timing of traffic signals to maximize the speed of transit vehicle flow, the most advanced form of this being the control of traffic signals by approaching buses or streetcars. Also, at congested entrance ramps to freeways and other facilities, transit vehicles can be given priority over other traffic by simply routing them on exclusive lanes around that traffic.

Lest one get the impression that priority treatment of public carrier vehicles is exclusive to urban transit, it should be noted that a similar situation exists at many airports with respect to common carriers versus private (general) aviation. At many airports, use during peak traffic periods is restricted to commercial aircraft, scheduled and charter vehicles; general aviation aircraft are prohibited.

In situations where road congestion is a problem, it is obvious that any means that will decrease the number of vehicles operated during congested periods will increase the average speed and level of service. In most instances it is undesirable to reduce the total number of persons traveling, which would disrupt patterns of living and the economy. Thus the problem can be viewed as reducing vehicle volume while retaining passenger volume at the same level. One means to accomplish this is to increase the average occupancy of private automobiles, as with car pools, in which two or more persons share the same automobile in making their trips. Typically this involves a driver picking up others and then traveling to a common destination (although some passengers could be dropped off on the way to the driver's destination), and the reverse for the trip in the other direction. In practice, a car pool vehicle is considered to be any automobile in which the number of persons is greater than one, or alternatively, greater than the average number occupying private automobiles, typically in the vicinity of 1.2 to 1.8 persons per automobile during the peak weekday travel hours. This sharing reduces costs per person, but increases travel time and compromises freedom to choose departure and arrival times.

The creation of a high-occupancy vehicle trip from a number of trips with fewer persons per car can be encouraged in a variety of ways. These include preferential parking locations and reduced tolls and parking rates (Dupree and Pratt, 1973). Many firms as well as municipalities and regional planning agencies are trying to encourage car pools by providing a central clearing house for information on when and where persons who wish to share rides are traveling.

Closely related is van pooling, in which a van, seating eight or nine persons, is substituted for the automobile, so that a larger number of persons share the same vehicle, thereby increasing the capacity further. Often an employer will purchase the van, charging its users an amount equal to its costs, and arrange for the assignment of driving responsibilities to various persons in the van pool. Sometimes the driver of the van is paid.

In a few cities, taxicab regulations have been altered to allow a taxi to operate as a for-hire car or van pool, meaning that it can carry more than one party at the same time, often with the passengers traveling from origins (and to destinations) which are nearby but not necessarily the same. Thus the taxi becomes similar to a van pool and also to a dial-a-bus.

Another encouragement is allowing these nontransit high-occupancy vehicles to use exclusive lanes which are reserved for buses or other transit vehicles. This is generally advantageous only when transit vehicles do not stop at stations in the running lanes, and hence it is limited to such cases. This is being experimented with in a few cities. Preliminary results indicate that a sizable number of such vehicles can be allowed on exclusive busways or bus lanes without impairing transit vehicle performance.

While carpools and other types of shared-ride transport are only beginning, there are some examples of very impressive usage. In Minneapolis-St. Paul, for example, over 1000 persons go to one factory in a van pool. Preliminary results of experiments with car and van pooling in two smaller cities—Knoxville and Portland—indicate that commuter vehicle-miles to many places of employment have been reduced by 5 to 40 percent (Pratsch, 1975, p. 55).

Reducing Travel Peaks

Another very important means of improving the flow of traffic is by reducing the severity of peaking of travel demand and hence of traffic volume. Peaks result primarily from the close proximity of beginning and ending times for work in many high-activity centers, particularly central business districts. In Fig. 17-5*a* are shown the volume of traffic demanded during peak periods and the effect on quality of service on roadways. With a flattening of the peak in volume, achieved by spreading the interval of time over which the peak period occurs, the maximum volume can be reduced and hence the level of service can be improved substantially, as indicated in Fig. 17-5*b*. A number of experiments with the staggering of working hours have had considerable success in reducing peak demands on roads and transit lines and in improving the level of service (Dupree and Pratt, 1973). Perhaps the most widely publicized example is that in lower Manhattan, New York City, where the Port of New York Authority and the Downtown-Lower Manhattan Association, an association of business firms, encouraged businesses to stagger work hours up to 30 min before or after the usual time to reduce peak demands on the subway (O'Malley, 1971). Peak volumes on rapid transit lines for short periods of 10 to 20 min were observed to drop by about 10 to 25 percent, with increases in volumes in less crowded periods on either side of the peak. The majority of employees and supervisors liked the change to an earlier working period, but most did not prefer the later working period to the usual period.

Figure 17-5 **Effects of shifting some peak period road volume to nearby periods. *(a)* Original condition with substantial peaking of traffic. *(b)* Modified condition with shifting of peak traffic volume to nearby periods.**

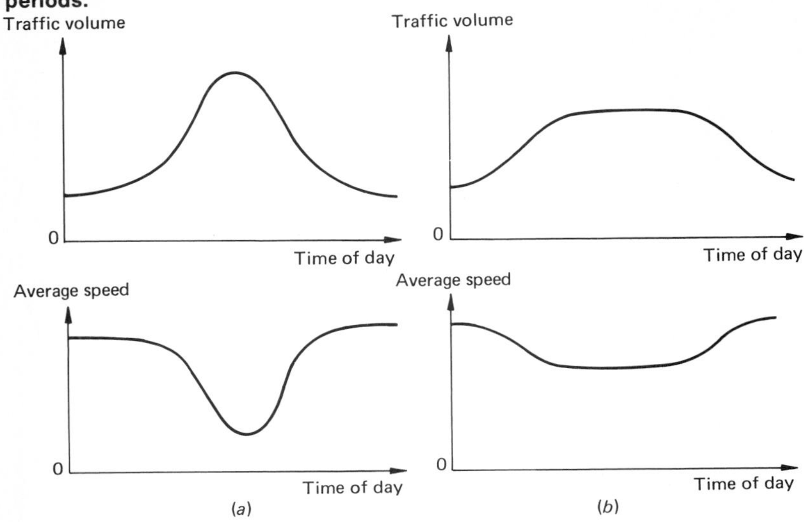

There are also indirect benefits of staggering work hours in the reduction in the cost of providing transportation services by reduced vehicle seat requirements (and perhaps reduced numbers of drivers or operators, reduced size of terminal facilities, etc.) in the case of public transit and reductions in the number of lanes required to meet peak flow volumes in the case of highways. There are various means of encouraging the staggering of work hours. At the present time emphasis has been given to convincing employers that the staggering of work hours will benefit the region as a whole and probably benefit their employees sufficiently to overcome any negative effects. In addition, facilities could be priced so as to encourage a staggering, by charging prices during peak periods that reflect the increased marginal cost of providing adequate capacity during those periods in comparison to periods of lower volume. While transit lines often have peak period fares different from base period fares, usually the period to which the higher fare applies is so long, often three or more hours, that it does not encourage substantial staggering of work hours. However, there is certainly no technological or other reason why transit fares and tolls on road facilities could not be varied over shorter intervals. Peak period pricing deliberately designed to encourage use of the facility during low-volume periods (in contrast to the peak) is practiced at a few airports and seems to be quite effective, particularly in discouraging aircraft with only a few passengers from operating into and out of the facilities at peak periods (Eckert, 1972).

Congestion Tolls

Clearly related to reducing peak flows is the use of congestion tolls—tolls or user charges which tranfer to the users the cost which they impose on other users due to their use of a facility. Any one user imposes a cost on others in increased transit time due to her or his presence on the facility. Yet in general users react only to their own costs (often perceived as travel time, as we noted in Chap. 11) and do not take into account added cost (including time) they impose on others by increasing average cost of travel on a facility with increasing volume, a relationship which seems to hold for almost all types of transport facilities—links, interchanges, and terminals of all modes.

This concept can be illustrated by an example, the same section of freeway used to illustrate the increasing road user cost curve in Chap. 11. The freeway is in rolling terrain, with three lanes in each direction, 8 percent trucks and 15 percent buses, and a design adjustment factor of 0.95. Figure 11-9 presented the relationships between average travel time per mile (slowness) and volume, and this is repeated in Fig. 17-6*a*.

Given a relationship between average travel time and volume as in Fig. 17-6*a*, which we shall denote by t(*q*), the total time to all users is

$$T(q) = q\bar{t}(q) \tag{17-1}$$

where $\bar{t}(q)$ = average travel time per unit distance-volume relationship
$T(q)$ = total travel time of all users per unit distance
q = volume

This is plotted in Fig. 17-6*b*.

The marginal time cost can be defined from this using the definition of

Figure 17-6 **Average, total, and marginal travel times on a road facility. *(a)* Average travel time for one vehicle. *(b)* Total travel time summed over all vehicles. *(c)* Marginal travel time of all vehicles due to one additional vehicle. Note: All flow measures are in one direction on a 1-mi section of freeway which has a 60 mi/h average highway speed, a 0.95 adjustment factor, 8 percent trucks, 15 percent buses, and is in rolling terrain. Capacity is 3,537 vehicles/h.**

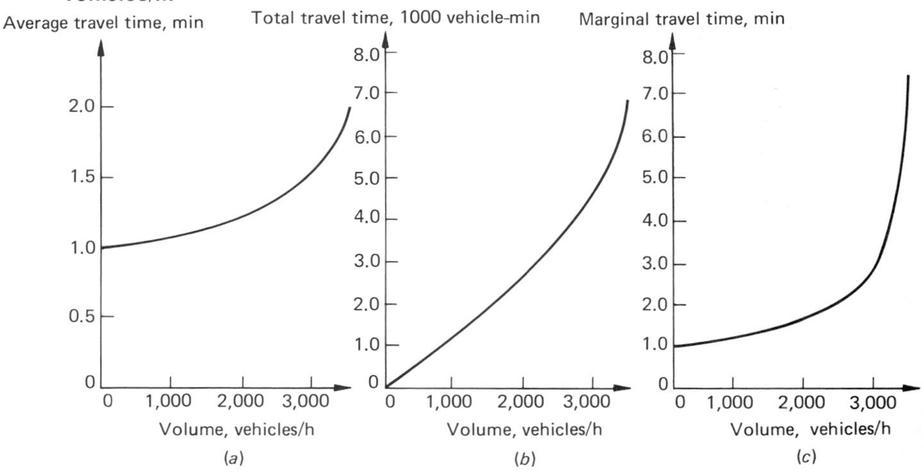

marginal cost obtained in Eq. (9-4). The marginal time cost per additional road user will therefore be

$$M(q) = \frac{d[T(q)]}{dq} \tag{17-2}$$

By graphical differentiation [taking the slope of the T(q) curve], this marginal time cost can be determined for the example road, and it is plotted in Fig. 17-6*c*.

As can be seen, the marginal cost starts out equal to the average cost at zero volume, as would be expected. However, as the volume increases the marginal cost of each individual user increases and becomes greater than the average cost. The reason for this is simply that the individual new (or marginal) user incurs a travel time cost equal to the average travel time at that volume and also imposes on all other users a small (but in some cases significant) additional travel time. The sum of these two components yields the (total) marginal cost at each volume. The significance of this is simply that each individual road user presumably behaves in response to the travel time that he actually experienced, approximately the average travel time. However, the additional cost to society associated with the use is equal to the marginal cost, which is therefore far greater than the perceived or average cost. Thus the user of the facility is not aware of the total time cost associated with the use of the facility, but rather only with the portion directly experienced.

It has often been proposed that road users should be charged in some manner to reflect the additional time cost they impose upon other road users. These are proposed as dollar tolls or prices which equal the value of the time each additional user forces others to experience. Although the logic is appealing, there are a number of problems with the concept. For one, it means that in effect the totality of users of the facility would be charged twice for some of the cost they incur, once in the form of the actual travel time they experience, and secondly in the form of a dollar charge or congestion price imposed to reflect the marginal time cost of the last (the marginal) user. And, of course, one could not simply turn this congestion tool revenue back to the users directly, for this transaction would cancel out the effect of the congestion toll. Also, the mechanism to establish the amount and to collect these tolls would have to be set up, possibly at great expense. However, the use of congestion tolls could result in changes in the routing of traffic and, perhaps more importantly, reductions in the amount of traffic. Users whose trips do not produce benefits at least equal to the marginal cost to society of the trips—the cost they experience plus the tolls reflecting costs they impose on others—would not make the trip or would shift it to another period. Thus congestion tolls might achieve a more economically efficient transportation system.

At the present time, however, congestion tolls are probably used in very few places in the world, if any. With increasing road congestion their use may be experimented with and possibly adopted in certain situations. They are also being vigorously proposed for other congested facilities, particularly airports, for the same reasons (for example, see Eckert, 1972).

Traffic Engineering Measures

Changes in traffic flow at a more localized level can also have very substantial benefits with respect to improved quality of service and increased capacity

(Pontier et al., 1971). One of the most effective means for improving the quality of service of road traffic is through the coordination of cycle lengths and offsets among traffic signals at nearby intersections. We saw in Chap. 5 how signal progression along a single street could substantially increase the speed of traffic, doubling of the average speed of traffic flow being common (for example, see Fig. 5-17). To achieve coordinated or progression timing of signals on two-way streets it may be necessary to close certain streets which make it difficult or impossible to achieve a smooth progression at reasonable speeds in both directions. Alternatively, those streets may remain open for traffic, but with traffic entering the main road from them being required to turn right only, perhaps with the intersection being controlled by a stop sign rather than a traffic signal.

Another type of signalization, which is used on freeway entrance ramps, is termed ramp metering. A traffic signal is placed at the entrance to the freeway, and this signal is controlled so that it allows only one vehicle to enter during each green phase in periods of high freeway volumes. The cycle time is adjusted so that the total volume of vehicles entering the freeway can be accepted by the freeway lane entered without flow being disrupted to the point where there is danger that speed will drop to the lower half of the speed-volume curve (see Fig. 5-11 again). In many installations of ramp metering, volume on the freeway lanes is measured continuously, and ramp metering rates (volumes) are adjusted automatically to ensure reasonably high freeway speeds. During low-volume periods, the ramp signal is set at green, and the ramp operates as usual.

Reallocation of traffic use of the various lanes within a street can also have beneficial effects. Particularly noteworthy are the gains from elimination of parking and possible gains from changing two-way streets to one-way streets. The substantial improvements in possible capacity can be illustrated with an example.

Consider a pair of parallel streets in a congested downtown area. These streets are each 50 ft wide, containing three traffic lanes in each direction, with parking allowed on both sides. Automobiles comprise most of the traffic on these streets, since trucks load and unload in alleys behind the stores and businesses, and transit service is provided on a nearby transit mall. The capacity of these streets can be estimated from Fig. 5-18. With a 25-ft approach width and parking on both sides, the capacity is 1100 vehicles/h of green time on each approach. With 60 percent effective green time in each cycle, the total capacity in each direction on the pair of parallel streets is

$$2(1100)(0.60) = 1320 \text{ vehicles/h}$$

If both streets are made one-way, the one-way capacity, with parking on both sides, is approximately 2200 vehicles/h of green time on each of the streets, yielding the same total capacity as with two-way operation.

This capacity can be substantially increased by eliminating parking (retaining two-way operation). This increases the capacity on each approach to 1700 vehicles/h of green time, yielding a 55 percent increase in traffic capacity. On the other hand, with one-way operation and all parking prohibited, the capacity increases to approximately 6000 vehicles/h of green time, yielding a capacity for the pair of streets of 3600 vehicles/h in each direction, almost three times the capacity of the original two-way operation with parking on both sides, and 76 percent more capacity than with two-way operation and parking eliminated. Even with parking allowed on one side (the right-hand side), the capacity increases to 3400 vehicles/h of green time, which yields the same capacity as two-way operation without parking.

Since in business districts on-street short-term parking attracts customers, the retention of some on-street parking usually is favored by businesses. In this example, the on-street parking could be allowed on one side, with one-way operation of the street, and still yield the same capacity as the situation of two-way operation without any parking at all. The beneficial effect of these increases in capacity on the speed of traffic flow and overall level of service is evident from Fig. 5-17, which indicates that at high volume-to-capacity ratios the average speed of flow on arterial streets decreases very rapidly with increasing volume.

Figure 17-7 **Some typical road traffic channelization markings. *(a)* Offset lane lines continued through intersection and crosswalk lines, and stop limit lines. *(b)* Turn lane lines, crosswalk lines and stop limit lines.** [**From U.S. Department of Transportation (1971), p. 192.**]

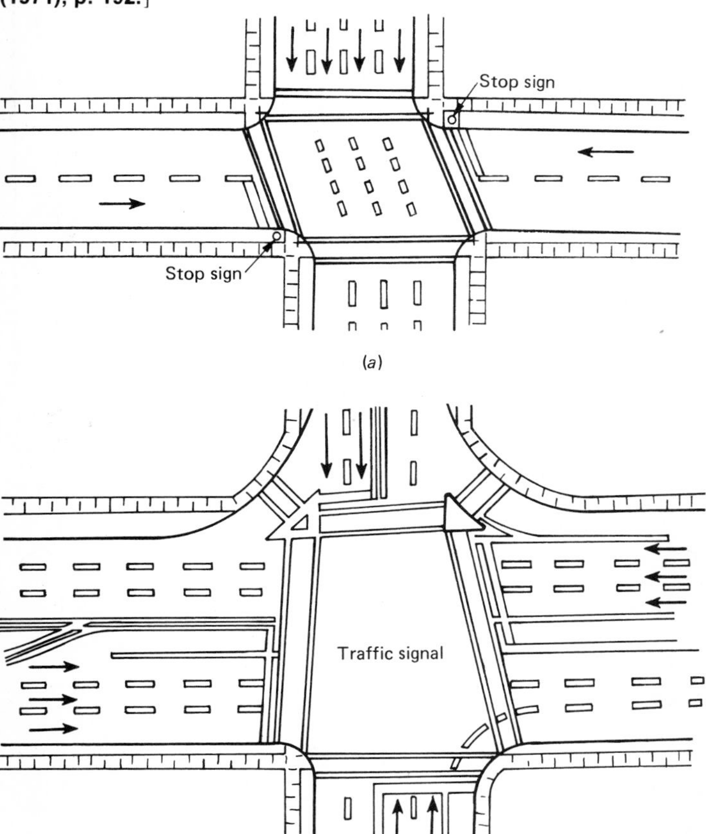

Also, the provision of separate channels of flow for various maneuvers through intersections can be extremely helpful in providing smoother traffic flow. Often such channelization can be provided simply by painting appropriate lines on the street surface, rather than by building special curbs and barriers. Typical forms of channelization include the designation of specific lanes for traffic turning right, turning left, and traveling straight through the intersection. At very complicated intersections where there may be more than three distinct maneuvers, the provision of additional lanes can be particularly helpful in increasing capacity and speed of traffic. Some typical examples of types of channelization are shown in Fig. 17-7. The basic principle is to provide for as many nonconflicting traffic movements as possible during any one phase of the signal cycle. In this manner, the number of phases can be minimized, and hence the time available for any one phase increased and the capacity of the intersection increased. By appropriately choosing the number of lanes assigned to any particular maneuver, the capacity of flow for that maneuver can be made adequate for the total volume of traffic desiring to make that maneuver.

A ROAD TRAFFIC MODEL FOR MAJOR ACTIVITY CENTERS

Traffic characteristics of selected areas such as central business districts are often analyzed by relatively simple, idealized models. Although such models are simplifications of reality, they can be useful guides in road traffic and transit planning. Below is presented an adaptation and generalization of a model developed in Europe (Leibbrand, 1970, pp. 121–136). The model is designed to take into account the major factors which influence the traffic-generating characteristics of major activity centers and the capacity of the transportation system to serve that traffic. It permits a rapid assessment of transportation alternatives for the area.

The model applies to a square central business district (or other area), in which there is a regular grid pattern of streets and square blocks, as shown in Fig. 17-8. All of the streets are assumed to be two-way. However, both of these assumptions could be altered and a new model derived with ease.

If there are $N - 1$ blocks on each side of the square CBD, then there are N streets feeding each side of the CBD. If the average capacity of each of these streets is Q_a vehicles/(h)(lane) and each street has n_a lanes available for general use in each direction, then the total volume of vehicles leaving or entering this area must be less than or equal to the road capacity, as below:

$$q \leqslant 4\, N n_a Q_a \tag{17-3}$$

where q = volume of private vehicle road traffic leaving (entering) CBD
N = number of streets on each side of CBD
n_a = number of lanes for private vehicle use in each direction on each street
Q_a = vehicle capacity per lane of private vehicle use lanes, vehicles/h

Since we shall treat transit-only lanes separately, these variables relate only to private vehicles and lanes for them, such vehicles including automobiles, trucks, and taxicabs. Also, the capacity of parking facilities in the area will

Figure 17-8 Layout of central business district in idealized road traffic analysis model.

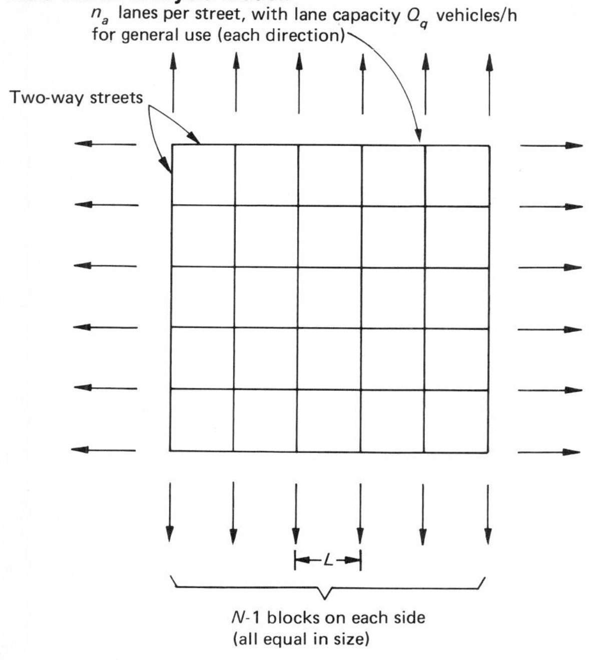

limit the number of automobiles which can be stored in the area at any one time, as

$$s \leq U \tag{17-4}$$

where s = maximum number of vehicles parked at any one time
U = number of parking spaces

The total amount of traffic entering or leaving the area during the peak period will consist of workers in the area, visitors, and some additional traffic such as trucks and cruising taxis. The average volume of workers entering or leaving the area during the peak period will be the total number of daytime employees in the area divided by the duration of the peak period:

$$p_w = \frac{\alpha_w w W'}{P} \tag{17-5}$$

where W' = number of workers entering (or leaving) the area in a day, persons/day
P = duration of one peak period, h
p_w = average volume of workers entering (or leaving) the area in peak period, persons/h
α_w = fraction of workers entering (or leaving) the area in peak period, dimensionless

The number of visitors traveling during the peak period will depend upon the temporal distribution of visitors entering and leaving the area throughout the day. If α_v is the fraction who travel in the peak direction during the peak period, then the volume of visitors will be

$$p_v = \frac{\alpha_v V}{P} \tag{17-6}$$

where V = number of visitors entering (or leaving) the area in a day, persons
p_v = average volume of visitors entering (or leaving) the area in peak period, persons/h
α_v = fraction of visitors entering (or leaving) the area in peak period, dimensionless

Given an average automobile occupancy for these two types of travelers and the fraction of travelers via public transit, the total volume of vehicle travel in the peak direction during the peak period will be

$$q = \frac{\alpha_w W(1 - R_w)}{E_w P} + \frac{\alpha_v V(1 - R_v)}{E_v P} + J \tag{17-7}$$

where E_w = average auto occupancy of workers in peak period, persons/auto
E_v = average auto occupancy of visitors in peak period, persons/auto
R_w = fraction of workers using transit in peak period, dimensionless
R_v = fraction of visitors using transit in peak period, dimensionless
J = volume of other vehicles, e.g., trucks, taxis, etc., in peak hour traffic stream, vehicles/h

Thus it is apparent that the total auto travel will be substantially influenced by the peaking of traffic, the average vehicle occupancy, and the percent of travelers via public transport.

The number of cars parking at any one time during the day will be influenced by essentially the same variables. It will be assumed that all the workers who work during the normal midday period leave their cars parked in the city during that entire period, rather than using them on business. Also, the duration of visitor parking must be included, and here it is assumed that visitors will generally remain for a period less than the duration of a work day, resulting in a ratio of visitors' automobiles entering the area to parking spaces required for them which is greater than unity, this ratio being termed the turnover of parking spaces for visitors. It will be further assumed that other traffic besides visitors and workers in the area will not require any significant number of parking spaces. Hence it will be ignored. Thus the total number of parking spaces required is given by the equation

$$s = \frac{\alpha_w W(1 - R_w)}{E_w} + \frac{V(1 - R_v)}{E_v T_v} \tag{17-8}$$

where T_v = turnover rate for visitor parking

These relationships can be used to judge the adequacy of the road system in accommodating expected future levels of traffic and also in specifying the minimum amount of traffic which the public transport system must accommodate in order for the area to experience road traffic within the capacity level of that system. An example may clarify this use. Consider a central business district which has a working population of 300,000 persons, of which 90 percent work the usual daytime period and travel during the peak periods.

Of those, 20 percent use public transport and 80 percent use automobiles, with an average of 1.4 persons/auto. 150,000 visitors come to the area each day, 95 percent in automobiles, with an average auto occupancy of 3.5 persons/auto. The peak period lasts 1.3 h, and during this period 20 percent of the visitors travel in the peak direction. The parking turnover rate for visitors is 3.0. Trucks and taxis add 10,000 vehicles/h to the peak traffic volume. This square CBD has eight blocks on each side (yielding nine streets in each direction), and the arterial streets facing each block consist of three lanes in each direction, with a capacity of approximately 750 vehicles/h. There are 40,000 parking spaces in the central area. From these data, Eq. (17-8) can be used to calculate the maximum possible flow of auto travel into or out of the region per hour, which is

$$q = \frac{0.9(300{,}000)(0.8)}{1.4(1.3)} + \frac{0.2(150{,}000)(0.95)}{3.5(1.3)} + 10{,}000$$
$$= 118{,}681.3 + 6263.7 + 10{,}000$$
$$= 134{,}945 \text{ vehicles/h}$$
$$4Nn_aQ_a = 4(9)(3)(750) = 81{,}000 \text{ vehicles/h}$$

Clearly the road system cannot accommodate this flow; the use of the public transit system or of other high-occupancy vehicles (e.g., car pools), or both, must be increased. Calculations reveal that a doubling of transit usage is still unfeasible. However, the combination of this transit usage and an increase of automobile occupancy to 2.0 persons/auto for work trips is feasible; it yields 78,242 vehicles/h.

The parking situation is similarly infeasible for the original conditions:

$$s = \frac{0.9(300{,}000)(0.8)}{1.4} + \frac{150{,}000(0.95)}{3.0(3.5)}$$
$$= 167{,}857 \text{ automobiles}$$

This is far greater than the number of spaces. With the reduced automobile usage the parking capacity remains inadequate, since

$$s = 93{,}857 \text{ automobiles}$$

Parking capacity is clearly the major constraint on auto use at the present time. To accommodate approximately 95,000 automobiles requires an increase in the number of parking spaces by about 140 percent. If the current average number of floors devoted to parking on plots of ground allocated to parking is 1.5, in order to expand parking capacity without diverting additional land in the central area to parking the average ratio of floor area to land area devoted to parking must be increased to about 3.6. Alternatively, more land could be devoted to parking, but this presumably would decrease traffic-generating activities or amenities such as parks and hence would adversely affect the economy and quality of the area.

Closely related to these problems of total traffic into and out of an area is the portion of the land within that area devoted to various land uses. For traffic and transportation purposes, these uses are divided by Leibbrand into the following categories: (1) traffic-generating areas, (2) traffic-serving areas, and (3) neutral areas such as parks and other public places. A number of convenient equations relating land area to traffic generation and road capacity can be derived for this square area. If the average lane width is L_w, the number of lanes devoted to auto traffic on each street is n_a, the number of lanes devoted to transit use n_t (buses and streetcars often preempt a lane even if it is not allocated solely to them), and the width of sidewalks on each side is K, the street width will be

$$b = 2L_w(n_a + n_t) + 2K \tag{17-9}$$

where b = total street width
L_w = width of one lane
n_a = number of lanes for automobiles
n_t = number of lanes for transit
K = width of one sidewalk

If one side of this square area of length is L, the total area devoted to roads will be $2NbL$ minus the land area devoted to intersections, which otherwise would be counted twice in this expression, that area being b^2N^2. Also, there will be additional land devoted to traffic purposes, such as alleys, pull-out ways, etc., which Liebbrand calculated as typically adding 30 percent to the land area calculated by considering streets alone. Thus the total land area devoted to roads will be

$$A_r = (2NbL - b^2N^2)\eta \tag{17-10}$$

where A_r = land area devoted to roads
η = factor to account for additional load for alleys, etc., dimensionless

The land area devoted to parking facilities can be calculated very easily once the area per car, H, which includes not only the parking area itself but also lanes, ramps, and pay areas, and the average number of floors per parking garage or lot (the plot ratio as defined above) S_p is known. The equation for land devoted to parking facilities is

$$A_p = \frac{H}{S_p} s \tag{17-11}$$

where A_p = land area devoted to parking
H = area required per automobile (including ramps, etc.)
S_p = average number of floors per parking facility (plot ratio)

In addition to these traffic-serving areas, there will be neutral land with area designated as A_n. Also, traffic-generating land can be related to the floor area of such generators according to the following definitional equation:

$$A_g = \frac{F_g}{S_g} \tag{17-12}$$

where A_g = land area devoted to traffic-generating uses
F_g = floor area devoted to traffic-generating uses
S_g = average number of floors per traffic generator (plot ratio)

Thus the total land area in the CBD will be

$$A = A_r + A_p + A_n + A_g \tag{17-13}$$

This area must equal L^2. As is made apparent by Eq. (17-13), any increase in land area devoted to parking or roadway facilities must be achieved with a decrease in the land devoted to neutral or traffic-generating activities. If the capacity of parking or road facilities must be increased, then this can be done only by either increasing the land area devoted to these activities or by increasing the plot ratio of road facilities, i.e., increasing the capacity of each lane. As we indicated above, it is possible by traffic engineering means to increase road capacity substantially above that which would be obtained on a street operated as a two-way throughfare with parking on both sides. But if any very substantial increase in capacity is required, particularly in areas where streets are narrow and land available for parking is severely limited, it is obvious that it can be achieved only by increasing the capacity and usage of the transit system. Even this requires taking lanes out of general use, but the gain in capacity on the transit-only lanes generally more than offsets the decrease in capacity of those lanes devoted to auto traffic.

CARRIER OPERATIONS PLANNING

Because carriers (as we have defined them) provide the vehicles in which passengers or freight are to be transported and also control the operation of those vehicles, they are faced with a wider range of operations planning options than facility operators typically have. Options available to a carrier include the following:

Points to be served

Routes to be operated to connect those points

Schedule on those routes, including vehicle type, accommodations, and capacity

Pricing

Advertising

Special services (e.g., tour packages of passenger carriers or combined warehousing and transport functions for freight carriers)

Some carriers are able to decide unilaterally on almost all these characteristics of the services to be provided by them, while as we noted in Chap. 9, many are regulated. In this section we shall explore the bases for decisions with respect to these options.

Conceptual Basis

The structure of the problem-solving process presented in Chap. 1 can be used to structure the evaluation of various alternatives facing passenger carrier decision makers. A fairly complete, but certainly not exhaustive, structuring for a passenger carrier is shown in Fig. 17-9. For purposes of constructing this figure, it was assumed that the carrier operated a service with a fixed route, fixed schedules, and fixed fares. Also, it was assumed that the major objective of the carrier is to maximize its profit, although of course in reality both private firms and publicly owned agencies have many different goals, which might include such factors as maximizing their share of the market and

Figure 17-9 **Process of selecting best operations plan and fare policy of an urban public transport system.**

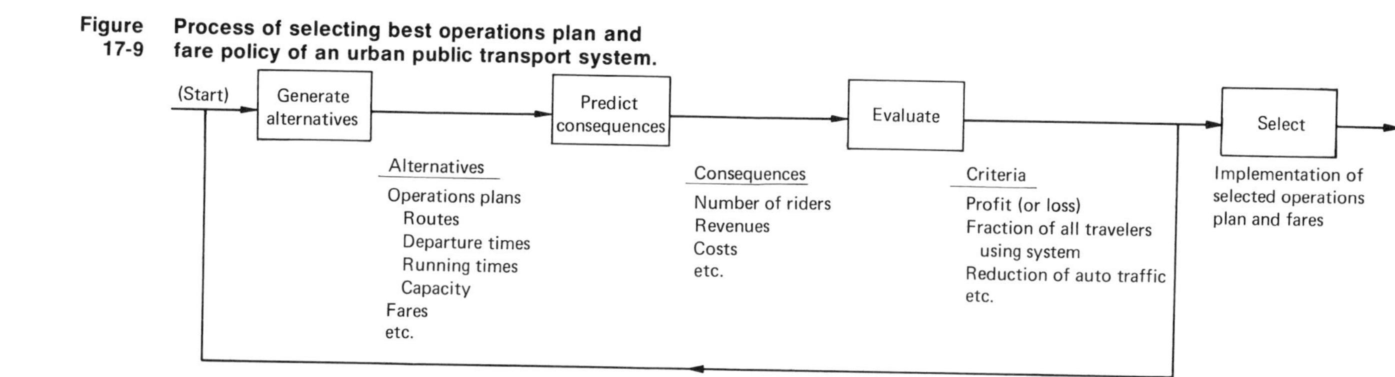

maximizing passengers carried subject to meeting minimal levels of profit (or not exceeding a certain financial loss to be made up by subsidies from government).

The roles of the various models which have been discussed in previous chapters are evident from this figure. Models of the technology of transportation and the associated cost-estimating relationships could be used to help specify the various alternatives and then to predict the capacity, level of service, and cost which would result from selecting any particular one. Models for estimating the demand for transportation (including changes in population and land use patterns if periods in the distant future are of interest) could be used to estimate the quantity of passenger traffic demanded and revenue resulting from any particular alternative. For purposes of evaluation, the use of a criterion such as profit is fairly straightforward, and in cases where the profit over an extended period were to be considered, discounting as discussed in Chap. 9 would be applied. If the sole objective were to maximize profit, then the problem of selection is simplified, since a single measure of the relative worth of any alternative is obtained.

To illustrate how this type of process would work, we might use an example for a very small transport system. Consider a carrier operating a service between two points, with a given price. The network is given, in both location and in the technology of transportation to be used (including a particular vehicle of fixed passenger capacity), resulting in a specification of the running time of vehicles between the two points served and in the access time and access cost of travelers (at any given origin or destination) to and from the end points. Thus the major option open to the management of this carrier, which will affect revenues and costs, is the frequency with which vehicles will depart from each of the two end points. To simplify this option, we shall assume that the frequency is the same in both directions.

The effect of variations in service frequency on the demand for travel in this system is indicated in Fig. 17-10*a*. For illustrative purposes, the quantity demanded has been shown to increase rapidly with increases in frequency in the vicinity of low values, the increase in quantity demanded decreasing with increasing frequency as the market penetration of this mode increases, as would be expected. With the fixed technology, vehicle size is fixed, and hence the total capacity resulting from any given level of frequency is proportional to the frequency, as indicated by line *OB*. As can be seen from the drawing, at frequency values between *C* and *D* the quantity demanded at any given frequency is greater than the capacity of vehicles to accommodate it, so some travelers must be turned away. Thus the actual traffic to be accommodated on the system at any given frequency would follow the capacity curve between frequencies *C* and *D* and the demand curve elsewhere.

Figure 17-10*b* presents the revenue and cost relationships for this same situation. At a constant fare, the revenue is simply the product of the constant price times the traffic on the system at any given frequency level, which results in the curve *OR*. The cost of providing service via this carrier is given by the line *NP*, which because of its nonzero intercept on the cost axis indicates that fixed costs of service equal to *ON* have been included. Profit is, of course, the total revenue minus the total cost. As can be seen from the figure, this profit is negative at low frequencies, becomes zero at *OS*, and then increases to reach a maximum at a frequency of *OK*, where the revenue is *OM*

Figure 17-10 Selection of departure frequency yielding maximum profit for a passenger carrier. ***(a)*** **Capacity and demand.** ***(b)*** **Revenue and cost.**

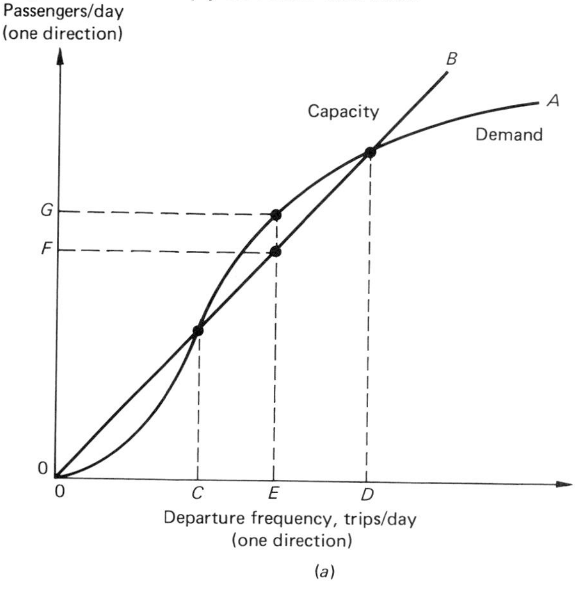

(*a*)

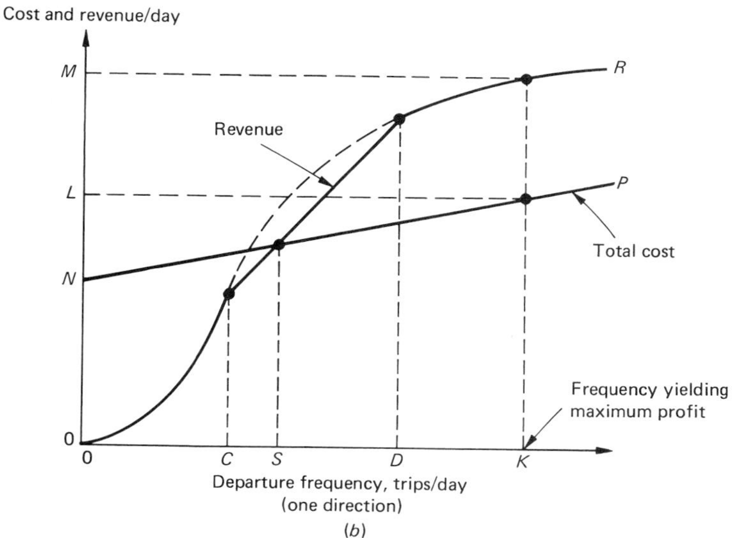

(*b*)

and the cost *OL,* for a profit equal to *LM.* At levels of frequency above this, the cost increases more rapidly than the revenue, which results in a decrease in profit. Thus the carrier would presumably choose to operate at frequency *OK.*

This same approach could be used to address more complex questions. For example, the carrier may have available vehicles of different types, each with different capacity, speed, and cost characteristics. The selection of the proper vehicle type then becomes a problem. Each type of vehicle might be analyzed as indicated along the lines above, resulting in an estimate of the optimal frequency at which that vehicle type should be operated and the maximum profit. This problem can be further complicated by considering a mixture of vehicle types operating simultaneously on the same route, as is often found in air networks. Also, these vehicles might differ with respect to their cruise speeds and hence the times required to travel between the various end points. In this case, the effect of different travel times on the quantity demanded would have to be considered, meaning that each vehicle type would have a different demand curve.

The same general procedure could be used to explore options with respect to price and speed of operation. It may be possible to purchase vehicles with different power-to-weight ratios or to redesign the terminals and way links connecting them to yield different running times, as in the case of land transport systems. Also, terminals may be relocated to reduce traveler access time. In these instances, all of these different options could be analyzed within this basic framework. Of course, when the number of variables (options) is increased, a graphical procedure is no longer appropriate. In such cases, the analysis must be performed by numerical consideration of the various alternatives, including the interaction among choices of the various types of options. However, the basic structure of the problem remains the same, that of attempting to find that particular combination of choices of these various types which will yield the maximum profit. Finally, it should be noted that this same approach can be applied to freight carriers, although in this case the range of options probably would expand even further, to include different vehicle types for different commodities and questions of whether to operate a scheduled service or a demand-responsive service.

A Transit Improvement Example

This example will illustrate how many of the methods presented in earlier chapters can be used in operations planning problems. The problem is one of deciding among a few improvements in an urban bus service, these alternatives falling within the field of transportation system management. The problem has been deliberately made small, so that the points would not be obscured by the computations.

The problem involves a bus service connecting a new high-rise residential development with the central business district 2 mi away. The existing bus service operates on the streets in mixed traffic, with an average speed of 12 mi/h (including stops but not layover time at each terminal), a fare of 35 cents (the fare throughout the city), and a headway of 10 min throughout the 16-h operating day. The buses serve only residents of this development, and operate nonstop between it and the CBD.

Three related proposals for improvement have been made. One is to ban parking in the arterial street used by these buses and to turn one lane in each direction into an exclusive bus lane, permitting average speeds to be increased to 25 mi/h. There is ample parking on side streets and in parking lots. The second is to reduce off-peak fares to 20 cents. Finally, any headway of 2, 5, or 10 min can be selected for the revised service. The management limits the headway options to three values that riders will be able to remember easily. All of these are to be analyzed from the standpoint of maximizing riders subject to revenues meeting all costs, the objective of this publicly owned transit line. This is, it might be added, the expressed objective of the London transit system and other publicly owned carriers (Quarmby, 1973).

The current weekday service accommodates an average of 280 passengers/h in the peak direction during the four peak hours and an average of 46 passengers/h otherwise, for a total revenue averaging $842.80/day. The bus line estimates its costs by the following formula, analogous to Eq. (9-16) for rail transit:

$$TC = \$20.346b + \$81.06d_f + \$0.564m_b \tag{17-14}$$

where TC = total cost/day
b = number of buses required
m_b = bus-mi/day
d_f = number of paid driver-days/day

The bus cost is based upon a $46,000 vehicle seating 52 and standing 25, with a 12-year life, a 10 percent interest rate, and an 8 percent maintenance reserve. On this route, with a 5-min layover at each terminal, the minimum cycle time is

$$t_m = \frac{4 \text{ mi}}{12 \text{ mi/h}} 60 \text{ min/h} + 10 \text{ min} = 30 \text{ min}$$

and hence by Eq. (8-13) the number of buses required is

$$b = \left\langle \frac{t_m}{h^p} \right\rangle = \left\langle \frac{30}{10} \right\rangle = 3 \text{ buses}$$

Assuming each driver works 8 h, 6 driver-days are needed each day. The number of bus-miles per day is

$$m_b = 4 \text{ mi} \left(16 \frac{\text{h}}{\text{day}} \right) \left(\frac{\text{h}}{10 \text{ min}} \right) \left(\frac{60 \text{ min}}{\text{h}} \right) = 384 \text{ bus-mi/day}$$

Hence the total costs is

$$TC = \$20.346(3) + \$81.06(6) + \$0.564(384) = \$763.97$$

for a route profit of $78.83/day.

With exclusive bus lanes, the service would have a reduced cycle time:

$$t_m = \frac{4}{25} 60 + 10 = 19.6 \text{ min}$$

Hence the number of buses required at each possible headway would be as in Table 17-1. The corresponding bus-miles per day and paid driver-days per day are also given there, separately for the peak and the nonpeak periods. In the peak period section of Table 17-1 only the items required to operate the peak service are given, and in the nonpeak section the additional items required for the nonpeak service are given. Then the corresponding costs are presented, based on Eq. (17-14).

In Table 17-2 the expected passenger traffic on the route for each of the headway and fare alternatives is given. This was calculated based on the knowledge that the total volume between this residential area and the CBD is

Table 17-1 **Bus Service Costs in Reserved Lane Problem**

h, Headway, min	*b*, Buses	m_b, bus-mi/day	*d*, driver-days/day	c, Cost, $
Peak period service				
2	10	480	10	1284.78
5	4	192	4	513.91
10	2	96	2	256.96
Nonpeak period service				
2	—	1440	10	1622.76
5	—	576	4	649.10
10	—	288	2	324.55

1000 persons/h during the peak period in the peak direction, and that the one-direction volume averages 200 persons/h otherwise, based on a 16-h day (7:00 A.M. to 11:00 P.M.). The modal split model presented in Chap. 10 (Fig. 10-6) was used, for economic class 5. The automobile travel time was composed of a 25 mi/h speed on the arterials, 1.5 min parking and a 0.25 mi access walk at 3 mi/h at the CBD, and a very short walk (0.5 min) at the residential end, for a total time of 11.8 min. Auto costs were calculated at 10 cents/vehicle-mi, with an average of 1.5 persons/vehicle. None of these factors varied between periods, although CBD parking was $2.00/vehicle for all-day parkers, $1.00 on average otherwise. Transit access was assumed to be 0.25 mi at each end, at 3 mi/h, and wait times of one-half the headway were used.

The modal split model can be applied directly with these data. An example

Table 17-2 **Bus Service Patronage in Reserved Lane Problem**

h, Headway, min	Travel time ratio	Percent via transit	Volume, passengers/h	Revenue, $/period
Peak period (4 h, 35 cent fare)				
2	1.35	60†	600	840.00
5	1.47	55†	550	770.00
10	1.69	41†	410	574.00
Nonpeak period (35 cent fare)				
2	1.35	43‡	86	842.80
5	1.47	41‡	82	803.60
10	1.69	35‡	70	686.00
Nonpeak period (25 cent fare)				
2	1.35	53‡	106	742.00
5	1.47	50‡	100	700.00
10	1.69	44‡	88	616.00

† From Fig. 10-6, with L_2, CR_1, and EC_5.
‡ From Fig. 10-6, with L_2, CR_2, and EC_5.

will illustrate its use, the example being for the 35 cent fare, peak hour work trips, and a 5-min headway:

h = headway = 5 min and w = average waiting time = 2.5 min

$$\text{TTR} = \text{travel time ratio} = \frac{(14.8 + 2.5)\ \text{min}}{11.8\ \text{min}} = 1.47$$

$$L = \text{service ratio} = \frac{(5 + 2.5 + 5)\ \text{min}}{(5 + 0.75)\ \text{min}} = 2.17$$

Hence we use curve L_2.

$$\text{CR} = \text{cost ratio} = \frac{\$0.35}{\dfrac{\$0.10 \times 2}{1.5} + \dfrac{\$2.00}{2(1.5)}} = 0.4375$$

Hence curves for CR_1 apply.

From the curves of income level EC_5 for work trips, the percentage of trips via transit is read to be 55 percent, resulting in 550 trips/h in the peak direction via transit. The revenue is also given in that table.

Table 17-3 summarizes the passengers carried per day, the revenue, and the cost for all combinations of price and level of service (headway). After eliminating all deficit-producing options, the one that maximizes riders is that with a 5-min headway throughout the day with a ridership of 5000 passengers/day and a profit of $306.99. This represents an increase of 108 percent over the 2408 riders in the existing service, a very substantial gain!

Note that this gain is due to a combination of operating changes: the use of an exclusive lane to speed transit vehicles, a decrease in headways, and a reduction in nonpeak period fares. None of these alone would have resulted in such a gain. This illustrates the importance of considering operations changes in addition to any one change that might be under consideration, which might have been the reserved bus lane in this example.

Table 17-3 Bus Service Patronage and Profits in Reserved Lane Problem

Peak headway, min	Nonpeak headway, min	Fare, cents	Passengers	Revenue, $	Cost, $	Profit, $
2	2	35	4808	1682.40	2907.54	−1225.14
2	2	25	5368	1582.00	2907.54	−1325.54
2	5	35	4696	1643.60	1933.88	−290.28
2	5	25	5200	1540.00	1933.88	−393.88
5	5	35	4496	1573.60	1163.01	410.59
5	5	25	5000	1470.00	1163.01	306.99
5	10	35	4160	1456.00	838.46	617.54
5	10	25	4664	1386.00	838.46	547.54
10	10	35	3600	1260.00	581.51	678.49
10	10	25	4104	1190.00	581.51	608.49

Some Practical Approaches

The procedures outlined above for making managerial decisions regarding schedules, fares, etc., all depend upon having available information on the demand curve for the service and on the costs of the various options. Also, in some cases it is essential to consider the reaction of competitors, whether it be from other carriers or from private transport (e.g., automobile) on roads that may become less congested if traffic is attracted to the carrier. In many instances, information on the demand and also on the reaction of competitors is either unavailable or available but not in the form of models which are applicable over the entire range of options.

An example of a limited information situation is illustrated by the Kraft-SARC intercity demand model presented in Chap. 10. That model contained estimates of the elasticity of demand for transportation via various intercity modes based on the assumption that the elasticity was constant over the entire possible range of prices or other characteristics of services, which implies that the quantity of travel demanded on a mode would be infinite at zero price. However, even though it is clear the parameter estimates do not apply over the entire ranges of price and level-of-service variables, such a demand model can be used to identify the most promising directions for changes in prices and level of service. In many cases knowing the most attractive direction of change is very helpful if it is combined with the judgment of persons knowledgeable in the problem area who can make educated guesses as to the optimal size of any changes.

The use which was made of the Kraft-SARC model in the planning for improved transportation service in the Boston-New York-Washington (Northeast) Corridor of the United States illustrates this. In the late 1960s it was known that intercity travel in this corridor would increase substantially in the next decades, taxing the capacity of existing highways and airports. It was felt that some type of improved surface common carrier transport, perhaps a high-speed rail service, would help to alleviate these conditions and improve the overall performance of the transportation system in the corridor. Two considerations were of paramount importance: (1) reducing road and airport congestion and (2) creating a new surface transport service which would attract sufficient patronage to be reasonably viable, that is, would require a minimum of operating subsidy and initial government expenditure. The near-term options included only changes in rail and bus service, since other more advanced technologies would require considerable time for research and development before an implementable system became available.

Inspection of Table 10-1 reveals the reasons why a high-speed rail service, with a price higher than the conventional (slower) rail service was adopted (to minimize the subsidy required). Considering business travel first, it was assumed in model development, as seems reasonable, that the effect of changes in bus costs or bus time on air demand would be negligible, since the characteristics of these two modes differ so greatly. However, the demand model revealed that while rail travel time had a considerable influence on the demand for air travel, the associated elasticity being close to unity (0.973); rail cost apparently had virtually no effect on air travel. Similarly, the influence of rail time and rail cost on the demand for auto travel was fairly substantial, again with elasticities close to unity. In contrast, the effect of bus cost and time on auto travel for business purposes was also minimal, again probably indicating that bus service was considered inappropriate by most business personnel for travel purposes.

Rail also appeared to be the most logical choice for improvements which would reduce the demand for personal (nonbusiness) travel via air and auto. Although rail cost does not influence air travel, the elasticity of rail time with respect to air travel is nearly unity. Also, the elasticity of auto travel for personal purposes with respect to rail travel time is approximately 0.458, although the influence of rail cost on auto travel seems quite small. In contrast, the influence of bus cost on auto travel is nil and the elasticity of auto travel with respect to bus time is close to zero.

For these reasons (and others, such as those related to the environment), it seemed most appropriate to attempt to reduce rail travel time substantially in the corridor to attract business travelers from air and auto travel. This was done in the southern half of the corridor (the most easily upgraded) by introducing a high-speed reserved-seat Metroliner service, which reduced running time between New York and Washington from 3 h 40 min or more on the previous conventional trains to about 3 h, with the new trains operating on headways of 1 h in each direction during the day. Fares on the new service are considerably higher than those on the conventional trains, but again the demand model elasticities revealed that this would not hinder the attraction of business travelers. To cater to travelers who are sensitive to price, conventional train service was retained, on a somewhat less frequent schedule than the new Metroliner service, with the fares remaining at the old levels. The effect of the experiment has been to approximately double intercity rail travel in the area served, and the revenues of the new Metroliner trains are approximately covering operating costs, although the surplus is not yet sufficient to pay the entire cost of the new rolling stock and line upgrading. Further improvements to rail service in the corridor are planned.

Conclusions

The application of quantitative methods for estimating demand to transportation carrier management and operating problems is just beginning. While the models for estimating passenger demand will have to be refined considerably to account for the full range of management options, to include not only major choices such as differential peak and off-peak fares but also smaller items such as various levels of passenger amenities. Parallel efforts are underway with respect to freight. Such tools are being developed and undoubtedly will find their way into standard management practice within the next decade. Similarly, cost models are still relatively crude compared to many of the decisions which management must make, which include not only decisions about different types of operating plans but also decisions regarding very specific aspects of work rules and labor contracts and day-to-day operating options which require much more detailed models than are now available. Thus the use of these tools to assist in the management of transportation systems is in an embryonic state and much fruitful research and development remains to be done.

MAINTENANCE

No discussion of the management of transportation systems would be complete without consideration of maintenance. All of the physical components of a transportation system must be maintained if they are to perform properly over their service lives, and a considerable part of the cost of transport is due to maintenance. For instance, railroads typically spend about 20 percent of their total operating cost on track and structure maintenance, and another 20 percent on locomotive and car maintenance (Association of American Railroads, 1976, pp. 15–16). Recently about 25 percent of the total expenditures by all levels of government roads in the United States has been on maintenance (U.S. Department of Transportation, 1973, p. 98). Airlines and bus lines typically spend about 15 percent of their total operating costs on maintenance (National Association of Motor Bus Owners, 1976, p. 24, and Air Transport Association of America, 1974, p. 25).

Maintenance considerations have become increasingly important in the United States in recent years, because so many parts of the transportation system are in financial difficulty and unable to maintain facilities at standards necessary for adequate service. The increase in railroad accidents is well known, as are the increased costs of train operation due to extremely slow operation on many lines. Many states are now finding it difficult to maintain roads properly, with the prospect of rapidly escalating expenditures to catch up with maintenance deferred from previous years.

Types of Maintenance

There are many different approaches to maintenance. One approach, preventive maintenance, in which a component is serviced in order to prevent failure while in use, is widely practiced, particularly if failure leads to unsafe situations, such as on aircraft. Alternatively, maintenance can be undertaken only when a failure is observed to have occurred. This is commonly used in maintenance of local streets, for example. Closely related to these approaches are options on inspection, which can be done on an ad hoc basis, whenever it can be conveniently scheduled, or on a regular basis, the timing dependent upon actual or expected use of the component (e.g., once every 300 flying hours or once every 30 days, respectively). Again the choice depends upon the circumstances.

Many transportation facilities and carriers are regulated with respect to safety considerations, including many that affect maintenance. Airlines are the most heavily regulated in this regard, for obvious reasons. Not only are the aircraft checked, but crews are also checked for fitness regularly (including some checks prior to each flight). Also regulated are equipment necessary on vehicles (e.g., life vests on aircraft operated over large bodies of water), vehicle performance (e.g., vehicle must be able to stop within a certain distance from a specified speed), vehicle condition, and work and rest periods of operators (e.g., railroad crews cannot be on duty more than 12 continuous hours). The exact form of these regulations and the manner of enforcement varies widely by mode.

Survivor Curves

The variation in maintenance practices among modes is so great that it is difficult to find many general principles or methods. However, one generally applicable concept is that of survivor curves, which can be used to help determine service lives of components, to define maintenance requirements,

Figure 17-11 **General form of survivor and related curves and an example for road pavements.** ***(a)*** **Typical form of survivor curve and curves derived from it.** [***From Winfrey (1969, p. 208).***] ***(b)*** **Survivor data and fitted curve for concrete pavement.** [***From Winfrey (1969, p. 221).***]

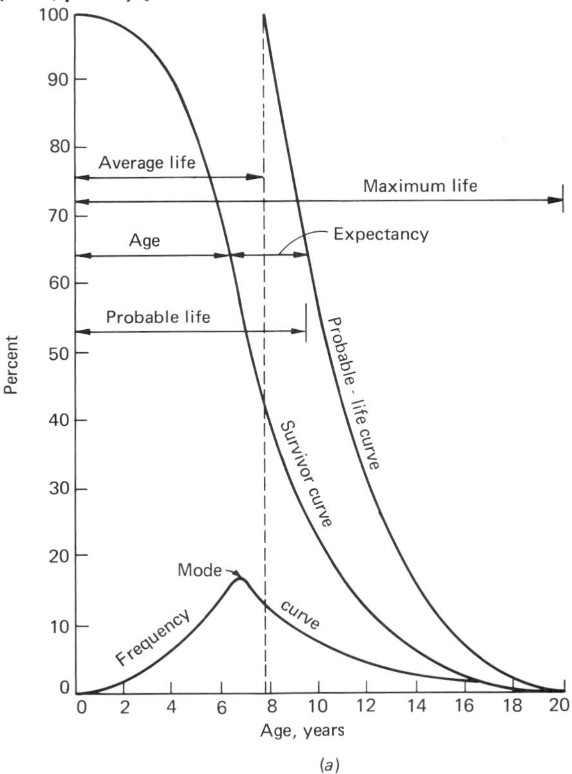

(a)

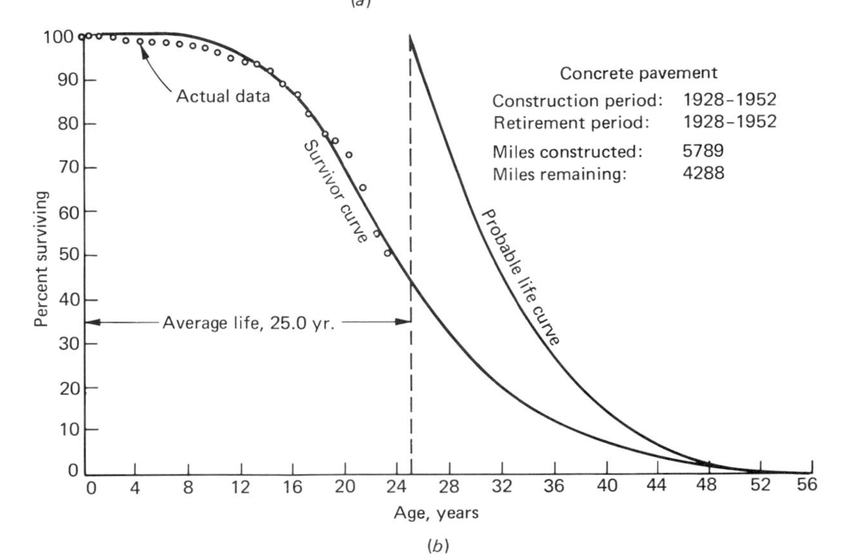

(b)

and to project maintenance or repair expenditures. We shall describe these briefly, using a road example.

There are many forms of survivor curves, but a form that is very useful in transport is the original group type of survivor curve. This curve presents the percentage of an original group of like components which survive to various years in the future. In Fig. 17-11*a* the lower curve indicates the number of units retired in each year, and hence can be used to predict the number of replacements of an original group necessary each year in the future. This curve is simply the difference between the number of units in service in one year (in the life of the items) from the next, and hence can be derived from the survivor curve. Finally, the probable life curve gives the expected years of future life for a unit at any given actual age.

To be used to predict service lives, replacements, and hence maintenance requirements for components, a specific survivor curve must be developed for it. For many types of components, standard curves have been developed. An example is presented in Fig. 17-11*b*, where data for a period of 23 years were fitted with a standard survivor curve to predict future survivors. All miles of pavements are expected to be retired and hence require replacement about 54 years after construction, and by 40 years only 6 percent of the original pavement miles will be serviceable. However, those pavement miles which do last 40 years can be expected to last for 4 more years, on the average. The exact number of miles expected to be retired each year can be constructed from the survivor curve as described above. For example, between ages of 32 and 36 years, 8 percent of the mileage is expected to need replacement. This information can be used to project maintenance expenditures.

INTEGRATED OPERATIONS PLANNING AND DESIGN OF A SYSTEM: A TRANSIT EXAMPLE

Most work in the design of transportation facilities, vehicles, and other system components is constrained by considerations of compatibility with an existing system. This is primarily because in most areas a transportation system already exists, and it is to be extended to serve new areas. However, there are situations where an entire system is to be designed, and where the limitations imposed by compatibility with existing systems are minimal. In this section we shall treat the design of a system with many options, an urban rail passenger transit system. This example will illustrate the very close relationship between certain aspects of planning (particularly level of service and volume of traffic), design of fixed facilities and vehicles, and operations plans.

Even though the vehicle design, track (including gauge), and control systems for railways are often selected so as to make use of standard equipment, there are many design options on any given route. These include the location and spacing of stations, the length of trains to be operated, speeds, maximum gradients and curvature, the manner of fare collection, the schedule plan (i.e., all-local, skip-stop, etc.), and train departure frequency. The entire system can be tailored to the specific tasks it is to perform. The choices will determine the capacity and level of service of the system, as well as its costs, and the degree to which the system can be modified in the future to changed travel demands.

The Example Route

For this example, we will assume that standard transit vehicles will be used, with the associated limitations imposed upon route characteristics such as the provision of high-level platforms. We will focus on other design options, which typically arise in practice, such as the length of train to be accommodated, the manner of fare collection, and the number of running tracks provided at various points along the line. This example will make use of the rail transit line example presented earlier in Chaps. 8 and 9, in which a line 15 mi long was considered. In that original example, the line was double-tracked throughout, each track being used in one direction, and stations were located every half-mile. A capacity of 8000 passengers/h in the peak flow direction was required during the peak period, which consists of two 2-h periods each weekday. Also, a capacity of 1143 passengers/h was required in other periods and directions. This led to the design of the system for the operation of five-car trains during the peak period, each car accommodating 100 passengers (52 seated), with trains being operated on a headway of 3.75 min. During the other periods, trains of three cars were operated, each car seating 52 passengers, with trains being operated on a headway of 8 min. This system required 120 cars. Each day it requires 36 train-crew-days, 24 operating on a straight 8-h shift and 12 operating a swing shift consisting of just the two peak periods, 60 station-agent-days, and operation of 17,700 car-mi. Using costs presented in Chap. 9 and the formulas for relating system design and operations plans to costs, the following cost model was derived:

$$TC = \$1839.36L + \$111.04c + \$0.982m_c + \$134.11d_c + \$41.29d_a \qquad (9\text{-}16)$$

where TC = total cost, \$/day
L = one-way route length
c = cars required
m_c = car-mi operated/day
d_c = train-crew-days paid/day
d_a = station-agent-days paid/day

For this particular line, this resulted in a total cost per day of \$65,601.96, of which the total operating cost (including the fixed daily cost per mile of track maintenance) was \$25,161.51, 38.4 percent of total cost. Thus 61.6 percent of the total cost was fixed.

Zone Service

An alternative system design which would change the quality of service as well as the cost is a zone type of operation. In order to do this, with such low headways and still reduced vehicle running time it is necessary to provide an additional track on each direction in the section over which the express running is to be provided. All trains will serve all stations in the central business district, which is 2 mi long; and hence if the remainder of the line is divided into two zones (with a common station on the zone boundary), the additional double track for express service will be required on 6.5 mi of route. Since the trains will not stop on that section, no additional station costs will be incurred. The additional fixed costs per day of providing 6.5 mi of double track and of maintaining that track will be

$$6.5 \text{ mi } (\$1258.10) = \$8177.65$$

There will also be changes in the operating costs. If we assume that the zone express type service is offered only during the peak periods, one operating option would be to continue to operate trains at the same headway into the CBD, with half of these trains being locals serving the inner zone and the CBD and the other half being expresses connecting the CBD via the new express tracks with the outer zone. From Fig. 4-20, assuming a maximum speed of 50 mi/h and station dwell times of 20 s, the round trip times of these trains can be determined. Each train requires 24 s and 1400 ft to accelerate to 50 mi/h and 26 s and 900 ft to decelerate, yielding a total of 50 s for 2300 ft plus 10 s for the remaining distance between stations 0.5 mi apart. Adding 15 s for dwell time yields 75 s per half-mile, with 7.5 min added at each end of the route for layover. Thus the time required for a round trip, in the notation of Chap. 8, is

$$t_{m(\text{local})} = 2(2.0 + 6.5)\text{mi}\left(\frac{75\text{ s}}{0.5\text{ mi}}\,\frac{\text{h}}{3600\text{ s}}\right) + 7.5\text{ min}\left(\frac{\text{h}}{60\text{ min}}\right) = 0.83\text{ h}$$

$$t_{m(\text{express})} = 2(2.0 + 6.5 + 0.5)\text{mi}\left(\frac{75\text{ s}}{0.5\text{ mi}}\,\frac{\text{h}}{3600\text{ s}}\right) + 7.5\text{ min}\left(\frac{\text{h}}{60\text{ min}}\right) + \frac{6.0\text{ mi}}{50\text{ mi/h}} = 0.99\text{ h}$$

Thus the fleet is reduced from 120 to 75 cars, which reduces the cost of vehicle ownership per day by

45($111.04) = $4996.80

Also, the reduction in trains will reduce the number of swing crews required, the reduction equaling the number of trains, nine. Hence crew costs will be reduced by

9($134.11) = $1206.99

Similarly, vehicle-miles will be reduced for the one-half of the trains now operated as inner-zone locals in the peak period. The reduction reflects the new round trip length during peak periods of 8.5 mi for 16 trips each peak period. The corresponding reduction in costs will be

2(16)(2)(5)(15.0 mi − 8.5 mi)($0.982) = $2042.56

Thus the daily operating and fleet ownership costs will be reduced by $8246.35.

The difference between the added costs of additional track and maintenance and the reduction in operating costs results in a savings of $68.69/day. This is a small amount compared to the total cost, about 2 percent, but the gains from the zone-type service are substantial reductions in the travel time for travelers between the outer zone and the CBD, the running time between the end stations being reduced from 37.5 to 29.7 min. Offsetting this slightly is the increase in average waiting time during the peak period from 1.875 to 3.75 min, although this might be reduced somewhat by issuing time tables to the public. Another important feature of this change in system design and operations is that the operating cost per day is reduced by $3249.55, a 12.9 percent reduction from the original value. This is particularly important to the operating agency, because usually it is responsible only for operating and maintenance costs, the capital costs of initial construction and equipment purchases being paid for by local, state, and federal governments.

It is interesting to inquire further as to whether a change in the base period operations plan to take advantage of the possibility of zone service would also be advantageous. Since the original all-local headway in the base period was 8 min, a simple splitting of trains between the two zones would result in a headway of 16 min, probably too lengthy for most transit operations. Hence we shall consider the operation of zone services with two-car (instead of three-car) trains, every 10 min in each zone, resulting in a headway of trains entering or leaving the CBD of 5 min. On many rapid transit lines when trains of two cars or less are operated, it is customary to collect fares on the trains, except at high-volume stations; in this case we will assume that the train guard collects all fares except those of passengers entering the system at the CBD stations, where volumes are liable to be too great for one person to handle.

This change in base period operations results in no additional capital costs, since the additional trackage was already included in the costs calculated above for the peak period. This revised operation requires 11 train crews to operate the service during the base period, as calculated below using Eq. (8-18):

$$n = \left\langle \frac{1}{h} t_m \right\rangle$$

$$n = \left\langle \frac{0.83 \text{ h}}{0.167 \text{ h}} \right\rangle + \left\langle \frac{0.99 \text{ h}}{0.167 \text{ h}} \right\rangle = \langle 4.97 \rangle + \langle 5.93 \rangle = 11 \text{ trains (or crews)}$$

Thus this operation with the revised service throughout the day requires 15 crews in service during each of the peak periods and 11 crews in service at any one time during the base period (two shifts), for a total crew requirement of 22 full shift crews and 4 swing shift crews, 26 crew-days/day. The service involving zone express trains during the peak period and all-stop trains at other times required 15 crews in service during the peak period and 12 crews in service during the base period, for a total of 27 crew-days/day. Thus there is a saving of 1 crew, which results from the institution of zone service in the base period at slightly longer headways than the all-stop service which it replaces. This results in a daily saving of \$134.11. Also, the elimination of station agents at 25 stations during 12 h of the day (out of 16 h)—assuming that they can be employed for fractions of a day—as is common on many transit lines—results in the saving of 37.5 station-agent-days/day:

$$37.5(\$41.29) = \$1548.38$$

Finally there is a savings in car-miles. The all-local base service called for three-car trains on an 8-min headway, resulting in

$$2(15 \text{ mi}) \left(\frac{3 \text{ cars}}{\text{train}} \right) \left(\frac{\text{train}}{8 \text{ min}} \frac{60 \text{ min}}{\text{h}} \frac{12 \text{ h}}{\text{day}} \right) = 8100 \text{ car-mi/day}$$

The new service calls for

$$2(8.5 \text{ mi} + 15 \text{ mi}) \left(\frac{2 \text{ cars}}{\text{train}} \right) \left(\frac{\text{train}}{10 \text{ min}} \frac{60 \text{ min}}{\text{h}} \frac{12 \text{ h}}{\text{day}} \right) = 6768 \text{ car-mi/day}$$

thus resulting in a daily cost saving of

1332($0.982) = $1308.02

The various base period changes result in a total saving of $2990.51/day.

The combined savings from the changes in the peak and the base period add up to $3059.20/day in total cost; and the reduction in operating cost is $6240.06/day, a reduction of 24.8 percent from the original operating cost. This clearly indicates that there are substantial trade-offs between operating costs and capital costs in the design and operation of transit systems. Furthermore, changes in operations which can substantially improve service for some travelers while maintaining essentially the same for others can often be made with a reduction in total costs!

More Frequent Trains

A sceptical person may wonder whether the slight degradation in service to travelers in the inner zone might be the real cause for these savings, thinking that if that service had to be maintained with the same frequency as with the all-stop service, the savings might disappear. That improvements in the quality of service for all passengers do not necessarily result in an increase in either initial costs or operating costs can be illustrated by considering a further change in the operation. Specifically, we will examine the costs of this same line, with four tracks in the middle zone to allow zone service, but with trains operated at higher frequencies (or lower headways) than in the initial case. Specifically, during the peak period trains would be three cars in length, operating on a headway in each zone of 4.50 min (for a headway in the CBD of 2.25 min), each train having the usual two-person crew and with fares collected on the train at outer stations and at the station in each of the five stations in the CBD. In the base period, trains will consist of one car and will be operated on a headway of 5.33 min in each of the two zonal services (for a headway of 2.67 min in the CBD), these trains being operated by a single operator, who also collects fares and who is paid at the premium rate of $81.06/day (in contrast to the $71.06/day for the operator of a train who does not collect fares). It might be noted that single-car trains on which operators collect fares are operated on the Chicago and Cleveland rapid transit systems, among others.

The calculation of cost for this system is rather straightforward. The only changes in addition to those already discussed are those due to the reduction in size of platforms at stations. In this case it is assumed that a station designed for a maximum of three cars/train will cost $700,000, in contrast to $1,000,000 assumed for the five-car train station design. Using the costs outlined above, the expression for total cost becomes

$$\begin{aligned} TC = \$1664.98L + \$1258.10L_e + \$111.04n_c + \$0.982m_c \\ + \$81.06d_f + \$63.05d_g + \$71.06d_m + \$41.29d_a \end{aligned} \quad (17\text{-}15)$$

where L_e = route length of express tracks, mi
d_f = operator-days with operator collecting fares
d_m = operator-days without collecting fares
d_g = train-guard-days (operates doors and collects fares)

The values of the various variables are as follows:

$L = 15.0$ mi
$L_e = 6.5$ mi

$$n_c = 3\left\langle\frac{1}{0.075}\,0.83\right\rangle + 3\left\langle\frac{1}{0.075}\,0.99\right\rangle = 3\langle 11.1\rangle + 3\langle 13.2\rangle = 72 \text{ cars}$$

$$\begin{aligned} m_c &= 3(27)(8.5)(2)(2) + 3(27)(15)(2) + 1(135)(8.5)(2) + 1(135)(15)(2) \\ &= 11{,}529 \text{ car-mi/day} \\ d_f &= 40 \text{ operator-days/day} \\ d_m &= 4 \text{ operator-days/day} \\ d_g &= 24 \text{ guard-days/day} \\ d_a &= 2.5 \text{ station-agent-days/day} \end{aligned}$$

The total cost is $59,998.04/day, $5603.91 less than the total cost per day of the system in which longer, less frequent trains are operated in all-stop service, an 8.5 percent reduction in total cost. Also, the operating cost is less, $18,807.28/day in contrast to the $25,161.51/day cost for the all-stop service. Thus, surprising as it may seem, a rail transit line can be designed so as to provide more frequent and faster service for passengers and cost less than the typical double-track all-stop design which might be the only design considered in many instances!

Skip-Stop Services

Another operating option is that of skip-stop service, in which only a fraction of the trains stop at certain stations, usually one out of every two, although one in three has been proposed for some lines. Different trains are assigned to different skip-stop stations, so that the running time of all trains remains equal, thus requiring only one track in each direction. On this example line, we will operate an *A* train and a *B* train, both stopping at all five CBD stations, but dividing all other stations equally, except for the last station, where both stop, and two additional shared stops where passengers can transfer between *A* and *B* trains. Train frequencies will be the same as for the zonal service above, with three-car trains in the peak and one-car trains otherwise.

Equation (17-15) can be used to calculate the costs. The differences with the zonal service are the absence of a four-track section and the round trip times and distances of both trains being 1.25 h and 15.0 mi, respectively. The time is longer than that for the zonal express because of speed restrictions past station platforms where passengers are standing, and two additional common stops so that passengers can change trains to go from an *A* station to a *B* station or vice versa. Hence,

$$n_c = 2(3)\left\langle\frac{1}{0.075}\,1.251\right\rangle = 2(3)\langle 16.7\rangle = 102 \text{ cars}$$

$$\begin{aligned} m_c &= 2(3)(27)(15.0)(2)(2) + (2)(135)(15.0)(2) = 17{,}820 \text{ car-mi/day} \\ d_f &= 58 \text{ operator-days/day} \\ d_m &= 5 \text{ operator-days/day} \\ d_g &= 34 \text{ guard-days/day} \\ d_a &= 2.5 \text{ station-agent-days/day} \end{aligned}$$

The total cost is $61,000.50, about equal to (precisely, 1.7 percent more than) the zone service with the same headways, and 7.0 percent less than the original all-stop service. With this scheme, service is more uniform in speed between the CBD stations and the other stations than with the zone operations plan, but some nearby stations are not connected except by backtracking (although very short distance trips are not usually made via rapid transit, on-street bus or streetcar service often being preferred).

Some General Comments

Before leaving this section, we point out that it would be incorrect to generalize from this particular example to all rail transit situations and assume that provision of express and local tracks and operation of very frequent zonal train services will necessarily reduce costs and improve the level of service for all passengers. No such generalizations are possible because each line and situation is unique. In some cases a double-track line with all-stop service will be most appropriate, while in others a double-track line with skip-stop service will be best, and in others (as the example above indicates) a four-track line with zone service will be most advantageous. The important point is that it is the responsibility of the design engineer or planner to identify and evaluate these various options. Just because the cost of constructing the track of a rail transit line (or, in general, any guideway) may seem to be tremendous, one should not assume that the provision of extra tracks for express or nonstop service will necessarily increase total cost, since their provision may permit the reduction of other costs. Similarly, it is erroneous to assume that the operation of larger vehicles at less frequent headways will necessarily reduce system operating costs, because by operating smaller, more frequent trains (or in general, smaller vehicles), it may be possible to institute other changes in the operations plan which will not only improve service for all passengers but also result in the reduction of the number of vehicles, operators, and vehicle-miles required, thereby reducing the total operating costs.

Transit Vehicle Design

The various operating plans used in the examples above impose different requirements on the transit vehicle, in that case the rail cars. First, the trains were assumed to be capable of acceleration and deceleration following the curves in Fig. 4-20, which imposes particular horsepower needs on the electric motors on the cars and similar needs for the braking system. The methods of vehicle performance simulation presented in Chap. 4 can be used to test any particular car design against such requirements.

Another requirement is that each car be able to accommodate 52 passengers seated and 48 standees. A possible design for this, illustrated in Fig. 17-12*a*, yields a relatively high ratio of seats to standee places for rapid transit cars. However, in the operating plans that included single-car trains on which the operators collect the fares, the interior design must be modified so that passengers pass the operator upon entering (or leaving). One design for this is shown in Fig. 17-12*b*, this arrangement using essentially the same car body as in Part *a* and having doors in the end of the car, so that guards are able to pass from one car to another. Only the front doors are used for entry and the rear doors for exit, just as in a bus, with operator fare collection. An alternative design for large capacity is in Part *c*, with few seats (along the walls, facing inward), which results in a higher total passenger capacity but a larger fraction standing than design *a* or *b*. Another arrangement for operator fare collection is in Fig. 17-12*d*, this being typical of streetcars or light rail vehicles, and similar to bus designs. Since this vehicle is designed

Figure 17-12 **Some alternative designs of rail transit cars for different service conditions. *(a)* Rapid transit car with two-abreast seats on both sides of aisle and relatively little standee room. *(b)* Rapid transit car as in *(a)* but designed for fare collection by motorman. *(c)* Rapid transit car with staggered doors and longitudinal seating for high-volume routes with many standees. *(d)* Bus or streetcar for single direction movement. *(e)* Bidirectional rail car for low- or high-level platforms.**

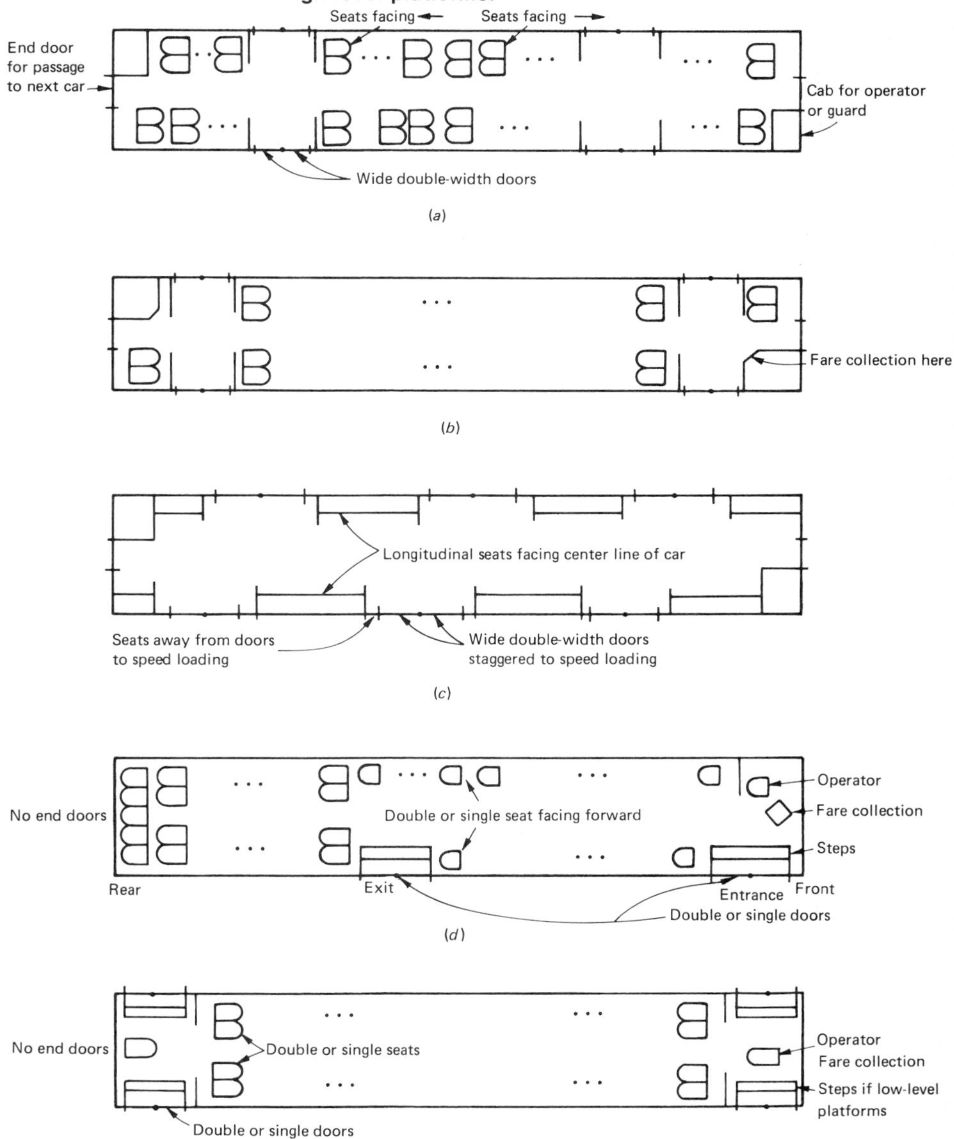

to be operated in one direction only it will require facilities for turning it around at each end of the line. Doors can be placed on both sides and operator consoles placed at both ends for bidirectional operation, as in Part *e*. Any of these designs can include a relatively large area free of seats near doors for the use of persons on wheelchairs.

There are numerous other design options. One is between third rail and overhead wire electric power distribution, the latter being preferred where there are road crossings at grade. Also, stations may have high-level platforms (at car floor level) or low-level platforms, requiring steps on the car. For high volumes of passengers boarding and alighting, the high-level platforms are usually used, to reduce dwell time and accidents. For low-volume operations, either can be used or both can be used on the same line with a varying-height step arrangement that moves to floor level at high level platforms. However, a major advantage of the high-level platform is that persons in wheelchairs can use the system without assistance, provided of course that there are ramps (or elevators) between the platforms, fare-collection area, and sidewalks, and that any doorways and turnstiles have adequate clearance.

Also, there are options in vehicle acceleration and deceleration rates and in maximum speeds. As we noted in Chap. 4, the gains from increasing any of these depends upon the spacing of stops contemplated, i.e., upon the operations plan and the design of the route which specifies station locations and scheduling options.

Options in Transportation System Design and Operations

The examples in this section have revealed that the range of options in the design and operation of transit systems are extremely large. The same is true for almost all transportation systems. It is therefore essential that these design and operations options be considered by engineers and planners in the development of any new system, and that they also be reconsidered by management as traffic and other conditions change. Only in this manner can the capacity and service provided by a transportation system be altered to provide the best combination of service and price to the public.

SUMMARY

1 The management of a transportation system is clearly as important as long-range planning or design of new facilities. Unless the system is adequately managed and operations are modified as demands and other requirements change, it will not serve its function well, any more than it would if new facilities are poorly planned or designed. Persons with a background in transportation system engineering (or planning) are being drawn into management, because the principles of this field are as necessary in effective management of transportation systems as in planning or design.

2 Management falls naturally into two categories, one concerned with facilities and the other with carrier services.

3 Facility management is concerned primarily with controlling the flow of vehicles or other traffic using the facilities. The major techniques are direct flow controls (e.g., traffic signals), routing, preferential treatment of certain classes of users, modifying demands (e.g., spreading peak flows by staggering work hours) and pricing. A relatively new concept is transportation system management, coordinated management of all facilities and carrier services in an urban area, to make the best use of the existing facilities.

4 The managements of carriers face an extremely large number of options with respect to pricing and operations plans, although often the range of choice is limited by regulatory agencies. These choices include whether or not to operate on fixed routes, whether to adhere to a fixed schedule or operate on demand, what size vehicles to operate, what types of traffic to accept (especially in freight transport), and what prices to charge.

5 All of these options for both facility managements and carrier managements can be addressed using the system planning process and the methods of cost estimation, demand estimation, etc., covered in previous chapters. However, often operational changes are relatively small and models are insensitive to the changes, resulting in the need to exercise judgment in making decisions. Fortunately, these changes can be treated as experiments which can be modified if the desired result is not achieved, in marked contrast to new facilities, which often cannot be altered except at great expense. However, since judgment and intuition are often misleading, it is important to analyze alternatives as carefully and rigorously as possible.

PROBLEMS

17-1

A 70 mi/h freeway with four lanes in each direction experiences very unbalanced traffic during the weekday peak periods. The traffic consists of 5 percent trucks and 6 percent buses in the peak hour, with a peak hour factor of 0.77. It has a design adjustment factor of 0.90. Volumes are 6000 vehicles/h in the peak direction and 2000 vehicle/h in the other. *(a)* What are the current average speeds in each direction? *(b)* It is proposed that one lane in the reverse peak direction be taken and used for buses and other high-occupancy vehicles (those with three or more persons), the latter being 6 percent of total volume. Assuming all buses and HOVs use the special lane, what would the resulting speeds be for vehicles (1) on the four regular peak direction lanes, (2) on the special HOV lane, and (3) on the three remaining reverse peak lanes? On a 5-mi trip on the freeway, what would be the corresponding changes in travel time?

17-2

A one-way arterial street has four 10-ft-wide lanes, with no parking allowed and right turns prohibited. What is the capacity in persons per hour with all automobile usage, and 1.5 persons/auto? What is the total capacity with one lane reserved exclusively for buses, with an average of 60 persons/bus? Green time is 50 percent and bus headway is 1.5 min.

17-3
Show on a figure similar to Fig. 17-10 the carrier frequency that maximizes the number of passengers using the system subject to the revenue being greater than or equal to total expenditures. For this, you will need to extend the revenue curve and the cost curve, and these should cross at a high frequency.

17-4
The demand curve faced by an airline connecting two cities is approximated by the formula

$$d = 650 \frac{f}{10 + f}$$

and its costs are given by the following equations for two different aircraft types, type 1 (100 seats) and type 2 (125 seats), respectively:

$$c_1 = 12{,}500 + 5000f$$
$$c_2 = 15{,}000 + 6000f$$

where d = daily passengers, one direction
f = flight frequency, per day, one direction
c_i = total cost, $ per day, sum of both directions

The fare is set by a regulatory agency at $100, in each direction. Demand is equal in both directions, and it is assumed that the frequency of flights will be identical in both directions. What is the optimal frequency of flights for each aircraft type? Which aircraft should be used? Assume the carrier wishes to maximize profit. Plot the revenue and cost curves for both aircraft.

17-5
Apply Leibbrand's idealized model of major activity center traffic to your city. What is the maximum passenger volume that can be accommodated entirely by autos? Compare this with the actual employee and shopper volumes into the area. If one lane on each arterial were devoted exclusively to transit, what would be the person capacity of the system? Note: To the extent possible, it is best to use actual street widths, auto occupancy, and other data for the city being studied, but some data may not be available. In such cases, use the values assumed in the example problem in the chapter. Clearly identify the source of all data.

17-6
Compare the ease with which two different types of urban public transport systems could be adapted to various types of changes through changes in operations (more vehicles can be purchased, but no new transit-only links, such as subways, constructed). The two types are (1) a system consisting of a few radial rail rapid transit lines, with parking at outlying stations and feeder bus service, and (2) an all-bus system operating on streets and on a few radial busways (one lane in each direction with parallel lanes at stations, including a four-lane busway through the CBD) and parking at outlying busway stations. The possible changes are (1) an increased number of passengers on existing routes, (2) a demand for service to newly developed outlying major activity centers, (3) a need for increased service from central city neighborhoods (not on rail line or busway) to outlying industrial areas (also not on the rail line or

busway), and (4) a fuel shortage in which use of the automobile was curtailed drastically. Assume any additional conditions you must to answer the question. Include in the comparison probable changes in level of service and in costs.

17-7
A state highway department is planning future expenditures for highway reconstruction. Estimate the amount of money needed to reconstruct a particular class of two-lane roads. The average reconstruction cost is estimated as $100,000/lane-mi in 1977, and this value is expected to grow at a rate of 4 percent per year for the foreseeable future. The mileage built in each year of construction is

1950: 507 mi
1951: 356 mi
1952: 25 mi
1953: 129 mi
1960: 190 mi

These roads require reconstruction as indicated in Fig. 17-11*b*. What is the cost of rebuilding in 1978 all lane-miles that have deteriorated through 1977?

REFERENCES

Air Transport Association of America (1974), "Air Transport," Washington, DC.

Association of American Railroads (1976), "Yearbook of Railroad Facts," Washington, DC.

Dupree, John H., and Richard H. Pratt (1973), "Low Cost Transportation Alternatives," 3 vols., U.S. Department of Transportation, Washington, DC.

Eckert, Ross D. (1972), "Airports and Congestion," American Enterprise Institute for Public Policy Research, Washington, DC.

Leibbrand, Kurt (1970), "Transportation and Town Planning," MIT Press, Cambridge, MA, trans. by Nigel Seymer from Kurt Leibbrand (1964), "Verkehr und Städtebau," Birkhäuser Verlag, Basel, Switzerland.

Levinson, Herbert S., William F. Hoey, David B. Sanders, and F. Houston Wynn (1973), Wilbur Smith and Associates, "Bus Use of Highways: State of the Art," National Cooperative Highway Research Program Report No. 143. Highway Research Board, Washington, DC.

National Association of Motor Bus Owners (1976), "1926–1976: One Half Century of Service to America," Washington, DC.

O'Malley, Brenden (1971), Staggered Work-Hour Project in Lower Manhattan, *Highway Research Record,* No. 348, pp. 152–165.

Pontier, Walter E., Paul W. Miller, and Walter H. Kraft (1971), "Optimizing Flow on Existing Street Networks," National Cooperative Highway Research Program Report No. 113, Highway Research Board, Washington, DC.

Pratsch, Lew (1975), Knoxville and Portland: Two Successful Commuter Pooling Programs, in Sandra Rosenbloom, ed., "Paratransit," Transportation Research Board Special Report 165, pp. 55–62.

Quarmby, David A. (1973), Implementing Research in a Transit Operating Agency, "Proceedings of the First International Conference on Transportation Research," College of Europe and Transportation Research Forum, pp. 875–881.

Remak, Roberta, and Sandra Rosenbloom (1976), "Peak-Period Traffic Congestion: Options for Current Programs," National Cooperative Highway Research Program Report No. 169, Transportation Research Board, Washington, DC.

"Transportation System Management" (1977), Transportation Research Board Special Report 172, Washington, DC.

U.S. Department of Transportation, Federal Highway Administration (1973), "Highway Statistics," Washington, DC.

U.S. Department of Transportation, Federal Highway Administration (1971), "Manual or Uniform Traffic Control Devices," Washington, DC.

Winfrey, Robley (1969), "Economic Analysis for Highways," International Textbook, Scranton, PA. Copyright © 1969 by International Textbook Co. Figures reproduced by permission of the publisher.

Concluding Remarks

18.

The field which we have called transportation engineering and planning has developed very recently, mainly in the last 30 years, and is still changing rapidly. This field emerged primarily as a result of the recognition that there is a common set of principles and methods which can be used in the engineering and planning of transportation systems. This commonality of principles and methods applies whether one is concentrating on long-range planning, design of particular facilities and other components, or the management and operation of an existing system. Nevertheless, there are unique features of engineering and planning peculiar to each mode and context (e.g., urban areas versus rural areas). Hence the field of transportation engineering and planning does not replace more traditional, narrower modal areas of study, but rather complements them. It serves as both a useful background to more specific studies of particular modes or problem contexts and as an area of study and practice in its own right.

A closely related but distinct development in this field is the increasing awareness of the need for a comprehensive treatment of transportation problems regardless of the context. From this perspective, all modes of transportation and all types of facilities are viewed as serving the overall objective of meeting the transportation requirements of the region under study, subject to economic, environmental, and other restrictions. It is through this comprehensive approach that it is hoped that we may begin to make real progress in solving long-standing transportation problems such as congestion and inadequate levels of service within rapidly growing urban areas and provision of adequate capacity for the growing freight and passenger traffic within developing countries and in international trade. This comprehensive view should not be interpreted as implying a single monolithic decision-making structure in transportation, for comprehensive planning can be undertaken in a number of different ways. The planning may be of a highly centralized sort, specifying precisely what facilities and services should be provided, or alternatively it may be directed toward specifying the types of policies and regulations necessary for private enterprise to provide the requisite transportation facilities and services. Hence the field of transportation systems engineering (or planning) is applicable regardless of a nation's economic organization.

Another important recent development in this field (as well as in many other fields of engineering and planning) is the increase in concern for effects of changes in the system on human activities and the environment external to the system. In the language of the economist, this is a concern for externalities. Instances of rapidly deteriorating natural environments and extremely undesirable social effects of major new technological systems have underscored the need to consider these effects on a level equal to the purely transportation impacts of changes in the system. As a result these impacts are playing an increasingly important role in transportation decision making and in government planning as well as in the decisions of private carriers.

While many of the methods of transportation system engineering and planning originally were developed for comprehensive public sector planning, it is increasingly recognized that most of these are relevant to decision making within the context of organizations with much more limited responsibilities, such as private firms and public agencies responsible for the operation of particular facilities or carriers. There is now renewed concern for the development of effective tools for management of transportation systems, and this concern will undoubtedly lead to the adaptation and development of additional methods specifically tailored to deal with problems in the operation and maintenance of transportation systems.

While this discussion has pointed out some current trends and possible future directions for the field as a whole, in rather abstract and philosophical terms, it is perhaps also important to consider some of the current trends within particular fields of transportation and related areas.

TRENDS IN TRANSPORTATION

Before looking into the future, it is instructive to look at the past to attempt to identify factors which apparently have influenced the path which led to our current transportation system. These seem to be somewhat different for local and long-distance transportation, so each of these will be discussed separately. The focus will be primarily upon developed nations such as the United States, although developing nations will be referred to with regard to current trends.

Local Transportation

Local transportation has changed very dramatically in the past century. Within this period the bulk of the population has shifted from rural areas to urban areas, with the associated high densities of development and concentrations of large numbers of persons. At first, urban areas were limited in physical size, and hence in population, by the necessity of walking for almost all trips, but the advent of the streetcar (originally horse-drawn, later electrically powered) and the rapid transit railway enabled persons to travel longer distance to work, and hence larger land areas could be encompassed with a single urban area. People in most industrialized nations concentrated themselves in large cities. Hence with the advent of technologies which enabled economical mechanized movement of large numbers of persons, the populations of urban centers grew and cities spread out to cover large land areas with much of the population commuting fairly long distances via some type of public carrier.

The advent of the automobile and the paved road (and to a much lesser extent the bus) added further options in the relative locations of residences, employment, shopping centers, etc., allowing persons to travel in any direction throughout the metropolitan area, not just along rail transit routes. This meant the further expansion of the populated area, and perhaps more significantly, the migration of much industrial and commercial activity to points outside of the central core of cities. The ubiquitous paved road network, with arterial streets and freeways designed for high speed and high capacity, has given considerable mobility to the large portion of the population which has access to automobiles.

The initial response to these most recent changes of rapid growth in both total population and in land area and the dispersal of activities and travel patterns was to attempt to expand road capacity to accommodate the growing amount of traffic. This was done primarily with freeway facilities, which are much cheaper per unit of capacity and allow higher speeds than arterial streets. But the reliance on the automobile has led to serious congestion problems in more densely developed areas of cities and a decline in use of public transit services, with attendant reductions in the quality and extent of such services; and the outward migration of residences and economic activities has created serious social problems for older central cities. In particular, the negative effects of freeways in consuming large amounts of land and in degrading the natural environment in their vicinities has led to a drastic reduction in new freeway construction in most urban areas. Highway building has been replaced to a large extent by efforts to improve public transport, most visibly in the form of new subways and similar rapid transit facilities designed for relatively long distance travel, and less visibly but equally important, improvements in local bus and streetcar services.

While the increase in trips from relatively newly developed suburbs to the central business districts of large cities makes sensible the provision of improved public transit service for such travelers, there is now considerable questioning of the wisdom of concentrating the bulk of urban public transit investment funds in such regional systems, since they typically serve only the very small fraction of all trips within an urban area which are oriented to the CBD. Also, there is now much questioning of the wisdom of trying to extend and improve local bus service as a means of providing an alternative to the automobile for short-distance trips, since these systems seem to be expensive to operate and so far at least have not made great inroads into auto travel. Many areas are experimenting with innovative forms of transport which might provide local transit service economically throughout the urbanized region. Some are low-density areas, where conventional large-vehicle transit services seem to be exceedingly costly. Innovations such as shared taxicab services and dial-a-bus and jitney (small-bus) services are being explored. In some cases these services are available to the general public, while in others they are tailored to the needs of particular groups, especially those who due to physical handicaps do not have access to automobiles and conventional transit.

Another recent development is the attempt to accommodate auto travel in a manner which minimizes negative environmental effects. Many urban areas are experimenting with incentives for the formation of car pools. Reserved lanes, which operate at speeds greater than other lanes, reduced tolls, and convenient mechanisms for identifying others with whom one might form a car pool, are all being tried. Another potentially very significant option is the creation of so-called auto-free zones, areas in which no automobiles (except emergency vehicles) are allowed, with the provision of appropriate parking facilities surrounding such areas. Much greater emphasis is now being given to better management of traffic flow on major arteries, with greater attention to proper timing of traffic signals, enforcement of restrictions such as no parking and no stopping, and reversal of the direction of some lanes to provide directional capacity at the time it is needed. Often these essentially traffic engineering measures are instituted and coordinated with transit improvement programs, to help reduce the quantity of auto travel as well as to improve the flow of that which remains. Finally, much attention is being devoted to reducing pollution emissions of motor vehicles and reducing their fuel consumption, substantial strides having been made in both of these areas with the promise of further gains.

It is of course very difficult to predict what changes might lie in the future, but nevertheless one can identify a number of directions of change which are fairly likely to occur. Certainly the most profound changes historically in urban transportation have resulted from new technologies and simultaneous changes in land use patterns which created travel demands for which the new technology was appropriate. An example is the emergence of the automobile as the dominant form of urban travel in a period when (1) urban populations were growing with most new land development occurring at the fringes of existing urbanized areas, (2) as a result trip origins and destinations became more dispersed, making much travel by transit difficult or impossible, and (3) trips lengths increased to the point where much travel had to be made by vehicle rather than by walking. Hence one must consider possible future changes in land use patterns and the associated demands for travel in assessing future changes in urban transportation and the appropriateness of possible new technologies.

At the present time, the very large urban areas which grew very rapidly in recent decades seem to be experiencing very low growth rates (or even negative rates). This seems to have occurred even before it became popular to limit growth to minimize the negative effects of rapid growth and change. While the populations of developed nations seem to be remaining urban in character, the locus of growth seems to be shifting toward smaller cities. In these areas perhaps transport as well as other important public services will be more manageable, and at least the experience with various policies in the already large urban centers can be used to avoid mistakes in these newly emerging areas.

In growing areas much probably can be done to minimize the length of trips and the need to employ vehicular transport and hence to reduce the total amount of resources consumed for transport. This might be accomplished by keeping the cost of transportation relatively high and planning land use so that most trip purposes can be satisfied through walking or short vehicular trips. And even in older areas that are not growing rapidly, policies can be instituted to reorganize land uses and travel patterns in a gradual way, in the direction of reducing dependence upon vehicular means of transport, particularly the more expensive forms. Of course, it is difficult to speculate on the political feasibility of such land use and transport pricing policies.

There also are many efforts to develop new transportation technologies which will make the moving of large as well as small volumes of persons over varying distances more attractive and less costly (in a total resource sense) than with existing technologies. Both the automobile-road system or private transport and public transport are the subjects of such technological development efforts. Although the number of specific public transit technologies being developed is extremely large, most have two characteristics in common. One is that they employ automated vehicles so that operating costs will be low, particularly with very short headways (perhaps as small as 1 or 2 min) envisioned to reduce waiting time to a minimum. Secondly, they employ guideways that are inexpensive to construct and maintain, these often being elevated and of small cross section, so that rather comprehensive networks can be implemented. The exact forms of these technologies vary considerably with intended use.

Illustrations of many types of new transit technologies appear in Fig. 18-1. So-called people movers are designed to handle large volumes of travelers over short distances within major activity centers, such as central business districts. Others, similar to automated rapid transit lines but with smaller vehicles, are designed for general service throughout an urban area, including long- and short-distance travel and high-volume as well as low-volume routes. One extreme is the so-called personal rapid transit concept, in which very small vehicles would operate nonstop from a traveler's origin station to destination station. The intent is to attempt to replicate the convenience and directness of automobile travel while obtaining the advantages of a fixed guideway and automatic control to reduce the number of channels required for any given volume of traffic. While none of these systems are in extensive urban service at the present time, some have already seen limited application (such as at airports and amusement parks) and in the not too distant future may be ready for general transit use. And it must be remembered that automated but otherwise conventional rail rapid transit lines have been successfully operated for many years. However, it remains to be seen whether these new systems will be available with the reliability and cheapness expected and whether or not the elevated guideways will receive the general public acceptance necessary for any widespread implementation.

A major problem with the development of new urban transportation technologies seems to be that many of the so-called new technologies are the results of random inventions of new mechanical devices rather than the results of conscious efforts to develop new means of transport designed to meet needs. Thus these new technologies may embody certain technical features, such as a novel provision for vehicle mobility (e.g., air cushion) or a novel propulsion system (e.g., a linear motor), which may not offer any real advantages and may yield some disadvantages compared to conventional transport technology. Another problem is that various levels of government in western nations where these technological developments are occurring have not come forward to provide the funding necessary to develop fully operational prototypes of these systems. While the government may be the logical

(a)

(b)

(c)

(d)

Figure 18-1
New forms of urban transport. ***(a)*** **Interior of people mover vehicle used at airports. [*Courtesy of Westinghouse Electric Corp.*]** ***(b)*** **Fully-automated rapid transit train. [*Courtesy of Westinghouse Electric Corp.*]** ***(c)*** **Personal rapid transit vehicles at multiple-berth terminal. [*Courtesy of Boeing Aerospace Co.*]** ***(d)*** **Passenger conveyor showing low-speed entrance and exit and high-speed main section (see Chap. 6 for description). [*Courtesy of Dunlop Limited*]** ***(e)*** **Electrically-powered automobile. [*Courtesy of Sebring Vanguard, Inc.*]**

(e)

organization to assume the risk in such technological development, most of its funding has been directed to conventional technologies, with the meager funds that have been devoted to the development of new technologies usually so scattered that no one option can be developed to the point of effective testing of the concept. Also, the public has been led to expect the development of entirely new technologies to the point of being practical, operational systems within a few years, even though such expectations are quite unrealistic. When operation systems were not forthcoming in such short periods, considerable distrust of all new transport technologies was created. For all of these reasons, the process of technological development in urban transportation will probably be much slower than it would otherwise be, but this may provide time to realign priorities for new technology to meet actual needs.

There are also many innovative concepts that involve modifications of existing technology, but which may produce major benefits. Most of these have been discussed in previous chapters, but a brief listing is useful. The automation of rail rapid transit is included here, as is the coordination of schedules of rail and feeder bus lines so travelers are not excessively delayed at transfer points. The operation of buses on exclusive roadways, perhaps to continue these journeys on public streets in low-density residential neighborhoods, to provide fast, nontransfer service, is an important idea, although to date this has been tried only to a limited extent. Similar in concept is the operation of streetcars on exclusive rights-of-way in congested areas, termed limited tram lines or light rail lines. Articulated and double-deck buses can increase capacity and reduce operating costs of bus lines, and electric power can reduce pollution and noise.

There are also prospects for potentially significant technological developments in automotive transportation. Already mentioned are the reduction in pollution emissions and fuel consumption rates for conventional vehicles. These two objectives might also be achieved by the development of practical electric cars or cars propelled by other means. Small autos propelled by electricity from batteries are now available (Fig. 18-1*e*). Another major problem with the existing automobile-road system is the large number of lanes required to move large volumes of traffic. It may be reduced by automation of vehicle control, permitting the operation of platoons of closely spaced automobiles on freeways. There would be automatic entry and exit of vehicles from moving platoons.

Autos would be guided electronically, following cables buried in the pavement. Deviations would be corrected by automatic steering. Speed would also be controlled. Another approach is to have the autos guided mechanically, as on a railroad. Many design concepts for such a "dual-mode" system, as it is termed, have been developed to the prototype stage. They simplify lateral control, but require additional wheels and appropriate guideway. Speed is automatically controlled, and entrance and exit from the guideways is limited to particular points, as in the automated highway concept.

Until very recently, the movement of freight within urban areas was almost entirely ignored by regional planning agencies. However, as we noted in Chap. 2, the total resources expended, measured in dollars, on urban goods movement roughly equals that of urban person movement, so this field is by no means insignificant. Most freight moving entirely within a single urban area is carried by truck at the present time, with the exception of fluids and gases moved in pipelines of various types. The costs of truck movement have been increasing very rapidly, particularly in larger urban areas, primarily due to congestion and problems of productivity in terminals. One possible solution is the consolidation of goods movement within urban areas to reduce the number of truck trips required to move the same amount of freight, although it is unclear whether the gains in trucking costs effected by this scheme would be offset by greater costs to the shipper, who might experience less prompt and reliable service. Another possibility is to move freight on a transport system separate from the truck-roadway system, such as in small capsules or containers (e.g., about 10 to 15 ft long and 5 ft in diameter) moving suspended by air pressure in an underground pipeline. Alternatively, the capsules could have wheels guided by rails or the inner surface of the pipe or tunnel. The technology exists for these to be entirely automated with respect to vehicle speeds, routing, etc. Shippers and receivers of freight could have terminals in their own buildings, where containers could be loaded and unloaded and future destinations encoded on them, or they might choose to use neighborhood terminals. Research now underway to determine the best physical form for such a system may lead to a demonstration so that the costs and benefits can be assessed.

Long-Distance Transport

Long-distance transport encompasses many different types of transport, including intercity transport, much transport which is purely rural, transport connecting rural and urban areas, and most international transport.

Long-distance passenger and freight transport has also been growing very rapidly in the last century, partly for the same reasons as the growth in urban transport. The growth in passenger transport is to a large extent due to increased personal income and leisure time and the development of cheaper and faster transport services. Also, earlier retirement has undoubtedly contributed to the increase in tourism and cruise trips. Business travel has mushroomed in recent decades, probably due largely to the advent of air transport and to a lesser extent faster ground transport (particularly rail), which has permitted many business trips to be made in one day which previously would have required many days of unproductive travel.

Freight transport (as measured in ton-miles) has grown less dramatically for the most part, probably because with a larger population concentrated in a single area it is possible to produce goods and services at more locations with economically sized facilities and thereby reduce the distances over which goods must be transported. Simultaneously, there has been a reduction in weight of many commodities shipped, primarily through technological advances in design and manufacturing, as illustrated by the substitution of lightweight metals and plastics for steel and the substitution of transistorized electronic parts for the much heavier tube-based components, to mention two examples. Also, there is the substitution of high-voltage electric transmission from mine-mouth generating stations for movement of coal from mines to distant electric power plants, which reduces ton-miles of freight. However, certain aspects of goods transport are growing very rapidly, including the movement of fuels (primarily international) and some raw materials.

There has also been much technological change in goods transportation.

Perhaps most notable are pipelines for the transmission of liquids and gases and large multi-unit intercity trucks and high-speed roadways. Also in recent decades there has been a resurgence of water transport of bulk commodities between points which are also connected by land carriers. Large trains of barges are proving to be a very economical means of transport, although they are slower than most other means. In most areas railroads still retain a very sizable fraction of intercity movements (often greater in ton-miles than any other mode), with the rail share of all freight tending to focus on long-distance high-volume movements, for which they are better suited than other carriers. Of great significance for international movements has been the development of containers, which have yielded substantial reductions both in costs and in time required to transfer freight between land vehicles and ocean-going ships.

In the United States and Canada, the bulk of long-distance passenger travel is now accommodated by the automobile, except for very long distance travel (over, say, 500 mi), where aircraft is widely used. Buses also provide an important service, serving almost all communities and providing a common carrier service which is much less expensive than air service. Between large centers of population on short (under about 300 mi) heavily traveled routes, rail service remains important and on many routes it will undoubtedly be improved to play an even greater role. Throughout the world long-distance water transport and rail transport have declined noticeably in their market shares and often in absolute passenger volumes, being replaced primarily by air travel for travelers who can afford to keep travel time to a minimum, although cruise ships and to a lesser extent cruise-type rail services still provide an important, primarily recreational, form of travel. In areas outside of North America, rail or bus travel represents a much larger fraction of all trips, and usually air traffic is limited by high fares.

Intercity transport is not devoid of problems, although in most nations it seems to be treated with less priority than urban problems, except in situations where intercity transport must be improved to achieve the economic potential of a nation. One problem with intercity road transport is that within urban areas the intercity movements are subjected to the same congestion and contribute to the same environmental problems as purely urban movements. Rail passenger systems seem to be plagued with high deficits which are increasingly difficult to justify in terms of benefits, whether these be in reducing congestion on other modes, reducing negative environmental impacts, or reducing fuel consumption, although these benefits may increase in the future.

In many instances rail freight costs are increasing, at a rate exceeding increases in revenues, leading to problems in continuation of essential services without subsidies. However, it may be true that in most of these situations rail prices are simply set too low, perhaps for political reasons (where governments control rail rates), but more likely because the prices of competitors reflect subsidization. It is alleged, for example, that trucks are not charged their full share of road costs, and the absence of user costs on inland waterways in the United States clearly indicates subsidization of that mode.

Although there are environmental problems with all forms of intercity transport, most seem to be manageable. One possible exception is air transport, where there are serious problems with noise and air pollution in the vicinity of major airports. Another exception is the severe problem with water and beach pollution resulting from accidents in which vessels lose their cargoes, this being especially damaging in the case of ships carrying oil and chemical products.

There continues to be much technological development in intercity transport, particularly with respect to passenger transport. Although extensive and relatively inexpensive air transport is scarcely three decades old, that mode has already experienced one major technological change, in the replacement of propellor-driven aircraft with jet aircraft, which has both increased speeds and reduced costs. Soon there may be a third major change, the advent of supersonic aircraft, that is, aircraft which can travel at greater than the speed of sound (Fig. 18-2*a*). There are potentially serious environmental problems with such aircraft, mainly due to noise they generate when passing through the sound barrier, which entails possible injury to humans and animals as well as damage to structures. However, if such aircraft are made economically and environmentally feasible, they could substantially reduce the time required to travel extremely long distances, and thereby shrink the effective size of the earth and help induce greater travel between very distant points.

Due to congestion in airports and airways at large metropolitan areas and especially in densely developed urban corridors, there is renewed interest in finding a substitute for air travel. In the past two decades there has been considerable interest in developing various forms of high-speed ground transportation. Already implemented in many densely developed corridors are high-speed trains, usually electrified and operating on tracks which are completely free of road crossings and other intrusions. One example of such a high-speed service is the Metroliner service between New York City and Washington, which also serves numerous intermediate cities along the 225-mi route. Trains are currently limited to a top speed of about 110 mi/h, which yields an average speed of approximately 85 mi/h due to track curvature, some road crossings at grade, and other limitations, although in the near future it is expected that maximum speeds of the order of 160 mi/h and average speeds in excess of 100 mi/h can be instituted. A photograph of one of these trains appears in Fig. 18-2*b*. Similar services already instituted in Europe and Japan have had considerable success in attracting travelers from other modes as well as in inducing increased travel between the cities served. Since there are limits to how fast wheeled vehicles can travel, much experimentation is underway with alternative systems for providing for the mobility of vehicles. These include air cushion support, with the vehicle being guided in somewhat the same way that a railroad train is, and vehicles which are separated from the guideway by magnetic levitation. New forms of propulsion must be used for these vehicles. Major possibilities under exploration include linear induction motors (motors in which the rotor is in effect the vehicle and the stator the track or guideway on which the vehicle operates) and the various forms of conventional propellor and jet propulsion as used for aircraft. Also, there continues to be much research and development of aircraft which can take off and land vertically (VTOL) or on very short runways (STOL). The latter type of aircraft (STOL) is particularly promising but still has largely unsolved problems in the noise generated during take-off and landing. All of these are illustrated in Fig. 18-2*c* through *f*.

Figure 18-2 **New forms of high-speed intercity transport.** ***(a)*** **Supersonic aircraft.** [***Courtesy of British Airways***] ***(b)*** **High-speed Metroliner train.** [***Courtesy of Amtrak***] ***(c)*** **Tracked air cushion vehicle.** [***Courtesy of Ste. Bertin & Cie***] ***(d)*** **Magnetically levitated vehicle.** [***Courtesy of Krauss-Maffei***] ***(e)*** **Experimental vertical take-off and landing (VTOL) aircraft.** [***Courtesy of NASA***] ***(f)*** **Experimental short take-off and landing (STOL) aircraft.** [***Courtesy of NASA***] ***(g)*** **Automobile and passenger carrying train.** [***Courtesy of auto-train™ corp.***]

Since much of the time (and often the expenditures) required for any intercity trip is consumed by the portion of the trip between the origin or destination and the terminal of the intercity carrier, much attention has been devoted to improving this segment of the trip. One approach is to coordinate the intercity system with an intra-urban transportation system which will exist to serve other travel. Examples of this include connecting airports with the rest of a city's rapid transit system by special airport-to-CBD rapid transit lines and providing special freeways to take travelers between airports and various locations along the intra-urban freeway system. Since so many origins and destinations within an urban area are easily reached via automobile, there is much interest in developing intercity transportation systems which will carry passengers as well as their automobiles. In one possible scheme for this travelers would drive their automobiles onto special railroad trains designed to operate between outlying stations strategically located along intra-urban freeways and arterial routes. Such services already exist, in the form illustrated in Fig. 18-2*g*. On appropriate track, such trains could be operated at Metroliner speeds; and they could be designed for rapid loading.

Another option would be to employ an automatic highway, identical in form to that discussed previously, which would relieve intercity travelers of long hours of driving and perhaps increase the average speed and possible range of travel in one's own automobile. Another possibility is, of course, to provide rental cars at the nonhome end of the trip, although this in practice turns out to be more expensive than most travelers seem to be willing to accept.

Another range of options for reducing access and terminal transfer times includes providing intercity terminals or access points throughout an urban area, to minimize the distance and other costs of access travel. This can be done with almost any transport technology, although it raises questions regarding the nature of connections between the various terminals in one city and the many terminals in another. One possibility is to offer direct nonstop services between each pair of terminals or to have each intercity vehicle stop at many terminals in each city, the latter obviously slowing the service. Alternatively, passengers might be placed in small vehicles at each of these terminals, with vehicles being routed to a central terminal where they would be joined together to operate as a single unit on the intercity link to the other city, where the reverse operation would take place. Clearly many options exist with respect to improvements in intercity travel, and only further research and actual demonstrations will indicate under what conditions each of these might be appropriate.

There is much less consideration of radically new transport technologies for goods movement than for person movement, although much experimentation is continuing with evolutionary changes. There are also attempts to adapt various types of high-speed ground passenger transport systems such as those described above to the carriage of goods. The evolutionary changes seem to be largely in the area of providing integrated multimodal services which may use each individual mode of transport for that portion of a movement for which it is most appropriate. For example, one might envision a containerized movement which involves truck movement from factory of origin to a central rail terminal at which the container is transferred to an intercity train, with rapid rail movement to the destination city, followed by truck movement to the final destination. The technology for such a system already exists, and there are numerous technological options, such as the speed and frequency of trains and the range of commodities to be accommodated in containers. There are many other possible intermodal technologies, such as modifying the characteristics of trucks and their trailers as well as railways so that the same vehicles can be moved over the highways in access movements and still combined into trains directly for line-haul movement. Barriers to multimodal service in almost all nations are present in institutional and regulatory constraints, which preclude any single firm such as a railway from initiating a comprehensive service of this sort (although some rail-truck coordination is usually possible). Until these institutional barriers to intermodal services are reduced, it is unlikely that such a freight system will evolve to any significantly greater extent than the current piggyback and containership services.

The other major new development in intercity transport is broadening the range of commodities carried by pipelines from just liquids and gases to some solids. Already coal and other bulk commodities have been carried in a liquid slurry, and it is technologically possible to ship solid freight in pipelines in containers or capsules, as described earlier, in the discussion of urban transport.

Some General Considerations

Transcending these particular types of transport are a number of general considerations which will radically influence the amount and form of future transport. Perhaps the most discussed among these is the possibility of substituting communication for travel. Rapid advances are being made in telecommunications which may make videophones sufficiently inexpensive to be used widely in business and perhaps even in private homes. This would presumably reduce the demand for transport. Currently business meetings are being held among widely dispersed persons via videophone in a few companies, which has eliminated the need for travel to a central point. And in principle many persons could work at their residences, communicating with others via videophone rather than in person. The development of paper copy transmitters will further this possibility. How far such substitution can go is impossible to predict now, but the possibility of significant substitution exists.

Also affecting the amount of travel and goods movement is the availability of the resources necessary for transport. Many current projections suggest that by the year 2000 the world supply of oil will be insufficient to support a continuation of current uses, among which transport accounts for about half in the United States today. Given the reliance of all modes on oil, changes will be necessary. It is possible to shift many modes to electric propulsion, with electricity being generated by coal or other nonoil sources. The technology for this substitution exists for railways (including rail transit) and buses in the form of transmission lines from which vehicles collect the power. The same technology can be used for any land vehicle which follows a fixed route, trucks being so powered in the Soviet Union, for example. Electric vehicles powered by fast-recharging or easily replaced batteries form another option. Research is currently underway on hydrogen as a vehicle fuel, which has the very attractive features of possibly being generated by solar energy, allowing easy storage on the vehicle, and its burning yielding no undesirable pollutants. Whether or not it will be practical remains to be seen. However, we must accelerate the task of developing practical nonoil-based propulsion systems if they are to be available in the next decade or two.

There are many other resources which existing transport systems consume in excessive amounts (Harris, 1968). One is land. Although in principle all ground modes could be placed in tunnels, in practice this is too expensive for all but a few routes—with existing transport technology and existing tunneling technology. Another problem is the friction between transport links and adjoining activities. Water, sewerage, and power supply are all accomplished with links that intrude only minimally into adjoining activities, but land transport links still interfere directly with activities in their immediate vicinity—through acting as a barrier, creating pollution, and causing accidents. Another general problem with existing technologies is the difficulty of transfer from one system to another, especially for passengers (containers and new bulk materials handling equipment having eased the problem for freight). All this suggests that there is indeed a need for a new transport technology—but a technology that is responsive to actual requirements for transport and to the need for compatibility with the environment, rather than simply another new way of moving things.

When one considers the many very valuable resources used in transport: oil, land in densely developed areas, and degradation of the natural environment, it is surprising that prices for transport are often set well below the average costs or long-run marginal costs. Prices for almost all products and services, housing, food, etc., are set above average costs, with the difference being sufficient to provide tax revenues to support general public services such as schools, welfare, fire and police protection, and the military forces. In contrast, much transport is subsidized, some aspects very heavily. A few examples based on experience in the United States illustrate the point. Strong evidence exists that road user taxes in total are insufficient to cover all road costs (construction and maintenance) and usually do not yield surpluses which can be used by government for general purposes (exceptions being toll facilities in the larger metropolitan areas). Urban public transit services now generally do not even cover operating costs with revenues, much less contribute to the capital cost. The air transport control system is subsidized by the federal government, and various levels of government subsidize most airports. Water carriers enjoy free use of the waterway system. Intercity rail passenger service covers about one-half its cost from revenues. This leaves pipelines and freight railroads as the only major sectors of transport that generate revenues sufficient to cover costs and create a surplus for general public purposes (taxes), and many railroads are having great difficulty continuing in this manner, some already requiring federal financial assistance.

Given the scarcity and value of many of the resources used by transport, one naturally wonders about the wisdom of current pricing and subsidization policies. Some transport may have its prices set low for social welfare reasons, such as public transit in low-income or high-unemployment areas. But that argument surely does not apply to air travel or to commuter services designed for suburbanites traveling to work in central cities. There may be substantial external benefits to cheap transport in particular instances, but these should be identified and substantiated as a justification of the policy.

While subsidizing all forms of transport might restore a modal balance, it destroys the balance with all other sectors of the economy and would lead to even more transport being consumed, and with it more resources such as land, oil, metals, etc. This issue seems to be one of the most important surrounding the future of transport and its institutions, public and private.

There is another important resource used in transport—people. For over 100 years the productivity, output per unit of input, of labor has increased, first by mechanization and more recently by automation, which has in the aggregate been responsible for much of the increase in our material standard of living. Undoubtedly further increases in productivity can be obtained in transport, such as through more automation and elimination of jobs (management or labor) no longer needed; and in many cases this would reduce the cost of transport. Yet such reductions in employment, in transportation as well as in other sectors of the economy, are adding to already difficult unemployment problems in many nations. It is possible, of course, for reductions in transport costs (through reduction in labor required or other means) to help stimulate other sectors of the economy and thereby increase employment in those sectors, but the net effect would have to be carefully determined in each instance. Thus automation, increases in labor productivity, and reductions in labor costs can no longer be considered as unambiguously positive effects, and employment impacts of technological change will have to be considered carefully along with environmental and other impacts external to the transport system.

No discussion of future transport is complete without reference to the spatial pattern of human activities, for it gives rise to the need for transport. For the past century, approximately, population has shifted to urban areas, where now about 72 percent of all persons in the United States live. Much of this population is concentrated in a few very large metropolitan areas, where some of the most difficult transport problems are found. For future planning, a critical question is whether or not this spatial pattern of population and activities will continue. There are strong indications that it is changing, at least in the United States; most growth is in smaller metropolitan areas in the South and West. Many older large cities have experienced absolute declines in population, and growth in their suburbs is small. Perhaps we are entering a period different from the past in the pattern of population migration and in the size of settlements and hence different in its transport requirements.

Extending this consideration further, one may ponder future patterns of human activity, not on a global scale but a galactic scale. Will humanity attempt to establish colonies on the moon or other planets? If so, future transport requirements will be radically different from today's.

CONCLUSION

One thing is certain: The future will be different from the past. Individual preferences, social movements, changes in religious thought and political beliefs, decisions by governments and industry, scientific discoveries, technological developments, and a myriad of other unknowns, will all influence transport requirements and capabilities of the future. Hopefully the principles and techniques presented in these chapters, as augmented and refined by future research and experience, will serve us well in meeting the challenges of the future.

REFERENCES

Harris, Britton (1968), Goals for Urban Transportation, *Transportation Research,* **2**(3): 249–252.

Morlok, Edward K. (1976), Scientific Research in Transportation: The Need for New Research Directions and New Support Policies, *Transportation Research,* **10**(6): 391–394.

English and Standard International Units

Appendix A

ENGLISH UNITS AND SI EQUIVALENTS

Length

1 in = 2.5400 cm
1 ft = 0.30480 m
1 mi = 1.6093 km

Area

1 in^2 = 6.4516 cm^2
1 ft^2 = 0.092903 m^2
1 mi^2 = 2.5899 km^2
1 acre = 4046.9 m^2

Volume (liquid)

1 gal = 3.7854 l

Volume (dry)

1 ft^3 = 0.02832 m^3
1 yd^3 = 0.76456 m^3

Pressure

lb/in^2 = 6894.8 N/m^2

Weight density

1 lb/ft^3 = 157.09 N/m^3

Angle

1° = 0.017453 rad

Temperature

$t_K = \frac{5}{9}(t_F + 459.67)$
t_K = Kelvin
t_F = Fahrenheit

Acceleration

1 ft/s^2 = 0.30480 m/s^2

Speed

1 mi/h = 1.6093 km/h
1 ft/s = 0.30480 m/s

Energy

1 Btu = 1055.1 J
1 ft-lb = 1.3558 J

Force

1 lb = 4.4482 N
1 cwt = 444.82 N
1 ton = 8896.4 N

SI PREFIXES

mega	10^6
kilo	10^3
hecto	10^2
deka	10
deci	10^{-1}
centi	10^{-2}
milli	10^{-3}
micro	10^{-6}

TRADITIONAL METRIC AND SI EQUIVALENTS

Force

1 kg = 9.8067 N
1 ton (metric) = 1000.0 kg·f
= 9806.7 N

Area

1 hectare = 10,000 m^2

Temperature

$t_K = t_C + 273.15$
t_K = Kelvin
t_C = Celsius

Volume

1 l = 1 m^3

ENGLISH UNIT EQUIVALENTS

Length

1 ft = 12 in
1 yd = 3 ft
1 mi = 5280 ft
1 nmi = 6080.2 ft

Area

1 acre = 43,560 ft^2
1 section = 1 mi^2 = 640 acres

Volume (dry)

1 qt = 67.2 in^3
1 pk = 8 qt = 537.6 in^3
1 bu = 4 pk = 2150.4 in^3

Volume (liquid)

1 gal = 231 in^3
1 gal = 4 qt = 8 pt
1 bbl = 31.5 gal
1 bbl (petroleum) = 42 gal

Angle

1° = 60′ = 3600″

Power

1 hp = 550 ft-lb

Force

1 ton = 2000 lb
(sometimes called short ton)
1 cwt = 100 lb
1 kip = 1000 lb
1 long ton = 2240 lb

ABBREVIATIONS

bbl	barrel
Btu	British thermal unit
bu	bushel
cm	centimeter
cwt	hundredweight
ft	foot
gal	gallon
h	hour
hp	horsepower
in	inch
J	joule
kg	kilogram
km	kilometer
l	liter
lb	pound
m	meter
mi	mile
N	newton
nmi	nautical mile
pk	peck
pt	pint
qt	quart
rad	radian
s	second
t_C	degrees, Celsius
t_F	degrees, Fahrenheit
t_K	degrees, Kelvin
yd	yard
°	degree
′	minute
″	second

For further information on units, see Meyer, Stuart L. (1975), "Data Analysis for Scientists and Engineers," Wiley, New York, App. I.

Densities of Common Commodities

Appendix B

Commodity	Weight density	
	lb/ft^3	lb/gal
Alcohol		6.6–6.8
Beer, crated bottles	32	
Benzene		7.5
Charcoal	18–35	
Cigarettes, cartons	19	
Coal, lumps	50–84	
Coffee, bagged	36	
Concrete, dry	110–115	
Concrete mix, wet	130–140	
Dresses, boxed	11	
Earth, dry loose	70–80	
Earth and gravel	95–110	
Electrical appliances	7–10	
Electrical machinery	35–70	
Food, packaged	14–18	
Fruits	30–40	
Fuel oil		6.9–8.0
Gasoline		6.0
Gravel	100–120	
Household furnishings	4–8	
Kerosene		6.6–6.9
Leather goods	7–10	
Lubrication oil		7.0–8.0
Meats, bulk	40–50	

Commodity	Weight density	
	lb/ft^3	lb/gal
Milk		8.5–8.7
Oats	25	
Oil, crude		7.0
Oranges, crated	37	
Petroleum		8.0
Refrigerators, crated	13	
Rock, crushed	85–105	
Sand, dry	90–115	
Sound equipment	3–40	
Steel sheets, bundled	250–450	
Water		8.35
Wheat, bulk	45–55	
Wood chips	11–14	

Index